Tissue Culture Clonal Propagation

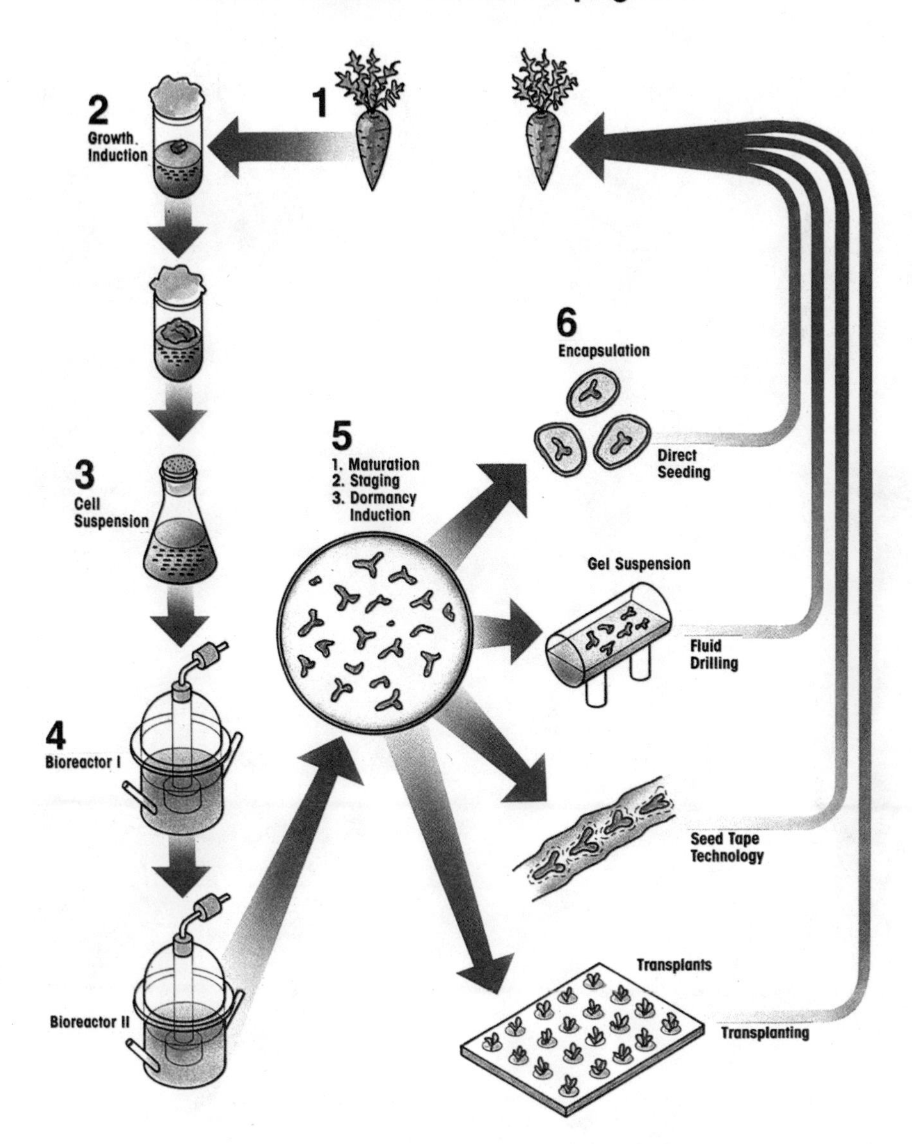

HANDBOOK OF PLANT CELL CULTURE

Published

Volume 1, Techniques for Propagation and Breeding
Editors: David A. Evans, William R. Sharp, Philip V. Ammirato, Yasuyuki Yamada

In Preparation (available Summer, 1984)

Volume 3, Crop Species
Editors: Philip V. Ammirato, David A. Evans, William R. Sharp, Yasuyuki Yamada

HANDBOOK OF PLANT CELL CULTURE,
Volume 2
Crop Species

Editors

William R. Sharp
DNA Plant Technology Corporation

David A. Evans
DNA Plant Technology Corporation

Philip V. Ammirato
Barnard College, Columbia University
DNA Plant Technology Corporation

Yasuyuki Yamada
Kyoto University

Macmillan Publishing Company
A Division of Macmillan, Inc.
NEW YORK

Collier Macmillan Publishers
LONDON

Macmillan Publishing Co.
866 Third Avenue, New York, N. Y. 10022

Collier Macmillan Canada, Inc.

Printed in the United States of America

printing number
2 3 4 5 6 7 8 9 10

Library of Congress Cataloging in Publication Data
Main entry under title:

Handbook of plant cell culture.

Includes bibliographies and indexes.
Contents: v. 1. Techniques for propagation and breeding / David Evans ... [et al.] -- v. 2. Crop species / William R. Sharp ... [et al.]
1. Plant cell culture--Collected works. I. Evans, David, 1952- . II. Sharp, William R., 1936-
QK725.H23 1983 631.5'2 82-73774
ISBN 0-02-949230-0

Contents

Section V. Root and Tuber Crops

Section VI. Tropical and Subtropical Fruits

Section VII. Temperate Fruits

Section VIII. Fiber and Wood

Section IX. Extractable Products

Preface

The tools of plant cell culture are increasingly being applied to a wide range of biotechnology ventures and in particular to the propagation and genetic improvement of crop plants. For this, the approaches and methodologies must be specifically adapted to the differing problems and potentialities of each crop and to the differing responses of plants that may be herbaceous or woody, dicotyledonous or monocotyledonous, annual or perennial, inbred or highly heterozygous. It is the application of plant cell culture techniques to the improvement of specific crop plants that is the subject of Volumes 2 and 3 of this series.

The list of plants that play an important part in agribusiness is longer than one might initially surmise. The selection of those to be included in these volumes reflects a necessary amalgam of several factors—the plants chosen must be of recognized economic importance, they need to have been successfully employed in cell culture research, and a key investigator had to be available and willing to contribute. Some important crops, then, are not here because these elements did not come together. And it is to be expected in any area where technology is only just being applied that there will be varying degrees of experience and success. This can be seen in the varying lengths of the presentations. The plants finally selected do represent major crops where cell culture methodology has been demonstrably applied.

In addition, in each volume, several general chapters provide overviews of topics that are particularly relevant to the subject at hand. In this volume, we have included a discussion of major food and energy crops, their relative economic importance, trends in production and trade, and indications of looming problems, including a potential diminution of food supplies with an increasing shift to energy crops. A second chapter summarizes current cell culture methods available for crop improvement and sets the stage for the subsequent discussions of individual crop plants. The reader is directed to Volume 1 in this series for a detailed exposition of these techniques.

As in our first volume, we are pleased to have an introductory essay by a distinguished scientist. Melvin Calvin, Nobel Prize recipient for his work in the elucidation of the path of carbon assimilation in photosynthesis, discusses the potential use of plants as a direct source of oil and other hydrocarbons. His search for plants that can grow on land not used for food production is particularly important.

The centerpiece of this volume, as for the entire series, is practical methodology, and we have again included for each chapter a major section with acutal protocols, whether as recipes, tables, charts, or narratives. To introduce the specific crop, the history and economic importance are presented, culminating in a discussion of important breeding and propagation problems, areas in which cell culture methods may be particularly applied. A critical review of the literature summarizes past and current cell culture work, and a discussion of future prospects details where we may expect the technology to go. Again, key references have been highlighted and references are given with full citation.

As before, our goal is to provide a comprehensive and practical compilation that students and scientists, academicians and businessmen alike will find informative and useful in both understanding current strategies and in extending scientific frontiers.

We are indebted to our editors at Macmillan, Sarah Greene and Frances Tindall, for the necessary support in bridging the gap from concept to book. Janis Bravo again served as editorial assistant, the crucial link from us to both authors and editors and for which she deserves our sincere appreciation. Lastly, we offer our deep thanks to our many authors who provided manuscripts on time and ushered them through the various publication stages with despatch and good cheer.

CONTRIBUTORS

A. Allavena	Instituto Sperimentale per l'Orticoltura, Montanaso L. (Milano), Italy
P.S. Baenziger	Field Crops Laboratory, USDA, Beltsville, MD
S. Biondi	Instituto Botanico, Universita di Bologna, Bologna, Italy
J.E. Bravo	DNA Plant Technology Corporation, Cinnaminson, NJ
M. Calvin	Department of Chemistry and Lawrence Berkeley Laboratory, University of California, Berkeley, CA
Chen Zhenghua	Institute of Genetics, Academia Sinica, Beijing, China
G.B. Collins	Department of Agronomy, University of Kentucky, Lexington, KY
S.S. Cronauer	Department of Biochemistry, State University of New York at Stony Brook, Stony Brook, NY
D.J. Durzan	Department of Pomology, University of California, Davis, CA
D.A. Evans	DNA Plant Technology Corporation, Cinnaminson, NJ
C.E. Flick	DNA Plant Technology Corporation, Cinnaminson, NJ
A.W. Galston	Department of Biology, Yale University, New Haven, CT
J.H.M. Henderson	Division of Natural Sciences, Tuskegee Institute, Tuskegee, AL
R. Kaur-Sawhney	Department of Biology, Yale University, New Haven, CT
P.J. King	Friedrich Miescher-Institut, Basel, Switzerland
A.D. Krikorian	Department of Biochemistry, State University of New York at Stony Brook, Stony Brook, NY
W.R. Krul	Department of Plant and Soil Science, University of Rhode Island, Kingston, RI
M.D. Lazar	Cell Culture and Nitrogen Fixation Laboratory, USDA, Beltsville, MD
R.E. Litz	Institute of Food and Agricultural Sciences, University of Florida, Homestead, FL
M.-C. Liu	Department of Plant Breeding, Taiwan Sugar Research Institute, Tainan, Taiwan
R.L. Meyer	Department of Agricultural Economics and Rural Sociology, Ohio State University, Columbus, OH
G.H. Mowbray	Montbray Wine Cellars, Ltd., Westminster, MD
T.J. Orton	Agrigenetics Corporation, Boulder, CO
G.C. Phillips	Department of Horticulture, New Mexico State University, Las Cruces, NM
B.R. Phills	Tuskegee Institute, Tuskegee, AL
N. Rask	Department of Agricultural Economics and Rural Sociology, Ohio State University, Columbus, OH
G. Reuther	Institut für Botanik, Geisenheim, Federal Republic of Germany

W.M. Roca	Genetic Resources Unit, Centro Internacional de Agricultura Tropical (CIAT), Cali, Colombia
G.W. Schaeffer	Cell Culture and Nitrogen Fixation Laboratory, USDA, Beltsville, MD
W.R. Sharp	DNA Plant Technology Corporation, Cinnaminson, NJ
K. Shimamoto	Friedrich Miescher-Institut, Basel, Switzerland
T.A. Thorpe	Department of Biology, University of Calgary, Calgary, Alberta, Canada
B. Tisserat	Fruit and Vegetable Chemistry Laboratory, USDA, Pasadena, CA
B.T. Whatley	Tuskegee Institute, Tuskegee, AL
R.H. Zimmerman	Fruit Laboratory, USDA, Beltsville, MD

"It's a genetic engineer. He wants to know if he can patent designer beans."

SPECIAL ESSAY:

Oil from Plants

M. Calvin

Growing oil, my subject for this essay, means just that. Up until now, we have been mining oil, the photosynthetic product of several hundred million years ago. The question now frequently arises about new sources of ancient photosynthetically produced materials, and there is no longer a tenable positive answer. Alternatives to mining oil must be found.

An immediate need for alternate energy sources is emphasized by the history of productivity of drilling rates for United States domestic petroleum from 1945 to the present time. It has been demonstrated that the amount of oil found per foot of well drilled is falling, from 35 barrels in 1945 to less than half that amount in 1975. Also, the rate of discovery is falling, and the energy cost of drilling and extracting oil is rising (Hall and Cleveland, 1981). Somewhere near the year 2000 the energy cost of finding and extracting a barrel of oil will exceed the energy content of that barrel of oil.

CARBON DIOXIDE PROBLEM

One alternative to oil that has been suggested is coal (also a product of ancient photosynthesis), huge supplies of which are available in the United States, the Soviet Union, Europe, and China, but this also has some problems. About 20-30 years ago we were admonished to restrict the use of coal because of environmental problems such as the destruction of land by strip mining, hazards to miners, and dangers posed by such effluents of burning coal as acid rain and carcinogenic hydrocarbons. We transformed our coal burning power plants into clean (gas) or low sulfur (oil) burning power plants to eliminate or reduce environmental hazards. As a result of the oil embargo in 1973, there has been a return to the use of coal for power plants, especially by indirect combustion (coal conversion to gas or liquid hydrocarbons) rather than by older combustion methods. When we suggest that coal in any form can be used as an alternative to burning oil, it should be remembered that environmental constraints will prevent its use over an extended period.

The combustion of fossil carbon in any form, but particularly coal, has created a problem, especially over the last 100 years and in an accelerated form in the last 20 years. Carbon that has been stored in the ground for several hundred million years and is suddenly used generates excess carbon dioxide. It is not possible to get the heat/energy values from the combustion of fossil carbon (oil or coal) without the production of CO_2. As a result, there has been a 7% increase in atmospheric CO_2 levels over the past 20 years, even from burning mostly oil, which has approximately two atoms of hydrogen for every atom of carbon. When coal is burned, however, there is less than one atom of hydrogen for one atom of carbon, with the result that roughly twice as much carbon dioxide per million BTUs is created than from the burning of oil.

It is possible to observe the annual increase in the CO_2 levels in the atmosphere from the data in Fig. 1, which is taken from a station at the top of Mauna Loa in Hawaii. This site was chosen because of

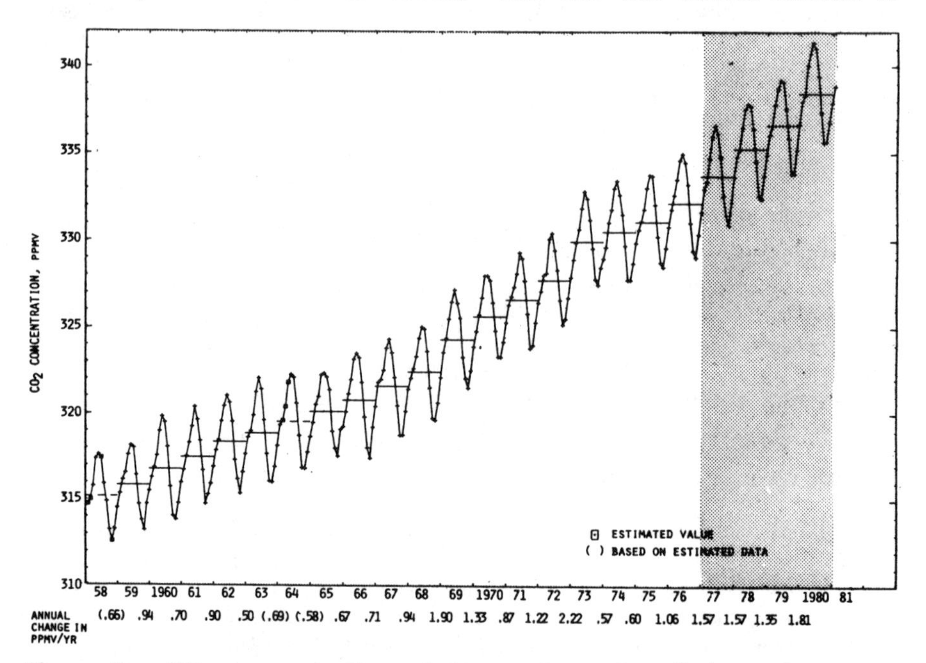

Figure 1. CO_2 concentration at Mauna Loa, Hawaii (data from C.D. Keeling et al., Scripps Institution of Oceanography).

its distance from urban and natural disturbances, so that the data obtained is representative of the actual atmospheric situation uncontaminated by manmade pollutants. The annual changes of the CO_2 levels in the atmosphere are clearly observable, with a rise in the winter and drop in the summer. Notice that the reduction in the summer is never as great as the increased concentration in the winter, resulting in a constant net CO_2 increase since 1958, from 315 to over 330 ppm.

We can actually extrapolate backward by another type of measurement, the ^{13}C content of tree rings that were laid down in 1860. From the ^{13}C deficiency we can estimate the total carbon dioxide that was in the atmosphere in 1860; it turns out to be about 290 ppm for that time. So, from 1860 to 1980 there has been an increase of about 15% atmospheric CO_2, and of that amount, half has been in the period 1958-1981. There has been a slight decrease in worldwide carbon-based fuel usage since the oil embargo period, but that is a relatively small perturbation on the overall rise of the CO_2 concentration.

Why is the CO_2 level important? Carbon dioxide is a peculiar gas. It is transparent to visible light. Approximately 99% of visible light is converted to heat when it strikes the surface of the earth after passing through the atmospheric CO_2 blanket. This heat is irradiated back into space, but the carbon dioxide is opaque to infrared light and absorbs it, re-reflecting some of it back down to the earth's surface. The CO_2 blanket thus acts as a one-way valve, letting heat into the surface of the earth but not allowing the heat to escape again into space, with a result that the earth's surface temperature rises (Smagorinsky, 1982; Macdonald, 1982).

Effect of Rising World Temperature

Let's discuss for a moment what the rising temperature of the earth may be expected to be and what the possible economic and social costs might be as a result. Estimates have been made by various means and the data show rising temperatures as a result of synthetic fuel use might be as great as 4 C, a very large change in the average global temperature of today. Even if we begin to use nonfossil replacement fuels, we still have the inertia of the CO_2 rise that has already begun and will continue. There is no way to stop the temperature rise that has already begun, and it will take 50-100 years before the increase levels off or even begins to fall again. We can be sure that if such temperature rises take place, there will be profound effects on agriculture, on human distribution, and on human societies that will have to adjust to a relatively very rapid change.

Several years ago the National Academy of Sciences organized a study on the effects of the increased CO_2 concentration in the atmosphere, and in 1982 they updated their findings (Smagorinsky, 1982). The 1982 report stated that the increase in temperature of 3 C postulated in 1979, which was predicted to accompany a doubling of atmospheric CO_2 concentration, needs "no substantial revision" at this time. The potential of increasing concentrations of atmospheric CO_2 to produce climatic effects is substantial, and the scientific uncertainties that surround the nature and extent of these effects require worldwide monitoring of temperature, ocean cycling, and other factors.

Is it possible to detect any consequence of the 15% rise in the CO_2 level that has occurred in the last 100 years? Can we extract this rise from the "noise" of the annual fluctuation in time and space of the temperature and weather in general? A number of efforts have been made to find "early warning signals" large enough to be believable but not so serious as to cause agricultural, economic, and social damage. If we wait until the evidence is unambiguously clear, until the temperature actually rises, it will be too late to alter our living patterns to depart from the inertial curve. We need some early warning signals to inform us about whether the temperature rise is happening and the order of magnitude of its occurrence (Hansen et al., 1981; Bryan et al., 1982). I have collected three or four different signals, and although they are independent observations, when taken together they appear to be unambiguous evidence that the heating effect is already beginning to show.

The simplest is the temperature history of some locality. The temperature change data for the Northern Hemisphere from 1880 to 1981 are shown in Fig. 2, indicating that 1981 was a peak year for warm weather. The rise in temperature from 1880 to 1980 is substantial, even allowing for the large "noise" level from year to year. This alone is not a good early warning signal because the annual fluctuations are too great to distinguish a trend. However, evidence has been accumulating that the ocean surface temperature is also increasing as a result of increased atmospheric CO_2 concentration, and an extrapolation has been made as to the time-dependent response of the ocean temperature

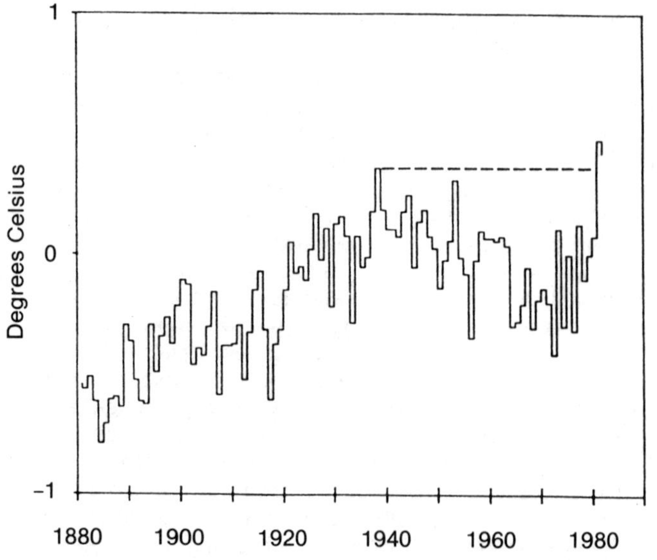

Figure 2. Northern Hemisphere temperature change (modified after P.D. Jones in Climate Monitor, 1982).

to a possible instantaneous doubling of CO_2 concentration (Fig. 3; Smagorinsky, 1982).

A more sophisticated piece of evidence of the rising global temperature concerns the polar ice caps. The accumulation of ice on the polar ice caps is an integrated effect unlike the fluctuations in annual temperature. Last year there was a satellite photograph of the South Polar Ice Cap, and that photograph, together with surface measurements by the U.S. and Soviet navies, has provided evidence that the extent of the South Polar Ice Cap has decreased in the last 100 years. In the last 7 years it has fallen from about 12-13 million km^2 to near 10 million km^2; in other words, 2 million km^2 of ice have disappeared (melted) from the South Polar Ice Cap. From temperature distribution data on the ice cap it is possible to state that the thickness of the remaining ice cap has also decreased (Kukla and Gavin, 1981). One can calculate how much ice has melted, and therefore, how much "new" water has come into the oceans. We can then deduce that the sea level should have risen and done so more rapidly in the last 20 years than before. The mean global sea level trend based on tide gauge measurements indicates that from 1800 to 1940 the rise was about 1 mm/year, while from 1940 to 1980 the rise has been 2-3 mm/year. That, I think, is unambiguous evidence, together with the satellite data and the temperature information for the Northern Hemisphere, that the heating effect (greenhouse effect) of the CO_2 increase is indeed occurring and will have its consequence on agriculture patterns throughout the globe in various way. Unfortunately, we do not as yet have good enough meteorological models to predict microclimatic changes (Schlesinger, 1982), i.e., to predict where the heat will be greatest, where rainfall will be increased or decreased, or where the agricultural patterns of the globe will change.

To give some clue as to what the surface change in the sea level might become we can examine a period about 50,000 years ago when the temperature was high and there was less polar ice. We find evi-

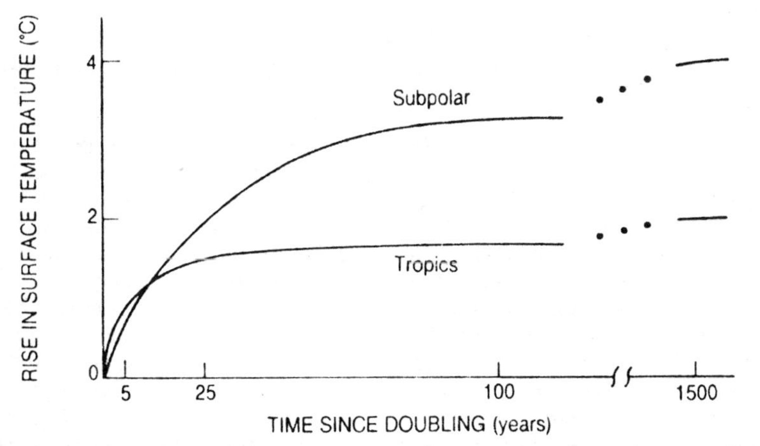

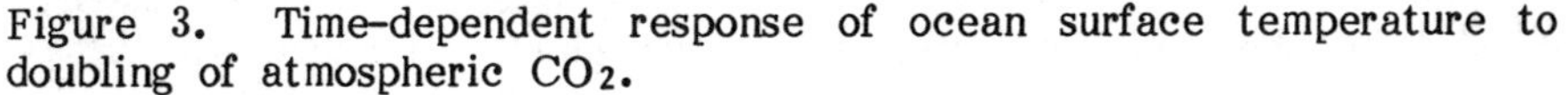
Figure 3. Time-dependent response of ocean surface temperature to doubling of atmospheric CO_2.

dence, at least on the eastern coast of the United States, that is unambiguous. It has been determined that most of Florida was totally under water, or marshland, and a good bit of the eastern coast was flooded or marshy from New Jersey to Florida. At that time the Antarctic ice covered less area than it does today. We currently are going in that direction once more, with the decreasing South Polar Ice Cap, increasing global temperatures, and other results of increased atmospheric CO_2 concentration.

What is the alternative? We cannot burn coal very long. If we do, we are headed toward major temperature changes. I believe that coal should not be used at all for environmental reasons, but the temperature change effects alone can and should limit its use.

PLANTS AS ENERGY SOURCES

We are left with only one alternative, and that is some way of harvesting the sun on an annual basis and not simply using the sun's harvest that has been stored in the ground for several hundred million years. As a first alternative, we must learn to harvest the sun annually, which means growing plants that will produce the kinds of materials we need (Calvin, 1983). We are already doing that today on a large scale in the form of forestry activities, but these are minor compared to what must be done to meet the fuel and material needs of the next generation. The fact that we can use green plants as a source of almost all the necessary fuel and materials is established through the photosynthetic carbon cycle itself.

Sugarcane—Puerto Rico

In some countries, energy agricultural practices have already begun. About ten years ago, just before the oil embargo, the Puerto Rican government constructed a very large power plant on the southern coast, intended to be fueled by oil. That became impractical, but Puerto Rico found an alternative to oil. It is possible to grow sugarcane there on a large scale, and research activities led the Puerto Rican government to encourage the planting of a special type of sugarcane, called "energy" cane. Normally, sugarcane is bred to reduce the fiber content and raise the sugar content so that processing of the cane for sugar can be economic. However, in Puerto Rico they realized that the fiber from the cane could be used to run the power plant. They therefore returned to an examination of some old clones that were poor in sugar and rich in fiber. These were planted experimentally a few years ago, leading to the development of energy cane (Fig. 4). The energy cane contains only about 8-9% sugar. Actually, the sugar produced per acre from the energy cane is about the same as in the standard cane (Gran Cultura), which is about 14-15%, but it contains about three times as much fiber (bagasse) which is the important part of the cane for fueling the power plants. By utilizing energy cane Puerto Rico now has both sugar and fiber (fuel) (Alexander, 1980).

Figure 4. Energy cane, Puerto Rico (Photo: Gene Elle Calvin).

Sugarcane—Brazil

When we first began looking for energy alternatives back in 1974, I knew there existed plants that were commercially producing hydrocarbons on a large scale. These were the *Hevea* rubber trees, mostly in Malaysia and Indonesia, but native to the Amazonian forest in Brazil, having been transferred from Brazil through Kew Gardens in London to the Southeast Asian plantations where they grow today. If *Hevea* is native to Brazil, belonging to the family Euphorbiaceae and making hydrocarbons, might there not be other members of that family, or even of that genus, which would produce hydrocarbons of lower molecular weight that might be useful as crude oil?

It was in Brazil on a trip in 1974 that we learned about the sugarcane industry and how it can be a self-contained, energy-efficient process. Brazil grows more sugarcane than any country in the world, and in 1974 they produced 10 million tons of raw sugar and 400 million liters of fermentation alcohol from the molasses of that sugar. As we traveled around Brazil looking for other species of Euphorbiaceae, we saw mostly sugarcane, and we visited mill after mill, finding that

Brazilians were harvesting the cane under conditions of maximum sucrose production.

If the Brazilians wanted to use alcohol as a possible source of liquid fuel and materials, the cane should be harvested for maximum fermentables rather than maximum sucrose. It was easy to determine that if the Brazilians waited for the sugarcane to ripen (for the glucose and fructose phosphate to dimerize to sucrose) some of the fermentables would be lost, because the dimerization is an energy-consuming process. However, if they harvested the sugarcane at the point of maximum fermentables, that is, a little earlier, they should have a higher alcohol yield. Indeed, the Brazilians now harvest a few weeks earlier and get 10–15% more alcohol per hectare.

In 1975 the Brazilian government made a decision to go into the alcohol-for-fuel program, what they call the Pro-Alcohol program. This encouraged the growing of cane for fermentables, which meant modifying the growing procedures, modifying the mill equipment to use the sugar juice directly, etc. The sugar mills themselves became self-supporting in energy, in that they had enough bagasse from the cane to generate the steam to run the crushers, to run the fermentation process, to distill the alcohol, and to fuel the trucks that collected the cane. The products from these mills were alcohol and excess electricity.

The Brazilians are now reaching the stage where they can use the excess electricity from the sugarcane operation to produce nitrogen fertilizer which is required for the soil of cane plantations. Also, potassium and phosphorus in the ash from the furnaces are available for these soils. The excess electricity from burning the bagasse can be used to electrolyze water, to make hydrogen, and to build small ammonia generators. The sugar plantations of Brazil therefore will be the first completely self-contained "energy plantations," which not only produce sugar, but also produce liquid fuels (alcohol) and petrochemicals from the fermentation process.

HYDROCARBON-PRODUCING PLANTS

Our trips to Brazil and elsewhere (Malaysia, Australia, Spain) were taken for the purpose of finding plants that could produce hydrocarbons of low molecular weight as alternative fuel and materials sources (Calvin, 1983). We continued to look for species in the family Euphorbiaceae, and we found a very large number of the genus *Euphorbia*, having roughly 2000 species, all of which produce a latex (similar to *Hevea* latex) containing one-third hydrocarbon, one-fifth protein, and water as the balance. Not only are there 2000 different species of *Euphorbia* growing worldwide and under all kinds of conditions and growth habits (tiny plants to large trees), but these growth habits and circumstances can be selected for the kind of land and climate that is available (McLaughlin et al., 1983).

One of the best candidates for hydrocarbon production is *E. lathyris* (commonly called the gopher plant), and plantations of this species have been developed in various parts of the world, specifically in the southwestern United States (Johnson and Hinman, 1978), California (Calvin, 1979), and Spain (Fig. 5). There are currently plantations of *E. lathyris* in other Mediterranean countries, Africa, the Canary Islands, and Australia (Coffey and Halloran, 1981). The reason for our choice of *E. laythyris* was its universal distribution throughout the United States, growing as far north as Montana, in the Southwest, and even on the East Coast. It requires only 12-15 inches of water annually and can grow in land that is not suitable for food production. The reason that *E. lathyris* might be one of the best candidates for an alternate energy crop comes from several factors. First, it is possible to grow this plant using irrigation on semiarid land not suitable for food or fiber production. Second, the plant grows in a 5-7 month period to size for harvest. Third, the extraction process is standard for the chemical industry. Also, in addition to the oil, the plant contains a substantial quantity of sugars, fermentable to alcohol. It seems that an annual yield of 6-10 barrels of oil per acre is achievable using wild seed. Agronomic and genetic improvements in the plant, of course, would increase that yield.

Because *E. lathyris* has been studied more than any other candidate for hydrocarbon production, data is available for such factors as genetic variability, irrigation requirements, process development, and analyses of the final products. There is no question that in spite of its widespread distribution *E. lathyris* may not be the best candidate for arid lands, as the water requirement is too high and the resulting products not economically competitive with crude oil. Efforts should be made to learn whether specialty chemicals from *E. lathyris* could be used for feedstock development in addition to those fractions which are similar to crude oil, thereby making growing *E. lathyris* more economically competitive and useful (Kingsolver, 1982).

Another species, *E. tirucalli*, a perennial that is common in Brazil, Africa, and Israel and grows prolifically in the southwestern United States, has been cultivated for its oil-producing quality, but it requires more water than *E. lathyris*. At the present time, *E. tirucalli* plantations are being developed in Okinawa and Thailand (Fig. 6), the latter effort under the direction of a Japanese chemical company.

There is another plant family, the Asclepiadaceae (milkweeds), which is widely distributed throughout the world. We have seen great areas of some of these plants (*Calotropis procera*) in Puerto Rico, and they are also being developed in the United States by a private firm as a possible candidate for liquid fuel production (Adams, 1982). Just recently I have learned of the existence of plantations of *C. procera* in Australia as well (Fig. 7). The planting of several thousand hectares is being sponsored by the Australian government to determine whether or not this plant would be a suitable candidate for fuel and materials production (L.R. Williams, personal communication). The chemistry of *Calotropis* and *E. lathyris* is similar, as are the extraction process and the various transformations involved.

The processing sequence to recover terpenoids and sugars from *E. lathyris* (Fig. 8) is almost the same for *E. tirucalli* and *C. procera* (Nemethy et al., 1981). This extraction procedure was worked out in the laboratory and has not been used on a pilot plant scale. The

Figure 5. *Euphorbia lathyris*, Spain.

Figure 6. *Euphorbia tirucalli*, Thailand (Yuko Chem. Co., Osaka, Japan).

process sheet is calculated for 1000 dry tons/day of *E. lathyris* material, which would yield 80 tons of crude oil and 200 tons of fermentable sugars. This in turn could produce 100 tons of alcohol. About 500 tons of bagasse is used to run the process, with a resulting 200 tons of bagasse available to distill the alcohol. The fermentation alcohol, which is a byproduct of processing the dried material from *E. lathyris*, is, of course, a starting point of an entire petrochemical industry.

The whole process is self-contained, as in the sugarcane situation, but the economics of processing *E. lathyris* and the products recovered from that effort are more conjectural. In 1980 I made calculations for a small miniplant, and the price that we would have to receive for the oil and alcohol to make the process economically feasible would be $40/barrel for the oil and $50/barrel for the alcohol. This type of return would make the process economic, including capital costs for construction of the extraction plant and payments to the farmer for growing the *E. lathyris* (which would be approximately $300–500/acre).

Studies of *E. lathyris* have shown that while it is possible to crack the material from this plant, as is done with crude oil, it might be more economic to determine whether this species and its products contain material more useful for chemical feedstock production, i.e., specialty chemicals that would not compete with the cheaper crude oil

Figure 7. *Calotropis procera*, Australia (Lyall Williams).

and that would have a unique situation in world markets (Kingsolver, 1982).

The crude oil from the *E. lathyris* has been converted, using special zeolite catalysts developed by the Mobil Corporation, to a suite of products similar to those from the standard catalytic cracking of petroleum (Weisz et al., 1979). These include olefins, paraffin, aromatics, and nonaromatics. This information confirms the desirability of the products of *E. lathyris* as possible raw materials to substitute for crude oil. Since these results were obtained, as a result of additional data on processing and cracking the material from *E. lathyris*, it now appears that a price of $100/barrel for the oil would be more realistic. This is still within reason, considering that it is only 2.5 times more than the 1982 price per barrel for crude oil from OPEC.

Isoprenoid Formation in "Oil" from Plants and Trees

The chemicals in plants that constitute the black oil are mostly triterpenes (C_{30}), sterols, and sterol esters. It is for this reason that the cracking pattern makes such desirable products. The biosynthetic

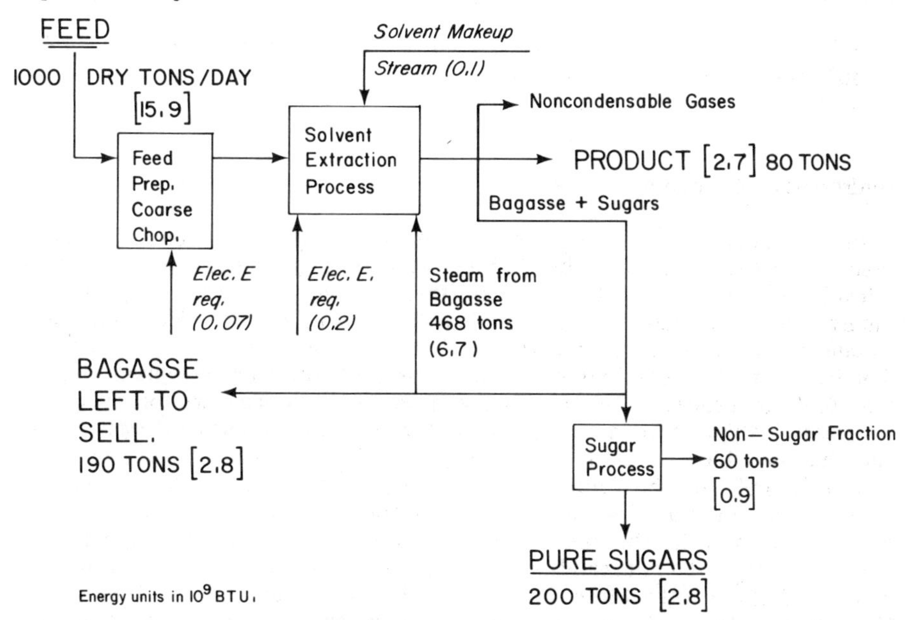

Figure 8. Conceptual processing sequence to recover terpenoids and sugars from *Euphorbia lathyris.*

route to the terpenes in the plant has not been worked out in great detail, but the route in *E. lathyris* is probably similar to that for rubber biosynthesis, except that the end products from the *E. lathyris* are lower molecular weight compounds. The biosynthetic route is from sugar via the glycolytic cycle to pyruvate, which is then built up to mevalonic acid and goes on to give isopentenylpyrophosphate (IPP). The IPP goes further to polymerize into a variety of isoprenoids. Normally, in *E. lathyris* the material goes on through the isoprenoid biosynthetic pathway to squalene (C_{30}), which is then folded up to make the C_{30} terpenoid alcohols. The latter constitute the greater percentage of the oil. The actual route is sugar to pyruvate to acetyl phosphate to mevalonic acid and on to dimethylallylpyrophosphate (DMAP) and isopentenylpyrophosphate (IPP). The isomeric C_5 compound (DMAP) drops off a pyrophosphate to create a carbonium ion, which takes an electron pair from the double bond of IPP to form a five-carbon to five-carbon link. The resulting carbonium ion then drops off the proton to give an identical structure (allypyrophosphate structure) which can go on, add another C_5 to make a C_{15} compound, then add another five-carbon piece to make a C_{20} compound. This process can continue on to polyisoprenes, essentially to rubber.

However, if the plant contains a cyclase enzyme, the carbonium ion can take an internal double bond and close at a cyclic C_{10} compound instead of adding another five-carbon group. In the case of *E. lathyris* and *E. tirucalli*, two C_{15} compounds come together to form squalene, a C_{30} open chain with no pyrophosphate which then folds up to form the triterpenes, closed C_{30} compounds, mostly sterols.

Would it be possible to improve the yield of the C_{15} compounds in the plants (which are the desired substances) if the process could be stopped at the C_{15} step? This would mean that no cracking operation would have to be performed on the materials since the C_{15} cyclic sesquiterpenes can be used directly as diesel fuel without further processing.

Hydrocarbon-Producing Trees

In our search for hydrocarbon-producing plants, we encountered a tree in Brazil that can be tapped and that produces sesquiterpenes (diesel oil) (Calvin, 1983). This tree belongs to the Leguminosae family, and there are several species growing in Brazil. We found one, *Copaifera langsdorfii*, growing in the Botanical Garden in Rio de Janeiro, but that particular species does not produce as much diesel-like fuel as others. In the Ducke Forest in Manaus we observed *C. multijuga*, which produces approximately 20 liters of sesquiterpene fuel material per tree in 24 hr every 6 months. The tree is harvested by drilling a hole in the trunk about 3 feet from the ground; the hole is about 2 cm in diameter and goes into the heartwood of the tree. A pipe is inserted in the hole and the oil drains out of the pipe into a bucket. After the material is "harvested," the hole is plugged with a bung, and six months later the tree will produce another 20 liters from the same hole. The oil comes not from the cambium, as does the rubber latex in *Hevea*, but from the 1-2-mm-diameter vertical pores in the heartwood.

The material from *C. multijuga* is a sesquiterpene that can be put directly into an automobile without any further processing or refining. An experimental plantation of *C. multijuga* has been started at the Amazon Research Institute at Manaus to show that the sesquiterpenes were "clean" enough to be used directly in automobile engines. They are also studying agronomic practices in connection with this species, with the hope of improving the yield, or perhaps being able to use more than one hole in each tree for diesel fuel production.

The oil from trees of the *Copaifera* species, called copaiba oil, has been analyzed by gas chromatography. Three main products are copaene, bergamotene, and carophyllene, all C_{30} compounds and all cyclic compounds with C_{15} (Wenninger et al., 1967).

The biosynthetic method by which the diesel oil from *Copaifera* is made is the same as that used by the *E. lathyris* up to the C_{15} step. The *Copaifera* cyclizes the C_{15} farnesyl pyrophosphate, that is, drops the phosphorus off to give the cyclic C_{15} compounds. One type of enzyme is responsible for the difference in the two end products of *C. multijuga* and *E. lathyris* and that is the farnesyl pyrophosphate cyclase enzyme. With *E. lathyris* this compound is dimerized while with *Copaifera* the material is cyclized, with many cyclic C_{15} products as a result.

Another family, the Pittosporaceae, contains trees that produce fruit (nuts) rich in terpenes. The terpenes rather than seed oils (glycerides) seem more desirable as fuel candidates. One of the first members of the Pittosporaceae which came to my attention is native to the Philippines and produces a fruit (nut) that is rich enough in terpenes to be used for illumination. This is *Pittosporum resiniferum* and the oil from the "petroleum nut" has been analyzed and found to contain a simple mixture of terpenes (Nemethy and Calvin, 1983).

With *P. resiniferum*, we can assume an average 10% oil yield and a maximum annual yield of 18 kg fruit/tree at a high density planting (2 m x 2 m). A maximum annual oil yield/acre would be 10 barrels. With smaller fruits the annual yield would be approximately 6.5 barrels/acre. The figure of 6.5 barrels/acre per year is the same as the yield for palm oil which is the best oil yield so far obtained for seed crops. Therefore, the *P. resiniferum* appears to be a very desirable fuel crop if high-yielding trees can be selected and planted at high density.

We then examined the fruit of a species of *Pittosporum* which is native to California (*P. undulatum*). When the fruits were extracted they were found to have the same terpenes (C_{10} compounds) as the ones native to the Philippines (Noble, 1979). Because this genus does grow prolifically in the Philippines and even in such places as California, these trees may turn out to be good candidates for fuel (terpene) production.

A comparison of biosynthetic routes might be of interest at this point. In the case of *E. lathyris*, the sequence is all the way to C_{15} compounds followed by dimerization to C_{30} materials. In the *Copaifera* (diesel tree), the sequence goes all the way to C_{15} followed by cyclization to the sesquiterpenes. For the *Pittosporum*, however, the route is to the C_{10} compounds, followed by cyclization to create monoterpenes in the fruit.

GENETIC ENGINEERING

The question now arises as to whether it is possible to transfer the gene for the production of sesquiterpenes from *C. multijuga* (or other *Copaifera* species) to such plants as *E. lathyris* (Calvin, 1983). It is not yet possible to grow *Copaifera* in the United States, and therefore it would be desirable to transfer the characteristics of sesquiterpene production in that species to another plant species that can be grown on an annual basis under existing conditions. There is only one enzyme, farnesyl pyrophosphate cyclase, involved in moving the C_{15} pyrophosphate to a cyclic C_{15} in *Copaifera*. This route would be preferable to having the compounds go all the way to C_{30} as is the normal route in *Euphorbia*. In other words, a single gene transplant from a donor cell of the *Copaifera* to the acceptor cell of a plant such as *E. lathyris* would be required. However, it is necessary to find a donor cell that has the genes for the desired enzyme, take out the messenger RNA, make a copy of the DNA, insert the copy DNA into a plasmid, clone the plasmid in *E. coli* and then, by means of the plasmids or other vector, insert the gene into a selected plant. Eventually, the piece of genetic information can be integrated into the nuclear gene of the transformed cell. This has been accomplished with bacteria but not yet with higher plants.

The regeneration of whole plants has not yet been accomplished through genetic engineering, although some plants, such as tobacco and carrot, have been regenerated from single cells. To accomplish whole plant regeneration by means of genetic engineering techniques it is necessary to have the two plant species available. *Copaifera*, as it does not grow in the United States, must be cultivated in greenhouse conditions. We managed to obtain some seeds of various species of *Copaifera* (*C. langsdorfii*, *C. lucens*, *C. multijuga*, and *C. officinalis*) and have tried to germinate them without too much success. We do have, however, some plant materials that we hope will provide us with the donor cells we need to place the enzymatic information into *E. lathyris*.

To accomplish gene insertion into the acceptor plant, however, it is necessary to prepare a plant cell culture. We have been able to take the callus tissue from *E. lathyris* and separate it into protoplasts. However, the protoplasts do not reproduce very well because they tend to aggregate. We have, however, separated single protoplasts from the callus and they do divide and form new callus. We have also regenerated whole plantlets from single protoplasts through such callus.

One alternative method to introduce new genetic material into plant protoplasts would be to use the technique of cell fusion; that is, take a cell that has the desired gene and fuse it with a cell that does not, then select for the fusion products (Redenbaugh et al., 1982). Eventually, there should be a cell into which the desired gene has been integrated. We have used a cell selection system and measured the process by flow cytometry. We took two sets of protoplasts from *E. lathyris*, staining one of them with rhodamine and one with fluorescein, both of which fluoresce in different regions of the spectrum. The two cell lines were mixed and a fusion agent was added so that the fused cells could be selected. The selection was done via flow cytometry, a method that passes the cells in front of a laser beam (Redenbaugh et al., 1982). Two photomultipliers on each side of the laser beam on the flow cytometer examine light of different wavelengths. As the cells go by, a charge is placed on them only if they have both types of fluorescence. That charge allows separation of the charged from the uncharged (fused from unfused) cells.

The next step in the process is to regenerate plants from the fused protoplasts and to examine the regenerated shoots to see if any of them make cyclic C_{15} compounds. Then we can begin to optimize the process of producing new plant material that synthesizes the sesquiterpene compounds for fuel and material uses.

CONCLUSION

The use of plants for fuel is not a new idea. Two different efforts were made over 40 years ago in this direction. The first was an attempt by the Italians in Ethiopia to use a member of the Euphorbiaceae, *E. abyssinica*, as a source of fuels. These plants grew prolifically throughout Ethiopia and the Italians built an extraction plant for the oils from these plants. However, time did not permit the development of this idea and the Italians abandoned the effort when they left Ethiopia in 1938. They had hoped to use this plant species, which contained oil, as the basis of a petroleum industry (Frick, 1983). In addition, the French in Morocco in 1940 used another species, *E. resiniferum*, and succeeded in obtaining oil from wild plants. Their yield was 3 metric tons of oil/hectare from wild plants (de Steinheil, 1941); they did not continue this effort, however.

It is clear that the use of plants to create hydrocarbon-like materials as a substitute for petroleum will become more important, especially in some of the less developed areas of the world that have land unsuitable for food production. Various efforts have been made—in Spain, Okinawa, Thailand, and Australia—toward this end, and attempts are underway to improve agronomic yields, develop small scale extraction plants, and learn more about the composition of the oil itself, using species of plants that heretofore have not been used for large-scale cultivation. In the United States efforts along these lines in the Southwest and California have shown that it would be possible to produce approximately 6-8 barrels of oil/acre annually on semiarid land.

What is now needed is a commitment by the agricultural community to an "energy" agriculture that would have long-term benefits for our country. However, it is my feeling that we will have to be shown by others that this type of development is feasible, before we begin to use this most important resource, the annually renewable green plant, for our own benefit.

ACKNOWLEDGMENT

The work described in this paper was supported, in part, by the Assistant Secretary for Conservation and Renewable Energy, Office of Renewable Energy, Biomass Energy Technologies Division, U.S. Department of Energy under Contract No. DE-AC03-76SF00098.

REFERENCES

Adams, R.P. 1982. Production of Liquid Fuels and Chemical Feedstocks from Milkweed. Institute of Gas Technology, Miami.

Alexander, A.G. 1980. The energy cane concept for molasses and boiler fuel. In: Symposium Proceedings on Fuels and Feedstocks from Tropical Biomass, San Juan, Puerto Rico.

Bryan, K., Komro, F.G., Manabe, S., and Spelman, M.J. 1982. Transient climate response to increasing atmospheric carbon dioxide. Science 125:57.

Calvin, M. 1979. Petroleum plantations for fuels and materials. BioScience 29:533.

______ 1983. New sources for fuel and materials. Science 219:24.

Coffey, S.D. and Halloran, G.M. 1981. Euphorbia: perspective and problems. In: Proceedings of National Conference on Fuels from Crops, Melbourne, Australia.

de Steinheil, P. 1941. L'Euphorbe resinifere, plante à caoutchouc et resine vernis. Rev. Gen. Caoutch. 18:55.

Frick, G.A. 1983. A new brand of gasoline. Cac. Succulent J. 10:60.

Hall, C.S. and Cleveland, C.J. 1981. Petroleum drilling and production in the United States: Yield per effort and net energy analysis. Science 211:576.

Hansen, J., Johnson, D., Lacis, A., Lebedeff, S., Lee, P., Rind, D., and Russell, C. 1981. Climate impact of increasing atmospheric carbon dioxide. Science 213:957.

Johnson, J.D. and Hinman, C.W. 1978. Oils and rubber from arid land plants. Science 208:460.

Kingsolver, B.E. 1982. *Euphorbia lathyris* reconsidered: Its potential as an energy crop for arid lands. Biomass 2:281.

Kukla, G. and Gavin, J. 1981. Summer ice and carbon dioxide. Science 124:497.

Macdonald, G.J. (ed.) 1982. The Long-Term Impacts of Increasing Atmospheric Carbon Dioxide Levels. Ballinger Pub., Cambridge, Massachusetts.

McLaughlin, S.P., Kingsolver, B.E., and Hoffmann, J.J. 1983. Bioenergy production in arid lands. Econ. Bot. (in press).

Nemethy, E.K. and Calvin, M. 1983. Terpenes from Pittosporaceae. Phytochemistry (in press).

Nemethy, E.K., Otvos, J.W., and Calvin, M. 1981. Natural production of high energy liquid fuels from plants. In: Fuels from Biomass (D.L. Klass and G.H. Emert, eds.) pp. 405-419. Ann Arbor Science Pub., Ann Arbor, Michigan.

Noble, B.F. 1979. More secrets of the petroleum nut. Canopy 4:6.

Redenbaugh, K., Ruzin, S., Bassham, J.A., and Bartholomew, J.C. 1982. Characterization and separation of plant protoplasts via flow cytometry and cell sorting. Z. Pflanzenphysiol. 107:65.

Schlesinger, M.E. 1982. CO_2-Induced Climatic Warming: A Review of Model Research and Prospectus for First Detectability. Report No. 36, Climatic Research Institute, Oregon State University, Corvallis.

Smagorinsky, J. 1982. Carbon Dioxide and Climate: A Second Assessment. National Academy of Sciences, Washington, D.C.

Weisz, P.B., Haag, W.O., and Rodewald, P.G. 1979. Catalytic production of high grade fuel (gasoline) from biomass compounds by shape-selective catalysis. Science 206:57.

Wenninger, J.A., Yates, R.L., and Dolinsky, M. 1967. Sesquiterpene hydrocarbons of commercial copaibe and American cedarwood oils. J. Am. Oil Chem. Soc. 50:1304.

SECTION I
Overview

CHAPTER 1
Major Food and Energy Crops: Trends and Prospects

R.L. Meyer and *N. Rask*

This chapter provides economic insights into issues associated with consumption, production, and distribution of major world food and energy crops. The chapter includes information that identifies the major food crops, shows trends in production and yields, identifies crops and geographic regions lagging in production and yields, charts recent trade patterns, and discusses crops and regions where competition between food and energy crops already occurs or where it may develop. Systematic data are not readily available for all crops. Furthermore, detailed data and analysis of individual crops would detract from the central objective of this chapter. Therefore, we have chosen to include crops that have greatest value in terms of current production and importance in human diets, and sometimes aggregate them when they appear to be reasonably good substitutions for one another. Some crops we have not included are undoubtedly important in isolated cases, especially when produced and consumed on the farm. They do not, however, play an important role in meeting total consumption needs at present, nor are they likely to do so in the near future. Since they are of lesser importance in the marketplace, it is likely they will continue to be less well researched, and whatever potential they may have will continue to be underestimated.

OVERVIEW OF THE WORLD FOOD SITUATION

The post-World War II world food situation can be characterized as a continuation of long-term trends mixed with periods of marked change and uncertainty, especially in the 1970s, with unfavorable weather, political upheaval, and policy changes, including sharp petroleum price increases. This section highlights and summarizes some of the key features of the situation. Subsequent sections will treat selected issues in more detail.

The period 1950–1980 has been one of strong and steady growth in world food production and consumption. The compound annual growth rate in per capita production for the entire period was 0.85 (O'Brien, 1981, p. 4). By the early 1970s, world food intake per capita had increased to 108% of the minimum FAO standard for normal activity and good health. At no time in history have so many people been so secure in their food supply. Paradoxically, there have never been so many poor, malnourished, and hungry people as there are in the world today, because of the uneven distribution of food, income, and agricultural resources among and within countries. These divergent trends, and the social and political problems they imply, are a source of great concern and debate.

The growth in agricultural production has been broad based. Worldwide production grew at a compound annual rate of 2.8% from 1950 to 1980 (O'Brien, 1981, p. 11). Developing nations and centrally planned economies actually performed better than developed economies during this period. A disturbing note, however, was that the production growth rate during the 1972-1980 subperiod declined to 2.3%. This decline in growth rate was observed in most regions, with the exception of the United States which brought reserve land back into production. Over one-third of the postwar gain in world food production has been due to expansion in resources, particularly arable land, committed to production. The remaining two-thirds was due to productivity gains brought about through technological advances in plant varieties and improved supplies of chemical fertilizers and other inputs. The pattern of productivity change in food production has been mixed. Some developing regions have performed as well as developed regions. Others have not. Wide weather fluctuations make it difficult to assess recent yield trends accurately, but it appears that productivity growth has slowed in the 1970s and is subject to substantial variation from year to year.

The relatively inexpensive gains in production through expansion in area are now becoming limited. Projected area expansion in the early 1980s is expected to be less than half the average increases during the postwar period. Future expansion in many countries will occur on fragile land, risking erosion and environmental damage. Technologically induced gains through increased use of fertilizer, pesticides, and irrigation will be more expensive in the future as a result of higher energy costs. In addition, some land that is now reserved exclusively for food production will be devoted to energy crops, thereby putting further pressure on food supplies. These energy-related impacts underscore the need for research on productivity-increasing technology, especially technology that is energy efficient.

The developing countries represent the areas of greatest concern regarding food production. They have the highest rate of population growth, much of their food production is consumed on the farm and in rural areas, their people have the lowest incomes and the least reserves with which to withstand adversity, and they have the least foreign exchange needed to buy food to compensate for short- or long-term production shortfalls. Asia and Latin America have generally done quite well. Sub-Saharan Africa is the only region in the world where per capita food production actually declined in the past two decades (Christensen et al., 1981). The 1978 index of per capita food production in Asia and Latin America had risen to about 115% of the 1961-1965 average, but had fallen in sub-Saharan Africa to less than 80%. Of the 38 low-income countries classified by the World Bank, 33 had less than the average daily per capita calorie requirement in 1977 (World Bank Development Report, 1980, p. 152); 22 of the 33 were African countries. When the uneven distribution of food within a country is considered, 50% or more of these populations may receive insufficient food.

A most striking feature of the world food economy concerns food trade. Because of its production problems, Africa has turned increasingly to world markets. The volume of its food imports has risen substantially, but because of price increases, the value of cereal imports increased a whopping sixfold from 1970 to 1978 (Christensen et al., 1981). Poor countries must increasingly compete with demand for food from middle- and high-income countries, where demand also outstrips domestic production but income is high enough so they can turn to imported supplies. More and more countries have shifted from net food exporters to net food importers. The United States has emerged as the largest single food exporter. Today almost half of total world cereal exports originate in the United States. A large portion of these exports go to Western Europe and Japan. Almost three-fourths of total world agricultural imports go to Europe and Asia.

High population birth rates in many countries assure a continual increase in the number of people to feed. Preferences for animal protein assure a continual increase in the demand for meat and the associated feedgrains as incomes rise around the world. The combined effect of greater populations and incomes will press hard on future food supplies. Furthermore, unconventional demand for food and food production resources is emerging in the form of alcohol and biomass use for energy. At the same time excess capacity has been wrung out of U.S. agriculture, and many other countries are facing the limit of their agricultural frontier. Thus the general peaking out of yield increases is a matter of deep concern because productivity increases must be the most important source of future growth in food supply. With this general summary, we now turn to some of the details.

FOOD CONSUMPTION PATTERNS

A discussion of world food crops must begin with an analysis of consumption patterns which determine those crops considered most important by the consumer. The average per capita world calorie intake per day in the mid 1970s was about 2600 (Table 1). North America had the highest level with over 3500, while Africa and Asia were in the

Table 1. Per Capita Calorie Consumption per Day by Region, 1975-1977

REGION	SOURCE		
	Total	Vegetable	Animal
World	2590	2149	441
Africa	2308	2140	167
Asia	2276	2077	200
North America	3519	2207	1312
Latin America	2563	2107	456
Europe	3410	2315	1095
USSR	3443	2505	938
Oceania	3204	2026	1177
Developed ME[a]	3329	2265	1063
Developing ME[a]	2219	2032	187
Centrally planned	2721	2259	461

Source: FAO Production Yearbook, 1979.

[a]Market economics.

2200-2300 range. Developed economies on average consumed about 3300 calories, while developing economies consumed only 2200. Worldwide, over 80% of the calories consumed are obtained from vegetable sources. The data in Table 1 show the well-known relation that the proportion of calories from animal sources increases as total caloric intake rises. Of the calories consumed in developed regions, 30-40% came from animal products compared to only 8% for developing regions.

Cereals are the most important single source of both calories and protein in the world (Table 2). Wheat and rice are the most important cereals. Together they provide almost 40% of total calories and over 30% of the protein. Maize is more important as a feed grain than as a direct source of food. Millet and sorghum are particularly important for human consumption in China, India, and parts of Africa. Roots and tubers are a staple food for many populations. The category of pulses, nuts, and oilseeds is particularly important as a supplementary source of protein for people with limited access to animal protein.

The distinction between food crops and animal feed crops varies by country. Wheat is used primarily as food in most countries, but it is an important livestock feed in the USSR and some European countries. Cereals also play an important role because they can be fairly easily stored and transported. Roots, tubers, vegetables, and fruits, on the other hand, tend to be bulky and perishable. Thus they tend to be consumed close to where they are produced. Only high-income countries have sophisticated methods for processing, transporting, and storing these commodities.

Table 2. Relative Importance of Food Groups in Average World Daily per Capita Intake

CATEGORY	AVERAGE WORLD DIET	
	Calories (%)	Proteins (%)
Cereals	52	47
Wheat	(19)	(20)
Rice	(19)	(13)
Maize	(6)	(6)
Millet and sorghum	(5)	(5)
Others	(3)	(3)
Roots and tubers	8	5
Sugar and sugar products	9	<1
Pulses, nuts, and oilseeds	5	12
Vegetables and fruits	4	4
Fats and oils	8	<1
Animal products	14	32
Total	100	100

Source: University of California, 1974, p. 36.

ALLOCATION OF LAND

The amount of land devoted to various crops reflects consumption requirements and market demand. Farmers allocate their land on the basis of expected returns, either in terms of value of product sold or proportion of household consumption needs met. The smaller the landholding and the poorer the farmer, the more important will be the consumption objective. Economists frequently argue that subsistence farmers use a safety-first approach to management: The household first attempts to satisfy its consumption requirement before producing for the market.

It is also clear, however, that area planted to crops and crop mix are related to agronomic requirements of crops and the quantity and quality of resources available to the farmer. For example, farmers with access to irrigation water produce irrigated rice. Farmers in the dry highlands produce wheat. Potatoes predominate in poor soils with cool temperatures. Thus farm households adjust their production and household consumption patterns to their production conditions. Changes in both occur, albeit at a fairly slow rate, when new crops emerge which are adaptable to previously inhospitable environments. Therefore data on area planted or harvested are related to farmers' perception of value of production for those crops that appear suited to their production conditions. Diets and consumption patterns are shaped accordingly.

Table 3 reports the worldwide distribution of major categories of crops. Over 700 million hectares of cereals were harvested in 1979. This area represented about one-half of the total arable land and land in permanent crops for that year. The second most important category was oil crops, utilizing about 10% of the land. Pulses represented about 5% and roots and tubers about 3.5%. The other categories represented less than 2% each.

Table 3. Area Harvested for Major Crops in the World, 1979

CATEGORY OF CROPS[a]	AREA HARVESTED (1000 ha)	PERCENT
Cereals	726,175	50.1
Oil crops	144,359	10.0
Pulses	71,355	4.9
Roots and tubers	49,833	3.4
Vegetables and melons	23,202	1.6
Sugar crops	22,077	1.5
Coffee and tea	16,483	1.1
Tobacco and fiber	9,449	0.7
Other crops, unharvested and unused land	385,929	26.7
Total arable land and land in permanent crops[b]	1,448,862	100.0

Source: Calculated from data in FAO Production Yearbook, 1980.

[a] These categories include the following crops:

Cereals:	Wheat, rice, barley, maize, rye, oats, millet, sorghum
Oil crops:	Soybeans, groundnuts, castor beans, sunflower seed, rapeseed, sesame seed, linseed, safflower seed, seed cotton
Pulses:	Dry beans, dry broad beans, dry peas, chickpeas, lentils
Roots and tubers:	Potatoes, sweet potatoes, cassava
Vegetables and melons:	Cabbages, artichokes, tomatoes, cauliflower, pumpkins, cucumbers, eggplants, chilies and peppers, dry onions, garlic, green beans, green peas, carrots, watermelons, cantaloupes, grapes
Sugar:	Sugarcane and beets
Coffee and tea:	Green coffee, cocoa beans, tea
Tobacco and fiber:	Tobacco, flax, hemp, jute, sisal, cotton.

[b] This value includes all crops, land planted but not harvested, and idle land.

Important regional differences exist in the allocation of land (Table 4). Overall, the crops included in these categories represent a larger proportion of total area reported for developing countries than for developed countries. Cereals occupy about 55% of the area in developing countries compared to 45% for developed countries. This trend also exists for pulses, roots, and tubers. Together these three categories represent relatively more important foodstuffs in lower-income countries. Consumers shift toward other foods as incomes rise and animal protein becomes more important than vegetable protein in the typical diet. Africa and Asia show a relatively higher proportion of land in pulses, roots, and tubers and this is consistent with what is know about diets in those regions. About two-thirds of the area in Asia is devoted to cereals. This region is the most intensively cropped, and double and triple cropping are found in many irrigated areas. The Western Hemisphere and Oceania are at the other end of the distribution where only 30–40% of the land is in cereals.

Oil crops are the second most important category in all regions except for Europe. They are particularly important in the Western Hemisphere and Asia. The composition of crops in this category, however, varies widely. Soybeans represent over half the area in oilseeds for developed countries as a group and about 70% of the area in North America alone, but only about 25% in developing countries. At the other extreme soybeans represent only about 3% of the area in Africa. Groundnuts, on the other hand, represent only 2% of the area in developed countries, but 20% for developing countries as a group and almost 50% of cultivated land area in Africa.

Vegetables and melons represent a heterogeneous category and the data are probably more suspect because so much production is consumed on the farm. A significantly larger proportion of land in Europe is dedicated to this category, in large part because grapes alone represent 70% of the total.

Sugar crops are somewhat more important for developing countries than developed countries but the crop mix is quite different. About 95% of the sugar area in developing countries is in the form of cane, while 90% in developed countries is in the form of beets. Asia and Latin America reported about the same amount of total area in sugar crops but the proportion of total crop area is much less in Asia.

Coffee, cocoa, and tea are tropical crops; therefore they are important only in developing countries. There is specialization by region. Latin America reported almost 60% of the total world coffee area, Africa reported over 70% of the cocoa bean area, and Asia reported 85% of the total area in tea.

The tobacco and fiber crops are also somewhat concentrated in their worldwide distribution. Over half of the area harvested in these crops was in Asia, which accounted for about half the tobacco and hemp and 95% of the total jute area. The USSR reported about 70% of the flax, while Latin America and Africa each reported about half the sisal area.

TRENDS IN PRODUCTION AND YIELDS OF SELECTED CROPS

Data in Tables 5 and 6 report trends in crop production and yields. The data in the two tables need to be analyzed together because crop

Table 4. Proportion of Land Area Devoted to Major Crops by Region, 1979

CATEGORY OF CROPS[a]	REGION								
	United States & Canada	Europe	Oceania	USSR	Africa	Asia	Latin America	All Developed Countries[b]	All Developing Countries[c]
Cereals	38.9	49.8	34.5	52.3	39.6	68.6	29.9	45.3	54.2
Oil crops	15.5	3.0	1.1	4.2	7.0	12.1	16.3	8.5	11.2
Pulses	1.0	2.0	0.4	2.2	6.4	9.5	4.4	1.3	8.0
Roots and tubers	0.4	4.2	0.6	3.0	6.8	4.2	3.1	2.1	4.6
Vegetables and melons	0.4	6.1	0.3	1.3	0.8	1.5	1.4	2.0	1.2
Sugar crops	0.2	2.6	0.7	1.6	0.5	1.4	5.2	1.3	1.7
Coffee and tea	<0.1	<0.1	0.3	<0.1	3.7	0.5	5.7	<0.1	2.1
Tobacco and fiber	0.1	0.6	<0.1	0.6	0.4	1.1	0.8	0.4	0.9
Other	43.5	31.7	62.1	34.8	35.6	12.0	33.2	39.1	16.1

Source: FAO Production Yearbook, 1979.

[a]Categories as in Table 1.

[b]Includes Canada, United States, Europe, USSR, Australia, New Zealand, Israel, Japan, South Africa.

[c]All other countries not included in .

yields are increasingly important in explaining total production as it becomes clear that more and more countries are reaching the limit of their good agricultural land. Consider the example of wheat, the most important crop in terms of world crop area. The total area in wheat worldwide expanded from 173 to 205 million hectares from 1950 to 1960, an increase of 18.3% (University of California Food Task Force, 1974 p. 63). The percentage increase for the period 1960–1970 was only 6.3% and from 1970 to 1980 the increase was 13.9%. For rice, the respective percentages were 16.8, 10.9, and 11.1. For maize, they were 17.9, 3.1, and 13.4. This decline in the rate of expansion in area during the decades of the 1950s, 1960s, and 1970s suggests that a larger proportion of the increase in future production must come from increased yields even though some areas will be able to expand area cropped, in part by increasing double and triple cropping where irrigation, fertility, and other factors are favorable.

Tables 5 and 6 present data for 10 important crops from 1950 to 1980 for the same regions as reported in Table 4. Unfortunately the data did not report the division of developed and developing countries for 1950 and 1960. Another limitation is that the data for 1980 are just for that year, while data for 1950, 1960, and 1970 represent averages for 3 or 4 years around those dates. It is possible that 1980 was an unusual year for a certain crop or region and, therefore, may not represent well what occured in the decade of the 1970s.

World wheat production increased by more than 150% from 1950 to 1980, rising from about 170 to more that 440 million metric tons. The rate of increase was particularly high in Asia which produced about 30% of the world's wheat in 1980, surpassing all other regions. This was due to a 40% increase in area and an even better yield performance. Average world yields almost doubled rising from 990 kg per hectare to almost 1900 kg from 1950 to 1980. This yield increase represented three-fourths of the total increase in world wheat production. Average yields in developed countries currently appear to be about 500 kg higher than in developing countries. Yields in Europe have grown most rapidly over this 30-year period and in 1980 were almost double the world average. Oceania, Africa, and Latin America are the only three areas where yields did not double in this 30-year period. Current yields in Africa are just a bit more than 50% of the world average.

Total rice production in this period grew at a rate only slightly slower than wheat, rising from 167 to 400 million metric tons. However, area and yield increases were even more evenly matched than with wheat. Area increased 40%, while average yields increased 70% from 1630 kg to 2750. Asia produces over 90% of the world's rice so it greatly influences the world data. Asian yields doubled from 1950 to 1980, but the rate of increase in the last decade appeared to slow down compared to the first two decades. This pattern of yield increase also existed in other regions but average yields are 50%–75% higher. If the yield of developing countries represents a reasonable yield target, Asia still has a long way to go to reach its potential. The regions of Africa and Latin America have much less area but also have made only modest progress in raising yields.

Table 5. Production Trends by Geographic Region for Major Crops (1,000,000 Metric Tons)

COMMODITY	YEAR[a]	REGION									
		World	United States and Canada	Europe	Oceania	USSR	Africa	Asia	Latin America	All Developed Countries[b]	All Developing Countries[c]
Wheat	1950	171.2	44.5	41.2	5.3	35.8	4.6	37.3	8.0	NA	NA
	1960	254.3	48.4	59.4	8.5	64.2	6.3	55.9	11.6	NA	NA
	1970	329.3	53.9	72.9	9.4	92.8	8.0	80.3	12.0	231.2	98.1
	1980	444.5	83.6	98.5	11.1	98.1	8.6	130.0	14.6	293.6	150.9
Rice	1950	167.5	1.9	1.3	<0.1	0.2	3.6	158.0	4.6	NA	NA
	1960	253.1	3.1	1.5	0.2	0.4	5.7	233.2	9.0	NA	NA
	1970	311.7	4.0	1.8	0.3	1.3	7.3	286.1	11.8	23.6	288.1
	1980	399.8	6.6	1.8	0.7	2.8	8.4	362.9	16.5	24.0	375.8
Maize	1950	139.9	74.3	12.9	0.1	5.8	9.2	21.8	15.6	NA	NA
	1960	216.4	95.5	25.7	0.2	13.1	16.0	37.8	28.1	NA	NA
	1970	293.8	125.1	38.6	0.3	10.0	22.1	60.7	46.9	180.7	113.0
	1980	392.2	174.3	53.5	0.3	9.7	27.2	83.1	44.0	248.1	144.1
Millet and sorghum	1950	47.4	3.9	0.1	0.1	1.8	12.7	28.2	1.0	NA	NA
	1960	74.6	13.9	0.2	0.3	2.7	19.2	35.9	2.5	NA	NA
	1970[d]	29.4	<0.1	<0.1	<0.1	2.5	9.7	17.0	0.2	2.6	26.8
	1980[d]	28.9	<0.1	<0.1	<0.1	2.0	10.2	16.5	0.2	2.1	26.8
Potatoes	1950	247.6	12.5	130.2	0.6	80.2	1.3	18.1	4.7	NA	NA
	1960	282.6	14.5	138.3	0.8	81.6	2.0	38.1	4.6	NA	NA
	1970	276.0	16.8	126.7	1.1	93.7	2.9	25.3	9.4	242.5	33.5
	1980	225.7	16.2	91.5	1.2	66.9	4.9	34.4	10.6	179.9	45.8

Sugarcane	1950	278.1	13.2	0.3	7.6	—	15.3	89.1	153.2	NA	NA
	1960	475.0	19.6	0.4	14.9	—	30.8	193.4	215.8	NA	NA
	1970	573.7	21.4	0.5	20.6	—	47.1	223.3	260.7	56.5	517.2
	1980	730.7	25.6	0.3	28.5	—	57.2	256.9	362.2	66.3	664.4
Cassava	1950	50.6	—	—	<0.1	—	25.2	9.7	15.6	—	NA
	1960	75.0	—	—	0.1	—	30.7	18.5	25.8	—	NA
	1970	96.8	—	—	0.2	—	38.4	22.9	35.2	—	96.8
	1980	122.1	—	—	0.2	—	46.8	43.6	31.6	—	122.1
Soybeans	1950	15.9	7.4	<0.1	—	0.2	<0.1	8.3	<0.1	NA	NA
	1960	32.5	19.7	<0.1	—	0.4	<0.1	11.8	0.5	NA	NA
	1970	44.8	31.4	0.1	<0.1	0.5	<0.1	12.6	2.0	32.2	12.6
	1980	83.5	50.2	0.6	<0.1	0.5	0.3	12.0	19.9	51.6	31.9
Dry beans	1950	6.8	0.8	0.8	<0.1	<0.1	0.7	2.6	1.9	NA	NA
	1960	9.8	0.9	0.9	<0.1	<0.1	0.7	3.9	3.2	NA	NA
	1970	12.4	0.9	0.8	<0.1	<0.1	1.1	5.6	3.9	2.0	10.4
	1980	14.7	1.3	0.7	<0.1	<0.1	1.3	7.3	4.1	2.3	12.4
Tomatoes	1950	12.3	4.0	3.9	0.1	<0.1	0.7	1.2	0.9	NA	NA
	1960	22.1	5.5	8.2	0.2	2.5	1.8	2.5	1.7	NA	NA
	1970	34.8	5.9	10.9	0.2	4.2	3.2	7.5	3.0	22.3	12.4
	1980	50.2	7.2	14.6	0.3	6.2	5.1	11.9	4.9	29.8	20.4

Source: FAO Production Yearbooks, 1971 and 1980.

[a]1950 = 1948-1952; 1960 = 1961-1965; 1970 = 1969-1971; 1980 = 1980

[b]Includes Canada, United States, Europe, USSR, Australia, New Zealand, Israel, Japan, South Africa.

[c]All other countries not included in [b].

[d]Millet only.

Table 6. Yield Trends by Geographic Region for Major Crops (100 kg/ha)

COMMODITY	YEAR[a]	REGION									
		World	United States and Canada	Europe	Oceania	USSR	Africa	Asia	Latin America	All Developed Countries[b]	All Developing Countries[c]
Wheat	1950	9.9	11.6	14.7	11.3	8.4	7.0	8.2	10.5	NA	NA
	1960	12.1	15.8	20.8	12.4	9.6	8.2	9.0	14.2	NA	NA
	1970	15.8	20.5	26.3	12.0	14.2	9.7	12.4	13.8	17.9	12.3
	1980	18.7	21.0	37.6	9.6	15.9	10.2	16.3	14.6	20.8	15.7
Rice	1950	16.3	25.6	43.0	31.1	14.5	12.9	14.4	17.0	NA	NA
	1960	20.5	43.7	46.5	46.4	24.6	18.0	17.9	17.2	NA	NA
	1970	23.8	50.9	46.1	58.4	35.7	18.3	24.1	17.2	52.0	22.8
	1980	27.5	49.4	48.9	48.1	42.2	17.1	28.0	19.8	49.4	26.7
Maize	1950	15.8	24.9	12.4	18.0	13.1	8.6	8.6	10.8	NA	NA
	1960	21.7	41.7	22.8	21.0	22.3	11.2	11.1	12.2	NA	NA
	1970	25.4	51.6	33.7	29.1	27.6	11.7	19.3	14.8	40.4	16.0
	1980	30.0	57.1	44.8	46.3	27.7	12.2	22.4	17.2	47.7	18.3
Millet and sorghum	1950	5.1	12.6	9.1	13.0	4.5	5.7	4.0	8.7	NA	NA
	1960	7.2	28.3	16.2	14.1	7.0	7.4	4.8	13.8	NA	NA
	1970[d]	6.7	—	13.0	9.5	8.8	6.3	6.6	10.5	8.9	6.5
	1980[d]	6.7	—	17.3	11.8	6.8	6.0	7.1	10.3	6.9	6.6
Potatoes	1950	109.0	155.0	137.0	102.0	94.0	57.0	80.0	59.0	NA	NA
	1960	119.0	216.0	160.0	158.0	95.0	36.0	95.0	73.0	NA	NA
	1970	137.4	242.7	176.7	203.9	116.9	79.5	94.8	84.8	150.2	85.0
	1980	125.2	281.3	161.7	260.0	96.5	82.7	110.1	96.8	134.4	98.6

Sugarcane	1950	422.0	780.0	644.0	584.0	—	552.0	348.0	443.0	NA	NA
	1960	492.0	895.0	693.0	694.0	—	624.0	447.0	484.0	NA	NA
	1970	539.8	—	721.3	747.2	—	677.5	499.1	545.2	813.6	520.7
	1980	556.7	—	670.0	789.6	—	649.2	488.5	585.7	784.1	541.1
Cassava	1950	74.0	—	—	83.0	—	62.0	70.0	111.0	NA	NA
	1960	84.0	—	—	112.0	—	67.0	82.0	127.0	NA	NA
	1970	88.9	—	—	109.7	—	66.4	92.9	135.1	—	88.9
	1980	87.7	—	—	110.7	—	64.1	113.4	114.0	—	87.7
Soybeans	1950	10.0	14.3	5.7	—	4.3	4.9	7.3	12.0	NA	NA
	1960	11.5	16.3	7.5	10.0	4.7	6.7	7.3	11.5	NA	NA
	1970	13.9	18.3	10.8	11.1	6.1	4.1	8.6	12.8	17.6	9.0
	1980	15.9	18.1	11.8	15.8	5.4	9.5	10.3	17.4	17.6	13.7
Dry beans	1950	4.4	11.9	2.2	10.0	5.4	5.3	3.1	5.6	NA	NA
	1960	4.5	14.4	2.2	10.0	13.0	6.0	3.3	5.7	NA	NA
	1970	5.2	14.2	3.6	3.2	21.3	5.8	4.5	6.1	6.3	5.1
	1980	5.6	16.1	4.4	7.0	18.0	6.1	5.1	5.4	9.0	5.2
Tomatoes	1950	131.0	150.0	165.0	153.0	—	135.0	115.0	85.0	NA	NA
	1960	182.0	285.0	214.0	239.0	137.0	136.0	138.0	117.0	NA	NA
	1970	179.7	339.1	247.1	241.0	143.4	128.8	132.2	142.0	236.3	125.6
	1980	209.9	424.5	307.0	247.8	157.7	136.8	168.8	186.7	274.6	156.1

Source: FAO Production Yearbooks, 1971 and 1980.

[a]1950 = 1948–1952; 1960 = 1961–1965; 1970 = 1969–1971; 1980 = 1980.

[b]Includes Canada, United States, Europe, USSR, Australia, New Zealand, Israel, Japan, South Africa.

[c]All other countries not included in [b].

[d]Millet only.

Maize ranks third after wheat and rice in total cropped area. Total maize production grew even faster, however, as it almost tripled in this 30-year period. The U.S. and Canadian share of the total production has fallen from 53 to 44%. One of the fastest growing areas is Asia where production increased almost four-fold. As is the case with wheat, world average maize yields doubled during this period. But the yield gap between developed and developing countries is even greater with rice, and appears to be increasing. Average yields are approaching 5000 kg in developed countries but have not yet reached 2000 kg in developing countries. Once again, the situation is worst in Africa where yields rose by 40% but are still only 40% of the world average of 3000 kg. Asia and Latin America have roughly the same area in maize, but the rate of increase has been faster in Asia and 1980 yields were 500 kg higher.

It is useful to contrast the pattern noted for maize with that of wheat and rice. Wheat and rice are the world's two most important food grains and demand for them is determined largely by population increase and the rise in income experienced by populations that use them as their major source of calories. Maize, on the other hand, is used primarily as a feed grain for livestock and poultry. Since the production of meat is an inefficient use of grains (to produce 1 lb of meat may require from 4 to 15 lb of grain), a small rise in animal product consumption may require a substantial increase in feed grain production. Therefore as incomes rise and the demand for livestock products grows, the demand for feed grains rises proportionately much faster than the demand for grain for human consumption. The rapid expansion in maize production during the 1970s represents a response to this demand for feed grains.

The data on millet and sorghum are combined in 1950 and 1960, but millet is reported alone for 1970 and 1980. Most of the production occurs in Africa and Asia. The combined yields greatly increased in the first decade, but there was no gain in the millet yield in the last decade. Once again, average yields were highest in developed countries and Africa registered the lowest yields.

Of the world's potatoes, 70-80% are grown in Europe and the USSR. All of the developing regions have increased their area in potatoes but this increase has been more than offset by declines in the traditional developed country producing areas. Therefore total production peaked in 1960, and then began to decline. Yields have been fairly stagnant in the USSR, while increasing about 18% in Europe from 1950 to 1980. The yield gap between developed and developing countries is large but seems to be closing. All three of the developing country regions raised yields by at least 40%. Yields in the United States, Canada, and Oceania are about double the world average suggesting considerable unexploited yield potential. However, because potatoes are bulky, they will be consumed locally and many regions do not have an appropriately cool growing climate.

Asia and Latin America are important to analyze for sugarcane as they have about 90% of the world's cane area. This is another crop that is expanding rapidly in Asia where total production has almost tripled. It more than tripled in Africa as well but that region started the period with a small total production. Both Asia and Latin America have increased average yields by over 30%, but Latin American yields are still 20% higher. Yields in Africa are surprisingly good compared to the other crops noted above. The yields of the developed countries suggest a significant yield potential still to be exploited by developing countries.

Cassava was included in this analysis for two reasons. First, it is an important food and feed crop in some regions and plays an important role on small farms with poor production conditions for other crops. Second, as will be noted in the energy section, it has a potential role in alcohol production. World production has more than doubled. Essentially all of it is produced in developing regions. Average yields have been stable for the last two decades so almost all the production increases have been due to expanded area. The reported yield data by country for the decade of the 1970s show a high of about 20,000 kg per hectare to a low of under 5000 kg. Relatively little research has been conducted on this crop so its yield potential is not well understood. Thailand produced about 10% of the world's production in 1980 due to a four-fold production increase during the previous decade. It has a well-established, modern processing industry for the exportation of cassava chips, pellets, and flour. Brazil produces about 20% of the world's production, but its production fell during the 1970s.

Historically about 90% of the world's soybeans have been produced in the United States and China, with the United States alone accounting for 70%. Production has begun to spread to other countries but no case is more dramatic than Brazil where production grew tenfold during the 1970s. In 1980, the share of total production of the United States, Brazil, and China was 59, 18, and 12%, respectively. Soybean yields worldwide have gone up by over one-half, but research has not yet provided the breakthrough needed to expand yields sharply. Yields in the United States and Canada increased by only one-fourth to 1800 kg in these three decades. Latin American yields, led by Brazil, are close behind. Yields in Africa and Asia have increased but have only reached 1000 kgs.

Production of dry beans worldwide has more than doubled. About half the production is in Asia and another quarter in Latin America. It is another disappointing crop in terms of yield levels and trends even though it has not been researched as heavily as some other crops. Three-quarters of the increased output from 1950 to 1970 was accounted for by increased area (University of California Food Task Force, 1974, p. 67). World yield averages have increased only about one-fourth. There have been no real scientific breakthroughs in yields of dry beans. Yields have increased significantly only in the USSR on a small production area. Yields are still low in the important producing areas of Asia and Latin America.

Tomatoes are one of the most important crops in the vegetable category and represented 10% of the total land in that category in 1979. Production increased four-fold these past three decades. Europe is the largest single producing area followed by Asia. Major technological advances have been made, especially with mechanized production in the United States. United States and Canadian yields were double the world average in 1980. Developing country yields still lag by a wide margin even though Latin America more than doubled yields in these 30 years. Africa experienced essentially no yield change during the period, while Asian yields improved by about one-half. For this crop, there is clearly a technological frontier for many countries to exploit.

The yield and production situation of these 10 crops presents a mixed picture. To summarize, it appears that excellent progress has been made with the three important cereals of wheat, rice, and maize. Production expansion and yield increases have been good to excellent over the past three decades. The most important concern is that the growth rate of yields may have slowed down. Potato production is actually falling and relatively little progress has been made in increasing yields. Sugarcane and cassava are potentially important energy crops, but, while sugarcane is expanding in both area and yields, cassava is essentially stagnant. Soybean production has expanded rapidly but there has been no dramatic breakthrough in yields. Dry beans are much like cassava in that they are produced mainly in developing countries and little progress has been made in increasing yields. Tomato yields have shot up in the United States but yields in most of the rest of the world have lagged.

In most crops there is a substantial difference in the performance of developing and developed countries. Average yields in developed countries are higher. The gap is large for all crops, increasing for maize, dry beans, and tomatoes, but decreasing for rice, millet-sorghum, potatoes, sugarcane, and soybeans. Asia and Latin America have made good progress in increasing yields for a number of crops. Africa lags behind and frequently average yields are less than half of developed countries. Yields in developed countries suggest a large unexploited potential yet to be realized in developing countries, especially in the production of rice, maize, soybeans, and tomatoes.

AGRICULTURAL TRADE

The production of many crops tends to be region- and even country-specific as noted above. Furthermore, because of differences in agricultural resources and production and demand trends, substantial food surpluses develop in one region or country, while deficits emerge elsewhere. The mechanism for distributing goods and balancing some of the surpluses and deficits is agricultural trade, which takes the form of market transactions and subsidized food aid.

Agricultural trade, however, is not an automatic process. Indeed, the free flow of agricultural products among countries is restricted and constrained by several factors. First, it is important when studying trade patterns to recognize that food security is an important national objective. Thus most countries prefer to produce a major portion of their own food supply often at a considerable economic sacrifice. Even major industrial countries such as Japan that are poor in agricultural resources strive to produce a significant portion of their food needs. Some countries will accept substantial production inefficiencies in order to achieve the security of reasonable levels of food self-sufficiency for the major food commodities. Subsidization of wheat production in Brazil, sugar in the United States, and a host of agricultural commodities in Western Europe are examples. Still other countries, because of the nature of their population size, low income levels, and lack of foreign exchange cannot enter world markets in a significant way and are forced to rely almost exclusively on domestic food production. China and India, among the major countries, are examples where food needs must be met chiefly by domestic production.

There are a series of situations in the world, however, that create the desirability or necessity for countries to specialize in the production of specific agricultural commodities, to produce these commodities in excess of their needs, and to use trade as a mechanism to distribute and exchange them. Economists refer to the law of comparative advantage to explain why production specialization and trade occur. Simply stated, it is to everyone's benefit to specialize in what they do best and then exchange the excess production to achieve distribution. Comparative advantage in the production of a specific commodity for a country or region is achieved through a combination of superior technology and quality, kind, and cost of production resources, including climate. For example, in the Midwest of the United States, fertile soil, favorable summer rainfall, level topography, and high levels of production technology combine to produce a substantial comparative advantage in the production of grains. Similar conditions exist in Argentina on a smaller scale. Also, to facilitate trade, a sophisticated marketing system of procurement, transportation, storage, and processing is needed along with accepted grades and standards. Cereals are well adapted to world market conditions and trade, while bulky and perishable food items such as tubers are not.

The above conditions that give rise to production specialization and trade are more easily met in developed countries, especially the conditions of production technology, marketing systems, and foreign exchange. As we will see later, much of the world's agricultural trade occurs between developed countries. In developing countries, agricultural development often proceeds on two levels. A few export crops such as coffee, sugar, and tropical fruits exhibit substantially greater sophistication of production and marketing technologies than is apparent for domestic food crops. This advanced development of export crops is fostered by the requirements of participating in international markets. Often specialization is limited chiefly to one or two export crops. Fluctuating international prices then leave these countries vulnerable to yearly variations in their export earnings. Developed countries with a greater variety of export commodities are generally less vulnerable to price fluctuations in any one commodity. With these general conditions in view, we proceed next to a discussion of specific agricultural trade patterns.

AGRICULTURAL TRADE PATTERNS

The conditions that give rise to agricultural trade, as noted above, are more prevalent in the developed world. Conversely, most people live in developing countries. In fact, the one-fourth of the world's population living in developed countries supplies over two-thirds of the value of agricultural products exported (Table 7). The developed world provides an even larger market, and accounts for almost four-fifths of agricultural imports. This means that developing countries are net exporters to the developed world. They provide about a third of the agricultural exports, but receive less than a fourth of the imports. The composition of this trade is varied. Cereals, which account for the major portion of world diets, show a deficit trade pattern for developing countries. Eighty-five percent of cereal trade originates in developed countries, while 45% of the imports go to developing countries, an amount three times as large as their contribution to cereal trade. The traditional tropical crops such as coffee show an opposite pattern: most of the trade originates in developing countries and flows to the developed world. A more even pattern is evident with trade in livestock products. However, since diets high in animal protein and much of the livestock production occurs in developed areas, the trade in livestock products is concentrated there as well. For example, 85% of the meat trade is between developed countries and much of this occurs within the European community.

Table 7. Trade Flows of Selected Commodities between Developed and Developing Areas, 1979

	DEVELOPED COUNTRIES		DEVELOPING COUNTRIES	
COMMODITY	Exports from (%)	Imports to (%)	Exports from (%)	Imports to (%)
All agricultural products	69	77	31	23
Cereals	84	55	16	45
Meat	86	85	14	15
Sugar	29	73	71	27
Coffee	7	94	93	6

Source: Calculated from data in FAO Production Yearbook, 1979.

Within both developed and the developing country groups, however, there are regions and countries that show significant variations from the above generalities. Europe, for example, contributes 40% of agricultural exports but, with over one-half of world agricultural imports, is in a strongly negative trade position within the developed region (Table 8). The USSR and Japan also show a negative trade balance. Conversely, North America and Oceania show very positive trade balances. Thus within the developed world, Europe, the USSR, and Japan in Asia are largely responsible for the overall negative trade balance.

Table 8. Value of Agricultural Trade by Region and Major Net Exporting Countries, 1979

	ALL AGRICULTURAL PRODUCTS					CEREALS	
	Exports ($ billion)	%	Imports ($ billion)	%	Net Trade ($ billion)	Exports ($ billion)	%
Africa	12.9	7	10.5	5	+ 2.4	0.4	1
N. America	41.8	21	21.9	10	+19.9	16.6	54
L. America	27.0	14	10.1	5	+16.9	1.9	6
Asia	24.2	12	46.6	21	-22.4	3.1	10
Europe	80.3	40	115.6	52	-35.3	7.0	23
Oceania	10.3	5	1.5	1	+ 8.8	1.3	4
USSR	2.8	1	13.6	6	-10.8	0.7	2
World total	199.3	100	219.8	100	-20.5	31.0	100
Major Net Exporting Countries							
United States	36.2	18	17.8	8	+18.4	14.2	46
Australia	6.8	3	0.8	-	+ 6.0	1.3	4
Argentina	5.4	3	0.3	-	+ 5.1	1.6	5
Brazil	6.8	3	2.4	1	+ 4.4	-	-
Netherlands	14.5	7	10.5	5	+ 4.0	0.5	2
New Zealand	2.9	1	0.3	-	+ 2.6	-	-
Malaysia	3.7	2	1.1	1	+ 2.6	-	-
Cuba	3.3	2	0.7	-	+ 2.6	-	-
Denmark	4.7	2	2.2	1	+ 2.5	0.2	1
Thailand	2.7	1	0.4	-	+ 2.3	1.0	3
Colombia	2.5	1	0.4	-	+ 2.1	-	-
Canada	5.6	3	4.1	2	+ 1.5	2.5	8
France	15.2	8	13.9	6	+ 1.3	3.5	11
Turkey	1.4	1	0.1	-	+ 1.3	0.1	-

Source: FAO Trade Yearbook, 1979.

Among developing regions, Latin America shows a strong positive trade balance built principally on coffee, sugar, and other tropical speciality crop exports, though grains and oilseeds from Argentina and Brazil are an important component. Africa has a slightly positive trade balance but, as noted earlier, sub-Sahara Africa has expanded food imports rapidly. As with Latin America, tropical speciality crops are important export earners for Africa, while cereals are important imports.

Among the major exporting countries, the United States stands out in several respects. It is by far the largest contributor to world agricultural trade, accounting for 18% of overall agricultural exports and a dominating 46% of cereal exports. It is truly a surplus agricultural nation, having net agricultural exports equivalent to almost 10% of total world trade. Australia and Argentina are the next most important countries in terms of net trade and are important cereal exporters as well. Five countries—the United States, Canada, Australia, Argentina, and France—account for three-fourths of all cereal exports. This is the result of a long-term trend in grain production and use that has seen most of the major regions of the world shift from surplus to deficit grain production, while North America and particularly the United States has emerged as the only major residual supplier of grains (Table 9). During the 1970s, the USSR and, to a lesser extent, China, entered world grain markets to stabilize their fluctuating production levels and to up-grade diets. Rising income levels in other Asian countries, such as South Korea, and in the oil exporting countries have also contributed to the rising grain deficits in Asia. As we will see later, major alcohol programs in the United States, France, Thailand, and Brazil could have a serious impact on the availability of grain and oilseed supplies in world markets.

MAJOR EXPORT CROPS

Wheat, maize, and rice are the principal world cereal crops as noted above. They are also the principal export cereals. Approximately 20% of wheat and maize production is traded internationally, while only 3% of rice production is committed to trade (Table 10). In the past 10 years, there has been a small increase in the percentage of wheat production exported, the percentage of rice exports has remained constant, and maize exports as a percent of production have doubled since 1970. While the production of these crops is distributed widely, only a few countries contribute significantly to exports. For wheat, over 90% of the trade originates in developed countries, with the United States, Canada, and France supplying two-thirds of wheat exports. More than 50% of the wheat production in these countries is exported as contrasted to the 20% average for all countries. Imports are widely distributed, however, with the USSR and China at 11% each, and Japan, Egypt, and Italy, with 7%, 5%, and 4%, respectively, account for almost 40% of all wheat imports.

The United States is dominant in maize production and trade. While maize production is widely distributed throughout the world, the United States supplies over three-fourths of the exports. In terms of world

Table 9. The Changing Pattern of World Cereal Trade, Net Imports and Exports by Region (in Million Metric Tons)

REGION	1934-38	1960	1975	1980
North America	+ 5	+39	+84	+137
Latin America	+ 9	0	- 3	- 13
Western Europe	-24	-25	-20	- 6
Eastern Europe and USSR	+ 5	0	-20	- 47
Africa	+ 1	- 2	- 8	- 11
Asia	+ 2	-17	-45	- 73
Australia and New Zealand	+ 3	+ 6	+12	+ 13

Source: Brown, 1974; FAO Food Outlook, 1981; and FAO Trade Yearbook, 1979.

Table 10. Exports of Major Crops as a Percentage of Total Production, 1970-1979

MAJOR CROP	EXPORTS AS PERCENTAGE OF WORLD PRODUCTION 1970	EXPORTS AS PERCENTAGE OF WORLD PRODUCTION 1979	MAJOR EXPORTING COUNTRIES	PERCENTAGE OF EXPORTS 1979
Wheat	17	19	United States	42
			Canada	15
			France	11
Rice	3	3	Thailand	23
			United States	19
Maize	11	20	United States	77
			Argentina	8
Soybeans	37	41	United States	72
			Brazil	16
			Argentina	8
Sugar	26	26	Cuba	38
			France	9
			Brazil	7
Coffee	82	77	Colombia	16
			Brazil	16

Source: Calculated from data in FAO Production Yearbook, 1980 and FAO Trade Yearbooks, 1975, 1979.

demand, maize shows the largest increase among the cereals, with trade more than doubling during the 1970s. Maize importing countries are principally developed nations. Western Europe, the USSR, and Japan with 38%, 16%, and 14%, respectively, together account for 68% of the maize imports.

Rice, an important food staple in developing countries, is not traded in significant quantities. The major producing countries consume almost all of their production domestically. Thailand and the United States, while accounting for only 6% of world rice production, provide over 40% of rice exports. Though export quantities are small they are distributed widely throughout the world.

Soybean trade is very country-specific and dominated by the United States (72%). In recent years, Brazil and Argentina have advanced to take 16% and 8%, respectively, of the soybean market. These three countries in the western hemisphere are responsible for virtually all soybean exports. Western Europe at 58% and Japan at 17% are the principal importers.

Sugar exports flow predominately from developing to developed countries. Latin America, with Cuba contributing 38% and Brazil 7%, accounts for over 50% of sugar exports. The USSR (33%), the United States (10%), and Japan (7%) account for one-half of the sugar imports. Western Europe is self-sufficient in sugar production.

Coffee is another tropical crop that flows principally to developed countries. North America and Western Europe import 83%, while Latin America exports 58% of the coffee trade. Within Latin America, Colombia and Brazil each represent 16% of total exports.

MAJOR ENERGY CROPS

As gaps between food supplies and demand have emerged in recent years in various countries and regions, international trade has channeled supplies from surplus areas to deficit ones. Over time, more and more countries have become net food deficit countries, increasingly dependent on foreign surpluses to meet their needs. Now some of the surplus countries are considering using part of their production for energy rather than to meet food needs. The potential for substantial changes in food trade availability and costs is now apparent.

The other side of the energy issue concerns the role of energy in the production, processing, and transporting of food products. Historically, energy has been a major input in the production of food. Fertilizer, irrigation, mechanization, and pesticides are chiefly responsible for much of the recent gain in agricultural productivity, especially in developed countries. With low energy prices, the use of these inputs has proven to be very profitable. Further, even at low energy prices food crops in most situations have been too expensive to be used as a fuel source. This is now changing. The rapid rise in energy prices during the 1970s and the relative scarcity of domestic energy resources in many countries have stimulated the consideration of using agricultural products as energy feed stocks. Since oil is the principle energy source appearing in world trade, the emphasis is on finding domestic substitute liquid fuels to replace dependence on uncertain and costly oil imports.

There are important trade-offs for each country to consider in determining whether or not to use food crops as food or as a source of energy. As noted earlier, the agricultural frontier in most countries is closing and additional production must now come chiefly from gains in productivity on fixed land area. The increased costs of energy inputs make these gains more expensive. Where agricultural expansion is limited, increased food costs are the probable price to pay for attaining greater energy independence through the use of agricultural energy feed stocks. Each country has a unique set of food-energy resource capabilities and needs that will dictate to what extent agriculture becomes a supplier as well as a user of energy. Countries that are surplus producers of agricultural commodities and major importers of oil are likely early candidates for major alcohol production programs (Fig. 1). Brazil, the United States, France, and Thailand are major food exporting countries that are now actively pursuing alcohol programs.

Four basic energy feed stocks are available to provide direct substitution for oil-based liquid fuels. Alcohol, which has assumed a leading role as a gasoline additive or substitute, can be produced from three general raw material sources: (a) sugar-bearing materials (such as sugarcane, molasses, and sweet sorghum), (b) starches (such as grains and root crops, including cassava), and (c) celluloses (such as wood and agricultural residues). Vegetable oils that can substitute for diesel are also being investigated, though at present they appear to be less economical than alcohol (Peterson et al., 1981).

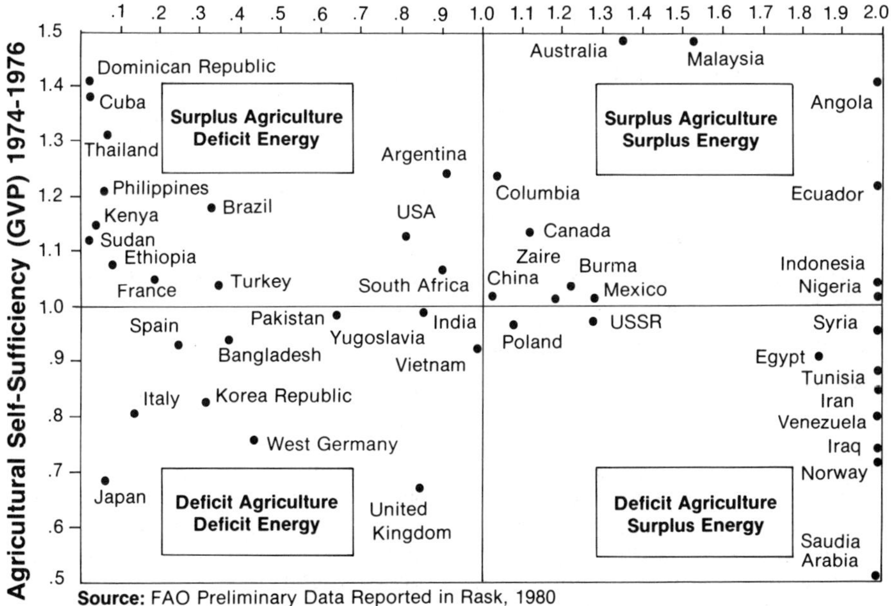

Figure 1. Energy and agricultural self-sufficiency.

Among the alcohol feedstocks, sugarcane is the leader (World Bank, 1980a). It has several distinct advantages. Production of alcohol per acre is high, partly as a result of production technology research (Table 11). The crushed stalk (bagasse) provides the necessary fuel for process heat; thus energy input costs do not rise as rapidly and energy balances are favorable. Finally, the processing technology and capital goods industry are available. Brazil, the leading fuel alcohol producer, relies almost exclusively on sugarcane as the energy feedstock. Among the disadvantages of sugarcane are the need for prime land and adequate moisture, a need which forces direct competition with important food crops for land use. Also the current scale technologies require concentrations of several thousands hectares of sugarcane production within a few kilometers of each distillery. This has important considerations for land use and land ownership concentration. Finally, sugarcane is generally a 3-6 month crop resulting in an uneven flow of alcohol throughout the year.

Table 11. Ethanol Yields of Main Biomass Raw Materials

RAW MATERIAL	ETHANOL YIELD (liters/ton)	RAW MATERIAL YIELD[a] (ton/ha)	ETHANOL YIELD (liters/ha/yr)
Sugarcane	70	50.0	3500
Molasses	280	NA	NA
Cassava	180	12.0	2160
		(20.0)[b]	(3600)[b]
Sweet sorghum	86	35.0[c]	3010[c]
Sweet potatoes	125	15.0	1875
Babassu	80	2.5	200
Corn	370	6.0	2220
Wood	160	20.0	3200

Source: World Bank, 1980.

[a]Based on current average yields in Brazil, except for corn which is based on the average in the United States.

[b]Potential with improved technology.

[c]Tons of stalks/hectare/crop. Two crops per year may be possible in some locations.

Among the starches, cassava in tropical countries and corn and other grains in temperate climates are the principal feedstocks. Cassava has many advantages. It is a hardy crop that can withstand adverse soil conditions and moisture stress. Thus geographically it can be produced under a much wider range of agro-climatic conditions than sugarcane. In many areas it can be harvested year round or chipped and dried for later use. It contains less water per ton than sugarcane, allowing for larger procurement distances. Yields of alcohol per hectare are somewhat lower than sugarcane though potential yields are not well known since little production research has been directed toward cassava. It has traditionally been a subsistence food crop produced on small farms. Recent research work suggests that substantial yield increases are possible. The starch must be changed to sugar before fermentation, therefore capital and processing costs are a little higher. A separate energy source must be provided. In most cases this energy will come from wood plantations, which in turn will increase the effective land area needed to produce a given quantity of alcohol from cassava.

In order to diversify alcohol production to different geographical areas, Brazil is currently putting increased emphasis on cassava as an important energy feedstock for future alcohol production. Thailand is considering sugarcane and cassava in combination to make better use of surplus bagasse as a heat source and to stretch distillery use over more months of the year. Thailand is currently the major exporter of cassava products for animal feed and is considering alcohol as an alternative use for cassava production due to a restricted export market.

In temperate climate agriculture, grains are the primary energy feedstocks. The United States is the leading fuel alcohol producer from maize. Grains, and maize in particular, present a more complicated situation. The production of alcohol from grains results in one of several by-products that are important to the economic viability of alcohol production. The principal by-product is a high-protein livestock feed that can partially substitute for other protein supplements in livestock feed rations. Alcohol yields per hectare of cereals are lower than those of sugarcane, but the added value of the by-product feed partially offsets this disadvantage. An outside heat source is needed for processing. Processing plants need to be large to gain production efficiencies, and maize is purchased in the market. Thus production costs are variable depending on the price of maize. The potentially large increase in the demand for maize will further affect both the level and stability of maize prices.

Major alcohol programs will require that substantial areas of land be removed from food crop production in order to provide the necessary energy feedstocks. Brazil is currently producing over 4 billion liters of alcohol annually. This is an amount equal to about one-fourth of their gasoline needs and requires 3% of the 1980 cropland. To give some indication of the magnitude involved, complete substitution for all imported petroleum and petroleum products in Brazil (including gasoline, diesel, and fuel oil) would require 40% of current crop land. In the United States, a gasohol program involving a 10% alcohol, 90% gasoline mixture would require 6% of crop land or 27% of maize crop land. Studies attempting to show the expected impact of major alcohol programs on land use, crop prices, and trade have demonstrated the high return that could be expected from successful output-increasing research for agricultural crops (Ott, 1981). If agricultural resources are to provide both fuel and food, increased agricultural research is not only a necessary prerequisite but an investment with a high potential payoff.

SUMMARY AND CONCLUSIONS

The post-World War II period has been one of major improvements in the world food situation. The production of many food crops has expanded sharply. Yields have also improved for many crops. Unexploited productive potential appears to exist in many of the developing countries thereby giving promise of some yield increases in the future. An elaborate world trade system has been developed which channels some of the production from surplus areas, especially in the United States, to deficit areas which are chiefly found in Japan, USSR, and Western Europe. More people have assurance of a reasonably adequate food supply today than at any other time in history.

There are serious food problems even today, however, and the long-term prospects are doubtful. First, the uneven distribution of agricultural production, agricultural resources, and disposable income among and within countries create great differences in food availability and consumption. Trade is an effective mechanism to bridge the gap between production and consumption for populations that have income and can compete in the international food market. Poor people, by definition, cannot rely on the market system to fulfill their needs. Thus millions of people overeat, while millions go hungry. Second, several nations have moved from being net food surplus countries to net food deficit ones. Thus more and more people are becoming dependent on the surpluses of a few countries. It is not at all clear how long these surpluses will be able to satisfy world needs. A major drought in the United States, for example, would send a tremendous shock into a world food system that has few stored surpluses and depends so heavily on U.S. cereal overproduction. Third, the world food market in the 1970s was able to respond to country and regional deficits by calling up greater production in countries such as Brazil and the United States where underutilized land existed. Now much of the good land is in production in these two countries, as well as in many other countries, so there are few easy ways to quickly achieve another major increase in food production. Fourth, average crop yields are a major concern. Yield increase must be the source of major improvements in future food production. Yet the rate of increase in yields for many crops including wheat, rice, and maize appears to have declined. No major breakthroughs have been realized for soybeans and dry beans. The yield gap between developing and developed nations is difficult to close when due to slow adoption of new technology by farms, and even more difficult when due to inferior production conditions of soil, water, and climate. Furthermore, the higher yields of developed areas are due in large part to modern inputs which have experienced sharp price increases because of the rise in petroleum prices. Fifth, the petroleum crisis has introduced a new unexpected demand for agricultural crops to produce energy. Countries with surplus food and deficit energy, like the United States, Brazil, Thailand, France, and the Philippines, have the option of converting food to energy. But when they do so, less food enters the world market. Price increases are the natural result. Therefore, their own populations will gain but food importing countries will lose.

This combination of positive and negative trends creates great uncertainty over the future of the world food economy. One thing is certain however: Everyone would gain with technological improvements in the production, processing, and distribution of food crops. A strategy of agricultural research to create such technology has the additional advantage of being more politically acceptable than one requiring more sacrifice by the rich to help the poor. Most food is still produced in the country where it is consumed. Therefore the clear objective should be to produce more in the developing countries where food deficits are most aggravated.

A three-part agricultural research strategy is required. First, research is needed to determine the reasons for the yield gap among and within countries. Even more effort is needed to adopt production technology to meet local needs. More research is required on crops important in tropical diets, such as cassava and dry beans, which have been largely ignored until recently. Second, research will continue to be needed in food surplus countries so that they can choose the option of producing energy from food crops, while simultaneously providing food for the world market. Third, crops which now are largely ignored in major research programs must be systematically scrutinized for their potential in making greater contributions to the world food supply.

A major worldwide agricultural research strategy may appear improbable at a time when country after country is struggling with problems of energy supplies, inflation, and unemployment. But failure to implement such a strategy will mean higher food prices as more people compete for scarce food, environmental chaos as more marginal land is brought into production, and even more malnutrition and hunger for the poor.

ACKNOWLEDGMENTS

We note with appreciation the assistance of Marvenia Howard in preparing some of the tables that appear in this chapter and the assistance of Barbara Lee and Jill Loar in preparing the manuscript.

REFERENCES

Brown, L.R. 1974. By Bread Alone. Praeger, New York.

Christensen, C., Donnen, A., Horenstein, N., Tryor, S., Riley, P., Shatouri, S., and Steiner, H. 1981. Food Problems and Prospects in Sub-Saharan Africa. USDA, Washington, D.C.

FAO Food Outlook. 1981. No. 11/12. Rome, Italy.

FAO Production Yearbooks. 1971, 1979, and 1980. Rome, Italy.

FAO Trade Yearbooks. 1975 and 1979. Rome, Italy.

O'Brien, P.M. 1981. Global prospects for agriculture. In: Agricultural-Food Policy Review, Perspectives for the 1980s, pp. 2-26. USDA, Washington, D.C.

Ott, S.L. 1981. The Economics of Producing Fuel Grade Alcohol from Corn in Western Ohio. Unpublished Ph.D. dissertation. Ohio State University, Columbus, Ohio.

Peterson, C.L., Auld, D.L., Thomas, V.M., Withers, R.V., Smith, S.M., and Vettis, D.L. 1981. Vegetable Oil as an Agricultural Fuel for the

Pacific Northwest. Bulletin No. 598. Agricultural Experiment Station. University of Idaho, Moscow, Idaho.
Rask, N. 1980. Biomass: Its Utilization as Food and/or Fuel. ESO 764. Department of Agricultural Economics and Rural Sociology, The Ohio State University, Columbus, Ohio.
University of California Food Task Force. July, 1974. A Hungry World: The Challenge to Agriculture. University of California, Berkeley.
World Bank. 1980a. Alcohol Production from Biomass in the Developing Countries. World Bank, Washington, D.C.
World Bank. 1980b. World Development Report, 1980. World Bank, Washington, D.C.
World Bank. 1981. Accelerated Development in Sub-Saharan Africa. World Bank, Washington, D.C.

CHAPTER 2

Cell Culture Methods for Crop Improvement

D.A. Evans, W.R. Sharp, and *J.E. Bravo*

Plant cell culture techniques have tremendous potential for crop improvement. This collection of techniques can be directed toward production of identical plants or to induce variability. Plants can be propagated from numerous explants including leaf sections, anthers, meristems, or even isolated single cells and protoplasts. Alternatively, callus or liquid cell suspension cultures can be established. When integrated with conventional crop improvement programs, techniques of cell culture could prove useful through (1) propagation in vitro, (2) meristem culture for virus elimination, (3) secondary product synthesis, (4) production of haploid plants from cultured anthers, and (5) development of new varieties via cellular or molecular genetics. The application of these methods to several major crop species is detailed in subsequent chapters of this volume. In this chapter we present an overview of these five areas of tissue culture. Basic procedures are outlined and the application of these methods to crop improvement is described (Table 1).

LARGE-SCALE PLANT REGENERATION

Innumerable plant species can be propagated in vitro (Murashige, 1974, 1978). In most cases clonal propagation is achieved by placing sterilized shoot tips or axillary buds onto a culture medium that is sufficient to induce formation of multiple buds. Shoot tip clonal propagation has already been used to commercially propagate via cloning a large number of marketable ornamentals (cf. Oglesby, 1978). In con-

Table 1. Progress in Application of Cell Culture Techniques to Crop Species

CROP	EXPLANTS USED FOR PLANT REGENERATION	RECOVERY OF HAPLOIDS	RECOVERY OF PLANTS FROM PROTOPLASTS
CEREALS			
Maize	Immature scutellum	Yes	No
Oats	Immature embryo	No	No
Wheat	Embryo, rachis, scutellum	Yes	No
Forage grasses	Several	Yes	Yes
Millets			
Pennisetum	Several	No	Yes
Panicum	Several	No	Yes
Sorghum	Immature embryo	No	No
Eleusine	Mesocotyl	No	No
Paspalum	Mesocotyl	No	No
Rice	Several	Yes	No
LEGUMES			
Phaseolus	Primary leaf	No	No
Clover	Several	Yes	Yes
Alfalfa	Several	No	Yes
Peanut	Mesocotyl, epicotyl	Yes	No
VEGETABLES			
Asparagus	Several	Yes	Yes
Celery	Petiole, leaf	No	No
Cole crops	Several	Yes	Yes
Tomato	Several	Yes (dihaploid)	Yes

ROOT AND TUBER			
Cassava	Cotyledon	No	Yes
Sweet potato	Stem, root	Yes	Yes
Potato	Several	Yes	Yes
Yams	Embryo, axillary bud	No	No
TROPICAL/SUBTROPICAL			
Banana	Shoot apex	No	No
Papaya	Several	Yes	No
Citrus	Ovules	No	Yes
Pineapple	Several	No	No
TEMPERATE FRUITS			
Apple	Several	No	No
Grape	Several	Yes	No
Blueberry	Several	No	No
Stonefruits	Several	Yes	No
Strawberry	Meristem tip	No	No
FIBER AND WOOD			
Conifers	Several	No	No
Cotton	Cotyledon, hypocotyl	No	No
Hardwoods	Several	Yes	No
EXTRACTABLE PRODUCTS			
Date palm	Several	No	No
Hevea	Stem	Yes	No
Sugarcane	Leaf, meristem	Yes	No
Tobacco	Several	Yes	Yes
Cacao	No	No	No
Coffee	Several	Yes	No
Oil palm	Leaf	No	No

trast to shoot tip propagation which faithfully produces clones, evidence has accumulated suggesting that regeneration of plants from callus, leaf explants, or plant protoplasts does not result in production of clones (Skirvin, 1978; Shepard, 1982). Regeneration of plants from long-term callus culture is associated with in vitro chromosome instability and recovery of aneuploid plants (Sacristan, 1971). Regeneration of plants directly from leaf explants has resulted in morphological variation reported in lettuce and tomato (Sibi, 1978, 1981). The observation that in both potato (Shepard, 1982) and sugarcane (Larkin and Scowcroft, 1981) regenerates varied in response to diseases, suggests that production of this variability in regenerated plants could prove agriculturally useful. Hence large-scale plant regeneration may have agricultural application for either clonal multiplication or production of novel genetic variability.

Shoot Apex Propagation

Certain species of ornamentals or parental breeding lines are often difficult or time consuming to propagate using conventional vegetative methods. Additionally, unique mutants or variants are often difficult or impossible to propagate using conventional sexual methods. In either case mass clonal propagation could be commercially useful. Clonal propagation requires the production of genetically identical plants. To insure stability, tissue cultures are usually initiated from organized structures and placed onto a simplified culture medium (Murashige, 1974). The standard method for plant propagation using shoot tip culture has been described. The apical shoot tip or axillary bud is surface sterilized and placed onto a simplified culture medium that will promote formation of additional shoots. Subsequently, multiple shoots are separated and subcultured to a rooting medium. Plantlets are then transplanted to soil or peat pots and transferred to the greenhouse or field.

Murashige (1974) has described a three-stage procedure that normally requires alteration of culture medium or growth conditions between stages. Stage I pertains to the establishment of tissue in vitro. Stage II is most important as it requires production of multiple shoots. Stage III must result in root formation and conditioning of propagules prior to transfer to the greenhouse. High light intensity is important in Stage III (Murashige, 1978). In many species, the media and culture conditions are not altered between Stage I and Stage II (Murashige, 1978).

While cell culture has already been demonstrated to be commercially important for propagation of ornamentals, cell culture may also be useful to propagate certain crop plants. (1) In plants with seed viability problems, breeding or parent lines could be regenerated in vitro. Somatic embryos produced in tissue culture could be directly planted into the field using systems already developed for pregerminated seed or transplants. (2) Mutated plants could be propagated in vitro. Albino mutants or other genetic lines useful in cellular genetics can be maintained in vitro. (3) Parents used in a breeding program may be propagated in vitro. Male sterile mutants have been used to produce commercial F_1 hybrids in a number of crop species. Normally, restorer lines must be used to maintain and propagate the male sterile. This procedure requires a series of sexual crosses with subsequent selection among all progeny for the male sterile types. Unless genetic markers are used, the selection must be completed on mature plants. This selection is particularly time consuming when the male sterile parent line is under nuclear genetic control. In some species, it is possible to maintain male sterile parent lines and preserve the male sterile character using in vitro propagation to allow for consistent hybrid seed production. (4) When advanced breeding lines have been identified, it is important to propagate as many seeds or plants as is possible to permit variety release. At this point tissue culture may be used in conjunction with other asexual methods to propagate clonal lines for seed production. Such an approach was used in Ecuador to establish elite clones of pyrethrum (*Chrysanthemum cinerariaefolium*) for commercial production (Levy, 1981). (5) In some cases cultivated crops, particularly cash crops where each individual plant is valuable, may be cloned using tissue culture. For example, varieties for greenhouse crop production may be propagated in vitro using shoot apices.

In crops where male sterility is unavailable, production of hybrid seed is expensive. In crosses between divergent varieties, hybrid seed cannot be propagated sexually without genetic segregation in the F_1 population. Genetic segregation in the F_1 population is inversely correlated with relatedness of the parental lines. Hybrid plants could be propagated using large-scale in vitro propagation. Such hybrids may be cheaper to produce than hybrids originating from seed produced by hand pollination. A business analysis of the economics of hybrid production should consider the cost of transplanting regenerated hybrid plants to the field as opposed to directly seeding the field with F_1 hybrid seed. Unless plants can be regenerated very inexpensively or are in a species in which transplants are commonly used, in vitro propagation of F_1 hybrids is not profitable.

Organogenesis and Embryogenesis

Plant regeneration from explants or callus may proceed via organogenesis or embryogenesis (Evans et al., 1981b). During organogenesis, separate culture media are usually used for shoot, then root formation. Generally, cytokinin is necessary for shoot formation, while an active auxin or reduction in the cytokinin concentration is required in a successful rooting medium. Embryogenesis is observed most frequently among umbelliferous species, although a number of nonumbelliferous species undergo somatic embryogenesis (Vol. 1, Chap. 3). During embryogenesis, root and shoot develop simultaneously on the same culture medium. The standard stages of embryo formation (globular, heart, then torpedo stage) can be detected. Although a complete analysis has not been presented in the literature, shoot apex culture is usually associated with genetic uniformity of regenerated plants, while organogenesis, particularly from leaf cells and protoplasts, is usually associated with genetic variability. Large populations of plants regenerated via embryogenesis have not been examined in the field, however preliminary reports suggest that these plants are also variable (e.g., Chap. 9; Sharp and Lipschutz, unpublished, both for celery). While several crop plants can be regenerated in vitro, regeneration is still limited among the legumes and cereals. In most species, regeneration is limited to a few explants. The explant source does not, though, predispose a culture to the embryogenic vs. organogenic mode of development. MS culture medium has been successfully used in most species. Certain legumes and cereals may require modifications of the MS salts to satisfy nutrient requirements (Schenk and Hildebrandt, 1972; Phillips and Collins, 1979).

SOMACLONAL GENETIC VARIABILITY AMONG REGENERATED PLANTS. Phenotypic variation has been reported in a number of plant species regenerated using tissue culture. Such genetic variability may be agriculturally useful when integrated into existing breeding programs. The phenotypic variability recovered in regenerated plants reflects either preexisting cellular genetic differences or tissue culture-induced variability. Unfortunately no crucial experiments to accurately distinguish between these two factors have been reported. Skirvin and Janick (1976) systematically compared plants regenerated from the callus of five cultivars of *Pelargonium* spp. Plants obtained from geranium stem cuttings in vivo were uniform, whereas plants from in vivo root and petiole cuttings and plants regenerated from callus were quite variable. This suggests that at least some of the variability is preexisting in the donor explants. Changes were observed when regenerated plants were compared to parent plants for plant and organ size, leaf and flower morphology, essential oil constituents, fasciation, pubescence, and anthocyanin pigmentation. It is likely that both preexisting genetic and tissue culture-induced variability has been recovered in these geranium lines.

Syono and Furuya (1972) recovered phenotypic variability, particularly abnormal floral morphology, in plants regenerated from long-term cell cultures. Long-term cell cultures do result in tissue culture-induced variability with regard to chromosome number in both callus and in plants regenerated from callus (Sacristan, 1971). This results in regeneration of aneuploid plants that are commercially useless in sexually propagated species. On the other hand aneuploidy may not interfere with productivity of asexually propagated crops such as sugarcane and potato.

Plants regenerated from callus of sugarcane have been examined in detail. A wider range of chromosome numbers (2n = 71 - 300) has been observed in plants regenerated from sugarcane callus (Heinz et al., 1977) than in parent lines. These clones of regenerated sugarcane were examined for disease resistance. Resistance to three diseases (eye-spot disease, Fiji disease, and downy mildew) was observed in some plants regenerated from a susceptible sugarcane clone (Heinz et al., 1977). One regenerated line was simultaneously resistant to both Fiji disease and downy mildew. These lines are being incorporated into a conventional breeding program.

More recently, plants regenerated from mesophyll protoplasts of potato have been analyzed. Although detailed genetic analysis has not been published, it is apparent that regenerated clones are highly variable. There is evidence that some of the variability is due to chromosomal rearrangements (Shepard, 1982). As in sugarcane, lines have been identified that are resistant to diseases (late blight and early blight), to which the parent line, Russet Burbank, is susceptible (Shepard, 1982).

Application of these techniques to sexually propagated crop species has not been examined in detail. Sibi (1978, 1981) has suggested that heritable variability has been recovered in plants regenerated from callus of lettuce and tomato without concomitant alteration of the chromosome number. However, detailed cytological analysis was not presented. No new crop varieties have yet been produced using the culture-induced genetic variability, although this method will probably produce new varieties shortly. Evans and Sharp (1983) have demonstrated a genetic basis to somaclonal variation in tomato.

TRANSFER OF SOMATIC EMBRYOS TO THE FIELD. Techniques have been developed which suggest that direct transfer of regenerated plants to the field may be economically feasible. These techniques relate to transfer of pregerminated seed to the field using fluid drilling and would be most easily applicable to species capable of somatic embryogenesis in vitro. Fluid drilling methods applied to pregerminated seed have already been developed for a number of crops with satisfactory results (Bryan et al., 1978). These crops include tomato, lettuce, cucumber, cabbage, okra, squash, broccoli, turnip, carrot, corn, and celery. Somatic embryos are available in carrot and celery. Fluid drilling of pregerminated seeds involves: (1) seed germination in optimum controlled environmental conditions, (2) mixing selected germinated seeds with a gel-type medium to protect and suspend the seeds for uniform distribution, and (3) sowing the seeds and gel using a fluid drill seeder (Bryan et al., 1978). Fluid drilling allows for early uniform plant development which can be reflected in an earlier harvest (Currah, 1977).

The fluid drilling technology appears to be ideal for a commercial operation dependent upon in vitro propagation to produce embryos or an artificial seed for sowing. The gel used in fluid drilling is comparable to agar, as it can be fortified with growth regulators, plant nutrients, pesticides, and other plant protection agents. Bryan et al. (1978) have listed some of the gels and fluid drilling machines that have been used successfully for sowing germinated seeds.

High frequency somatic embryogenesis offers the best opportunity to develop a system for fluid drilling. Unfortunately, substantial research must be accomplished pertaining to synchronization of embryo development and perfection of large-scale separation procedures for collecting synchronous embryos before an artificial seed operation can be established. Sieves or glass beads may be used for the separation and collection of uniform embryos (Warren and Fowler, 1977). Thereafter the embryos should be maintained in an arrested state of development as an artificial seed until the time of sowing. The arrested embryo development could be achieved using cold storage or mitotic inhibitors. Alternatively, mechanized transplanting methods may be used if transplants are normally grown in the field (Rogers, 1982). These techniques may be appropriate for several vegetable crops.

A summary of commercial applications of large-scale plant regeneration is presented in Table 2. Cost of propagating plants may be of greater importance for clonal propagation, as the regenerated plant is the commercial product. Propagation cost has been estimated at 13-25 cents per plant for some ornamentals (Oglesby, 1978) and 11.6 cents per plant for broccoli hybrids propagated in vitro (Anderson et al., 1977). This cost is mostly labor and has no doubt escalated in the past 5 years. In production of novel genetic variability by organogenesis, the final product may be a new variety that is sold as seed. The released variety may be removed several generations from the original regenerated plant. The cost of in vitro regeneration would be negligible if the new variety was particularly valuable.

Table 2. Procedures for Commercial Use of Large-Scale Propagation

	CLONAL PROPAGATION	GENETIC VARIABILITY
Application	1. Multiply existing ornamentals cheaply	1. Alter crop varieties
	2. Maintain unique parents	2. Incorporate disease resistance
	3. Multiply hybrid plants	3. Produce new ornamentals
Explant source	Apical meristem	Leaf explants
	Axillary buds	Callus
		Protoplasts
Culture medium	MS propagation medium	High cytokinin in medium
Sample species where useful	Bromeliads	Sugarcane
	Orchids	Potato
	Ferns	Lettuce
	Male sterile parents	Geranium

MERISTEM CULTURE FOR VIRUS ELIMINATION

It has been observed in a large number of plant species that the concentration of infective viruses is low in the apical meristem of a plant. This reduced titer of viruses reflects a combination of factors. The lack of vascular differentiation in the meristem impairs intercellular movement of viruses and the active metabolism of mitotic cells precludes viral infection. In certain crops or ornamental species, it has been economically useful to develop in vitro techniques to produce virus-eradicated plants. This eradication is accomplished by regenerating a plant from a virus-free explant. When an organized structure such as the apical dome is cultured, plant regeneration represents the enlargement of a preexisting shoot followed by root induction. Genetic stability as measured by phenotypic and chromosomal stability is usually preserved in plants recovered from cultured meristems. Alternatively, plants may be regenerated from callus. Although plants regenerated from callus are often virus-free, it should be noted that chromosomal abnormalities are frequently observed among callus-derived plants. It has been demonstrated that populations of callus-mediated regenerates are not as uniform as regenerates from preorganized shoot apices. Consequently, meristem culture is used most frequently to obtain virus-eradicated plants.

Techniques of Meristem Culture

As the explants to be cultured must be virus-free, young rapidly growing shoot tips are usually used. Meristems are often obtained from presterilized young plants or shoot cultures maintained in vitro (Kartha et al., 1979). Although apical meristems that are protectively covered with young leaf primordia are usually sterile, the exposed leaves are usually surface-sterilized prior to excision of the meristem. Sterilization is usually accomplished with 75% ethanol or with diluted sodium hypochlorite. Precise sterilization procedures should be ascertained prior to commercial propagation as the shoot apex of some species is sensitive to the chemicals used for sterilization.

Meristems are detached from the shoot tip in a sterile environment using a dissecting microscope. This procedure may prove difficult in species with abundant hairs on the young leaf primordia. The apical dome is placed onto solidified culture medium that is sufficient to insure development of the preorganized shoot apex and to permit root development. For most species, a simplified hormoneless culture medium is adequate. White's culture medium was used in many early publications, but most recent publications used a medium with higher salt concentrations, such as MS. Premixed MS medium is commercially available. Addition of an auxin, IAA or NAA, may be necessary to induce root development. In some species it is possible to achieve shoot and root development on the same culture medium, although most published protocols suggest sequential media for shoot, then root, development. The importance of different light and temperature regimes in meristem culture has been discussed in Vol. 1, Chap. 5.

Numerous factors influence the ability to eliminate viruses through tissue culture. Both the location and size of the bud are related to presence of infecting viruses in the explant. In order to take advantage of virus-free cells, the smallest possible meristem explant should be cultured. On the other hand, small explants do not regenerate or produce roots as well as large meristems. These two factors must be balanced to identify optimal explant size. Additionally, terminal buds respond better than lateral buds.

Heat treatment is commonly used to inactivate the virus prior to culturing shoot tip explants. In many cases, using 34-35 C will result in inactivation of the virus without deleterious effects on the plant. About one-half of the known plant viruses, particularly small spherical viruses, can be inactivated with thermotherapy (Wang and Hu, 1980). Heat treatment during plant growth can result in successful use of larger meristem explants than from unheated plants (Mullin et al., 1974). In some species, heat treatment is essential for successful virus elimination. In chrysanthemum, 98.5% of progeny from heat-treated plants were virus-free, while only 15% of apices from room temperature grown plants were virus-free (Hollings et al., 1972). If plants are damaged by continuous heat treatment, fluctuating temperatures may be used (Walkey, 1976). Heat treatment also can be applied during tissue culture of the meristem. Walkey and Cooper (1975) cultured meristems of *Nicotiana rustica* at 32 C which resulted in elimination of some viruses. TMV, though, appeared to be heat resistant.

Applications of Meristem Culture

Viruses have been eliminated using meristem culture from a number of economical species. These species include the elimination of mottle virus from strawberry, potato virus X from potato, mosaic virus from cassava, and cauliflower mosaic virus from cauliflower (Quak, 1977). Putative virus-free stock plants must be tested to insure virus-free status. Virus infection can be assayed using serological methods (Wang and Hu, 1980), examination in the electron microscope, or by infecting indicator plants. If suitable indicator plants are available, infectivity tests are extremely sensitive assays for virus.

Virus-free plants produced by meristem culture have resulted in significant increases in yield when compared to virus-infected plants (Walkey and Cooper, 1975). Results have been particularly favorable in potato (Gregorini and Lorenzi, 1974), where yield of virus-free potatoes is 60% higher than that of infected plants. As expected, field-grown plants often become reinfected with virus. Nuclear stock plants should be maintained to reproduce virus-free plants. In vitro maintenance of some stock plants is advisable, although more costly than greenhouse maintenance. It should be noted that meristems can be revived following freeze preservation in liquid nitrogen (Kartha et al., 1979).

Plants regenerated from callus may also be free of viral infection. Callus tissue is similar to meristematic tissue as cell mitosis is rapid and vascular differentiation is incomplete. In callus or cell suspension cultures, viral infection may be completely eliminated via repeated subculture. Virus-free plants have been regenerated from callus in tobacco, geranium, Gladiolus, and potatoes (Wang and Hu, 1980). Although callus tissue results in virus-free plants, chromosome abnormalities have been documented in plants regenerated from callus tissue (Sacristan, 1971).

Plants regenerated from uninfected leaf tissue may also be used to recover virus-free plants. Murakishi and Carlson (1976) regenerated tobacco plants from dark green uninfected islands of a TMV infected leaf. Approximately 50% of regenerated plants were virus-free. Similarly, Shepard (1975) reported that 7.5% of 4140 plants regenerated from potato-virus-X-infected tobacco leaves were virus-free.

SECONDARY PRODUCT SYNTHESIS

Most proposed benefits of plant tissue culture require regeneration of a genetically modified or cloned plant to achieve successful application of cell culture methodology. It has been proposed, though, that cultured plant cells may produce useful chemicals. While plant cells cannot compete with bacterial or yeast batch cell culture for cost of production, certain unique compounds, synthesized exclusively by plant cells, may have commercial value. While microorganisms are used commercially for production of chemicals, several differences exist between plant cells and microbes grown in culture. (1) Microorganisms are faster growing than plant cells. (2) Plant cells accumulate products within the cytoplasm, while bacterial cells normally excrete products. (3) Plant cells tend to produce useful chemicals in stationary phase rather than in log phase. (4) Techniques of genetic manipulation are more advanced with microbial systems than with plant cells, so that superproducing microbes can be isolated easily.

Several conditions should be manipulated to optimize the yield of desired products. Plant cells can be grown in batch cultures or in fermenters. However, several problems will no doubt be encountered when cell suspension cultures are scaled up to fermenters. It is necessary to ascertain optimal cell harvest time, stability of the culture, and reproducibility of chemical production.

Selection methods should be used to identify stable, high-producing lines. It has already been demonstrated that some compounds may be present in cultured cells at a higher concentration than in plant tissue. Those compounds already identified in higher concentration include the anthraquinones, rosmarinic acid, glutathione, and ubiquinone-10 (see review by Zenk, 1978). In certain cultures, expensive precursors or growth regulators must be added to the plant cells to produce commercially feasible yields. Selection of variant cell lines that accumulate very large concentrations of desired compounds has been reported (Tabata and Hiraoka, 1976). Selection for improved production may be based on physiological or genetic methods. Most researchers have attempted to select for production based on alterations of culture medium, environmental growth conditions, and harvest time. Physiological selections may be epigenetic and therefore more difficult to control. Selection should probably be combined with a mutagen treatment to increase the probability that variants are due to genetic lesions, rather than epigenetic variation. Mutants isolated in this method may result in maintenance of more stable cell lines. For example, it may be possible to select for hormone autotrophic growth that will result in lower cost of culture medium. Similarly, photoautotrophic cells could be grown using CO_2 as a carbon source rather than glucose and sucrose. Selection for resistance to amino acid analogs or other toxic analogs could result in mutated cells that overproduce amino acids or other useful chemicals. Once mutant lines are selected it is important to maintain stability of these lines. Yamada and Fujita (Vol. 1, Chap. 23) have described selection and cloning methods to produce stable cell lines capable of overproduction of useful chemicals.

A detailed cost analysis of secondary product synthesis is necessary for plants. It is important to consider the cost of the final product per kg. Unless the cost of the chemical is high, it will be impossible to use cell cultures to produce a commercially competitive product. Cost evaluation should also consider demand of the product, opportunities and value of producing a product with improved purity, and government regulations on distribution. Finally, it should be emphasized that if there is any possibility that a microbe can make the product, plant cell biosynthesis will not be economical. At present only the production of expensive drug compounds is currently commercially feasible (Zenk, 1978), although work in Japan is proceeding to identify new potentially useful compounds produced by plant cell culture.

HAPLOID PLANT PRODUCTION

Haploid plants are important in plant breeding as haploids greatly reduce the time for obtaining homozygous lines following diploidization in allogamous populations. Doubled haploids, when integrated into a conventional breeding program, have resulted in rapid development of new varieties. Kasha and Reinbergs (1980) have reported that use of doubled haploids reduced the time for release of a new variety of barley from 12 years to 5 years. Haploid plants can also be used for genetic analysis. Burk (1970) has demonstrated that single gene heterozygous diploid plants produce a 1:1 ratio of the two haploid parental types following anther culture. In tobacco, haploids have been used to analyze the number of genes controlling disease resistance (Nakata and Kurihara, 1972) and nicotine production (Nakamura et al., 1974), as well as to aid in analysis of mutants induced in vitro (Chaleff, 1981) and of somatic hybrids (Schieder, 1978).

Techniques of Haploid Plant Production

Haploid plants have been obtained in crop species using conventional breeding methods, anther culture, or pollen culture. As the haploid plants produced by each method seem equally desirable, the method which produces the highest percentage of haploids should be used.

Haploid plants arise frequently following interspecific hybridization and haploids produced by hybridization have been integrated into breeding programs for barley, potato, and tobacco. In each of these species, haploids are produced following fertilization of a cultivated crop with a wild relative. In barley, haploids are produced following fertilization of *Hordeum vulgare* with *H. bulbosum*, as *H. bulbosum* chromosomes are preferentially eliminated. Recently, the barley system has been extended to produce haploids in closely related cereals, as additional species have been used as female parent. In potato, pollination of *Solanum tuberosum* with wild species pollen, particularly *S. phureja*, results in parthenogenic development of *S. tuberosum*. Potato doubled haploids and haploids can be used to aid in the introduction of wild species genes into the tetraploid cultivated potato. Haploid tobacco has been produced following pollination of cultivated tobacco with *Nicotiana africana* (Burk et al., 1979). Each of these methods results in production of maternal haploids of the cultivated crop. On the other hand, the indeterminate gametophyte (ig) gene can be used in corn to produce haploids with unique mixtures of nuclear and cytoplasmic DNA. Both paternal and maternal haploids can be produced using the ig gene in heterozygous condition (Kermicle, 1969). This system may be modified in the future to aid in transfer of cytoplasmic DNA between breeding lines without transfer of nuclear genes.

Haploid tissue and plants can be produced in higher frequency in many species using anther culture. Haploid plants were first produced in *Datura innoxia* (Guha and Maheshwari, 1964). For each species examined, the best developmental stage of cultured anthers must be ascertained to optimize the frequency of haploid production. First-pollen-grain mitosis is optimal for tobacco and *Datura*, while an earlier stage is required for tomato and a later stage is required for *Brassica* (Sunderland, 1974). In some species, the culture medium does not appear to be as important as anther stage in determining the conditions for haploid production as the nutrient requirements for cultured anthers may be quite simple. On the other hand, in some species the response of cultured anthers is medium-dependent (e.g., rice varieties). Numerous other factors affect the degree of success in anther culture. The growth environment and age of the donor plant are important. Pretreatment of the anthers after removal from the plant, but prior to culturing in vitro, may be essential. Both cold shock (Nitsch, 1974) and heat treatment (Keller and Armstrong, 1979) have been proposed to increase the number of haploid plants produced per cultured anther.

Haploid plants have been recovered either via androgenesis or by plant regeneration from microspore-derived haploid cells following callus formation. From a recent review (Sharp et al., 1983) it is evident that most species capable of haploid production proceed via direct androgenesis. Taxonomic differences are apparent though, between androgenesis and callus-mediated regeneration. The majority of solanaceous species capable of producing haploid plants undergo direct embryogenesis, while the majority of graminaceous species undergo callus-mediated haploid plant regeneration (Sharp et al., 1983). The crop species capable of haploid plant production from cultured anthers include: rapeseed, tobacco, rice, rye, potato, corn, asparagus, pepper, strawberry, rubber, barley, sweet potato, wheat, grapes, and clover.

As anthers are a mixed population of diploid and haploid cells, all capable of cell reproduction, great effort has been directed toward isolated pollen culture. Pollen can be liberated by squashing anthers in liquid culture medium and subsequently collected via filtration. Isolated pollen can be cultured on complex medium or using nurse culture (reviewed by Nitsch, 1971). Glutamine and serine appear to be important additives to pollen culture medium.

Applications of Haploids

Haploids are integrated into a breeding program as homozygous diploid lines. Diploidized haploid lines can be produced using a colchicine treatment or alternately may be produced using leaf midrib culture (Kasperbauer and Collins, 1972). Selection from a population of doubled haploids, particularly an F_1 population, can result in rapid fixation of desired characteristics. Griffing (1975) has statistically demonstrated that use of doubled haploids, when rapidly obtained, is more efficient than conventional selection.

Doubled haploids have been incorporated into breeding programs to produce new cultivated varieties of barley (Kasha and Reinbergs, 1980) and tobacco (Collins and Genovesi, 1982). Although not used as extensively in other crop species, doubled haploids have also been used in asparagus, *Brassica*, rice, corn, barley, wheat, rye, and potato. In asparagus, anther culture of mature male (XY) plants can be used when chromosomes are doubled to produce supermales (YY). The supermales, when crossed with females (XX), have been very useful in establishment of high-yielding, uniform all-male (XY) lines of asparagus (Dore, 1974). Doubled haploid techniques are being utilized in *Brassica* for breeding for lowered glucosinolate content (Nitzsche and Wenzel, 1977; Wenzel et al., 1977). Current varietal development in rice (Yin et al., 1976) and peppers (Pochard and Vaulx, 1971) also involves doubled haploids. In corn, doubled haploids are being utilized for analytical breeding. Because of the genetic uniformity of doubled haploids when compared to regular diploid lines, doubled haploids of inbred lines are used for evaluation of the breeding lines for parents of F_1 hybrids. Following the analysis, F_1 hybrids are constructed with more confidence (Nitzsche and Wenzel, 1977).

Doubled haploids also can be used in genetic analysis to study the inheritance of qualitative and quantitative traits. Because haploids and doubled haploids exhibit the genotype of the gamete, ratios obtained are less complex if doubled haploids rather than diploids are used for genetic analysis (Collins and Legg, 1980). Doubled haploids have been utilized for the genetic analysis of several quantitative traits in tobacco. In one study, biparental progenies of doubled haploid plants were used to estimate genetic variance for morphological, agronomic, and chemical traits in a population of burley tobacco (Legg and Collins, 1974). Procedures for using doubled haploids to estimate additive and additive x additive genetic variances have also been outlined (Choo et al., 1979). Haploids are useful in cellular mutagenesis programs as recessive alleles are expressed in haploids while they may be undetected in diploids. Anther culture also can be used to manipulate chromosome number in somatic hybrids (Schieder, 1978; Evans et al., 1982) and to generate nullisomics useful for genetic analysis (Moore and Collins, 1982).

CELLULAR AND MOLECULAR GENETICS

Techniques developed for the plating of single cells and for the selection of desired genotypes at the cellular level, suggest that cellular genetics can be used to produce genetically modified plants. It has been suggested that by using cell suspension culture, large numbers of cells can be screened for the isolation of agriculturally useful variants

(Parke and Carlson, 1980). The underlying requirement for successful agricultural application of cellular genetics is that variation identified in cultured cells continues to be expressed in regenerated plants. Variation can be introduced into cultured cells via mutagenesis, protoplast fusion, or uptake of exogenous DNA. Variant plants must be introduced or integrated into a conventional breeding program to result in development of new varieties. Although cultivated varieties have not yet been released using cellular genetics, this area of research is perhaps the most promising research area for future development of new varieties.

Cultured plant cells have been used for isolation of mutants in vitro with subsequent plant regeneration. This area has been reviewed recently (Chaleff, 1981). Often, mutants have been induced for resistance to chemicals. Variant cell lines resistant to amino acid analogs have been used to understand biochemical pathways of amino acid biosynthesis. Unfortunately, most amino acid analog-resistant lines have not been genetically characterized. These lines, though, are often quite stable, even when cultures are reinitiated from regenerated plants (Flick et al., 1981). Other resistant mutants have been used to delineate biochemical pathways, including lysine-threonine resistance in corn (Hibberd et al., 1980), chlorate resistance in tobacco (Muller and Grafe, 1978), and bromodeoxyuridine resistance in soybean (Ohyama, 1976).

In addition to analysis of biochemical pathways, certain resistance mutants may be directly applicable for crop improvement. Herbicide resistance could be used to develop efficient weed control programs in certain species. Cell lines have been used to select for resistance to 2,4-D, paraquat, and picloram. Picloram-resistant tobacco mutants isolated in vitro, have been genetically characterized (Chaleff, 1980). Pathotoxin resistance could be used to develop varieties resistant to crop pathogens. Lines have been developed with resistance to *Phytophtora* toxin in potato (Behnke, 1980) and to *Helminthosporium* toxin in maize (Gengenbach et al., 1977). The *Helminthosporium* resistance is under maternal control. Resistance to antibiotics may have unique application to crop improvement. As resistance to antibiotics has been demonstrated to be encoded in cytoplasmic DNA, antibiotic resistance may be used to mark unique plant cytoplasms. The ability to distinguish cytoplasms may prove useful in somatic hybridization research particularly in efforts to transfer cytoplasmic male sterility or other organelle-encoded traits.

Two other classes of mutants reported in the literature less frequently are temperature-sensitive/lethal (ts) mutants and auxotropic mutants. In each case these mutants are probably recessive, so that the development of haploid cell lines is particularly important. Unfortunately, when haploid lines have been used, plants recovered have not retained the haploid karotype. For example, the ts mutant isolated from haploid *N. tabacum* (n = 24), contained 37 chromosomes (Malberg, 1980), and mutants isolated from haploid *D. innoxia* (n = 12) could not be regenerated (King et al., 1980).

Protoplast fusion has often been suggested as a means to develop unique hybrid plants impossible to achieve via conventional sexual hybridization. While any two plant cells can be fused using a modified-

PEG method, creation of unique somatic hybrid plants is limited by the ability to regenerate plants from isolated plant protoplasts.

Protoplasts are obtained by enzymatically digesting plant cell walls with cellulase and pectinase dissolved in a concentrated sugar solution. When the enzyme is removed, protoplasts from two species can be mixed, then fused using polyethylene glycol or other means, and rinsed and cultured in appropriate medium.

Following the fusion treatment, the protoplasts regenerate cell walls and undergo mitosis. Fusion products must be distinguished within a mixture of parental cells and homokaryotic and heterokaryotic fusion products. Usually genetic complementation, particularly correction of albino mutants, has been used to visually identify fusion products (see review by Evans, 1982). Albino mutants have been used to visually identify fusion products in *Datura, Daucus, Nicotiana,* and *Petunia.* Use of in vitro isolated mutants for amino acid analog resistance and chlorate resistance has been proposed, though not yet extensively used, to identify somatic hybrids. Interspecific somatic hybrid plants have been produced in the following genera: *Datura, Daucus, Nicotiana, Petunia,* and *Solanum.* Recently, intergeneric plants have been reported in four species combinations. In each case only aneuploid plants were recovered and the gross morphological abnormalities were reported for each combination. These intergeneric combinations include: *Solanum tuberosum + Lycopersicon esculentum* (Melchers et al., 1978), *Atropa belladonna + Datura innoxia* (Krumbiegel and Schieder, 1979), *Daucus carota + Aegopodium podagraria* (Dudits et al., 1980), *Arabidopsis thaliana + Brassica campestris* (Gleba and Hoffmann, 1980), and *Nicotiana chinensis + Atropa belladonna* (Gleba et al., 1982). Unfortunately, none of these fusion products has resulted in crop improvement, as the plants produced in each of these fusion experiments are abnormal and sterile.

As distant hybrids produced by protoplast fusion have not resulted in crop improvement, somatic hybridization should be critically analyzed as an aid to crop improvement. The primary limitation to application of protoplast technology is certainly the inability to regenerate plants from protoplasts. Numerous hybrids could be proposed in cereal and legume programs, but very little success has been reported in plant regeneration from these important crops. This limitation currently precludes application of this technology to cereals and legumes.

Protoplast fusion can be used to produce unique nuclear-cytoplasmic combinations not possible using conventional breeding (Gleba, 1979). As information on cytoplasmic genetics accumulates, greater value may be achieved from intra- and interspecies fusion products. Protoplast fusion has been suggested as a means to transfer cytoplasmically controlled male sterility (e.g., Izhar and Power, 1979) between breeding lines.

We have developed techniques for the production of unique somatic hybrids in the genus *Nicotiana* (Evans et al., 1981a). Production of somatic hybrids has been directed toward production of agriculturally useful somatic hybrids. Protoplasts from leaves of *N. nesophila* were fused with cell suspension culture protoplasts from an albino mutant of cultivated *N. tabacum.* Hybrid plants produced by protoplast fusion were distinguished as light green shoots and were verified as somatic hybrids using morphological, chromosomal, isoenzymic, and genetic criteria. Stable incorporation of disease resistance was tested using resistance to tobacco mosaic virus. These somatic hybrids should be resistant to numerous diseases of cultivated tobacco, and seed from backcrosses (BC) is being evaluated for disease response. Useful material will be incorporated into a breeding program to develop resistant varieties.

Numerous attempts have been made to use exogenous DNA to induce transformation in higher plants. While early reports were directed toward injection of exogenous bacterial or plant DNA—into seeds of isolated pollen grains—difficulty in analysis of the fate of the exogenous DNA has hampered the widespread application of this method of DNA feeding. The success of these whole plant experiments has been critically reviewed (Kleinhofs and Behki, 1977). More recently, interest has accumulated in the transformation of cultured plant cells or perhaps more appropriately, isolated plant protoplasts (Lurquin and Kado, 1977). The successful transfer of DNA may require using naturally ineffective DNA from a plant pathogen. Use of the *Agrobacterium tumefaciens* Ti-plasmid and cauliflower mosaic virus has been proposed. As exogenous DNA is available for digestion by host cell endonucleases, exogenous DNA could be encapsulated in liposomes to afford protection (Matthews et al., 1979). Following DNA uptake, expression and stable inheritance of the foreign DNA is required.

CONCLUDING REMARKS

The potential agricultural application of plant cell culture to crop improvement is astonishing. Cell culture represents a collection of techniques, that when carefully integrated into an existing breeding program, should result in development of unique germplasm. In addition, as techniques of cellular genetics continue to develop, the possibility exists to alter the nature of future crop species. Even if only the less technical methods of meristem culture and plant regeneration for cloning and genetic variability are utilized, plant cell culture should have a tremendous impact on crop improvement in the near future. The hope is that the remaining chapters in this volume will aid the reader in establishing guidelines for application of cell culture for crop improvement of specific crops.

REFERENCES

Ammirato, P.V. 1983. Embryogenesis. In: Handbook of Plant Cell Culture, Vol. 1 (D.A. Evans, W.R. Sharp, P.V. Ammirato, and Y. Yamada, eds.) pp. 82-123. Macmillan, New York

Anderson, W.C., Meagher, G.W., and Nelson, A.G. 1977. Cost of propagating broccoli plants through tissue culture. HortScience 12:543-544.

Behnke, M. 1980. Selection of dihaploid potato callus for resistance to the culture filtrate of *Fusarium oxysporum*. Z. Pflanzenzuecht. 85: 254-258.

Bingham, E.T., Hurley, L.V., Kaatz, D.M., and Saunders, J.W. 1975. Breeding alfalfa which regenerates from callus tissue in culture. Crop Sci. 15:719-721.

Bryan, H.H., Stall, W.M., Gray, D., and Richmond, N.S. 1978. Fluid drilling of pre-germinated seeds for vegetable gardening. Proc. Fla. State Hortic. Soc. 91:88-90.

Burk, L.G. 1970. Green and light-yellow haploid seedlings from anthers of sulfur tobacco. J. Hered. 61:279.

______, Gerstel, D.U., and Wernsman, E.A. 1979. Maternal haploids of *Nicotiana tabacum* L. from seed. Science 206:585.

Chaleff, R.S. 1980. Further characterization of picloram-tolerant mutants of *Nicotiana tabacum*. Theor. Appl. Genet. 58:91-95.

______ 1981. Genetics of Higher Plants—Applications of Cell Culture. Cambridge Univ. Press, Cambridge.

Choo, T.M., Christie, B.R., and Reinbergs, E. 1979. Doubled haploids for estimating genetic variances and a scheme for population improvement in self-pollinating crops. Theor. Appl. Genet. 54:267-271.

Collins, G.B. and Genovesi, A.D. 1982. Anther culture and its application to crop improvement. In: Application of Plant Cell and Tissue Culture to Agriculture and Industry (D.T. Tomes, B.E. Ellis, P.M. Harney, K.J. Kasha, and R.L. Peterson, eds.) pp. 1-24. Univ. of Guelph Press, Guelph, Ontario.

Collins, G.B. and Legg, P.D. 1980. Recent advances in the genetic applications of haploidy in *Nicotiana*. In: The Plant Genome (D.R. Davies and D.A. Hopewood, eds.) pp. 197-213. The John Innes Charity, Norwich, England.

Currah, I.E. 1977. Fluid Drilling Research. National Vegetable Res. Station Report. Wellesbourne, Warwick, England.

Dore, C. 1974. Production de plantes homozygotes males et femelles a partir d'antheres d'asperge cultivees in vitro (*Asparagus officinalis* L.). C.R. Acad. Sci. Paris 278:2135-2138.

Dudits, D., Fejer, O., Hadlaczky, G., Koncz, C., Lazar, G.B., and Horvath, G. 1980. Intergeneric gene transfer mediated by plant protoplast fusion. Mol. Gen. Genet. 179:283-288.

Evans, D.A. 1982. Plant regeneration and genetic analysis of somatic hybrid plants. In: Plants Regenerated from Tissue Culture (L. Earle and Y. Demarly, eds.) pp. 303-323. Praeger Press, New York.

______ and Sharp, W.R. 1983. Single gene mutations in tomato plants regenerated from tissue culture. Science 221:949-951.

______, Flick, C.E., and Jensen, R.A. 1981a. Incorporation of disease resistance into sexually incompatible somatic hybrids of the genus *Nicotiana*. Science 219:907-909.

______, Sharp, W.R., and Flick, C.E. 1981b. Plant regeneration from cell culture. In: Horticultural Reviews, Vol. 3 (J. Janick, ed.) pp. 211-311. AVI Press, Westport, Connecticut.

______, Flick, C.E., Kut, S.A., and Reed, S.M. 1982. Comparison of *Nicotiana tabacum* and *Nicotiana nesophila* hybrids produced by ovule culture and protoplast fusion. Theor. Appl. Genet. 62:193-198.

Flick, C.E., Jensen, R.A., and Evans, D.A. 1981. Isolation, protoplast culture, and plant regeneration of PFP-resistant variants of *N. tabacum* (Su/Su). Z. Pflanzenphysiol. 103:239-245.

Gengenbach, B.G., Green, C.E., and Donovan, C.M. 1977. Inheritance of selected pathotoxin resistance in maize plants regenerated from cell cultures. Proc. Natl. Acad. Sci. U.S.A. 74:5113-5117.

Gleba, Yu.Yu. 1979. Nonchromosomal inheritance in higher plants as studied by somatic cell hybridization. In: Plant Cell and Tissue Culture: Principles and Applications (W.R. Sharp, P.O. Larsen, E.F. Paddock, and V. Raghavan, eds.) pp. 775-788. Ohio State Univ. Press, Columbus.

______ and Hoffmann, F. 1980. "Arabidobrassica": A novel plant obtained by protoplast fusion. Planta 149:112-117.

______, Momot, V.P., Cherup, N.N., and Skarzynskaya, M.V. 1982. Intertribal hybrid cell lines of *Atropa belladonna* (x) *Nicotiana chinensis*. Theor. Appl. Genet. 62:75-79.

Gregorini, G. and Lorenzi, R. 1974. Meristem-tip culture of potato plants as a method of improving productivity. Potato Res. 17:24-33.

Griffing, B. 1975. Efficiency change due to use of doubled-haploids in recurrent selection methods. Theor. Appl. Genet. 46:367-385.

Guha, S. and Maheshwari, S.C. 1964. In vitro production of embryos from anther of *Datura*. Nature 204:497.

Heinz, D.J., Krishnamurthi, M., Nickell, L.G., and Maretzki, A. 1977. Cell, tissue, and organ culture in sugarcane. In: Plant Cell, Tissue and Organ Culture (J. Reinert and Y.P.S. Bajaj, eds.) pp. 3-17. Springer-Verlag, Berlin.

Hibberd, K.A., Walters, T., Green, C.E., and Gengenbach, B.G. 1980. Selection and characterization of a feedback-insensitive tissue culture of maize. Planta 148:183-187.

Hollings, M., Stone, O.M., and Dale, W.T. 1972. Tomato ringspot virus in *Pelargonium* in England. Plant Pathol. (London) 21:46-47.

Hu, C.Y. and Wang, P.J. 1983. Meristem and shoot tip cultures. In: Handbook of Plant Cell Culture, Vol. 1 (D.A. Evans, W.R. Sharp., P.V. Ammirato, and Y. Yamada, eds.). Macmillan, New York.

Izhar, S. and Power, J.B. 1979. Somatic hybridization in *Petunia*: A male sterile hybrid. Plant Sci. Lett. 14:49-55.

Kartha, K.K., Leung, N.L., and Gamborg, O.L. 1979. Freeze-preservation of pea meristems in liquid nitrogen and subsequent plant regeneration. Plant Sci. Lett. 15:7-15.

Kasha, K.J. and Reinbergs, E. 1980. Achievements with haploid in barley research and breeding. In: The Plant Genome (D.R. Davies and D.A. Hapwood, eds.) pp. 215-230. The John Innes Charity, Norwich.

Kasperbauer, M.J. and Collins, G.B. 1972. Reconstitution of diploid from leaf tissue of anther-derived haploids in tobacco. Crop Sci. 12: 98-101.

Keller, W.A. and Armstrong, K.C. 1979. Stimulation of embryogenesis and haploid production in *Brassica campestris* anther cultures by elevated temperature treatments. Theor. Appl. Genet. 55:65-67.

Kermicle, J.L. 1969. Androgenesis conditioned by a mutation in maize. Science 166:1422-1424.

King, J., Horsch, R.B., and Savage, A.D. 1980. Partial characterization of two stable auxotrophic cell strains of *Datura innoxia* Mill. Planta 149:480-484.

Kleinhofs, A. and Behki, R. 1977. Prospects for plant genome modification by nonconventional methods. Annu. Rev. Genet. 11:79-101.

Krumbiegel, G. and Schieder, O. 1979. Selection of somatic hybrids after fusion of protoplasts from *Datura innoxia* Mill. and *Atropa belladonna* L. Planta 145:371-375.

Larkin, P.J. and Scowcroft, W.R. 1981. Somaclonal variation—a novel source of variability from cell cultures for plant improvement. Theor. Appl. Genet. 60:197-214.

Legg, P.D. and Collins, G.B. 1974. Genetic variances in a random-intercrossed population of burley tobacco. Crop Sci. 14:805-808.

Levy, L.W. 1981. A large-scale application of tissue culture: The mass propagation of pyrethrum clones in Ecuador. Environ. Exp. Bot. 21:389-395.

Lurquin, P.F. and Kado, C.I. 1977. *Escherichia coli* plasmid pBR313 insertion into plant protoplasts and into their nuclei. Mol. Gen. Genet. 154:113-121.

Malmberg, R.L. 1980. Biochemical, cellular, and developmental characterization of a temperature-sensitive mutant of *Nicotiana tabacum* and its second site revertant. Cell 22:603-609.

Matthews, B., Dray, S., Widholm, J., and Ostro, M. 1979. Liposome-mediated transfer of bacterial RNA into carrot protoplasts. Planta 145:37-44.

Melchers, G., Sacristan, M.D., and Holder, A.A. 1978. Somatic hybrid plants of potato and tomato regenerated from fused protoplasts. Carlsburg Res. Comm. 43:203-218.

Moore, G.A. and Collins, G.B. 1982. Identification of aneuploids in *Nicotiana tabacum* by isozyme banding patterns. Biochem. Genet. 20: 555-568.

Muller, A.J. and Grafe, R. 1978. Isolation and characterization of cell lines of *Nicotiana tabacum* lacking nitrate reductase. Mol. Gen. Genet. 161:67-76.

Mullin, R.H., Smith, S.H., Frazier, N.W., Schlegel, D.E., and McCall, S.R. 1974. Meristem protoplast frees strawberries of mild yellow edge, pallidosis and mottle diseases. Phytopathology 64:1425-1429.

Murakishi, H.H. and Carlson, P.S. 1976. Regeneration of virus-free plants from dark green islands of tobacco mosaic infected tobacco leaves. Phytopathology 66:931-932.

Murashige, T. 1974. Plant propagation through tissue culture. Annu. Rev. Plant Physiol. 25:135-166.

______ 1978. The impact of tissue culture on agriculture. In: Frontiers in Plant Tissue Culture (T.A. Thorpe, ed.) pp. 15-26. Univ. of Calgary Press, Calgary.

Nakamura, A., Yamada, T., Kadotani, N., Itagaki, R., and Oka, M. 1974. Studies on the haploid method of breeding in tobacco. Sabrao J. 6:107-131.

Nakata, K. and Kurihara, T. 1972. Competition among pollen grains for haploid tobacco plant formation by anther culture. II. Analysis with resistance to tobacco virus (TMV) and wildfire diseases, leaf color, and leaf base shape characters. Japan. J. Breed. 22:92-98.

Nitsch, C. 1974. Pollen culture: A new technique for mass production of haploid and homozygous plants. In: Haploids in Higher Plants, Advances and Potential (K.J. Kasha, ed.) pp. 123-135. Univ. of Guelph Press, Ontario, Canada.

Nitsch, J.P. 1971. La production in vitro d'embryons haploides: Resultats et perspectives. Colloq. Int. C.N.R.S. 193:282-294.

Nitzsche, W. and Wenzel, G. 1977. Haploids in plant breeding. Verlag Paul Parey, Berlin.

Oglesby, R.P. 1978. Tissue cultures of ornamentals and flowers: Problems and perspectives. In: Propagation of Higher Plants through Tissue Culture (K.W. Hughes, R. Henke, and M. Constantin, eds.) pp. 59-61. U.S. Dept. Energy, Washington, D.C.

Ohyama, K. 1976. A basis for bromodeoxyuridine resistance in plant cells. Environ. Exp. Bot. 16:209-216.

Parke, D. and Carlson, P.S. 1980. Somatic cell genetics of higher plants: Appraising the application of bacterial systems to higher plant cells cultured in vitro. In: Physiological Genetics (J.G. Scandalios, ed.) pp. 196-237. Academic Press, New York.

Phillips, G.C. and Collins, G.B. 1979. In vitro tissue culture of selected legumes and plant regeneration from callus culture of red clover. Crop Sci. 19:59-64.

Pochard, E. and Dumas de Vaulx, R. 1971. La monoploidie chez le piment (*Capsicum annuum* L.). Realisation pratique d'un cycle de selection accelere par passage a l'etat monoploide en trisieme generation. Z. Pflanzenzuecht. 65:23-46.

Quak, F. 1977. Meristem culture and virus-free plants. In: Plant, Cell, Tissue and Organ Culture (J. Reinert and Y.P.S. Bajaj, eds.) pp. 598-615. Springer-Verlag, Berlin.

Rogers, H.T. 1982. Plugging away at labor-less transplanting. Am. Veg. Grower 30(5):8-10.

Sacristan, M.D. 1971. Karyotypic changes in callus cultures from haploid and diploid plants of *Crepis capillaris* (L.) Wallr. Chromosoma 33:273-283.

Schenk, R.U. and Hildebrandt, A.C. 1972. Medium and techniques for induction and growth of monocotyledonous and dicotyledonous plant cell cultures. Can. J. Bot. 50:199-204.

Schieder, O. 1978. Genetic evidence for the hybrid nature of somatic hybrids from *Datura innoxia* Mill. Planta 141:333-334.

Sharp, W.R., Reed, S.M., and Evans, D.A. 1983. Production and application of haploid plants. In: Novel Techniques in Plant Breeding (P.B. Vose and S. Blixt, eds.) Pergamon Press, Oxford (in press).

Shepard, J.M. 1975. Regeneration of plants from protoplasts of potato virus x-infected tobacco leaves. Virology 66:492-501.

______ 1982. The regeneration of potato plants from leaf cell protoplasts. Sci. Am. 246:154-166.

Sibi, M. 1978. Multiplication conforme, non conforme. Le Selectionneur Francais 26:9-18.

______ 1981. Variants epigeniques et cultures in vitro chez *Lycopersicon esculentum* L. In: Application de la culture in vitro a L'Amelioration des Plantes Potageres (C. Dove, ed.) pp. 179-185. Institut National de la Recherche Agronomique, Versailles, France.

Skirvin, R.M. 1978. Natural and induced variation in tissue culture. Euphytica 27:241-266.

Sunderland, N. 1974. Anther culture as a means of haploid induction. In: Haploids in Higher Plants--Advances and Potential (K.J. Kasha, ed.) pp. 91-122. Univ. Guelph, Ontario, Canada.

Syono, K. and Furuya, T. 1972. Studies on plant tissue culture 18. Abnormal flower formation of tobacco plants regenerated from callus cultures. Bot. Mag. (Tokyo) 85:273-284.

Tabata, M. and Hiraoka, N. 1976. Variation of alkaloid production in *Nicotiana rustica* callus cultures. Physiol. Plant. 38:19-23.

Walkey, D.G.A. 1976. High temperature inactivation of cucumber and alfalfa mosaic virus in *Nicotiana rustica* cultures. Ann. Appl. Biol. 84:183-192.

______ and Cooper, V.C. 1975. Effect of temperature on virus eradication and growth of infected tissue cultures. Ann. Appl. Biol. 80:185-190.

Wang, P.J. and Hu, C.Y. 1980. Regeneration of virus-free plants through in vitro culture. In: Advances in Biochemical Engineering, Vol. 18 (A. Fiechter, ed.) pp. 61-99. Springer-Verlag, New York.

Warren, G.S. and Fowler, M.W. 1977. A physical method for the separation of various stages in the embryogenesis of carrot cell cultures. Plant Sci. Lett. 9:71-76.

Wenzel, G., Hoffmann, F., and Thomas, E. 1977. Anther culture as a breeding tool in rape. I. Ploidy level and phenotype of androgenetic plants. Z. Pflanzenzuecht. 78:149-155.

Yamada, Y. and Fujita, Y. 1983. Production of useful compounds in culture. In: Handbook of Plant Cell Culture, Vol. 1 (D.A. Evans, W.R. Sharp, P.V. Ammirato, and Y. Yamada, eds.) pp. 717-728. Macmillan, New York.

Yin, K.C., Hsu, C., Chu, C.Y., Pi, F.Y., Wang, S.T., Liu, T.Y., Chu, C.C., Wang, C.C., and Sun, C.S. 1976. A study of the new cultivar of rice raised by haploid breeding method. Sci. Sin. 19:227-242.

Zenk, M.H. 1978. The impact of plant cell culture on industry. In: Frontiers of Plant Tissue Culture 1978 (T.A. Thorpe, ed.) pp. 1-13. Univ. of Calgary Press, Calgary.

SECTION II
Cereals

CHAPTER 3
Maize

P. King and *K. Shimamoto*

Maize (*Zea mays* L.) is a domestic annual species in the Maydeae tribe of the grass family, Gramineae. The plant has commonly a single sturdy stem bearing long, broad, alternate leaves with occasional shorter basal tillers (Fig. 1). The monoeceious trait of maize is unusual among the major cereals and, apart from the obvious advantages to the geneticist for easy pollination control, the separation of the sexes allowed a high degree of specialization and adaptation, particularly of the female inflorescence, during the domestication of the crop. The ear terminates a lateral branch (usually just one) and can produce from 400 to 1000 ovules. The male inflorescence (the tassel) terminating the main stem produces about 10,000 anthers, and clouds of pollen (about 2×10^7 grains per plant) are released by air currents agitating the tassel. The pollen:ovule ratio of 20,000:1 insures efficient pollination including a high frequency of cross pollination. The duration of the life cycle is, like many characters, highly variable but is typically ca. 100 days (Kiesselbach, 1949).

The exact ancestry of maize is still somewhat controversial. The "teosinte theory" is perhaps the most generally acceptable now, namely, that maize arose either directly from teosinte or from hybridization of teosinte and some unknown species of the same tribe (Andropogeae). A somewhat opposing view suggests that pod corn, with each kernel individually enclosed in long glumes, is the ancestral form and that introgression with teosinte, a proposed derivative of maize, has contributed to the evolution of modern maize. The controversy revolves around the age of the many-ranked spikelets/condensed internode habit of maize which retains the ear within a blanket of many husk leaves thus

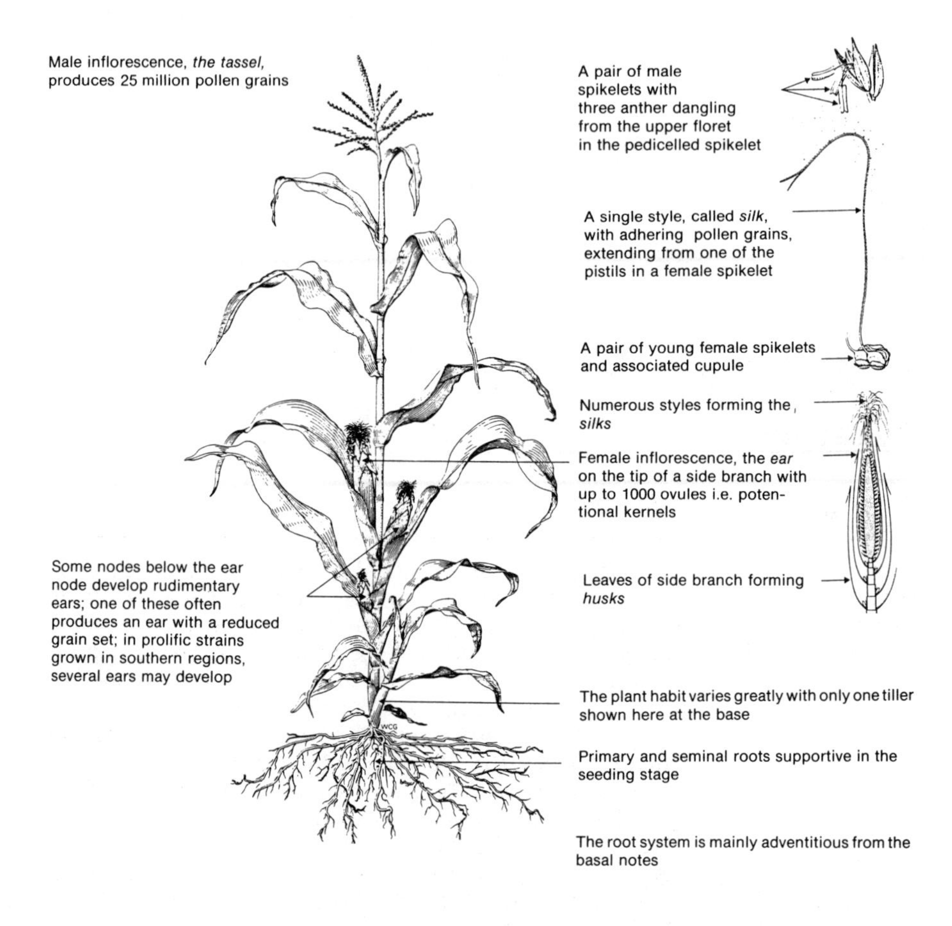

Figure 1. The morphology of maize. (Reproduced from "MAIZE"—a technical monograph published by Ciba-Geigy AG, Basel, with the permission of Dr. W.C. Galinat.)

preventing seed dispersal and making the modern species incapable of surviving in the wild (Galinat, 1977; Beadle, 1978).

Maize is characterized by a high degree of genetic heterozygosity built up by persistent out-crossing. As the cultivation of early maize spread to different geographical regions there was a rapid evolution of many races adapted to a wide variety of growing conditions. Thus, today there is a bewildering complex of natural geographic races as well as a classification into pop, flint, floury, dent, and sweet based mainly upon the nature of the endosperm of the kernels. The most productive race, the USA Corn Belt dent, is a hybrid of northern flints and southern dents. Inbreeding experiments begun in the early 1900s have produced a massive array of inbred lines from the original eighteenth-century open-pollinated dents. During this inbreeding phase, as the genetic variability in maize was brought under control, the variability was also systematically studied, with the result that more is known about the cytology and genetics of maize than any other plant species. Approximately 350 loci relating to determination of colors, chemical constituents, and development have been mapped on the 10 chromosomes. Another 600 loci have been identified but not yet completely defined (Coe and Neuffer, 1977).

Despite the wide spectrum of genetic variability available in maize and close relatives, only a very small fraction is represented in the present commercially favored hybrids; the inbred parents of the major Corn Belt hybrids trace back either to Lancaster Sure Crop or the Iowa Stiff Stalk Synthetic. Only six hybrids account for more than 70% of the maize crop in the U.S. corn belt and there is growing concern about the vulnerability of such a narrow-based germplasm. About 385 million tons of maize is produced per year on more than 120 million hectares of moist, fertile land in the temperate, subtropical, and higher tropical regions. Whilst both the production and total area is less than either of wheat or rice, the yield of maize per unit area is higher than any other food crop. Apart from the manufacture of mixed feed, maize products include corn starch, maltodextrins, corn oil and corn syrups of the large milling industry, and well-known products of the fermentation and distilling industries (Steele, 1978; Watson, 1977).

Breeding objectives in maize fall into three general groups:

1. Yield. As yet it has not been possible to identify many single genetic loci and phenotypes which have a simple impact on yield. Yield is by definition a quantitative trait and may be viewed as controlled by many genes, each with a small additive effect. The modern approach to increased yield aims at the increase of this additive variance in breeding populations by recurrent selection and the evaluation of hybrid combinations following elaborate statistical genetic models.
2. Environment. Most of the environmental stresses experienced by a maize plant (high temperature, cold, drought, flooding, salt) are basically water stresses and ways to improve the water economy of the plant are of great interest.
3. Disease and insect damage. Worldwide, about 9% of maize output is lost due to pathogenic microorganisms. However, some diseases can cause losses in excess of 50% in particular outbreaks. Among the worst are the stalk rots, caused by a combination of several different fungi and bacteria. Resistance to southern corn leaf blight (*Helminthosporium maydis*) or some other method of combat-

ing the disease would alleviate the difficulties caused by the disastrous sensitivity of maize hybrids containing Texas male-sterile cytoplasm to new races of the pathogen. In the United States (the major maize producer) there are about 25 insect species persistently causing economic losses, most of them soil insects. The European corn borer, the western and northern rootworms, and the corn leaf aphid are the primary, consistent pests. In some cases breeding for resistance has reduced infestation but control relies heavily on insecticide application (Sprague, 1977; Walden, 1978).

Maize is not only of vital economic importance but it is also a plant full of potential for plant geneticists and, at the moment, a challenge to many plant cell biologists who would like to see the useful aspects of tissue culture applied to plant breeding.

EVOLUTION OF MAIZE TISSUE CULTURE

The history of maize tissue culture probably begins with the attempts of La Rue to continuously culture maize endosperm (La Rue, 1947). La Rue (1949) eventually established long-term cultures and Black Mexican Sweet and to a lesser degree other sweet corns became the standard for use in many different types of studies (Straus and La Rue, 1954; Straus, 1958). Attempts to produce starchy endosperm cultures first succeeded in 1970 when Shannon and Batey (1973) obtained cultures of the dent inbreds A636 and R168. Reddy and Peterson (1977) subsequently obtained waxy endosperm cultures by backcrossing wx-m8 into R168. These successes indicate genetic variation for tissue culture response in maize and its genetic basis was first studied by Tabata and Motoyoshi (1965). Unfortunately we still have no idea why these particular genotypes alone respond. As subsequent studies have shown, the commonly applied culture conditions play a minor role compared to choice of the right genotype in improving a particular response to maize tissue culture.

A program begun in 1964 at the National Chemical Laboratory, Poona, India resulted in continuously propagated tissue cultures of several important cereals including maize (Mascarenhas et al., 1969). The cultures were initiated from seedling shoot sections placed on a wide range of culture media. Subsequent observations published in 1975 (Mascarenhas et al., 1975a,b; Hendre et al., 1975) gave the first clues to the unhappy results that are still obtained when cultures are established from differentiated cereal tissues: (a) Cultures consist only of proliferating roots and callus cell lines rarely appear; those that do appear are not morphogenic. (b) It is, therefore, not possible to obtain suspension cultures by simply immersing the initial culture in liquid medium. (c) There are differences in the growth and behavior of tissue cultures from different cereals with maize being the least satisfactory. (d) Shoots do not appear from cultures unless the initial explant contained the shoot apex. A similar picture emerges from the other reports of maize tissue culture at this time in which various explants were used: immature tassel meristems (Linsmaier-Bednar and Bednar, 1972), macerated, imbibed mature embryos (Gresshof and Doy, 1973), seedling stem sections (Sheridan, 1974), mature embryos, and stem apices (Green et al., 1974). The nature of typical maize cultures was first critically examined by Mott and Cure (1978), Cure and Mott (1978), and King et al. (1978).

In 1975, two reports appeared dealing with two new extremes of maize tissue culture. First, Sheridan (1975) established a finely dispersed liquid suspension culture from Black Mexican Sweet beginning with seedling stem sections. This genotype was the only 1 out of 14 tested which responded in this way and such cultures have been repeatedly isolated in other laboratories from Black Mexican Sweet. This type of culture is suitable for plating experiments and yields dividing protoplasts (see below) but is not morphogenic. Similar cultures have in the meantime been obtained from B73 (Potrykus et al., 1979a), F71 (Bartkowiak, 1981), and G155 x C103 (Polikarpochkina et al., 1979). Second, a routine method for the production of maize tissue cultures from which plants could be regenerated was reported by Green and Phillips (1975). The cultures arose from the scutellum of immature embryos of a specific age and could be subcultured for many months without loss of regenerative ability. Once again particular genotypes (e.g., the inbred A188) out of several tested produced morphogenic cultures, and subsequent testing of several hundred different maize lines (King and Potrykus, unpublished) has revealed a complex gradation of response. Histological examination of such maize cultures (Springer et al., 1979) and studies of other cereals (e.g., *Sorghum* by Dunstan et al., 1980) suggest that the cultures are a complex mixture of adventitiously proliferating embryos, shoots, and roots. The process of plant regeneration from these cultures bears little resemblance to the classical system of plant regeneration in tobacco, in which a nondifferentiated cell population gives rise to organized meristems following a change in hormone concentration in the medium.

Despite their complexity and the fact that many of the plants regenerated from such cultures were already present as multicellular structures, this type of culture has been used with success in selecting for resistance mutants in corn (Gengenbach et al., 1977; Brettell et al., 1980a; Hibberd et al., 1980). Nevertheless, the search for a morphogenic cereal culture of the "tobacco type" has continued. Rapidly growing, undifferentiated cultures which grow as a fine suspension in liquid medium but which differentiate to plantlets via somatic embryogenesis have been reported in *Pennisetum* (Vasil and Vasil, 1981) and rice (Wakizuka, Lazar, Wernicke, and Potrykus, unpublished). Again maize has proved the most troublesome, but recently P. Gordon (Pfizer Genetics) and E. Green (University of Minnesota) have observations as yet unpublished of this tissue culture type. Ziyi et al. (1981) have also obtained an embryogenic, haploid culture from maize anthers.

The establishment of cultures such as those just described is important in the context of cereal protoplasts. Induction of sustained divisions in protoplasts isolated from the tissues of a cereal plant (maize) has been reported only once (Potrykus et al., 1977). But division of protoplasts from cell cultures of the Black Mexican Sweet type described above is routine (e.g., Potrykus, et al., 1979a; Chourey and Zurawski, 1981). The aim in many laboratories now is to achieve protoplast division using fine embryogenic suspension cultures.

The production of haploid tissues in maize presents no special problems. Several cytogenetic systems with suitable markers are available (see general details in King et al., 1978) and methods have been described for isolating morphogenic, haploid cultures from immature embryos (Dhaliwal and King, 1978; Rhodes and Green, 1981). The yields from maize anther culture have also increased considerably over the last ten years but unfortunately only through screening for responsive genotypes. Neither the factors in the anthers nor the medium that determines the response are yet understood. For example, Ting et al. (1981) found ca. 17% of anthers from the Chinese hybrid Dan-San 91 produced calli and embryoids, with more than 50% of the calli regenerating haploid plants.

PROTOCOLS

The best way to learn and adopt a tissue culture method is to study the original paper and visit the author. The authors of the present review use maize tissue culture techniques, but, of course, not all of them. What follows is a guide to the currently available techniques, each with a brief outline of the method to give some idea of the degree of difficulty encountered.

Plant Propagation and Pollination

Maize is not an experimental plant that can be grown routinely all over the world. Not suprisingly, Corn Belt lines grow better in Bloomington, Indiana, than in Basel. Thus field plantings during the summer for experiments bigger than are possible in a greenhouse must be planned carefully to achieve optimum pollination and development. The protocol for propagation follows.

1. Plant extra seed to allow for poor germination.
2. When the season is too short, try preparing seedlings in the greenhouse for later transfer to the field. One result of poor conditions is a breakdown in coordination between silk emergence and anther ripening. Plant several times at intervals, especially when crosses between lines of unknown flowering time are contemplated.
3. If seed ripening is slow, try one of two solutions: take out the embryos and germinate them in vitro to begin the next generation or dig up important plants after seed set and transfer into the greenhouse.

Maize is also not the ideal plant for greenhouse work, as it requires high light intensity (ca. 50,000 lux) and high temperatures, and grows very poorly in pots. Crossing experiments can be done on pot-grown material although there is often poor tassel growth and failure of ear emergence. To produce normally flowering plants the greenhouse should be designed for growing plants in the ground. A good extra crop in Europe can be produced by sowing in a heated but unlighted greenhouse in early February. Such plants which flower at the beginning of April are often superior to a summer field planting when foreign lines are involved.

The protocol for pollination follows.

1. Allow plenty of room between plants when planting for access during pollination time.
2. Cap ears with waterproof bags a few days before silks emerge. Use only the first ear unless the second ear is absolutely necessary.
3. When tassels of pollen parents start to flower (usually they shed pollen for a week), cut off the top 3 cm of bagged ears. (Avoid cutting too deep into the cob). Silks form a uniform "bouquet" 1 or 2 days after cutting and at this stage collect pollen.
4. To collect pollen cover whole tassels with waterproof bags in the morning and collect pollen in the afternoon by shaking the bagged tassels. Use pollen immediately. Alternatively tassels are bagged on the previous day and collected and used for pollinations on the following morning.
5. Cover pollinated ears with bags to avoid furthur pollen contamination. Write pollination dates on the bag.

In Vitro Pollination and Direct Ovule Pollination

1. Cover maize ears with paper bags prior to silk emergence. Ideally the plants should be grown in isolation and the tassels removed prior to anther dehiscence.
2. After silk emergence remove ear from the plant and immediately cut away protruding silks. Remove husk layers one at a time, thoroughly wiping the last layer with ethanol (90% w/v).
3. Remove the last husk layer under sterile conditions and trim the silks to a length of 1-2 cm.
4. Cut the ear into transverse sections 2-3 caryopses thick (Fig. 2a). Slice each section radially to give sections bearing 4-6 caryopses (Fig. 2b) and trim off rachis tissue to make a flat base (Fig. 2c).
5. Place the sections rachis down onto agar-solidified culture medium in 9-cm-diameter petri dishes with the silks overlapping in the center (Fig. 2d). The composition of culture medium is not very critical. The medium of Green and Phillips (1975) minus hormones and with sucrose increased to 5% w/v has been used successfully (Dhaliwal and King, 1978).
6. Hold freshly dehiscing anthers over the centrally arranged silks and tease with a needle to release pollen. No parts of the anther must touch or fall into the petri dish. At least 80% of the dishes remain sterile using this technique and it is not essen-

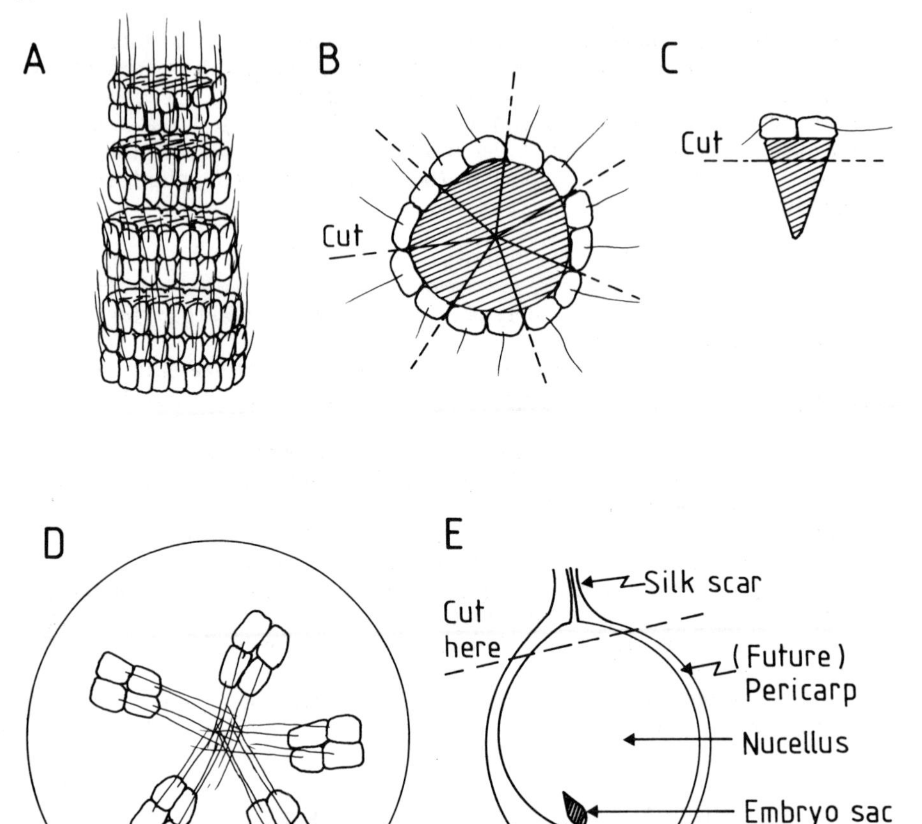

Figure 2. Stages in the preparation of cob sectors for in vitro pollination and caryopsis culture.

tial to trail the silks over the edge of the petri dish for pollination as previously described (Gengenbach, 1977)

7. Incubate the dishes in the dark at 25-30 C.
8. Mature seeds develop at a high frequency (at least 50%).
9. To pollinate maize ovules directly, cut through the ovary wall below the insertion of the silk and remove a cap of tissue to expose the nucellus (Fig. 2e). Apply pollen as described above.
10. Culture media effective for direct maize ovule pollination are B'' (Norstog, 1973) and Green and Phillips (1975) with 400 mg/l CH and 28.9 μM GA.

Initiation of Tissue Cultures from Different Explants

SEEDLING SHOOT AND ROOT. Shoot or root explants of seedlings produce cultures of the "rooting type" relatively easily irrespective of the genotype. However, the establishment of friable cell cultures is possible with only a limited number of genotypes (see earlier section, Evolution of Maize Tissue Culture). The protocol for initiation of cell cultures from seedling shoots and roots follows.

1. Surface-sterilize dry seeds with sodium hypochlorite (2.5% v/v plus a few drops of detergent) for 15 min, shaking vigorously, and rinse three times with sterile distilled water. (Mercuric chloride at 0.01% w/v is also effective.)
2. Shake surface-sterilized seeds at room temperature in sterile distilled water (just enough to cover the seeds) overnight.
3. Repeat Step 1.
4. Place individual seeds on hormone-free culture medium (ca. 5 ml) in 150 x 25 mm glass tubes.
5. Allow to germinate in the light for 5-6 days until seedlings reach the 3-leaf stage.
6. For shoot explants, excise 1-cm segments above the first seedling node (Fig. 3) and slice into disks 1 mm thick.
7. Place 5-10 disks onto culture medium (ca. 10 ml) in a 6-cm diameter petri dish.
8. For root explants cut 1-2-cm segments of primary root and place 5-10 segments onto culture medium. The response may decrease toward the root tip in some genotypes.
9. Incubate dishes in the light (ca. 2000 lux) at 25 C. The commonly used culture medium is Murashige and Skoog (1962) often as modified by Green and Phillips (1975).

IMMATURE EMBRYO. The protocol for the initiation of cell cultures from immature embryos follows.

1. Carry out a controlled pollination (see earlier instructions for pollination).
2. The time required for development of embryos of a size suitable for culturing varies with the genotype and growing conditions.
3. Examine the development of embryos on a pollinated ear by carefully folding down the husks (but do not detach) or opening them up around a longitudinal cut through all layers. Remove individual caryopses by levering them with a blunt spatula. With practice it is possible to remove embryos (see Fig. 4b) rapidly at the plant and determine their size. Embryos of the correct size (ca. 0.5-1.5 mm) are clearly visible.
4. Remove suitable caryopses as described above without damaging the pericarp and rub off the adhering glumes.

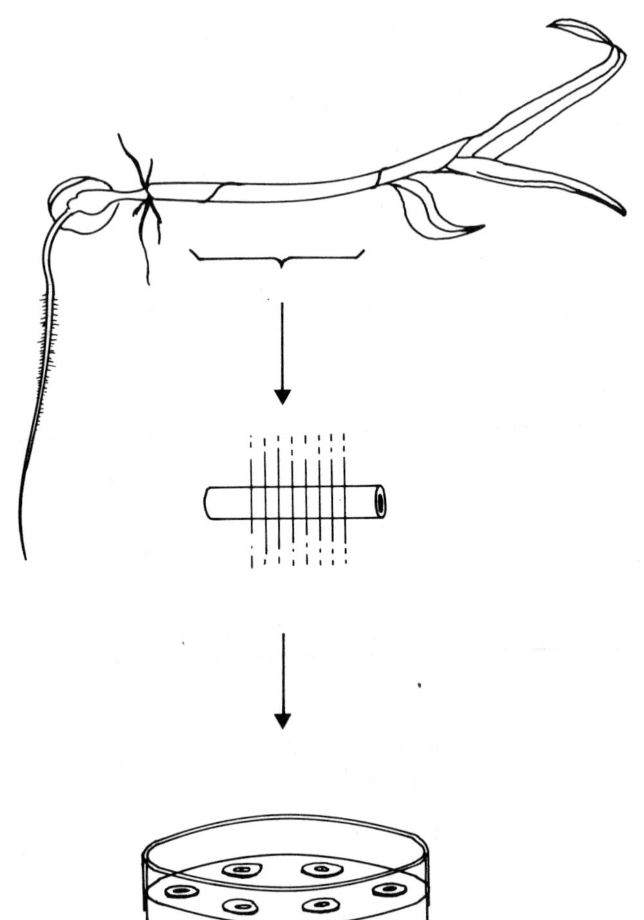

Figure 3. Initiation of tissue cultures from seedling shoot explants.

5. Surface-sterilize the caryopses by shaking vigorously in sodium hypochlorite solution (2.5% w/v plus a few drops of detergent) for 10 min. Remove sterilant and wash three times with sterile, double-glass distilled water. Waste solutions are most easily removed under sterile conditions by sucking them off via a sterile pipette attached through a liquid trap to a vacuum (see Fig. 5).
6. Keep the sterile caryopses moist in the sterilizing container and transfer one at a time to the lid of a sterile petri dish for dissection. First attempts are best carried out under a low-power

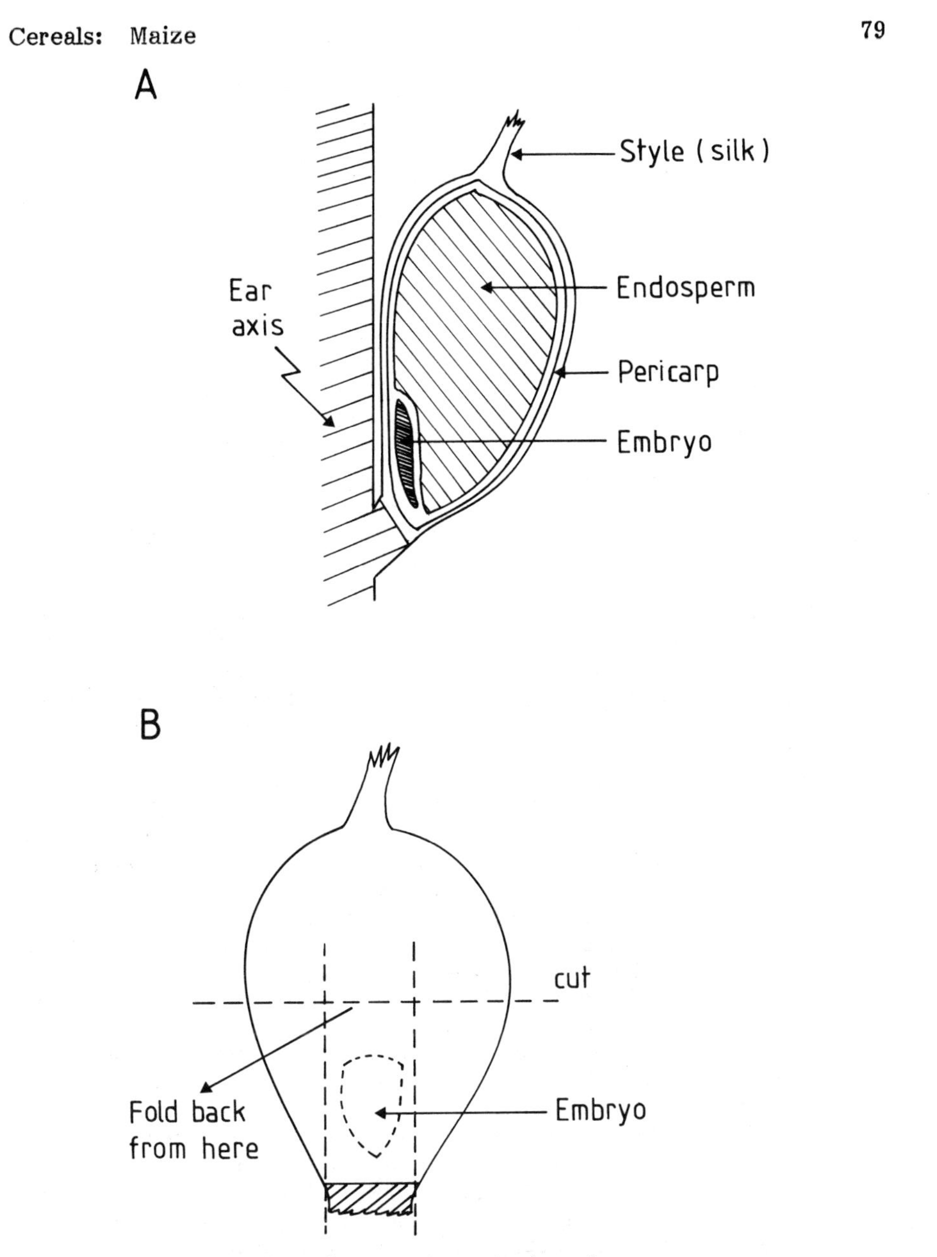

Figure 4. The location of the embryo in immature caryopses.

dissecting microscope wiped over with 70% ethanol and placed on the laminar air flow bench. Place a short strip of Parafilm under the dish to prevent sliding.

7. The embryo lies on the (at this stage) jellylike endosperm towards the base of the caryopsis, just underneath the flattened surface which was pressed against the ear axis (see Fig. 4a). Place the

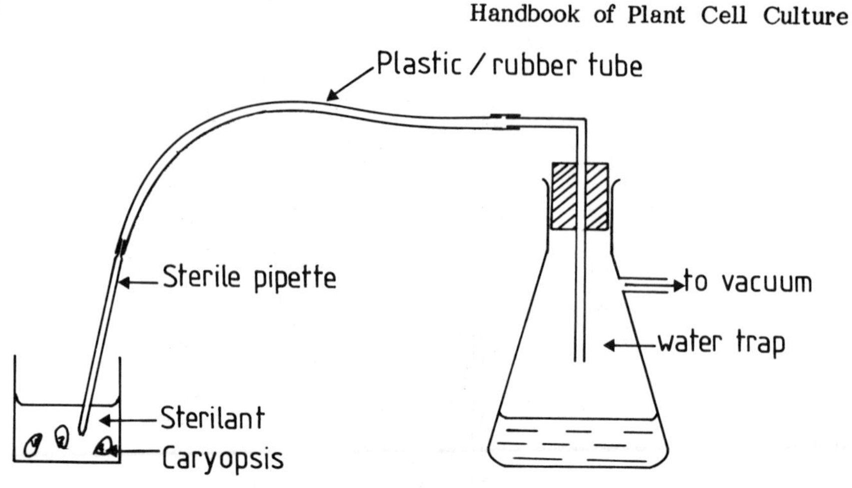

Figure 5. A simple apparatus for removing sterilant and wash solutions when sterilizing explants.

caryopsis with the rounded side down and secure by stabbing with a pair of needle-point forceps. By means of two parallel longitudinal and one transverse cut with a fine-pointed scalpel (see Fig. 4b) remove a flap of pericarp just above the embryo. Remove the embryo on the point of the scalpel and transfer to culture medium.

8. To induce proliferation of the scutellum place the embryo on an agar-solidified medium with the embryonic axis (on the flattened face) downwards. The most commonly used culture initiation medium is Murashige and Skoog (1962) medium as modified by Green and Phillips (1975). Embryos of a similar size will germinate on this medium without 2,4-D and the method can be used to shorten the generation time of maize. Illuminate the cultures with a standard incubator light source at 5000–10,000 lux.
9. The response of the scutellum varies considerably with the genotype (see earlier section, Evolution of Maize Tissue Culture) as does the optimum embryo size. In the best cases a complex tissue culture many times the mass of the original embryo is produced within 2–3 weeks.
10. Maintain the cultures by transfer of small selected pieces to fresh medium at 2-week intervals. Shoot buds or embryos present in the culture will mature and produce small plantlets when the 2,4-D is removed stepwise, e.g., 9.0 μM to 0.9 μM to zero. The plantlets, which frequently tiller, normally root on the culture medium and can be transferred to individual culture dishes on agar medium. Wash the agar from the roots of plantlets 4–5 cm high and transfer to light, moist soil. Cover the plant with a glass beaker during the first week and remove in stages.

IMMATURE INFLORESCENCE. There have been several reports of plant regeneration from cultured inflorescences of cereals including

Triticum (Dudits et al., 1975), *Sorghum* (Brettell et al., 1981), and *Zea mays* (Rice et al., 1979). It is interesting and potentially very useful that, at least in sorghum, plants are produced by somatic embryogenesis. As this has not been demonstrated widely for maize and in general the response is unsatisfactory (R. Brettell, personal communication) there is no recommended method. The procedure briefly outlined below is meant to encourage rather than lead.

1. Strip away the outer leaves of plants bearing up to 40-mm long unemerged tassels. Swab the last layers of young leaves with ethanol (90% v/v) and remove the inflorescence under aseptic conditions.
2. Cut the rachis into ca. 5-mm segments and place on culture medium in 9-cm diameter petri dishes. Use MS medium as modified by Green and Phillips (1975). The subsequent further modifications by Brettell et al. (1981) for culture of *Sorghum* inflorescences should also be examined.
3. The optimal response depends upon the stage of maturity of the inflorescence.
4. The best responding explants produce a proliferating, scutellum-like tissue, parts of which resemble normal zygotic embryos with a prominent root and shoot axis. However, the embryo-like structures rapidly give rise to a complex mass of shoots and roots. Individual plantlets can be separated and cultured to maturity as described earlier in the Immature Embryo section.

YOUNG LEAVES AND LEAF PRIMORDIA. The demonstration by Wernicke and Brettell (1980) that leaf primordia of sorghum give rise to somatic embryos is in stark contrast to the complete absence of any in vitro response by *mature* cereal leaf blades. Young leaves would be a practical source of embryogenic cultures normally only obtained from immature embryos (see earlier discussion of this topic) and although there is as yet no working protocol for maize-producing morphogenic cultures the following procedure can be used to examine the potential of maize genotypes of special interest.

1. Prepare sterile seedlings by sterilizing seed (see earlier section, Initiation of Tissue Cultures from Different Explants) and germinating on agar-solidified culture medium minus hormones in glass tubes (150 mm x 25 mm). Incubate at 25-27 C under 2000-5000 lux.
2. Cut through the shoot just above the apical meristem and slice the upper part into 1-2-mm sections. Unroll each leaf (under a stereo low-power microscope) and place on agar-solidified medium (see previous section).
3. A proliferating tissue culture may arise from the young leaf pieces. On the youngest leaves, before the differentiation between leaf sheath and leaf blade, all sections should produce a culture. On older leaves cultures never arise from the leaf blade.
4. The desired response from the very youngest leaves is the initiation of adventitious bud formation and somatic embryogenesis (see two previous sections) but this depends upon the genotype used.

ENDOSPERM CULTURE. Genotype and developmental stage of the endosperm are the most important factors affecting response of explants. Pollination of plant materials should be performed carefully and pollination dates should be recorded.

1. Harvest 7-10-day postpollination ears and bring them into the laboratory with husks intact.
2. Remove all husks and silks carefully by hand so as not to damage kernels or give unnecessary infection and cut ears transversely into two.
3. Surface-sterilize stripped ears in 20% calcium hypochlorite for 20 min in a large, sterile beaker and rinse three times with sterile distilled water.
4. All operations after this step are performed asceptically in a laminar air-flow cabinet.
5. Holding the ends of an ear with the fingers, cut off the tops of the kernels with a scalpel, 10-20 kernels at a time.
6. Scoop out endosperm carefully with forceps making sure that maternal tissue is not attached to endosperm.
7. Place four endosperms on Straus medium (10 ml) in 6-cm-diameter petri dishes and seal them with Parafilm.
8. Incubate in the dark at 25 C.

Protoplast Isolation and Culture

FROM MAIZE PLANTS. Despite many years of intensive attempts to isolate and induce divisions in protoplasts from cereal plant tissues, there is only one clearly positive report (Potrykus et al., 1977) which has not proved repeatable. Thus there is no established method. The essential points of the successful experiment with maize stem protoplasts are given below. It is surely worth repeating.

1. Remove all leaves from mature maize plants just before tassel emergence (ca. 8 weeks).
2. Surface-sterilize the stem ($HgCl_2$ 0.01% w/v with a few drops of detergent) for 10 min, wash in sterile distilled water and place sections in a small volume of culture medium (P2/mod, Potrykus et al., 1976) in sterile petri dishes.
3. Slice disks (0.3 mm thick) of stem into the culture medium and leave 30 min.
4. Incubate sliced tissue (ca. 2 g per 30 ml solution) in the following enzyme solution: cellulase (Onozuka R10) 0.2% w/v, pectinol fest (Roehm, Damstadt) 0.2% w/v in P2/mod, pH 5.8 in the dark at 12 C followed by 6 h at 32 C.
5. Agitate the dishes gently and filter the mixture through a sieve (100 μm).
6. Centrifuge the protoplasts (5 min at 20 x g) after diluting the solution 1:2 with $CaCl_2$ (0.24 M).
7. Wash the sedimented protoplasts with a mixture of 2 vol. 0.24 M $CaCl_2$ to 1 vol. of P2/mod and centrifuge again.

8. Remove debris by collecting the protoplasts onto a cushion of 0.54 M sucrose (5 vols.) with P2/mod (1 vol.) at 40 x g. Repeat this collection twice.
9. Wash once in 0.24 M $CaCl_2$ (2 vol.) plus P2/mod (1 vol.) by centrifuging at 20 x g and resuspend in P2/mod.
10. All solutions should be adjusted to 650 mOs/kg H_2O and pH 5.8.

The protoplasts were cultured in a wide range of conditions using the MDA technique of Potrykus et al. (1979b). Divisions were recorded in 50% of the combinations tested, with the best result (callus formation from, ca., 5% of the original protoplasts) at a population density of 5×10^3 ml^{-1}, 23.0 μM 2,4-D, 12.3 μM indole-3-butyric acid, 24 hr at 12 C, 14 days at 22 C, and further culture at 27 C, always in the dark. After 35 days the medium osmolality was reduced to 360 mOs/kg H_2O. The cultures resulting from this experiment still exist, growing in suspension with a doubling time of ca. 20 hr, but are not morphogenic.

FROM MAIZE CELL CULTURES. In maize, as in some other cereals and many dicotyledon species, protoplasts isolated from a particular type of cell culture will readily divide and re-form continuously proliferating cultures. The source cell cultures are characterized by a short doubling time; homogeneous, small, highly cytoplasmic cells; the absence of organized tissues and organs; and high friability growing well in suspension culture (see earlier section, Evolution of Maize Tissue Culture). The occurrence of such cultures in cereals is less frequent than in dicots. The following protocol was found useful with a cell line from maize inbred B73.

1. Culture the cells regularly in suspension in a medium with the same osmolality as the enzyme solution subsequently used. The best time in the growth cycle for protoplast isolation must be determined for each culture. Success is greatest with long-established cultures.
2. Collect the cells by centrifugation, wash in fresh medium and suspend in the enzyme solution: cellulase (Onozuka R10, Kinki Yakult, Nishinomiya, Japan) 1% w/v, Pectinol fest (Roehm, Damstadt, FRG) 0.5% w/v, hemicellulase (*Rhizopus*, Sigma) 0.5% w/v, mannitol 0.2 M, $CaCl_2$ 80 mM, MES buffer 0.5%, osmolality of 440 mOs/kg H_2O, pH 6.0.
3. Swirl gently on a reciprocal shaker for 1 hr at 24 C.
4. Sediment (ca. 100 x g), remove the enzyme mixture and replace at a 1/10 dilution (maintaining the osmolality constant by adjustment of the $CaCl_2$ concentration).
5. Incubate with occasional gentle swirling motion for 1-2 hr. Eighty to ninety percent of the cells are converted to protoplasts.
6. Filter through a sieve (50 μm) and wash protoplasts by repeated centrifugations (5 min at 50 x g) with an osmotic solution [0.5 M mannitol (3 vol.), 0.2 M $CaCl_2$ H_2O (1 vol.), 0.5% MES pH 5.8].

7. Transfer to culture medium maintaining a constant osmolality, either in liquid thin-layer cultures or soft agar plating. Ca. 30% of the protoplasts divide at least once and ca. 10% undergo sustained division.
8. There is no single recommended culture medium. Out of 27 common culture media tested with the B73 maize cell line (Potrykus et al., 1979a), division occurred in 12 of them. There was an optimum 2,4-D concentration (9.0 µM) and an optimum protoplast density (5-9 x 10^4 protoplasts per ml).
9. Extensive modifications of the isolation conditions and culture conditions did not significantly improve the results.

Organ Culture

EMBRYO. The isolation technique for maize immature embryos is given earlier. Many different culture media have been used for maize embryo culture over the last 62 years and the benefits of any particular medium supplement are difficult to discern from studying the literature. Loescher and Mock (1974) found a medium consisting of MS salts plus thiamine, 175 mM sucrose, and 1% agar satisfactory for culturing maize embryos as small as 0.1 mm in length. Addition of complex supplements, e.g., CW, yeast extract, and CH had no advantages. The "Barley Medium II" described by Norstog (1973) for culturing very small (0.2 mm) barley embryos also supports the growth of immature maize embryos (Dhaliwal and King, 1978). There are differences between species in response to embryo culture media and it would not be surprising to find variation among maize lines in this respect. The two media quoted above make a good starting point.

CARYOPSIS. This technique is applicable for any genotype and can be used for in vitro pollination investigation (see also earlier section, In Vitro Pollination and Direct Ovule Pollination) of plant to seed translocation of nutrients (Shimamoto and Nelson, 1981) and physiological study of grain filling during seed development.

1. Harvest 2-7-day postpollination ears and bring them into the laboratory with husks intact.
2. Carefully remove all husks and silks by hand or forceps in a laminar air-flow hood.
3. Surface-sterilize ears in 20% calcium hypochlorite for 20 min, rinse three times with sterile distilled water, and transfer onto a sterile petri dish.
4. Cut ears transversely by cuts between two double rows of caryopses across the inflorescence axis.
5. Cut each section of the cob into wedge-shaped blocks containing 4-10 caryopses (2 x 2 to 2 x 5) (Fig. 2).
6. Place two to three blocks with the cob tissue down on 50-ml hormone-free culture medium in a 125 ml Erlenmeyer flask or a 9-cm-diameter petri dish.

7. Incubate in the dark at 25 C.

The basic medium is not critical, but see Gengenbach, 1977.

AXILLARY BUD. Vegetative propagation of a tillering grass species like maize would not seem to pose any problems. However, dent maize lines usually produce none or just one or two basal tillers. Further, even in high tillering lines most tillers are committed to produce a terminal female inflorescence. Although it is difficult to imagine a need for vegetative propagation in a normal maize genetics program, it may be useful, for example, in propagating haploid plants. Fowler and Peterson (1978) found that a maize tiller could be genetically different to the main stem and that the tiller trait was heritable. The sexual committment of maize axillary buds is itself an interesting phenomenon.

1. Cut the stem at the base of 20-day seedlings, wash, and surface-sterilize 15 min in sodium hypochlorite (6% v/v) and rinse in sterile water.
2. Remove all leaves except the youngest primordia to produce a short (10–20 mm) stem bearing five-six axillary bud primordia.
3. Culture this stem explant for ca. 10 days on a medium made up from mineral salts of Murashige and Skoog (1962), vitamins according to White (1943), myo-inositol (555 mM), sucrose (87.6 mM), adenine sulphate dihydrate (784 μM), monobasic sodium phosphate (1.42 mM), KIN (13.9 μM), IAA (5.7 μM) or IBA (5.4 μM), and agar (8 g/l), pH 5.6.
4. Remove the developed axillary buds and place on rooting medium (the same basic medium omitting adenine sulphate, sodium phosphate, IAA, and KIN, and adding NAA (27.0 μM).
5. After ca. 15 days transfer to soil. Most of the buds will root and survive the transfer. Some of the lowest buds grow up into plants bearing a tassel and lateral ear but increasingly higher buds produce more and more female florets in the tassel. The highest buds produce only a terminal female inflorescence.

ANTHER CULTURE. Since the first successful production of pollen-derived maize plants was reported in 1975 (Research Group No. 401, 1975) the low frequency of response has been a major obstacle for the practical use of this technique for haploid production. However, a recent report with improved results (Ting et al., 1981) should now stimulate more research, particularly into the biological factors causing a high response in some genotypes.

1. Select healthy plants that have young tassels of proper developmental stages, cut them at the base, and bring them into the laboratory.
2. Remove the leaves covering tassels and cut out whole tassels.
3. Excise a few spikelets from the top and bottom of the central spike and also from branch tassels in order to examine stages of pollen development.

4. Assess the stage of pollen development by squashing anthers in aceto-carmine solution. Pollen of mononucleate stage appear to give the most satisfactory results (Brettell et al., 1981). Note the gradient of pollen development along the tassel.
5. Surface-sterilize the tassel in 20% calcium hypochlorite (ca. 5 min) in a sterile glass cylinder, rinse three times with sterile water, and place in a large sterile petri dish.
6. Remove a branch of the tassel and place in a sterile dish.
7. Pick up individual spikelets and open them by gently pushing the bases of glumes with forceps.
8. Place 10–20 anthers on 10 ml medium in 6-cm-diameter petri dishes and incubate at 25 C in the dark.

ROOT. Culture of organized roots can be used to study nutrient requirements of plants (Street, 1957). While cereal roots are difficult to culture one successful protocol is listed below.

1. To surface-sterilize maize seeds place the seeds (ca. 10 seed per 20 ml sterilant) in a clean screw-capped container with mercuric chloride solution (0.01% w/v plus a few drops of detergent). Shake vigorously for 20 min. Remove sterilant using the vacuum line (see Immature Embryo section). Wash three times with sterile water.
2. Add a small volume of sterile water and shake gently overnight in an unsealed container. Remove water and repeat sterilization procedure. The overnight incubation and second sterilization may not be necessary with some seed batches.
3. Arrange the seeds around the edge of a sterile petri dish containing two layers of sterile filter paper moistened with sterile distilled water. Place the seeds embryo upwards with the radicle pointing to the center of the dish.
4. Incubate in unsealed dishes in the dark in a moist environment at ca. 30 C for 2-3 days.
5. Cut ca. 10-mm lengths from the tip of the terminal roots. This may be done quantitatively by first transferring ca. 20-mm lengths to a sterile petri dish placed upon a sheet of millimeter graph paper. Avoid any damage to the first 1-2 mm of the tip.
6. Transfer the tips in pairs to 20 ml of White's root culture medium containing 0.12 M glucose and CH in 100-ml Erlenmeyer flasks. Cap the flasks with cotton-wool stoppers or sterile aluminum foil and shake at ca. 110 rpm on an orbital shaker at 25-30 C and ca. 2000 lux.
7. The cultures can be multiplied by alternating "tip" and "sector" cultures at 2-3-week intervals (see Fig. 6).
8. Cereal roots are in general difficult to culture. The cultures "stale" after a few passages and growth ceases. There is considerable variation among different genotypes.

FUTURE PROSPECTS

The attraction of plant cell culture which is often inferred but rarely realized is that of working with populations of *single, independent,*

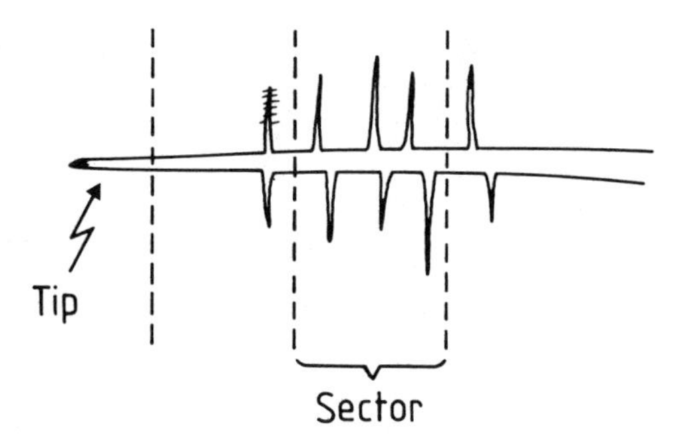

Figure 6. Tip and sector explants from cultured roots.

totipotent cells. Up until 1980, cereal cultures were as far away from this ideal as one could imagine (King et al., 1978; Thomas et al., 1979). Although the prospects for isolating totipotent protoplasts directly from the plant have not noticeably improved, there are encouraging signs of totipotent cultures which might be reduced to the single-cell level. This would immediately open up many of the much-discussed possibilities for genetic manipulation of maize, particularly if it were in the form of protoplast culture. As it is, the presently available techniques are only suitable, for example, for the isolation of pesticide or pathogen/toxin resistance (Gengenbach et al., 1977) or for resistant mutants showing amino acid overproduction (Hibberd et al., 1980). General variability arising amongst regenerated plants might also prove useful. More directly practical techniques are available for plant breeding in the form of anther culture and the rescue of embryos following wide crosses.

ACKNOWLEDGMENT

The authors are indebted to Nguyen van Hai for the drawings and to colleagues at the FMI for helpful discussions.

KEY REFERENCES

Dhaliwal, H.S. and King, P.J. 1978. Direct pollination of *Zea mays* ovules in vitro with *Z. mays*, *Z. mexicana* and *Sorghum bicolor* pollen. Theor. Appl. Genet. 53:43-46.

Green, C.E. and Phillips, R.L. 1975. Plant regeneration from tissue cultures of maize. Crop Sci. 15:417-421.

Raman, K., Walden, D.B., and Greyson, R.I. 1980. Propagation of *Zea mays* L. by shoot tip culture: A feasibility study. Ann. Bot. (London) 45:183-189.

Street, H.E. 1957. Excised root culture. Biol. Rev. (Cambridge) 32: 117-155.

Ting, Y.C., Yu, M., and Zheng, W.-Z. 1981. Improved anther culture of maize (*Zea mays*). Plant Sci. Lett. 23:139-145.

Wernicke, W. and Brettell, R.I.S. 1980. Somatic embryogenesis from sorghum leaves. Nature (London) 282:138-139.

REFERENCES

Bartkowiak, E. 1981. Tissue culture of maize: Selection of friable callus lines. Plant Cell Reports 1:52-55.

Beadle, G.W. 1978. Teosinte and the origin of maize. In: Maize Breeding and Genetics (D.B. Walden, ed.) pp. 113-128. John Wiley and Sons, Toronto.

Brettell, R.I.S., Ingram, D.S., and Thomas, E. 1980a. Selection of maize tissue cultures resistant to *Drechslera (Helminthosporium) maydis* T-toxin. In: Tissue Culture Methods for Plant Pathologists (D.S. Ingram and J.P. Helgeson, eds.) pp. 233-237. Blackwell Scientific Pub., Oxford.

Brettell, R.I.S., Wernicke, W., and Thomas, E. 1980b. Embryogenesis from cultured immature inflorescences of *Sorghum bicolor*. Protoplasma 104:141-148.

Brettell, R.I.S., Thomas, E., and Wernicke, W. 1981. Production of haploid maize plants by anther culture. Maydica XXVI:101-111.

Chourey, P.S. and Zurawski, D.B. 1981. Callus formation from protoplasts of a maize cell culture. Theor. Appl. Genet. 59:341-344.

Coe, E.H. and Neuffer, M.G. 1977. The genetics of corn. In: Corn and Corn Improvement (G.F. Sprague, ed.) pp. 111-224. American Society of Agronomy, Madison, Wisconsin.

Cure, W.W. and Mott, R.L. 1978. A comparative anatomical study of organogenesis in cultured tissues of maize, wheat and oats. Physiol. Plant. 42:91-96.

Dudits, D., Nemet, G., and Haydu, Z. 1975. Study of callus growth and organ formation in wheat (*Triticum aestivum*) tissue cultures. Can. J. Bot. 53:957-963.

Dunstan, D.I., Short, K.C., and Thomas, E. 1980. The anatomy of secondary morphogenesis in cultured scutellum tissues of *Sorghum bicolor*. Protoplasma 97:251-260.

Fowler, R.G. and Peterson, P.A. 1978. An altered state of specific *EN* regulatory element induced in a maize tiller. Genetics 90:761-782.

Galinat, W.C. 1977. The origin of corn. In: Corn and Corn Improvement (G.F. Sprague, ed.) pp. 1-48. American Society of Agronomy, Madison, Wisconsin.

Gengenbach, B.G. 1977. Development of maize caryopses resulting from in vitro pollination. Planta 134:91-93.

______, Green, C.E., and Donovan, C.M. 1977. Inheritance of selected pathotoxin resistance in maize plants regenerated from cell cultures. Proc. Natl. Acad. Sci. U.S.A. 74:5113-5117.

Green, C.E., Phillips, R.L., and Kleese, R.A. 1974. Tissue cultures of maize (*Zea mays* L.): Initiation, maintenance and organic growth factors. Crop Sci. 14:54-58.

Gresshoff, P.M. and Doy, C.H. 1973. *Zea mays*: Methods for diploid callus culture and the subsequent differentiation of various plant structures. Aust. J. Biol. Sci. 26:505–508.

Hendre, R.R., Mascarenhas, A.F., Pathak, M., and Jagannathan, V. 1975. Tissue cultures of maize, wheat and sorghum: Part II—Growth and nutrition of callus cultures. Indian J. Exp. Biol. 13:108–111.

Hibberd, K.A., Walter, T., Green, C.E., and Gengenbach, B.G. 1980. Selection and characterization of a feedback-insensitive tissue culture of maize. Planta 148:183–187.

Kiesselbach, T.A. 1980. The Structure and Reproduction of Corn. Univ. of Nebraska Press, Lincoln.

King, P.J., Potrykus, I., and Thomas, E. 1978. In vitro genetics of cereals: Problems and perspectives. Physiol. Veg. 16:381–399.

La Rue, C.D. 1947. Growth and regeneration of the endosperm of maize in culture. Am. J. Bot. 34:585–586.

______ 1949. Cultures of the endosperm of maize. Am. J. Bot. 36: 798.

Linsmaier-Bednar, E.M. and Bednar, T.W. 1972. Light and hormonal control of root formation in *Zea mays* callus cultures. Dev. Growth Differ. 14:165–174.

Loescher, W.H. and Mock, J.J. 1974. A nutrient medium for in vitro cultivation of maize and sorghum embryos. Maize Genet. Coop. Newsletter 48:64–66.

Mascarenhas, A.F., Hendre, R.R., Rao Seetharama, B., and Jagannathan, V. 1969. Tissue cultures of maize, wheat, jowar and rice. Indian J. Exp. Biol. 7:65–67.

Mascarenhas, A.F., Pathak, M., Hendre, R.R., and Jagannathan, V. 1975a. Tissue cultures of maize, wheat, rice and sorghum: Part I—Initiation of viable callus and root cultures. Indian J. Exp. Biol. 13: 103–107.

Mascarenhas, A.F., Pathak, M., Hendre, R.R., Ghugale, D.D., and Jagannathan, V. 1975b. Tissue cultures of maize, wheat, rice and sorghum: Part IV—Studies of organ differentiation in tissue cultures of maize, wheat and rice. Indian J. Exp. Biol. 13:116–119.

Mott, R.L. and Cure, W.W. 1978. Anatomy of maize tissue cultures. Physiol. Plant. 42:139–145.

Murashige, T. and Skoog, F. 1962. A revised medium for rapid growth and bioassays with tobacco tissue cultures. Physiol. Plant. 15:473–496.

Norstog, K. 1973. New synthetic medium for the culture of premature barley embryos. In Vitro 8:307–308.

Polikarpochkina, R.T., Gamburg, K.Z., and Kharkin, E.E. 1979. Cell-suspension culture of maize (*Zea mays* L.). Z. Pflanzenphysiol. 95:57–67.

Potrykus, I., Harms, C.T., and Loerz, H. 1976. Problems in culturing cereal protoplasts. In: Cell Genetics in Higher Plants (D. Dudits, G.L. Farkas, and P. Maliga, eds.) pp. 129–140. Akademiai Kiado, Budapest, Hungary.

______, and Thomas, E. 1977. Callus formation from stem protoplasts of corn (*Zea mays* L.). Mol. Gen. Genet. 156:347–350.

_____ 1979a. Callus formation from cell culture protoplasts of corn (*Zea mays* L.). Theor. Appl. Genet. 54:209-214.

_____ 1979b. Multiple-drop-array (MDA) technique for the large-scale testing of culture media variations in hanging micro-drop cultures of single cell systems. I. The technique. Plant Sci. Lett. 14:231-235.

Reddy, A.R. and Peterson, P.A. 1977. Callus initiation from waxy endosperm of various genotypes of maize. Maydica XXII:125-130.

Research Group No. 401, Genetics Institute Academia Sinica, Beijing, China. 1975. Primary study on induction of pollen plants of *Zea mays*. Acta Genet. Sinica 2:138-142.

Rhodes, C. and Green, C.E. 1981. An in vitro detection system for monoploid maize tissue cultures. Maize Genet. Coop. Newsletter 55: 93-94.

Rice, T.B., Reid, R.K., and Gordon, P.N. 1979. Propagation of higher plants through tissue culture. Proceedings of the University of Tennessee Symposium, p. 262.

Shannon, J.C. and Batey, J.W. 1973. Inbred and hybrid effects on establishment of in vitro cultures of *Zea mays* L. endosperm. Crop Sci. 13:491-493.

Sheridan, W.F. 1974. Tissue culture of maize. I. Callus induction and growth. Physiol. Plant. 33:151-156.

_____ 1975. Growth of corn cells in culture. J. Cell Biol. 67:396a.

Shimamoto, K. and Nelson, O.E. 1981. Movement of ^{14}C-compounds from maternal tissue into maize seeds grown in vitro. Plant Physiol. 67:429-432.

Sprague, G.F. 1977. Corn and Corn Improvement. American Society of Agronomy, Madison, Wisconsin. 774 pp.

Springer, W.D., Green, C.E., and Kohn, K.A. 1979. A histological examination of tissue culture initiation from immature embryos of maize. Protoplasma 101:269-281.

Steele, L. 1978. The hybrid corn industry in the United States. In: Maize Breeding and Genetics (D.B. Walden, ed.) pp. 29-40. John Wiley and Sons, Toronto.

Straus, J. 1958. Spontaneous changes in corn endosperm tissue cultures. Science 128:537-538.

_____ 1960. Maize endosperm tissue growth in vitro. III. Development of a synthetic medium. Am. J. Bot. 47:641-647.

_____ and La Rue, C.D. 1954. Maize endosperm tissue grown in vitro. I. Culture requirements. Am. J. Bot. 41:687-694.

Tabata, M. and Motoyoshi, F. 1965. Hereditary control of callus formation in maize endosperm cultured in vitro. Japan. J. Genet. 40:343-355.

Thomas, E., King, P.J., and Potrykus, I. 1979. Improvement of crop plants via single cells in vitro—an assessment. Z. Pflanzenzuecht. 82:1-30.

Vasil, V. and Vasil, I.K. 1981. Somatic embryogenesis and plant regeneration from suspension cultures of pearl millet (*Pennisetum americanum*). Ann. Bot. (London) 47:667-678.

Walden, D.B. 1978. Maize Breeding and Genetics. John Wiley and Sons, Toronto. 794 pp.

Watson, S.A. 1977. Industrial utilization of corn. In: Corn and Corn Improvement (G.F. Sprague, ed.) pp. 721-764. American Society of Agronomy, Madison, Wisconsin.
White, P.R. 1943. A Handbook of Plant Tissue Culture. J. Cattell, Lancaster, Pennsylvania.
Ziyi, C., Caiyne, G., and Jianjun, H. 1981. A study of embryogenesis in pollen callus of maize (*Zea mays* L.). Acta Genet. Sin. 8:269–274.

CHAPTER 4
Oats

R. Kaur-Sawhney and *A.W. Galston*

The common cultivated oat (*Avena sativa* L.) is the principal modern derivative of the common wild oat *Avena fatua* (Metcalfe and Elkins, 1980). Ancient records, though scarce, indicate that oat cultivation originated in Asia Minor and then spread into Europe and other regions favorable for oat culture (Baum, 1977). The records further suggest that the Romans (250 BC–50 AD) and Greeks (400 BC–200 BC) were familiar with the genus *Avena*.

Oats comprise one of the important cereal crops, with about 130 million acres under cultivation around the world. They are one of the standard feeds for domestic animals and also are used for human consumption. The whole grain contains as much as 12% protein and fair amounts of the major vitamins. Approximately 40% of the total crop is grown in the USSR (Table 1) while the United States, Canada, China, and West Germany are the other primary producers (Metcalfe and Elkins, 1980). Although oats are best suited for cool, moist climates, they can be grown successfully in a wide range of climatic conditions because of the large number of varieties available. There are as many as 5000 varieties and strains of oats recorded in the files of the USDA (1980), of which only a few improved and selected varieties have been adopted by farmers on recommendations by agricultural agencies.

The chromosome complements of the genus fall into 3 groups, all based on n = 7; the basic numbers seen are 14 (diploid), 28 (tetraploid), and 42 (hexaploid). The latter is most common in commercial oats.

With increasing world population and demands for high-protein foods, a great deal of attention is being given to improvement, yield, and

Table 1. The Distribution of Oat Production in the World[a]

COUNTRY	TOTAL ACREAGE (%)
USSR	39.3
United States	18.1
Canada	7.9
China	4.0
West Germany	3.9
Others	26.8

[a]Source: Metcalfe and Elkins, 1980.

quality of cereal plants, including oats. In this endeavor, protoplast and cell culture techniques could be important. These techniques are not only powerful tools for studying the control of growth, development, and differentiation in plants, but they are also potentially valuable in the development of genetically modified crops. Before genetic engineering can be applied to crop improvement, plant regeneration from single cells or small tissue explants must be achieved. Plant regeneration can be used to obtain useful variants, induced in vitro via protoplast fusion, mutagenesis and selection, or DNA uptake. Such techniques, if successful, can eventuate in enormous economic gains through improved crop yield and quality.

Earlier demonstrations of the regeneration of entire plants from single plant cells (Muir et al., 1954; Steward et al., 1964; Halperin and Wetherell, 1964) and protoplasts (Takebe et al., 1971) have led to successful regeneration of plants of many species. These include haploid plants from microspores and somatic hybrids through protoplast fusion in vitro (reviews by Schieder and Vasil, 1980; Vasil, 1980; Flores et al., 1981). Improvements in these techniques provide the basis for the current widespread interest in their use in biotechnology.

Despite a great deal of progress, mainly with dicots, successful tissue culture of two major families of food plants, the Gramineae and the Leguminosae, has been hindered by slow growth rates in vitro, and by the refractory behavior of their protoplasts. Until recently, neither cell division from protoplasts nor organ regeneration from calli of protoplast origin had been achieved in either group. This chapter examines the state of the art in technological developments, as well as the problems and potential of protoplast and cell culture of the oat plant.

PROTOPLAST CULTURE

Protoplasts are uniquely suited for experiments in plant growth and development, especially where attempted genetic modification is involved. The following are some of the most important advantages of these naked cells: (1) Large numbers of intact, uniform, viable protoplasts can be extracted from most parts of the plant and to a lesser extent from cell suspensions and calli. (2) When cultured on suitable media, protoplasts may regenerate a cell wall, divide to form a callus, and regenerate an entire plant. (3) Protoplasts can be fused and the fused products can be cultured to yield hybrids indistinguishable from those sexually produced. Some novel hybrids, such as between tomato and potato, can also be obtained through protoplast fusion. (4) The absence of a cell wall permits protoplasts to take up, by pinocytosis, externally supplied particles such as nuclei, chloroplasts, chromosomes, and virus particles. These properties have kindled interest in protoplasts as an important new route for producing novel types of plants. Various aspects of this process have been reviewed recently (Galston, 1978; Potrykus, 1980; I.K. Vasil and V. Vasil, 1980; Flores et al., 1981).

Totipotency of single protoplasts was first demonstrated with tobacco mesophyll protoplasts in 1971 (Takebe et al., 1971). Subsequently, protoplasts from many plant species, belonging mostly to the Solanaceae, have been cultured successfully, leading to the regeneration of complete plants (Shepard et al., 1980). In contrast, the culture of cereal protoplasts has extreme difficulties (Brenneman and Galston, 1975; Potrykus et al., 1976; Cocking, 1978; King et al., 1978; Pental and Gunckel, 1979; Flores et al., 1981). Protoplasts can be isolated from actively growing cell and tissue cultures and from many parts of most cereal plants, but it is only from the leaves that high yields of protoplasts can be obtained. These protoplasts, when cultured in a suitable medium, generally regenerate a cell wall, but then senesce rapidly and usually do not divide to form a callus (Potrykus et al., 1976; Galston, 1978). Attempts to induce division in cereal protoplasts have largely failed, despite innumerable variations in plant material and culture conditions (Brenneman and Galston, 1975; Potrykus et al., 1976). Only a few examples of sustained division in cereal protoplasts have been reported; these are listed in Table 2, and have been discussed by Flores et al. (1981). Recently, true regeneration from a cereal protoplast culture has been reported in *Pennisetum americanum* (Vasil and Vasil, 1981). Protoplasts from an embryo-derived suspension culture were able to proceed all the way to an intact plant. It is not yet known whether this phenomenon will turn out to be general for cereals.

Clearly, the improvement of cereals by protoplast technology must await development of reproducible methods of plant regeneration. With this in view, our laboratory undertook, about 9 years ago, detailed investigations of the culture of oat mesophyll protoplasts. We have discussed the methods of isolation and culture of protoplasts in several reports (Brenneman and Galston, 1975; Galston et al., 1977; Kaur-Sawhney et al., 1980). In addition, protoplast technology in relation to crop plants has been reviewed recently (Bhojwani et al., 1977; King et al., 1978; Pental and Gunckel, 1979; Potrykus, 1980). The discussion below will thus concern only the most important aspects of the culture of oat mesophyll protoplasts.

Table 2. Species of Cereal Plants in which Sustained Division of Cultured Protoplasts has been Reported

SPECIES	SOURCE OF PROTOPLASTS	REFERENCE
Hordeum vulgare	Callus	Koblitz, 1976
Oryza sativa	Haploid callus	Academia Sinica, 1975; Tsai et al., 1978
	Leaf sheath	Deka & Sen, 1976
Pennisetum americanum	Embryogenic suspension cultures	V. Vasil & I.K. Vasil, 1980
Triticum monococcum	Suspension cultures	Nemet & Dudits, 1977
Zea mays	Stem	Potrykus et al., 1977
	Callus from stem protoplast	Potrykus et al., 1979

Protocol for Isolation and Culture

The first leaf of each 5- to 7-day-old seedling of *Avena sativa* L. (cv. Garry, Lodi, and Victory), grown in controlled-growth rooms (Brenneman and Galston, 1975), was used for protoplast isolation and culture. All postharvest operations were performed aseptically in a laminar air-flow cabinet.

1. Sterilize leaves by immersion in 70% ethanol for 2 min. Rinse leaves with distilled sterile water (3 x), dip into 10% commercial bleach containing Tween-20 (1 drop/10 ml) for 5 min, and finally rinse with distilled sterile water (5 x).
2. Remove the lower epidermis by peeling with a curved pair of fine forceps. Float leaves, stripped-side down, on 0.5% (w/v) Cellulysin (B grade, Calbiochem) in B5 medium (Gamborg et al., 1968) or in 1 mM phosphate buffer (pH 5.8) with 0.4 M sorbitol or 0.6 M mannitol for 2 hr at 30 ± 1 C in a 15 x 100 mm petri dish. These treatments cause the release of ca. 10^6 protoplasts/g fresh wt. Exposure to enzyme must occur shortly after peeling, otherwise a resistance to protoplast release by Cellulysin develops (Kaur-Sawhney et al., 1976).
3. Remove protoplast suspension from petri dish by gentle aspiration with a Pasteur pipet. Discharge gently into a centrifuge tube and seal with Parafilm.

4. Centrifuge at 50 x g for 5 min. Wash protoplast pellet several times by gentle resuspension in medium and recentrifugation, or by passing centrifugally through a sucrose (0.6 M) cushion.
5. Resuspend the protoplasts in B5 medium with 2,4-D, NAA, BA, and other additives at a concentration of ca., 3 x 10^5 protoplasts/ml (Brenneman and Galston, 1975; Kaur-Sawhney et al., 1980).
6. Suspend protoplasts in 2 ml of B5 medium and mix with equal volumes of B5 medium containing 1.2% Ionagar (Colab Laboratories, Inc., Illinois) at a temperature of 45 C. Pour suspensions into glass petri dishes. Alternatively, culture the protoplast suspension as a thin layer or as 20 µl droplets in 15 x 60 mm petri dishes, 6–8 droplets/dish.
7. Seal petri dishes with Parafilm.
8. Incubate in moist chambers at 22 C in diffuse light or in darkness.

Oat mesophyll protoplasts, under these conditions, like most other cereal protoplasts, are labile and lyse easily. The majority of protoplasts do regenerate a cell wall, which is patchy and irregular when viewed as Calcofluor White fluorescence (Brenneman and Galston, 1975; Galston et al., 1977; Kinnersley et al., 1978). Such protoplasts show only limited and sporadic cell division (Brenneman and Galston, 1975; Galston et al., 1977; Kaur-Sawhney et al., 1980). Thus far, sustained cell division has not been achieved from mesophyll protoplasts of any cereal on any medium (Bhojwani et al., 1977; Galston, 1978; Potrykus, 1980; V. Vasil and I.K. Vasil, 1980).

Reasons for Lack of Cell Division in Oat Protoplasts

The lack of sustained cell division in cereal protoplasts, including oats, may derive either from their genetic makeup, from injury caused during protoplast isolation, or from yet unrevealed nutrient requirements.

GENETIC FACTORS. Failure of cell division in protoplasts from cereal plants ranging from wild to highly inbred forms, and over a variety of culture media, suggests that lack of cell division is due to conditions specified by the genetic nature of the parent tissue (Potrykus et al., 1976). The slow growth rates of cereal callus tissues and the low frequency of plantlet formation compared with dicotyledonous plants support this view (Bhojwani et al., 1977; King et al., 1978). Success in inducing callus from cereal leaf has been attained, thus far, only in one instance (Saalbach and Koblitz, 1978): with barley leaves. Results obtained in our laboratory (Kaur-Sawhney et al., 1980) with oat mesophyll protoplasts suggest that lack of division may be due to a block in DNA synthesis, the protoplasts being locked into the G_0-G_1 phase. Removal of the block in nuclear division by fusion of cereal protoplasts with dicotyledonous protoplasts (Constabel et al., 1975) further supports the hypothesis that this lack of mitotic activity is due to genetic factors.

INJURY. Protoplasts cannot be isolated without several kinds of injury. Leaves must first be excised from the plant and cut into segments. Then the epidermis of each segment must be peeled off and the segments plasmolyzed by floating them on a hypertonic medium. Finally, they are exposed to somewhat impure enzymes. These procedures must cause varying degrees of injury to the tissue and can lead to rapid senescence, manifested by increases in RNase and protease activities, decreases in rates of RNA and protein synthesis, and loss of chlorophyll in leaf tissue (Thimann, 1980; Kaur-Sawhney and Galston, 1979) and in protoplasts (Fuchs and Galston, 1976; Altman et al., 1977; Galston et al., 1978). Obviously cereal leaves are more sensitive to injury than those of many dicotyledonous plants, in which mesophyll protoplasts can show consistent cell division when cultured on appropriate media.

Plasmolysis prior to protoplast isolation causes water stress, which in turn may produce increased levels of abscisic acid and ethylene (Gepstein and Thimann, 1980). This is known to inhibit cell division (Cocking et al., 1974; Eriksson et al., 1974). Severe plasmolysis is also known to rupture plasmodesmata, resulting in leakage of RNase into the medium, which may then lyse protoplasts (Galston et al., 1978). Plasmolysis also alters the transmembrane electrical potential from strongly negative to slightly positive (Racusen et al., 1977). Such protoplasts do not regenerate a uniform cell wall as determined by Calcofluor staining (Kinnersley et al., 1978). It is only after recovery and reversion to a negative transmembrane potential that tobacco protoplasts are able to proceed to cell wall regeneration and cell division.

Improvement in Oat Protoplast Culture

Oat mesophyll protoplasts are labile upon isolation and usually lyse within 24 hr of incubation. Their ability to incorporate labeled precursors into protein and RNA remains constant for a few hours and declines rapidly thereafter (Fuchs and Galston, 1976). Furthermore, the protoplasts are incapable of DNA synthesis, since labeled thymidine is not incorporated into authentic DNA hydrolyzable by DNase (Kaur-Sawhney et al., 1980). This observation together with the rare mitotic activity observed in protoplasts suggests that they are blocked in the G_0-G_1 phase of the cell cycle (Kaur-Sawhney et al., 1980). Such protoplasts, which senesce rapidly in culture, can be stabilized somewhat by pretreatment of the leaves prior to protoplast isolation with known senescence inhibitors such as kinetin and cycloheximide. These agents also decrease the rise in RNase and protease activities and increase net RNA and protein synthesis in isolated protoplasts (Kaur-Sawhney et al., 1976; Kaur-Sawhney et al., 1977). More recently we have found that the naturally occurring polyamines (putrescine, cadaverine, spermidine, spermine) and their precursors, L-arginine and L-lysine, stabilize protoplasts against lysis and increase their mitotic activity (Galston and Kaur-Sawhney, 1980).

Protoplasts obtained from oat leaf segments pretreated with 1-10 mM polyamine or isolated in medium containing polyamines showed elevated rates of macromolecular synthesis (Kaur-Sawhney et al., 1977; Altman et al., 1977). Some authentic DNA synthesis was observed in the presence of 1 mM cadaverine or spermidine in aseptic microdrop cultures; spermidine also triggered some mitosis, leading to an increased incidence of binucleates (Kaur-Sawhney et al., 1980). Such binucleate protoplasts failed to undergo cytokinesis and did not develop further.

Animal cells in culture frequently show a close relationship between polyamine content and ability to divide (Heby, 1981). In plant systems too, polyamines can play an important role in anabolic cell activity and prevention of senescence (Galston and Kaur-Sawhney, 1980; Altman and Bachrach, 1981). Endogenous levels of polyamines and their key biosynthetic enzymes decline with increasing age of oat leaves and with senescence of excised leaves (Kaur-Sawhney et al., 1982). Although polyamines are potent senescence inhibitors and appear to be promising tools for improving oat protoplast culture, the optimum conditions for establishing cell division and polyamine biosynthesis in oat mesophyll protoplasts remain to be established. We have recently found that cereal leaves exposed to 0.6 M mannitol respond by a rapid and dramatic rise in the titer of putrescine (1,4-diaminobutane; Flores and Galston, 1982). This rise is due to the increased synthesis of the enzyme arginine decarboxylase (ADC; EC 4.1.1.19). We do not know whether putrescine formation is related to the expression of the injury syndrome, its prevention, or neither.

TISSUE CULTURE

Callus Formation

Successful production of callus from tissue explants and suspension cultures of monocotyledonous plants was achieved much later than in dicotyledonous plants. During the past several years, however, rapid progress has been made in the induction of callus formation and plant regeneration from explants of some cereal crops. The earlier work in cereal tissue culture includes callus formation from maize endosperm (Straus, 1954), rye endosperm (Norstog, 1956), rye embryo (Carew and Schwarting, 1958), and corn stems and roots (Mascarenhas et al., 1965; Hendre et al., 1972; Sheridan, 1975). Among cereals, it is perhaps rice cultured on Murashige-Skoog (MS) medium (Murashige and Skoog, 1962) with 10-50 µM 2,4-D that forms callus most easily from young explants. Shoot regeneration has been reported from almost all parts of the rice plant (see reviews by Pental and Gunckel, 1979; Evans et al., 1981), but occurs only when 2,4-D is removed from the medium. Callus growth has also been achieved in both wild and cultivated species of wheat; successful callus induction has been reported from the rachis and embryos of seven species of *Triticum* including diploid (*T. monococcum*, *T. speltoides*, and *T. tauschii*), tetraploid (*T. timopheevii* and *T. turgidum*), and hexaploid (*T. aestivum*) species (Gosch-Wackerle et al., 1979). Other cereals in which callus formation has been successful include barley (Gamborg and Eveleigh, 1968; Schenk and Hildebrandt, 1972), sorghum (Masteller and Holden, 1970; Gamborg et al., 1977; Thomas et al., 1977), and millet (V. Vasil and Vasil, 1981).

In oats, callus induction from explants and suspension cultures has been limited. To date in oats we have not been able to apply satisfactorily the tissue culture techniques that have been successful for callus formation in other cereals. Some of the earliest work on callus induction in oats was done by Webster (1966) and by Carter et al. (1967). In both investigations callus was induced from germinating seeds, with maximum growth obtained on a medium containing 2,4-D. Webster (1966) was able to maintain a firm callus through several subcultures using Heller's medium (Carew and Schwarting, 1958) supplemented with IAA, NAA, and 2,4-D. Carter et al. (1967) obtained viable callus when seeds were grown on LS medium (Linsmaier and Skoog, 1965) containing a complex mixture of auxins, and kinetin. They concluded that induction of callus is dependent on auxin (IAA and 2,4-D) but not on KIN concentration. In fact, high concentrations of KIN suppressed callus growth, and cytokinins seem generally less effective with cereals than with other plants. Concentrations of IAA required for callus initiation generally were much larger than the concentrations of 2,4-D, maximum growth of callus being obtained with 22.5-225 µM 2,4-D. In our laboratory (Brenneman and Galston, 1975), callus formation was induced when explants from root and hypocotyl tissue of young seedlings of Garry and Victory oats were grown on LS medium containing high auxin concentration (22.5 µM 2,4-D). When the callus was excised from the original explant and repeatedly subcultured, roots could be induced on a medium with low levels of auxins (IAA or NAA) or when supplements of cytokinins (BAP, IBA, or KIN) were used. Bright green nodules and meristemoids without root formation occurred only on a modified LS medium, roots were formed instead of green nodules. Anatomical study of the nodules revealed tracheids as well as meristematically active cells, but with little orderly and functional arrangement of vascular tissue. Some typical roots developed, but even stemlike nodules failed to develop into typical shoots.

Organogenesis and Regeneration

Although callus induction has been achieved from explants of most cereals, organ differentiation and plant regeneration is limited to a few species (Evans et al., 1981). We have summarized conditions under which callus induction and plant regeneration have been successful in Table 3. In general, immature embryos are best for callus initiation and subsequent plant regeneration (Green and Phillips, 1975); however, in some instances regeneration has been obtained from callus derived from rachises, shoot apices, and roots. The type and age of the explant are always important for determination of callus induction.

For maximum callus induction and regeneration from immature embryos, the scutellum side must be oriented upwards; with more mature seedlings, downwards orientation is better. The reason for this is obscure. To produce a totipotent callus, a suitable explant is generally cultured in MS medium containing 2,4-D; NAA and 2iP are generally added for increased callus growth (Green and Phillips, 1975). Transfer of the growing callus to medium without 2,4-D is usually sufficient to induce root and sometimes shoot regeneration. Successful

Table 3. Induction of Callus and Plant Regeneration in Oats (*A. sativa*)

CULTIVAR	MEDIUM	GROWTH REGULATORS (μM)		EXPLANT	REFERENCE
		Callus	Organogenesis		
Sun II	Heller's	23.0 2,4-D 11.0 IAA 134.3 NAA	---	Seed	Webster, 1966
Victory	LS	23.0–230.0 2,4-D or IAA	no 2,4-D	Seed	Carter et al., 1967
Garry	LS	23.0 2,4-D	---	Hypocotyl	Brenneman & Galston, 1975
Victory (25 genotypes)	B5 or MS	2.3-13.6 2,4-D	no auxin	Embryo	Cummings et al., 1976
Flamingskrone, Arnold, Tiger	SH	2.3-9.0 2,4-D	300 NAA, 10 IAA, 1.5 BA	Hypocotyl	Loerz et al., 1976

organogenesis and plant regeneration has been obtained, using modifications of the above medium, in rice (Nishi et al., 1968), wheat (Shimada et al., 1969), maize (Green and Phillips, 1975), barley (Chin and Scott, 1977), sorghum (Masteller and Holden, 1970), and millet (Rangan, 1976; V. Vasil and Vasil, 1981).

In oats, Cummings et al. (1976) were perhaps the first to achieve initiation of callus and regeneration of plants from immature embryos by culturing the explants from a number of genotypes on B5 or MS medium containing 2.3-13.5 μM of 2,4-D. The cultures were maintained by subculturing on B5 medium with 4.5 μM 2,4-D. Callus from the variety Lodi was maintained for 1 1/2 yr through 13 subcultures without losing its ability to regenerate plants. For organ differentiation, B5 medium without 2,4-D was used and 90% of the regenerated plants, transferred from culture medium to sterilized soil, survived. Most plants grew vigorously and produced seeds varying from sterility to high viability. Although most of the regenerated plants appeared normal, some phenotypic and genotypic variability was observed. The variability ranged from occasional albinos to variegated plants to differences in awns and type of headings. In some regenerated plants, chromosomal instability was also observed. Callus capable of regenerating plants was also obtained from germinating embryos and excised apical meristems grown on B5 medium containing 2,4-D.

Loerz et al. (1976) were able to induce callus from hypocotyls of the three varieties of *Avena sativa* (Flamingskrone, Arnold, and Tiger) by culturing on SH medium with 2,4-D 2.3 μM. By increasing the 2,4-D concentration to 9.0 μM, they could increase the rate of callus growth. Organogenesis was induced in the 8th subculture by a proper balance of auxin and cytokinin hormones. Roots were initiated only with the combinations IAA-kinetin or NAA-BA-GA_3, and roots and shoots with NAA 3.0 μM, IAA 10.0 μM, and BA 1.5 μM. More than 50% of the calli produced shoots, ranging from 1 to 5 shoots per callus when cultured in 4000 lux, 16:8 hr, light:dark photoperiod. Calli with shoots, but no roots, developed roots when transferred to hormone-free MS medium. Recently (Galston et al., 1977), we have succeeded in inducing vigorous callus organogenesis from the scutellum of mature embryos using a slight modification of the technique of Cummings et al (1976). We have also recently succeeded in efficiently regenerating plants from callus derived from lateral buds of tillers.

Protocol for Callus Formation

1. Sterilize husked seeds of oat (*Avena sativa* L. cv. Lodi) by immersion in commercial bleach: H_2O (1:2 v/v) for 15 min. Rinse several times in distilled sterile water, blot, and place on moist filter paper in 25 x 100 mm petri dishes.
2. Incubate dishes in darkness at 24 C in an incubator for ca., 18 hr.
3. Dissect intact embryos from the imbibed seeds with sterile instruments. Place embryos, scutellum-side down, on callus-inducing media (MS or B5) containing 2.3-23.0 μM 2,4-D.

4. Incubate cultures at 23 C in diffuse light in controlled growth cabinets. Callus formation is obtained in about 6 weeks.
5. Subculture into fresh medium (2x) at intervals of 4 weeks.
6. Transfer to B5 medium containing 2.3 µM of 2,4-D. This treatment results in initiation of multiple green centers of organization with cylindrical stemlike properties. On excision, with some calli, these centers differentiate into roots and shoots.
7. Transplant plantlets into vermiculite containing a mineral nutrient solution (1.2 g/l Hyponex, Copley Chemical Co., Copley, Ohio) in controlled-condition growth rooms at 23 C with 18 hr daylight (ca. 12,000 lux). The plants developed into normal-appearing vegetative oat plants. It is not clear if age of embryo, size of scutellum, or concentration of 2,4-D in the culture initiation medium are critical for plant regeneration.

These recent experiments have demonstrated that callus induction and plant regeneration are attainable in oats and other cereals. However, neither the optimum conditions of medium and environment nor chemical factors involved in successful regeneration of plants are yet fully understood in any species.

OVERVIEW: PROBLEMS AND PROSPECTS

While a number of successful attempts at whole plant regeneration from cereal tissue cultures have recently been reported, such regeneration has been sporadic and somewhat unpredictable. Despite attempts with many species, there has been only one instance (*Pennisetum*) in which protoplasts from suspension cell cultures of immature embryos have resulted in complete plant regeneration.

We have tentatively attributed the lack of cell division in cereal protoplasts to genetic factors and/or the sensitivity of the protoplasts to injury or osmotic stress during their isolation. This inference is based upon the relative ease of cell division in protoplasts from other plant species under similar experimental conditions. Despite the relative lack of success with cereals, substantial advances have been made in determinations of the various metabolic events accompanying cell senescence which may prevent cell division. These events include decreases in rates of RNA and protein synthesis and increases in RNase and protease activities. Treatments with specific senescence inhibitors, such as polyamines and cytokinins, have resulted in stabilizing protoplasts against lysis and have prevented the rise in RNase and protease activities and the decrease in rates of macromolecular synthesis. Polyamines have also induced at least one round of mitosis after some stimulation of DNA synthesis. Unhappily such protoplasts do not undergo cytokinesis, and thus remain binucleate. Thus it appears that polyamines can facilitate the G_1-S transition, but no more. The limited success in cell division and plant regeneration from protoplasts in cereals must await a more complete understanding of the metabolic processes needed for optimum cell division and their control.

In spite of a number of reports of successful plant regeneration from cereal explants, regeneration has been extremely difficult and inconsistent at best. Indeed, anatomical study of organogenesis in cultured tissues of oats and other cereals by Cure and Mott (1978) indicates that the incidence of shoot regeneration in cultures may be due to adventitious bud formation on roots rather than controlled de novo organogenesis from undifferentiated callus. Besides, the ability of only certain explants (Table 3) from maize, wheat, and sorghum (Evans et al., 1981) and of only certain genotypes of oats to organize (Cummings et al., 1976) emphasizes our lack of understanding of the basis of regeneration in cereals.

Yet another phenomenon which may be a source of serious concern in the attempts to produce pure cell lines for crop improvement in cereals involves chromosomal variations. Although there has been some success in the regeneration of haploid plants from anther culture of cereals, especially rice, success from oats has been negligible. This is especially unfortunate, since the importance of haploid plants as a source of useful mutations for crop improvement is an obvious attraction. The technology of producing haploid plants from anther culture has its own difficulties--among them the production of many variants, including albinos. For useful application of tissue culture technology to cereal improvement these problems need much more careful study and understanding.

KEY REFERENCES

Brenneman, F.N. and Galston, A.W. 1975. Experiments on cultivation of protoplasts and calli of agriculturally important plants. 1. Oats (*Avena sativa* L.). Biochem. Physiol. Pflanz. 168:453-471.

Cummings, D.P., Green, C.E., and Stuthman, D.D. 1976. Callus induction and plant regeneration in oats. Crop Sci. 16:465-470.

Flores, H.E., Kaur-Sawhney, R., and Galston, A.W. 1981. Protoplasts as vehicles for plant propagation and improvement. In: Advances in Cell Culture (K. Maramorosch, ed.) pp. 241-279. Academic Press, New York.

Galston, A.W. and Kaur-Sawhney, R. 1980. Polyamines and plant cells. What's New Plant Physiol. 11:5-8.

Loerz, H., Harms, C.T., and Potrykus, I. 1976. Regeneration of plants from callus in *Avena sativa* L. Z. Pflanzenzuecht. 77:257-259.

REFERENCES

Academia Sinica. 1975. Isolation and culture of rice protoplasts. Sci. Sin. 18:779-784.

Altman, A. and Bachrach, U. 1981. Involvement of polyamines in plant growth and senescence. In: Advances in Polyamine Reserach Vol. 3

(C.M. Caldarera, V. Zappia, and U. Bachrach, eds.) pp. 365-375. Raven Press, New York.

Altman, A., Kaur-Sawhney, R., and Galston, A.W. 1977. Stabilization of oat leaf protoplasts through polyamine-mediated inhibition of senescence. Plant Physiol. 60:570-574.

Baum, B.R. 1977. Oats: Wild and Cultivated. Thorn Press, Ottawa, Canada.

Bhojwani, S.S., Evans, P.K., and Cocking, E.C. 1977. Protoplast technology in relation to crop plants: Progress and problems. Euphytica 26:343-360.

Carew, D.P. and Schwarting, A.E. 1958. Production of rye embryo callus. Bot. Gaz. (Chicago) 119:237-239.

Carter, O., Yamada, Y., and Takahashi, E. 1967. Tissue culture of oats. Nature 214:1029-1030.

Chin, J.C. and Scott, K.J. 1977. Studies on the formation of roots and shoots in wheat callus cultures. Ann. Bot. 41:473-481.

Cocking, E.C. 1978. Protoplast culture and somatic hybridization. In: Prod. Symp. Plant Tissue Culture, pp. 225-263. Science Press, Peking.

______, Power, J.B., Evans, P.K., Safwat, F., Frearson, E.M., Hayward, C., Berry, S.F., and George, D. 1974. Naturally occurring differential drug sensitivity of cultured plant protoplasts. Plant Sci. Lett. 3: 341-350.

Constabel, F., Dudits, D., Gamborg, O.L., and Kao, K.N. 1975. Nuclear fusion in intergeneric heterokaryons. Can. J. Bot. 53:2092-2095.

Cure, W.W. and Mott, R.L. 1978. Comparative anatomical study of organogenesis in cultured tissue of maize, wheat and oats. Physiol. Plant. 42:91-96.

Deka, P.C. and Sen, S.K. 1976. Differentiation in calli originated from isolated protoplasts of rice (*Oryza sativa* L.) through plating technique. Mol. Gen. Genet. 145:239-243.

Eriksson, T., Bonnett, H., Glimelius, K., and Wallin, A. 1974. Technical advances in protoplast isolation, culture and fusion. In: Tissue Culture and Plant Science (H.E. Street, ed.) pp. 213-231. Academic Press, New York.

Evans, D.A., Sharp, W.R., and Flick, C.E. 1981. Growth and behavior of cell cultures: Embryogenesis and organogenesis. In: Plant Tissue Culture. Methods and Applications in Agriculture (T.A. Thorpe, ed.) pp. 45-114. Academic Press, New York.

Flores, H.E. and Galston, A.W. 1982. Polyamines and plant stress: Activation of putrescine biosynthesis by osmotic shock. Science 217: 1259-1261.

Fuchs, Y. and Galston, A.W. 1976. Macromolecular synthesis in oat leaf protoplasts. Plant Cell Physiol. 17:475-482.

Galston, A.W. 1978. The use of protoplasts in plant propagation and improvement. In: Propagation of Higher Plants through Tissue Culture (K.H. Hughes, R. Henke, and M. Constantin, eds.) pp. 200-212. Tech. Info. Center, U.S. Dept. of Energy, Washington, D.C.

______, Adams, W.R., Jr., Brenneman, F., Fuchs, Y., Rancillac, M., Kaur-Sawhney, R., and Staskawicz, B. 1977. Opportunities and obstacles in the culture of cereal protoplasts and calli. In: Molecular Genetic Modification of Eucaryotes (I. Rubenstein, R.L. Phillips, C.E. Green, and R.J. Desnick, eds.) pp. 13-42. Academic Press, New York.

______, Altman, A., and Kaur-Sawhney, R. 1978. Polyamines, ribonuclease and the improvement of oat leaf protoplasts. Plant Sci. Lett. 11: 69-79.

Gamborg, O.L. and Eveleigh, D.E. 1968. Culture methods and detection of glucanases in suspension cultures of wheat and barley. Can. J. Biochem. 46:417-421.

Gamborg, O.L., Miller, R.A., and Ojima, K. 1968. Nutrient requirements of suspension cultures of soybean root cells. Exp. Cell Res. 50:151-158.

Gamborg, O.L., Shyluk, J.P., Brar, D.S., and Constabel, F. 1977. Morphogenesis and plant regeneration from callus of immature embryos of sorghum. Plant Sci. Lett. 10:67-74.

Gepstein, S. and Thimann, K.V. 1980. Changes in abscisic acid content of oat leaves during senescence. Proc. Natl. Acad. Sci. U.S.A. 77: 2050-2053.

Gosch-Wackerle, G., Avivi, L., and Galun, E.Z. 1979. Induction, culture and differentiation of callus from immature rachises, seeds and embryo of *Triticum*. Z. Pflanzenphysiol. 91:267-278.

Green, C.E. and Phillips, R.L. 1975. Plant regeneration from tissue cultures of maize. Crop Sci. 15:417-421.

Halperin, W. and Wetherell, D.F. 1964. Adventive embryony in tissue culture of wild carrot, *Daucus carota*. Am. J. Bot. 57:274-283.

Heby, O. 1981. Role of polyamines in the control of cell proliferation and differentiation. Differentiation 19:1-20.

Hendre, R.R., Mascarenhas, A.F., Pathak, M., Seetarama Rao, B., and Jagannathan, V. 1972. Studies on tissue cultures of maize, wheat, rice and sorghum. Biochem. J. 128:27.

Kaur-Sawhney, R. and Galston, A.W. 1979. Interaction of polyamines and light on biochemical processes involved in leaf senescence. Plant Cell Environ. 2:189-196.

Kaur-Sawhney, R., Rancillac, M., Staskawicz, B., Adams, W.R., Jr., and Galston, A.W. 1976. Effect of cycloheximide and kinetin on yield, integrity and metabolic activity of oat leaf protoplasts. Plant Sci. Lett. 7:57-67.

Kaur-Sawhney, R., Adams, W.R., Jr., Tsang, J., and Galston, A.W. 1977. Leaf pretreatment with senescence retardants as a basis for oat protoplast improvement. Plant Cell Physiol. 18:1309-1317.

Kaur-Sawhney, R., Flores, H.E., and Galston, A.W. 1980. Polyamine-induced DNA synthesis and mitosis in oat leaf protoplasts. Plant Physiol. 65:368-371.

Kaur-Sawhney, R., Shih, L., Flores, H.E., and Galston, A.W. 1982. The relation of polyamine synthesis and titer to aging and senescence in oat leaves. Plant Physiol. 69:405-410.

King, P.J., Potrykus, I., and Thomas, E. 1978. In vitro genetics of cereals: Problems and perspectives. Physiol. Veg. 16:381-399.

Kinnersley, A.M., Racusen, R.H., and Galston, A.W. 1978. A comparison of regenerated cell walls in tobacco and cereal protoplasts. Planta 139:155-158.

Koblitz, H. 1976. Isolierung und Kultivierung von Protoplasten aus Calluskulturen der Gerste. (English summary.) Biochem. Physiol. Pflanzen. 170:287-293.

Linsmaier, E.M. and Skoog, F. 1965. Organic growth factor requirements of tobacco tissue cultures. Physiol. Plant. 18:100-127.

Mascarenhas, A.F., Sayagaver, B.M., and Jagannathan, V. 1965. Studies on the growth of callus cultures of *Zea mays*. In: Tissue Culture (C.V. Ramakrishna, ed.) pp. 283-292. Dr. W. Junk, Pub., The Hague, Netherlands.

Masteller, V.J. and Holden, D.J. 1970. The growth of and organ formation from callus tissue of sorghum. Plant Physiol. 45:363-364.

Metcalfe, D.S. and Elkins, D.M. 1980. Crop Production. Macmillan Pub., New York.

Muir, W.H., Hildebrandt, A.C., and Riker, A.J. 1954. Plant tissue cultures produced from single cells. Science 119:877-878.

Murashige, T. and Skoog, F.K. 1962. A revised medium for rapid growth and bioassays with tobacco tissue cultures. Physiol. Plant. 15:473-497.

Nemet, G. and Dudits, D. 1977. In: Use of Tissue Cultures in Plant Breeding (J. Novak, ed.) pp. 145-163. Czech. Acad. Sci., Prague.

Nishi, T., Yamada, Y., and Takahashi, E. 1968. Organ redifferentiation and plant restoration in rice callus. Nature 219:508-509.

Norstog, K. 1956. Growth of rye-grass endosperm in vitro. Bot. Gaz. (Chicago) 117:253-259.

Pental, D. and Gunckel, J.E. 1979. Cereals. In: Plant, Cell and Tissue Culture (W.R. Sharp, P.O. Larsen, E.F. Paddock, and V. Raghavan, eds.) pp. 633-709. Ohio State Univ. Press, Columbus, Ohio.

Potrykus, I. 1980. The old problem of protoplast culture: Cereals. In: Advances in Protoplast Research (L. Ferenczy, G.L. Farkas, and G. Lazar, eds.) pp. 243-254. Akademiai Kiado, Budapest, Hungary.

______, Harms, C.T., and Loerz, H. 1976. Problems in culturing cereal protoplasts. In: Cell Genetics in Higher Plants (D. Dudits, C.L. Farkas, and P. Maliga, eds.) pp. 129-140. Akademiai Kiado, Budapest, Hungary.

______, Harms, C.T., Loerz, H., and Thomas, E. 1977. Callus formation from stem protoplasts of corn (*Zea mays* L.). Mol. Gen. Genet. 156:347-350.

______, Harms, C.T., and Loerz, H. 1979. Callus formation from cell culture protoplasts of corn (*Zea mays* L.). Theor. Appl. Genet. 54: 209-214.

Racusen, R.H., Kinnersley, A.M., and Galston, A.W. 1977. Osmotically induced changes in electrical properties of plant protoplast membranes. Science 198:405-406.

Rangan, T.S. 1976. Growth and plantlet regeneration in tissue cultures of some Indian millets: *Paspalum scrobiculatum* L., *Eleusine coracana* Gaertn., *Pennisetum typhoideum* Pers. Z. Pflanzenphysiol. 78:208-216.

Saalback, O. and Koblitz, H. 1978. Attempts to initiate callus formation from barley leaves. Plant Sci. Lett. 13:165-169.

Schenk, R.U. and Hildebrandt, A.C. 1972. Medium and techniques for induction and growth of monocotyledonous and dicotyledonous cell cultures. Can. J. Bot. 50:199-204.

Schieder, O. and Vasil, I.K. 1980. Protoplast fusion and somatic hybridization. In: Perspectives in Plant Cell and Tissue Culture, Supplement 11B (I.K. Vasil, ed.) pp. 21-42. Academic Press, New York.

Shepard, J.F., Bidney, D., and Shahin, E. 1980. Potato protoplasts in crop improvement. Science 208:14-17.

Sheridan, W.F. 1975. Tissue culture of maize. I. Callus induction and growth. Physiol. Plant. 33:151-156.

Shimada, T., Sasakuam, T., and Tsunewaki, K. 1969. In vitro culture of wheat tissues. I. Callus formation, organ redifferentiation, and single cell culture. Can. J. Genet. Cytol. 11:294-304.

Steward, F.C., Mapes, M.O., Kent, A.E., and Holsten, R.D. 1964. Growth and development of cultured plant cells. Science 143:20-27.

Straus, J. 1954. Maize endosperm tissue grown in vitro. II. Morphology and cytology. Am. J. Bot. 41:833-839.

Takebe, I., Labib, G., and Melchers, G. 1971. Naturwissenschaften 58: 318-320.

Thimann, K.V. 1980. The senescence of leaves. In: Senescence in Plants (K.V. Thimann, ed.) pp. 85-115. CRC Press, Boca Raton, Florida.

Thomas, E., King, P.J., and Potrykus, I. 1977. Shoot and embryolike formation from cultured tissue of *Sorghum bicolor*. Naturwissenschaften 64:587.

Tsai, C., Chien, Y., Chou, Y., and Wu, S. 1978. A further study on the isolation of the rice *Oryza sativa* L. protoplasts. Acta Bot. Sin. (Peking) 20:97-103.

U.S.D.A. 1980. Agricultural Statistics. U.S. Gov. Printing Office, Washington, D.C.

Vasil, I.K. 1980. Androgenetic haploids. In: Perspectives in Plant Cell and Tissue Culture, Supplement 11A (I.K. Vasil, ed.) pp. 195-217.

______ and Vasil, V. 1980. Isolation and culture of protoplasts. In: Perspectives in Plant Cell and Tissue Culture, Supplement 11B (I.K. Vasil, ed.) pp. 1-16.

Vasil, V. and Vasil, I.K. 1980. Isolation and culture of cereal protoplasts. II. Embryogenesis and plantlet formation from protoplasts of *Pennisetum americanum*. Theor. Appl. Genet. 56:97-99.

______ 1981. Somatic embryogenesis and plant regeneration from tissue cultures of *Pennisetum americanum* and *P. americanum* x *P. purpureum* hybrid. Am. J. Bot. 68:860-864.

Webster, J.W. 1966. Production of oat callus and its susceptibility to a plant parasitic nematode. Nature (London) 212:1472.

CHAPTER 5
Wheat

G.W. Schaeffer, M.D. Lazar, and *P.S. Baenziger*

HISTORY OF THE CROP

Wheat is a very diverse and ancient crop. The cultivated wheats include einkorn (*Triticum monoccocum* L., 2x = 14), emmer [*T. diccocum* (Shank) Schuebl., 4x = 28], durum (*T. turgidum* L., 4x = 28) (also called *T. durum* Desf.), and common or bread (*T. aestivum* L. em. Thell, 6x = 42) wheats (Feldman and Sears, 1980). *T. aestivum* is the most widely grown wheat. Club wheat (*T. compectum* Host, 6x = 42) and other diploid, tetraploid, and hexaploid wheat species have been cultivated at some time (Briggle, 1980).

Despite extensive research, no one knows where or when wheat originated. Wheat evolved from wild grasses, most likely in the Near East near the Tigris-Euphrates river basin (Briggle, 1980). Wheat may have been gathered by man as early as 15,000-10,000 B.C. As early as 2400 B.C., the estimated yields of emmer in Mesopotamia were 3672 liters/ha (Harlan, 1975a). By 2100 B.C. wheat had become a minor crop, possibly attributable to increased salinization of the soil (Jacobsen and Adam, 1958). The spread of wheat cultivation into Europe, Asia, and Africa resulted in more intense selection pressures on wheat with subsequent increases in the variation among wheats. Not all of the wheat species spread to the same regions. Ethiopia is a center of diversity for tetraploid wheats, but not for hexaploid wheats. Harlan (1975b) postulated that the tetraploids spread to Ethiopia before hexaploids had evolved. Similarly, einkorn apparently never spread into

Egypt, Ethiopia, or India but did extend into western Europe (Harlan, 1975a). *T. aestivum* had reached China and Japan before 1000 B.C. The great geographic diversity and isolation of common wheat allowed divergent evolution. Harlan described common wheat as being oligocentric (having many centers of diversity).

ECONOMIC IMPORTANCE

Today wheat is grown throughout the world and is the most widely adapted of the cereal crops. Table 1 presents the hectarage and total production of wheat for 1959, 1969, and 1979. Wheat is a major crop in every continent with the exception of Antarctica. In a recent description of wheat production, Allan (1980) concluded the best wheat-growing areas are between 30 and 50 north latitude and 25 and 4 south latitude. The crop generally is grown without irrigation and 85% is grown in areas where rainfall is less than 900 mm annually. Winter wheat is grown in areas (the Balkan countries and southern USSR, China, western Europe, and most of the United States except for the northern plains) where it can survive the winter. Spring wheat is grown in those areas (Argentina, Canadian prairies, northern United States, and northern and central USSR) where winters are too severe for winter wheat to survive and in areas (Australia, Brazil, India, Mexico, and the southwestern United States) where winters are too mild for winter wheat to vernalize.

Wheat is the chief food for one-third of the world's population and it provides more nourishment for more people than any other food crop (Allan, 1980; Inglett, 1974). For human consumption, wheat has diverse uses and is marketed by quality classes. Hard winter and spring common wheats are used to make bread, farina, and rolls. Hard wheat classes grow best in temperate climates with fairly low rainfall. Kernels have a vitreous appearance and a hard texture (thus giving them their name). Protein content of hard wheat ranges between 11 and 17%. Soft wheats are used to make cakes, crackers, cookies, and other assorted pastries. Soft wheats are generally grown in areas where there is abundant moisture and relatively mild winters. Most soft wheats are sown in the fall and have a relatively low kernel protein content of 6-11%. Durum wheat is processed into semolina, a coarser material than standard wheat flour and is used to make pasta (Finney and Yamazaki, 1967). Durum wheat kernels normally have a hard or vitreous appearance and have a kernel protein content similar to the hard common wheats. While millers and bakers have developed processing techniques which can alter the chemical nature of wheat flour, it is recognized that good quality wheat is desirable for all flours. Reviews of the wheat processing industry are numerous (Finney and Yamazaki, 1967; Ford and Kingswood, 1981).

Wheat is an important animal feed crop and accounts for approximately 10% of all of the grain fed to livestock. Wheat also is used for pasture, hay, and silage (Allan, 1980).

Table 1. Wheat Area and Production for 1959, 1969, and 1979 by Continent[a]

	AREA (1000 ha)			PRODUCTION (1000 MT)		
	1959	1969	1979	1959	1969	1979
North America	31,740	30,117	36,372	43,273	60,364	77,441
United States	21,467	19,253	25,172	30,703	39,705	57,535
South America	6,688	7,626	9,621	7,620	9,975	11,794
Europe	29,384	28,074	24,433	55,656	73,089	82,782
USSR (Europe and Asia)	63,563	66,327	57,500	62,596	76,600	86,000
Asia	57,785	64,693	77,582	51,302	66,867	119,219
Africa	7,360	7,964	8,646	5,307	7,089	9,354
Oceania	4,724	9,909	11,862	5,455	11,322	16,337
Total	201,247	214,705	226,016	220,312	291,866	402,927
Rice (Total)	113,442	98,751	104,198	213,654	195,082	368,562

[a]Agricultural Statistics 1960, 1970, and 1980.

Breeding and Crop Improvement Problems

Wheat is a self-pollinated crop with little natural cross pollination (usually less than 1%) although some weather conditions may enhance cross pollination. The breeding methodologies for self-pollinated crops have been thoroughly described elsewhere (Allard, 1960; Simmonds, 1979) as have the specific methods of wheat breeding (Lelley, 1976; Schmidt, 1980; Heyne and Smith, 1967; Patterson and Allan, 1981). All breeding procedures rely upon the successful exploitation of genetic variation. Wheat is a self-pollinated crop and since selfing will lead to homozygosity, wheat breeders have used controlled hybridization to exploit genetic variation. Allan (1980) described methods for wheat hybridization. Irradiation and chemical mutagenesis have also been used as methods for introducing variation. Genetic resources for a crop such as wheat are extensive. (Presently the USDA World Collection of Small Grains contains 37,000 different wheats and wild relatives.) Great care is taken to choose the parents for hybridization. Parent selection is guided by the breeding objective.

Problems of crop improvement are as diverse as the ecological areas in which wheat is grown. The breeder's objective is to overcome these problems. It would be impossible to discuss all of the problems in detail, however, breeding objectives can be grouped into the following categories: (a) breeding for improved agronomic type, (b) breeding for improved market quality, and (c) breeding for improved stress resistance.

The objective of breeding for improved agronomic type is to increase the amount of grain produced per unit area or the economic return per unit of investment (Schmidt, 1980). Yield can be improved by increasing one or more components of yield: number of spikes per unit area, number of kernels per spike, or kernel weight. The proper balance among these components will give the best yield. Production practices and climate can affect the components. Seeding rate will affect the number of spikes per unit area and so will moisture supply. Unfortunately the increase in one component is often offset by decreases in other components. Concurrent with yield increases, better straw strength and lodging resistance need to be developed. Lodging can severely increase harvest losses and reduce kernel quality. Breeders are also trying to improve responsiveness to fertilizer applications. Increased levels of available nitrogen simultaneously increase both plant growth and lodging. Cultivars are very desirable that can effectively respond to higher nitrogen fertility by increasing yield without increasing lodging. In areas where a second crop is planted immediately following wheat harvest (known as double cropping), early maturity of wheat provides a longer growing season for the second crop.

Breeding for market quality improves wheat utilization. As mentioned previously, each end product requires flours having special characteristics (Finney and Yamazaki, 1967; Ford and Kingswood, 1981). It is the breeder's objective to develop a high-yielding wheat having good functional quality. Improved nutritional quality is also an important objective in wheat breeding. Elevation of lysine and protein in wheat is an objective of some breeding programs. Lysine is the most limiting amino acid in wheat protein for human nutrition (Mattern et al., 1975).

Development of wheats that can withstand various stresses and still have high yield at harvest is a major goal. Stress may be either abiotic or biotic. Examples of abiotic stresses are temperature or moisture extremes and mineral stresses caused by acid, alkaline soils, or saline soils (Jung, 1978; Wright, 1976). Among biotic stresses are disease epiphytotics, insect infestations, and weed infestations.

Both types of stress can severely reduce yield. A severe winter can destroy a local wheat crop. Some abiotic soil stresses can be tempered by the addition of soil amendments, but these are often too costly. Soil amendments tend to stay only in the upper soil levels necessitating exploitation of genetic variation. Cultivars differ widely for tolerance to cold, moisture levels, soil pH, and salinity (Jung, 1978; Wright, 1976).

Because biotic stresses are dynamic and ever-changing, wheat is constantly vulnerable to pathogens with new virulences (Horsfall, 1972). Breeders are constantly developing cultivars for changing conditions. For example, Kilpatrick (1975) summarized the changes of leaf, stem, and stripe rust (*Puccinia recondita* f. sp. tritici Rob. ex. Desm., *Puccinia graminis* Pers. f. sp. tritica Eriks. and E. Henn., and *Puccinia striiformis* West., respectively) in 26 countries and 10 states (United States) and found that on the average a shift in leaf, stem, and stripe rust virulence occurred every 5.6, 5.3, and 5.5 years, respectively. These are average values. In some areas the virulence changes were more frequent. Since it requires approximately 10 years to develop an average cultivar, it often takes longer to develop a cultivar than it can be maintained as a disease-free cultivar. While rusts are the most important wheat diseases and can be found wherever wheat is grown (Loegering et al., 1967), there are many other important diseases and insect pests of wheat (Dahns, 1967; Weise, 1977). For example, in Maryland where the rusts are of little importance, powdery mildew (*Erysiphe graminis* DC. f. sp. tritici E. Marchal) reduced yield by 34% in a susceptible cultivar (Johnson et al., 1979). Phytotoxins have been associated with the pathogenicity of *Septoria nodorum* (Berk.) Berk. and *Helminthosporium sorokinianum* Sacc. (previously reported as *H. sativum*) (Kent and Strobel, 1975; Pringle, 1977).

The number of diseases and pests of wheat is very large, therefore several breeding and production techniques have been developed to alleviate those stresses (see Jenkyn and Plumb, 1981, for recent reviews). Besides using major single-gene resistance (which has the drawback that a change in the pathogen's virulence can easily overcome the host resistance), pathologists and breeders have used race-nonspecific resistance gene(s) giving resistance to a spectrum of pathological races and gene pyramiding, i.e., adding more than one gene for resistance. Multigenic resistance has the advantage that more than one gene in the pathogen must be altered before it becomes pathogenic on the new host line. Multilines (mixtures of lines that have a common genetic background but different specific genes for resistance), and blends have also been proposed as an efficient way to exploit genetic diversity.

It should be clear to the reader that wheat breeders and other researchers must frequently adapt multiple objectives for which they utilize both qualitative and quantitative genetics.

LITERATURE REVIEW

Callus and cell suspension cultures have been established from somatic tissues of several *Triticum* species as well as many cultivars of *T. aestivum*. Tissues amenable to such culture include root tips, root sections, stem nodes, intercalary meristems, shoot tip meristems, rachis sections, mature and immature excised embryos, and whole seeds. In vitro culture of isolated wheat roots was reported as early as 1951 (Burstrom, 1951). The development of culture methodology for callus and cell suspensions came later with the reports from Trione et al. (1968), Gamborg and Eveleigh (1968), and Shimada et al. (1969).

Several medium formulations have proven successful in culture of wheat somatic tissues (Table 2). Early formulations frequently made use of complex components such as casein hydrolysate (Gamborg and Eveleigh, 1968) or coconut milk (Shimada et al., 1969). Synthetic auxinlike compounds used in the absence of complex additives, however, produced adequate callus growth from a variety of explant sources and genotypes. This success may relate to the potency of such compounds and/or their slow metabolic turnover. Especially popular is 2,4-D, though others have been found useful as either supplements or substitutes for 2,4-D, including CPA (Schenk and Hildebrandt, 1972), 2,4,5-T (Dudits et al., 1975; Bhojwani and Hayward, 1977); 4-amino-3,5,6-trichloropicolinic acid (picloram) (Collins et al., 1978); NAA (Schaeffer et al., 1979); and 3,5-dichloro-anisic acid (dicamba) and 4-chloro-2-oxobenzo-thiazolin-3-yl acetic acid (benazolin) (Dudits et al., 1975). Cytokinins generally inhibit callus proliferation (Schenk and Hildebrandt, 1972; Dudits et al., 1975). Though a variety of other organic additives have been examined, no convincing evidence has been produced indicating any effect upon rate of callus or cell suspension culture proliferation.

In contrast to the general inconsequence of many changes in composition of cell culture media (strong hormones excepted), it is clear that the genotype of the cultured tissue significantly affects the rate of cell proliferation. In addition to variation in growth rate of cultures among the species *Triticum monococcum*, *T. dicoccum*, *T. longissimum*, *T. speltoides*, *T. tauschii*, *T. turgidum*, *T. timopheevi*, and *T. aestivum* found by Shimada et al. (1969) and Gosch-Wackerle et al. (1979), data from several laboratories point to significant variation among cultivars within *T. aestivum* (Trione et al., 1968; O'Hara and Street, 1978; Schaeffer et al., 1979; Sears and Deckard, 1982). Data on rate of callus proliferation from mature embryo explants have indicated that the varieties Baart and GWO-1809 perform particularly well (Trione et al., 1968; Schaeffer et al., 1979) (Fig. 1). Although these two varieties were not examined, the recent work of Sears and Deckard (1982) indicated that such differences may not exist when immature embryos are used as explants, at least with respect to callus

Table 2. Basal Medium Formulations useful for In Vitro Culture of Wheat Somatic Tissues

COMPONENT	BASAL MEDIUM (mg/l)			
	B5	MS[a]	SH	T[a]
KNO_3	2500.0	1900.0	2500.0	1000.0
NH_4NO_3	—	—	—	—
KCl	—	—	—	65.0
$(NH_4)_2SO_4$	134.0	—	—	—
$NH_4H_2PO_4$	—	—	300.0	—
$Na_2H_2PO_4H_2O$	150.0	—	—	165.0
KH_2PO_4	—	170.0	—	—
$Ca(NO_3)_2 \cdot 4H_2O$	—	—	—	260.0
$CaCl_2 \cdot 2H_2O$	150.0	440.0	200.0	—
$MgSO_4 \cdot 7H_2O$	250.0	370.0	400.0	360.0
$MnSO_4 \cdot H_2O$	10.0	22.3	10.0	5.0
$FeSO_4 \cdot 7H_2O$	27.8	27.8	15.0	27.8
Na_2EDTA	37.2	37.2	20.0	37.2
H_3BO_3	3.0	6.2	5.0	3.0
$ZnSO_4 \cdot 7H_2O$	3.0	8.6	1.0	3.0
$Na_2MoO_4 \cdot 2H_2O$	0.25	0.25	0.1	0.25
KI	0.75	0.83	1.0	—
$CuSO_4 \cdot 5H_2O$	0.25	0.025	0.2	0.025
$CoCl_2 \cdot 6H_2O$	0.25	0.025	0.1	0.025
Inositol	100.0	100.0	1000.0	100.0
Glycine	—	2.0	—	10.0
Thiamine HCl	10.0	0.1	5.0	10.0
Pyridoxine HCl	1.0	0.5	0.5	1.0
Nicotinic acid	1.0	0.5	5.0	5.0
Ascorbic acid	—	—	—	10.0
Ca pantothenate	—	—	—	0.25
L-glutamine	—	—	—	50.0
Sucrose	20,000.0	30,000.0	30,000.0	20,000.0
2,4-D	2.0	—	0.5	—
CPA	—	—	2.0	—
KIN	—	—	0.1	—
pH	5.5	5.8	5.9	5.5
Reference:	Gamborg et al., 1967	Murashige & Skoog, 1962	Schenk & Hildebrandt, 1972	Dudits et al., 1975

[a]Hormones optional and/or variable.

initiation. They found that over 90% of immature embryos cultured from each of seven varieties produced callus, though they did not report relative growth rates. Additionally, several studies have been

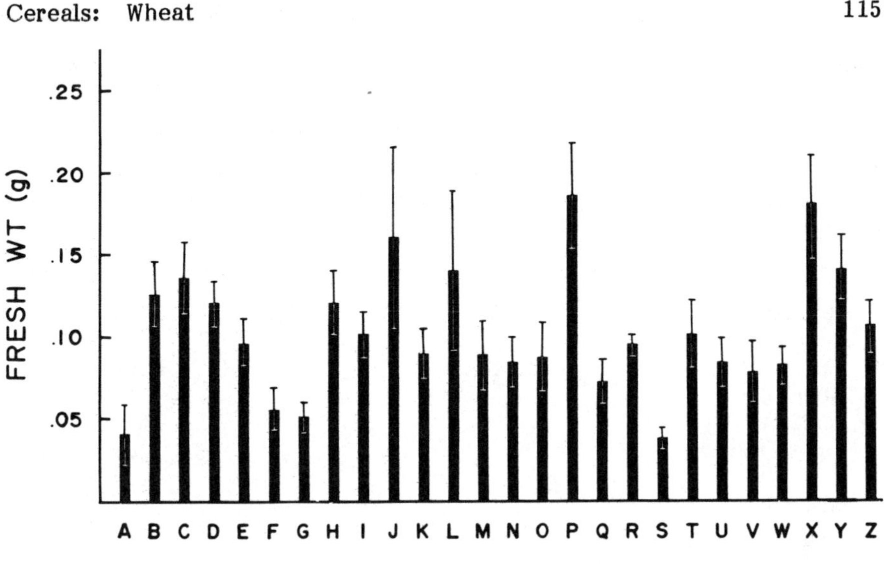

Figure 1. Response of embryos of *Triticum aestivum* L. genotype to in vitro culture. Growth response is measured as fresh weight from 26 lines after a 21-day growth period on the callus proliferation medium (Table 4) after three passages on the same medium. The data are presented as mean ± standard error for each variety or cultivar (LSD = 0.06 g). Letters represent the cultivars: A = Arthur 71; B = Coker 68-15; C = McNair 1003; D = McNair 3003; E = Chris; F = Stoddard; G = Olaf; H = Fielder; I = Ticonderoga; J = Kitt, K = Coker 747; L = Coker 75-6 (Southern Bell); M = Downy; N = Holley; O = Potomac; P = Baart; Q = Chancellor; R = Tecumseh; S = Centurk; T = Blueboy; U = Polk; V = Waldron; W = Era; X = GWO1809; Y = Profit 75; Z = Protor.

reported concerning rate of callus proliferation in several ditelocentric lines of the variety Chinese Spring (Baroncelli et al., 1978; Gosch-Wackerle et al., 1979; Khoteleva and Ermishin, 1980). While the results of the various studies are somewhat conflicting, all show a reduction in callus proliferation in the absence of the long arm of chromosome 1A. The importance of the A genome might have been guessed since *T. monococcum* (AA) callus proliferates rapidly (Shimida et al., 1969; Gosch-Wackerle et al., 1979).

While root differentiation from wheat somatic cell cultures has been easily inducible via classic reduction or elimination of exogenous auxin (Shimada et al., 1969; Gosch-Wackerle et al., 1979; O'Hara and Street, 1978; Chin and Scott, 1977) or by addition of the antiauxin TIBA (Dudits et al ., 1975), shoot differentiation by either organogenic or embryogenic pathways has been both rare and unpredictable. Explants of specific cell types, such as scutellar tissue, result in calli capable of producing plantlets after relatively long culture periods. Reports of shoot differentiation from somatic cell culture are summarized in Table 3. Many of the cultures from which shoots were obtained were derived from embryo explants. This makes the claim of de novo shoot meristem

Table 3. Status of Shoot Regeneration from Somatic Cell Cultures of Wheat

SPECIES/ VARIETY	GREEN SPOTS[a]	NUMBER OF SUB-CULTURES	PLOIDY OF REGENER-ANTS	REFER-ENCE
T. aestivum Chinese Spring	—[b]	1	Euploid	Shimada et al., 1969
T. aestivum Tobari 66	—	Several	—	Dudits et al., 1975
T. aestivum Chinese Spring	—	≥1	—	Bhojwani & Hayward, 1977
T. aestivum Mengari	+	1	—	Chin & Scott, 1977
T. aestivum Maris Ranger	—	≤2	—	O'Hara & Street, 1978
T. durum 10 varieties	—	1	Most euploid many aneuploid	Bennici & D'Amato, 1978
T. aestivum Chinese Spring	+	≤3	—	Shimada, 1978
and Salmon	+	≤4	Euploid	
T. aestivum Chinese Spring	+	≤6	Euploid	Shimada & Yamada, 1979
and Salmon	+	≤6	Euploid	
T. aestivum Chinese Spring	—	≤4	—	Gosch-Wackerle et al., 1979
T. longissimum	—	1	—	
T. aestivum 26 varieties	+	≤4	—	Schaeffer et al., 1979
T. durum	—	1	Many aneuploid	D'Amato et al., 1980

[a]Designates visible macroscopic centers of organization present preceding shoot differentiation.

[b]Not discussed.

formation suspect and difficult to interpret. Some authors report the differentiation of shoots from rachis segments (Chin and Scott, 1977; Gosch-Wackerle et al., 1979), though others report difficulty in obtaining plants from this source (Dudits et al., 1975; O'Hara and Street, 1978). The production of multiple shoots from single callus and the production of shoots from cultures after several passages, while not definitive evidence of de novo meristem initiation, are interesting phenomena which could facilitate the use of such cultures in genetic selection experiments.

Generally, ploidy of callus and regenerated plants has not been determined. It remains to be determined whether differences in ploidy levels underlie differences in differentiation capacity. Certainly alterations in medium composition have not produced any repeatable increases in the capacity of wheat somatic cell cultures to differentiate shoots. Recent work (Lazar, 1981) has indicated that although shoot differentiation occured at low frequency, a significant relationship existed between source cultivar and differentiation frequency and that significant genotype x medium interactions also existed. Considering experience with other cereals (e.g., Green et al., 1974) and with microspore-derived wheat culture (e.g., Bullock, 1980), it is becoming increasingly clear that genetic factors are very important for shoot production in vitro. Thus with respect to both the proliferation and the regeneration of somatic cell cultures, the importance of genetic factors in wheat cannot be overemphasized. The use of responding genotypes is currently a prerequisite to the study of the influence of environmental factors on growth and differentiation of wheat cultures. Eventually it may be possible to identify and locate genetic factors which can be transferred between important cultivars.

A technique with great import for the potential genetic manipulation of wheat is protoplast culture. Unfortunately several efforts in this area have resulted in little progress. Protoplasts have been isolated with little difficulty from leaf mesophyll (Evans et al., 1972; Vasil and Vasil, 1974; Okuno and Furusawa, 1977), cell suspension culture (Szabados et al., 1981), and seedling epicotyl (De la Roche et al., 1977). No reports, however, indicate that any such protoplasts either regenerate a cell wall or undergo mitosis. There is only sparse evidence that wheat protoplasts even maintain metabolic integrity after isolation (De la Roche et al., 1977). In this respect wheat proved even more difficult than other cereal genera, several of which do reform cell walls after protoplast isolation (Potrykus et al., 1976), though subsequent mitosis has been reported only in maize (Chourey and Zurawski, 1981). It should be noted that in wheat no reports exist of successful regeneration of plants from leaf or cell suspension sources. Further, several attempts at induction of callus growth from leaf segments have failed (e.g., O'Hara and Street, 1978). Perhaps greater success at protoplast culture could be expected utilizing tissue sources of relatively high regenerative capacity, e.g., anther cultures of certain genotypes (see below).

To date, the most successful in vitro methodologies with wheat utilize not sporophytic, but gametophytic tissues, most notably anthers and isolated microspores. The application of these techniques to wheat species was first reported by Fujii (1970) and by Kimata and Sakamoto (1972), though application to *T. aestivum* and great refinements of the techniques have been produced since (e.g., Ouyang et al., 1973; Picard and De Buyser, 1973; Chuang et al., 1978; Schaefer et al., 1979).

The development of the wheat anther culture methodology has proceeded along lines similar to the development of somatic cell culture techniques, though anther culture has met with greater success and, therefore, more rapid exploration. The major practical differences between the culture of anthers as opposed to somatic tissues are that anther culture (a) may result in the predictable de novo formation of shoots and whole plants and (b) results in the production of largely haploid and/or dihaploid callus cultures and plants.

While the success of wheat anther cultures does not depend upon the additon of undefined components to the medium (Chu, 1978), best rates of anther callus formation have been obtained utilizing a potato extract (Ouyang et al., 1973; Schaeffer et al., 1979). Since under the best of conditions no more than 10% of anthers plated produced callus, the use of this extract has proved beneficial (Chuang et al., 1978; De Buyser and Henry, 1980; Schaeffer et al., 1979). Subsequent cell proliferation and plant regeneration have generally been adequate utilizing completely defined media (Chu et al.,1973; Picard and De Buyser, 1973; Schaeffer et al., 1979). As for somatic cell culture, 2,4-D and other synthetic auxins have been useful for initiation and maintenance of callus from anther cultures (Chuang et al., 1978). Removal of the auxins from the culture medium, though not required, has enhanced shoot regeneration. Some evidence suggests that the addition of glutamine may aid in both callus maintenance and shoot regeneration (Schaeffer et al., 1979; De Buyser and Henry, 1980; Henry and De Buyser, 1981).

Another factor that has been shown important to the successful production of callus from microspores during anther culture is the stage of development of the microspores. Although callus has been obtained from anthers having microspores as young as the tetrad stage (Fujii, 1970) and as old as binucleate stage (authors unpublished results), the optimum callus production has consistently occurred when microspores were in the mid- to late-uninucleate stages (Wang et al., 1973; Chuang et al., 1978). Further, relationships have been established between the natural occurrence of anomalous pollen grains and the development of microspore-derived callus during anther culture (Chou, 1980; Amssa et al., 1980). It has been suggested that anomalies which encourage embryogenic development may be promoted by environmental modifications, such as cold shock, prior to the first pollen mitosis (Amssa et al., 1980). It seems likely that embryonic development in wheat anthers is analogous to the well-analyzed pathways of barley microspores (Sunderland et al., 1979; Sunderland and Evans, 1980; Xu et al., 1981) in which multinucleate microspores, the frequency of which is enhanced by cold treatment, precede embryo development.

As in the case of wheat somatic cell culture, one of the factors most critical to the success of wheat anther culture is genotype. Several authors have found differences among wheat species in the capacity of excised anthers to produce haploid calli and ultimately haploid plants (Fujii, 1970; Kimata and Sakamoto, 1972; Chu, 1978). Within *T. aestivum*, differences among varieties have been firmly established (Picard et al., 1978; Schaeffer et al., 1979; De Buyser and Henry, 1979b), though all genotypes examined thus far have produced anther calli at some frequency. High-responding varieties, including the commonly cultivated lines, Centurk and Chris, produce calli for 8 to 10% of cultured anthers. Differences also exist among varieties with respect to the relative frequency of green albino plantlets produced from microspore-derived calli. Work in one laboratory suggests that production of dihaploid plants constitutes a selection device for frequency of androgenic haploid production and that anther culture of dihaploid plants therefore results in higher frequencies of callus production than anther culture of the parent variety (Picard and De Buyser, 1977; De Buyser and Henry, 1980). Our own work, however, indicates that such improvement of callusing frequency is itself dependent upon the choices of parental variety as well as various environmental factors (unpublished results). Further, it is clear that successful selection within an inbred variety requires that either intracultivar variation exists for androgenic capacity or that the anther culture technique induces relatively high mutation rates. There is evidence that mutations occur during embryogenesis (Bullock, 1980) though it is unlikely it is the primary mechanism for androgenesis. In any case it would be expected that only some dihaploid plants from any one variety might have higher androgenic capacity than the variety. Anther callus production has also been studied in the variety Chinese Spring and its aneuploids (Shimada and Makino, 1975). These authors suggest that genes may exist on the long arm of the chromosome 4A which inhibit anther callus initiation, though the small number of anthers cultured from each aneuploid must make these interpretations tentative.

PROTOCOLS

Success of wheat cell cultures is influenced by the choice of appropriate genotype(s), careful selection of the explant, proper environmental treatment of the tissue both before and after excision, and selection of a suitable medium for callus induction. The protocols described below (Fig. 2) are presented as examples of the current state of the art. There are some differences among laboratories but we consider these differences incidental to the primary factors listed above. Nonetheless, the readers are advised to review all procedures for details useful to each experimental situation.

Seed Germination

Winter wheats germinate and must be vernalized for a period of 4-8 weeks (varies dependent upon variety) at temperatures ranging from 1 to 7 C. Spring wheats may be germinated and grown without vernalization after harvest. If seedling tissues are to be used as explants for callus initiation, aseptic germination is helpful. Surface sterilization of the mature seed can be accomplished satisfactorily by treating with 95% ethanol for 1-4 min, followed by 20% commercial bleach for 10-20 min, $HgCl_2$ for 0-60 sec, and at least 3 rinses with sterile, distilled water. Variation exists among seed lots for amount and type of surface contaminants. Also, varieties differ in their susceptibility to sterilization treatments. Therefore, each seed lot requires some initial testing in order to establish exact requirements. The sterilized seed

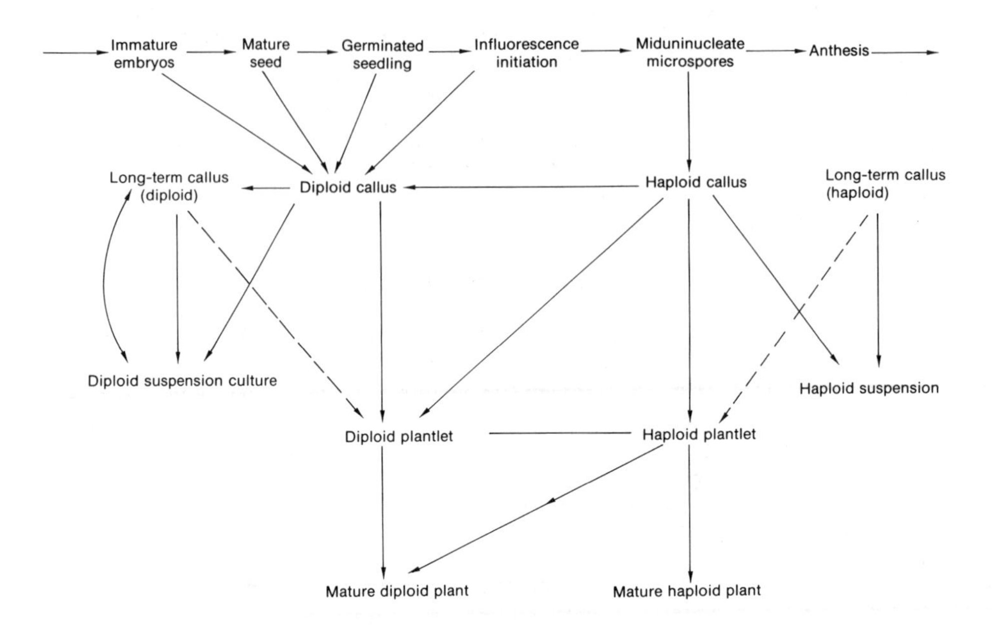

Figure 2. Protocols for wheat tissue culture.

can be germinated on sterilized filter paper in petri plates or on a low-salt, low-hormone medium such as the rooting medium described in Table 4 with agar added.

Plant Growth to Jointing Stage

Wheat can be either daylight insensitive or a long-day plant. Most vigorous growth of the plant occurs if allowed to grow for relatively short daylength for 4-6 weeks after germination or vernalization. This allows significant tillering to occur prior to floral induction. As growth to this stage under asepsis may be difficult, mature vegetative tissues used for callus initiation (e.g., root) generally require surface sterilization. Treatments generally must be less severe than those used for seeds, since growing tissues are more susceptible to injury. One must anticipate, however, that not all explants will be fully sterilized.

Growth of Inflorescence

As the wheat head grows within the leaf sheath or boot, microspores within each anther develop fairly synchronously so that at the midboot stage, when the flag leaf is fully developed and a pronounced swelling exists within the boot, the microspores are almost all post-meiotic, most in the uniculeate stage (Fig. 3a,b,c). It is at this stage that

Table 4. Media for Wheat Tissue and Cell Culture (μM)

Callus proliferation[a] (Schaeffer et al., 1979)
5.4 NAA
4.5 2,4-D
0.49 2iP
8.0 g/l agar
Anther culture (Research Group 301, 1973)
0.26 M sucrose
6.8 2,4-D
2.3 KIN
100 $FeSO_4 \cdot 7H_2O$
110 Na_2EDTA
8.0 g/l agar
Potato extract (see text)
Differentiation[a] (Schaeffer et al., 1979)
5.7 IAA
4.7 KIN
8.0 g/l agar
Plantlet rooting[b]
29.0 IAA
0.47 KIN

[a]These media contain MS basal salts (available prepackaged from K.C. Biological, Inc. and Grand Island Biological Co.), 0.03 mM glycine, 4.1 μM nicotinic acid, 2.4 μM pyridoxine·HCl, 0.3 μM thiamine HCl, 0.55 mM myo-inositol, 1 mM glutamine, and 0.087 M sucrose.

[b]Basal medium is as above except that MS salts are supplied at one-half strength.

anthers should be excised for culture. The inflorescence in the compact leaf sheath is sterilized with 20% commercial bleach for 15 min and rinsed 3 times with sterile, distilled water. The inflorescence is then removed aseptically from the leaf sheath and placed into a 10-cm petri dish containing several drops of sterile water to maintain high humidity. After the spikelets are removed, the rachis can also be used, without further sterilization, as an explant source for callus initiation.

Development of Immature Embryo

After anthesis the grain begins to form on the head, passing through 3 stages termed milk, soft dough, and hard dough. Roughly 9 days after anthesis, in the late milk stage, the embryo becomes visible to

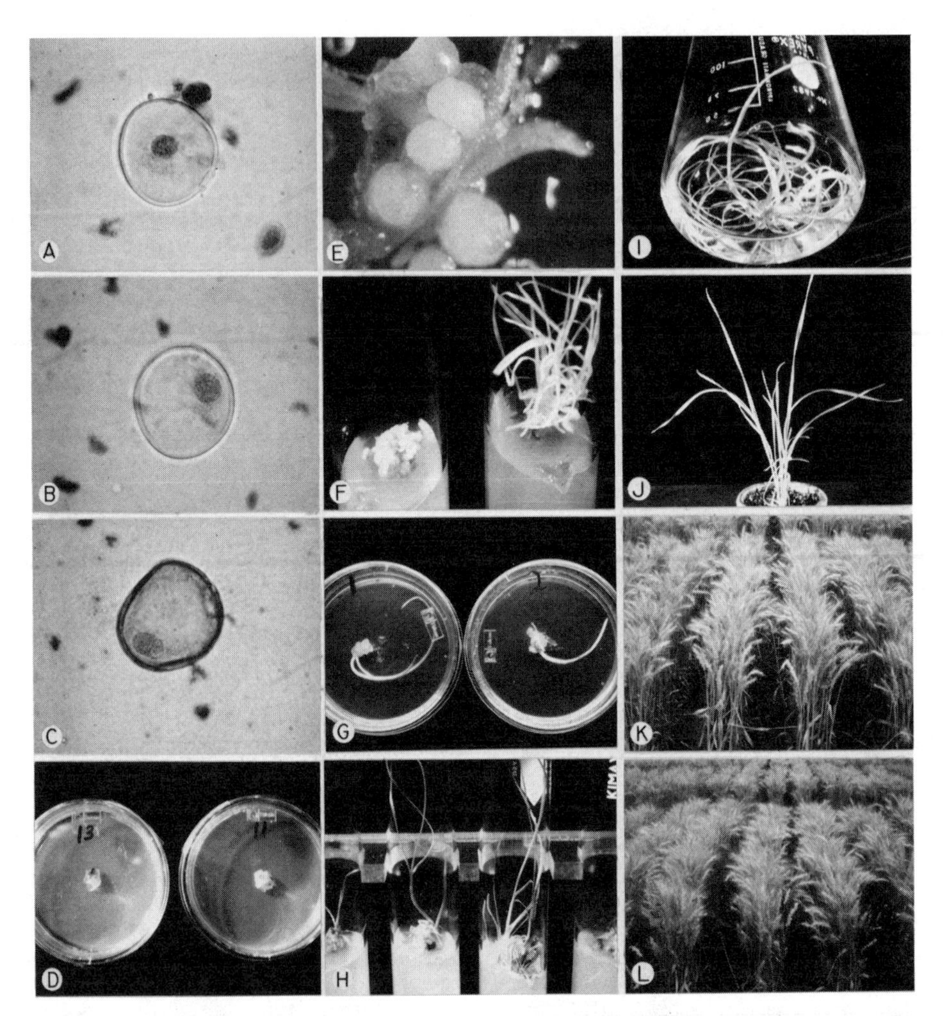

Figure 3. Stages in the development of wheat grown in vitro. A-C: Wheat microspores. A. The early uninucleate stage. B. Middle uninucleate stage, optimal for initiation of callus cultures. C. Late uninucleate stage. D. Embryo-derived callus in second culture passage grown on callus proliferation medium (Table 4). E. Embryoidal clusters emerging from an anther (cv. Kitt) grown on anther culture medium (Table 4). F. Anther-derived callus (left) after six culture passages (cv. Downy) growing on callus proliferation medium and albino plantlet (right) derived from anther culture (cv. Downy) after two passages, shown growing on regeneration medium (Table 4). G. Embryo-derived calli (cv. GWO1809) undergoing shoot differentiation on regeneration medium. H. Green shoots developing from anther cultures on regenera-

tion medium (1-r cvs. Potomac, Kitt, Centurk, Downy). I. Anther-culture-derived plantlet (cv. Downy) grown in rooting medium (Table 4). J. Anther-culture-derived plant after transfer to soil (cv. Downy). K. Field plot of a dihaploid line derived from anther culture of cv. Kitt (two center rows). L. Field plot of cv. Kitt (two center rows).

the eye as a pale yellow disk 1-2 mm in diameter within the white endosperm. From this stage to maturity the embryo can be excised for callus initiation. In general, younger plants can undergo more culture passages after callus initiation without losing totipotency than can older explants. Surface sterilization of the developing grain is similar to sterilization of the mature seed, though ethanol treatment can be only 1-2 min. Also, an additional treatment of the excised embryo with 10% commercial bleach for 1-3 min is recommended.

Callus Initiation from Somatic Tissues

As indicated above callus cultures may be initiated from a variety of explant sources. Growth rates and growth patterns of calli vary among explant sources as well as among genotypes (see Fig. 1). Variation dependent upon medium composition can be observed but is less dramatic. Generally, the callus proliferation medium outlined in Table 4 is adequate for all explant sources (see Fig. 3d).

Callus Initiation from Anther Cultures

Wheat anthers are an excellent explant source for the production of haploid cultures since the microspores can be induced to form callus to the exclusion of diploid anther tissues. The inflorescence is subdivided into 2-3 groups of spikelets and placed into small plastic petri dishes (25 mm diameter) containing several drops of water. The needles and scalpels are kept moist to help the anthers adhere to the tools and prevent electrostatic repulsion. The anthers are excised from the florets under a dissecting scope and transferred to 25 mm x 150 mm test tubes containing 20 ml of medium or to plastic petri dishes, 10 cm in diameter, containing 20 ml of medium, usually a potato medium (Research Group 301, 1976; Chuang et al., 1978). The potato (*Solanum tuberosum* L. variety Kenebec) extract is prepared by boiling diced debudded potatoes for 30 min. The material is then squeezed through several layers of cheesecloth to remove insoluble residue and the liquid portion used for the anther culture medium (Table 4). Anthers are plated at the rate of 1-10 anther/ml of medium. In liquid float cultures even higher numbers of anthers give optimum conditions in cereals (Xu et al., 1981). The individual containers are wrapped in foil and placed into a refrigerator at 4-7 C for at least 72 hr. The optimum cold treatment may be as long as 10-14 days. After the cold temperature shock, the containers are transferred to 25 C with continuous light at 15 $\mu E\ M^{-2}\ sec^{-1}$ for 28-35 days. The excised anthers are kept in the dark for at least 8 days, or the duration of the cold temperature treatment if it exceeds 8 days, by keeping the test tubes wrapped in foil. Callus consisting of one or more proembryonic structures grows along the interior surface of the anther wall (Fig. 3e). Callus can be lifted from the anthers during early stages of embryo development thereby assuring the regeneration of single plants from individual growth centers. Some U.S. varieties have a relatively high frequency of callus-producing anthers (5-15%), including the spring varieties Kitt and Chris and the winter varieties Centurk and Downy.

Long-term Callus Growth

Regardless of explant source, wheat callus may be subcultured routinely using callus proliferation medium (Table 4). Differentiation capacity declines with each passage and reaches zero within 3-7 subcultures. The competence for leaf and shoot development is often lost before the capacity to form roots is lost. Rooting with the callus proliferation medium is at first spontaneous. Rooting can be reduced or eliminated by increasing the auxin concentration, though the capacity of the callus to produce shoots is also reduced by this treatment. The use of higher auxin concentration (e.g., 45.0 μM 2,4-D) produces a very friable callus suitable for initiation of liquid suspension cultures (Fig. 3f).

Suspension Culture

Wheat suspension cultures can be initiated from either newly initiated or long-term calli by finely chopping calli in a small volume (e.g., 1 g callus per 2 ml) of liquid medium. The callus proliferation medium (Table 4) may be used though such cultures are very slow growing and tend to differentiate into roots unless continuously chopped for the first few weeks of culture. The use of higher auxin concentrations (45-90 μM 2,4-D or 11.7-23.5 μM 2,4,5-T) reduces root formation and eventually induces more rapid growth of small cell aggregates. The cultures are best initiated in small petri plates (e.g., 60 mm diameter), adding 2 ml of fresh medium every 3-4 days (and chopping up roots or large calli). These can be kept on a gyrotary shaker at about 70 rpm for the first 2-3 weeks. At the end of that time, a milky suspension should form which can be separated from macroscopic calli, subcultured into a 125-ml Erlenmeyer flask with an equal volume of fresh medium, and placed on a gyrotary shaker at about 120 rpm. Using a high-auxin medium, the cultures will double in density every 3-4 weeks initially. After several culture passages the doubling time may be as short as 1 week.

Reinitiation of Callus from Suspension Cultures

Replating small suspended cell aggregates on the same medium in agar results in the formation of macroscopic colonies within 14 days. The resultant callus generally has no capacity to initiate shoots. Elimination of auxin from the medium sometimes allows root initiation.

Diploidization of Haploid Calli

Approximately 10–15% of the calli recovered from anther culture are spontaneously diploid (i.e., dihaploid). These callus masses can differentiate diploid shoots and subsequently produce fertile, dihaploid plants.

Plantlet Regeneration

Utilizing young ($\leq$6 subcultures) diploid or haploid calli, shoot differentiation may occur spontaneously on the callus proliferation medium, but can be increased two- to three-fold by transfer of callus to a differentiation medium outlined in Table 4. Initiation of new shoots occurs in 4-8 weeks after transfer (Figs. 3f, 3g, 3h). Regeneration frequency is affected by the choice of parent genotype. Also, rate of production of albino shoots (Fig. 3f) is related to parent genotype. Frequently wheat plantlets grown on agar medium do not support healthy root systems even though the remainder of the plantlet appear vigorous. Healthy roots may be developed by growing the plantlets partially or fully differentiated in liquid culture aerated by shaking at 100–120 rpm on a gyrotary shaker, using the rooting medium (Table 4). A growth period of 7-10 days with 15 μE $M^{-2}sec^{-1}$ continuous light produces plants with clear actively growing root tips (Fig. 3i). These root tips are suitable for cytological studies because the root tip cells are slightly enlarged and thin walled, which facilitates staining and microscope examination. The liquid medium may be used to increase tillering and the tillers can be separated for vegetative propagation of precious material or for the multiplication of infertile materials.

Diploidization of Haploid Plants

The liquid rooting medium may be used to add colchicine to the plant under controlled conditions. This procedure is effective for chromosome doubling, but was inconsistent in its inhibitory effect upon subsequent growth. This may be due in part to differences in vigor of the plants in liquid culture partially conditioned by the previous tissue culture history. Colchicine can also be used to induce doubling just prior to transfer of the rooted plantlets to soil, treating the whole root system for 4 hr with a 0.1% (w/v) aqueous colchicine solution. The addition of 2% (v/v) DMSO, 0.26 μM GA_3, and a few drops of surfactant such as Tween-20 is helpful. The roots must be washed thoroughly before potting.

Development of Regenerant Plants

After root formation, with or without chromosome doubling, plants can be transferred to soil or to a soil/peat mixture (Fig. 3j). It is recommended that these plants be maintained under conditions of high relative humidity (80%) for the first 1-2 weeks after transfer because they are susceptible to dessication. Subsequent plant growth may be atypical in chlorophyll development, tillering rate, and final height. Also, partial infertility is not unusual. Such traits probably relate to prior culture history, especially auxin treatment, as they generally do not persist to the next generation (Fig. 3k,l).

APPLICATIONS

Basic Science

GENETIC STUDIES. The development of a wide range of genetic marker stocks in wheat is a goal of great importance for classical genetics and breeding as well as for application of new technologies. It has been hoped that selection for specific biochemical markers could be accomplished through the use of cell suspension cultures as has been done in some other species (Handro, 1981). The cultivated wheats, however, are hexaploid and therefore carry substantial gene duplication. Another barrier is that explants from most commercial cultivars of wheat grow slowly in vitro. Further, shoot differentiation from suspension cultures has not been possible thus far, and we have neither the experience nor precedents to anticipate the biophysical or biochemical events initiated during in vitro culture which can change the whole plant phenotype.

In spite of such limitations, some utilization of wheat tissue and cell cultures for genetic studies, including the identification and localization of marker genes, has been possible. Since, as noted earlier, growth and development of wheat in vitro is strongly influenced by genotype, genes governing in vitro growth may themselves be useful markers. For example, callus growth-promotion genes have been located on chromosome 1A (Khoteleva and Ermishin, 1980; Baroncelli et al., 1978) and callus inhibitor genes on chromosome 4A (Shimada and Makino, 1975). Such genetic factors have been found not to correlate with similar quantitative genetic factors affecting seedling growth (Baroncelli et al., 1978). Similar work utilizing ditelocentric lines indicated that the absence of the 5B chromosome, which contains a locus controlling mitotic diploidization (Okamota, 1957; Riley and Chapman, 1958), results in a high degree of aneuploidy in callus cultures grown on a medium containing 4.5 μM 2,4-D as compared to the euploid (Asami et al., 1975).

PHYSIOLOGICAL STUDIES. Cell and tissue cultures can also serve as a system suitable for studying an array of problems in cellular physiology and biochemistry in the absence of the many complex interactions among tissues and organs found in whole plants. Wheat cell cultures have been utilized, for example, in analyzing patterns of cell physiological development with respect to glucanases (Gamborg and Eveleigh, 1968) and esterases (Nakai and Shimada, 1975). Quite a number of reports have indicated use of wheat cell cultures to analyze effects of alterations of the chemical composition of the medium on metabolic pathways, especially those of nitrogen assimilation (Gamborg,

1970; Gamborg and Shyluk, 1970; Bayley et al., 1972; Fukunaga et al., 1978). All of these studies were conducted using cell cultures of several species, though wheat was the only cereal examined. Wheat cells were found to possess some similarities and some differences compared to the other species with respect to various aspects of nitrogen assimilation. Additionally, some authors have made use of in vitro wheat cultures in studies of physiological effects of various growth regulating compounds, including IAA, NAA, 2,4-D, KIN, ADE, CH (Nakai and Shimada, 1975), ε-aminocaproic acid, and lysine (Taira and Larter, 1977a,b).

Breeding and Crop Improvement

DIHAPLOID BREEDING. Tissue culture and associated techniques are becoming an important aid in crop improvement. For example, the theoretical efficiencies of dihaploid breeding have been described by Griffing (1975). He concluded that the greatest difficulty with dihaploid breeding was that dihaploids were hard to obtain in sufficient numbers for plant breeders to utilize effectively. With the advent of various haploid-producing methods, this objection is less relevant. In 1979, the first dihaploid barley cultivar, Mingo, was released. When wheat anther culture or other haploid-producing techniques are perfected, dihaploid breeding should become a useful breeding method greatly decreasing the number of generations required to achieve homozygosity in segregating populations. Dihaploid breeding should be particularly useful in the winter wheat regions since single-seed descent breeding, the conventional breeding technique most similar in objective to dihaploid breeding, is very time consuming since winter wheats require vernalization.

Currently, androgenic frequency in wheat is approximately 10%. This raises several questions regarding application, among them the issue of incomplete gametic representation. However, as noted above, evidence exists that androgenic capacity is a heritable trait (Picard et al., 1978; De Buyser and Henry, 1979b) which may in fact be selected for by successive cycles of anther culture (Rives and Picard, 1977). Conceivably, by such genetic assimilation (Waddington, 1961), androgenesis-favoring alleles may be fixed resulting in maximum androgenic frequency. Insufficient work has been done on determination of gametic representation in currently available dihaploid populations, though it has been shown recently that such populations can contain unexpected variability (Bullock, 1980). If gametic representation is adequate, such populations derived from F_1 plants could be used for early and precise analysis of segregants.

The production of haploid wheat should facilitate mutagenesis and selection programs. The resulting reduction in genetic redundancy would expose many 'masked' alleles for selection. Additionally, this technique removes a major problem in classical mutation breeding in that, after diploidization, all alterations of agronomic importance both dominant and recessive can be observed among the segregants, not just those initially selected for. Mutation breeding at the diploid level requires generations of selfing to allow all of the mutations to become homozygous and expressed. Biochemical screening techniques utilizing haploid cell cultures should be effective in selecting plant types that are resistant to mineral and salinity stresses, certain herbicide and insecticide stresses, disease toxins, and perhaps temperature and osmotic shocks. Such selection should also allow altered metabolic pathway selection which could improve forage and grain quality. However, in the absence of suitable plant regeneration from such cultures, mutagenesis and selection should be possible directly at the level of haploid plants or plantlets. Very direct selection, for example, for pathogen resistance has been successfully applied in other systems using very rapid passage through callus cultures, without need for cell suspension cultures or protoplasts (Gengenbach and Green 1975; Deaton et al., 1981). It may also be that the process of dihaploid production may eliminate the need for mutagenic treatment, if sufficient variability is created by that process.

WIDE HYBRIDIZATION. Conventional breeding methods have relied on sexual crosses to transfer genomes or genes into a species. As early as 1901, Carleton proposed the transfer of genes for rust resistance from durum and emmer into common wheats. McFadden and Sears (1947) describe their efforts and problems encountered in genomic transfer via crossing. Triticale is one example of a successful sexual genome transfer. McFadden and Sears (1947) also reviewed wild species' gene introgressions and proposed that the hexaploid wheat genomes be resynthesized by wild species crosses. The value of this technique was to introgress the important wild species' genes into a hexaploid wheat which could then be readily crossed with our common wheats. Wild relatives of wheat have always been and continue to be an important genetic resource (Feldman and Sears, 1981). The gene Lr 9 for leaf rust resistance came from *Aegilops umbellulata* and is used in common wheat (Solinman et al., 1963). Schultz-Schaeffer and McNeal (1977) recently described the anticipated development of an *Agrotriticum* by crossing of *Triticum turgidum* L. var. durum with *Agropyron intermedium*. Still, wheat and its relatives cannot all be intercrossed by conventional methods. The use of bridge species is not uncommon. One can easily foresee the utilization of immature embryo culture (e.g., Sears and Deckard, 1982) for rescue of hybrids that normally do not survive to form a mature embryo. This technique has been used successfully in *Nicotiana* (Lloyd, 1975; Reed and Collins, 1980), for example, as well as with many other crops.

WHEAT AND THE NEW BIOTECHNOLOGIES. Many potential applications can be envisioned for technologies not currently available in wheat. These are of sufficient importance to merit discussion. Protoplast fusion techniques could be useful in transferring nuclear and cytoplasmic genes. Male fertile wheats could be readily converted into cytoplasmically male sterile wheats through somatic cell fusion and cybrid formation. This development would be useful for hybrid wheat programs (Aviv et al., 1980). Protoplast fusion would also allow new genetic combinations to be formed. While interspecific crosses are certainly not uncommon, protoplast fusion could make these techniques more efficient. New crops may also be formed, though it must be recognized that any new crop normally takes many years of development before it is commercially acceptable.

Protoplast technologies and gene vector systems may allow specific genes to be transferred. The transfer of single genes would be a boon to the rapid development of multilines (conventional methods require back-crossing the required gene from the donor parent into the recurrent parental type), and for gene pyramiding. Gene vector systems using whole chromosomes or large portions of chromosomes may allow gene movement in functional and functioning units. Recent work regarding the formation of "mini-microspores" (Hao et al., 1980) and nuclear fragmentation (D'Amato et al., 1980) in wheat may allow recovery of 'prepackaged' groups of chromosomes or even single chromosomes. Additionally, it may prove possible to supply foreign DNA to microspores or to egg cells at the time of fertilization.

Finally, plant and tissue cloning techniques would allow rapid propagation of important plant types. The potential uses of this involve hybrid testing programs, in which one needs large quantities of cytoplasmic male sterile plants, and the initial seed increase cycle after cultivar release. The time required for the development of certified seed for release to commercial growers is several years. The limiting step in this increase procedure is usually the first increase since only a small quantity of seed can be planted. If that initial planting could be augmented by clonally propagated material, the seed harvest in the following generations could be greatly increased.

SUMMARY

Wheat is the world's most important food crop. It is little wonder that the germplasm available to plant breeders is extensive and that classical genetics and cytogenetics of wheat are among the best understood of all crops. Even so, our knowledge about the macromolecular and cellular aspects required for in vitro culture of wheat is less than many crops, particularly tobacco and the other Solanaceae. Thus an obvious challenge for molecular geneticists and biochemists is to solve basic genetic, biochemical, and developmental problems in wheat and systematically develop in vitro technologies for faster and more predictable changes of the components of the wheat genome(s).

Tissue culture of wheat is not well established, growth rates are slow, and plants can be regenerated from calli with only limited in vitro histories. No plants have been derived from calli in culture for more than a year or calli which have undergone more than seven subtransfers. Anther culture, however, and the derivation of dihaploids is an in vitro system that functions well and can be used by plant breeders and developmentalists at the 'present state of the art.' Nonetheless, progress would be much faster if selection pressure could be applied at the microspore stage and calli resistant to known physical or biochemical pressures developed from individual microspores. Many such challenges and much promise remain for the future.

KEY REFERENCES

Chuang, C.C., Ouyang, T.W., Hsu, C., Chou, S.M., and Ching, C.K. 1978. A set of potato media for wheat anther culture. In: Proceedings of Symposium on Plant Tissue Culture, Science Press, Peking.

De Buyser, J. and Henry, Y. 1980. Induction of haploid and diploid plants through in vitro anther culture of haploid wheat (n = 3x = 21). Theor. Appl. Genet. 57:57-58.

Inglett, G.E. 1974. Wheat in Perspective. In: Wheat: Production and Utilization (G.E. Inglett, ed.) pp. 1-7. AVI Press, Westport, Connecticut.

Schaeffer, G.W., Baenziger, P.S., and Worley, J. 1979. Haploid plant development from anthers and in vitro embryo culture of wheat. Crop Sci. 19:697-702.

Schenk, R.U. and Hildebrandt, A.C. 1972. Medium and techniques for induction and growth of monocotyledonous and dicotyledonous plant cell cultures. Can. J. Bot. 50:199-204.

REFERENCES

Allan, R.E. 1980. Wheat. In: Hybridization of Crop Plants (W.R. Fehr and H.H. Hadley, eds.) pp. 709-720. American Society of Agronomy and Crop Science Society of America, Publishers, Madison, Wisconsin.

Allard, R.W. 1960. Principles of Plant Breeding. John Wiley and Sons, New York.

Amssa, M., De Buyser, J., and Henry, Y. 1980. Origine des plantes obenues par culture in vitro d'anthèrs de Ble fendre (*Triticum aestivum* L.): Influence du pretraitment au froid et de la culture in vitro sur le doublement. C. R. Acad. Sci. Series D 290:1095-1097.

Asami, H., Shimada, T., Inomata, N., and Okamoto, M. 1975. Chromosome constitution in cultured callus cells from four aneuploid lines of the homeologous group 5 of *Triticum aestivum*. Japan. J. Genet. 50: 283-289.

Aviv, D., Fluhr, R., Edelman, M., and Galun, E. 1980. Progeny analysis of the interspecific hybrids: *Nicotiana tabacum* (cms) + *Nicotiana sylvestris* with respect to nuclear and chloroplast markers. Theor. Appl. Genet. 56:145-150.

Baroncelli, S., Buiatti, M., Bennici, A., Foroughi-Wehr, G., Mix, B., Gaul, H., Tagliasacchi, A.M., Loiero, M., and Giorgi, B. 1978. Genetic control of in vitro and in vivo growth in hexaploid wheat. I. Behavior of ditelocentric lines. Z. Pflanzenzuecht. 80:109-116.

Bayley, J.M., King, J., and Gamborg, O.L. 1972. The effect of the source of inorganic nitrogen on growth and enzymes of nitrogen assimilation in soybean and wheat cells in suspension cultures. Planta 105:15-24.

Bennici, A. and D'Amato, F. 1978. In vitro regeneration of durum wheat plants. I. Chromosome number of regenerated plantlets. Z. Pflanzenzuecht. 81:305-311.

Bhojwani, S.S. and Hayward, C. 1977. Some observations and comments of tissue culture of wheat. Z. Pflanzenphysiol. 85:341-347.

Briggle, L.W. 1980. Origin and botany of wheat. In: Wheat (E. Haflinger, ed.) pp. 6-13. Ciba-Geigy, Basel, Switzerland.

Bullock, W.P. 1980. The capacity for in vitro androgenesis in wheat and polyhaploid resistance to *Erysiphe graminis* tritici. M.S. Thesis, Univ. of Maryland, College Park, Maryland.

Burstrom, H. 1951. Studies on the growth and metabolism of roots. V. Cell elongation and dry matter content. Physiol. Plant. 4:199-208.

Carleton, M.A. 1901. Emmer: A grain for the semi-arid regions. U.S.D.A. Farmers Bulletin 139.

Chin, J.C. and Scott, K.J. 1977. Studies on the formation of roots and shoots in wheat callus cultures. Ann. Bot. 41:473-481.

Chou, J.Y. 1980. Pollen dimorphism and its relation to the formation of pollen embryos in anther culture of wheat (*Triticum aestivum*). Acta Bot. Sin. 22:117-121.

Chourey, P.S. and Zurawski, D.B. 1981. Callus formation from protoplasts of a maize cell culture. Theor. Appl. Genet. 59:341-344.

Chu, C.C. 1978. The N-6 medium and its application to anther culture of cereal crops. Proceedings of Symposium on Plant Tissue Culture, Science Press, Peking.

______, Wang, C.C., Sun, C.S., Chien, N.F., Yin, K.C., and Hsu, C. 1973. Investigations on the induction and morphogenesis of wheat (*Triticum vulgare*) pollen plants. Acta Bot. Sin. 15:1-11.

Chuang, C.C., Ouyang, T.W., Hsu, C., Chou, S.M., and Ching, C.K. 1978. A set of potato media for wheat anther culture. Proceedings of Symposium on Plant Tissue Culture, Science Press, Peking.

Collins, G.B., Vian, W.E., and Phillips, G.C. 1978. Use of 4-amino-3,5,6-trichloropicotinic acid as an auxin source in plant tissue cultures. Crop Sci. 18:286-288.

Dahns, R.G. 1967. Insects attacking wheat. In: Wheat and Wheat Improvement (K.S. Quesenberry, ed.) pp. 411-444. American Society of Agronomy, Madison, Wisconsin.

D'Amato, F., Bennici, A., Cionini, P.G., Baroncelli, S., and Lupi, M.C. 1980. Nuclear fragmentation followed by mitosis as a mechanism for wide chromosome number variation in tissue cultures: Its implications for plant regeneration. In: Plant Cell Cultures: Results and Perspectives (F. Sala, B. Parisi, R. Cella, and O. Ciferri, eds.) pp. 67-72. Elsevier, Amsterdam.

Deaton, W.R., Keyes, G.J., and Collins, G.B. 1981. In vitro expression of resistance and susceptibility to black shank in *Nicotiana*. Agron. Abs. 73:59.

De Buyser, J. and Henry, Y. 1979a. Comparaison de differents milieux utilises encluture d'anthères in vitro chez le Ble tendre. Can. J. Bot. 58:997-1000.

______ 1979b. Androgenese sur des Bles tendres en cours de selection. 1. L'obtention des plantes in vitro. Z. Pflanzenzuecht. 83:49-56.

_____ 1980. Induction of haploid and diploid plants through in vitro anther culture of haploid wheat (n = 3x = 21). Theor. Appl. Genet. 57:57-58.

De la Roche, A.I., Keller, W.A., Singh, J., and Siminovitch, D. 1977. Isolation of protoplasts from unhardened and hardened tissues of winter rye and wheat. Can. J. Bot. 55:1181-1185.

Dudits, D., Nemet, G., and Haydu, Z. 1975. Study of callus growth and organ formation in wheat (*Triticum aestivum*) tissue cultures. Can. J. Bot. 53:957-963.

Evans, P.K., Keates, A.G., and Cocking, E.C. 1972. Isolation of protoplasts from cereal leaves. Planta 104:178-181.

Feldman, M. and Sears, E.R. 1981. The wild gene resources of wheat. Sci. Am. 244:102-113.

Finney, K.F. and Yamazaki, W.T. 1967. Quality of hard, soft and durum wheats. In: Wheat and Wheat Improvement (K.S. Quesenberry, ed.) pp. 471-503. American Society of Agronomy, Madison, Wisconsin.

Ford, M. and Kingswood, K.W. 1981. Milling in the European economic community. In: Soft Wheat: Production, Breeding, Milling, and Uses (W.T. Yamazaki and C.T. Greenwood, eds.) pp. 129-167. American Association of Cereal Chemists, St. Paul, Minnesota.

Fujii, T. 1970. Callus formation in wheat anthers. Wheat Inf. Serv. 31:1-2.

Fukunaga, Y., King, J., and Child, J.J. 1978. The differential effects of TCA-cycle acids on the growth of plant cells cultured in liquid media containing various nitrogen sources. Planta 139:199-202.

Gamborg, O.L. 1970. The effects of amino acids and ammonium on the growth of plant cells in suspension culture. Plant Physiol. 45:372-375.

_____ and Eveleigh, D.E. 1968. Culture methods and detection of glucanases in suspension cultures of wheat and barley. Can. J. Biochem. 416:417-421.

_____ and Shyluk, J.P. 1970. The culture of plant cells with ammonium salts as the sole nitrogen source. Plant Physiol. 45:598-600.

_____, Miller, R.A., and Ojima, K. 1968. Nutrient requirements of suspension cultures of soybean root cells. Exp. Cell Res. 50:151-158.

Gengenbach, B.G. and Green, C.E. 1975. Selection of T-cytoplasm maize callus cultures resistant to *Helminthosporium maydis* race T pathotoxin. Crop Sci. 15:645-649.

Gosch-Wackerle, G., Aviv, L., and Galun, E. 1979. Induction, culture and differentiation of callus from immature rachises, seeds and embryos of *Triticum*. Z. Pflanzenphysiol. 91:267-278.

Green, C.E., Phillips, R.L., and Kleese, R.A. 1974. Tissue cultures of maize (*Zea mays* L.): Initiation, maintenance and organic growth factors. Crop Sci. 14:54-58.

Griffing, B. 1975. Efficiency changes due to use of doubled haploids in recurrent selection schemes. Theor. Appl. Genet. 46:367-386.

Handro, W. 1981. Mutagenesis and in vitro selection. In: Plant Tissue Culture (T.A. Thorpe, ed.), Academic Press, New York.

Hao, S., He, M.Y., Xu, Z.Y., Zou, M.Q., Hu, H., Xi, Z.Y., and Ouyang, J.W. 1980. Formation of 'mini-microspores' in anther derived haploids of *Triticum aestivum* and its possible uses. In: Plant Cell Cultures:

Results and Perspectives (F. Sala, B. Parisi, R. Cella, and O. Ciferri, eds.) pp. 389-392. Elsevier, Amsterdam.

Harlan, J.R. 1975a. Crops and Man. American Society of Agronomy and Crop Science Society of America, Madison, Wisconsin.

______ 1975b. Geographic patterns of variation in some cultivated plants. J. Hered. 66:184-191.

Henry, Y. and De Buyser, J. 1981. Float culture of wheat anthers. Theor. Appl. Genet. 60:77-79.

Heyne, E.G. and Smith, G.S. 1967. Wheat breeding. In: Wheat and Wheat Improvement (K.S. Quesenberry, ed.) pp. 269-306. American Society of Agronomy, Madison, Wisconsin.

Horsfall, J.G. 1972. Genetic Vulnerability of Major Crops. Nat. Acad. Sci., Washington, D.C.

Inglett, G.E. 1974. Wheat in perspective. In: Wheat: Production and Utilization (G.E. Inglett, ed.) pp. 1-7. AVI Press, Westport, Connecticut.

Jacobsen, T. and Adams, R.M. 1958. Salt and silt in ancient Mesopotamian agriculture. Science 128:1251-1258.

Jenkyn, J.F. and Plumb, R.T., eds. 1981. Strategies for the Control of Cereal Disease, Blackwell Scientific Pub., Boston, Massachusetts.

Johnson, J.W., Baenziger, P.S., Yamazaki, W.T., and Smith, R.T. 1979. Effects of powdery mildew on yield and quality of isogenic lines of 'Chancellor' wheat. Crop Sci. 19:349-352.

Jung, G.A., ed. 1978. Crop Tolerance to Sub-optimum Land Conditions. American Society of Agronomy, Crop Science Society of American, and Soil Science Society of America, Madison, Wisconsin.

Kent, S.S., and Strobel, G.A. 1975. Phytotoxin from *Septoria nodarum*. Trans. Br. Mycol. Soc. 67:354-358.

Khoteleva, L.V. and Ermishin, A.P. 1980. Callus formation in in vitro culture of ditelocentric lines of bread wheat, variety Chinese Spring, dependent upon the composition of the nutrient medium. Dokl. Akad. Nauk BSSR 24:186-187.

Kilpatrick, R.A. 1975. New Wheat Cultivars and Longevity of Rust Resistance, 1971-75. USDA, Agricultural Research Service, Publication No. NE-64.

Kimata, M. and Sakamoto, S. 1972. Production of haploid albino plants of *Aegilops* by anther culture. Japan. J. Genet. 47:61-63.

Lazar, M.D. 1981. Studies of somatic cell culture of wheats (*Triticum* spp.). Agron. Abs. 73:65.

Lloyd, R. 1975. Tissue culture as a means of circumventing lethality in an interspecific hybrid. Tob. Sci. 19:4-6.

Loegering, W.T., Johnston, C.O., and Hendrix, J. 1967. Wheat rusts. In: Wheat and Wheat Improvement (K.S. Quesenberry, ed.) pp. 307-336. American Society of Agronomy, Madison, Wisconsin.

Loving, H.H. and Brenneis, L.J. 1981. Soft wheat uses in the United States. In: Soft Wheat: Production, Breeding, Milling, and Uses. (W.T. Yamazaki and C.T. Greenwood, eds.). American Association of Cereal Chemists, St. Paul, Minnesota.

Mattern, P.J., Johnson, V.A., Stroike, J.E., Schmidt, J.W., Klepper, L., and Ulmer, R.L. 1975. Status of protein quality improvement in wheat. In: High Quality Protein Maize (L.F. Bauman, ed.) pp. 387-397. Dowden, Hutchinson and Ross, Stroudsbury, Pennsylvania.

McFadden, E.S. and Sears, E.R. 1947. The genome approach in radical wheat breeding. J. Am. Soc. Agron. 39:1011-1026.

Murashige, T. and Skoog, F. 1962. A revised medium for rapid growth and bioassay with tobacco tissue cultures. Physiol. Plant. 15:473-497.

Nakai, Y. and Shimada, T. 1975. In vitro culture of wheat tissues. II. Morphological, cytological, and enzymatic variations induced in wheat callus by growth regulators, adenine sulfate, and casein hydrolysate. Japan. J. Genet. 50:19-31.

O'Hara, J.F. and Street, H.E. 1978. Wheat callus culture: The initiation, growth, and organogenesis of callus derived from various explant sources. Ann. Bot. 42:1029-1038.

Okamoto, M. 1957. Asynaptic effect of chromosome V. Wheat Inf. Serv. 5:6.

Okuno, T. and Furusawa, I. 1977. A simple method for the isolation of intact mesophyll protoplasts from cereal plants. Plant Cell Physiol. 18:1357-1362.

Ouyang, J.W., Hu, H., Chuang, C.C., and Tseng, C.C. 1973. Induction of pollen plants from anthers of *Triticum aestivum* L. cultured in vitro. Sci. Sin. 16:79-95.

Patterson, F.L. and Allan, R.E. 1981. Soft wheat breeding in the United States. In: Soft Wheat: Production, Breeding, Milling, and Uses (W.T. Yamazaki and C.T. Greenwood, eds.). American Association of Cereal Chemists, St. Paul, Minnesota.

Picard, E. and De Buyser, J. 1973. Obtention de plantules haploides de *Triticum aestivum* L. a partir de cultures d'anthèrs in vitro. C.R. Acad. Sci. 277:1463-1466.

______ 1977. High production of embryoids in anther culture of pollen derived homozygous spring wheats. Ann. Amelior. Plan. 27:483-488.

_____, and Henry, Y. 1978. Technique de production d'haploides de Ble par culture d'anthères in vitro. Le Selectionneur Francais 26:25-37.

Potrykus, I., Harms, C.T., and Lorz, H. 1976. Problems in culturing cereal protoplasts. In: Cell Genetics in Higher plants (D. Dudits, G.L. Farkas, and P. Maliga, eds.) pp. 129-140. Academiai Kiado, Budapest, Hungary.

Pringle, R.B. 1977. Role of toxins in etiology of root rot disease of wheat. Can. J. Bot. 55:1801-1806.

Reed, S.M. and Collins, G.B. 1980. Chromosome pairing and black shank resistance in three *Nicotiana* interspecific hybrids. J. Hered. 71:423-426.

Research Group 301. 1976. A sharp increase of the frequency of pollen plant induction in wheat with potato medium. Acta Genet. Sin. 3: 30-31.

Riley, R. and Chapman, V. 1958. Genetic control of the cytologically diploid behavior of hexaploid wheat. Nature (London) 182:713-715.

Rives, M. and Picard, E. 1977. A case of genetic assimilation: Selection through androgenesis or parthenogenesis of haploid producing systems (an hypothesis). Ann. Amelior. Plan. 27:489-491.

Schaeffer, G.W., Baenziger, P.S., and Worley, J. 1979. Haploid plant development from anthers and in vitro embryo culture of wheat. Crop Sci. 19:697-702.

Schenk, R.U. and Hildebrandt, A.C. 1972. Medium and techniques for induction and growth of monocotyledonous and dicotyledonous plant cell cultures. Can. J. Bot. 50:199-204.

Schmidt, J.W. 1980. Genetics and breeding in wheat. In: Wheat (E. Haflinger, ed.) pp. 14-18. Ciba-Geigy, Basel, Switzerland.

Schultz-Schaeffer, J. and McNeal, F.H. 1977. Alien chromosome addition in wheat. Crop Sci. 17:891-896.

Sears, R.G. and Deckard, E.L. 1982. Tissue culture variability in wheat: Callus induction and plant regeneration. Crop Sci. 22:546-550.

Shimada, T. 1978. Plant regeneration from the callus induced from wheat embryo. Japan. J. Genet. 53:371-374.

______ and Makino, T. 1975. In vitro culture of wheat. III. Anther culture of the A genome aneuploids in common wheat. Theor. Appl. Genet. 46:407-410.

______ and Yamada, Y. 1979. Wheat plants regenerated from embryo cell cultures. Japan. J. Genet. 54:379-385.

______, Sasakuma, T., and Tsunewaki, K. 1969. In vitro culture of wheat tissues. I. Callus formation, organ redifferentiation and single cell culture. Can. J. Genet. Cytol. 11:294-304.

Simmonds, N.W. 1979. Principles of Crop Improvement. Longman, New York.

Solinman, A.S., Heyne, E.G., and Johnston, C.O. 1963. Resistance to leaf rust in wheat derived from Chinese *Aegilops umbellulata* translocation lines. Crop Sci. 3:254-256.

Sunderland, N. and Evans, L.J. 1980. Multicellular pollen formation in cultured barley anthers. II. The A, B, and C pathways. J. Exp. Bot. 31:501-514.

Sunderland, N., Roberts, M., Evans, L.J., and Wildon, D.C. 1979. Multicellular pollen formation in cultured barley anthers. I. Independent division of the generative and vegetative cells. J. Exp. Bot. 30: 1133-1144.

Szabados, L., Hadlaczky, G., and Dudits, D. 1981. Uptake of isolated plant chromosomes by plant protoplasts. Planta 151:141-145.

Taira, T. and Larter, E.N. 1977a. Effects of ε-amino-n-caproic acid and L-lysine on the development of hybrid embryos of triticale (x *Triticosecale*). Can. J. Bot. 55:2330-2334.

Trione, E.J., Jones, L.E., and Metzger, R.J. 1968. In vitro culture of somatic wheat callus tissue. Am. J. Bot. 55:529-531.

USDA. 1960. Agricultural Statistics. U.S. Govt. Printing Office, Washington, D.C.

______ 1970. Agricultural Statistics. U.S. Govt. Printing Office, Washington, D.C.

______ 1980. Agricultural Statistics. U.S. Govt. Printing Office, Washington, D.C.

Vasil, V. and Vasil, I.K. 1974. Regeneration of tobacco and petunia plants from protoplasts and culture of corn protoplasts. In Vitro 10: 83-96.

Waddington, C.H. 1961. Genetic assimilation. Adv. Genet. 10:257-293.

Wang, C.C., Chu, C.C., Sun, C.S., Wu, S.H., Yin, K.C., and Hsu, C. 1973. The androgenesis in wheat (*Triticum aestivum*) anthers cultured in vitro. Sci. Sin. 16:218-222.

Wiese, M.V. 1977. Compendium of Wheat Diseases. American Phytopathological Society, St. Paul, Minnesota.
Wright, M.J., ed. 1976. Plant Adaptation to Mineral Stress in Problem Soils. Special publication of Cornell University Agricultural Experiment Station, Ithaca, New York.
Xu, Z.H., Huang, B., and Sunderland, N. 1981. Culture of barley anthers in conditioned media. J. Exp. Bot. 32:767-778.

SECTION III
Legumes

CHAPTER 6
Beans *(Phaseolus)*

A. Allavena

The genus *Phaseolus* belongs to the Papilionaceae family, the Papilionatae subfamily, and the Phaseoleae tribe (Tonzig, 1968). The latter has been divided into five subtribes based on morphological traits: Erythryninae, Glycininae, Galactiinae, Cajaninae, and Phaseolinae. The genus *Phaseolus* has more than 180 species (Colin, 1966) of which 126 originated in the Americas, 54 in Asia and East Africa, 2 in Australia, and 1 in Europe. Mexico is the center of diversity with 70 species (Colin, 1966).

HISTORY OF THE CROP

The area of greatest diversity for the common bean is the western belt of Mexico and Guatemala. In these regions, beans grow spontaneously as wild species (*P. vulgaris* ssp. *aborigineus*). Frequently they utilize Teosinte as a natural support. Archeological remains of bean seeds have been found close to a town which was 7000 years old. While there is documented evidence of the evolutionary changes in corn, no such evidence exists for beans.

Wild bean plants are annual, or rarely, short-lived perennial, climbing vines that flower and fruit in the first year. Most of the plants have small seeds; however, large-seeded types exist. Bush forms have not been found among wild bean populations. Their short stature makes them unfit to compete in the thicket communities (Gentry, 1969). Domesticated varieties have bigger seeds, larger pods, thicker stems, and either bush or vined habit (Kaplan, 1965).

The common bean was introduced into Europe during the sixteenth century by the Spanish and Portuguese and reached England in 1954. The Spanish also introduced it to Africa and other parts of the Old World. The species is now widely cultivated in many parts of the tropics and subtropics, and throughout the temperate regions (Purseglove, 1969).

At least nine species are economically important, four of these come from the New World while five are from the Old World. The regions of cultivation and uses of these nine species are summarized in Tables 1 and 2.

ECONOMIC IMPORTANCE

Yield and production data for dry and green beans are summarized in Table 3. Developing countries are the primary producers of dry beans while developed countries produce more green beans. China, India, and Brazil are the three major producers of dry beans. The dry beans are distinguished on the basis of seed-coat color, seed size, and seed shape. The most important types are: Navy (pea), Great Northern, Pinto and Red Kidney (light- and dark-seeded types), Baby Lima, Big Lima, Blackeye, Small White, Small Red, Pink, Cramberry, Yelloweye, White Marrow, White Kidney, Cannellino, Black Turtle, and Borlotto. Dry beans are an excellent protein source with 22-23% protein content. Beans have been developed with 30% protein. The percentage of protein varies according to the variety, crop year, the date of planting and harvesting, and the availability of nitrogen in the soil. Bean protein generally contains low methionine and cysteine and therefore bean diets have to be supplemented with other foods such as eggs, milk, meat, and cereals to achieve a balanced diet. Dry beans contain toxic substances which are partially inactivated by cooking. Other compounds result in flatulence and reduce protein digestibility. Improvement against such compounds is one of the main goals of breeding. The other principle bean seed components are carbohydrates (58%), fat (1.6%), fiber (4%), and ash (3.5%). While dry beans are a good source of thiamine and niacin, they are low in riboflavin and very low in vitamin A, C, and B_{12}.

The three leading producers of green beans are China, Italy, and Egypt. Climbing cultivars of French green beans are popular in home gardens since they produce pods over a long period of time. On the other hand, bush cultivars are more often grown on a commercial scale. Green beans used for direct consumption are generally bigger in size with a flat or oval shape. Flat-podded cultivars are also used for slicing. Shorter, round-podded cultivars with white seeds are preferred for canning. The suitability of green beans for quick-freezing is evaluated by taking into account skin peeling, appearance, color, firmness, and taste of the product after cooking (Crivelli et al., 1979).

Table 1. Old World Cultivated Species of *Phaseolus*

SPECIES	COMMON NAME	FARMING AREA	CROPS
P. aconitifolius Jacq. syn. *P. trilobus* Ait.	Mat or moth bean	India, Texas, California	Dry seeds, green pods, fodder, green manure
P. mungo L. syn. *Vigna mungo* (L.) Hepper	Black gram, urd	India	Green pods, dry seeds, green manure
P. aureus Roxb. syn. *P. radiatus* L., *Vigna aurea* (Roxb) Hepper, *Vigna radiata* (L.) Wilczek	Green or golden gram, mung	India, United States	Dry beans, green pods, bean sprouts, hay, green manure
P. angularis (Willd) Wight syn. *Vigna angularis* (Willd) Ohwi and Ohashi	Adzuki bean	Japan, China	Dry seeds, herbage
P. calcaratus Roxb.	Rice bean	Southeastern Asia	Dry beans, green pods, fodder

Table 2. New World Cultivated Species of *Phaseolus*

SPECIES	COMMON NAME	FARMING AREA	CROPS
P. vulgaris L.	Common bean	Throughout the cultivated land	Green pods, dry and fresh seeds
P. coccineus L. syn. *P. multiflorus* Willd.	Scarlet runner bean	Central America, tropics, Europe (higher altitudes)	Dry seeds, fleshy tubers
P. lunatus L.	Lima or sieva beans	United States, Africa, southern Europe	Dry seeds, green beans
P. acutifolius Gray var. latifolius Freem	Tepary bean	Arizona, northwestern Mexico, various African lands	Dry seeds

Table 3. Yield and Production of Green and Dry Beans (Based on FAO Production Yearbook, 1979)

CONTINENT OR COUNTRY	DRY BEANS		GREEN BEANS	
	Yield[a]	Production[b]	Yield[a]	Production[b]
World	580	14781	6457	2527
Africa	584	1259	7515	291
Egypt	1923	13	9398	234
North America	875	2378	4835	248
Canada	1809	69	5360	52
Mexico	666	1056	3053	29
USA	1633	937	5165	164
South America	571	2814	3698	116
Argentina	1160	232	5000	3
Brazil	519	2187	-	-
Chile	1057	116	5997	81
Asia	536	7545	6210	636
China	929	4000	9857	345
India	289	2400	2105	40
Japan	1632	155	8407	95
Syria	1687	11	8519	46
Thailand	622	260	2455	41
Turkey	1557	165	-	-
Europe	449	699	7473	1189
Belgium/Lux	3030	3	17336	51
Germany/France	3482	36	10774	53
Italy	1700	85	8112	322
Netherlands	3011	11	10000	64
Romania	110	78	3385	44
Spain	729	102	8826	203
UK	-	-	9179	113
Developed Countries	838	2005	7008	1584
Developing Countries	553	12776	5705	943

[a]Yield in kilograms per hectare.

[b]Production times 1000 metric tons.

BREEDING METHODS

Phaseolus vulgaris is the main cultivated species in this genus. A great deal of breeding work has been carried out in North America, South America, and Europe on the common bean both as a pulse and as a vegetable. The common bean is a self-fertile, autogamous species, with usually less than 1% out-crossing. Most breeding programs are therefore aimed to obtain pure breeding lines. This can be done quite easily considering the relatively short reproductive cycle. In a single year, four generations can be evaluated in Italy, one in the field and three in the greenhouse. The field generation is suitable for evaluation of both agronomic characteristics and disease resistance, while the greenhouse generations provide an opportunity to select for disease resistance in a controlled environment. The low coefficient of multiplication of bean has led breeders to work with small progenies, a factor which delays variety development.

Depending on the breeding objective, several methods of breeding may be applicable to beans. (1) Intraspecific crossing followed by progeny selection is the most common breeding method when selection is aimed at improved yield, quality, earliness, and plant structure. (2) Crossing among selected sib-progenies possessing favorable traits, followed by selection, is very effective for accumulating additive genes. For example, intercrossing enhances the accumulation of additive genes for resistance into lines quite similar to the susceptible cultivated variety. When working with wild, disease-resistant parents distantly related to cultivated varieties, selection for resistance in the F_3 generation is suggested to prevent loss of variability. This is due to the increased probability of finding new recombinant types during meiosis in F_2 plants. (3) Back-crossing is used when the goal is to introduce one or a few major genes for resistance or other Mendelian traits (glossy leaves, plant markers) into varieties with good performance. Multiple cycles of backcross and selection are appropriate when polygenic resistance has to be transferred from wild accessions to cultivated varieties. This is done after a suitable number of backcross lines have been selected that express resistance and good agronomic traits. (4) Because of the short life cycle of beans, even the single-seed descent (SSD) breeding method should be evaluated. SSD would be useful for selection of additive resistance genes when both parents are cultivated species. Strong selection pressures in an early generation could lead to selection for vertical resistance instead of nonspecific resistance. Recurrent selection has been suggested by Bliss (1977) for improving diverse populations of *Phaseolus vulgaris*, particularly with regard to pod quality, plant type, and root-rot tolerance in snap beans, and seed yield and protein content in dry beans. Since three generations can be grown each year, one cycle of recurrent selection can be completed per year. A satisfactory estimation of the amount of genetic gain can be obtained after evaluating first- and second-cycle selections. Selected plants can then be advanced by single-seed descent.

Hybridization of Beans

Hand pollination with and without emasculation of the buds has been described by Buishand (1956). The proper stage for emasculation and hybridization is about one day before anthesis. To verify cross pollination, plants with one or more genetic markers should be used as female parents. Hybridization is best accomplished in the spring, under greenhouse conditions, where high humidity can be maintained.

Notwithstanding hybridization barriers, normal growth and fertility have been reported in progeny of interspecific hybrids between Old and New World *Phaseolus* species (Smartt, 1970; Dana and Das, 1974; Biswas and Dana, 1975a,b, 1976; Braak and Kooistra, 1975; Alvarez et al., 1981).

Growth regulator treatment or embryo culture has been used to recover interspecific hybrids. Al-Yasiri and Coyne (1964) found that naphthalene acetamide alone or mixed with potassium gibberellate showed a remarkable effect in delaying pod abscission in *P. vulgaris* x *P. acutifolius*. Embryos excised from treated pods were viable and grew in vitro. N-m-tolyphthalamic acid applied to the pedicel of the pollinated flower produced pod set when *P. lunatus* was pollinated with *P. coccineus* (Honma and Heeckt, 1958).

When strong interspecific sterility is encountered, it is helpful to cross one or both parents with a related intermediate species. Progenies of both parents obtained in this fashion can sometimes be intercrossed more easily. Difficulty in making crosses between Old World and New World groups of species is explained by considering that Old World species are more closely related to the genus *Vigna* than to New World *Phaseolus* spp. McComb (1975) pointed out that, with one possible exception, none of the intergeneric crosses listed among legumes produced hybrids. Abdel Aziz et al. (1979) crossed *Phaseolus vulgaris* (cv. Ideal Market) with *Vigna sinensis* (cv. Black Eye). The seed set was higher when *P. vulgaris* was used as the female parent, but the hybrids were completely sterile. The reciprocal cross gave a hybrid which was partially fertile. The cross between *P. vulgaris* as female parent and *Cajanus cajan* or *Lablab purpureus* as male parent produced sterile hybrids.

Interspecific hybridization could contribute significantly to the transfer of several traits to *Phaseolus vulgaris*. Some of these would be the resistance to drought, low-temperature tolerance, disease resistance (bean yellow mosaic virus, *Pseudomonas phaseolicola*, root rot), male sterility, hypogeal seed germination, desirable morphological traits, increased protein content, and improved amino acid composition present in *P. coccineus*, *P. acutifolius*, *P. lunatus*, and *P. ritensis*.

Introgression from *P. vulgaris* to *P. coccineus* has been useful for converting the climbing *P. coccineus* variety to a bush type (Campion et al., 1980). Other characteristics which would be desirable for introduction into *P. coccineus* by interspecific hybridization would be short inflorescence, neutral photoperiod, and self-fertilization. It is a general concept that introgression in natural populations occurs primarily via repeated back-crossing of progeny to one or both of the parental species. Wall (1970) made an analysis of the effect of three mating systems upon introgression of a quantitatively determined morphological character, hypogeal germination, from *P. coccineus* into *P. vulgaris*, an epigeal species. The mating systems evaluated measured gene flow through male gametes, and female gametes when a sib mating generation was produced between backcross generations. The results demonstrated that the mating system significantly affects gene flow between partially isolated species. Of the mating systems tested, sib mating together with back-crossing appears to achieve the best balance from genome recombination. Although a prerequisite for introgression, genome recombination has deleterious effects which may lead to hybrid breakdown.

Mutagenesis

Being a self-pollinated crop, beans are amenable to mutation breeding. Among the different types of radiation used on dormant seeds, neutrons are preferable since they are repeatable. Among the tested chemicals, the highest mutation rates were obtained using ethyl- and methyl methane sulphonate, ethyleneimine, and nitrosomethyl urea (Blixt and Gottschalk, 1975).

Using gamma rays, Rubaihayo (1975) isolated lines with improved seed yield and protein yield. Sarafi (1973) identified mutants with altered protein content and seed size after irradiation with gamma rays and neutrons. The most promising mutant has 12.8% larger seeds and 14.8% more protein. As compared to controls, there was a positive correlation (r = 0.703) between percentage of protein and weight of seeds of mutants.

Moh and Alan (1964) isolated a dwarf mutant in the progeny of irradiated beans. Segregation studies have revealed that the dwarf allele is recessive and controlled by a single gene. As the mutant is normalized by GA treatment, Moh and Alan (1967) postulated that the dwarf allele blocks the biosynthetic pathway for the production of gibberellinlike substances necessary for normal growth.

Variants for seed-coat color have also been identified. A white-seeded variety of the common bean was developed by Motto et al. in 1978. Many other authors (Moh, 1969; Hussein and Disouki, 1976) have observed mutation for seed color after treatments with mutagens. Swarup and Gill (1968) treated dry seeds of the variety Wax Podded with X-rays. Eleven green-podded mutants were obtained in an M_2 generation originated from 300 M_1 seeds.

Leaf mutants were induced with X-rays by Moh (1968), Moh and Nanne (1969), and Motto et al. (1979). Mutants with altered leaf area, leaf weight, and dry matter have been isolated. Mutants with resistance to the golden mosaic virus were isolated by Tulmann Neto et al. (1977) using gamma radiation and chemical mutagens. The mutants and their progeny showed milder mosaic symptoms and a tendency for faster recovery.

BREEDING GOALS AND RELATED PROBLEMS

Many characteristics of beans can be improved by breeding. Some of these, such as yield, protein quality and quantity, digestibility, toxic and antinutritional content, resistance to diseases and pests, and resistance to adverse environmental conditions are very important, and breeding programs should be concerned with one or more of them. Characteristics such as pollution tolerance, with particular reference to herbicide residue, adaptability to mechanical harvesting, and processing are also important.

Yield Improvement

The best yields obtained with present cultivars of dry beans are relatively low when compared to most other field crops. Both physiological and morphological characteristics of the bean plant are believed to play major and interdependent roles in determining yield. Denis and Adams (1978) tried to identify patterns of morphological characteristics in a set of cultivars related to yield. It was concluded that an approximation of a high-yielding bean ideotype would be a relatively large plant with numerous nodes, leaves, and a canopy facilitating light interception.

Coyne (1969) found low heritability for total yield and yield components in F_3 progeny of a cross between PI 165078 and GN 1140. No progress was achieved through selection because of the environmental effect on the expression of these traits making it difficult to identify genetically superior individuals. It was suggested that bulk population breeding would be more efficient than the pedigree method in selecting for higher yields in populations derived from this cross. Physiological factors which were found to be associated with high yield by other investigations may not be useful in the identification of genotypically superior plants segregating in the field because these traits are quantitatively inherited and strongly influenced by the environment. It seems more appropriate to identify varieties which possess one or more of these desirable physiological characteristics and intercross them to obtain favorable recombinants for different yield components. A bulk breeding method could then be adopted.

The effect of bulk breeding on yield and its components (seed number, pod number, seed/pod, seed weight) was studied by Hamblin (1977) for two high-yielding and two low-yielding crosses. Bulk breeding did not change the population mean of the high-yielding crosses but there was a steady increase for populations derived from low-yielding crosses.

Also, variance as a determinant of the potential of the cross was studied. It was observed that for characters with variable expressivity, including grain yield, progeny variance did not differ significantly from parent variance. It was concluded that interplant competition was the environmental factor responsible for this effect (Hamblin, 1977). Considering that yield and fitness are positively associated, bulk breeding would be more efficient in terms of resource utilization than pedigree selection. Single-plant selection should be delayed until late generations following bulk breeding. Modified pedigree and single-seed descent selection methods are also suggested by Evans and Gridley (1979) when breeding for characters involving additive and low heritable characters.

Selection procedures that maximize recombination, such as recurrent crossing between selected genotypes, have received little attention in legume breeding. In Cambridge, a recurrent selection program has been carried out to simultaneously select for high yield and high protein content. In the first selection cycle, hybrid populations derived from intercrosses between single F_3 plants could be selected that were higher in both yield and protein as compared to the parents. These populations also outyielded F_5 progenies derived from F_3 single-plant selections (Evans and Gridley, 1979).

Protein Quality

Three main aspects are involved with protein improvement in beans: yield/unit area, amino acid profile, and digestibility. A combination of high yield and high protein content would obviously maximize protein production per unit area.

Kelly and Bliss (1975a) studied the inheritance of protein content and its correlation with yield. The low negative correlation ($r = -0.30$) between seed yield and percentage of protein suggested that selection should be made initially for high yield. Within high-yielding families the plants with the highest protein percentage should be further selected and intermated.

No protein of a single vegetable species is able to provide all the amino acids in the right concentration for human diets. Fortunately not all plant proteins are deficient in the same amino acids, so that mixtures of different plant proteins provide balanced amino acid composition. Cereal-pulse mixtures have been predominant in vegetarian cultures or in cultures where animals are not a significant source of meat (Smartt et al., 1975).

Bressani (1975) grew rats with diets containing an equal amount of protein derived from maize and dry bean in different proportions. He found that a maximum protein value was obtained when 50% of the protein in the diet was derived from bean and 50% from maize corresponding to a maize:bean ratio of 2.6:1. Lysine is the limiting amino acid in maize while methionine is deficient in beans. It has been estimated by nutritionists that beans need a doubling in tryptophan content and a threefold increase in the sulfur amino acids to be able to supply up to 24% of dietary protein (Evans and Gridley, 1979).

Kelly and Bliss (1975a) reported a positive correlation ($r = 0.33$) between percentage of protein and available methionine as percentage of protein indicating that both characteristics could be improved simultaneously. Since heritability for percentage protein, percentage available methionine, and available methionine as a percentage of protein are not low, and gene action appears to be largely additive, genetic improvement can be expected in breeding programs using strains having high levels of protein and available methionine. It is usually reported that methionine is the main limiting amino acid of legumes. However, it is important to consider the content of both sulfur amino acids, cysteine and methionine, since methionine can be derived from cysteine. Kelly and Bliss (1975b) calculated the correlation coefficients between percentage available methionine and each of percentage total methionine and the protein efficiency ratio (PER) of four bean lines. They reported a highly significant correlation between PER and available methionine. On the other hand, Porter et al. (1974) found a highly significant correlation between total sulfur as a percentage of crude protein and the sum of methionine and cysteine as a percent of protein. A significant correlation of PER with each of the above two variables was observed while correlations of PER with methionine or cysteine alone were not significant. These data apparently contrast with that of Kelly and Bliss (1975b). A statistically significant negative correlation between the percentage of vicillin and available methionine was also detected. As methionine is an indicator of protein quality, this characteristic could be useful in breeding programs to reduce vicillin content and increase bean protein utilization.

Antinutritional Factors

Common beans contain a large number of antinutritional factors such as hemagglutinins, trypsin inhibitors, protease inhibitors, amylase inhibitors, metal binding constituents, alcohol-insoluble flatulence factors, cyanogenic glycosides, antivitamin factors, and other unidentified growth inhibitors. Cooking and other processing is effective in destroying most toxic constituents. Breeding programs for nutritional value should also take into account factors that affect digestibility and the toxic constituents of beans.

Jaffe (1975) described different types of hemagglutinins based on their activities against different animal red blood cells. Prolonged boiling or autoclaving of presoaked bean seeds destroys agglutinating activity. Often not all hemagglutinin present in bean and cereal flour mixtures is destroyed. This causes diarrhea and other signs of toxicity in humans who eat similar food. All bean hemagglutinins are glycoproteins differing in molecular weight and sugar component. Quantitative differences in agglutinating power and toxicity among varieties were observed. The distribution of hemagglutinin-positive seed in a segregating population indicated this trait is controlled by a single dominant gene (Brucher et al., 1969). Hamblin and Kent (1973) suggested that the possible role of hemagglutinins in beans is to bind Rhizobium to the root at sites suitable for the infection of plant by bacteria.

Pusztai (1966) isolated from the kidney bean a trypsin-inhibiting protein containing more than 14% cysteine which was relatively poor in aromatic amino acids and valine. This protein closely resembles a trypsin inhibitor that was previously isolated from lima bean. Amylase inhibitors are also present in most kidney beans but are destroyed by proper heating. Little is known about their physiological importance (Jaffe and Vega, 1968).

Several other undesirable compounds are present in beans. Metal-binding constituents lead to a decrease of some trace metals such as zinc, manganese, copper, and iron. For example, phytic acid may interfere with iron absorption and the flatulence-producing factor; it is not destroyed by heating. Raw kidney beans are believed to contain an antagonist of vitamin E an excess of which results in liver necrosis in rats, and muscular dystrophy and low levels of plasma tocopherol in chickens. The antivitamin E action of raw kidney beans can be eliminated partially be heat treatment (Liener, 1975). Cyanogenic glycosides are present in *P. lunatus*. Many cases of human toxicity have occurred even with cooked beans. In spite of a great deal of research, very little is known about the flatulence factors in beans. They are under genetic control and the possibility exists for reducing these by genetic selection.

Resistance to Diseases and Insect Pests

Zaumeyer and Thomas (1957) described 38 diseases that attack dry and French beans; of these diseases, 21 are caused by fungi, 8 by bacteria, and 9 by viruses. Numerous pests (nematodes, insects, and spiders) are also known. The severity of some diseases and the cost and inefficiency of chemical control has stimulated breeders to develop resistant varieties. The most important bean diseases are summarized in Table 4. Accessions of noncultivated species and collections have been identified with resistance to each of these diseases.

Table 4. Major Bean Diseases

DISEASE	SPECIES	SOURCE OF RESISTANCE	REFERENCE
Anthracnose	*Colletotrichum lindemuthianum*	*P. vulgaris, P. coccineus, P. formosus*	Hubbeling, 1977
Rust	*Uromyces phaseoli*	*P. vulgaris*	Coyne & Schuster, 1975
Root rot	*Fusarium solani*	*P. coccineus*	Boomstra et al., 1977
	Phytium ultimum	*P. vulgaris*	York et al., 1977
	Rhizoctonia solani	*P. vulgaris*	Deakin & Dukes, 1975
BCMV	Bean common mosaic virus	*P. coccineus*	Dickson & Natti, 1968
Halo blight	*Pseudomonas phaseolicola*	*P. acutifolius*	Coyne et al., 1971
Common blight	*Xanthomonas phaseoli*	*P. vulgaris, P. acutifolius, P. coccineus*	Coyne & Schuster, 1974

CELL CULTURE LITERATURE

A review on legume cell culture with emphasis on beans has been published by Boulter and Crocomo (1979). This review deals with recent developments in bean cell culture.

Bean callus and cell cultures have been established on different media with both low- and high-salt concentration, as shown in the work of Veliky and Martin (1970), using media of White (1963), Murashige and Skoog (1962), and Gamborg et al. (1968). Liau and Boll (1970) developed a solid medium for callus cultures derived from explants of roots, hypocotyls, and cotyledons of bush bean seedlings, and a liquid medium for growth of cell suspension cultures established from these callus cultures. Different media have been used that vary in growth factors, organic substances, and salt composition and concentration. The concentration of mineral salts for cell suspension cultures was lower than for callus cultures. CW and other organic substances were required for optimum growth. There were no obvious differences in yield of callus derived from roots, hypocotyls, or cotyledons. Few differences in yield were obtained among cell suspension cultures from roots, hypocotyls, and cotyledons; cells from roots gave the highest yield in dry weight.

Liau and Boll (1971) recorded change in cell numbers, cell size, fresh weight, and dry weight during the growth cycle of a serially subcultured cell suspension culture derived from hypocotyls of the bush bean (var. Contender). MS mineral salts at one-fourth the strength and organic substances such as those described by Liau and Boll (1970) were used in the medium. Growth factors added were 2,4-D (9 μM) and KIN (3 μM). In the 12-day growth cycle there was a lag phase not exceeding 48 hr in duration, throughout which conditions necessary for exponential growth developed. During the exponential phase the cell numbers increased rapidly and clumps formed. However, this was not accompanied directly by increase in cell size and dry weight. Increase in dry weight was delayed by about 2 days and cell enlargement started only when the logarithmic phase was more or less completed. Cell seperation was not obvious until about day 12, but by about day 16 the cells were almost completely free. Seven-day-old cells repeatedly subcultured at 5-day intervals could be maintained in a state of continuous cell division. The time for the average mitosis was less than 24 hr.

The primary cell forms most frequently reported to occur in batch-propagated cell cultures, were found in bean cell cultures. No growth pattern leading to definite embryolike structures, either isolated or in cell clumps, was observed. Early stages of clump formation from short filaments appeared as young embryos with cell clusters at one end of the filament. However, such clusters developed into clumps and not into organized embryos. The formation of a clump at one end of a filament of cells indicated that physiological polarity existed in these filaments. The fact that clump formation could also occur at both ends, in the middle, or all along a filament of cells means that no consistent polarity was observed. This lack of consistent polarity with the medium used here accounts for the absence of embryolike forms. Appropriate manipulation of the medium can be used to induce organization and embryogenesis in the cell cultures.

Growth Factors Requirement

Mok and Mok (1977) evaluated the response to auxins of seven genotypes belonging to three species: *P. vulgaris*, *P. lunatus*, and *P. acutifolius*. MS basal synthetic medium was used. Callus growth of *Phaseolus* sp. could be effectively promoted by the synthetic auxin Picloram. The range of Picloram concentration favorable for callus growth was broad in most genotypes (1.25-40 μM) and the quality of cells cultured on suitable Picloram levels was good. The auxin 2,4-D also enhances callus growth, however the range of optimal concentration was markedly narrower than that of Picloram. Optimum concentrations of 2,4-D for tissue of four genotypes tested were between 1.25 and 20 μM. Tissue growth of the remaining three genotypes was inhibited at concentrations higher than 1.25 μM. NAA induced callus growth when supplied at relatively high concentrations (10-40 μM). IAA was ineffective in supporting callus growth as browning of the tissue was observed at an early stage.

Mok et al. (1980) studied differences in the cytokinin requirement of callus cultures of *P. vulgaris* and *P. lunatus*. Picloram (2.5 μM) and KIN (5 μM) were included in the MS basal medium used for callus initiation. Tests for cytokinin autonomy were performed after a passage on a medium containing 5 μM KIN. Of 10 *P. vulgaris* genotypes, 1 required cytokinin for callus growth, 6 exhibited moderate growth on cytokinin-free medium and 3 grew well in the absence of cytokinin. Six genotypes of *P. lunatus* out of 10 were cytokinin-dependent and 4 displayed irregular callus growth on cytokinin-free medium. The behavior of bean callus tissues was independent of the tissue of origin and the time in culture.

Inheritance of the cytokinin requirement in *Phaseolus vulgaris* tissue cultures studied in F_1, F_2, and BC_1 tissue from crosses between strictly dependent and independent genotypes results indicated that the cytokinin requirement may be regulated by one set of alleles.

M.C. Mok et al. (1978) investigated the activities of eight cytokinins in promoting callus growth in two *Phaseolus* genotypes. The feature which contributes to the major genotypic difference in cytokinin structure-activity relationships was the presence or absence of a double bond at the 2,3 position of the isoprenoid N_6 side chain. In *P. lunatus* var. Kingston, trans-zeatin and 2iP are either equally active or more active than the cytokinins bearing the corresponding saturated side chains. In *P. vulgaris* var. Great Northern the presence of a double bond in the N_6 side chain of trans-zeatin and 2iP results in a dramatic reduction of activity relative to dihydrozeatin and 2iP. Cytokinin activity does not appear to be influenced when 2,4-D was supplied in place of Picloram as an auxin source. Dihydrozeatin occurs naturally as a free base and possibly as a ribonucleoside in fruits of *P. vulgaris* var. Pinto (Krasnuk et al., 1971).

Callus growth on hormone-free medium containing a high concentration of sucrose (12%) was reported by Bennici et al. (1976) from proximal portion of the *Phaseolus coccineus* embryo suspensor. Callus formation was initiated by amitosis. Amitosis was followed by mitosis of both nuclei resulting from previous amitosis and nuclei of euploid cells preexisting in the suspensor. The integrity of the embryo-suspensor complex is essential for callus formation. Callus formation via amitosis followed by mitosis is peculiar to tissue culture grown in vitro on media lacking KIN or on media with an unbalanced auxin:cytokinin ratio (D'Amato, 1977). Nuclear fragmentation followed by mitosis is a mechanism for wide chromosome number variation (D'Amato et al., 1980).

Biotransformation and Metabolite Production

In studies involving the cultivation of isolated plant cells and their metabolism, actively growing cultures are required. Veliky and Martin (1970) developed a fermenter for cultivation of bean cells. Bertola and Klis (1979) modified a commercially available bacterial fermenter to allow continuous cultivation of glucose-limited bean cells at a defined pH. Continuous cultures of substrate-limited plant cells are useful for studying metabolism because the cells grow in a solution of constant composition at a constant rate. Cells can thus be kept indefinitely in a particular metabolic rate.

Veliky (1972) and Veliky and Barber (1975) found that cell suspension cultures of *Phaseolus vulgaris* possess a biosynthetic potential for transforming tryptophan into the alkaloids harman and norharman. Cell culture establishment from the roots of a plant that is not known to produce alkaloid in vivo will produce harman and norharman when grown in vitro in the presence of trytophan. Hence bean cell cultures could be used for the production of useful metabolites.

Mante and Boll (1975, 1978) studied the effect of CW, 2,4-D, and KIN on extracellular polysaccharide production in cell culture of *Phaseolus vulgaris* cv. Contender. Factors in CW had relatively little effect on the natural sugar composition of the extracellular polysaccharide in comparison with a completely synthetic medium. More extracellular polysaccharide was produced in the presence of 2,4-D alone than in the presence of KIN alone. However, production was least in the control medium. When only KIN was present in the medium there was little extracellular polysaccharide production during the stationary phase. This may indicate that extracellular polysaccharide produced after cell division was influenced more by 2,4-D than KIN.

Mante and Boll (1976) made an analysis of the effect of the culture cycle phases on the single extracellular polysaccharide constituent (cold-water-soluble pectin, EDTA-soluble pectin, and neutral polysaccharide). The neutral sugar present in the pectin fraction showed considerable fluctuation during the culture cycle as compared with rather small changes in the concentrations of the sugar present in the neutral polysaccharide.

Isozyme patterns of *Phaseolus vulgaris* plant material and callus induced from plant tissue have been investigated by Bassiri and Carlson (1978). Although patterns of enzyme systems such as peroxidase, malate dehydrogenase, and acid phosphatase were specific for individual organs, a great uniformity of isozyme patterns was observed in rapidly growing callus cultures initiated from different plant parts. Callus cultures assume a distinct isozyme pattern independent of the tissue of origin. This is of importance to crop improvement since many genetic selections and rapid assays for agronomic traits in vitro require the cellular expression of tissue-specific traits.

Arnison and Boll (1974) performed electrophoretic analyses of peroxidase, polyphenol oxidase, catalase, malate and glutamate dehydrogenases, esterase and leucine aminopeptidase in root, hypocotyl, and cotyledon callus cultures derived from a single seedling. Enzyme patterns changed during the culture cycle and several isozymes appeared only at a certain time. The isoenzymatic patterns of the three cultures were very similar but persistent differences between them were observed.

In similar work performed with extracts of root, hypocotyl, and cotyledon suspension cultures, Arnison and Boll (1975) found considerable differences in content and activity of isoenzyme bands of the enzymes acid phosphatase, peroxidase, and malate dehydrogenase. The authors concluded that differentiation in specialized tissues may involve irreversible changes.

Arnison and Boll's (1975) results are in contrast with those of Bassiri and Carlson (1978); the divergence may reflect the different genetic material used and the different conditions in which the studies were carried out.

The effect of growth factors such as 2,4-D and KIN on isoenzyme patterns of various enzymes has also been studied by Arnison and Boll (1978). Differences in isozyme patterns of cells grown with and without growth regulators were predominantly quantitative. However, certain minor isozymes were only detected in cultures grown in the presence/absence of growth regulators.

Activity of phenylalanine ammonia lyase (PAL), caffeate-o-methyl transferase, and peroxidase was investigated after the transfer of bean callus from a maintenance medium with 2,4-D to an induction medium with NAA and KIN (Haddon and Northcote, 1976a). The peak of maximum PAL activity in callus grown on maintenance medium remained constant at less than half the concentration on induction medium between 6 and 35 days after transfer. The activity of caffeate-o-methyl transferase in the callus rose between 3 and 9 days after transfer to either induction or maintenance medium and was reduced to its initial level after 15 days on maintenance medium. In induction medium the activity remained constant between 9 and 15 days then decreased slowly and remained significantly above maintenance medium for at least 5 weeks. Peroxidase activity remained constant for callus grown on maintenance medium but fell rapidly in callus transferred to induction medium. The increase of PAL activity occurred when ABA (30 μM) was added to the induction medium.

Morphogenesis

Only a small number of reports have appeared on morphogenesis from bean tissue culture. This probably reflects a large number of unsuccessful experiments. Crocomo et al. (1976a) cultured sections of bean leaves on media containing Veliky and Martins' (1970) salt and vitamin, 10 g/l agar, IAA 11.2 μM, NAA 5.4 μM, and KIN 0.9 μM. Extract from bean seeds soaked for 2 hr in tap water prior to homogenation was added to the medium at a concentration of one-fourth seed/ml. Two plantlets were induced in a test of 9 cultures.

Somatic embryogenesis has been observed by Rosetti and Allavena (unpublished). Breeding line P263, selected for resistance to BCMV at our laboratory, was used in the preliminary experiment. Primary leaves of 8-day-old seedlings were cultured in 120 ml bottles containing 15 ml

B5 basal medium plus 18 μM 2,4-D. After 20 days callus was transferred into 10 ml liquid medium and placed on a shaker. The secondary medium contained B5 basal salts, sugar, and vitamins with the following growth factors: 0.53 μM NAA, 2.3 μM KIN, 0.29 μM GA_3, and 0.11 μM ABA. All cultures were maintained in diffuse light, 16:8 hr light-dark cycle, at a temperature of 25 ± 0.5 C. After 10 to 12 days of culture on the secondary medium a few embryos developed in some bottles (Fig. 1). No mature plants have yet been obtained. We will continue to repeat these experiments to confirm our previous observation. Because degenerating embryolike structures have also been observed after culturing bean tissue on other media the precise ratio between GA_3 and ABA should be further identified for bean somatic embryogenesis (cf. Vol. 1, Chap. 3).

Peters et al. (1976) studied the effect of caffeine and nicotine on callus growth and root morphogenesis of *Phaseolus vulgaris* var. Bico de Ouro tissue culture. Nicotine was tested on four media: basal medium (BM) (modified Veliky and Martins, plus IAA 11.5 μM and KIN 1 μM), BM minus KIN, BM minus IAA and KIN. On BM, nicotine was inhibitory to root induction. On BM minus KIN, nicotine was unable to promote callus growth but was associated with root induction. On BM minus both growth factors nicotine was unable to stimulate callus growth or root morphogenesis. Caffeine was antagonistic to both callus growth and root morphogenesis.

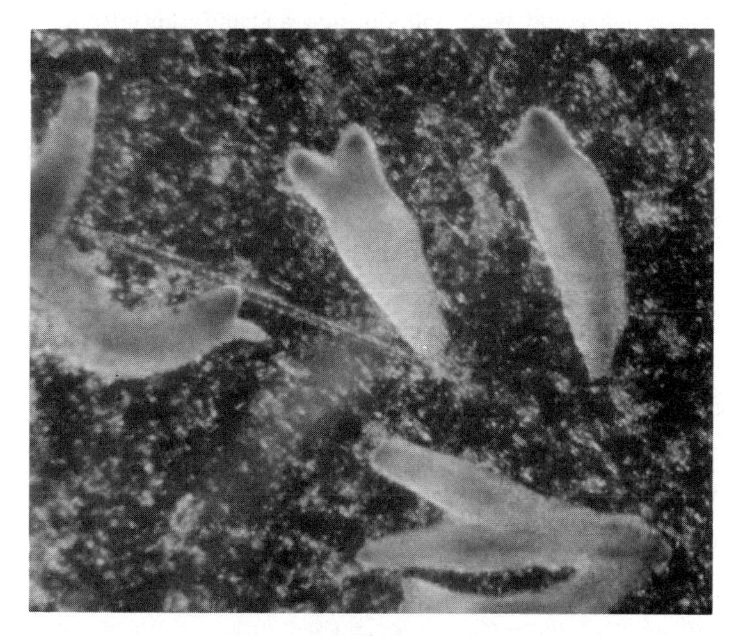

Figure 1. Somatic embryos developed after transfer of bean calli grown on B5 agar medium containing 18 μM 2,4-D to a B5 liquid medium with the following growth factors: 0.53 μM NAA, 2.3 μM KIN, 0.29 μM GA_3, and 0.11 μM ABA.

Tonin et al. (1982) studied the effect of amino acids on bean cell culture. A medium containing a modified MS (nutrient solution) with 4.6 μM KIN, 28.5 μM, and three amino acids—arginine (0.34 mM), aspartic acid (0.37 mM), and cysteine (0.08 mM)—promoted optimal root morphogenesis. The same medium supplemented with 0.44 mM glutamic acid, 0.33 mM glycine, and 0.06 mM histidine was inhibitory to both growth and morphogenesis. Glutamic acid promoted callus growth at low concentration (3.4 μM) and was inhibitory at a high concentration (0.44 mM). Crocomo et al. (1976b) discovered a stimulatory effect of ZEA (0.2-0.45 μM) on root morphogenesis.

Jeffs and Northcote (1967) found patches of differentiation in calli induced on medium supplemented with 2,4-D, and with wedges containing IAA and sucrose or certain other glucosil disaccharides such as trehalose and maltose. Nodules containing xylem at the center and phloem at the perimeter with a cambium-like layer between them were found when IAA and sucrose were present in the wedges or when wedges containing IAA were replaced by wedges containing only sucrose. If sucrose was applied initially, followed by IAA, only traces of xylem were induced. IAA-induced mitosis and young cells could be further stimulated to differentiate by sucrose which induced cell enlargement and secondary thickening in the presence of residual amounts of IAA. When sucrose was applied in absence of IAA it was rapidly used by cells for metabolism and for storage as starch. KIN alone in the wedge stimulated cell divisions; in the presence of IAA and sucrose it enhanced both cell division and differentiation.

Root and vascular nodule formation has been induced from bean callus initiated on a defined maintenance medium supplemented with 9 μM of 2,4-D and .058 M sucrose after transfer to an induction medium with 5.4 μM NAA, 1 μM KIN, and 0.088 M sucrose (Haddon and Northcote, 1975). Calli have been transferred 3-6 times on maintenance medium before being transferred to induction medium. Root differentiation was lost at an earlier stage of subculture in maintenance medium (fourth subculture) than was nodule formation (sixth subculture). PAL and 1-3 glucan synthetase activity have been used as biochemical markers for vascular differentiation.

NAA and KIN interact in induction medium (Bevan and Northcote, 1979a). NAA was required at an optimal concentration of 5.4 μM 2 days prior to the increase in PAL activity. KIN caused a rapid stimulation of the rate of PAL induction and the total amount of PAL was induced in the concentration range of 0.5 to 2 μM of KIN. The inhibitory effect of 10.7 μM NAA on PAL induction was overcome by an increased concentration of KIN.

The effect of GA_3 and ABA on tissue and cell differentiation has been investigated by Haddon and Northcote (1976b). Both types of differentiation were inhibited by the addition of ABA to the medium at concentrations greater than 1 μM. Inhibition appeared complete and irreversible after 9 days of culture on medium containing 30 μM ABA. This inhibition was not affected by the addition of GA_3. Once differentiation had been initiated, subsequent transfer to medium with ABA did not inhibit the normal development of nodules.

Application of GA_3 stimulated organized development at a concentration of 30 to 45 µM in the presence of a suitable ratio of auxin to cytokinin. In the absence of KIN, root formation is also stimulated at a concentration of 0.1 to 30 µM. Roots appeared in the presence of GA_3 also at an auxin:cytokinin ratio of 50:1 which did not normally induce differentiation. The effect of GA_3 on the formation of vascular elements is dependent upon the concentration used. GA_3 in induction medium increased the lag phase before the nodules were formed. At 45 µM concentration, GA_3 inhibited the rate of cell differentiation as well as delaying its initiation.

The morphogenetic potential of the tissue of origin did not persist in the callus derived from it (Haddon and Northcote, 1976c). Transfer of callus from different plant parts to induction medium after four subcultures on maintenance medium induced the formation of the same number of vascular nodules. Only callus derived from anther culture formed a significantly greater amount of vascular tissue.

When CW was added to maintenance medium, the number of transfers after which roots and vascular nodules could be induced was increased. CW increased the proportion of diploid cells which were necessary for differentiation. KIN was unable to replace CW as loss of potential had been partially prevented but the ploidy of the culture was not affected. A component of CW different from KIN apparently stimulates the division of developed cells. Other factors such as the cell-cell interaction are important in determining the ability of cultures to form vascular tissue.

The loss of morphogenetic potential in bean suspension cultures has been investigated by measuring the amounts of PAL activity (Bevan and Northcote, 1979b). Tissue has been grown in two types of medium: the first supplemented with 2,4-D as the only growth medium, and the second supplemented with 2,4-D and CW. When cells were grown in medium with 2,4-D for a period of 5-10 subcultures and samples were transferred to induction medium with 5.4 µM NAA and 0.93 µM KIN at various intervals, the amount of PAL activity and the number of xylem elements induced progressively declined. Cells grown in the presence of CW did not lose the ability to induce PAL or xylem elements. Apparently cell selection nor genetic changes were the cause of the loss of inducibility. The induced changes seem to be connected with a reversible change in the cells' response to the hormones added to the induction medium. One way this might occur, since the cells produce endogenous growth factors, is that the amounts of thesse and their proportions could vary during continued growth and with different types of media. Thus the amounts of auxin and KIN within the cell may determine PAL concentration and these amounts depend on both exogenous and endogenous hormones. Therefore it is possible that the cells which have lost their inducibility would respond to different exogenous concentrations of auxin and cytokinin. However, other changes in the cell could be envisaged whereby they no longer respond to any concentrations of exogenous growth hormones.

Androgenesis and Haploid Production

Callus from bean anther culture was first described by Haddon and Northcote (1976c). The medium contained the salts and vitamins of Gamborg et al. (1968), and 0.058 M sucrose was supplemented with 9 µM 2,4-D, alone or in combination with 1.9 µM KIN or 20% CW (v/v). Callus was produced after 2 months in culture from anthers containing mono- and binucleate pollen grains. In the presence of KIN, 20% of the anthers initiated callus. Callus was similar in appearance and growth rate to those coming from other plant parts, but retained the potential for vascular differentiation for a longer period of time. No chromosomal analysis has been done.

Peters et al. (1977) observed callus development from anthers grown on Veliky and Martins medium supplemented with 4.5 µM 2,4-D alone, or with 5.4 µM NAA, 11.4 µM IAA, and 0.9 µM KIN. Two temperatures were tested: 25 and 30 C. About the same percentage of anthers produced callus at these temperatures (57 and 61%, respectively) on media with or without 2,4-D, but the number of callus obtained on 2,4-D medium was almost double. Thirty degrees Celsius was inhibitory to callus induction and growth. Cytological analysis revealed that approximately the same amount of haploid and diploid cells were present, with less than 3% polyploids observed.

Cytological analysis of callus cells from bean suspensors grown in hormone-free medium (Bennici et al., 1976) revealed a great variability in chromosome number. About 28% of the cells had 11 chromosomes but there was no clear evidence that these cells were true haploids as precise karyotype analysis was impossible. Amitosis (nuclear fragmentation) followed by mitosis, which is the pattern of division of cells grown on a medium lacking cytokinin or with an unbalanced ratio of auxins to cytokinin, could be the cause of the chromosome variability (Alpi et al., 1975).

Micropropagation and Meristem Tip Culture

Micropropagation has been achieved by Allavena and Sharp (1981) culturing in vitro apical buds of two bean breeding lines. Multiple and adventitious shoots developed at high frequency when the basal medium of Murashige and Skoog (1962) was supplemented with 20-40 µM of BA in combination with 2-50 µM IAA or 2-5 µM NAA (Fig. 2). Further shoot bud development occurred when small clusters of shoots were subcultured on the same medium. Root formation and longer shoots were obtained after transferring shoot clusters to B5 basal medium with 0.45 µM BA (Fig. 3). Regenerated plantlets transferred into soil were able to produce flowers and seeds (Fig. 4).

Preliminary results of meristem culture in *P. vulgaris* were reported by Kueneman (1976). Approximately 0.5 mm apical dome of 4-day imbibed seeds were aseptically dissected out and placed on B5 basal solid medium. Data indicate that 1 µM NAA promotes, while 0.2 µM BA inhibits, root initiation.

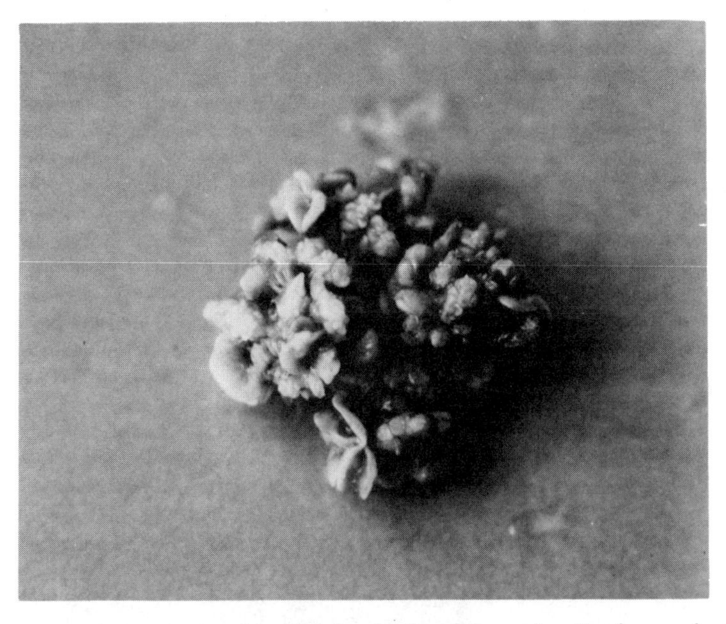

Figure 2. Development of multiple buds after subculturing apical buds onto MS basal medium supplemented with 2 µM NAA and 20 µM BA.

Kartha et al. (1981) studied the regeneration potential of shoot apical meristems of some legumes. Common bean cv. Dwarf Green Stringless seeds, were surface-sterilized and germinated on moist cotton contained in a sterile 110-ml glass. Four to 6 days after germination shoot apical meristems with subjacent tissue measuring 0.4-0.5 mm in length and containing a pair of leaf primordia were aseptically isolated and cultured in small test tubes containing 2.5 ml nutrient medium composed of MS mineral salts, 0.088 M sucrose, vitamins as found in B5 medium, and 8 g/l Difco bacto agar. BA and NAA were added in various combinations and concentrations. Cultures were maintained at 26 C for a 16-hr photoperiod at 7500 lux. Meristems differentiated into vigorously growing plants on basal medium alone or when cultured on medium supplemented with 1 µM NAA as the only hormone. Regeneration of whole plants did not occur when the meristems were cultured with various concentrations and combinations of BA and NAA. Multiple buds regenerated on medium containing 1-10 µM BA, or BA and NAA at a concentration of 5 and 10 µM, respectively. On medium with BA alone, maximum differentiation (15-30 buds per meristem) occurred at 10 µM of BA. At 0.1 µM BA, meristems regenerated directly into single shoots 2-3 cm long. Twenty days after transfer to a medium containing BA, NAA, and GA_3 at concentrations of 1, 0.1, and 0.1 µM, respectively, multiple shoots elongated to about 2.5 cm in height without differentiating roots. Rooting was successfully induced by culturing the elongated shoots on half-strength MS medium supplemented with 1 µM IAA. Plantlets thus obtained also were transferred to soil and grown to maturity.

Figure 3. Bean plantlet obtained from a small cluster of multiple buds ready to be transferred to soil.

Embryo Culture

A review on applied aspects of embryo culture has been presented by Raghavan (1977). Studies on bean embryo culture have been carried out to overcome hybrid inviability and to understand seed biology.

Honma (1955) grew 14- to 24-day-old interspecific embryos (*P. vulgaris* x *P. acutifolius*), collecting them when the pods showed signs of abortion. Liquid White medium was satisfactory for root and shoot development when the sucrose concentration was 0.12 M. Plantlets were transferred to medium with a reduced sugar concentration before transplanting into potting soil. Full fertile plants were obtained. CW, nicotinic acid, and vitamin B_1, when added, did not improve the efficiency of the medium.

Braak and Kooistra (1975) used embryo culture to grow hybrid embryos of the cross *P. vulgaris* x *P. ritensis* with the aim of transfer-

Figure 4. Mature and fertile plant regenerated from in vitro multiple shoot culture.

ring the tolerance to low temperatures carried by the latter species. Media of Tomaszewski (Kroh, 1962) and White were used. Very young embryos grew better on White medium with 0.088 M sucrose; older embryos grew faster on 0.014 M sucrose medium. According to Honma (1955), growth of hybrid embryos was most rapid on Tomaszewski solid medium with 0.12 M sucrose concentration. Addition of 1 g/l casein hydrolysate accelerates the growth of embryos measuring up to 0.7 mm (without cotyledons) and retards the growth of embryos longer than 0.7 mm.

Alvarez et al. (1981) made the following crosses within the section Euphaseolus of the genus *Phaseolus* with the help of embryo cultures: *P. coccineus* x *P. acutifolius*; *P. coccineus* x *P. vulgaris*; *P. vulgaris* x *P. acutifolius*; F_2 (*P. vulgaris* x *P. coccineus*) x *P. acutifolius*.

D.W.S. Mok et al. (1978) studied the effects of glutamine and GA_3 on immature embryos. Basal medium consisting of MS inorganic salts was supplemented with the following organic substances: sucrose (0.088 M), myo-inositol (0.55 M), thiamine HCl (3.0 µM), nicotinic acid (40.6 µM), and pyridoxine HCl (2.4 µM). Glutamine at 0.07 mM or 0.7 mM, and GA_3 at 0.1 µM were supplied. Addition of glutamine to the culture medium had a beneficial effect on hybrid embryo survival especially when the embryo size was very small (less than 0.3 mm). GA_3 did not improve survival. Embryo culture techniques have also been applied to obtain hybrids among Old World bean species (Sawa, 1973; Ahn and Hartmann, 1978).

Protoplasts

High yields of protoplasts were obtained from leaf tissue of *Phaseolus vulgaris* var. Pinto by Pelcher et al. (1974). Protoplasts were obtained from both primary and trifoliate leaves. Plants coming directly from the greenhouse gave preparations with a considerable amount of debris. Prior maintenance of the plants for 24 hr or longer at 23-25 C, 45-48% relative humidity, and 200-400 lux led to high yields of protoplasts. Protoplasts were obtained from leaf sections in an enzyme mixture consisting of 0.3 M mannitol, 0.25% desalted Cellulase, and 0.25% Pectinase. A pH of 7 had a stabilizing effect on protoplast production and the final yield of viable protoplasts was significantly higher than that at pH 5 or 6.

After separation from debris, protoplasts were collected by centrifugation, washed three times, and suspended in B5 medium supplemented with 0.3 M mannitol, 0.1 M glucose, 0.1 g/l casein hydrolyzate, 6.8 µM $CaCl_2 \cdot 2H_2O$, 3.3 mM ribose, 3.3 mM xylose, 0.7 mM hydroxyl-L-proline, 4.5 µM 2,4-D, and 9.3 µM KIN. In freshly prepared protoplasts the nucleus was located close to the plasmalemma and the chloroplasts appeared to cluster around the nucleus. Twenty-four to 28 hr after isolation the nucleus assumed a more central position and the chloroplasts began to be distributed by cyclosis throughout the cytoplasm. Staining indicated the presence of cell wall material being deposited around the protoplasts within 24-48 hr after isolation. After 2 weeks of culture an equal volume of fresh medium was added to each droplet. Fresh medium was added at weekly intervals. After 1 month, cell suspensions were transferred to agar plates on B5 medium containing 4.5 µM 2,4-D and 9.3 µM KIN. After callus formation, KIN could be omitted.

PROTOCOLS

Morphogenesis from Leaf Explants of *Phaseolus vulgaris* (Crocomo et al., 1976a)

1. Surface-sterilize leaves from 2-week-old plants in 20% hypochlorite for 15 min. Rinse twice in distilled water.
2. Cut sterilized leaves into 5 mm squares and inoculate onto 10 ml of nutrient medium in a test tube.
3. Incubate cultures under constant conditions at 25 C with 12 hr photoperiod (200 foot candles).
4. First regenerated shoots are observed within 6 weeks.
   Nutrient Medium:
   - 67-V salts and vitamins (Veliky and Martin, 1970)
   - Bean seed extract: one-fourth seed/ml water (soaked 2 hr in aerated water, then homogenized)
   - IAA 11.0 µM
   - NAA 5.4 µM
   - KIN 0.9 µM
   - pH = 4.5; 1% agar.

Protoplast Isolation and Callus Formation in *Phaseolus vulgaris* (Pelcher et al., 1974)

1. Plants should be maintained at least 24 hr at 23–25 C, 45–48% relative humidity, and 200–400 lux before use.
2. Leaves of 9-day- to 5-week-old plants are sterilized in 10% commercial bleach containing 0.05% Tween 80 for 5 min followed by immersion in 70% ethanol for 2 min. Rinse twice in 0.3 M mannitol.
3. Cut leaves into 5 x 5 cm sections and remove lower epidermis using jewelers forceps.
4. Place peeled sections into 8 ml sterile 0.3 M mannitol, 0.25% desalted Driselase (Vol. 1, Chap. 4), and 0.25% Pectinase (Sigma Chem. Co.), at pH 5–7. Incubate in darkness for 12–18 hr at 23–25 C.
5. Separate protoplasts from debris by passing preparation through an 88-μM stainless steel sieve. Collect protoplasts by centrifugation at 100 x g for 3 min.
6. Wash protoplasts 3 times by centrifugation/resuspension.
7. Resuspend the washed protoplasts in B5 (Gamborg et al., 1968) medium containing 0.3 M mannitol, 0.1 M glucose, 0.1% casein hydrolysate, 6.8 μM $CaCl_2 \cdot 2H_2O$, 3.3 μM ribose, 3.3 μM xylose, 0.76 μM hydroxy-L-proline, 4.5 μM 2,4-D, and 9.3 μM KIN.
8. Culture the protoplasts in 50 μl droplets at a density of 5 x 10 protoplasts per ml.
9. Incubate protoplast preparation in a humidified chamber at 23–25 C. Cell wall regeneration should occur 24–48 hr after isolation.
10. Add fresh culture medium after 2 weeks in culture, then at weekly intervals.
11. After 3–4 weeks in culture, cells can be transferred to agar medium.

FUTURE PROSPECTS

Until now we have not set up a protocol to permit full plant regeneration through callus or cell suspension cultures. Factors controlling root morphogenesis are well known, but shoot morphogenesis and somatic embryogenesis are poorly understood. The two plants obtained by Crocomo et al. (1976a) and the few somatic embryos recovered (Rossetti and Allavena, unpublished) only suggest the directions for future research. More work has to be done with the goal of obtaining cultures capable of embryo formation which can be developed into mature plants.

Practical application has not yet been found for techniques such as protoplast fusion, androgenesis, and selection of nuclear and cytoplasmic mutants (e.g., resistance to diseases and nonregulated enzymes involved in biosynthesis of sulfur amino acids, tryptophan, and lysine). Techniques for vegetative propagation using cultured apical buds (Allavena and Sharp, 1982) or meristems (Kartha et al., 1981) allow the following applications: (1) induction of cytoplasmic and nuclear variation, (2) isolation of true breeding mutants from mosaic tissue, (3) vegetative multiplication of peculiar genotypes (male-sterile, self-incompatible), (4) production and propagation of pathogen-free material, and (5) cryopreservation of germplasm.

There are promising techniques for culturing young interspecific bean embryos. Many papers report the possibility of fertilization between bean species of the New World (*Euphaseolus*) or between species of the Old World. Crosses between the two groups of species seem more difficult. The application of growth factors to the female plant pods and the in vitro culture of immature embryos makes it possible to overcome some barriers of incompatibility. The primary objective of interspecific hybridization among *Phaseolus* species is to facilitate gene transfer from wild and uncultivated species to cultivated varieties. Resistance to disease, drought, cold weather; high protein content; altered amino acid composition; and plant architecture are some of the desirable traits that breeders wish to transfer. Considering the importance of beans in human nutrition and the potential of tissue culture methods for the genetic improvement of beans, a greater effort should be made in this field.

KEY REFERENCES

Boulter, D. and Crocomo, O.J. 1979. Plant cell culture implications: Legumes. In: Plant Cell and Tissue Culture Principles and Applications (W.R. Sharp, P.O. Larsen, E.F. Paddock, V. Raghavan, eds.) pp. 615-631. Ohio State University Press, Columbus.

Evans, A.M. and Gridley, H.E. 1979. Prospects for the improvement of protein and yield in food legumes. Current Adv. Plant Sci. 32:1-17.

Purseglove, J.W. 1969. Tropical crops Dicotyledons 1. Longmans, New York.

Zaumeyer, W.J. and Thomas, H.R. 1957. A monographic study of bean diseases and methods for their control. Technical Bulletin 868, U.S. Dept. Agric.

REFERENCES

Abdel Aziz, H.M., Ali, A.M., and Desouki, I.A.M. 1979. Success of some intergeneric crosses in the tribe *Phaseoleae*. Res. Bull., Fac. Agric. 1956:18.

Ahn, C.S. and Hartmann, R.W. 1978. Interspecific hybridization between mung bean (*Vigna radiata* (L.) Wilczek) and Adzuki bean (*V. angularis* (Willd.) Ohwi & Ohashi). J. Am. Soc. Hortic. Sci. 103:3-6.

Allavena, A. and Sharp, W.R. 1981. Adventitious shoots and plantlets development of *Phaseolus vulgaris* apical buds grown "in vitro." Proceedings XXV Meeting of the Italian Society for Agricultural Genetics.

Alpi, A., Togoni, F., and D'Amato, F. 1975. Growth regulator levels in embryo and suspensor of *Phaseolus coccineus* at two stages of development. Planta 127:153-162.

Alvarez, M.N., Ascher, P.D., and Davis, D.W. 1981. Interspecific hybridization in Euphaseolus through embryo-rescue. HortScience 16: 541-543.

Al-Yasiri, S. and Coyne, D.P. 1964. Effect of growth regulators in delaying pod abscission and embryo abortion in the interspecific cross *Phaseolus vulgaris* x *P. acutifolius*. Crop Sci. 4:433-435.

Ammirato, P.V. 1983. In: Handbook of Plant Cell Culture, Vol. 1 (D.A. Evans, W.R. Sharp, P.V. Ammirato, and Y. Yamada, eds.). Macmillan, New York.

Arnison, P.G. and Boll, W.G. 1974. Isoenzymes in cell cultures of bush bean (*Phaseolus vulgaris* cv. Contender): Isoenzymatic changes during the callus culture cycle and differences between stock cultures. Can. J. Bot. 52:2621-2629.

______ 1975. Isoenzymes in cell cultures of bush bean (*Phaseolus vulgaris* cv. Contender): Isoenzymatic differences between stock suspension cultures derived from a single seedling. Can. J. Bot. 53:261-271.

______ 1978. The effect of 2,4-D and kinetin on the activity and isoenzyme pattern of various enzymes in cotyledon cell suspension cultures of bush bean (*Phaseolus vulgaris* cv. Contender). Can. J. Bot. 56:2185-2195.

Bassiri, A. and Carlson, P.S. 1978. Isozyme patterns and differences in plant parts and their callus cultures in common bean. Crop Sci. 18: 955-958.

Bennici, A., Cionini, P.G., and D'Amato, F. 1976. Callus formation from the suspensor of *Phaseolus coccineus* in hormone-free medium: A cytological and DNA cytophotometric study. Protoplasma 89:251-261.

Bertola, M.A. and Klis, F.M. 1979. Continuous cultivation of glucose-limited bean cells (*Phaseolus vulgaris* L.) in a modified bacterial fermenter. J. Exp. Bot. 30:1223-1231.

Bevan, M. and Northcote, D.H. 1979a. The interaction of auxin and cytokinin in the induction of phenylalanine ammonia-lyase in suspension cultures of *Phaseolus vulgaris*. Planta 147:77-81.

______ 1979b. The loss of morphogenetic potential and induction of phenylalanine ammonia-lyase in suspension cultures of *Phaseolus vulgaris*. J. Cell Sci. 39:339-353.

Biswas, M.R. and Dana, S. 1975a. Black gram x rice bean cross. Cytologia 40:787-795.

______ 1975b. *Phaseolus aureus* x *P. lathyroides* cross. Nucleus (Boston) 18:81-85.

______ 1976. *Phaseolus aconitifolius* x *Phaseolus trilobus*. Indian J. Genet. Plant Breed. 36:125-131.

Bliss, F.A. 1977. Recurrent selection in beans. Bean Improv. Coop. 20:50-52.

Blixt, S. and Gottschalk, W. 1975. Mutation in the leguminosae. Agri Hort. Genet. 33:33-85.

Boomstra, A.G., Bliss, F.A., and Beebe, S.E. 1977. New sources of *Fusarium* root rot resistance in *Phaseolus vulgaris* L. J. Am. Soc. Hortic. Sci. 102:182-185.

Braak, J.P. and Kooistra, E. 1975. A successful cross between *Phaseolus vulgaris* L. and *Phaseolus ritensis* Jones with the aid of embryo culture. Euphytica 24:669-679.

Bressani, R. 1975. Legumes in human diets and how they might be improved. In: Nutritional Improvement of Food Legumes by Breeding (M. Milner, ed.) New York.

Brucher, O., Wecksler, M., Levy, A., Palozzo, A., and Jaffe, W.G. 1969. Comparison of phytohaemagglutinins in wild beans (*Phaseolus aborigineus*) and in common beans (*Phaseolus vulgaris*) and their inheritance. Phytochem. 8:1739-1743.

Buishand, T.J. 1956. The crossing of beans (*Phaseolus* spp.). Euphytica 5:41-50.

Campion, B., Fadda, A., and Soressi, G.P. 1980. Conversione a taglia nana del fagiolo di Spagna rampicante (*Phaseolus coccineus* L.). In: Risultati di un Quinquennino di Lavoro sul Miglioramento Genetico del Fagiolo da Granella. C.N.R. Atti Convegno di Cuneo 37-57.

Colin, S.M. 1966. Identificacion de las especies-mexicanas y cultivadas del genero *Phaseolus.* Colegio de Postgraduados, Escuela Nacional de Agricoltura n° 8, Chapingo, Mexico.

Coyne, D.P. 1969. Heritability and selection of yield components in beans. Bean Improv. coop. 12:14-15.

______ and Schuster, M.L. 1974. Breeding and genetic studies of tolerance to several bean (*Phaseolus vulgaris* L.) bacterial pathogens. Euphytica 23:651-656.

______ and Schuster, M.L. 1975. Genetic and breeding strategy for resistance to rust (*Uromyces phaseoli*) (Reben) Wint) in beans (*Phaseolus vulgaris* L.). Euphytica 24:795-803.

______, Schuster, M.L., and Gallegos, C.G. 1971. Inheritance and linkage of the halo blight systemic chlorosis and leaf watersoaked reaction in *Phaseolus vulgaris* variety crosses. Plant Dis. Rep. 55:203-207.

Crivelli, G., Abbo, A., Pizzocaro, F., and Allavena, A. 1979. Researches on the suitability of bean varieties to quick freezing. XV Intern. Cong. of Refriger. Venezia, Sept 23-29.

Crocomo, O.J., Sharp, W.R., and Peters, J.E. 1976a. Plantlet morphogenesis and the control of callus growth and root induction of *Phaseolus vulgaris* with the addition of a bean seed extract. Z. Pflanzenphysiol. 78:456-460.

Crocomo, O.J., Peters, J.E., and Sharp, W.R. 1976b. Interactions of phytohormones on the control of growth and root morphogenesis in cultured *Phaseolus vulgaris* leaf explants. Turrialba 26:232-236.

D'Amato, F. 1977. Cytogenetics of differentiation in tissue and cell cultures. In: Plant Cell, Tissue, and Organ Culture (J. Reinert and Y.P.S. Bajaj, eds.) Springer-Verlag, Berlin, New York.

______, Bennici, A., Cionini, P.G., Baroncelli, S., and Lupi, M.C. 1980. Nuclear fragmentation followed by mitosis as mechanism for wide chromosome number variation in tissue cultures: Its implications for plant regeneration. In: Plant Cell Cultures: Results and Perspectives (F. Sala, B. Parisi, R. Cella, and O. Cieferri, eds.) pp. 67-72. Elsevier/North Holland Biomedical Press, Netherlands.

Dana, S. and Das, N.D. 1974. Natural amphidiploidy in a *Phaseolus* hybrid. Sabrao J. 219-222.

Deakin, J.R. and Dukes, P.D. 1975. Breeding snap beans for resistance to diseases caused by *Rhizoctonia solani* Kuehn. HortScience 10:269-271.

Denis, J.C. and Adams, M.W. 1978. A factor analysis of plant variables related to yield in dry beans. I. Morphological traits. Crop Sci. 18: 74-78.

Dickson, M.H. and Natti, J.J. 1968. Inheritance of resistance of *Phaseolus vulgaris* to bean yellow mosaic virus. Phytopathology 58:1450.

Evans, D.A. and Bravo, J.E. 1983. Protoplast isolation and culture. In: Handbook of Plant Cell Culture, Vol. 1 (D.A. Evans, W.R. Sharp, P.V. Ammirato, and Y. Yamada, eds.). Macmillan, New York.

Gamborg, O.L., Miller, R.A., and Ojima, K. 1968. Nutrient requirements of suspension cultures of soybean root cells. Exp. Cell. Res. 50:151-158.

Gentry, H.S. 1969. Origin of the common bean, *Phaseolus vulgaris*. Econ. Bot. 23:55-69.

Haddon, L.E. and Northcote, D.H. 1975. Quantitative measurement of the course of bean callus differentiation. J. Cell Sci. 17:11-26.

______ 1976a. Correlation of the induction of various enzymes concerned with phenylpropanoid and lignin synthesis during differentiation of bean callus (*Phaseolus vulgaris* L.). Planta 128:255-262.

______ 1976b. The influence of gibberellic acid and abscisic acid on cell and tissue differentiation of bean callus. J. Cell Sci. 20:47-55.

______ 1976c. The effect of growth conditions and origin of tissue on the ploidy and morphogenetic potential of tissue cultures of bean (*Phaseolus vulgaris* L.). J. Exp. Bot. 27:1031-1051.

Hamblin, J. 1977. Plant breeding interpretations of the effects of bulk breeding on four populations of beans (*Phaseolus vulgaris* L.). Euphytica 26:157-168.

______ and Kent, S.P. 1973. Possible role of phytohaemagglutinin in *Phaseolus vulgaris* L. Nature (London) New Biol. 245:28-30.

Honma, S. 1955. A technique for artificial culturing of bean embryos. Proc. Amer. Soc. Hort. Sci. 65:405-408.

______ and Heeckt, O. 1958. Bean interspecific hybrid involving *Phaseolus coccineus* x *P. lunatus*. J. Am. Soc. Hortic. Sci. 72:360-364.

Hubbeling, N. 1977. The new jota race of *Colletotrichum lindemuthianum*. Bean Improv. Coop. 20:58.

Hussein, H.A.S. and Disouki, I.A.M. 1976. Mutation breeding experiments in *Phaseolus vulgaris* L. I. EMS and gamma-ray-induced seed coat colour mutants. Z. Pflanzenzuecht. 76:190-199.

Jaffe, W.G. 1975. Factors affecting the nutritional value of beans. In: Nutritional Improvement of Food Legumes by Breeding (M. Milner, ed.) Wiley-Interscience, New York.

______ and Vega, C.L. 1968. Heat-labile growth-inhibiting factors in beans (*Phaseolus vulgaris*). J. Nutr. 94:203.

Jeffs, R.A. and Northcote, D.H. 1967. The influence of indole-3 acetic acid and sugar on the pattern of induced differentiation in plant tissue culture. J. Cell Sci. 2:77-88.

Kaplan, L. 1965. Archeology and domestication in American *Phaseolus* (beans). Econ. Bot. 19:358-368.

Kartha, K.K., Pahl, K., Leung, N.L., and Mroginski, L.A. 1981. Plant regeneration from meristems of grain legumes: Soybean, cowpea, peanut, chickpea, and bean. Can. J. Bot. 59:1671-1679.

Kelly, J.D. and Bliss, F.A. 1975a. Heritability estimates of percentage seed protein and available methionine and correlations with yield in dry beans. Crop Sci. 15:753-757.

______ 1975b. Quality factors affecting the nutritive value of bean seed protein. Crop Sci. 15:757-762.

Krasnuk, M., Witham, F.H., and Tegley, J.R. 1971. Cytokinins extracted from Pinto bean fruit. Plant Physiol. 48:320-324.

Kroh, M. 1962. Vergleichende untersungen an *Phaseolus coccineus*. Selbstungen und Kreuzungen zwischen *Ph. vulgaris* und *Ph. coccineus*. Z. Pflanzenzuecht. 47:201-216.

Kueneman, E.A. 1976. Meristem culture in *Phaseolus vulgaris*. Bean Improv. Coop. 19:53-55.

Liau, D.F. and Boll, W.G. 1970. Callus and cell suspension culture of bush bean (*Phaseolus vulgaris*). Can. J. Bot. 48:1119-1130.

______ 1971. Growth, and patterns of growth and division, in cell suspension cultures of bush bean (*Phaseolus vulgaris* cv. Contender). Can. J. Bot. 49:1131-1139.

Liener, I. 1975. Antitryptic and other antinutritional factors in legumes. In: Nutritional Improvement of Food Legumes by Breeding (M. Milner, ed.) pp. New York.

Mante, S. and Boll, W.G. 1975. Comparison of growth and extracellular polysaccharide of cotyledon cell suspension cultures of bush bean (*Phaseolus vulgaris* cv. Contender) grown in coconut-milk medium and synthetic medium. Can. J. Bot. 53:1542-1548.

______ 1976. Changes in the amount and composition of fractions from extracellular polysaccharide during the culture cycle of cotyledon cell suspension culture of bush bean (*Phaseolus vulgaris* cv. Contender). Can. J. Bot. 54:198-201.

______ 1978. Effect of either 2,4-dichlorophenoxyacetic acid or kinetin on production and composition of various fractions from extracellular polysaccharides produced by cotyledon cell suspension cultures of bush bean (*Phaseolus vulgaris* cv. Contender). Can. J. Bot. 56:1816-1822.

McComb, J.A. 1975. Is intergeneric hybridization in the leguminosae possible? Euphytica 24:497-502.

Moh, C.C. 1968. Bean mutant induced by ionizing radiation. III. Wrinkled leaf. Turrialba. 18:181-182.

______ 1969. Seed-coat color changes induced by ethyl methanesulfonate in the common bean (*Phaseolus vulgaris* L.). Mutation Res. 7: 469-471.

______ and Alan, J.J. 1964. Bean mutant induced by ionizing radiation. I. Dwarf mutant. Turrialba 14:82-84.

______ and Alan, J.J. 1967. The response of a radiation-induced dwarf bean mutant to gibberellic acid. Turrialba 17:176-178.

______ and Nanne, H. 1969. Bean mutant induced by ionizing radiation. IV. "Pepper" mutant. Turrialba 19:292-293.

Mok, D.W.S., Mok, M.C., and Rabakoarihanta, A. 1978. Interspecific hybridization of *Phaseolus vulgaris* with *P. lunatus* and *P. acutifolius*. Theor. Appl. Genet. 52:209-215.

Mok, M.C. and Mok, D.W.S. 1977. Genotypic responses to auxins in tissue cultures of *Phaseolus*. Physiol. Plant. 40:261-264.

______, and Armstrong, D.J. 1978. Differential cytokinin structure-activity relationships in *Phaseolus*. Plant Physiol. 61:72-75.

______, Armstrong, D.J., Rabakoarihanta, A., and Kim, S.G. 1980. Cytokinin autonomy in tissue cultures of *Phaseolus*: A genotype-specific and heritable trait. Genetics 94:675-686.

Motto, M., Soressi, G.P., and Samamini, F. 1979. Growth analysis in a reduced leaf mutant of common bean (*Phaseolus vulgaris* L.). Euphytica 28:593-600.

Murashige, T. and Skoog, F. 1962. A revised medium for rapid growth and bioassays with tobacco tissue cultures. Physiol. Plant. 15:473-497.

Pelcher, L.E., Gamborg, O.L., and Kao, K.N. 1974. Bean mesophyll protoplasts: Production, culture and callus formation. Plant Sci. Lett. 3:107-111.

Peters, J.E., Crocomo, O.J., and Sharp, W.R. 1976. Effect of caffeine and nicotine on the callus growth and root morphogenesis of *Phaseolus vulgaris* tissue cultures. Turrialba 26:337-341.

______, Paddock, E.F., Tegenkamp, I., and Tegenkamp, T. 1977. Haploid callus cells from anthers of *Phaseolus vulgaris*. Phytomorphology 27:79-85.

Porter, W.M., Maner, J.H., Axtell, J.D., and Keim, W.F. 1974. Evaluation of the nutritive quality of grain legumes by an analysis for total sulfur. Crop Sci. 14:652-654.

Pusztai, A., 1966. The isolation of two proteins, glycoprotein I and a trypsin inhibitor, from the seeds of kidney bean (*Phaseolus vulgaris*). Biochem. J. 101:379-384.

Raghavan, V. 1977. Applied aspects of embryo culture. In: Plant Cell, Tissue, and Organ Culture (J. Reinert and Y.P.S. Bajaj, eds.) pp. 375-397. Springer-Verlag, Berlin.

Rubaihayo, P.R. 1975. The use of α-ray induced mutations in *Phaseolus vulgaris* L. Z. Pflanzenzuecht. 75:257-261.

Sarafi, A. 1973. Utilisation de rayons ionisants dans l'amelioration du haricot (*Phaseolus vulgaris* L.). Ann. Amélior. Plantes 23:77-81.

Sawa, M. 1973. On the interspecific hybridization between the Adzuki bean, *Phaseolus angularis* (Willd) W.F. Wight and the Green Gram, *Phaseolus radiatus* L. I. Crossing between a cultivar of the Green Gram and a Semi-Wild relative of the Adzuki bean, in Endemic name "Bakaso." Japan. J. Breed. 23:61-66.

Smartt, J. 1970. Interspecific hybridization between cultivated American species of the genus *Phaseolus*. Euphytica 19:480-489.

______, Winfield, P.J., and Williams, D. 1975. A strategy for the improvement of protein quality in pulses by breeding. Euphytica 24:447-451.

Swarup, V. and Gill, H.S. 1968. X-ray induced mutations in French bean. Indian J. Genet. Plant Breed. 28:44-58.

Tonin, G.S., Derbyshire, M.T.V.C., Sharp, W.R., and Crocomo, O.J. 1982. Amino acids in the callus growth and root morphogenesis of bean (*Phaseolus vulgaris*) leaves cultured in vitro.

Tonzig, S. 1968. Botanica, Vol. 2 (Casa Ed. Ambrosiana). Milano.

Tulmann Neto, A., Ando, A., and Costa, A.S. 1977. Attempts to induce mutants resistant or tolerant to golden mosaic virus in dry beans (*Phaseolus vulgaris*). In: Induced Mutations against Plant Diseases, pp. 281-290. Vienna.

Veliky, I.A. 1972. Synthesis of carboline alkaloids by plant cell cultures. Phytochem. 11:1405-1406.

______ and Barber, K.M. 1975. Biotransformation of tryptophan by *Phaseolus vulgaris* suspension culture. Lloydia 38:125-130.

______ and Martin, S.M. 1970. A fermenter for plant cell suspension cultures. Can. J. Microbiol. 16:223-226.

Wall, J.R. 1970. Experimental introgression in the genus *Phaseolus*. I. Effect of mating systems on interspecific gene flow. Evolution 24: 356-366.

White, P.R. 1963. The Cultivation of Animal and Plant Cells. Ronald Press, New York.

York, D.W., Dickson, M.H., and Abawi, G.S. 1977. Inheritance of resistance to seed decay and pre-emergence damping-off in snap beans caused by pythium ultimum. Plant Dis. Rep. 61:285-289.

CHAPTER 7

Red Clover and Other Forage Legumes

G.C. Phillips and *G.B. Collins*

HISTORY OF THE CROP

Red clover (*Trifolium pratense* L.) has a long and significant history of agricultural importance as a forage and soil improvement crop in temperate regions of the world. It is one of about 250 species of true clovers, several of which are important as forage legumes. Among these, white or ladino clover (*T. repens* L.) also is a leading crop in temperate regions of the world. Another group of forage legumes consists of the trefoils (*Lotus* spp.), among which birdsfoot trefoil (*L. corniculatus* L.) may be the most important. Trefoils are produced for pasture in tropical as well as temperate regions of the world. Significant progress has been made in recent years in the development and application of in vitro culture methods for the improvement of certain clovers and birdsfoot trefoil. This progress will be reviewed and state-of-the-art protocols will be presented for red clover.

Red clover was cultivated in southern Europe as early as the third century A.D. (Evans, 1976). It was introduced into northern Europe in the fifteenth century A.D. and into North America during the colonial period (Taylor and Smith, 1979). It replaced the fallow in cropping systems and was therefore an extremely important factor in the agricultural revolutions which occurred in both the Old and New Worlds (Fergus and Hollowell, 1960). Diploid cultivars are predominantly grown in the Western Hemisphere while autotetraploid cultivars are becoming prevalent in the Eastern Hemisphere (Taylor and Smith, 1979).

ECONOMIC IMPORTANCE

Red clover is most important as a hay crop (Fergus and Hollowell, 1960; Taylor, 1973; Taylor and Smith, 1979), yielding on the average one-three cuttings per year and as many as five cuttings under irrigation in the western United States. The number of cuttings obtained generally depends on the length of the growing season and moisture availability. When grazed, it may cause bloat (foam production) in livestock. This is a common problem with forage legumes. The bloat problem usually does not arise when the forage is wilted or grown with companion grasses or grains. Red clover or its mixtures can be ensiled following wilting. All classes of livestock and poultry perform well on red clover forage. Its feed value, digestibility, and palatability compare well with other forage legumes. It serves best in 3- and 4-year rotations as a hay and pasture crop and for renovation of old pastures. It is the easiest legume to establish in renovated or closely grazed sods. It is excellent for use as a rotation crop due to its soil improvement properties.

Although fed to livestock as forage, red clover is marketed to the general consumer in the form of meat, fiber, and dairy products (Taylor and Smith, 1979). Red clover is grown on about 5.4 million ha each year in the United States, with an annual hay yield of about 21.5 million MT. Yields could easily be twice this figure if improved cultivars and proper management practices were used exclusively. It is estimated that the annual economic worth of red clover hay is $1.4 billion in the United States alone. The pasture value of this crop is important. Seed crop values of red clover in the United States are about $25 million annually. A red clover crop returns to the soil between 125 and 200 kg of nitrogen per ha, thus greatly reducing the need for and cost of nitrogen fertilizer for subsequent crops.

BREEDING AND CROP IMPROVEMENT

The breeding and genetics of red clover have been thoroughly reviewed (Fergus and Hollowell, 1960; Taylor, 1973; Taylor and Smith, 1979). Red clover is self-incompatable by virtue of an oppositional S-allele gametophytic system. Severe inbreeding depression develops and intensifies as selfing is continued.

The factor which most limits the use of red clover as a pasture crop is its weak perenniality under field conditions. Longevity may be a complex character, dependent on genetic composition, diseases, pests, unfavorable environments, and other factors. Intraspecific variation has been utilized in breeding programs, but it has not increased longevity greatly. Interspecific hybridization with long-lived perennial clovers may be the best source of increased longevity. Conventional hybridization procedures have not produced such hybrids for evaluation.

Phosphorus is a limiting factor on most soils for maximum red clover productivity. Potassium and other nutrients are limiting in some areas. Increased nutrient efficiency is highly desirable, but whole plant evaluations are cumbersome and have not resulted in significant progress with red clover.

Table 1. Some of the Cultivated *Trifolium* L. Species[a]

TAXONOMIC SECTION OF THE GENUS	SPECIES	COMMON NAME OF CLOVER	HABIT	CHROMOSOME NUMBER (2n)
Lotoidea Zoh.	*repens* L.	White, ladino	Perennial	32
	nigrescens Viv.	Ball	Annual	16
	hybridum L.	Alsike	Perennial	16
	ambiguum Bieb.	Kura, caucasian	Perennial	16, 32, 48
	semipilosum Fres.	Kenya	Perennial	16
Trifolium Zoh.	*pratense* L.	Red	Perennial	14
	medium L.	Zigzag	Perennial	64-80
	incarnatum L.	Crimson	Annual	14
	pannonicum Jacq.	Hungarian	Perennial	130, 180[b]
	striatum L.	Striate, knotted	Annual	14
	hirtum All.	Rose	Annual	10
	lappaceum L.	Lappa, burdock	Annual	14[b], 16
	arvense L.	Rabbitfoot	Annual	14
	alexandrinum L.	Berseem, egyptian	Annual	16
Chronosemium (Ser.) Zoh.	*agrarium* L.	Hop	Annual	14
	campestre Shreb.	Large hop	Annual	14
	dubium Sibth.	Small hop, yellow suckling	Annual	14, 28, 32[b]
Tricocephalum Zoh.	*subterraneum* L.	Subterranean, sub	Annual	16
Vesicaria Crantz.	*fragiferum* L.	Strawberry	Perennial	16
	resupinatum L.	Persian	Annual	16
Mistyllus (Presl) Zoh.	*vesiculosum* Savi.	Arrowleaf	Annual	16

[a]Compiled from Zohary (1971, 1972), Whyte et al. (1953), Leffel (1973), Knight and Hoveland (1973), Evans (1976), Taylor and Smith (1979), Darlington and Wylie (1955), and Moore (1973).

[b]Infrequently reported chromosomal races.

Superior clones of red clover are normally propagated by crown cuttings. In the absence of readily available inbreeding, clonal propagation serves as the principal means of maintaining specific, desirable genotypes. One problem associated with vegetative clonal propagation is that virus infections disrupt the continued maintenance of synthetic cultivars by the debilitation or loss of irreplaceable genotypes.

Intraspecific variation has been utilized in breeding for disease resistance in red clover. Fungal and viral infections are much more severe than bacterial infections in red clover. Interspecific hybridization will offer the best source of resistance to many diseases, since limited sources of resistance have been identified within the species. Progress has been made by breeding for resistance to certain destructive insects and nematodes. Several insects carry viruses which are mechanically transmitted to the plants.

Adaptations to more extreme temperature and moisture conditions should extend the agronomic distribution of red clover.

Several other clovers are of agricultural importance as forages, as indicated in Table 1. Many clovers merit attention for specific adaptational characteristics (Evans, 1976; Leffel, 1973; Knight and Hoveland, 1973; Taylor, 1973). The genus includes both perennial and annual species. Generally annuals are self-compatible while perennials tend to be self-incompatible (Taylor et al., 1979). Base chromosome numbers in the genus include n = 5–8 (Taylor et al., 1979).

Because of its strong perennialism, white clover is much more important as a pasture crop than red clover, especially for renovation of permanent pastures (Leffel and Gibson, 1973; Evans, 1976). It is a self-compatible tetraploid which is well diploidized. Its production is limited by pests, extremes in temperature and moisture, salinity, and alkalinity. Little intraspecific variation exists for most of these characteristics. Introgression following interspecific hybridization is viewed as the best approach for improving white clover.

Birdsfoot trefoil is a self-incompatible tetraploid which is well diploidized (Seaney and Henson, 1970). It is an excellent permanent pasture legume also used for hay and silage production. It is well adapted to cooler climates and tolerates poor soils better than other leading forage legumes. Seedling establishment is the most limiting factor in its production. Weed competition is also a serious problem in establishment. It recovers slowly after harvest. Improved seed set and yield are needed. Its production is limited in warm, humid climates by disease problems. Birdsfoot trefoil offers the distinct advantage of being a nonbloating legume.

DEVELOPMENT OF TISSUE AND CELL CULTURE SYSTEMS

Technologies

RED CLOVER. Initial attempts to generate callus from red clover tissue and plants from callus were disappointing. Niizeki and Kita (1973) cultured 99 anthers from two cultivars on the basal media of Miller (1961) and Bourgin and Nitsch (1967) modified with various combinations of IAA, NAA, 2,4-D, BA, and GA. Callus formation occurred only in a single instance, on Miller medium containing 17 µM IAA and 6.6 µM BA. No morphogenetic development was observed. Microscopic evaluation indicated that the callus arose from somatic tissue.

Zakrzewski and Zakrzewska (1976) investigated the potential of using stem-derived callus of one cultivar as a propagation medium for the nematode, *Ditylenchus dipsaci* (Kuhn) Filipjve. Various combinations of 2,4-D, IAA, NAA, KIN, and corn milk were evaluated using a basal medium composed of the major salts of Gautheret (1955), minor salts modified from Burkholder and Nickell (1949), and organic components modified from Gautheret (1955). Callus growth was obtained using 9 µM 2,4-D and 15% corn milk. Very little growth occurred without the addition of corn milk. This suggests that undefined nutritional factors were missing from the basal medium.

Ranga Rao (1976) initiated callus from root, stem, and leaf tissues of one cultivar and two breeding lines of red clover cultured on a basal medium modified from Miller and supplemented with 9 µM 2,4-D and 10 µM KIN. Callus suspensions were grown in a liquid version of the same medium and inoculated with a symbiotic strain of *Rhizobium trifolii* Danegeard. Callus suspensions were subsequently cultured on liquid medium free of NO_3^-, 2,4-D, and KIN. In 6-8 weeks, 75% of the infected cultures produced roots regardless of the origin of the tissue. Uninoculated tissues did not undergo morphogenesis. It was concluded that hormones produced and released by the *Rhizobium* in infected cultures were responsible for root formation by the red clover callus.

Ahloowalia (1976) cultured seeds of one tetraploid cultivar on MS medium (Murashige and Skoog, 1962) supplemented with 7.5 µM 2,4-D, 37 µM IAA, and 10 µM KIN. Callus was produced from 63% of the seeds within 6 weeks. Somatic embryos at the globular, heart, and torpedo stages were recovered from callus cultured on half-strength MS medium free of growth regulators. No plants were regenerated.

Significant progress in the development of useful red clover callus and cell culture systems has been made just in the past 3-4 years. Beach and Smith (1979) established callus cultures from seedling hypocotyls and excised pistils of two cultivars on B5 medium (Gamborg et al., 1968) containing 10 µM each of 2,4-D, NAA, and KIN. Hypocotyls produced more callus than did pistils. Callus production was very poor in the absence of 2,4-D. Shoot buds formed on this medium in some cases. After 4-5 weeks callus was transferred to B5 medium containing twice the normal concentration of thiamine, 10 µM NAA, and 15 µM ADE. Numerous shoots developed, but few roots were formed. Rooting of shoots was accomplished on B5 medium containing the higher concentration of thiamine and 1 µM NAA. Growth regulator combinations including IAA, 2iP, or BA were inferior to those listed. The SH basal medium (Schenk and Hildebrandt, 1972) was a suitable substitute for B5 basal medium. MS and Blaydes' (1966) basal media did not support callus growth as well. Performance was not improved on B5 medium supplemented with CW or leaf extracts. Most of the regenerated plants were normal, fertile diploids. A few abnormal, sterile plants were recovered.

Phillips and Collins (1979a) investigated the growth of callus from seedling sections of five cultivars. Initial experiments evaluating the SH, B5, and MS basal media containing 2,4-D and KIN were not very productive. Visual ratings of various combinations of NAA, IAA, CPA, 2,4-D, PIC, KIN, 2iP, and BA were carried out. The combination of 0.25 μM PIC and 0.44 μM BA was optimal for callus initiation and cell proliferation. The use of PIC as an auxin in plant tissue cultures had been described previously (Collins et al., 1978).

Since other basal media were deemed unsatisfactory, an improved basal medium designated L2 was experimentally developed (Phillips and Collins, 1979a). The chemical forms and concentrations of individual components were visually rated. Major and minor inorganic elements and organic compounds were evaluated in that order. The final composition of the L2 medium contained a major salts formulation similar to that of the MS medium, with a reduction in the concentration of NH_4^+ and increased concentrations of PO_4^{3-}, K^+, Mg^{2+}, and Ca^{2+}. The minor salts formulation was similar to that of the SH medium, with adjustments in the concentrations of several salts. The organic formulation (vitamins and sucrose) was similar to that of Linsmaier and Skoog (1965), with increased concentrations of thiamine and myo-inositol. The addition of nicotinic acid was inhibitory to callus growth. Fresh weight increases were statistically evaluated in a final experiment comparing the L2, MS, SH, and Miller inorganic formulations using the growth regulators and organic formulation of the L2 medium. Although no statistical differences were detected between the L2 and MS inorganic formulations, the L2 formulation was superior to the others tried. However, the L2 formulation supported callus growth of 92/100 genotypes tested whereas the MS formulation supported callus growth of only 80/100 genotypes tested. It was concluded that the L2 medium was more broadly supportive of red clover genotypes in culture than other basal media. In tests with alfalfa (*Medicago sativa* L.) and soybean [*Glycine max* (L.) Merr.], the L2 inorganic formulation was statistically shown to be superior to the MS, SH, and Miller formulations, confirming that the L2 medium was more broadly supportive of certain legumes in culture.

Collins and Phillips (1982) routinely germinated seedlings to provide explants for callus production on a modification of the L2 medium containing only 10% of the major salt concentrations, 25% of the vitamin concentrations, 60% of the sucrose concentration, and 0.65% agar. Phillips and Collins (1979a) also obtained vigorous callus from both mature and immature vegetative and reproductive explant sources on the L2 medium. Plants were regenerated from callus cultures. The frequency of plant regeneration was dependent on the origin of the tissue and the source cultivar. Meristem-derived callus regenerated from 30 to 80% of the genotypes while nonmeristem-derived callus regenerated from only 1% of the genotypes. Anther-derived callus was infrequent but highly morphogenic. The cultivars Altaswede and Arlington produced more plants from callus than did Redman, Kenstar, and Tensas. Plants regenerated from callus cultured on various combinations of 2,4-D, NAA, CPA, PIC, BA, and KIN. Combinations including 2,4-D yielded more plants than a combination of PIC and BA.

Plants were recovered more efficiently when callus grown on media containing 2,4-D was transferred to a medium containing a low concentration of PIC and a high concentration of BA. Plants were rooted on a half-strength L2 medium supplemented with nicotinic acid and a vitaminlike analog, 3-aminopyridine. All regenerated plants were diploid and normal in appearance.

Cell suspension cultures of red clover were successfully obtained on a modified liquid version of the L2 medium (Phillips and Collins, 1979a). Rapid cell division rates were observed. Callus colonies were recovered from cell suspensions when recultured on solidified L2 medium. No plants were regenerated from suspension-derived callus in this study.

In a subsequent study, Phillips and Collins (1980) successfully regenerated plants from callus derived from a cell suspension culture. Three seedlings selected from one cultivar were sectioned and placed in SL2 liquid culture medium to initiate the cell suspension culture. The SL2 medium was modified from the L2 medium by reduction of the concentrations of certain major salts and all minor salts. The cell suspension was serially subcultured weekly on three growth regulator combinations (0.25 μM PIC and 0.44 μM BA, 0.9 μM 2,4-D and 25 nM PIC, or 0.45 μM 2,4-D and 15 μM ADE) for 6 months prior to initiating regeneration studies. Regeneration studies evaluated various combinations of IAA, NAA, PIC, 2,4-D, BA, and ADE in solidified L2 basal medium used to recover callus from three cell suspension lines. There were no statistical differences detected between growth regulator combinations used in cell suspension cultures for subsequent plant regeneration. No statistical differences were detected between cytokinins for plant regeneration from callus. Statistical differences were detected between auxins, 2,4-D being the best for plant regeneration. The concentration of 2,4-D was important, 23 nM-0.14 μM being statistically superior to other concentrations.

Based on numerical differences alone, it was concluded that the SL2 medium containing 0.25 μM PIC and 0.44 μM BA was the best medium for maintaining cell suspension cultures. L2 medium containing 45 nm 2,4-D and 15 μM ADE was optimal for initiating plant regeneration. Plants regenerated via somatic embryogenesis, as evidenced by the developmental pattern of regeneration. Cotyledons and a unifoliate leaf, as found in seedlings, emerged from the somatic embryos prior to the development of trifoliate leaves. Some somatic embryos developed complete shoots and roots on the induction medium. Shoot development of most somatic embryos was most efficient on a L2 medium containing a low concentration of PIC and a moderate concentration of BA. Most of the regenerated plants were normal, fertile diploids. Some of the plants were morphologically abnormal and/or sterile. About 10% of the regenerated plants were tetraploid, suggesting spontaneous doubling of the chromosomes during culture. It was determined that callus derived from other genotypes also responded to the regeneration procedure, although some callus required several subcultures onto induction medium for the initiation of somatic embryogenesis.

Phillips and Collins (1981) used these procedures for cell suspension culture and plant regeneration in a cellular selection scheme to study phosphorus efficiency. Two genotypes were exposed to progressively lower concentrations of phosphorus in the culture medium during serial subculture. The standard level of phosphorus was 1.8 mM. A 50% increase in cell-doubling rates occurred on medium containing 0.1 mM phosphorus. Cells were subsequently selected for sustained growth on 1.8 µM phosphorus for a period of 4 to 12 months. Cell-doubling rates of selected cells on the selective concentration of phosphorus were about three times as long as those of unselected cells. Selected cells of one genotype were returned to standard conditions and readily adapted to selective conditions again. Selected and unselected cells of both genotypes were analyzed for phosphorus and total nitrogen contents. Selected cells of one genotype when returned to standard growth conditions exhibited significantly higher concentrations of phosphorus and nitrogen compared to unselected cells. Plants were regenerated from this line. These regenerated plants also exhibited elevated concentrations of phosphorus and nitrogen. The phosphorus content of these plants was two to three times greater than that found previously in other red clover plants.

Keyes et al. (1980) performed genetic studies evaluating variation in callus cultures of red clover. Additive genetic variance was a significant source of variability for most in vitro traits evaluated, including rapid callus growth, colony vascularization, root initiation, chlorophyll production, and somatic embryogenesis. These traits were highly heritable. It was concluded that they should respond to breeding and selection toward the development of superior populations for use in tissue culture studies. The genetic variance for dominance was significant for only a few in vitro traits. In some cases, genetic variance for dominance was expressed on only one of the two regeneration media compared (Beach and Smith, 1979; Phillips and Collins, 1980). Maternal and cytoplasmic factors were important primarily during early subcultures. Additive genetic variance was highly correlated on the two regeneration media for 12 traits including somatic embryogenesis. It was concluded that common genes account for correlated performance on both regeneration media.

Other in vitro methods have been developed for red clover to use in special situations. Phillips and Collins (1979b) developed procedures for shoot meristem-tip culture to eliminate viruses from the 10 parent clones of the synthetically derived Kenstar cultivar. The meristematic dome of the first foliar primordium, measuring 0.1-0.4 mm in diameter, were excised from crown buds. Explants were placed on L2 basal medium with six combinations of PIC and BA. A combination of 16 nM PIC and 4.4 µM BA produced shoot growth and development. No growth was obtained in the absence of PIC. Higher concentrations of PIC yielded callus. Little growth was obtained in the absence of BA. Fewer shoots were obtained on 0.44 µM BA than on the higher concentration of BA. About 65% of the smaller explants yielded shoots on the optimal medium tested, while 80% of the larger explants produced shoots. However, 85% of the shoots derived from the smaller explants appeared symptom-free whereas only 50% of the shoots derived from the larger explants appeared symptom-free. Shoots were rooted on half-strength L2 basal medium containing full-strength KH_2PO_4 and iron, 8.5

μM nicotinic acid, 24 μM 3-aminopyridine, and 1.2 μM IAA. *Chenopodium amaranticolor* Coste & Reyn. plants were used as virus bioassay hosts. All parent clones were infected with one or more viruses, according to bioassays, while meristem-tip culture eliminated viruses from at least two-thirds of the plants.

Subsequent studies indicated that the specific concentrations of PIC and BA required varied with genotype (Collins and Phillips, 1982). The combination considered optimal for most genotypes was 12 nM PIC and 2.2 μM BA. This modification of the L2 medium was found to be suitable for rapid clonal propagation as well as virus elimination. Most genotypes produced shoots with no callus on this medium. Hundreds of shoots were produced within a few months using this medium. All plants generated by this procedure appeared comparable to the parent clones.

Cheyne and Dale (1980) cultured shoot tips measuring 0.2-0.7 mm in length from one diploid cultivar and one tetraploid cultivar. B5 or Blaydes basal media supplemented with 1.2 μM IAA and 0.9 μM 2iP yielded about 67% plant regeneration frequencies. However, some callus was produced at the base of the shoots. Although the shoot meristems were not imbedded in callus, the presence of callus in long-term propagation cultures is undesirable because of the greater genetic instability generally associated with callus cells. Media containing 2,4-D and KIN, or NAA, produced even more callus. Some shoot tip cultures were placed in cold storage (2-6 C) for 15-18 months which successfully regenerated plants when returned to standard conditions. Better than 80% of the plantlets survived transplantation to soil. No abnormalities were observed among more than 200 plants.

Skucinska and Miszke (1980) developed methods for clonal propagation from immature inflorescences. Contamination rates exceeded 70%. About 10% of the contaminant-free inflorescences regenerated shoots on MS medium containing 0.5 μM IAA and 22 μM BA. The coefficient of shoot number increase was about 2-4 per month. Rooting was achieved on medium free of growth regulators. Rooting was inhibited on medium containing 30 μM IAA. About 1000 plants were obtained in 8 months. All plants appeared to be normal. This method of propagation is comparable in plant production to shoot tip propagation. However, infection rates of these cultures were very high and the proportion of explants responding was low. The use of shoot tips has greater appeal for maintaining genetic integrity during clonal propagation since the organized shoot is already present, whereas the state of differentiation of inflorescences must be altered before shoots can be obtained.

Embryo culture techniques have received attention among red clover breeders, particularly as a means for obtaining new interspecific hybrids. Keim (1953) cultured immature embryos (8 days postpollination) of red clover on a simple nutrient agar medium (Randolph and Cox, 1943). Mature, normal plants were obtained. Reciprocal crosses between tetraploid red clover and zigzag clover (*T. medium* L.) failed to produce embryos for culture. Reciprocal crosses between diploid red clover and *T. alpestre* L. yielded only one embryo when *T. alpestre* was used as the female parent. The embryo was already degenerated and did not respond in culture.

Evans (1962) used similar cultural procedures as above in an attempt to rescue hybrid embryos of red clover crossed with zigzag clover, *T. rubens* L., *T. alpestre*, crimson clover (*T. incarnatum* L.), *T. lupinaster* L., and alsike clover (*T. hybridum* L.). In some cases both diploid and tetraploid forms of the parents were used. Embryos were available 8-15 days after pollination in the crosses of red clover with zigzag clover, *T. rubens*, *T. alpestre*, and alsike clover. Embryos developed abnormally on the culture medium and no hybrid plants were rescued. It was concluded that the cause of species cross-incompatibility was due to abnormal development of the endosperm and its inability to nourish the developing embryos. It was speculated that modification of the culture medium may improve embryo development.

Phillips et al. (1982) successfully rescued immature hybrid embryos of diploid red clover crossed with *T. sarosiense* Hazsl. (2n = 48) by in vitro culture. Embryos were excised, 12-19 days after pollination, from the female parent, *T. sarosiense*. The heart-staged hybrid embryos obtained were cultured for 8-14 days on L2 basal medium containing 0.365 M sucrose, 25 nM PIC, and 15 μM ADE. Other concentrations of PIC and sucrose or combinations with BA were suboptimal for embryo survival. Viable embryos were transferred to L2 basal medium containing the standard concentration of sucrose (73 mM), 4 nM PIC, and 0.66 μM BA where shoots developed directly from the embryos. Other concentrations of PIC and BA or combinations with ADE were suboptimal. Higher concentrations of PIC or BA promoted the loss of organization as evidenced by callus formation. Some embryos produced only callus even on the optimal medium. Plants were obtained from some callus by culture on the medium used to induce somatic embryogenesis, described in a previous paragraph. The medium used to promote shoot development directly from hybrid embryos was also suitable for promoting shoot development from somatic embryos. All hybrid shoots were rapidly propagated using the medium described earlier for propagation from shoot tips, with a 4- to 10-fold increase per month in shoot number. About 10% of the hybrid embryos were successfully rescued using these procedures.

No viable hybrid seeds were obtained from undisturbed interspecific crosses, confirming that in situ embryo abortion was the cause of reproductive isolation. Hybrid plants exhibited the leaf-mark trait inherited from the male parent, red clover, which was lacking in the female parent. Hybrid plants possessed the expected F_1 chromosome number of 31. Hybrid plants were essentially sterile due to early arrest of meiosis. Backcrosses of hybrid plants to the male parent produced a very few globular- and heart-shaped embryos. It was suggested that embryo rescue procedures could be used to advance the backcross generations.

Collins et al. (1981) used the same procedures for rescuing presumed hybrid embryos from crosses of red clover with zigzag clover and *T. alpestre*. Verifications of hybrid status have yet to be performed in these cases. These results, however, indicate the potential for using improved embryo rescue procedures for generating previously unavailable interspecific hybrids of red clover.

A summary of the key contributions toward the development of in vitro culture methods for red clover is presented in Table 2.

WHITE CLOVER. Pelletier and Pelletier (1971) cultured cotyledons from an unspecified number of seedlings from one white clover cultivar. Callus was grown on a medium composed of MS major salts, Heller minor salts (given in Gautheret, 1955), B vitamins, agar, 0.17 M glucose, 0.9 µM 2,4-D, and 0.5 µM KIN. Callus was transferred to medium containing 0.5 µM NAA, 0.5 µM KIN, and 10% CW after 6 weeks. Callus derived from a single cotyledon exhibited organization. This callus was subsequently propagated on the same medium with and without CW. Numerous plantlets were obtained and rooted on a dilute nutrient medium containing 0.56 M glucose and 50 nM NAA. Chromosome numbers were evaluated in 42 regenerated plants. Several plants were cytochimeric. Only 4 plants possessed the normal complement of 32 chromosomes; 4 plants, 30 chromosomes; and 2 plants, 62 chromosomes. Many plants possessed some proportion of unifoliate rather than trifoliate leaves, regardless of chromosome number. Other morphological alterations were observed, particularly among aneuploids. Questions have been raised about the origin of aneuploid and polyploid tissues in this case (Gresshoff, 1980).

Oswald et al. (1977a) cultured 2-day-old seedlings of one cultivar on modified Miller basal medium containing 2 µM vitamin E, 20 µM 2,4,5-T, and 0.5 µM KIN for the induction of callus. Callus was used to generate cell suspension cultures on a liquid version of this medium containing 2.3 µM 2,4-D and 0.5 µM KIN. Callus was recovered from cell suspensions using a solidified version of the last medium. Callus obtained in this manner was cultured on medium containing 2 µM 2,4-D and 4.6 µM KIN but lacking vitamin E. Of 50 cultures, 37 developed buds which produced shoots with roots within 3-4 weeks. Since entire plantlets were regenerated, it could be speculated that the mode of regeneration was by somatic embryogenesis. During these investigations, it was noted that four times the normal concentration of chelated iron favored callus growth, cell suspension growth was favored on twice the normal concentration of iron, and plant regeneration was favored on the normal concentration of iron. The presence of vitamin E promoted cell dispersion in cell suspension cultures. Plants were not transferred to soil.

Rupert and Seo (1977) obtained callus from hybrid embryos of kura clover (*T. ambiguum* Bieb.) by white clover crosses. Ovules were dissected from the female parent 7 days after pollination and cultured on MS medium containing 9 µM 2,4-D. After 10 days, the enlarged embryos were dissected from the ovules and cultured on the same medium to produce callus. Callus cultures regenerated plants for at least 3 years (Rupert et al., 1979a), presumably by culture on medium containing 1 µM NAA and 21 µM 2iP (Rupert et al., 1979b).

Parrott and Collins (1982) have successfully obtained callus and cell suspension cultures from white clover using the methods and media developed by Phillips and Collins (1979a, 1980) for red clover. Callus

Table 2. Key Contributions toward the Development of in vitro Cultural Techniques for Red Clover

CONTRIBUTION	BASAL MEDIUM	GROWTH REGULATORS (μM)	MATERIALS	REFERENCES
Development of L2 basal medium, callus proliferation	L2	0.25 PIC, 0.44 BA	Mature or immature, vegetative or reproductive	Phillips & Collins, 1979a
Cultivar variation for plant regeneration from callus	L2	Various combinations	Seedling sections, anthers, petiole sections	Phillips & Collins, 1979a
Plant regeneration from young callus by organogenesis	B5 or SH including 2x thiamine	Transfer from 10 2,4-D, 10 NAA, 10 KIN, to 10 NAA, 15 ADE for shoots, then transfer to 1 NAA for roots	Seedling hypocotyls, excised pistils	Beach & Smith, 1979
Virus elimination, rapid clonal propagation	L2	12 nM PIC, 2.2 BA	Shoot meristem tips, crown buds	Phillips & Collins, 1979b; Collins & Phillips, 1982
Rooting of shoots	RL (modified L2)	1.2 IAA	Shoot tips	Phillips & Collins, 1979b
Cell suspension culture	SL2 (modified L2)	0.25 PIC, 0.44 BA	Seedling sections, callus	Phillips & Collins, 1980
Induction of somatic embryogenesis in long-term cultures	L2	4.5 nM 2,4-D, 15 ADE	Cell suspensions, callus	Phillips & Collins, 1980

CONTRIBUTION	BASAL MEDIUM	GROWTH REGULATORS (μM)	MATERIALS	REFERENCES
Cold storage, rapid clonal propagation (diploid and tetraploid)	B5 or Blaydes	1.2 IAA, 0.9 2iP	Shoot tips	Cheyne & Dale, 1980
Rapid clonal propagation	MS	0.5 IAA, 22 BA	Inflorescences	Skucinska & Miszke, 1980
Correlation of additive genetic variance, genotype influence on plant regeneration	L2 and B5	Following Beach and Smith, 1979, and Phillips and Collins, 1980	Seedling sections, callus	Keyes et al., 1980
Immature embryo rescue	L2 with 0.365 M sucrose	25 nM PIC, 15 ADE	Heart-staged hybrid embryos	Phillips et al., 1982
Shoot development from embryos	L2	4 nM PIC, 0.66 BA	Hybrid embryos, somatic embryos	Phillips et al., 1982
Cellular selection suitable for regeneration of variant plants	SL2 with 1.8 μM phosphorus	0.25 PIC, 0.44 BA	Cell suspensions	Phillips & Collins, unpublished

derived from 20% of the tested genotypes produced roots. Cell suspensions grew more slowly than those of red clover and the cells tended to remain in larger aggregates. Callus recovered from cell suspensions grew more vigorously than the parent callus. This callus developed buds but no plants were regenerated. These results indicate the potential for modifying procedures developed for red clover to culture white clover.

Gresshoff (1980) cultured seeds and seedling sections of one cultivar on B5 or MS basal media supplemented with 0.5 μM KIN and either 4 or 10 μM 2,4-D. Callus induction was complete on the higher concentration of 2,4-D. The presence or absence of 0.1% CH was not critical. Some rhizogenesis was observed in callus during the first 2 months of growth. Cell suspension cultures were initiated from callus and grown on liquid B5 medium containing either 0.25 μM KIN, 2 μM 2,4-D, and 0.1% CH, or 50 nM KIN, 4 μM 2,4-D, and 0.05% CH. Callus was recovered from cell suspensions on the medium used for callus induction. Protoplasts were isolated from cell suspensions using 0.25 M mannitol, 0.25 M sorbitol, 2 mM $CaCl_2$, 5 mM KH_2PO_4, 2% cellulysin, 2% R10 cellulase, and 1% driselase, pH 5.8 to 6.0. Callus was obtained from protoplasts cultured on the callus medium.

Callus from all sources, up to 2 years in culture, produced buds on MS medium containing 2 μM 2iP and 0.5 μM IAA. Shoot development and multiplication occurred when budding callus was transferred to half-strength MS medium containing 2 μM 2iP. Rhizogenesis was induced from shoots, buds, and callus when cultured on MS medium containing 1 μM 2iP and 15 μM IAA. Plantlet hardening was obtained on half-strength B5 medium containing 0.1 μM 2iP and 0.5 μM IAA. Most plants subsequently survived transplantation to soil. Some plants exhibited variation in leaflet number (one, two, and three leaflets) but 13 plants evaluated cytologically possessed the normal chromosome number of 32.

Oswald et al. (1977b) studied herbicide tolerance in cell suspension cultures. The same materials and cultural procedures were used by these authors as described in a previous paragraph. Three phenoxy analogs were evaluated at concentrations of 18 μM 2,4-D, 38 μM 2,4,5-T, and 8 μM 4-(2,4-dichlorophenoxy)butyric acid (2,4-DB). At these concentrations, the herbicidal activity of the three analogs was comparable. Each analog was used for 5-day pretreatments in cell suspension cultures. Pretreated and control cultures were then evaluated for tolerance to all three analogs. Pretreated cells exhibited significantly less stress than control cells when exposed to the analogs. Pretreatment by each analog resulted in tolerance to treatment by all three analogs. The presence of vitamin E in the culture medium enhanced selection efficiency, particularly to 2,4-DB. Cells selected with 2,4-DB as the pretreatment pressure were more tolerant to all three analogs than were cells pretreated with 2,4-D or 2,4,5-T.

Variation in cultures was due to herbicide sensitivity during the initial exposure. The stress conditions exhibited by surviving cells became less pronounced during subsequent incubation, resulting in stabilized cultures. Stabilized cultures grew under treatments which were herbicidal to control cells, indicating that adaptation had occurred.

No data was presented on plants regenerated from adapted cell populations.

Barnett et al. (1975) developed stolon meristem-tip culture procedures for elimination of viruses from the six parent clones of Tillman white clover. Stolon tips, including one or two foliar primordia, were cultured on MS basal medium containing 0.15 M sucrose, 0.5% agar, and 0.06 nM IAA. About 9% of the stolon tips survived culture and grew to maturity. Nearly 60% of these plants were free of all viruses as determined by extensive host plant bioassays, whereas all parent clones contained at least one virus. This procedure was more efficient than the use of heat treatments for the elimination of viruses.

Cheyne and Dale (1980) cultured shoot tips from two cultivars of white clover using the same procedures as they used for clonal propagation and cold storage of red clover. Blaydes basal medium was superior to B5 medium. Plantlet yield was about 80% with some callus forming at the base of the shoots. Better than 90% of the plants survived transplantation to soil.

Richards and Rupert (1980) found that a modified MS medium containing 2.9 µM IAA and 4.5 µM 2iP encouraged shoot multiplication from cultured embryos. Some callus was produced. This procedure is comparable to that described in the previous paragraph.

Parrott and Collins (1982) used the procedures developed for red clover by Phillips and Collins (1979b) and Collins and Phillips (1982) to culture shoot tips from white clover seedlings. Rapid shoot multiplication was obtained with little or no callus.

Keim (1953) cultured immature embryos of white clover and successfully obtained mature plants using the same procedures as for red clover. At 15 days post-pollination embryos were excised and cultured from white clover by *T. nigrescens* Viv. crosses. Mature hybrid plants were obtained. This was not a new interspecific combination, but reciprocal hybrids were obtained whereas the original hybrid was produced only when white clover was used as the female parent.

Evans (1962) cultured immature embryos, excised 12-25 days after pollination, from crosses of white clover with *T. nigrescens*, alsike clover, *T. uniflorum* L., berseem clover (*T. alexandrinum* L.), subterranean clover (*T. subterraneum* L.), and *T. arvense* L. using the same procedures as described for red clover. Hybrid plants were grown from crosses of white clover with *T. nigrescens* and *T. uniflorum*. Embryos were recovered and cultured from the other crosses but in these cases were either abnormal, undifferentiated, or exhibited no growth. Some of the interspecific hybrids obtained represented new ploidy and reciprocal combinations.

Williams (1978) utilized transplanted nurse endosperm to culture heart-staged embryos, excised 14-16 days after pollination, from white clover by tetraploid kura clover crosses. The culture medium used contained major salts modified from White (1963), minor salts modified from MS, and organic composition modified from Kanta and Maheshwari (1963). No growth regulators were used. Nurse endosperm was dissected 8-9 days after pollination from normal tetraploid kura clover crosses. The normal kura clover embryo was removed and the hybrid embryo was placed in the endosperm which was then cultured on a filter-paper disk on the solidified culture medium. About 10% of the available hybrid embryos grew in culture to produce plantlets after 4 to 6 months. Hybridity was verified on the basis of unique chromosomes possessing satellites, morphological intermediacy, and electrophoretic banding patterns. This hybrid combination had not been previously obtained.

The use of the nurse endosperm technique to rescue interspecific hybrid embryos was reviewed by Williams and DeLautour (1980). The hybrid of kura clover and white clover was also obtained using normal white clover endosperm. Mature hybrid plants from white clover crossed with *T. uniflorum* have been produced. This hybrid is not a new interspecific combination and white clover endosperm was used in this case. The same culture medium as above or one containing organic compounds modified from Bourgin and Nitsch (1967) was used successfully. In the latter case, the medium contained 0.5 μM NAA or IAA. Plant production was more efficient using this medium. The media formulated by Gresshoff (1980) were less efficient but still yielded plants. The medium used was as important as the source of endosperm, which suggests that normal endosperm cannot provide all the necessary growth factors for the rescue of hybrid embryos.

Rupert and Evans (1980) obtained several interspecific hybrids utilizing defined culture media for embryo rescue. New plants were generated of the previously reported hybrid of white clover crossed with kura clover, *T. nigrescens*, *T. uniflorum*, and *T. isthomocarpum* Brot. In the latter case, the hybrid was originally obtained unilaterally while embryo culture rescued the reciprocal crosses. Apparently the same culture medium was used as described earlier (Rupert et al., 1979b). Embryos were cultured but not rescued from *T. michelianum* L. by white clover crosses.

Richards and Rupert (1980) developed procedures for in vitro pollination and culture of ovularies. Calyx lobes and pedicels had to be attached to the ovularies to obtain fertilization and embryo development in white clover. In kura clover by white clover crosses, embryological development in cultured ovularies with accessory floral parts closely paralleled in situ development. The presence of growth regulators in a modified MS medium had some effect but did not improve embryo development greatly.

A summary of the key developments in the in vitro culture of white clover is given in Table 3.

OTHER CLOVERS. Graham (1968) cultured seeds of subterranean clover on modified MS medium (Linsmaier and Skoog, 1965, supplemented with 4 μM nicotinic acid) containing 0.2 μM KIN and 1 μM 2,4-D to induce callus formation. Callus was maintained on the same medium. Cell suspension cultures were grown on a liquid version of the same medium. Cultures were inoculated with *Rhizobium trifolii*, which did not interfere with growth. No morphogenetic development was observed.

Schenk and Hildebrandt (1972) obtained callus growth from seedlings of alsike clover cultured on SH medium containing 2.2 μM 2,4-D, 11 μM CPA, and 0.5 μM KIN. No morphogenetic development was reported.

Table 3. Key Contributions in the Development of In Vitro Cultural Procedures for White Clover

CONTRIBUTION	BASAL MEDIUM	GROWTH REGULATORS (μM)	REFERENCES
Hybrid embryo rescue	Randolph and Cox, 1943	None	Keim, 1953; Evans, 1962
Virus elimination from stolon tips	MS	0.06 nM IAA	Barnett et al., 1975
Plant regeneration from callus and cell suspension cultures	Modified Miller (1961)	Transfer from 20 2,4,5-T, 0.5 KIN, to 2 2,4-D, 0.5 KIN, then transfer to 2 2,4-D, 4.6 KIN	Oswald et al., 1977a
Plant regeneration from hybrid embryo-derived callus	MS	Transfer from 9 2,4-D to 1 NAA, 21 2iP	Rupert & Seo, 1977; Rupert et al., 1979b
Hybrid embryo rescue using nurse endosperm	Modified from WH, MS, and Kanta and Maheshwari, 1963	None	Williams, 1979
Plant regeneration from long-term callus, cell suspension, and protoplast cultures	MS or B5	Transfer from 10 2,4-D, 0.5 KIN to 0.5 IAA, 2 2iP for buds; transfer to half-strength MS with 2 2iP for shoots; then transfer to MS with 15 IAA, 1 2iP for roots	Gresshoff, 1980

Table 3. Cont.

CONTRIBUTION	BASAL MEDIUM	GROWTH REGULATORS (μM)	REFERENCES
Cold storage, rapid clonal propagation	Blaydes (1966)	1.2 IAA, 0.9 2iP	Cheyne & Dale, 1980
Hybrid embryo rescue	(MS)[a]	(1 NAA, 21 2iP)[a]	Rupert & Evans, 1980
Ovulary culture with accessory floral parts	Modified MS	None or 5.7 IAA, 11 BA, 1.5 GA	Richards & Rupert, 1980
Rapid clonal propagation	L2	12 nM PIC, 2.2 BA	Parrott & Collins, 1982; Collins & Phillips, 1982

[a]Not specified; implied from previous reports.

Mokhtarzadeh and Constantin (1978) cultured seedling hypocotyl sections of berseem clover on MS medium containing various combinations of IAA, NAA, 2,4-D, 2iP, BA, and KIN. Optimal callus induction occurred on a combination of 5.5 μM NAA and 7.5 μM KIN. Berseem clover was very sensitive to 2,4-D. Callus was subsequently propagated on medium containing 11 μM NAA and 0.45 μM 2iP. Cell suspension cultures were grown from callus inoculated in liquid medium containing 11 μM NAA and 0.9 μM 2iP. Callus colonies were recovered from cell suspensions inoculated on a solidified version of the same medium. Immature anthers excised from one plant formed callus when cultured on medium containing 5.5 μM NAA, 0.45 μM 2,4-D, and 45 nM 2iP. Callus of all sources produced shoots when cultured on medium containing 2.7 μM NAA and 2.5 μM KIN. Certain other combinations of growth regulators yielded shoots less efficiently. Shoots were rooted on medium containing 5.7 μM IAA and 0.44 μM BA. Plants were established in the greenhouse. Cytological evaluation indicated that hypocotyl-derived callus and cell suspensions yielded diploid plants. Preliminary evaluation indicated that anther-derived plants were haploid. This has not been verified or reproduced.

Beach and Smith (1979) obtained callus from crimson clover in the same manner as from red clover. Regeneration of plants from callus was achieved in the same manner as for red clover except that transfer to rooting medium was not necessary. The recovery of entire plantlets suggests that crimson clover may regenerate by somatic embryogenesis.

Parrott and Collins (1982) cultured large hop clover (*T. campestre* Schreb.), subterranean clover, crimson clover, *T. alpestre*, zigzag clover, and *T. rubens* following procedures developed by Phillips and Collins (1979a,b, 1980) and Collins and Phillips (1982) for red clover. Large hop clover failed to respond to shoot tip culture while rapid clonal propagation was achieved with *T. alpestre*, crimson clover, *T. rubens*, and subterranean clover. The latter species multiplied most rapidly in shoot tip culture. Zigzag clover produced callus as well as shoots on the standard medium, which was corrected by using one-third the normal concentration of PIC as auxin. Subterranean clover rooted prolifically on the standard rooting medium. Rooting frequencies of the other species were more efficient using only water.

Callus of subterranean clover, large hop clover, and *T. alpestre* did not grow well, exhibiting varying degrees of necrosis. The L2 medium was suboptimal for these species. About 20% of the *T. alpestre* genotypes rooted before the callus turned brown. All callus of subterranean clover rooted. One callus sector of large hop clover survived and proliferated slowly. Zigzag clover and *T. rubens* produced callus which grew more vigorously than that of red clover, while callus of crimson clover grew more slowly. Some crimson clover callus developed buds and roots. Rapidly growing cell suspensions were initiated from callus of *T. alpestre*, crimson clover, zigzag clover, and *T. rubens*. Callus colonies were recovered from cell suspensions. *T. alpestre* and zigzag clover callus developed buds but did not regenerate plants. Cell suspension-derived callus of *T. rubens* regenerated plants by somatic embryogenesis, as evidenced by the developmental pattern of regeneration. One-half of the *T. rubens* genotypes tested regenerated plants. Plants were transplanted to soil. It was suggested that manipulation of the growth regulators may result in plant regeneration from the species which developed buds, namely zigzag clover, *T. alpestre*, and crimson clover. These results indicate the potential for modifying the procedures developed for red clover to culture other clover species.

Keim (1953) excised and cultured immature embryos 15 days after pollination from crosses of kura clover with alsike clover and *T. nigrescens*. The former hybrid was successfully rescued for the first time. Two ploidy combinations of this hybrid were obtained using kura clover as the female parent. Only one of these survived in the greenhouse and it did not flower. Hybridity was verified by chromosome number.

Evans (1962) obtained the hybrid of kura clover and alsike clover by culture of embryos excised 12-15 days after pollination. Four ploidy combinations and all reciprocals were rescued. Only one of the eight combinations flowered and it was fertile.

The same interspecific hybrid was obtained using the transplanted nurse endosperm technique (Williams, 1980; Williams and DeLautour, 1980). A new ploidy combination was obtained, but not the reciprocal. Kura clover, the female parent, was used to provide the nurse endosperm. These hybrid plants flowered.

Rupert and Seo (1977) obtained callus from immature hybrid embryos of kura clover crossed with alsike clover and *T. montanum* L. Plants were regenerated from the former hybrid and from pentaploid and triploid kura clover (Rupert et al., 1979a,b). Richards and Rupert (1980) investigated these materials in ovulary culture. Rupert and Evans (1980) obtained the hybrid of kura clover and alsike clover by embryo culture. The hybrid of alsike clover crossed to *T. michelianum* was not rescued although embryos were cultured. The hybrids of kura clover crossed with *T. montanum* and *T. occidentale* Coombe were successfully rescued for the first time.

Table 4 presents a summary of the cultural procedures used with the other clovers.

BIRDSFOOT TREFOIL. Niizeki and Grant (1971) cultured anthers from two cultivars and two other lines of birdsfoot trefoil and a related species, *L. caucasicus* Kupr. The basal media of Miller (1961), Bourgin and Nitsch (1967), and MS (Murashige and Skoog, 1962) were used with various combinations of IAA, 2,4-D, KIN, and BA. The most consistent callus formation occurred on Miller basal medium containing IAA and either KIN or BA. Shoot regeneration proceeded most consistently on 8.8 or 23 µM IAA with 6.6 or 17.6 µM BA. Root formation from callus resulted only when IAA was used in combination with KIN. Shoots were rooted optimally on medium free of growth regulators. Microscopic evaluation indicated that callus arose from somatic tissues. About 5% of the regenerated plants possessed doubled chromosome numbers while the remainder possessed the somatic chromosome number. Considerable variation in chromosome number was observed in the callus. Isolated microspores in culture failed to give rise to callus or plants, although some cell division was observed.

Table 4. Key Developments in the In Vitro Cultural Procedures for Other Clovers

SPECIES	DEVELOPMENT	BASAL MEDIUM	GROWTH REGULATORS (μM)	REFERENCE
T. alexandrinum	Plant regeneration from callus and cell suspension cultures	MS	Transfer from 11 NAA, 0.9 2iP to 2.7 NAA, 2.5 KIN	Mokhtarzadeh & Constantin, 1978
T. incarnatum	Plant regeneration from callus	B5 or SH	Transfer from 10 2,4-D, 10 NAA, 10 KIN to 10 NAA, 15 ADE	Beach & Smith, 1979
T. incarnatum, *T. subterraneum*, *T. medium*, *T. alpestre*, *T. rubens*	Rapid clonal propagation from shoot tips	Following Collins & Phillips, 1980		Parrot & Collins, 1982
T. incarnatum, *T. medium*, *T. alpestre*, *T. rubens*	Callus and cell suspension cultures	Following Phillips & Collins, 1979a, 1980		Parrot & Collins, 1982
T. rubens	Plant regeneration from callus and cell suspension cultures	Following Phillips & Collins, 1980		Parrott & Collins, 1982
T. ambiguum x *T. hybridum*	Hybrid embryo rescue	See Table 3	None	Keim, 1953; Evans, 1962; Williams, 1980

Table 4. Cont.

SPECIES	DEVELOPMENT	BASAL MEDIUM	GROWTH REGULATORS (μM)	REFERENCE
T. ambiguum, *T. ambiguum* x *T. hybridum*	Plant regeneration from embryo-derived callus	MS	Transfer from 0.9 2,4-D to 1 NAA, 21 2iP	Rupert et al., 1979a,b
T. ambiguum x *T. hybridum*, *T. ambiguum* x *T. montanum*, *T. ambiguum* x *T. occidentale*	Hybrid embryo rescue	(MS)[a]	(1 NAA, 21 2iP)[a]	Rupert & Evans, 1980

[a]Not specified; implied from previous reports.

Tomes (1976) cultured seedling hypocotyls from five cultivars and one germplasm release on a medium containing cytokinin and auxin. After 4-6 weeks, callus was transferred to medium containing cytokinin for one or two passages. Cultivar variation for morphogenesis was observed, in that plants regenerated from 60 to 98% of the genotypes from each cultivar. Callus derived from anthers, ovaries, and stems responded similarly.

Swanson and Tomes (1980a) established callus cultures from several genotypes of cv. Leo on B5 medium containing $CaCl_2$ and $MgSO_4$ concentrations as in MS medium and 4.5 µM 2,4-D. Fast- and slow-callus-growth types were selected and bulked from among nine genotypes. Unselected control callus was bulked from 15 randomly selected genotypes. Callus was subcultured onto media containing 4.5 µM or 0.18 mM 2,4-D and subcultured at varying intervals. Callus was used to generate cell suspension cultures grown on liquid versions of the medium containing 4.5 or 67.5 µM 2,4-D. Callus colonies were recovered from cell suspensions using the medium designed for callus induction. Shoot regeneration from all callus sources occurred on medium containing 0.44 µM BA. Shoots were rooted on medium free of growth regulators. Callus selected for fast growth was more homogeneous and friable than that selected for slow growth. Unselected control callus was intermediate in growth rate and morphology. Control callus regenerated shoots more efficiently and roots less efficiently than callus selected for fast growth.

Callus selected for fast growth was more tolerant to 0.18 mM 2,4-D than was the control callus after seven passages of 3-week duration. The mechanism for tolerance to 2,4-D appeared to be inducible, stable, and distinct from that reported in a previous in vivo study. Plants were regenerated from 2,4-D tolerant callus and from cell suspensions grown on both standard and elevated concentrations of 2,4-D. Regenerated plants were increased using meristem culture (to be described in a subsequent paragraph).

Swanson and Tomes (1980b) continued these studies with callus established as before from cv. Leo and a line found to be 2,4-D-tolerant in the field (T-68). Shoot and root regeneration were tested in 1-year-old callus on media free of growth regulators, with 0.22 µM BA alone, and BA in combination with 0.45 or 0.23 µM 2,4-D, 0.29 µM IAA, or 0.27 µM NAA. The two callus lines responded differently. The best shoot and root production, considering both callus lines, occurred with no growth regulators or BA alone. T-68 was slower to respond than Leo, but additional transfers to regeneration medium resulted in equivalent regeneration frequencies. No differentiation occurred on medium containing 4.5 µM 2,4-D while responses on lower concentrations of 2,4-D were intermediate compared to the absence of growth regulators. T-68 was less sensitive to reduced 2,4-D concentrations than was Leo.

Both callus lines were exposed to progressively higher concentrations of 2,4-D, up to 0.18 mM. The differential superior tolerance of T-68 occurred only during very early subcultures and at lower concentrations of 2,4-D. By the third transfer, T-68 did not exhibit superior 2,4-D tolerance when compared to Leo. Growth of both callus lines was suppressed on the highest concentration of 2,4-D after 1 month exposure, with no difference in the response of the two lines. Buds formed during this treatment which produced plants when transferred to medium containing 0.22 μM BA. Plants regenerated from Leo callus selected on high concentrations of 2,4-D possessed increased tolerance compared to Leo controls, but did not approach the tolerance exhibited by T-68 controls.

A positive relationship between the in vivo and in vitro responses of T-68 to 2,4-D was noted. It was suggested that tolerance in T-68 may be based on a mechanism for 2,4-D conjugation. While the presence of 90 μM to 0.18 mM 2,4-D was effective as a selection pressure for isolating in vitro 2,4-D tolerance in Leo, indications were that a selection pressure of about 65 μM 2,4-D may be more effective for isolating 2,4-D tolerance comparable to T-68 controls. Leo plants with improved 2,4-D tolerance were intercrossed to form a population for use in further studies.

Tomes (1979) cultured shoot meristem tips (about 0.3 mm in diameter) and stem nodes (5 mm in length) from 100 genotypes of Leo on B5 medium containing 0.22 μM BA and on medium free of growth regulators. Shoots proliferated more rapidly from node cultures than from meristem tips but shoot tips were easier to disinfect. Shoots proliferated about five times faster on medium containing BA. Callus formation was rare. Some shoot cultures were stored at 2-5 C for 1 month with no significant change in survival rate or mean shoot height. It was suggested that long-term cold storage would be feasible. Shoots were rooted on medium free of growth regulators. Up to 10-fold increases in shoot number were obtained monthly.

Bent (1962), Grant et al. (1962), DeNettancourt and Grant (1964), Phillips and Keim (1968), and Somaroo and Grant (1971) used embryo culture techniques with trefoils similar to those used by Keim (1953) and Evans (1962) with clovers. Hybrids of birdsfoot trefoil with narrowleaf trefoil (*L. tenuis* Waldst. & Kit.) and big trefoil (*L. pedunculatus* Cav.) were obtained. These were not new interspecific hybrids except in the latter case where a new reciprocal hybrid was obtained. Several hybrids were rescued among species related to birdsfoot trefoil, including narrowleaf trefoil, big trefoil, *L. alpinus* Schleich., *L. filicaulis* Dur., *L. frondosus* Freyn, *L. japonicus* (Reg.) Lars., *L. schoelleri* Schweinf., and *L. krylovii* Schischk. & Serg. Generally only embryos at or beyond the torpedo stage were amenable to culture. Many of these hybrids have been obtained without the aid of embryo culture, but the efficiency of hybrid production was sometimes improved using this technique.

DeLautour et al. (1978) used a modification of the nurse endosperm technique to rescue the hybrid of tetraploid narrowleaf trefoil with birdsfoot trefoil. Endosperm transplants were not required since the embryos were fully mature. Hybridity was verified by inheritance of a biochemical trait from the male parent. The hybrid was partially fertile. The reciprocal cross did not provide embryos for culture. Another hybrid, between narrowleaf trefoil and big trefoil, was rescued from heart- and torpedo-shaped embryos utilizing nurse endosperm. This hybrid was produced more efficiently than it was by Bent (1962).

A summary of the cultural procedures developed for birdsfoot trefoil is provided in Table 5.

Table 5. Key Developments in the In Vitro Cultural Procedures Used for Birdsfoot Trefoil

CONTRIBUTION	BASAL MEDIUM	GROWTH REGULATORS (μM)	REFERENCES
Plant regeneration from anther-derived callus	Miller (1961)	8.7 or 23 IAA, 6.6 or 17.6 BA	Niizeki & Grant, 1971
Plant regeneration from callus derived from seedling hypocotyls, anthers, ovaries and stems; cultivar variation for plant regeneration	Not given	Transfer from auxin and cytokinin (not specified) to cytokinin only	Tomes, 1976
Rapid clonal propagation from shoot meristem tips and stem nodes, cold storage	B5	0.22 BA	Tomes, 1979
Plant regeneration from long-term callus and cell suspension cultures, selection for improved 2,4-D tolerance	Modified B5	Transfer from 4.5 2,4-D (or higher concentrations) to 0.22 or 0.44 BA for shoots, then transfer to medium free of growth regulators for roots	Swanson & Tomes, 1980a,b

Current Applications to Crop Improvement

The most immediate application of tissue culture techniques to crop improvement among forage legumes is the use of shoot meristem-tip culture to eliminate viruses from elite clones used to produce synthetic cultivars. The parent clones of Tillman white clover (Barnett et al., 1975) and Kenstar red clover (Phillips and Collins, 1979b) have been freed of infection by this technique. The methods of Cheyne and Dale (1980) may improve the efficiency of virus-free plant production from white clover. The methods of Tomes (1979) are amenable to virus elimination from clones of birdsfoot trefoil. The methods of Phillips and Collins (1979b), and Collins and Phillips (1982), as implemented by Parrott and Collins (1982), make this technique available for subterranean, crimson, zigzag, and certain other clovers. The technique has limited application, since the resultant plants are not immune to virus. Virus elimination may have to be performed again on such materials due to reinfection, but this need should not arise for several additional years. Virus-free stocks are superior for seed production and continued maintenance of desirable synthetic cultivars.

Methods for rapid clonal propagation from shoot tips of diploid red clover were developed by Phillips and Collins (1979b) and Collins and Phillips (1982). Clonal propagation from cultures of immature inflorescences as developed by Skucinska and Miszke (1980) are less efficient to establish but result in a comparable number of plants. The technique is available for white clover (Cheyne and Dale, 1980) and birdsfoot trefoil (Tomes, 1979). The methods developed for diploid red clover were extended to white, subterranean, crimson, zigzag, and certain other clovers (Parrott and Collins, 1982). Larger shoot tips can be used than those required for virus elimination, which renders this application less laborious. Crown buds of red clover, stolon tips of white clover, or stem nodes of birdsfoot trefoil can be used rather than shoot tips. Although clonal propagation is normally achieved among forage legumes by means of cuttings, the tissue culture technique allows for faster rates of multiplication.

Cheyne and Dale (1980) developed methods for long-term cold storage of shoot cultures from red and white clovers. Tomes (1979) demonstrated the potential for long-term cold storage of birdsfoot trefoil shoot cultures. In both cases the standard shoot propagation procedure was used to generate the cultures which were then stored at 2 to 6 C. Indications are that long-term cold storage of shoot propagation cultures can be achieved with all forage legume species for which in vitro clonal propagation methods are available.

Phillips et al. (1982) developed methods for the rescue of heart-staged hybrid embryos of red clover prior to in situ abortion. A series of defined culture media were adjusted for the osmotic sensitivity of immature embryos and to prevent precocious germination of the radicle (first step). Subsequent cultural conditions promoted shoot germination and development (second step). Shoots were then multiplied (third step) and rooted (fourth step). Plant regeneration was also achieved from some embryos which produced only callus. Indications are that these methods will be useful for generating previously unavailable wide interspecific hybrids of red clover (Collins et al., 1981).

These hybrids should be of great importance for improving the longevity of red clover under field conditions. Rescuing the appropriate ploidy combinations and/or using embryo culture to advance backcross generations should result in hybrid-derived populations which possess a useful degree of fertility to facilitate introgression. Other useful traits, such as disease resistance, may be inherited from the long-lived perennial species.

The results obtained in red clover studies indicate that wide hybrids among forage legume species can be obtained using defined cultural media. The progress made by Rupert and Evans (1980) supports this view in studies with white clover and related species. The nurse endosperm technique used by Williams (1978, 1980) and DeLautour et al. (1978) has been useful for producing some hybrids among certain clovers and certain trefoils. However, this technique is more laborious and involves the undefined nutritional support of the endosperm. The comparatively simple media used by Keim (1953), Evans (1962), and others are probably only useful for producing hybrids among closely related species.

Phillips and Collins (1981) have used cellular selection methods to investigate phosphorus efficiency in red clover, the first study of its kind at the cellular level. In this preliminary study, plants derived from a single genotype were regenerated possessing forage concentrations of phosphorus considerably elevated compared to other red clover plants evaluated previously. These plants also possess elevated concentrations of nitrogen. It is yet to be seen whether these plants will continue to express these traits under field conditions and whether the traits will be transmitted to progeny. If subsequent studies are fruitful, such plants will be useful for improving the forage quality and fertilizer use efficiency in red clover. It would be desirable to use the cellular selection method to improve the phosphorus efficiency in several genotypes which then regenerate plants. This would provide a basis for developing populations with sufficient genetic variability among other traits to facilitate the transfer of improved forage quality traits into cultivars.

Swanson and Tomes (1980a,b) used selection methods at the callus tissue and cellular levels to select for improved 2,4-D tolerance in birdsfoot trefoil. A positive relationship was established in one genotype for in vivo and in vitro responses to 2,4-D. Additional plants were regenerated with some degree of improved 2,4-D tolerance. It is hopeful that a combination of in vitro cellular selection and in vivo phenotypic recurrent selection will result in base populations with significantly improved 2,4-D tolerance. New cultivars developed from such populations would greatly benefit the production of birdsfoot trefoil by effectively incorporating broadleaf-weed control and reducing competition during stand establishment.

Key Variables and Factors for Success

Probably the most important factor for successful in vitro culture of forage legumes is the genotype used. Phillips and Collins (1979a) and Keyes et al. (1980) demonstrated that red clover cultivars and genotypes respond differently in culture. Regeneration of red clover plants is very dependent on the genotype in culture (Keyes et al., 1980) and cultural conditions have been optimized for a large number of genotypes (Phillips and Collins, 1979a; Collins and Phillips, 1982).

The observations of Parrott and Collins (1982) indicate that genotypic differences are important in many of the clover species. Cultivar differences for plant regeneration was observed in birdsfoot trefoil (Tomes, 1976). In the case of clovers, regeneration frequencies among populations may be improved by breeding methods as has been done with alfalfa (Keyes et al., 1980). This is important where cellular selection methods are to be applied to cross-pollinating species.

The origin of the tissue used for establishing callus and cell suspension cultures is important (Phillips and Collins, 1979a; Beach and Smith, 1979). Immature (less organizationally differentiated) tissue sources generally respond more rapidly than mature tissue sources. Meristematic tissue sources tend to regenerate plants more efficiently than nonmeristematic tissue sources. Reproductive tissue sources sometimes produce cultures which regenerate plants better than explants from vegetative sources. However, cultures can be successfully established from various tissues (Phillips and Collins, 1979a; Tomes, 1976) and this flexibility can be important in establishing a culture from a unique plant.

Gresshoff (1980) noted that the clovers exhibit a certain degree of plasticity in their response to various cultural conditions. He also pointed out that there are similarities among the cultural conditions used by various workers. It can be argued further that the basal medium used is important for cultural success. In pioneering studies with tobacco (*Nicotiana tabacum* L.), Murashige and Skoog (1962) developed an improved salts medium which was experimentally optimized for callus growth. This work culminated with research by Linsmaier and Skoog (1965) in which it was shown that relatively few vitamins or organic supplements were required with the optimized salts formulation. The implication of this work is that as a greater number of vitamins and organic supplements are required for successful callus culture, the less optimal the salts formulation must be. Other important research along these lines was performed by Miller (1961), who reported an improved basal medium for soybean callus culture; Gamborg et al. (1968), who reported an improved medium for soybean cell culture; and Schenk and Hildebrandt (1972), who developed a medium suitable for culture of callus from a number of dicotyledonous and monocotyledonous species. The media developed by these workers have formed the basis of the cultural conditions used for clovers and trefoils. Many forage legume workers have observed that one or two basal media but not others were suitable for their purposes.

Phillips and Collins (1979a) developed an improved basal medium for the culture of callus from red clover and certain other legumes. This medium supported more rapid growth and was more broadly supportive of a number of red clover genotypes than other basal media tested. The hypothesis of Skoog and co-workers was supported in that relatively few vitamins and organic supplements were required with the optimized salts formulation. The use of an optimized basal medium was important in subsequent research with red clover (Collins and Phillips, 1982), in which cultural conditions were modified for cell suspension culture, meristem-tip culture, and embryo rescue.

Research with other clover species supports the claim that this improved medium and its modifications are generally useful (Parrott and Collins, 1982). Research performed with birdsfoot trefoil proceeded better using a modification of one of the standard basal media (Swanson and Tomes, 1980a,b). However, successful cultures can result from the use of other media. Although it is not clearly substantiated that improved or modified basal media are necessary for success there is a certain appeal in using the best cultural conditions available. This appeal pertains especially to those interested in breeding applications, since those interested in basic physiological or biochemical studies may wish to use standard media or a different approach.

The choice of the appropriate growth regulators at appropriate concentrations for a particular cultural response is critical, as has been clearly shown in the red clover studies (Collins and Phillips, 1982). This is particularly true with plant regeneration. The use of 2,4-D as the auxin is critical for the efficient induction of somatic embryogenesis and shoot buds. Somatic embryogenesis proceeds best when induced on a low concentration of 2,4-D in combination with a weakly active cytokinin, ADE (Phillips and Collins, 1980). Shoot bud induction involves a shift from a high concentration of 2,4-D in combination with NAA and KIN to a medium free of 2,4-D and containing NAA and ADE (Beach and Smith, 1979).

Callus and cell proliferation occur optimally using a relatively high concentration of PIC as the auxin, in combination with cytokinin (Phillips and Collins, 1979a). The maturation of heart-staged sexually derived embryos proceeds best using a moderate concentration of PIC with ADE (Phillips et al., 1982). The cytokinin requirement of immature sexual embryos parallels that of somatic embryos. Shoot development proceeds best from sexually derived and somatic embryos using a low concentration of PIC with a moderate concentration of BA as the cytokinin (Phillips and Collins, 1980; Phillips et al., 1982). Shoots proliferate most rapidly with minimal callus formation using low-moderate concentration of PIC with a high concentration of BA (Phillips and Collins, 1979b; Collins and Phillips, 1982). Root induction on shoots occurs best using IAA as the auxin (Phillips and Collins, 1979b).

The step-by-step approach taken during the development of cultural methods for red clover may account for the progress achieved. The first step was the identification of basic nutritional requirements for in vitro culture of callus (Phillips and Collins, 1979a). The second step was the development of methods for shoot growth and proliferation from meristem tips (Phillips and Collins, 1979b), and the induction of root formation on shoots was the next step (Phillips and Collins, 1979b). Then methods were developed for cell suspension culture (Phillips and Collins, 1980) and for regenerating plants from callus and cell suspension cultures via somatic embryogenesis (Phillips and Collins, 1980). Previously established methods were modified to facilitate the rescue of immature hybrid embryos by a gradual approach (Phillips et al., 1982). Finally, cellular selection methods were utilized with the recovery of variant plants (Phillips and Collins, 1981). Several of these methods were shown to be useful with a number of other clovers (Parrott and Collins, 1982). All the necessary groundwork is now laid for approaching protoplast and subcelluar technologies. The development of a complete in vitro system in a gradual fashion, perhaps using the red clover system as a model, may be a useful approach for achieving other successes with legume species.

PROTOCOLS

A flow chart summarizing the in vitro cultural capabilities and sequences for red clover is provided in Fig. 1. The compositions of the media used are provided in Tables 6 (basal media) and 7 (growth regulators). Detailed protocols are presented below. Operations are carried out using standard aseptic technique in a laminar air-flow cabinet.

Explant Preparation Including Seedling Germination

The protocol for seedling germination follows.

1. Seeds are scarified with medium-grade sandpaper.
2. Seeds are placed in a double layer of cheesecloth and tied with a short length of string to facilitate transfer through solutions.
3. Seeds are immersed in 95% ethanol for 5 min.
4. Seeds are transferred to a saturated solution of calcium hypochlorite for 15 min. Very clean seed lots may only require the less rigorous treatment of 40% commercial bleach (2% sodium hypochlorite) for 15 min.
5. Seeds are rinsed in sterile, deionized, distilled water for 5 min for three or more times.
6. Individual seeds are placed onto SGL medium. Cultures are maintained at 25 C under low- to moderate-intensity light.
7. Germination should occur within 1-2 weeks. Normal seedlings free of contamination are selected for explanting. Each seedling generally represents a different genotype.
8. Cotyledons, epicotyls, hypocotyls, and/or roots are excised and/or sectioned and placed individually onto the appropriate medium, e.g., onto L2 medium for callus formation.

The protocol for explanting from other plant parts follows.

1. Organs or tissues are collected fresh and kept moist until used. Materials are rinsed with plain or soapy water when necessary to remove large particles of dirt. For example, 1-3 cm of crown or shoot material is prepared for excision of meristem tips, florets are prepared for excision of pistils or embryos, etc.
2. Prepared materials are immersed in 70% ethanol for 1-2 min.
3. Materials are transferred to 40% commercial bleach (2% sodium hypochlorite) for 5-8 min. This solution may also contain a drop of soap. Some tissues, e.g., mature petioles collected from the

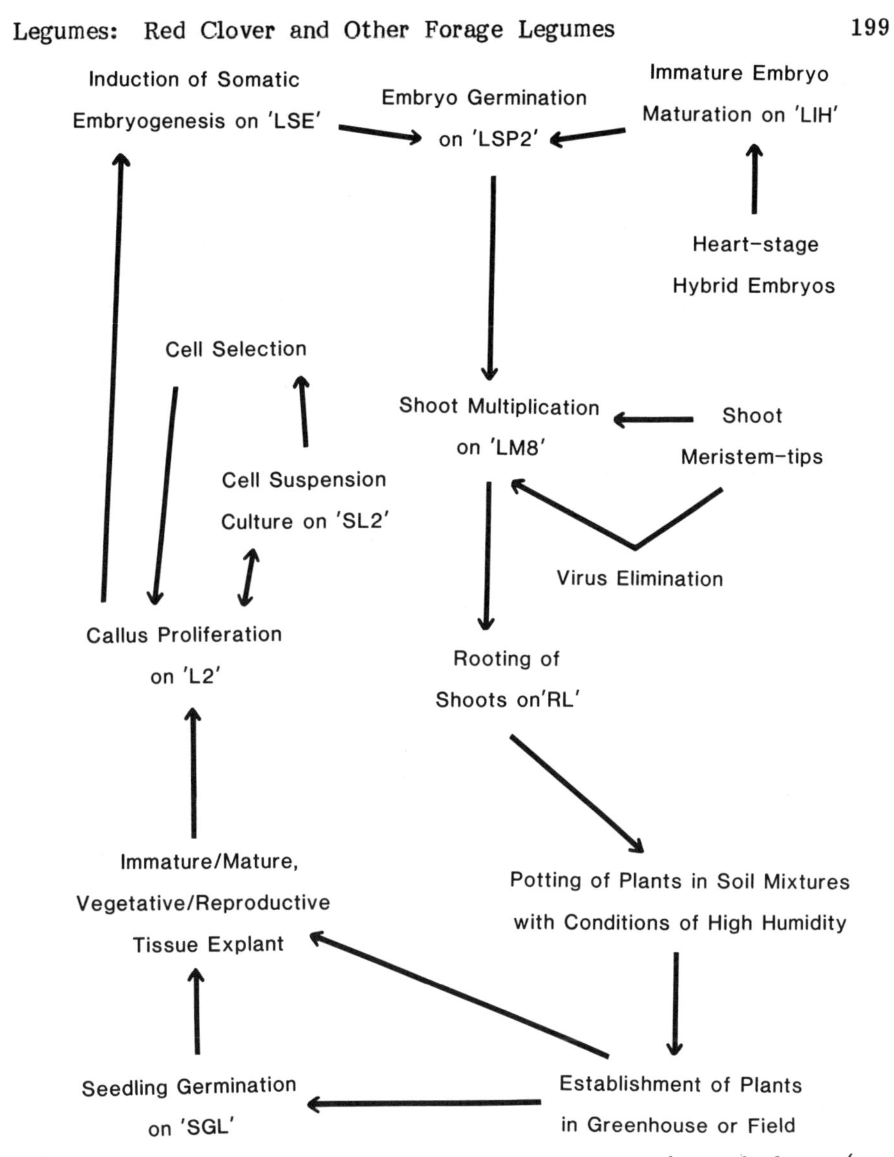

Figure 1. In vitro capabilities and media sequences for red clover (see Tables 6 and 7 for media composition).

field, may require more rigorous treatments for disinfection, such as a saturated solution of calcium hypochlorite for 5-8 min.

4. Materials are rinsed in sterile, deionized, distilled water for at least 5 min. Additional rinses are necessary if calcium hypochlorite is used in Step 3.
5. Target explants (e.g., meristem tips, embryos) are excised from the disinfected materials and individually placed on the appropriate medium. The aid of a dissecting scope may be necessary for excision of small explants.

Table 6. Compositions of the L2 Basal Nutrient Medium and Its Modifications Used for In Vitro Culture of Red Clover[a]

COMPONENT	BASAL MEDIUM			
	L2	SL2	RL	SGL
KNO_3	20.8 mM	20.8 mM	10.4 mM	2.1 mM
NH_4NO_3	12.5 mM	7.5 mM	6.25 mM	1.25 mM
KH_2PO_4	2.34 mM	1.8 mM	2.34 mM	0.23 mM
$MgSO_4$	1.8 mM	1.6 mM	0.9 mM	0.18 mM
$CaCl_2$	4.1 mM	3.75 mM	2 mM	0.4 mM
NaH_2PO_4	0.6 mM	—	0.3 mM	60 μM
$FeSO_4$·EDTA	90 μM	90 μM	90 μM	10 μM
$MnSO_4$	90 μM	80 μM	45 μM	—
H_3BO_3	82 μM	74 μM	41 μM	—
$ZnSO_4$	18 μM	16 μM	9 μM	—
KI	6 μM	5.4 μM	3 μM	—
Na_2MoO_4	1.7 μM	1.5 μM	0.85 μM	—
$CoCl_2$	0.42 μM	0.38 μM	0.21 μM	—
$CuSO_4$	0.4 μM	0.36 μM	0.2 μM	—
Myo-inositol	1.4 mM	1.4 mM	0.7 mM	0.35 mM
Thiamine·HCl	6 μM	6 μM	3 μM	1.5 μM
Pyridoxine·HCl	2.4 μM	2.4 μM	1.2 μM	0.6 μM
3-Aminopyridine	—	—	24 μM	—
Nicotinic acid	—	—	8.5 μM	—
Sucrose	73 mM	73 mM	44 mM	29 mM
Agar	0.8%[b]	—	0.65%	0.65%
pH 5.8-5.9				

After addition of growth regulators and adjusting pH, autoclave at 121 C (1.2-2 atm) for 18 min.

[a]See Table 7 for growth regulator compositions.

[b]Reduce to 0.3-0.6% for recovery of callus colonies from cell suspension cultures.

Shoot Meristem-Tip Culture and Clonal Propagation

The protocol for virus elimination follows.

1. Meristem tips, including the meristematic dome and one or two leaf primordia, are excised from disinfected materials and placed onto LM8 medium.
2. Cultures are maintained at 25 C under low-intensity light. One or more shoots should grow from about three-fourths of the explants within 4-6 weeks.
3. Shoot number increase and rooting are accomplished as described in the protocol for rapid clonal propagation below.

Table 7. Growth Regulator Compositions of the L2 Medium and Its Modifications Used for In Vitro Culture of Red Clover[a]

MEDIUM DESIG-NATION	USE	BASAL MEDIUM	AUXINS (μM)			CYTOKININS (μM)	
			PIC[b]	2,4-D[c]	IAA[b]	BA[d]	ADE[d]
SGL	Seedling germination	SGL	—	—	—	—	—
L2	Callus	L2	0.25	—	—	0.44	—
SL2	Cell suspension	SL2	0.25	—	—	0.44	—
LIH	Heart-staged hybrid embryos	L2 with 0.365 M sucrose	25 nM	—	—	—	15
LSE	Induction of somatic embryogenesis	L2	—	45 nM	—	—	15
LSP2	Shoot development from embryos (somatic and sexual)	L2	4 nM	—	—	0.66	—
LM8	Shoot multiplication	L2	12 nM	—	—	2.2	—
RL	Rooting of shoots	RL	—	—	1.2	—	—

[a]See Table 6 for compositions of basal media.

[b]Steamed into solution. The IAA stock solution is stored in the dark.

[c]Solubilized as a sodium salt. Calculations of molarity are based on the adjusted molecular weight of the salt.

[d]Solubilized in 1 N NaOH and brought to final stock solution volume with H_2O.

4. Host plant bioassays or serological tests should be performed to identify those plants which are free of viruses. About two-thirds of the plants should be virus-free.

The protocol for rapid clonal propagation follows.

1. Shoots or crown buds of suitable size are placed onto LM8 medium and maintained as described in Step 2 of the protocol for virus elimination.
2. A 4- to 10-fold increase in shoot number is obtained during each monthly subculture onto LM8 medium. Shoots proliferate most rapidly when a small cluster of shoots (three to six) is subcultured onto fresh medium. There is often a lag phase for shoot number increase when individual shoots are subcultured.
3. Some genotypes may produce some calli at the base of the shoots when cultured on LM8 medium. Subtle adjustments in the concentration of auxin and/or cytokinin may be necessary to obtain the optimal propagation rate with minimal callus formation.
4. Based on observations by other workers (Cheyne and Dale, 1980), shoot cultures on LM8 medium can probably be stored for at least 1 or 2 years at 2-6 C without appreciable loss of viability.
5. Shoots are rooted as described in the following pages (Rooting of Shoots).

Rescue of Immature Hybrid Embryos

1. Hand pollinations are performed in the greenhouse following the methods of Taylor (1980). Emasculations may or may not be required depending on the compatibility system of the female parent. For example, hand pollination of *T. sarosiense* with red clover pollen does not require emasculation since *T. sarosiense* is self-incompatible. Clonal identity of parental clones is important since interspecific combining abilities vary.
2. Florets are collected from the female parent 14-19 days after pollination in the case of *T. sarosiense* x *T. pratense*. The optimal collection time varies with the species combination involved.
3. Florets are disinfected as described earlier in Explant Preparation Including Seedling Germination.
4. Immature embryos are excised from each floret. Generally heart-staged embryos are obtained in the case of *T. sarosiense* x *T. pratense*.
5. Immature embryos are individually placed onto LIH medium for 8-14 days to prevent precocious germination of the radicle and to allow for embryo maturation. About 20% of the *T. sarosiense* x *T. pratense* embryos survived this step. Embryo growth response improves with the globular, heart, and torpedo stages, respectively. Cultures are maintained at 25 C under low-intensity light.
6. Prior to degeneration during Step 5, embryos are transferred to LSP2 medium to encourage shoot germination and development. About half of the *T. sarosiense* x *T. pratense* embryos produced shoots within 2-4 weeks.

7. Hybrid shoot numbers are increased as described earlier in Shoot Meristem-Tip Culture and Clonal Propagation. Clonal identity is important since each hybrid embryo represents a potentially different F_1 line.
8. Embryos which produce only callus during Step 6 are handled as described in Induction of Somatic Embryogenesis and Plant Regeneration (following) to obtain hybrid plants from callus.
9. Hybridity of rescued plants is verified by genetic morphological markers, chromosome numbers and/or karyotypes, and/or biochemical traits.

Callus Induction and Proliferation

1. Prepared explants are individually placed onto L2 medium.
2. Cultures are maintained at 25 C under low- or moderate-intensity light. Callus should form within 3-4 weeks.
3. Callus is maintained and proliferated by monthly subculture onto fresh L2 medium. Selection of the healthiest, most vigorous callus sectors allows for indefinite proliferation.
4. Callus colonies are recovered from suspension cultured cells by placing 1-3 ml of suspension onto L2 medium containing reduced concentrations of agar (see Table 6). These colonies are recovered within 4-8 weeks. Callus obtained in this manner are proliferated as described in Step 3 above.
5. Callus can be removed at any time and handled as described in Induction of Somatic Embryogenesis and Plant Regeneration (following). Callus retains its morphogenetic potential for 1-2 years. Plant regeneration frequencies depend on the genotype, the origin of the tissue, and the age of the culture.

Cell Suspension Culture

1. Selected tissue explants are inoculated directly into SL2 medium for cell proliferation and initiation of cell suspensions from sloughed cells. Cell suspensions are initiated more rapidly by placing small clumps of friable callus into SL2 medium. In the latter case, about 0.5 g of callus is inoculated into about 12 ml of medium.
2. Fresh SL2 medium is introduced weekly or at other desired intervals. For the first two culture intervals, dilution of the culture at 1:1 rates is desirable. Once a large population of single cells and small aggregates of cells are proliferating in the culture, about 5 ml of suspension is used to inoculate about 25 ml of fresh medium. Subculture dilution or inoculation rates will vary depending on the growth rate of the cells in culture and the desired population density of cells. Densities of about 1×10^6 cells/ml are routine.
3. Cultures are rotated at 110 rpm either in the dark or under low-intensity light. The standard culture temperature is 25 C.
4. Passing suspensions through 75 µm sieves or using narrow-bore pipets during serial subculture facilitates the maintenance of single

cells and small aggregates of cells in the cultures. This may not be required with all cultures.

5. Cell suspensions are maintained indefinitely using these procedures. Callus colonies are recovered as described earlier in Callus Induction and Proliferation. Morphogenic potential is maintained for 1 or 2 years when appropriate genotypes are used.
6. Cellular selection is practiced by introducing or removing appropriate components to or from the medium to establish selective conditions, or by providing appropriate environmental conditions during culture. It is recommended that serial subculture procedures be practiced during selection. This helps eliminate the possibility of recovering cross-fed cells which are not adapted to the selective conditions.

Induction of Somatic Embryogenesis and Plant Regeneration

1. Callus, including that recovered from cell suspension cultures, is placed onto LSE medium. Large rather than small clumps are preferable. Cultures are maintained at 25 C under low- to moderate-intensity light. A photoperiod may be desirable.
2. The appearance of somatic embryos from callus begins within 4 weeks. Additional somatic embryos are obtained by monthly subculture of the callus onto LSE medium. Somatic embryogenesis initiates from cultures of additional genotypes after repeated subculture onto LSE medium. Generally, somatic embryogenesis will not be obtained from a particular callus source if it has not responded within six subcultures.
3. Some embryos develop shoots during culture on LSE medium. Many embryos develop roots on LSE medium. Embryos which do not develop shoots on LSE medium are transferred in clusters or singly onto LSP2 medium. Repeated monthly subculture onto LSP2 medium may be required for shoot development from additional embryos.
4. Shoots derived from somatic embryos are increased in number as described earlier in Shoot Meristem-Tip Culture and Clonal Propagation. Fully developed somatic embryos (with both shoot and root) are alternately established in vivo as described in Establishment of Plants in Soil (following).
5. Plants are alternately regenerated following the methods of Beach and Smith (1979). Callus is placed onto medium containing 10 μM 2,4-D, 10 μM NAA, and 10 μM KIN for one or two monthly subcultures for shoot bud induction. Callus is then transferred to medium containing 10 μM NAA and 15 μM ADE for one or more monthly subcultures for the development of shoots from buds. B5, SH, or L2 basal media are suitable. Shoots obtained in this manner are rooted as described below.

Rooting of Shoots

1. Shoots of red clover obtained via in vitro methods are placed individually onto RL medium. Cultures are maintained at 25 C under low-intensity light with a photoperiod.

2. Roots appear within 2-4 weeks on many of these shoots. An additional 1-month culture on fresh RL medium encourages further root and plant development.
3. Shoots derived from some genotypes do not develop roots on RL medium. These shoots are placed in sterile tap water within a beaker. The beaker is covered with film (e.g., Parafilm) with several small holes in it. This allows for gas exchange while maintaining conditions of high humidity. These shoots usually develop roots within 1-2 months. The water may need replacement weekly or biweekly.
4. Rooted plants are established in vivo as described below.

Establishment of Plants in Soil

1. Plants obtained in vitro are freed of agar when appropriate. This is facilitated with the use of forceps and a gentle rinse in lukewarm water.
2. Plants are potted in a mixture of soil, peat, and prewashed vermiculite (1:1:1, by volume). Other soil mixtures can be used. Root tips must be planted in a downward orientation. Inoculation with *Rhizobium trifolii* is performed at this time when N_2-fixation by these plants is desired.
3. Conditions of high humidity must be provided for about 2 weeks. This is accomplished with a plastic tent or provided by an environmental cabinet or room. Mist benches may provide excess moisture.
4. Potted plants must be regularly watered. Care is taken to see that the plants and soil do not dry out, since these plants are especially susceptible to wilt during the establishment period. Care is also taken to assure that plants are not overwatered, since excess moisture will promote damping-off.
5. After about 2 weeks, most plants will adjust to normal greenhouse conditions. It is recommended that in-vitro-derived plants be well established in the greenhouse prior to transplanting to the field.

FUTURE PROSPECTS

Some of the earlier discussion was relevant to the potential for using the red clover in vitro system as a model for developing methods for other forage legumes.

Preliminary investigations indicate that certain in vitro and in vivo growth characteristics of red clover exhibit a positive correlation (Keyes et al., 1979). In particular, a potential was indicated for using the in vitro growth and morphogenesis of lines inbred for a single generation as an index of agronomic yield, maturity, and persistence. This potential, if realized, could greatly facilitate the introgression of longevity from long-lived perennial species into red clover, for example. Other such correlations could potentially be useful for in vitro screening of traits such as seedling vigor in the field.

Cell and tissue selection methods for disease resistance offer great potential for improving forage legumes. Diseases which are associated with toxin production are most amenable to cell selection via the use of concentrated toxin or pathogen culture filtrates. In vitro procedures may be useful for screening segregating populations for inherited resistance to disease. Preliminary investigations have been performed with several legume species and a positive correlation was established between in vitro and in vivo production of phytoalexins (Gustine and Moyer, 1979). Such research could be useful to the breeder if phytoalexins are shown to be important factors in disease resistance. This research demonstrates the potential for performing biochemical selection at the cellular level among forage legumes. Cellular selection methods could also be developed for overproduction of specific amino acids and improving legume tolerance to various stress conditions, such as extreme temperatures, extreme moisture situations, high-salt concentrations in irrigation water and soils, and the presence of heavy metals in soils.

Cryobiological methods could be useful among forage legumes as a means of long-term storage of genetic variability. The advantage of cryobiological methods over existing procedures for storage would be the considerably longer period of storage, thus necessitating less handling of the materials. Shoot tips may be the most desirable tissue to store in this manner.

In vitro methods of producing haploid plants could revolutionize forage legume breeding approaches. Advances in the methodologies for culturing microspores in isolation and achieving androgenesis may eventually be useful for haploid production among forage legumes. Anther or ovule cultures are alternatives which have yet to be productive among these crops for haploid production. The availability of haploid and dihaploid true-breeding lines would provide a means of efficient inbreeding and single- or double-cross hybrid cultivar production among the cross-pollinating species.

Methodologies more readily approachable among forage legumes include wide interspecific hybridization via somatic cell fusion of protoplasts. Methods for isolating and culturing protoplasts of white clover are available (Gresshoff, 1980). Such methods are likely to be developed for red clover in the near future, considering that all the basic technologies are available. Somatic cell fusion offers a potential alternative to embryo rescue as a means of obtaining new wide hybrid combinations. The development of protoplast technologies for forage legumes would provide the basis for investigating organelle and subgenome manipulations and molecular genetic engineering technologies in the future.

ACKNOWLEDGMENTS

The authors thank N.L. Taylor, Professor of Agronomy, University of Kentucky, for his critical reading of the manuscript and suggestions provided.

This paper is presented as a review article from the Agricultural Experiment Station, New Mexico State University, and paper no. 81-3-260, Agricultural Experiment Station, University of Kentucky. The paper is published with the approval of both Directors.

KEY REFERENCE

Collins, G.B. and Phillips, G.C. 1982. In vitro tissue culture and plant regeneration in *Trifolium pratense* L. In: Variability in Plants Regenerated from Tissue Culture (E. Earle and Y. Demarly, eds.) pp. 22-34. Praeger, New York.

REFERENCES

Ahloowalia, B.S. 1976. Tissue culture investigations with red clover. An Foras Taluntais Plant Science and Crop Husbandry Research Report, p. 29. An Foras Taluntais, Dublin.

Barnett, O.W., Gibson, P.B., and Seo, A. 1975. A comparison of heat treatment, cold treatment, and meristem-tip culture for obtaining virus-free plants of *Trifolium repens*. Plant Dis. Rep. 59:834-837.

Beach, K.H. and Smith, R.R. 1979. Plant regeneration from callus of red and crimson clover. Plant Sci. Lett. 16:231-237.

Bent, F.C. 1962. Interspecific hybridization in the genus *Lotus*. Can. J. Genet. Cytol. 4:151-159.

Blaydes, D.F. 1966. Interaction of kinetin and various inhibitors in the growth of soybean tissue. Physiol. Plant. 19:748-753.

Bourgin, J.P. and Nitsch, J.P. 1967. Obtention de *Nicotiana* haploides a partir d'etamines cultivees in vitro. Ann. Physiol. Veg. 9:377-382.

Burkholder, P.R. and Nickell, L.G. 1949. Atypical growth of plants. I. Cultivation of virus tumors of *Rumex* on nutrient agar. Bot. Gaz. 110:426-437.

Cheyne, V.A. and Dale, P.J. 1980. Shoot tip culture in forage legumes. Plant Sci. Lett. 19:303-309.

Collins, G.B., Vian, W.E., and Phillips, G.C. 1978. Use of 4-amino-3,5,6-trichloropicolinic acid as an auxin source in plant tissue cultures. Crop Sci. 18:286-288.

Collins, G.B., Taylor, N.L., and Phillips, G.C. 1981. Successful hybridization of red clover with perennial *Trifolium* species via embryo rescue. Proceedings of Fourteenth International Grasslands Congress. Westview Press, Boulder, Colorado.

Darlington, C.D. and Wylie, A.P. 1955. Chromosome Atlas of Flowering Plants, pp. 159-160. G. Allen and Unwin Ltd., London.

DeLautour, G., Jones, W.T., and Ross, M.D. 1978. Production of interspecific hybrids in *Lotus* aided by endosperm transplants. N.Z.J. Bot. 16:61-68.

DeNettancourt, D. and Grant, W.F. 1964. The cytogenetics of *Lotus* (*Leguminosae*). IV. Additional diploid species crosses. Can. J. Genet. Cytol. 6:29-36.

Evans, A.M. 1962. Species hybridization in *Trifolium*. I. Methods of overcoming species incompatibility. Euphytica 11:164-176.

______ 1976. Clovers; *Trifolium* spp. (*Leguminosae - Papilionatae*). In: Evolution of Crop Plants (N.W. Simmonds, ed.) pp. 175-179. Longman, London.

Fergus, E.N. and Hollowell, E.A. 1960. Red clover. Adv. Agron. 12: 365-436.

Gamborg, O.L., Miller, R.A., and Ojima, K. 1968. Nutrient requirements of suspension cultures of soybean root cells. Exp. Cell Res. 50:151-158.

Gautheret, R.J. 1955. The nutrition of plant tissue cultures. Ann. Rev. Plant Physiol. 6:433-484.

Graham, P.H. 1968. Growth of *Medicago sativa* L. and *Trifolium subterraneum* L. in callus and suspension culture. Phyton (Horn, Austria) 25:159-162.

Grant, W.F., Bullen, M.R., and DeNettancourt, D. 1962. The cytogenetics of *Lotus*. I. Embryo-cultured interspecific hybrids closely related to *L. corniculatus* L. Can. J. Genet. Cytol. 4:105-128.

Gresshoff, P.M. 1980. In vitro culture of white clover: Callus, suspension, protoplast culture, and plant regeneration. Bot. Gaz. 141:157-164.

Gustine, D.L. and Moyer, B.G. 1979. Use of tissue culture for studying phytoalexin biosynthesis in legumes and the possible role of phytoalexins in foliar disease resistance. Minutes of the Sixth *Trifolium* Conference, p. 89. University of Wisconsin, Madison, Wisconsin.

Kanta, K. and Maheshwari, P. 1963. Test-tube fertilization in some angiosperms. Phytomorphology 13:230-237.

Keim, W.F. 1953. Interspecific hybridization in *Trifolium* utilizing embryo culture techniques. Agron. J. 45:601-606.

Keyes, G.J., Collins, G.B., and Taylor, N.L. 1979. Genetics of red clover tissue cultures. Minutes of the Sixth *Trifolium* Conference, pp. 75-84. University of Wisconsin, Madison, Wisconsin.

______ 1980. Genetic variation in tissue cultures of red clover. Theor. Appl. Genet. 58:265-271.

Knight, W.E. and Hoveland, C.S. 1973. Crimson clover and arrowleaf clover. In: Forages: The Science of Grassland Agriculture (M.E. Heath, D.S. Metcalfe, R.F. Barnes, eds.) pp. 199-207. Iowa State University Press, Ames, Iowa.

Leffel, R.C. 1973. Other legumes. In: Forages: The Science of Grassland Agriculture (M.E. Heath, D.S. Metcalfe, R.F. Barnes, eds.) pp. 208-220. Iowa State University Press, Ames, Iowa.

______ and Gibson, P.B. 1973. White clover. In: Forages: The Science of Grassland Agriculture (M.E. Heath, D.S. Metcalfe, R.F. Barnes, eds.) pp. 167-176. Iowa State University Press, Ames, Iowa.

Linsmaier, E.M. and Skoog, F. 1965. Organic growth factor requirements of tobacco tissue cultures. Physiol. Plant. 18:100-127.

Miller, C.O. 1961. A kinetin-like compound in maize. Proc. Natl. Acad. Sci. U.S.A. 47:170-174.

Mokhtarzadeh, A. and Constantin, M.J. 1978. Plant regeneration from hypocotyl- and anther-derived callus of berseem clover. Crop Sci. 18:567-572.

Moore, R.J., ed. 1973. Index to Plant Chromosome Numbers 1967-1971, pp. 246-249. International Organization of Plant Biosystematists, Utrecht, the Netherlands.

Murashige, T. and Skoog, F. 1962. A revised medium for rapid growth and bioassays with tobacco tissue cultures. Physiol. Plant. 15:473-497.

Niizeki, M. and Grant, W.F. 1971. Callus, plantlet formation, and polyploidy from cultured anthers of *Lotus* and *Nicotiana*. Can. J. Bot. 49:2041-2051.

Niizeki, M. and Kita, F. 1973. Studies on plant cell and tissue culture. III. In vitro induction of callus from anther culture of forage crops. J. Fac. Agric. Hokkaido Univ. 57:293-300.

Oswald, T.H., Smith, A.E., and Phillips, D.V. 1977a. Callus and plantlet regeneration from cell cultures of ladino clover and soybean. Physiol. Plant. 39:129-134.

______ 1977b. Herbicide tolerance developed in cell suspension cultures of perennial white clover. Can. J. Bot. 55:1351-1358.

Parrott, W.A. and Collins, G.B. 1982. Callus and shoot-tip culture of eight *Trifolium* species in vitro with regeneration via somatic embryogenesis of *T. rubens*. Plant Sci. Lett. 28:189-194.

Pelletier, G. and Pelletier, A. 1971. Culture in vitro de tissus de trefle blanc (*Trifolium repens*): Variabilite des plantes regenerees. Ann. Amelior. Plant. 21:221-233.

Phillips, G.C. and Collins, G.B. 1979a. In vitro tissue culture of selected legumes and plant regeneration from callus cultures of red clover. Crop Sci. 19:59-64.

______ 1979b. Virus symptom-free plants of red clover using meristem culture. Crop Sci. 19:213-216.

______ 1980. Somatic embryogenesis from cell suspension cultures of red clover. Crop Sci. 20:323-326.

______ 1981. Growth and selection of red clover (*Trifolium pratense* L.) cells on low levels of phosphate. Agronomy Abstracts, p. 187. American Society of Agronomy, Madison, Wisconsin.

______, and Taylor, N.L. 1982. Interspecific hybridization of red clover (*Trifolium pratense* L.) with *T. sarosiense* Hazsl. using in vitro embryo rescue. Theor. Appl. Genet. 62:17-24.

Phillips, R.L. and Keim, W.F. 1968. Seed pod dehiscence in *Lotus* and interspecific hybridization involving *L. corniculatus* L. Crop Sci. 8: 18-21.

Randolph, L.F. and Cox, L.G. 1943. Factors influencing the germination of iris seed and the relation of inhibiting substances to embryo dormancy. Proc. Am. Soc. Hortic. Sci. 43:284-300.

Ranga Rao, V. 1976. Nitrogenase activity in *Rhizobium* associated with leguminous and non-leguminous tissue cultures. Plant Sci. Lett. 6:77-83.

Richards, K.W. and Rupert, E.A. 1980. In vitro fertilization and seed development in *Trifolium*. In Vitro (Rockville) 16:925-931.

Rupert, E.A. and Evans, P.T. 1980. Embryo development after interspecific cross-pollinations among species of *Trifolium*, section *Lotoidea*. Agronomy Abstracts, p. 68. American Society of Agronomy, Madison, Wisconsin.

Rupert, E.A. and Seo, A. 1977. Hybrid cell cultures from undifferentiated *Trifolium* embryos. Agronomy Abstracts, p. 69. American Society of Agronomy, Madison, Wisconsin.

Rupert, E.A., Barnett, O.W., and Camper, N.D. 1979a. Tissue culture of white clover species. Minutes of the Sixth *Trifolium* Conference, p. 74. University of Wisconsin, Madison, Wisconsin.

Rupert, E.A., Seo, A., and Richards, K.W. 1979b. *Trifolium* species hybrids obtained from embryo-callus tissue cultures. Agronomy Abstracts, p. 75. American Society of Agronomy, Madison, Wisconsin.

Schenk, R.U. and Hildebrandt, A.C. 1972. Medium and techniques for induction and growth of monocotyledonous and dicotyledonous plant cell cultures. Can. J. Bot. 50:199-204.

Seaney, R.R. and Henson, P.R. 1970. Birdsfoot trefoil. Adv. Agron. 22:119-157.

Skucinska, B. and Miszke, W. 1980. In vitro vegetative propagation of red clover. Z. Pflanzenzuecht. 85:328-331.

Somaroo, B.H. and Grant, W.F. 1971. Interspecific hybridization between diploid species of *Lotus* (*Leguminosae*). Genetica (The Hague) 42:353-367.

Swanson, E.B. and Tomes, D.T. 1980a. Plant regeneration from cell cultures of *Lotus corniculatus* and the selection and characterization of 2,4-D tolerant cell lines. Can. J. Bot. 58:1205-1209.

______ 1980b. In vitro responses of tolerant and susceptible lines of *Lotus corniculatus* L. to 2,4-D. Crop Sci. 20:792-795.

Taylor, N.L. 1973. Red clover and alsike clover. In: Forages: The Science of Grassland Agriculture (M.E. Heath, D.S. Metcalfe, R.F. Barnes, eds.) pp. 148-158. Iowa State University Press, Ames, Iowa.

______ 1980. Clovers. In: Hybridization of Crop Plants (W.R. Fehr and H.H. Hadley, eds.) pp. 261-272. American Society of Agronomy, Madison, Wisconsin.

______ and Smith, R.R. 1979. Red clover breeding and genetics. Adv. Agron. 31:125-154.

______, Quesenberry, K.H., and Anderson, M.K. 1979. Genetic system relationships in *Trifolium*. Econ. Bot. 33:431-441.

Tomes, D.T. 1976. Plant regeneration from callus cultures of *Lotus corniculatus*. Agronomy Abstracts, p.64. American Society of Agronomy, Madison, Wisconsin.

______ 1979. A tissue culture procedure for propagation and maintenance of *Lotus corniculatus* genotypes. Can. J. Bot. 57:137-140.

White, P.R. 1963. The Cultivation of Animal and Plant Cells, 2nd ed. Ronald Press, New York.

Whyte, R.O., Nilsson-Leissner, G., and Trumble, H.C. 1953. Legumes in agriculture. F.A.O. Agric. Stud. 21, 367 pp.

Williams, E. 1978. A hybrid between *Trifolium repens* and *T. ambiguum* obtained with the aid of embryo culture. N.Z.J. Bot. 16:499-506.

______ 1980. Hybrids between *Trifolium ambiguum* and *T. hybridum* obtained with the aid of embryo culture. N.Z.J. Bot. 18:252-258.

Williams, E.G. and DeLautour, G. 1980. The use of embryo culture with transplanted nurse endosperm for the production of interspecific hybrids in pasture legumes. Bot. Gaz. 141:252-257.

Zakrzewski, J. and Zakrzewska, E. 1976. Hodowla in vitro kallusa koniczyny czerwonej jako podloza do rozmnazania nicieni—*Ditylenchus dipsaci* (Kuhn). Hodowla Rosl. Aklim. Nasienn. 20:97-104.

Zohary, M. 1971. A revision of the species of *Trifolium* sect. *Trifolium* (*Leguminosae*). I. Introduction. Candollea 26:297-308.

______ 1972. A revision of the species of *Trifolium* sect. *Trifolium* (*Leguminosae*). II. Taxonomic treatment. Candollea 27:99-158.

SECTION IV
Vegetables

CHAPTER 8
Asparagus

G. Reuther

HISTORY OF THE CROP

Asparagus officinalis L. is an European-Sibirican continental plant related to the East Mediterranean vegetation (Meusel et al., 1965). As a crop it is native to the Orient and to the eastern parts of the Mediterranean genecenter. The Greeks introduced it into their motherland from the eastern nations and at a later date the Romans adapted its culture from the Greeks (Luzny, 1979). Cato (200 BC) reported that asparagus was a popular delicious vegetable in Imperial Rome, and the Roman troops are said to have introduced it to central Europe. An old chronicle of the town of Worms on the Rhine River reports that crusading troops brought asparagus seeds from Arabian countries to the Rhine valley around 1214. The first mention of asparagus as a vegetable is documented in a French monastery in the eleventh century. The first beds of it were established in Germany in 1567 in the vicinity of Stuttgart. Before the seventeenth century its growth was limited mainly to the gardens of castles and monasteries. Blanching of asparagus was introduced by the Dutch at the beginning of the nineteenth century. Activities in asparagus breeding first took place in France and Germany. According to Henslow (1911), until the seventeenth century ancient herbalists directed more attention to the medicinal properties of asparagus than to its use as a vegetable. The same author describes several areas of Europe with native *Asparagus officinalis*—for example, the coast of Wales, Cornwall and Dorset, and the southern parts of Russia and Poland—where the waste steppes are literally covered with it. It is suggested that the native forms of asparagus originated from cultivated types which then reverted to wild forms.

ECONOMIC IMPORTANCE

Asparagus is grown in temperate climates which are characterized by relatively high soil temperatures in early spring; its cultivation is also of economic importance under subtropical conditions. Growth of asparagus spears normally takes place in sandy soils with temperatures of 12 to 18 C. Green asparagus production is less dependent on the type of soil than white or blanched asparagus. In European countries (e.g., Germany, the Netherlands, France, and Italy), blanched asparagus is produced mainly for the fresh market whereas canned white asparagus is imported from overseas (e.g., Taiwan). In the United States and Canada only green asparagus is cultivated. In Europe, Italy, France, and to a smaller extent in the German Democratic Republic, production of green asparagus is relatively low compared with the white or blanched crop. In Europe, countries exporting fresh asparagus include France, Spain, and Italy. Taiwan is the main exporting country for processed asparagus in the world (Table 1).

Table 1. Asparagus Production in Different Countries (1979)[a,b]

LOCATION	AREA (ha)	YIELD (kg/ha)	TYPE
Belgium	100	3,000	White
Canada	1,459	2,740	Green
Chile	600	-	White
Denmark	200	1,150	White
France	18,500	2,540	White/green
German Fed. Rep.	3,300	3,190	White
German Dem. Rep.	4,103	-	White/green
Greece	400	2,500	
Hungary	380	2,650	White/green
Ireland	4	1,000	
Italy	5,340	5,640	White/green
Netherlands	2,400	3,880	White
South Africa	2,000	-	-
Spain	13,961	3,561	White
Taiwan	13,512	7,750	White/green
U.K.	300	900	
United States	33,663	2,210	Green

[a]The list of asparagus-growing countries is not complete; information from other asparagus-producing countries was not available.

[b]Sources: Eurostat, Crop Production 3 - 1981; Anuario de Estadistica de Espana; Secretaria General Tecnica; Taiwan Mushroom and Asparagus Research Committee; Department of Agriculture and Fisheries, Pretoria; Agricultural Statistic 1980, California Crop Reporting Service; Hungarofruct, Budapest; Facultad de Ciencias Agrarias, Santiago, Chile; Kaufmann, F. 1981, Arch. Gartenbau.

A great number of cultivars have been selected in the past according to the different climatic and soil conditions of the asparagus-growing areas and to the local demand of the market (Table 2). The older cultivars are dioecious, consisting of female and male plants. Several selections and crossings with other varieties have been developed with improved yield and quality from the old German cv. Ruhm Von Braunschweig, e.g., Schwetzinger Meisterschuss, Limburgia, Huchels, Eros, and some others. Selections exist in the United States, Canada, and Taiwan from the American cv. Mary Washington and the University of California types. As a result of the inbreeding of andromonoecious lines and of anther culture, pure male F_1 hybrids have recently been introduced into asparagus cultivation.

Table 2. Asparagus Varieties Cultivated in Various Countries

	Canada	Chile	France	German Fed. Rep.	German Dem. Rep.	Hungary	Italy	Netherlands	Spain	Taiwan	United States
Ruhm v. Braunschweig				X	X	X					
Schwetzinger Meisterschuss		X		X				X			
Huchels Leistungs-auslese				X							
Lucullus			X	X			X	X			
Eros					X						
Limburgia								X			
Limbras								X			
Minerve			X				X			X	
Diana			X								
Darbonne IV			X							X	
Precoce d'Argenteuil							X				
Violette d'Albenga							X				
Blanco de Navarra									X		
Mary Washington	X	X				X			X		X
Paradise											X
Robert Super											X
Rutgers Bacon											X
Viking 2 K	X										
Eden	X										X
U.C. 309										X	X
U.C. 711		X								X	X
U.C. 72	X										
U.C. 66		X									

BREEDING AND CROP IMPROVEMENT

Asparagus officinalis L. is a dioecious species with a sex ratio of 1:1; however, male plants have flower types that vary from staminate to hermaphroditic. The inheritance of sex in asparagus depends, as in many dioecious species, on sex chromosomes. Genetic experiments have demonstrated that asparagus females are homogametic (XX), males are heterogametic (XY), and hermaphroditic flowers arise on male (XY) plants. The males with hermaphroditic character are andromonoecious plants which can be cross- or self-fertilized. The identification of sex chromosomes by means of cytological methods has not been possible. One chromosome has been identified as the sex chromosome in triploid asparagus with the aid of Giemsa staining in trisomics of the progenies (Loeptien, 1979a,b). The degree of expression of the hermaphroditic flowers in andromonoecious individuals is determined hereditarily by a major factor on the Y chromosome and by modifying genes on the autosomes (Wricke, 1968, 1973; Loeptien, 1979a,b). It is assumed that andromonoecious plants have their origin in rare mutations that occur in a normally dioecious population that changes males to hermaphrodites (Wricke, 1979).

A classification of various male flower types and their possible distribution is described in Fig. 1 based on the findings of Beeskow (1967). In a population of male asparagus plants five different classes of the types I-IV can be distinguished in various frequencies: (a) pure male plants, type I, no variations; (b) andromonoecious plants, types I-IV, highest variation; (c) andromonoecious plants, types II, III, and IV, moderate variation; (d) andromonoecious plants, types III and IV, low variation; and (e) andromonoecious plants, type IV, no variation (hermaphroditic plants). The development of berries depends on the degree of andromonoecism and is highest in the hermaphroditic (type IV) plants.

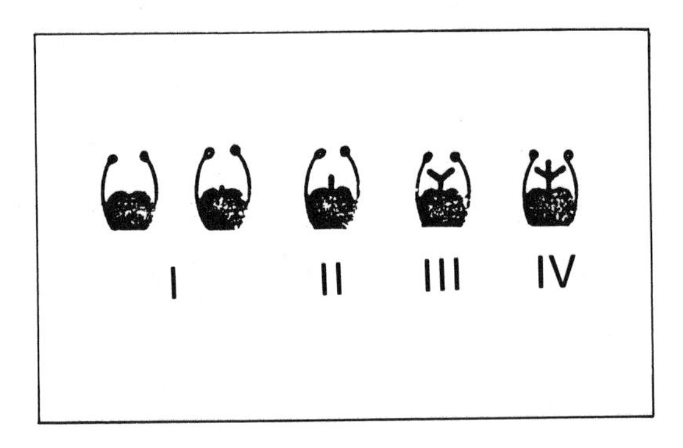

Figure 1. Scheme of flower types of male asparagus plants and classification of the degree of andromonoecism: type I, no style or small style initial on the ovary (pure male); type II, style without stigma; type III, style with two stigma; type IV, style with three stigma, completely hermaphroditic.

Asparagus has an average harvesting period of 10 years and also demands a great amount of financial and labor input before any economic return is realized. Therefore improvement of the cultivars by breeding has high priority. Important breeding aims are high yield and quality characterized by an increased number of spears of large diameter per plant, uniformity among the spears, low level of fiber in the spears, high resistance to disease, salt tolerance, climatic adaptability, and in white spears, a low tendency to discolor. If conventional breeding methods are used, the development of improved cultivars may require 10 years or more because of the generation time of the crop.

Breeding of Simple and Double Hybrids

In the past, improvement of asparagus cultivars has been achieved by cross-pollination of selected high-yielding male and female plants of the same variety or of different varieties with subsequent mass selection among the progeny. F_1 hybrids were tested for their combining ability in order to improve or to maintain heterosis. By using a polycross system the optimal combination of female and male genotypes could be identified. This method is demonstrated in Figure 2 with the breeding of cv. Limbras according to Boonen (personal communication, 1981).

Genotypes were selected based on combining ability after pollination of each female with all male plants. In 1974 a seed-bearer field was established with clones of the following combinations:

10 x 3 = Limbras 10
18 x 37 = Limbras 18
22 x 37 = Limbras 22
26 x 3 = Limbras 26

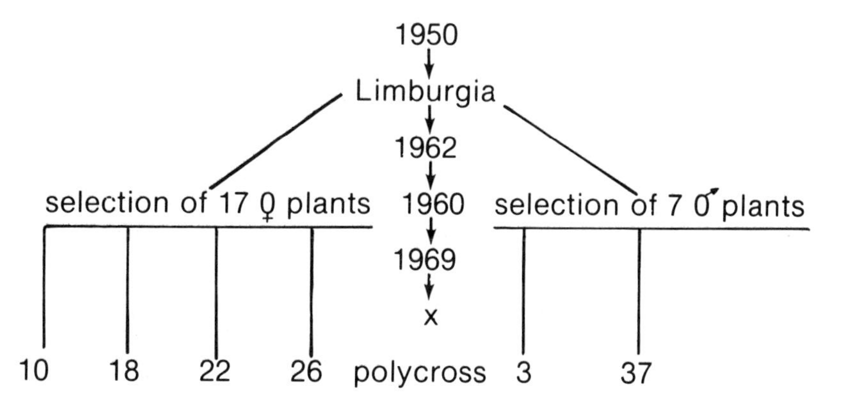

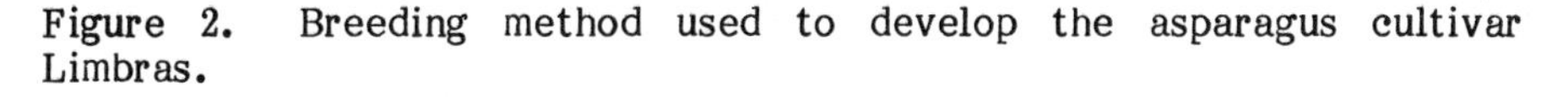

Figure 2. Breeding method used to develop the asparagus cultivar Limbras.

As seen in Figure 2 the time span for conventional breeding of new cultivars such as Limbras is extremely long.

Suggestions given by Corriols-Thevenin (1979) for reduction of the testing period led to a shorter method based on a test cycle of 5 years. This test cycle, used in the breeding of double hybrids, provides a high yield of seeds that is quickly produced by a greater number of female hybrid plants with a relatively low variability (Figs. 3 and 4). Using four progenitors in the first step, two pairs of plants

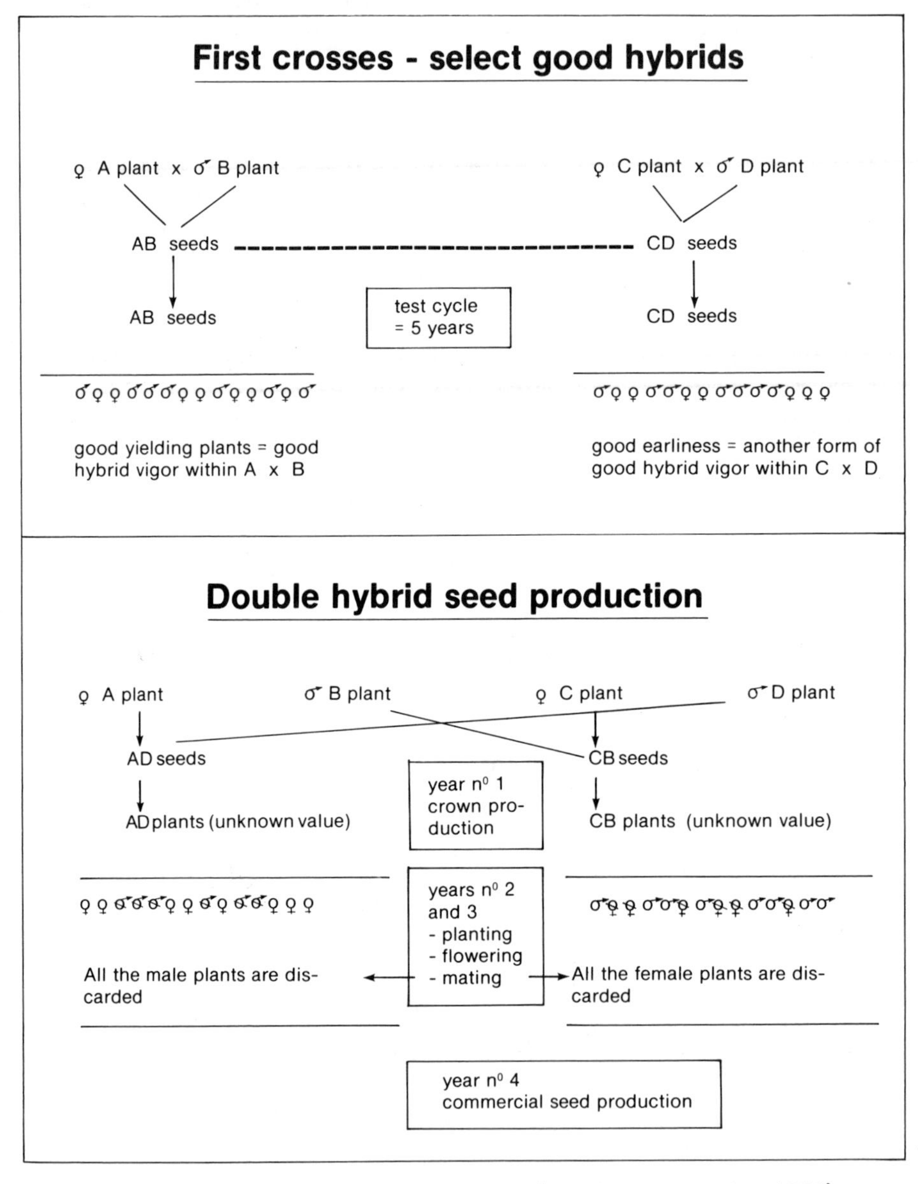

Figure 3. Double hybrid seed production (Corriols-Thevenin, 1979).

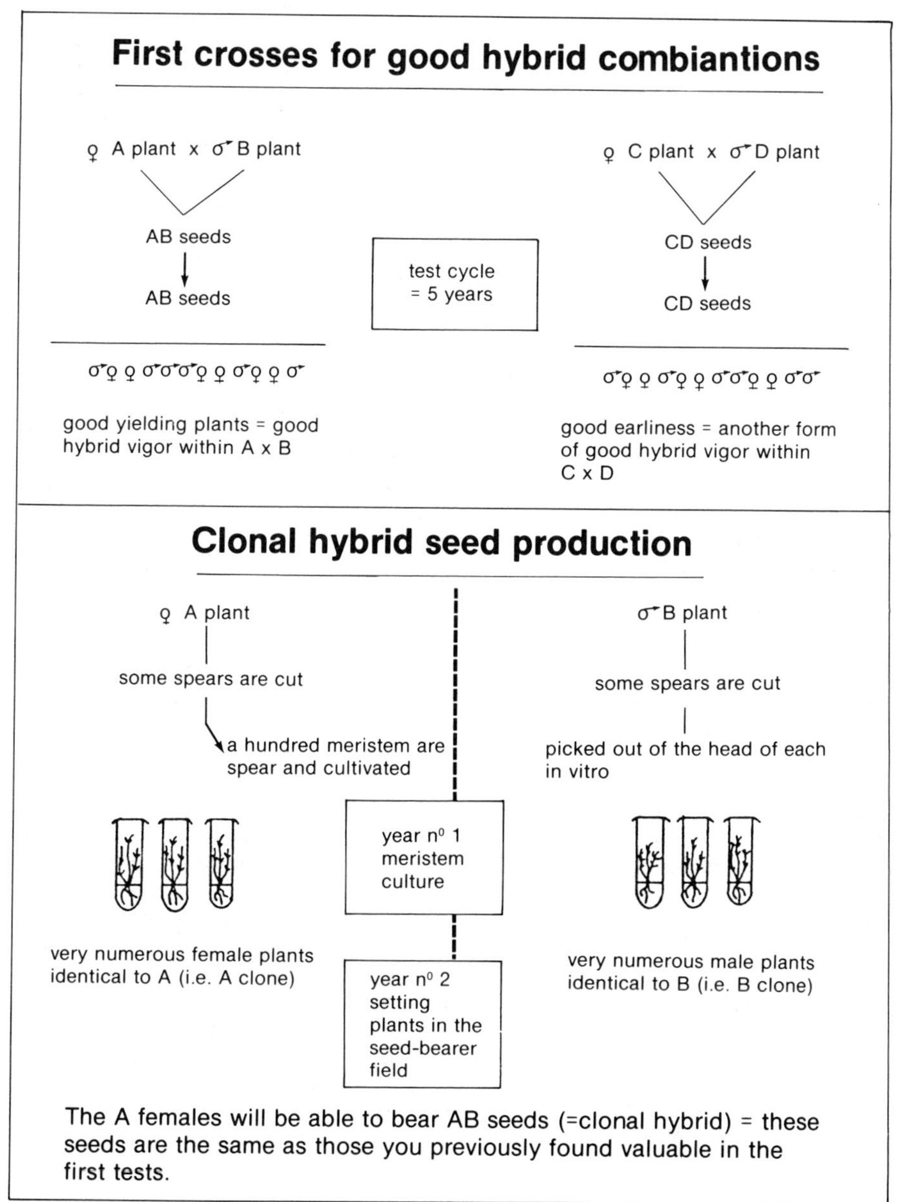

Figure 4. Clonal hybrid seed production by application of in vitro techniques for multiplication of parent plants (Corriols-Thevenin, 1979).

are crossed (A x B and C x D), selected, and tested for their hybrid value. For double-hybrid seed production, the genotype A female is pollinated with D male and C female with B male. The male plants from AD plants are discarded; similarly the female plants are discarded from the CB cross. After pollination in an isolated seed-bearer field, the seeds represent a double hybrid consisting of ADCB genotype.

Rapid introduction of newly estabished hybrids with proven combining ability for commercial use is still a problem because of their limited capability for vegetative propagation in addition to the long period of time required for testing. Division of the crowns of elder plants into smaller parts results in only a few cloned plants and also implicates possible problems with soilborne infectants. The most practical method of vegetative multiplication uses cell culture techniques. Clonal multiplication can be achieved by direct rooting of explanted meristems from the spear top as reported by Andreassen and Ellison (1967), Murashige et al. (1972), Morel (1972), Greiner (1974), and Doré (1975).

Breeding of All-Male F_1 Hybrids

Since higher yield is correlated with staminate plants, there have been efforts in asparagus breeding to produce all-male varieties. In conventional breeding methods this aim can be reached by pollination of andromonoecious plants which are cytogenetically male, but with hermaphroditic flowers. Such crosses produce inbred lines consisting of two Y chromosomes—the supermales. Pollination of female plants with pollen from the supermale plants results in all-male F_1 hybrids. The occurrence of andromonoecious plants is about 2% of the male plants (Weit and Stein, 1981) and reaches a value of 10–20% in some breeding material (Boehne, 1977). A stable equilibrium of andromonoecious plants only exists if there are males and andromonecious types, but no females, as shown by population genetic methods (Maynard-Smith, 1976).

The essential steps for the development of an all-male variety are summarized in Fig. 5.

After the identification of supermale (YY) genotypes, the combining ability in respect to yield and quality must be evaluated (Table 3). According to breeding experiences, the all-male F_1 hybrids differ from each other with respect to yield, quality, and disease resistance (Table 4). The F_1 hybrids in Table 4 are derived from the same pollinator in combination with different female genotypes. If a YY plant with a set of genes for yield and other valuable characters has been selected, limiting factors for a large-scale F_1-hybrid seed production are the inbreeding depression of the supermale parent plant and the failing of vegetative propagation by conventional methods. Therefore cloning of parent plants by cell culture is important for setting up a seed bearer field and for the production of a great amount of F_1-hybrid seeds in a short time (Fig. 4).

Haploids and Homozygous Lines in Asparagus Breeding

Since asparagus is a cross pollinated plant, it has a high degree of natural heterozygosity. In many cases it would be of considerable interest to produce completely homozygous plants by obtaining haploids and subsequently doubling the haploid chromosomes. F_1 hybrids could be produced from homozygous parents with demonstrated combining ability. In asparagus haploids have been recovered in two ways: first,

A.

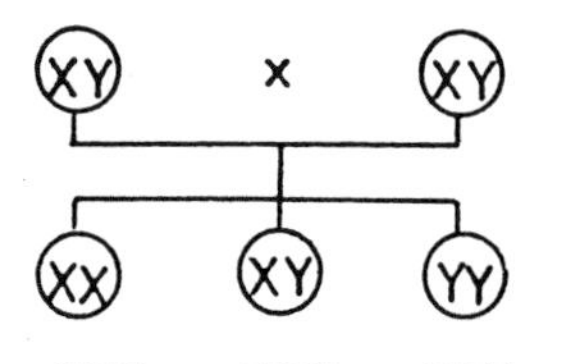

25% 50% 25%

Distinguish classes by testcrossing

B.

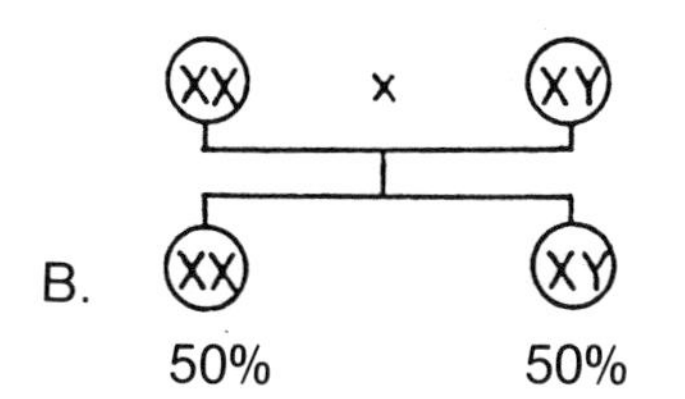

Normal cross

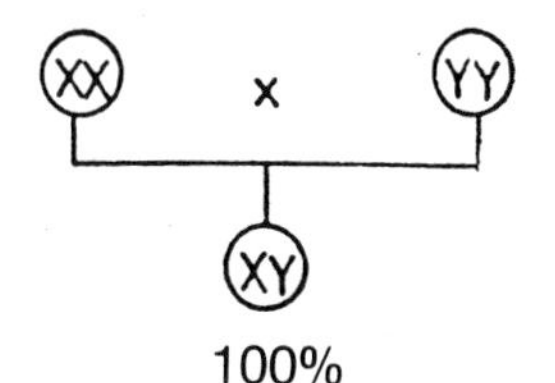

Uniform male progeny using supermale parent

Figure 5. Breeding scheme of supermale progenitors for all-male F_1 hybrid seed production. (a) Crossing of andromonoecious plants to produce all-male variety. (b) Cytogenetically different male plants can be distinguished only by test crossings.

by selection of polyembryonic seeds, and second, by anther culture (Corriols-Thevenin, 1979). Sometimes seeds are produced with two or more embryos instead of the usual one, and occasionally one of the supernumeraries is a haploid. A mean frequency of 0.23 haploids was found in 1000 seeds (Thevenin, 1968). After treatment with colchicine, fertile homozygous female plants are obtained. The more promising way of producing homozygous plants is to induce pollen or pollen mother cells to grow directly into haploid plants. If intact anthers are used at a precise stage of development under in vitro conditions, the microspores may develop haploid plantlets directly or via callus. In the case of asparagus this method has been applied with limited success for breeding supermale parent plants.

Cultivar trials with progenies derived from crossing heterozygous female, male, and supermale (YY) plants demonstrate the progress in yield of the male F_1 hybrids when compared with older cultivars and existing double hybrids (Table 4). Superior yields are possible using the all-male F_1 hybrids derived from an inbred supermale genotype with tested combining ability and different female progenitors. First attempts with all-male F_1 hybrids obtained by crossing diploid pure lines from anther culture are shown in Table 5. The data in Tables 4 and 5 demonstrate that homogenous male F_1 hybrids are not in all cases superior in yield and quality when compared with dioecious cultivars. The results indicate further great differences among the progenies from homozygous genitors developed by anther culture. Thus the development of supermale genitors by anther culture may not be the single

Table 3. Time Table for the Development of an All-Male Asparagus Variety[a]

TIME (years)	MALE PARENT PLANT	FEMALE PARENT PLANT
1st	Pollination of andromonoecious plants for obtaining YY males	Development of pure lines as female parent plants
2nd–4th	Cultivation of progenies up to flower stage	
4th	Test crossings of the males	
5th	Cultivation of the progenies of test crossings	
6th–8th	Analysis of the test crossings and identification of the supermale genotypes	
9th	Test crossings for combining ability of the YY males with female lines	
10th	Cultivation of the crossings for combining ability and testing of the seedlings	
11th–14th	Testing for vigor, yield, and quality and the occurrence of andromonoecious plants; cloning of the most suitable YY males by cell culture	Generative and/or vegetative propagation of special female lines and planting into the nursery
15th	Establishment of a seed-bearer field with tested YY male clones	

[a]Weit and Stein, 1981.

prerequisite for high-yielding all-male F_1 hybrids; they must be tested for their combining ability as well as for inbreeding lines with heterozygous or homozygous female genotypes. Homozygosity of the genitors is not principally an advantage for the breeding of high-yielding asparagus cultivars. On the other hand, anther culture provides shortening of the breeding cycle by selection of supermales with good vigor. Selfing of hermaphrodites creates a broader genetic base for F_1 hybrids which can consist of a mixture of four genotypic combinations and is therefore equivalent to a double-cross hybrid (Marks, 1979). An explanation for higher spear productivity of male asparagus plants is given by Fiala and Jolivet (1979) who demonstrated a sex-specific accumulation of reserves in the roots of 2- to 4-year-old plants.

Table 4. Yield of All-Male F_1 Hybrids, Dioecious Simple and Double Hybrids, and Older Asparagus Cultivars

CULTIVARS	HYBRID TYPES[c]	1979[b]		1980		1981	
		kg/ha	% of Lucullus Control	kg/ha	% of Lucullus Control	kg/ha	% of Lucullus Control
Lucullus	F_1 H	2670	100	4950	100	6070	100
Rekord	F_1 H	2900	109	5710	115	6830	112
Sieg	F_1 H	3020	113	5430	110	7070	116
Optima	F_1 H	3200	120	5820	118	8050	133
Schwetz	LC	2360	88	3790	77	5320	88
Huchels	LC	2720	102	4500	91	5870	97
Limbras	LC	2550	96	3970	80	4560	75
Desto	H	1780	67	2990	60	2910	48
Aneto	H	1350	51	3890	79	3940	65
Diana	DH	2190	82	3050	62	4550	75
Minerve	DH	3040	114	3910	79	4020	66
Laroc	DH	2790	105	3760	76	4420	73
Junon	DH	2620	98	4180	84	5480	90

[a] Hartmann, 1982.

[b] First harvest.

[c] Abbreviations: LC = local cultivars; H = hybrid; DH = double hybrid; F_1 H = all-male F_1 hybrid; Schwetz = Schwetzinger Meisterschuss.

Table 5. Mean Weight per Plot of Marketable Spears[a] and Mean Diameter of Spears over the Whole Harvesting Period (1981) of All-Male F_1 Hybrids Obtained from Homozygous Lines from Anther Culture Related to Progenies of Heterozygous Genitors[b]

CULTIVARS	HYBRID TYPE[c]	MEAN WEIGHT (g)[d]	MEAN DIAMETER (mm)[d]
Lucullus	F_1 H	799 a	14.3 b
Minerve	DH	637 a,b	15.6 a
Diane	DH	627 a-c	14.1 b,c
H 21	F_1 HAN	609 b-d	12.2 d-g
H 20	F_1 HAN	510 b-e	12.8 c-f
H 12	F_1 HAN	451 c-f	11.0 g
Md 10xVLG1-15	F_1 H	355 e-g	12.0 e-g
387 Gx14	F_1 H	307 f,g	12.9 b-f
H 17	F_1 HAN	286 f,g	11.0 g
Prec.d'Arg.	LC	270 f,g	13.5 b-d

[a]Diameter >7 mm.

[b]Falavigna, 1981.

[c]Abbreviations: F_1 HAN = all-male F_1 hybrid from homozygous lines obtained by anther culture. For others see Table 4.

[d]Values with a common letter are not significantly different at $p = 0.05$ according to Duncan's multiple range test. Data was collected for five plants per plot with five replications.

The content of carbohydrates at the beginning of the rest period in winter is higher in the roots of male plants, and the catabolism and anabolism of organic acids, free amino acids, and carbohydrates are faster. In seedlings the dry-weight accumulation in the fern cladophylls, stalks, rhizome, roots, and finally, in the whole plant differed significantly in favor of higher values in F_1 hybrids than in open-pollinated plants (Benson, 1979).

REVIEW OF THE LITERATURE

Considering the long time required for breeding new asparagus cultivars using conventional methods, rapid multiplication of tested parental lines is of high interest. Using conventional methods *Asparagus officinalis* has low regenerative potential. In the past, division of the crown into one or two smaller parts with subsequent regeneration of a whole plant was applied with limited success because injury to the roots provided a site for introduction of pests. Thus a very limited number of progenies could be obtained using this method. The development of aerial crowns on asparagus stems in potted plants kept under high moisture provides an alternate method of vegetative propagation (Yang and Clore, 1973, 1975). After enlargment of the aerial crowns, shoots and roots are formed and detached from the stem nodes or are separated artificially to form new plants. In cell culture aerial crowns have been observed occasionally, mostly following inhibited growth of the main or lateral shoots (Fig. 6). Propagation by division of individual crowns and induction of aerial crowns is not commercially feasible for production of genetically identical plants. Only the in vitro technique has been proved to be a suitable method for cloning special selections of parent plants or cultivars in large quantities. Complete plantlets have been obtained using explants from stem tips, from which meristems and shoot apices have been taken, and from lateral buds of field-grown ferns or white asparagus spears. Initial experiments were unsuccessful in establishing rooted cuttings outside aseptic conditions

Figure 6. Crowns emerging from callus on the base of expanding shoots (left); aerial crowns on fasciated shoots (middle and right).

(Andreassen and Ellison, 1965, 1967). Depending on the composition of the nutrient medium the explanted organ initials develop into shoots with subsequent root formation or to callus with various stages of dedifferentiation.

Propagation of Asparagus through Organ Culture without Callus Formation

For asexual multiplication, shoot apices (approximately 0.15 mm long) were excised from young lateral branches of a field-grown spear following sterilization (Murashige et al., 1972). The explant contains the meristematic apical dome and a few leaf primordia. Larger explants resulted in a higher frequency of survival and a greater tendency for multiple spears and callus, whereas smaller explants formed complete plantlets. The composition of the medium was as follows:

Inorganic constituents	MS
Organic constituents	
NAA	0.3 μM
KIN	0.49 μM
Thiamine·HCl	2.96 μM
Pyridoxine·HCl	0.024 mM
Nicotinic acid	0.04 mM
Myo-inositol	0.55 mM
Adenine sulfate·dihydrate	0.15 mM
Sucrose	0.07 M
Other supplements	
Difco Bacto malt extract	500 mg/l
$NaH_2PO_4 \cdot H_2O$	1.23 mM
Difco Bacto agar	6.0 g/l

Similar results with respect to the above-mentioned basal medium and the size of the shoot tip explant have been evaluated by several authors. The concentrations of the supplementary cytokinin and auxin compounds was varied from 0.3-1.4 μM for KIN and from 0.5-1.6 μM for NAA (Matsubara, 1973; Greiner, 1974; Yang, 1977). Variations in the composition of the mineral nutrients, especially a 1.66 higher content of KNO_3 (related to MS medium) influenced rooting of explanted meristems in the presence of 0.5 μM NAA (Tendille, and Lecerf, 1974; Tendille, 1974). Furthermore, rooting of shoots from shoot tip explants of various sizes was induced with 0.5 μM NAA in a KIN-free subculture. The vigor of adventitious shoots is dependent on their age which in turn affects rooting capacity. Best results were obtained with 4- and 5-month-old plantlets after a rooting period of 4-8 weeks (Yang and Clore, 1974a). The time required for developing asparagus stock plants transferred into soil from in vitro organ culture, calculated from the excision of the primary explant, varies from 6 to 8 months (Yang, 1977) or from 10 to 12 months (Tendille and Lecerf, 1974). After in vitro establishment of a stock plant, culture of its lateral buds can be used for further rapid vegetative propagation. The multiplication rate of subcultured buds depends on the vigor of the shoots and the position of the buds; more complete plantlets were obtained from basal and middle parts of the shoot (Yang, 1977). A constant temperature of 27 C is optimum for plant formation from excised shoot apices. The optimum light intensity for spear as well as root initiation was found to be 1000 lux at daily light periods of 16 hr (Hasegawa et al., 1973); shoot and root initiation occurred equally satisfactorily up to 3000 lux. Root formation is initiated adventitiously in a swollen mass of calluslike tissue at the base of the shoot apex explant whereas spear formation resulted from growth of the axillary bud.

Unsatisfactory survival rates of plantlets after transfer to soil gave rise to experiments for improving the consistency of the plantlets under in vitro conditions and horticultural techniques. It was suggested that rooted meristems cultured on a nutrient medium with sucrose are heterotrophic, unable to form crowns, and consequently are not adapted to horticultural conditions (Tendille and Lecerf, 1974). Morphological deficiencies were also cited by Hasegawa et al. (1973) as the cause of transplant loss. A pretransplant reculture step under high light intensities of 10,000 lux is recommended for better results because ferning, i.e., the differentiation of cladophylls on spears, is stimulated by high light intensity. There were direct correlations between light intensity, degree of ferning, and survival of transplanted plants. However, at 3000 lux, no significant reduction in ferning or survival was observed. Immediately following transfer of plants from test tubes to the greenhouse, the pots must be covered with a plastic sheet and kept at relatively low temperatures. Survival of transplants in Jiffy 7 peat pots under intermittent mist of approximately 90% was significantly greater than under 70% relative humiditiy (Yang and Clore, 1974b).

Shoot tip explants of a very small size that undergo direct development to complete plantlets without a callus phase are considered to provide genetic stability. According to Hasegawa et al. (1973) the chromosome numbers of cell culture-derived plants have been consistent with the original plants. Another aspect of application of the shoot tip method is the possibility of elimination of virus infections by excision of apical meristems with a size of less than 0.1 mm in length (Yang, 1979). For elimination of the asparagus virus AV-1, shoot tips of 0.1-0.3 mm can be used successfully, but for AV-2, smaller meristems without leaf primordia should be used.

Regeneration of Asparagus Plantlets from Callus Cultures Started with Organ Explants

Clonal multiplication can be achieved by direct rooting of explanted shoot tips as cuttings. This method is limited to the production of one plant per cutting, or of a few plants in the case of basal branching of the primary explant. Low multiplication rates have been correlated with low survival rate after transfer to soil. To enhance the number of vegetatively propagated asparagus plants, experiments for callus formation and regeneration of adventitious organs have been conducted. Primary explants from shoot tips, apical and lateral meristems of the spear, pith tissue of basal spear segments, and single cells of the cladophylls were induced to form callus by supplementing basal medium with 2,4-D and NAA in combination with different cytokinins. Callus cultures were maintained on solid and liquid medium and stimulated to undergo organogenesis by including appropriate concentrations of the hormones.

The first step of differentiation of highly vacuolated callus cells is their transformation into densely cytoplasmic, meristemoid, or embryogenic clumps of cells. Using 0.5-cm portions of etiolated shoots on a Galston-Loo medium with 27.0 µM NAA, Gorter (1965) induced a basal callus with root initals that developed into plantlets in liquid medium augmented with boric acid. Information on propagation efficiency was not given. Vigorously growing callus, derived from hypocotyls of sterile seedlings on basal medium of Linsmaier-Skoog (1965) (LS) with 4.5 µM 2,4-D and 1.5 µM KIN changed its growth pattern a short time after subculture into liquid medium of the same composition. Large numbers of dense cell aggregates developed into globular units interpreted as embryos. The number of these units increased with decreasing concentrations of 2,4-D (Wilmar and Hellendoorn, 1968). Beyond a size of 2 mm the embryos dedifferentiated again in suspension culture. Among the transferred plantlets some tetraploids were found. Spear slices freed of lateral buds form callus on the basal end of the disk within 4-6 weeks on MS medium supplemented with 2.7 µM NAA and 15% CW (Takatori et al., 1968). Within a spear, disks obtained from the region more distal to the apex produced better callus growth than those adjacent to the apex. Shoot formation was enhanced by addition of 50 mg/l ADE. All callus-derived plantlets were diploid in spite of the ploidy changes observed in the callus tissue. Ploidy levels of asparagus plants propagated by rooted cuttings and by callus culture have been compared (Malnassy and Ellison, 1970). It was observed that more plants derived from callus culture were tetraploid than those from rooted cuttings.

Replacement of 2,4-D by NAA, especially in combination with BA, favored root and shoot regeneration in callus from pith explants (Yakuwa et al., 1971a,b). The optimal concentrations of BA and NAA varied between 0.1 and 1 mg/1, whereas 10 mg/1 of one of the hormones reduced yield of plantlets. Hunault (1975) reported a distinct influence of the basal medium on callus proliferation. For example, LS enriched with thiamine, nitrate, and ammonium ions proved to be most effective compared with six other basal media. Harada (1973) described the effects of different auxins, cytokinins, and casein hybrolysate on morphogenetic responses of asparagus stem explants on MS medium for the formation of callus, root, shoot, and somatic embryos. Callus growth was stimulated more by 0.5-5.4 µM NAA than with 2,4-D; whereas ZEA, BA, and KIN were equally active. Adventitious roots developed slightly more with 5.4 µM NAA than with 2,4-D; the inhibitory effect on rooting was highest with 0.05 µM KIN, moderate with BA, and lowest with ZEA. From callus, NAA (0.5 µM) in combination with ZEA (4.6 µM) favored regeneration of shoot whereas BA and KIN were less effective; casein hydrolysate was stimulatory only for shoot development.

Histological studies of cultured stem pieces during callus induction initiated by 2,4-D in the light revealed that callus started from a subepidermal diffuse cambium with numerous tracheids (Hunault, 1973). Further development of the callus culture resulted in formation of numerous roots from the surface of the callus, but the roots had an atrophied morphology. The abnormal structure of adventitious roots promoted by 2,4-D and in some cases by NAA is responsible for the great losses of in vitro plantlets after transfer to soil. These atrophied roots degenerate in soil and do not retain the capacity for water and mineral supply absorption.

In order to improve the multiplication rate from callus cultures and to provide differentiation of adventitious organs without morphological deviations and functional disorders the propagation procedure was divided into two different steps (Reuther, 1977a,b, 1978, 1979). From excised apical and lateral apices of white and green spears on a solid LS basal medium supplememted with 5.4 µM NAA and 4.6–9.3 µM KIN, a proliferation phase was initiated after subculture (Fig. 7a). After callus formation and proliferation (Fig. 7b), callus clumps were transferred onto the same basal medium with 5.7 µM IAA and 0.4–4.4 µM BA or 0.5–4.9 µM 2iP. The majority of callus clumps developed shoots first (Fig. 7c), or simultanously developed shoots and roots (Fig. 7d), but isolated root formation was not observed. Callus-derived shoots formed roots rapidly on LS medium with 2.9 µM IAA (Fig. 7e). By replacement of NAA with IAA the high rooting tendency in callus is reduced and the shoot initials differentiate. With crown structures on the base of the shoots and roots very similar to those of seedlings the plantlets continued to grow under greenhouse conditions without mortality losses (Fig. 7f,g).

Development of Callus Cultures from Isolated Single Cells and Protoplasts

Mesophyll cells separated mechanically from cladophylls were plated on modified MS media with oligoelements (Nitsch, 1969), NAA, and BA (Jullien, 1973). Survival and mitotic activity depended on the concentration of cells. A high plating density (3.5 x 10^5 cells/ml) allows mitosis in up to 40% of the living cells that develop to very small colonies of approximately 10 cells each. With the reduction of cell concentration to about 5 x 10^4 cells/ml the percentage of division decreases to 5% but the residual colonies increase in size. Callus proliferation originating from single cells may be enhanced by the addition of NAA, BA, or ZEA. When mineral nitrogen, NO_3^- and NH_4^+, is omitted in the basal medium and replaced with glutamine (30 mM) survival is higher and the mitotic activity of the cell populations increases (Jullien and Guern, 1979). This phenomenon is related to the special nitrogen metabolism of asparagus cladophylls which lack nitrate reductase.

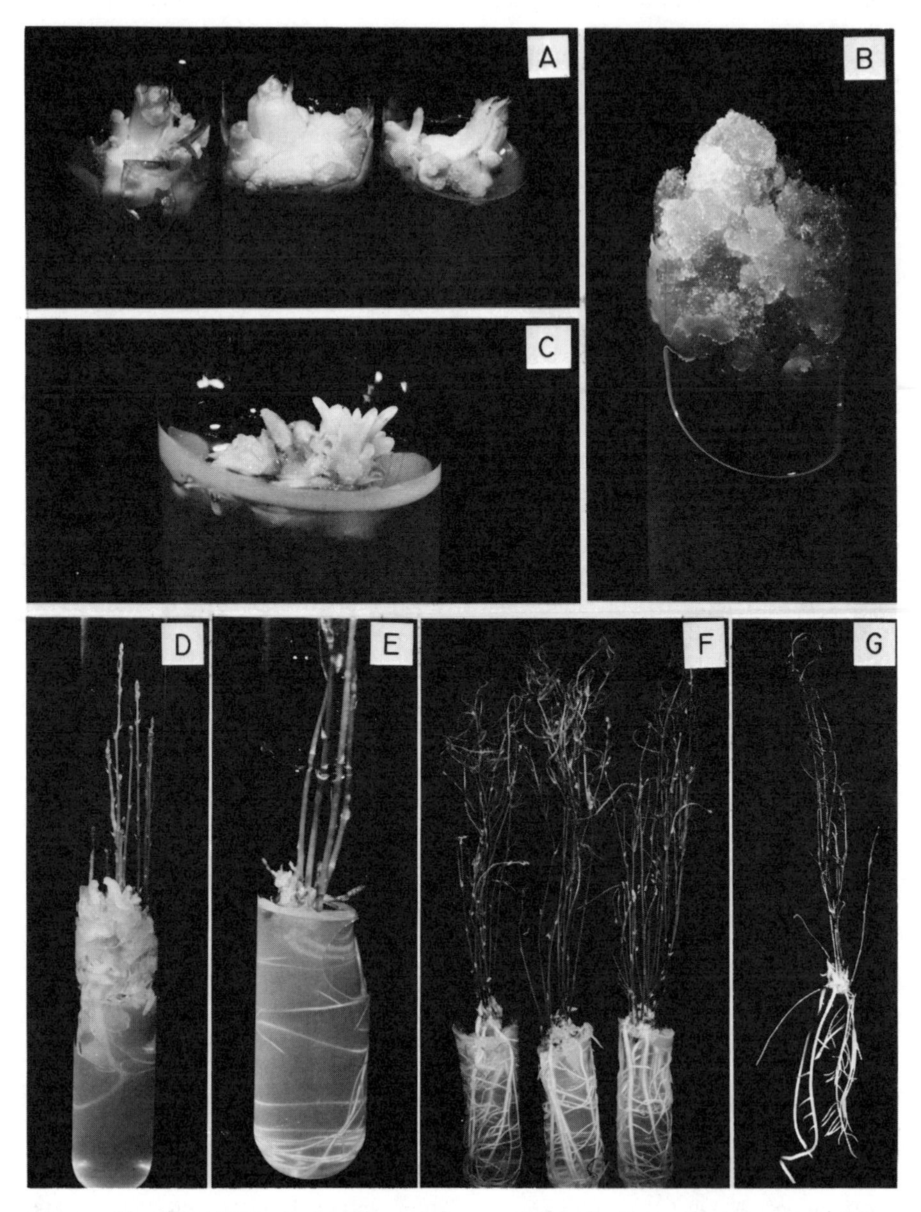

Figure 7. (a) Explanted apices of cv. Huchels 5 weeks on mod. LS + 5.4 μM NAA + 4.6 μM KIN; callus induced on the basal end of the explant in contact with the medium; shoot tips transversally enlarged and longitudinal growth inhibited. (b) Second callus subculture of cv. Schwetzinger Meisterschuss on mod. LS + 5.4 μM NAA + 4.6 μM KIN; proliferating callus of a high degree of differentiation. (c) Crowns

isolated from callus phase of a supermale genitor; 1 week on mod. LS + 5.7 μM IAA + 0.98 μM 2iP. (d) Organogenic callus of a supermale genotype with shoots, crown initials, roots; 8 weeks on mod. LS + 5.7 μM IAA + 0.44 μM BA. (e) Transferred shoots to mod. LS + 2.9 μM IAA, 8 weeks; formation of crown, expanding shoots and roots. (f) Completed supermales before planting into soil after 3 months of the transfer to the rooting medium. (g) Female callus-derived plant.

Organogenesis in callus derived from protoplasts of mesophyll cells of cladophylls was reported by Bui Dang Ha (1973). Completely undifferentiated callus was formed starting from a protoplast concentration of 3 to 5 x 10^5/ml on a liquid medium and subsequent transfer of the colonies of 1 mm diameter to a solid MS medium with microelements and vitamins according to Nitsch. Callus proliferation was promoted by addition of NAA, BA or ZEA. In the presence of ZEA and 2,4-D no shoot initiation was observed (Bui Dang Ha et al., 1975). The regeneration potency of the callus tissue was stimulated by combinations of IAA (0.5-2.5 μM) and BA (4 μM). Globular and banana-shaped units, described as somatic embryos, originated from callus on the basal medium lacking growth regulators (as a subculture step) after transfer from a medium supplemented with ADE (300 μM), NAA, and BA.

In Vitro Androgenesis in Asparagus Anther Culture

The reasons for the application of anther culture for the establishment of homozygous female and male parent plants for the breeding of all-male F_1 hybrid asparagus varieties were discussed above. Many attempts have been made to induce androgenesis in asparagus anthers based on a relatively simple method for the regeneration of haploids from explants of *Nicotiana* anthers. Anthers from flower buds of different stages of maturity were placed on a basal medium of the following composition: MS macroelements, microelements according to Heller without $FeCl_3$, 0.1 mM $FeSO_4$ ($7H_2O$), 0.11 mM Na_2EDTA, a mixture of B vitamins according to Morel, 0.058 M sucrose and 5 g agar (Pelletier et al., 1972). Histological studies revealed proliferation of haploid cells stimulated by addition of 0.5 μM NAA, 0.9-4.4 μM BA, and 4.5 μM 2,4-D. The frequency of anthers producing at least one androgenic callus varied from 4 to 20% depending on the genotype (Doré, 1973). Plantlets regenerated from anther callus contained a range of ploidy levels, i.e., n, 2n, 3n, 4n, and 6n (Doré, 1974). Spontaneous diploidization in originally haploid callus results in homozygous male and female plants. More detailed cytogenetic studies in cells of androgenic callus showed a high karyotypic instability (Cateland and Lambert, 1973). After 2 months the average ploidy level was 6n. Under the same conditions callus derived from somatic tissues had a lower grade of mixoploidy. In spite of mixoploidy of androgenic callus, adventitious buds were homoploid. In general, no haploid asparagus plants have been obtained via anther culture. In most cases diploids were derived from pollen via diploidization of callus. This can be verified by test-crossing the male progeny; in the case of uniformly male F_1 hybrids, the supermale character of the pollinator has been recovered (Doré and Corriols, 1980). Falavigna (1979) has observed variation in callus formation in relation to genotype. After culturing 95 genotypes only 7 yielded more than 3% callus. The regeneration capability of callus also could be related to specific anther donors. According to the ploidy properties, two different callus groups were distinguished. An 'isoploid callus' gives rise to plantlets with the same ploidy level in spite of the cytological heterogeneity of the callus cells; a 'mixoploid' callus gives rise to plantlets with different ploidy levels.

When anthers were cultured at the uninuclear stage of pollen on MS medium with 0.4 µM BA and 5.4 µM NAA, a high percentage of anther cultures produced callus (79.3%). On the other hand, anthers at mature stage formed callus at a considerably lower rate (3.9%) (Yakuwa et al., 1972). Diploid and tetraploid chromosome complements were observed in plants regenerated by this system. The development of callus from diploid anther tissue (i.e., anther wall), connective tissue and the cut end of the filament on MS medium substituted by various concentrations of sucrose and NAA must be taken into consideration (Hondelmann and Wilberg, 1973).

The success of asparagus anther culture is highly affected by genetic and environmental factors. Falavigna et al. (1982) examined nearly 98,000 cultured anthers, taken from 20 heterozygous genotypes and 9 pure lines, to determine the influence of size of the flower bud environmental conditions of anther donor plants, periods of cold treatment, and modifications of the medium on anther culutre.

Summary of the Literature Review

In the last 10 years significant progress has been made in vegetative propagation of asparagus by the application of tissue culture techniques. Direct rooting of stem tip cuttings and explanted meristems with cladophyll primordia was promoted with various concentrations of NAA. Cytokinins stimulated shoot growth from explanted initials. By means of explanting very small meristems, virus elimination was achieved. The survival of plantlets after transfer to soil was improved under high light intensities, by changing nitrate supply, and by horticultural management. The establishment of moderately growing callus cultures with high regenerative potential from organ explants provides a very suitable system for the differentiation and rapid multiplication of asparagus plantlets. The replacement of 2,4-D and NAA by IAA in combination with various cytokinins resulted in callus growth and root formation in combination with crown formation. Thus morphologically conditioned plantlets adapt easily to a greenhouse environment. The formation of callus-derived somatic embryos has been described in several papers usually based on morphological similarities with zygotic embryos. It seems in some cases shoot or root fragments were interpreted as embryos.

An important development was the regeneration of asparagus plantlets from callus originating from single cells and protoplasts of cladophyll mesophyll cells. This method offers future prospects for the induction of somatic mutations and screening for disease resistance or other factors in crop improvement. Considering the long breeding cycle in asparagus and the difficulties in obtaining supermale inbred lines, the development of homozygous supermale and superfemale plants from anther callus will shorten the breeding procedures for the development of male F_1 hybrids. But the high genetic instability of asparagus callus and especially of anther callus is still a problem.

PROTOCOLS

Cloning Techniques

Two different ways of cloning have been identified for asparagus: (1) direct one-step organogenesis or (2) callus-mediated plant regeneration. For the induction of organ development from primary explants such as anthers, single cells, and protoplasts, callus formation is an obligatory first step. Somatic organ explants also easily produce callus before adventitious organs are differentiated. On the other hand proliferating and differentiating callus is a very suitable material for achieving high multiplication rates. However, this method requires that the regeneration potency of callus strains obtained from different genotypes be maintained over a long period of time in culture.

The morphological and histological features of callus from organ explants in the course of dedifferentiation, proliferation, and organization, as well as the media related to each phase, are described according to Reuther (1977a,b, 1978, 1979) supplemented with previously unpublished data (Table 6). Steps one and two are proliferation phases whereas step three represents the organogenic phase. The process of callus formation should be achieved in NAA- and KIN-supplemented medium so that root formation is suppressed. Callus lines with high regenerative potential have been maintained for 5 years on IAA-BA media. The relatively low or moderate callus-proliferation capacity on IAA-BA medium may result in a gradual decrease in the growth rate. Therefore callus clumps should be transferred from time to time onto NAA-KIN medium in order to enhance callus proliferation. Depending on the genotypic origin of the callus culture, many subculture steps on NAA-KIN medium (more than 10) may decrease the regenerative potential. However, genotypes exist with more or less constant organogenic properties while in the NAA-KIN callus-proliferation medium. The tendency for callus formation from primary explants is highest in female and male plants and lowest in supermale plants. The demand for BA or 2iP for organogenesis (e.g., shoot initiation) varies among cultivars.

Histological investigations of proliferating callus revealed rounded, highly vacuolated cells with an extremely thin cytoplasmic film along the cell wall and wide intercellular spaces without reserve metabolites (Fig. 8a). The transformation of callus cells to cells with organogenic potential is indicated first by protein synthesis in the cytoplasm with subsequent secretion into the vacuole (Fig. 8b). The granules move into the center of the vacuole and fuse into larger, irregularly shaped

Table 6. Scheme for Asparagus Callus Cultures and Adventitious Organ Formation (after Reuther, 1978, 1979)

STEP	TREATMENT	RESPONSE
1	Shoot tips (1-2 mm) are excised from apical or lateral meristems of white or green spears and placed on modified LS medium[a] with 5.4 μM NAA and 4.6 μM KIN; 26 C, darkness for 4-6 weeks	Explants grow atypically and slowly, producing callus consisting of solid areas with some more friable sectors (Fig. 7a)
2	Tissues are subcultured onto medium of the same composition; 2-4 subcultures of 4-6 weeks each	There is increasing dedifferentiation and proliferation of the tissue, the extent dependent on genotype. Callus is friable (Fig. 7b)
3	Callus is transferred to mod. LS with 5.7 μM IAA and 0.44 μM BA or 2.5 μM 2iP; 26 C, light (2000-3000 lux) for 4-6 weeks	At lower growth rate, callus gradually differentiates to form shoot and crown initials, occasionally with root formation that may or may not connect to shoots (Fig. 7c,d)
4	Excised shoot and crown initials are placed on mod. LS with 2.9 μM IAA, 26 C, light for 4-6 weeks	Roots develop, shoots continue to grow with formation of cladophylls (Fig. 7e,f)
5	After removal of agar, plantlets are transferred to Jiffy 7 pots or mixture of peat (TKS_1) with sandy soil and kept in humid atmosphere for 3-4 weeks	Plants form consisting of numerous shoots with cladophylls, crown, and roots (Fig. 7f,g)

[a]LS = Linsmaier-Skoog (1965) modified by higher concentrations of KH_2PO_4 (1.84 mM) and KNO_3 (24.7 mM).

protein bodies. In the course of organ formation the protein droplets are metabolized. Metabolism is associated with subsequent mitotic activity (Fig. 8c). The next stages of organogenesis indicate irregular distribution of procambial initials and the appearance of shoot buds on the periphery of the callus (Fig. 8b and d). It is likely that the

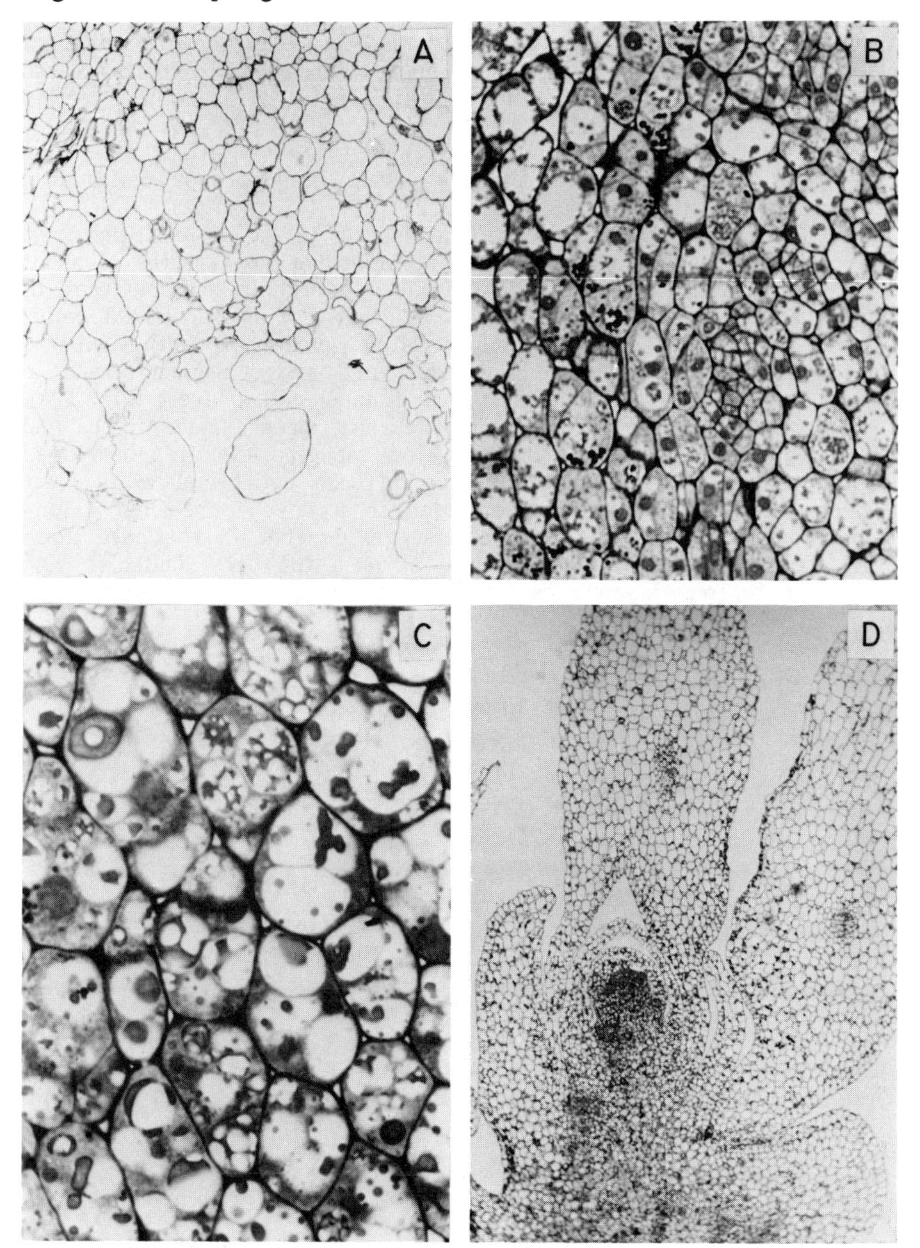

Figure 8. (a) Highly vacuolated cells from proliferating callus; mod. LS + 5.4 μM NAA + 4.6 μM KIN. (b) Differentiating callus developed on mod. LS + 5.7 μM IAA + 0.44 μM BA; (left) cells with increasing content of cytoplasm and cytoplasmic origin of protein droplets segregating through the tonoplast into the vacuole; in some cells starch deposition (black particles); in the central region meristemoid cells forming pro-

vascular bundles. (c) Irregularly shaped proteins showing different stages of decomposition with simultaneous increase of cytoplasm. (d) Longitudinal section through a shoot initial enclosed in a cavity formed by an enlarged cladophyll.

pattern of differentiation of storage tissue in asparagus callus is a prerequisite for organ development.

The histological structures of organs (shown in Fig. 7) emerging from the peripheral layers of the callus represent various features of shoot buds and somatic embryos. Two different organization patterns of shoots have been identified (Fig. 9). Originating from superficial callus layers, buds consist of an apical meristem, cladophyll initials in the axil of scales which enclose the apex, and a procambium strand in loose connection with tracheidal nodes in inner-callus layers (Fig. 10a). In general these buds grow rapidly, turning green under light, and white crown initials with limited growing capacity are often developed at their base (Fig. 10b). Histologically these cylindrical white units are composed of a shoot bud enclosed in a cave formed by a long tissue coat (Fig. 8d). It has been suggested that in the past these units were misinterpreted as embryos because they are similar in size and shape to somatic embryos. Somatic embryogenesis is characterized

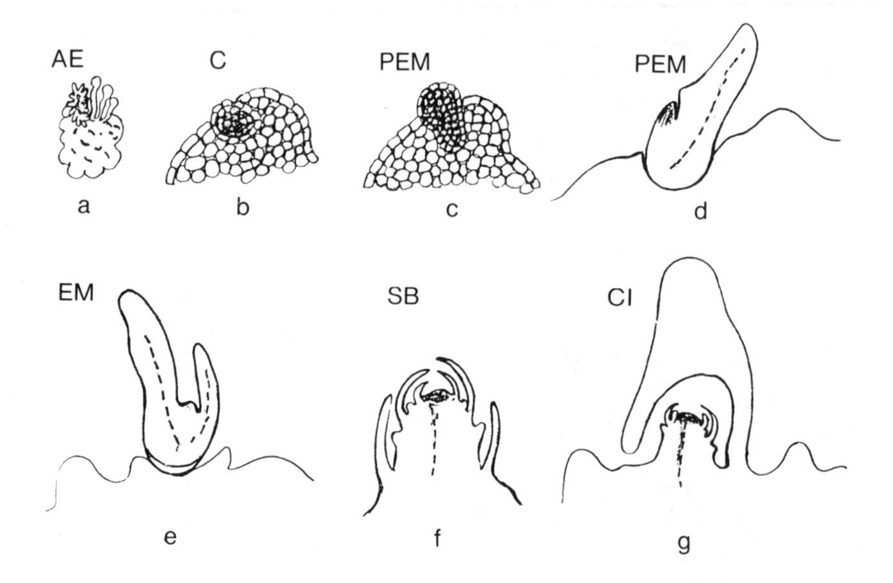

Figure 9. Scheme of differentiation patterns of adventitious organs originating from asparagus callus (a) primary culture starting with apex-explant (AE) with shoot and crown initials and callus, (b) callus differentiating meristemoid units; (c) proembryoid (PEM) in callus; (d) intermediate stage of proembryoid; (e) final stage of embryoid (EM); (f) shoot bud initials; (g) shoot bud initials enclosed in hypertrophically enlarged cladophyll interpreted as crown initials.

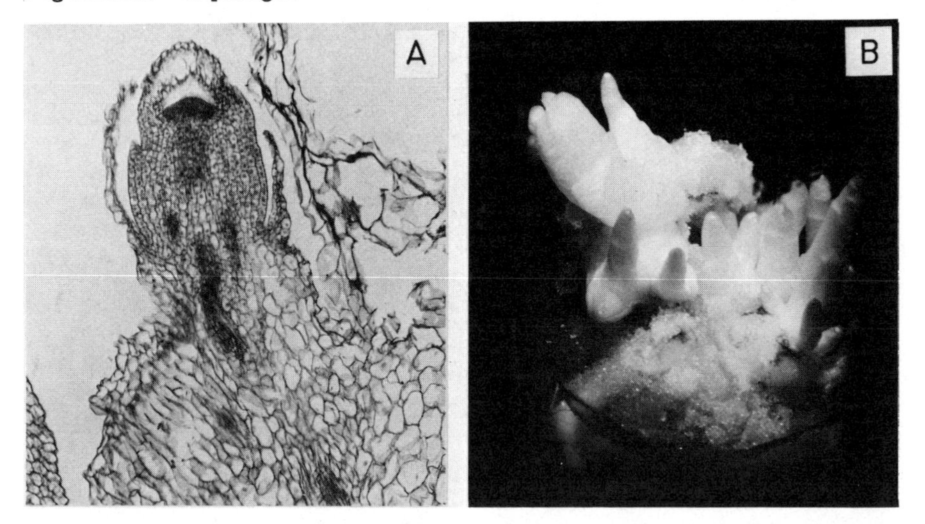

Figure 10. (a) Longitudinal section through a shoot bud of asparagus indicating apical meristem, scales and cladophyll initials, procambial strand; callus tissue consisting of vacuolated cells and tracheidal nodes; male genotype on modified LS + 5.7 μM IAA + 0.44 μM BA. (b) Numerous shoot initials differentiated in the second subculture step of callus on modified LS + 5.4 μM NAA + 4.6 μM KIN; cv. Lucullus. Shoots are of the same histological constitution as described in Fig. 3.

by segmentation into multicellular, globular, proembryonal cell complexes surrounded by a protoderm layer. The proembryos elongate to bipolar structures with shoot and root poles and separate more or less from the callus (Fig. 11a-c). Compared with the findings for somatic embryogenesis of other monocotyledonous plants such as iris (Reuther, 1977a,b), the asparagus embryos deviate from the structure of zygotic embryos.

A great number of progenies of inbred supermale genotypes with tested combining ability (pollinator of cv. Lucullus) and also female parent plants were propagated vegetatively from callus cultures. Chromosome counts in preparations of squashed root tip cells of greenhouse plants have shown no changes in ploidy levels, although abnormally enlarged callus cells probably of high polyploidy were found. The genetic identity of supermale and superfemale clones obtained by callus cultures and plants propagated directly from rooted shoot tip explants was compared with the original YY-inbred line and with the female progenitors (Greiner, 1979. From the original inbred line and each of 50 callus-derived and shoot tip-propagated supermale plants, 50 females were pollinated. No differences in seed productivity and development of the offspring were detected. In the field, callus-derived plants had improved vigor expressed by a greater number of shoots and a distinctly higher growth rate.

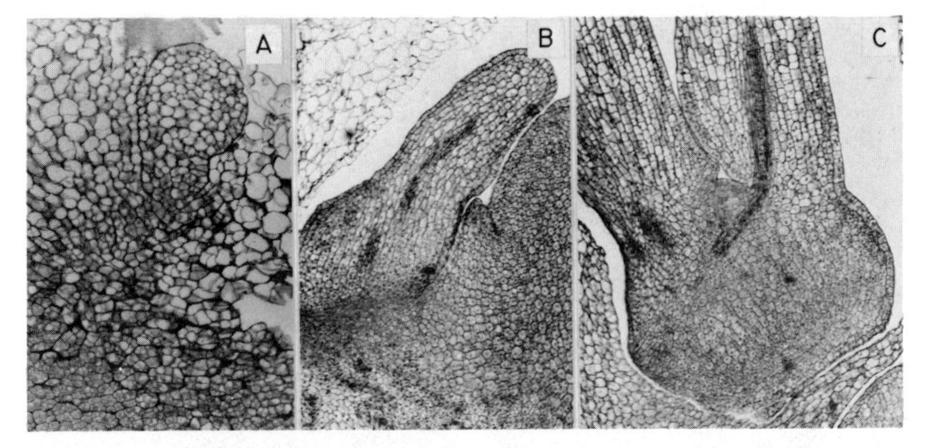

Figure 11. (a) Globular proembryoid differentiated in callus of a supermale on mod. LS + 5.7 μM IAA + 0.44 μM BA. (b) Later stage of somatic embryoid formation indicating cotyledonlike structure with procambium strand and lateral shoot pole initial. (c) Embryo with shoot pole, root pole, calyptra (abnormally flat) and deviating cotyledon; note the bipolar structure.

Homozygous Asparagus Lines from Anther Culture

The importance of homozygous supermale and superfemale progenitors for breeding all-male F_1-hybrid cultivars has been discussed. The success in androgenesis is markedly influenced not only by the appropriate composition of the medium but also by the developmental phase of the explanted anthers and the environmental conditions of the donor plants. According to Falavigna et al. (1982) the maximum androgenic potential is expressed in callus formation under the following conditions: (a) length of flower bud (uninucleate microspore) of 1.6–2.1 mm, (b) cold pretreatment of flower buds at 7 C for 10 days (dependent on genotype), (c) solid medium (on liquid medium only half of the anthers responded), (d) 0.12 mM sucrose, and (e) anthers taken from outdoor plants in the summer respond better than those from greenhouse plants.

Only 2–3% of 98,000 explanted anthers resulted in callus formation. Composition of the medium for anther callus formation (Pelletier et al., 1972; Doré, 1974) was as follows: MS macroelements, diluted twice; microelements without $FeCl_3$ according to Heller (1953); vitamins according to Morel and Wetmore (1951); 0.1 mM $FeSO_4$ ($7H_2O$); 0.11 mM Na_2EDTA; 0.058 mM sucrose (0.12 M, see Falavigna, 1981); 6 g/l agar-agar, pH 5.8; and 2.2 μM BA, 2.3 μM 2,4-D, and 0.5 μM NAA.

The small anther-derived callus should be transferred after 4 to 6 weeks to the same basal medium supplemented with 1.3 μM BA. Stems derived from callus are then excised and cut into pieces with a bud for further propagation and subsequent rooting. A subculture medium where the cuttings are grown to plantlets is used either without hormones (Falavigna, 1979) or with 5.4 μM NAA, 0.5 μM KIN, and 0.3 μM GA (Doré, 1975).

Cultures from Isolated Single Cells

The method was described by Jullien (1973) and Jullien et al. (1979). The cladophylls used for isolation of mesophyll cells were taken one week after blooming from 30-day-old spears grown in the greenhouse. A mechanical method was used for cell separation. Cladophylls were surface sterilized for 20 min with 7% Ca-hypochlorite, washed, and then ground in 30 ml of an 0.12 M sucrose solution. After washing three times with sucrose solution the cell suspension was diluted to 50 ml containing approximately 5 x 10^6 cell/ml.

A liquid medium is recommended consisting of MS macroelements, Nitsch (1969) microelements, and Nitsch and Nitsch (1965) organic supplements without glycine. The inorganic nitrogenous compounds were replaced with 30 mM glutamine, potassium was added as KCl; and the growth regulators NAA (5 µM) and BA (0.5 µM) were added. All media were sterilized by filtration through millipore filters. The final inoculation density was 2 x 10^5/ml in 25 ml of liquid medium in a 250 ml Erlenmeyer flask, agitated at a speed of 100 rpm in continuous fluorescent daylight (6000 erg $cm^{-2}s^{-1}$) with a temperature of 26 ± 1 C. Glutamine was vital for a better survival rate of isolated cells and earlier entry into mitotic activity. The age of the cladophylls affects the yield of living cells and the further development of the cell cultures.

FUTURE PROSPECTS

The genetic instability of asparagus callus, i.e., mixoploidy, is present mainly in callus originated from anthers, but is also observed in somatic explants. Genetic stability is a prerequisite for the cloning of progenitors with tested combining ability as well as for mass propagation of high-yielding genotypes selected from various cultivars. It has been demonstrated that callus strains maintain a high regeneration potential over numerous subculture steps, an asset which provides for the formation of several shoots in a test tube together with several crown initials. Changes in ploidy compared with the original genotype are related to the effects of 2,4-D and NAA in the media. It may be valuable to use IAA and other indole derivatives in combination with cytokinins to achieve greater genetic stability in callus cells. It is important to know how the genetic stability varies after many subcultures and to identify those conditions that stabilize anther derived callus. Numerous reports confirm the regeneration of diploid plants from mixoploid callus, so that euploid cells may be able to regenerate better than aneuploid cells. All-male F_1 hybrid cultivars are the most desired varieties for asparagus cultivation. The superiority in yield has already been demonstrated with progenies from inbred supermales. Direct regeneration of haploid asparagus plants from cultured anthers is rare and therefore the homozygosity of regenerated diploids must be confirmed by test crossings. Further attempts should be made for direct androgenic development of plantlets to improve the regeneration frequency from cultured anthers.

For a mutation breeding program, the genetic instability of callus may permit selection of genotypes with improved yield or disease resistance. Because cultivated asparagus is affected by several viral and fungal pathogens, selection for resistant genotypes should be continued by exposure of callus and in vitro plantlets to pathogen filtrates. Successful recovery of organogenic callus from isolated mesophyll single cells and protoplasts also encourages the establishment of a mutation breeding program in asparagus.

ACKNOWLEDGMENT

For providing data on asparagus production I am indebted to colleagues in different countries. I gratefully acknowledge the excellent assistance of Mrs. S. Hell.

KEY REFERENCES

Arbeitskreis Spargelbau in Suedwestdeutschland. 1974. Ingelheim-Geisenheim (H.D. Hartmann, ed.). Annual Report.

Asparagus Research. 1980–1981. Vegetable Crops Series; Dep. of Vegetable Crops, University of California, Davis. Annual Report.

Hung, L. 1975. Annotated Bibliography on Asparagus, p. 547. Publ. by Department of Horticulture, National Taiwan University, Taipei, Taiwan, Republic of China.

Reuther, G., ed. 1979. Proceedings of the 5th International Asparagus-Symposium, 314 pp. Eucarpia Section Vegetables, 1979 Geisenheim. Geisenheim Forschungsanstalt, Germany.

Thevenin, L., ed. 1973. 4eme Reunion sur la Selection de l'Asperge, Eucarpia Section Horticole, Versailles.

REFERENCES

Andreassen, D.C. and Ellison, J.H. 1965. Produce roots on asparagus cuttings. N.J. Agr. 47(3):7.

______ 1967. Root initiation of stem tip cuttings from mature asparagus plants. Proc. Am. Soc. Hortic. Sci. 90:158–162.

Beeskow, H. 1967. Untersuchungen ueber die Variabilitaet der Andromonoezie/Dioezie und ihre Korrelationen mit verschiedenen Ertragsfaktoren bei Spargel (Asparagus off. L.) unter besonderer Beruecksichtigung der Zuechtung rein maennlicher Sorten. Z. Pflanzenzuecht. 57:254–283.

Benson, B.L. 1979. Photosynthate partitioning in open-pollinated and F_1-hybrid asparagus cultivars. In: Proc. 5th International Asparagus Symposium (G. Reuther, ed.) pp. 60–73. Eucarpia Section Vegetables.

Bui Dang Ha, D. 1973. Formation de tiges, racines, embryoides et plantes entieres a partir des cals d'asperge provenant de protoplastes. In: 4eme Reunion sur la Selection de l'Asperge (T. Thevenin, ed.) pp. 155-162. Eucarpia Sect. Horticole, Versailles.

______, Norreel, B., and Masset, A. 1975. Regeneration of *Asparagus officinalis* L. through callus derived from protoplasts. J. Exp. Bot. 26:263-270.

Cateland, B. and Lambert, A. 1973. L'androgenese in vitro par culture d'antheres d'*Asparagus officinalis*. II. Les problems de polyploidie. In: 4eme Reunion sur la Selection de l'Asperge (L. Thevenin, ed.) pp. 183-189. Eucarpia Sect. Horticole, Versailles.

Corriols-Thevenin, L. 1979. Different methods in asparagus breeding. In: Proc. 5th International Asparagus Symposium (G. Reuther, ed.) pp. 8-20. Eucarpia Section Vegetables. Geisenheim Forschungsanstalt, Germany.

Doré, C. 1973. Androgenese in vitro par culture d'antheres d'*Asparagus officinalis*. I. Etat actuel des recherches. In: 4eme Reunion sur la Selection de l'Asperge (L. Thevenin, ed.) pp. 173-181. Eucarpia Sect. Horticole, Versailles.

______ 1974. Production de plantes homozygotes males et femelles a partir d'antheres d'asperge, cultivees in vitro (*Asparagus officinalis* L.). C.R. Acad. Sci. Paris 278:2135-2138.

______ 1975. La multiplication clonale de l'asperge (*Asparagus officinalis* L.) par culture in vitro: Son utilisation en selection. Ann. Amelior. Plantes 25:201-224.

______ and Corriols, L. 1980. Divers aspects de la culture in vitro chez l'asperge. In: Application de la Culture in Vitro a l'Amelioration des Plantes Potageres (L. Theuenin, ed.) pp. 82-85. Reunion Eucarpia, Section Legumes, Versailles.

Falavigna, A. 1979. Report on pure line production of *Asparagus officinalis* L. by in vitro culture in Italy. In: Proc. 5th International Asparagus Symposium (G. Reuther, ed.) pp. 91-99. Eucarpia Section Vegetables, Geisenheim Forshungsanstalt, Germany.

______ 1981. Miglioramento genetical dell'asparago di altedo. Istituto Spermentale per l'Orticultura di Salerno.

______, Chiapperini, E.E., and Soressi, G.P. 1982. Influenza de fattori genetici ed ambientali sull'androgenesi in vitro di asparago (*Asparagus officinalis* L.). Genetica Agraria 36:

Fiala, V. and Jolivet, E. 1979. Variation in biochemical composition of *Asparagus officinalis* L. Roots in relation with sex expression and age. In: Proc. 5th International Asparagus Symposium (G. Reuther, ed.) pp. 74-81. Eucarpia Section Vegetables, Geisenheim Forschungsanstalt, Germany.

Gorter, C.J. 1965. Vegetative propagation of *Asparagus officinalis* by cuttings. J. Hort. Sci. 40:177-179.

Greiner, H.D. 1974. Vegetative Vermehrung von Spargel (*Asparagus officinalis* L.) durch die Kultur von Sprosspitzen. Gartenbauwissenschafts 39:549-554.

______ 1979. Phenotypic features in sexually produced progenies of vegetatively propagated asparagus strains by tissue culture. In: Proc. 5th International Asparagus Symposium (G. Reuther, ed.) pp. 150-155. Eucarpia Section Vegetables. Geisenheim Forschungsanstalt, Germany.

Harada, H. 1973. Differentiation of shoots, roots and somatic embryos in asparagus tissue culture. 4eme Reunion sur la Selection de l'Asperege (L. Thevenin, ed.) pp. 163-172. Eucarpia Sect. Horticole, Versailles.

Hartmann, H.D. 1982. Arbeitskreis Spargelanbau in Suedwestdeutschland. 10. Jahresbericht, Ingelheim-Geisenheim.

Hasegawa, P.M., Murashige, T., and Takatori, F.H. 1973. Propagation of asparagus through shoot apex culture. II. Light and temperature requirements, transplantability of plants, and cyto-histological characteristics. J. Am. Soc. Hortic. Sci. 98:143-148.

Heller, R. 1953. Recherches sur la nutrition minerale des tissus vegetaux cultives in vitro. Ann. Sci. Nat. Bot. Biol. Veg. 14:1-223.

Henslow, G. 1911. The origin and history of our garden vegetables and their dietetic values. J. Roy. Hortic. Soc. 36:590-595.

Hondelmann, W. and Wilberg, B. 1973. Breeding all-male varieties of asparagus by utilization of anther and tissue culture. Z. Pflanzenzuecht. 69:19-24.

Hunault, G. 1973. Etude de l'histogenese au cours de la formation du cal sur des fragments de tiges d'Asperge (*Asparagus officinalis* L.) cultives in vitro. Rev. Cytol. Biol. Veg. 36:335-356.

______ 1975. Influence de differents milieux de culture sur la croissance de tissus d'Asperge (*Asparagus officinalis* L.) cultives in vitro. C.R. Acad. Sci. Paris 280:2661-2664.

Jullien, M. 1973. La culture in vitro de cellules separees du tissu foliaire d'*Asparagus officinalis*: De la cellule parenchymateuse au cal organogene. C.R. Acad. Sci. Paris Serie D 276:733-736.

______ and Guern, J. 1979. Induction de la division cellulaire et croissance des populations de cellules separees du parencyme foliare chez *Asparagus officinalis* L. Physiol. Veg. 17:445-456.

______, Rossini, L., and Guern, J. 1979. Some aspects of the induction of the first division and growth in cultures of mesophyll cells obtained from *Asparagus officinalis* L. In: Proc. 5th International Asparagus Symposium (G. Reuther, ed.) pp. 103-129. Eucarpia Section Vegetables. Geisenheim Forschungsanstalt, Germany.

Linsmaier, E.U. and Skoog, F. 1965. Organic growth factors requirements of tobacco tissue cultures. Physiol. Plant. 18:100-127.

Loeptien, H. 1979a. Identification of the sex chromosome pair in asparagus (*Asparagus officinalis* L.). Z. Pflanzenzuecht. 82:162-173.

______ 1979b. Untersuchungen zur Vererbung der Andromonoezie beim Spargel (*Asparagus officinalis* L.). Z. Pflanzenzuecht. 82:258-270.

Luzny, J. 1979. The history of asparagus as a vegetable, the tradition of its growing in Czechoslovakia (CSSR) and the prospect of its further propagation and breeding. In: Proc. 5th International Asparagus Symposium (G. Reuther, ed.) pp. 82-86. Eucarpia Section Vegetables. Geisenheim Forschungsanstalt, Germany.

Malnassy, P. and Ellison, J.H. 1970. Asparagus tetraploids from callus tissue. HortScience 5:444-445.

Marks, G.E. 1973. Selecting asparagus plants as sources of haploids. Euphytica 22:310-316.

______ 1979. Hermaphrodites: Do they have a role in asparagus breeding? In: Proc. 5th International Asparagus Symposium (G. Reuther, ed.) pp. 39-41. Geisenheim Forschungsanstalt, Germany.

Matsubara, S. 1973. Population effect in lateral bud culture of asparagus and promotion of root formation by transplanting. J. Japan. Soc. Hort. Sci. 42:142-146.

Maynard-Smith, J. 1976. The Evolution of Sex. Cambridge University Press, Cambridge, England.

Meusel, H., Jaeger, E., and Weinert, E. 1965. Vergleichende Chorologie der zentraleuropaeischen Flora. VEB Gustav Fischer Verlag, Jena.

Morel, G.M. 1972. Morphogenesis of stem apical meristem cultivated in vitro: Application to clonal propagation. Phytomorphology 22:265-277.

_____ and Wetmore, R.H. 1951. Fern callus tissue culture. Am. J. Bot. 38:141-143.

Murashige, T., Shabde, M.N., Hasegawa, P.M., Takatori, F.H., and Jones, J. B. 1972. Propagation of Asparagus through shoot apex culture. I. Nutrient medium for formation of plantlets. J. Am. Soc. Hortic. Sci. 97:158-161.

Nitsch, J.P. 1969. Experimental androgenesis in *Nicotiana*. Phytomorphology 19:389-404.

_____ and Nitsch, C. 1965. Neoformation de fleurs in vitro chez une espece de jours courts: *Plumbago indica* L. Ann. Physiol. Veg. 7: 251-256.

Pelletier, G., Raquin, Ch., and Simon, G. 1972. La culture in vitro d'antheres d'Asperge (*Asparagus officinalis*). C.R. Acad. Sci. Paris 274:848-851.

Reuther, G. 1977a. Embryoide Differenzierungsmuster im Kallus der Gattungen Iris und Asparagus. Ber. Dtsch. Bot. Ges. 90:417-437.

_____ 1977b. Adventitious organ formation and somatic embryogenesis in callus of Asparagus and Iris and its possible application. Acta Hort. 78:217-224.

_____ 1978. Verklonung von weiblichen und maennlichen Spargelstaemmen durch Gewebekultur. Gartenbauwissenschaft 43:1-10.

_____ 1979. Differentiation patterns in callus and cloning of different asparagus varieties. In: Proc. 5th International Asparagus Symposium (G. Reuther, ed.) pp. 130-145. Eucarpia Section Vegetables, Geisenheim Forschungsanstalt, Germany.

Takatori, F.H., Murashige, T., and Stillman, J.I. 1968. Vegetative propagation of Asparagus through tissue culture. HortScience 3:20-22.

Tendille, C. 1974. La multiplication vegetative de l'Asperge (*Asparagus officinalis* L.) influence des divers facteurs et en particulier de la nutrition minerale sur le developpement des meristemes d'Asperge et sur l'obtention d'Asperges adultes issues de ces meristemes. Acadamie d'Agriculture de France, pp. 1127-1180. Numéro d'ordre 83234 Alengon.

_____ and Lecerf, M. 1974. La multiplication vegetative de l'asperge (*Asparagus officinalis* L.). Action de divers facteurs, en particulier de la nutrition minerale, sur le developpement des meristemes d'asperge, sur la croissance des plantules issues de ces meristemes et sur la production de plantes adultes. Ann. Amelior. Plantes 24:269-282.

Thevenin, L. 1968. Les problems d'amelioration chez *Asparagus officinalis* L. II. Haploidie et amelioration. Ann. Amelior. Plantes 18: 327-365.

Wilmar, C. and Hellendoorn, M. 1968. Growth and morphogenesis of Asparagus cells cultured in vitro. Nature 217:369-370.

Wricke, G. 1968. Untersuchung zur Verebung des Geschlechts bei *Asparagus officinalis* L. I. Ein Majorfaktor fuer die Auspraegung des Andromonoeziegrades. Z. Pflanzenzuecht. 60:201-211.

______ 1979. Breeding research in *Asparagus officinalis*—Introductory remarks. In: Proc. 5th International Asparagus Symposium (G. Reuther, ed.) pp. 1-7. Eucarpia Section Vegetables, Geisenheim Forschungsanstalt, Germany.

Yakuwa, T., Harada, T., Saga, K., and Shiga, Y. 1971a. Studies on the morphogenesis of asparagus. I. Callus formation originating in the pith tissue of asparagus spears in tissue culture. J. Japan Soc. Hort. Sci. 40:230-236.

______ 1971b. Studies on the morphogenesis of asparagus. II. Effect of auxins and 6-benzyladenine on callus and organ formation of stem pieces cultured in vitro. Jap. Soc. Hort. Sci. 40:343-353.

Yakuwa, T., Harada, T., Inagaki, N., and Shiga, Y. 1972. Studies on the anther culture of horticultural crops. I. Callus and organ formation in anther culture of asparagus. J. Jap. Soc. Hortic. Sci. 41: 272-280.

Yang, H.J. 1977. Tissue culture technique developed for asparagus propagation. HortScience 12:140-141.

______ 1979. Mass production of virus-free *Asparagus officinalis* L. plants in tissue culture. In: Proc. 5th International Asparagus Symposium (G. Reuther, ed.) pp. 156-162. Eucarpia Section Vegetables, Geisenheim Forschungsanstalt, Germany.

______ and Clore, W.J. 1973. Aerial crowns in *Asparagus officinalis* L. HortScience 8:33.

______ and Clore, W.J. 1974a. Development of complete plantlets from moderately vigorous shoots of stock plants of asparagus in vitro. HortScience 9:138-140.

______ and Clore, W.J. 1974b. Improving the survival of aseptically-cultured asparagus plants in transplanting. HortScience 9:235-236.

______ and Clore, W.J. 1975. In vitro reproductiveness of asparagus stem segments with branch-shoots at a node. HortScience 10:411-412.

CHAPTER 9

Celery

T.J. Orton

HISTORY OF THE CROP

It is likely that the umbelliferous species *Apium graveolens* L. first arose in the marshes and streambeds of southeastern Europe and the Mediterranean coast of Asia Minor, possibly to Egypt. Although little is known about the domestication of *A. graveolens*, there is some evidence that its first use was as a medicine and spice by the Greeks some 3000 years ago (Simmonds, 1976).

Some time prior to 1700 A.D., domestication of *A. graveolens* with edible vegetative structures took place. Three distinct horticultural types are cultivated presently: var. Dulce (Mill.) Pers. (commonly referred to as celery), var. Rapaceum (Mill.) Gared. Beaup. (commonly referred to as celeriac), and var. Secalinum (Mill.). All three types are biennial and grow vegetatively as a rosette.

Enlarged, succulent petioles, the economic portion of celery, are consumed raw or cooked, and are most commonly found in salads, soups, and stir-fries. Domestication of celery probably took place in Western Europe and independently in China. A preferred characteristic of fresh market celery in Western Europe is blanched, or yellow petioles. This is often accomplished by covering green plants with soil several weeks before harvest. To avoid this costly procedure, self-blanching cultivars have been developed, primarily in Western Europe, where they are grown exclusively. Self-blanching cultivars were widely grown in North America until about 1930 when a disease known as Fusarium yellows became a major economic problem. Resistance to the causative pathogen was found in an Oriental cultivar with green petioles, and selections derived from this line presently dominate in the United States and Canada.

ECONOMIC IMPORTANCE

Enlarged taproot and stem tissues are the consumed portion of celeriac. Its main use is in soups and stews, although it can be eaten raw. Celeriac was probably domesticated in Eastern Europe, where it is cultivated almost exclusively at present.

Secalinum types are cultivated for their leafy foliage, which is used as a leafy base for salads. Plants of this type produce a large number of emaciated petioles and exhibit extensive loss of apical dominance. They occupy an extremely limited area in Europe. Little research has been reported specific to Secalinum cultivars, and they will not be addressed in this chapter.

Celery seed for spice is still a popular commodity. The seed is extremely easy to produce, and celery and celeriac cultivars (and expendable breeding lines) are commonly used. However, a small number of cultivars specific to seed spice production occupy an extremely small land area in North America, Europe, and the Orient. Those from Europe and North America are sometimes referred to as "smallage." These cultivars are mainly distinguished by their annual flowering behavior and resultant short generation times. While these types are not economically important, they have provided valuable genes for linkage studies and for probing the genetics of flowering behavior (Orton, unpublished).

A. graveolens crops are used exclusively for their value in flavor and texture enhancement. Nutritionally they provide only digestive fiber. Essential oils and steroids of possible exploitation occur in seeds, but these have not yet been pursued commercially (Table 1).

Table 1. Nutritional Comparison between Celery and Other Major Vegetable Crops[a]

	AMOUNT PER 100 g				
CROP	Water (%)	Energy (cal)	Protein (g)	Fat (g)	Carbohydrate (g)
Celery	94	17	0.9	0.1	3.9
Snap bean	90	32	1.9	0.2	7.1
Lettuce	96	13	0.9	0.1	2.9
Tomato	94	22	1.1	0.2	4.7

[a]After Lorenz and Maynard, 1980.

The geographical range on which *A. graveolens* can be successfully cultivated is relatively limited. Growth is optimal at approximately 20 to 22 C, and extremes are not well tolerated. Growth is extremely slow below 17 C, and chronic exposure to temperatures less than 14 C results in stem elongation and flowering when plants are returned to higher temperatures. Symptoms of stress, such as pithiness and leaf burn, occur on plants after growth in temperatures in excess of 25 C. Consequently, *A. graveolens* crops are generally restricted to mild temperate regions during summer and coastal subtropical areas during the winter.

Both celery and celeriac also have stringent water requirements. They grow best when water is readily available and show reduced growth rate in excessive or deficient water regimes. Inadequate water supply frequently culminates in pithiness. This characteristic probably developed in *A. graveolens* as a consequence of the plentiful water in marsh habitats where it evolved.

With respect to nutrition, celery and celeriac are classified as "heavy feeders." That is, they require relatively high levels of nitrogen (23 kg/ha), phosphorus (46 kg/ha P_2O_5), and potassium (46 kg/ha K_2O). Deficiencies of one or more of these major nutrients results in undesirable stress symptoms such as pithiness. A condition known as blackheart develops in celery plants grown under a calcium deficiency, while boron deficiency can result in "brown checking." A yellowing of older leaves can develop if magnesium is limited. For all of these disorders, significant differences have been noted among cultivars with respect to severity, suggesting that a genetic basis exists for resistance and susceptibility (Sims et al., 1977). With proper management, *A. graveolens* crops can be successfully cultivated in a wide range of soil types, provided that drainage is adequate.

Uniform stands of *A. graveolens* crops at optimum densities are extremely difficult to produce. Seeds characteristically show slow and variable germination, and small seed and seedling size results in high susceptibility to microenvironmental fluctuations. Consequently, direct seeding must be done at high density, and hand thinning is required to attain a uniform stand. Slow growth, particularly at early growth stages, results in prolonged time to maturity. Transplants have been used extensively to reduce the problems of excessive growth time and poor germination simultaneously.

This difficulty in cultivating *A. graveolens* crops and the problem of limited range may ultimately have conditioned the products' supply and demand. In the case of celery in the United States, growers tend to plant approximately the same area continually, depending on experience to produce a marketable crop. High cash values are balanced by the high production costs. The market price fluctuates somewhat, but the supply is highly consistent.

Reliable figures on the total economic value of *A. graveolens* crops are not available on a worldwide basis. Hence average data of celery production in the United States between 1975 and 1978 will be used as representative. The land area devoted to celery cultivation during this period was 13,525 ha per year, producing an average total yield of 746,103 MT (55.16 MT/ha). The total cash value was $146,912,000, or $10,862 per ha. This compares with $8066 per ha for tomatoes over the same period. Approximately 90% of this production was restricted to California (55%), Florida (25%), and Michigan (10%). These figures take total production into account, but most of the crop is used for fresh market and a small portion for canning.

GENETICS AND CROP IMPROVEMENT

Available germplasm for genetical studies and crop improvement is extremely limited. There has been little collection of wild plants in the center of origin of celery, and native habitats are being progressively destroyed. Collections of cultivars of recent origin are maintained by government ministries in a small number of countries (e.g., the United States, the U.S.S.R., the German Democratic Republic, and China). The U.S. collection consists of 34 celery and 45 celeriac cultivars (some smallage) of international scope, nearly all maintained at the Geneva, New York, repository.

Cytogenetically, *A. graveolens* behaves as a true diploid (2n = 2x = 22). The chromosomes are relatively large and centromere positions are usually discernible. The karyotype consists of seven long and two short acrocentrics, one metacentric, and one telocentric with no apparent satellites (Fig. 1). Spontaneous change in chromosome number by Robertsonian translocation has been observed (Sharma and Bhattacharya, 1959; Williams and Collin, 1976a).

Long generation times, lack of definitive phenotypes, difficulty in affecting controlled hybridization, and low seed:pollination ratio have all contributed to the lack of basic inheritance research in *A. graveolens*. Until recently, only one Mendelian gene had been characterized: conditional pithiness (Emsweller, 1933). Research in our laboratory has now characterized 16 additional nuclear genes, and seven of these have been tentatively mapped with respect to each other (Fig. 2) (Arus and Orton, 1983). Increased knowledge of the organization of the *A. graveolens* genome will ultimately facilitate plant breeding efforts, as it has in corn and tomato.

Celery and celeriac breeding programs exist mostly in universities and research institutions, and not in private concerns, because of the relatively low economic value of seed. However, a few seed companies

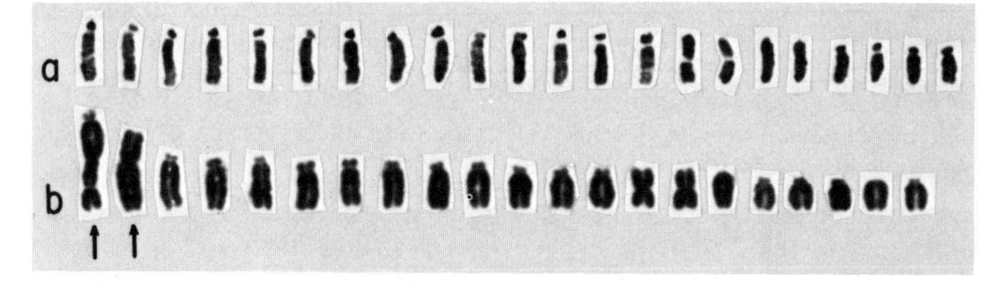

Figure 1. Karyotype of *A. graveolens* PI 169001 x cv. Tall Utah 52-70 R F_1 (a) native (root tip) and (b) suspension cell, prepared as described by Murata (1982). Arrows denote chromosomes with altered morphology as compared to the native karyotype.

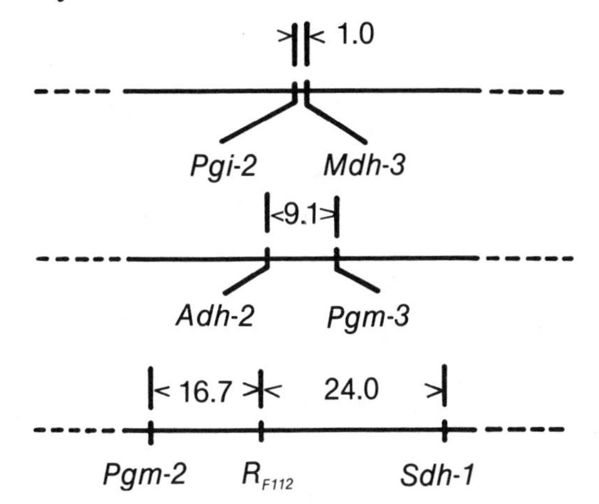

Unmapped: **Adh-1, Got-1, Aps-1, Pgi-3, Var-1, Var-2, Ms-8, Ms-143,* +*A-112*

**Adh-1* unlinked to *Adh-2*

+*A-112* unlinked to *Pgm-2*, R_{F112}, and *Sdh-1*

Figure 2. Summary of known genetic markers as of Jan. 1983 and their linkage relationships (Arus and Orton, 1983).

in Europe and North America have been active in this area. *A. graveolens* is considered an outcrosser with respect to preferred breeding methods, although genetic evidence suggests that it is not obligate (Orton and Arus, 1982; Arus and Orton, in preparation). Virtually all commercial cultivars have been developed by mass selection and random mating. The advantages of this technique are low cost and residual variability; the latter probably buffers against environmental stresses and permits a broad range of cultivation. Disadvantages include loss of uniformity, progressive attenuation of genetic base, and yield losses due to inbreeding depression.

The most important attributes of potential new commercial cultivars are earliness, stable high yield, uniformity, broad adaptation, and acceptable horticultural type (particularly with celery) for the intended market. Acceptable horticultural type in celery is the most difficult composite breeding objective. Components of horticultural type include petiole shape, thickness, succulence, ribbiness (external manifestation of vascular bundles), color, the manner in which petioles are borne on the rosette, overall height, and heart (crown and taproot) size. Nothing specific is known about the inheritance of variation for these characters; hence breeding has been by intuition.

A more or less extended period of vegetative growth is necessary to produce a marketable crop of either celery or celeriac (150–200 days from seed). Thus flowering is undesirable from a vegetable production standpoint, but absolutely essential for purposes of seed production.

Commercial cultivars represent a compromise between these opposing yet necessary attributes.

A great deal is known about the factors which mediate bolting (stem elongation) and flowering in *A. graveolens.* Accumulating exposure to temperatures less than 12 C (optimally greater than 5 C) for 4-6 weeks will result in bolting and subsequent flowering shortly after return to higher temperatures. Shorter exposures, marginal temperatures, high intervening temperatures, and long days during exposure all quantitatively and progressively reduce the incidence and individual symptomology of bolting (Thompson and Kelly, 1957). Breeding for slow-bolting varieties has been pursued to permit both earlier planting in temperate regions and winter production in coastal temperate and subtropical areas. Mass selection has achieved some success, but slower bolting seems to be related to undesirable slow vegetative growth (J.E. Welch, personal communication).

Celery and celeriac have a number of more or less pressing deficiencies with respect to environmental stress, particularly disease. *Fusarium oxysporum* f. sp. *apii* (Snyder & Hansen), the causitive agent of Fusarium yellows, presently is the most important disease, partly because no cultural techniques have been developed for its control (Opgenorth and Endo, 1979). The pathogen is soilborne and enters the host through the root system. The vascular system and surrounding tissues are slowly colonized while visible external symptoms progress from mild to severe chlorosis, wilt, and death. During the early 1900s, a virulent race of the organism caused progressively greater losses of celery in France and North America. In the case of North America, the problem was solved by switching from susceptible self-blanching types to resistant green cultivars (particularly cv. Utah). However, in the mid 1960s, a new race of the organism virulent on green celeries was observed in southern California (Hart and Endo, 1978). The distribution of this new race (race 2) has now spread to nearly all major production areas in North America. New cultivars with higher tolerance than unselected varieties have been identified, but these are not as horticulturally desirable and nonetheless sometimes show losses to the disease. Tolerance in one such selection has been preliminarily shown to be controlled by two genes with at least partially additive effects (Orton, 1981). Many celeriac cultivars, one Secalinum cultivar, California wild populations, and Chinese celery accessions show moderate-high resistance to race 2 as compared with standard green celery varieties (particularly Tall Utah 52-70R) (Opgenorth and Endo, 1979; Endo, 1981; Orton, 1981). Resistance in one celeriac line appears to be conditioned by a dominant allele at a single locus (Orton, 1981), and the back cross method is currently being used to move this allele into green celery.

Late blight, a foliar disease caused by *Septoria apiicola*, is particularly severe under conditions of high humidity or excessive rainfall. In mild cases, the older leaves exhibited circular black lesions up to 1 cm in diameter. As severity increases, lesions become denser and may occur on petioles and leaves of younger petioles. Although the morphology of the lesions strongly suggests that a pathotoxin is involved, preliminary assessments of this possibility were negative (Rappaport and Grogan, 1978). Effective control of late blight can readily be achieved with fungicides (e.g., anilazine, benomyl), but this is costly. Screening available *A. graveolens* germplasm for resistance to *S. apiicola* revealed no evidence of useful genetic variability (P.E. Hill, personal communication), although parsley (*Petroselinum crispum*) is apparently immune. Recently, celery and parsley have been successfully hybridized, and the resultant F_1 is partially fertile. F_2 progeny segregate for resistance and susceptibility to *S. apiicola*, suggesting that a single dominant resistance gene from parsley may be transferrable to celery (Honma and Lacy, 1980).

There are numerous other deficiencies of minor importance for which genetic solutions are not being widely pursued at present. Examples are susceptibility to early blight (*Cercospora apii*), Pink rot (*Sclerotina sclerotium*), Celery mosaic virus, nematodes, insects (*lepidopterous* larvae, leaf miners), nutrient deficiencies (calcium, magnesium, nitrogen, boron), drought, and salt. Pink rot is of particular interest because lesions are associated with high levels of psoralen derivatives, which are highly mutagenic due to their ability to intercalate DNA (T. Kosuge, personal communication).

A. graveolens would seem to be an exceptional candidate for improvement via F_1 hybrids. The species exhibits some inbreeding depression, but is relatively easy to self-pollinate, and wide crosses are strikingly heterotic (P.E. Hill, personal communication). It is reasonable to speculate that hybrids could increase uniformity and reduce the necessary transplant and field culture period. The difficulty in affecting controlled hybridization and low seed:pollination ratio necessitates a developmental genetic tool such as male sterility for purposes of seed production. Extensive searches for spontaneous male sterile plants in commercial celery seed production fields have been unsuccessful (P.E. Hill, personal communication). Cytoplasmic male sterility has been recently reported and used to produce a commercial F_1 celery hybrid (cv. Green Giant; Takii Seed Co.). Our observations of this hybrid under greenhouse and field conditions have convinced us that it generally exhibits a faster vegetative growth rate and relatively larger petioles than standard green cultivars. Insufficient observations were made to judge relative uniformity. Unfortunately the hybrid is horticulturally deficient for the North American fresh market, and is highly susceptible to *Fusarium oxysporum* f. sp. *apii* 'race 2' and is therefore not presently used in the United States.

Recessive alleles at two nuclear loci that result in male sterility when homozygous were recently discovered in the author's program (Orton and Rabin, unpublished). One of these (designated ms8) ultimately may be extremely useful for hybrid seed production and being the gene is transferred to an acceptable horticultural background. An inbreeding and selection scheme will subsequently be imposed to develop suitable parent lines.

REVIEW OF PREVIOUS RESEARCH

Callus Initiation

Little empirical information has been reported on methods of explant disinfestation. Most commonly, tissues were immersed in calcium or sodium hypochlorite solutions (1-10%) with a surfactant followed by several washes in sterile water. Browers (1981) recently reported experiments contrasting various disinfestation treatments, including 1-5% calcium hypochlorite with and without surfactant, 90% ethanol, and the absence or presence of constant agitation. A number of different explants were used, including taproot, apical shoot, petiole sections, and leaf disks. The optimal method was determined to be 2-5% calcium hypochlorite with surfactant under continuous agitation for 20 min.

A. graveolens responds in much the same way as does carrot with respect to optimal tissue and cell culture protocols. Chen (1976) and Williams and Collin (1976a) were the first to successfully culture and regenerate celery. Little progress has been made toward improving these techniques, except for recent successes in protoplast isolation and culture.

Conclusions about optimum techniques for callus initiation in *A. graveolens* are not possible since different cultivars, plants at different ages, and different explant tissues were used by different groups, and sometimes even by the same group, over time. Callus initiation in *A. graveolens* was first reported by Reinert et al. (1966), who used Murashige and Skoog (MS) basal medium, White's vitamins, 0.06 M sucrose, and 0.23 µM 2,4-D. This medium was subsequently shown to be comparatively ineffective for callus initiation (Williams and Collin, 1976a). Chen (1976) reported that young petiole sections of celery cv. Florimart produced callus after 3 weeks on MS basal medium containing 5.4 µM NAA and 0.47 µM KIN. It was noted, however, that 2,4-D in combination with KIN maximized initiation of rapidly proliferating "undifferentiated callus," which the author feared would not regenerate. Williams and Collin (1976a) also used young petiole explants, but of a different cultivar (Latham Blanching). Their callus initiation medium was similar to Chen's except that 2.3 µM 2,4-D was substituted in place of the NAA and KIN was fixed at 2.8 µM. If cytokinin was omitted from the medium, growth rate was slower, and root regeneration was diminished. They noted further that the original callus after continual subculture actually consisted of a mixture of undifferentiated parenchymal cells and embryoids. Similar observations were reported by Kandeel et al. (1981), who used petiole segments of the celeriac cv. Frigga.

Dormant axillary shoots found at the base of older petioles were determined to be the preferable explant of the celery cv. Tall Utah 52-70 R by Rappaport et al. (1980). Additionally, they used 9.1 µM 2,4-D and 4.4 µM BA for callus initiation. In further work, Fujii (1982) determined that BA is preferrable to both KIN and ZEA for callus initiation with this explant and cultivar, and that 9.1 µM 2,4-D also was preferrable to 2.3 µM. Cytological observations revealed only undifferentiated tissue; no somatic embryogenesis (Orton, unpublished histological observations). Using the celery cv. Tendercrisp and celeriac cvs. PI 169001, PI 171500, and PI 177266 (all Turkish), Browers (1981) found that a broad range of auxin (2,4-D)/cytokinin (KIN and BA) mixtures in MS basal medium promoted callus initiation from both leaf and petiole explants. The optimal levels appeared to be 6.8 µM 2,4-D and 2.8 µM KIN. No differences were noted among cultivars, and petiole explants were consistently better than leaves.

Little or no definitive experiments have been conducted to elucidate the effects of other attributes of the physical environment on callus initiation. Chen (1976) performed all experiments at 25 C in the dark, Williams and Collin (1976a) used 26 C and a 12-hr light-dark diurnal cycle, and Rappaport et al. (1980) and Fujii (1982) used continuous light at 25 C. This is curious, since the optimum growth for commercial celery production is approximately 20 C. An experiment conducted in the author's laboratory suggested that continuous light fostered the initiation of vigorous callus tissue more quickly and at a higher frequency than darkness (Orton, unpublished results).

While it is impossible to make unqualified judgements about the effects of such factors as genotype, plant age, explant source, and interactions of these factors with media with respect to callus initiation, some speculations can be made. Callus has been successfully initiated from a broad range of genotypes and plant ages of *A. graveolens*, and it seems likely that these factors do not bear significantly on success. Young or dormant tissues have been used exclusively, and it remains unknown whether mature tissues could be substituted. Likewise, callus has been successfully initiated from different explant tissues including petiole segments, leaf pieces, and dormant shoot tissues. Only slight quantitative differences were noted in comparative studies, and since no other adverse consequences related to explant source have been reported, the choice seems arbitrary. The optimal medium for callus initiation appears to be MS basal supplemented with 2.3-9.1 μM 2,4-D and 2.3 μM KIN. When using different cultivars or explants, BA may be more effective than KIN. When lower concentrations of 2,4-D are used, embryo formation is observed concomitantly with or without undifferentiated tissues, and particularly if the cytokinin is omitted. From the standpoint of maintaining genetic fidelity and the ability to regenerate from long-term cultures, maintaining a state of continual mixture of undifferentiated and "partially" differentiated tissue is probably desirable (i.e., lower 2,4-D and cytokinin levels).

Callus and Suspension Maintenance

Optimization of techniques for maintaining in vitro growth of callus and suspension cultures of *A. graveolens* has not been done exhaustively. For maintaining callus, Williams and Collin (1976a) employed a simple serial transfer technique and used the same medium and physical conditions as those for callus initiation. Fujii (1982), who used 9.1 μM 2,4-D to initiate callus, found that the growth rate of serially transferred callus was greater at lower auxin concentrations of 0.45-2.3 μM 2,4-D. Fujii (1982) determined further that 4.4 μM BA, while optimum for callus initiation, was toxic to continued growth of callus tissue. KIN at 0.47 μM was found to optimize growth rate established by callus tissue in MS medium supplemented with 2.3 μM 2,4-D.

Successful propagation of *A. graveolens* cells in liquid suspension culture was first reported by Williams and Collin (1976a). They introduced established callus tissue into 50 ml of liquid medium, of the same composition as that for initiation, in a 250-ml conical flask, and agitated it on a rotary shaker at 100 rpm. After 3 weeks, the resulting suspension consisted of a mixture of embryos, undifferentiated cell clusters, and single cells. Rappaport et al. (1980) adopted a similar procedure, but modified the medium by substituting 1.1 µM BA in the place of KIN. Cultures were shaken continuously on a rotary shaker (80 rpm) under continuous light at 25 C. Under these circumstances, their cultures doubled in cell volume every 5 days, and were constituted entirely of undifferentiated cells occurring singly or in aggregates.

Regeneration

Williams and Collin (1976a) noted spontaneous somatic embryogenesis in the course of callus initiation and maintenance. Moreover, root formation was observed from callusing petiole explants, suggesting that plant regeneration via organogenesis may also occur under some circumstances. Somatic embryos occuring in medium containing 2,4-D and KIN continued to differentiate to a stage where embryonic shoots and roots could be distinguished. Auxin:cytokinin ratios were critical to the initiation of somatic embryos. While progressively higher concentrations of KIN promoted embryogenesis, 2,4-D was clearly inhibitory. At 2,4-D concentrations in excess of 9.0 µM, only globular embryos were formed initially, and no embryo formation was seen after 2 weeks. After 10 weeks on high 2,4-D, the potential for regeneration was irreversibly lost (Al-Abta and Collin, 1978a). Spontaneous embryogenesis persisted in most cultures for over 2 years, while a few cultures seemed to have lost this ability, apparently due to differential responses to KIN.

Spontaneous embryogenesis in combination with undifferentiated growth also persisted when callus tissue was dispersed in agitated liquid medium, although younger somatic embryos seemed to predominate. When 2,4-D and KIN were omitted, further development of embryos was observed. Continued growth and development could be fostered by reducing mineral nutrient concentration by 50%, resulting in a dense mass of entangled plantlets (Williams and Collin, 1976a).

Chen's (1976) description of embryogenesis was not nearly as complete as that of Williams and Collin (1976a), but the following information can be abstracted: Callus tissue initiated on 5.4 µM NAA and 0.47 µM KIN exhibited some signs of regeneration, while that initiated on 4.5 µM 2,4-D and 4.7 µM KIN remained undifferentiated. In either case, embryogenesis could be induced by transferring callus to a medium containing 0.54 µM NAA and 13.9 µM KIN. However, embryogenesis diminished within 8 months after serial propagation of callus on medium containing 2,4-D. Further development beyond the globular stage was accomplished by plating embryos onto MS medium devoid of hormones.

Globular embryos were obtained directly from the surface of petiole explants of Chinese celery on MS medium with 2.3 µM 2,4-D, occurring concomitantly with undifferentiated callus growth (Zee and Wu, 1979). Development did not proceed further on this medium. When the 2,4-D was replaced with 2.8 µM KIN, torpedo-stage embryos were observed.

Subsequently, it was shown that somatic embryogenesis was taking place in cortical cells proximal to vascular bundles at the cut surface of the explant (Zee et al., 1979).

Rappaport et al. (1980) and Fujii (1982) reported the phenomenology of plant regeneration in callus and suspension cultures of celery cv. Tall Utah 52-70 R. They noted rapid plantlet formation from undifferentiated callus tissue after removal of 2,4-D from the culture medium. Development proceeded normally on MS basal medium containing 2.3 μM KIN, and regenerated plantlets could be transplanted directly to vermiculite in greenhouses. Cells and tissues in liquid suspension could be regenerated either directly or by plating onto solid medium. When cultures maintained on MS basal medium containing only 1.1 μM KIN, asynchronous embryogenesis was observed. After 4 weeks, mixtures of undifferentiated cells and tissues, embryos of diverse stages, and young green plantlets were present. Regeneration could also be achieved by introducing 1.5 ml of the culture to the surface of solid MS medium containing no hormones.

The suspension cultures of Rappaport et al. (1980) contained single cells and aggregates containing up to 10^5 cells, as determined by direct counting of squash preparations (Orton, unpublished observations). An experiment was conducted to determine the aggregate size class(es) from which somatic embryos originated. Although some regeneration was observed from small aggregates and single cells, by far the highest frequency occurred among the largest aggregates (Rappaport et al., 1980).

These independent reports, although using methods and germplasm which were not necessarily equivalent, are remarkably consistent. Callus tissue initiation on 2.3 μM 2,4-D occurs concomitantly with spontaneous somatic embryogenesis. The admixture is maintained after serial transfer to the same medium, but embryogenesis predominates if 2,4-D is removed. Callus tissue initiation on concentrations of 2,4-D greater than 4.5 μM does not occur concomitantly with somatic embryos, even if the callus is later transferred to 2.3 μM 2,4-D. However, embryogenesis can be induced by complete removal of auxins. The qualitative and quantitative impact of cytokinins on somatic embryogenesis is not significant. These relationships are summarized schematically in Fig. 3.

Like other species of the Umbelliferae (e.g., carrot, parsley), *A. graveolens* responds predictably, rapidly, and at high frequency with respect to exogenously induced developmental changes. Moreover, the impact of genotype on amenability to in vitro culture and regeneration appears to be insignificant. Hence it is not surprising that *A. graveolens* (particularly celery) is being explored as a model species for the application of in vitro culture technology to basic biology, crop improvement, and field production.

Haploids

Successful isolation of haploid callus tissue or whole plants has not been reported for any of the Umbelliferae. This is a consequence of

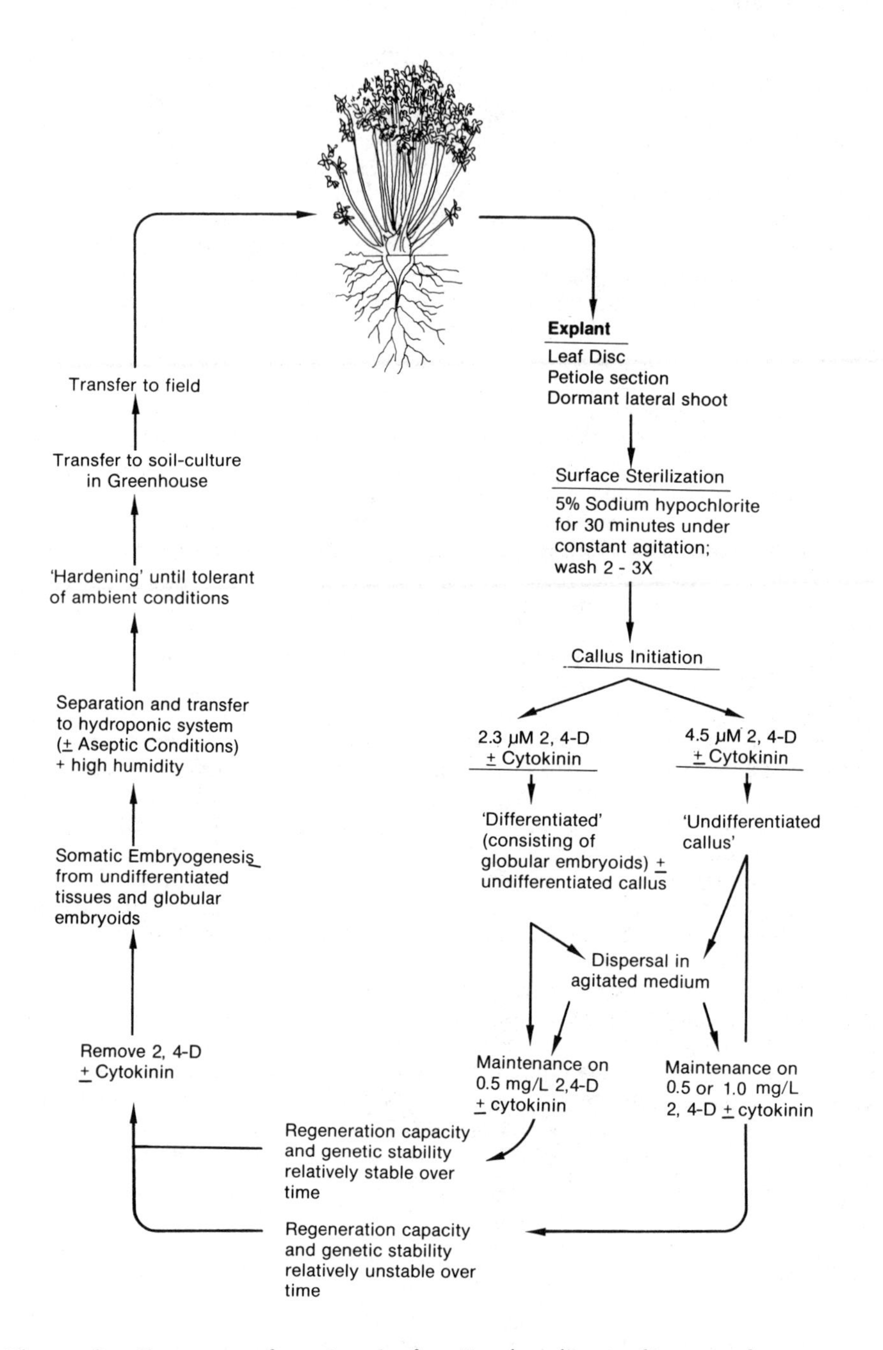

Figure 3. Summary of protocols for the in vitro culture and regeneration of somatic celery tissue.

logistical difficulties associated with small flower size which make the manipulation of gametophytes extremely difficult. The process of obtaining haploid plants of *A. graveolens* by selecting parthenogenic plants or by chromosome elimination following interspecific hybridization has not been explored. Cells with haploid chromosome numbers have occasionally been observed in vitro, but it is unknown whether these are true haploids. The possibilities of utilizing haploids in biochemical selection schemes, and for producing inbreds for possible use as hybrid parents are exciting, and research in this important area is badly needed.

APPLICATIONS OF CELERY CELL CULTURE

Basic Studies of Somatic Embryogenesis

Celery suspension cultures have been used to probe the effects of exogenous compounds on morphogenesis and associated endogenous biochemical changes. The process of somatic embryogenesis in *A. graveolens* is extremely similar to that established for carrot (McWilliam and Street, 1974). However, Williams and Collin (1976a) have pointed out that celery responds differently to hormones and tends to retain embryogenic potential for longer periods than does carrot.

Observations by Al-Abta and Collin (1978b) suggested that somatic embryos exhibited less internal differentiation than do zygotic embryos at comparable stages of gross development. Early development from cell determination to the globular embryo stage was found to be dependent on the presence of both 2,4-D and KIN in the external medium (Al-Abta and Collin, 1979). However, development beyond the globular stage was found to be inhibited by exogenous auxin or cytokinin.

Zee et al. (1979) observed three distinct early morphological events on celery petiole explants, cultured on 2.3 μM 2,4-D, which preceded globular embryos: (a) a lag period (0–4 days), (b) proliferation of cortical cells surrounding the vascular bundles and the formation of meristematic layers (4-10 days), and (c) the meristematic layers giving rise to proembryos. Unlike the account of Al-Abta and Collin (1979), Zee and Wu (1979) reported that globular embryos could be induced to develop further by adding 2.8 μM KIN while removing the auxin. Histological observations of torpedo-stage embryos suggested that subsequent polarity developed as a consequence of differential starch distribution. A similar sequence of morphological events was observed with cultured leaf explants (Zee and Wu, 1980). Preliminary comparative experiments showed both qualitative and quantitative differences in proteins present during all stages from the formation of proembryo derivative tissue (Zee et al., 1979) through torpedo-stage embryos (Zee and Wu, 1979).

Genetic Instability In Vitro

A great deal of interest has centered on the phenomenon of spontaneous genetic variability in cultured eukaryotic cells. In plants, circumstantial evidence has associated this variability with the appearance of new genetic entities among regenerated plants (both desirable and undesirable) and the loss of embryogenic or organogenic potential (Bayliss, 1980; Larkin and Scowcroft, 1981). The questions of central importance are: (a) What are the genetic and physiological factors which bear on the appearance of spontaneous genetic variability? and (b) What is the source of the apparent differences between differentiated and undifferentiated tissues with respect to genetic stability? Callus and cell cultures exhibit little or no external manifestation of variation; hence changes in chromosome number and structure have been the primary source of information. *Haplopappus gracilis* and *Crepis cappillaris* have therefore been studied extensively in this context by virtue of a low chromosome number and definitive structure. However, both species have proven to be somewhat more difficult to work with in vitro than carrot or tobacco, particularly with respect to regeneration. Tobacco is inappropriate for this work due to allopolyploidy and difficulty in discerning chromosome structures, and carrot, which has been used to determine the effects of 2,4-D on mitotic anomolies in vitro (Bayliss, 1973, 1975), is even more difficult to examine cytogenetically. By comparison, *A. graveolens* has a number of attributes which deem it highly appropriate as an organism for this research: it is diploid (2n = 22), has large well-defined chromosomes (Fig. 1), and is easy to work with in vitro. Moreover, it is a significant crop entity on which some genetic work has been done.

The standard karyotype of somatic tissues of seed-propagated *A. graveolens* contains seven pairs of long acrocentrics, two pairs of short acrocentrics, and one pair each of metacentrics and telocentrics (Fig. 1). Chromosome counts from root tips of celery cv. Tendercrisp and celeriac PI 169001 were invariably 2n = 22 (Browers and Orton, 1982a). Rare cases of reduction of chromosome number have been reported and appear to be a consequence of Robertsonian fusions (Sharma and Bhattacharya, 1959; Marks, 1978).

Williams and Collin (1976b) compared morphological features and cytology among seed propagated plants and regenerates from 9- and 15-month-old callus cultures of celery cv. Lathom Blanching. Growth rate and petiole architecture were similar among the three groups and no striking phenotypic variants were noted. Approximately 68% of all regenerates were 2n = 22, while 27% were 2n = 18, and the remainder 2n = 42 and 44. Chromosome numbers were determined in only 4 seed-propagated plants, but one of these was 2n = 18 (and the rest 2n = 22). It is reasonable to conclude that reduction of chromosome number in this cultivar occurs commonly in vivo with little or no drastic phenotype effect. Moreover, 2n = 18 regenerated plants were merely a reflection of this normal in vivo phenomenon, which is apparently transient and restricted to somatic tissues. Chromosome counts were insufficient among the seed-propagated plants to ascertain the source of the 4-5% polyploidy observed among regenerates. Presumably the genetic stability of these cultures was due to the perpetual organization of cultured cells as globular embryo structures.

In a comparative study of chromosome numbers in vitro among four diverse *A. graveolens* germplasm entries (cv. Tendercrisp, PI 169001, PI

171500, and PI 169001 x cv. Tall Utah 52-70 R F_1), Browers and Orton (1982a) found reproducible significant differences attributable to genotype. Cultured cells of PI 169001 were nearly 70% 2n = 22, while the remainder were mostly aneuploid in the range of 2n = 23 to 43. In contrast, only 45% of callus cells of the cv. Tendercrisp had 22 chromosomes. In separate experiments PI 171500 showed even more instability. Differences were also detected among cultures from different plants of the same germplasm, but never among cultures derived from the same plant (e.g., different explants). However, studies indicated changes in karyological makeup over time in culture (Browers, 1981). It was concluded that genotype and culture age were the major factors bearing on the appearance and persistance of karyologically variant cells, while physiological differences between explant tissues and random drift were not significant.

A modification of protocols commonly used in mammalian cytogenetics, utilizing protoplasts, was recently accomplished for plants (Murata, 1982). The technique produces excellent spreads of metaphase chromosomes with extremely well-preserved morphology. We have successfully used the technique to study chromosome structural changes in cultured *A. graveolens* cells (Murata and Orton, 1982). Ultimately we envision other applications such as differential stainings and in situ hybridization with cloned probe sequences.

The culture used for chromosome structural analyses was PI 169001 x cv. Tall Utah 52-70 R, an F_1 hybrid between a celeriac and celery that was maintained in liquid suspension for 21 months after initiation. Concomitant loss of morphologically "normal" chromosomes (mean 2.40 per cell) and gain of structurally altered chromosomes (mean 1.35 per cell) was observed (Fig. 1). Further, the loss of morphologically normal chromosomes was not random with respect to type (e.g., acrocentric, telocentric, etc.). Among presumptive eudiploid (2n = 22) cells, 56% exhibited clear evidence of the loss of chromosome segments, while the remaining 44% all had at least one structural change from the native karyotype.

A great deal of potential genetic variability goes undetected in analyses of chromosome number, structure, or DNA content. Since cultured cells and tissues do not exhibit morphological variation to exploit for genetic purposes, it will be necessary to detect variation due to mutation, recombination, insertions, and so forth by direct characterization of DNA sequences and gene products. One approach to this problem utilizing *A. graveolens* has been to construct plants heterozygous at a large number of known loci and to assess phenotypic changes in corresponding cultured tissues and regenerated plants. Certain of these phenotypes, isozyme bands, can be visualized in vitro (Orton, 1983). Single-cell clone cultures of one plant heterozygous at 4 loci, including *Pgm-2* and *Sdh-1*, showed the loss of expression of the fast-migrating *Pgm-2* electromorphon 24 of 100 single cell clones, while the heterozygous phenotype was always retained at *Sdh-1*. Of all the possible causes of this phenomenon, its properties strongly suggested that insertion sequences were responsible. However, when $Pgm\text{-}2^{FS}$ and $Pgm\text{-}2^{S}$ clones were contrasted karyologically, background variability was too great to draw valid conclusions.

Chromosomally variant cells which presumably arise in vitro are also transmitted to regenerated plants (Browers and Orton, 1982b). While a suspension culture of PI 169001 consisted of 80% presumptive diploid and 20% aneuploid or polyploid cells, root tip cells of regenerated plants were 85% diploid and 15% hypodiploid; no aneuploidy in excess of 2n = 22 or polyploidy was observed. Karyotype analyses have shown that hypodiploid cells in regenerates are true aneuploids, and do not correspond to any karyotype observed with PI 169001 in vivo. Further, hypodiploid cells were only observed in intraplant mixtures, and never as pure sectors or whole plants, suggesting either a multicellular origin of embryos or that spontaneous variability can occur during development. These observations are somewhat contradictory to those of Williams and Collin (1976b) who observed polyploidy and hypodiploidy (but equivalent to diploidy) in addition to diploidy, but always among and not within plants. The differences between these reports may be due to the developmental stage of the regenerates at the time chromosome counts were taken: Williams and Collin (1976b) used new lateral roots of 2-month-old plants while Browers and Orton (1982b) used primary roots of very young plantlets.

As noted previously for different plant groups (D'Amato, 1977), the degree of chromosomal variability generated in vitro seems to be negatively correlated with ability to regenerate plants of *A. graveolens*. PI 171500, in which aneuploid and polyploid cells quickly predominate after callus initiation, has been recalcitrant to induced embryogenesis (Browers and Orton, 1982b). Suspension cultures of PI 162001 x Tall Utah 52-70 R, which were used in the in vitro karyotype study cited above (Murata and Orton, 1982), yielded only abnormal organized growth at the time when this experiment was done.

Loss of expression of the *Pgm-2*^{F} electromorph in PI 169001 x Tall Utah 52-70 R (originally *Pgm-2*^{FS}) cultures also seems to be associated with a reduction in the ability to regenerate whole plants (Orton and Lassner, 1982). When mass-transferred callus cultures, known to consist of a mixture of *Pgm-2*^{FS} and *Pgm-2*^{S} tissues were transferred to a medium lacking 2,4-D, only the *Pgm-2*^{FS} phenotype was present among regenerates. Sublines of the *Pgm-2*^{S} phenotype regenerated sporadically, but regenerates retained the variant *Pgm-2* phenotype. Attempts to nurture *Pgm-2*^{S} regenerated plantlets to flowering status have thus far been unsuccessful. Embryos appeared to be normal until well past torpedo stage, but showed some morphological anomalies at the cotyledon stage. Structures corresponding to cotyledons were fused, and a large number of adventitious globular embryos arose on the apical surfaces of the plantlet; no conspicuous apical meristem was evident. Subsequently, plantlets become chlorotic and root growth ceased, while development of adventitious embryos continued. The culture was maintained in this manner for over 7 months in the absence of hormones. This "abnormal development syndrome" is not limited to callus cultures of *Pgm-2*^{S} phenotype, since similar observations have been made with *Pgm-2*^{FS} cultures. Further studies are in progress utilizing molecular biological tools in the hope of pinpointing the responsible genetic lesions more conclusively.

A. graveolens has proven to be highly appropriate material for these studies, and has already provided new insights into the phenomenology and underlying causes of spontaneous genetic variability in cultured somatic tissues. The full potential has not yet been realized, particularly with respect to the formulation of new techniques to resolve events at the molecular level. Ultimately *A. graveolens* may provide techniques and basic information of broad biological significance, such as the genetics of embryogenesis and the mechanism and impact (e.g., in tumor biology and evolution) of spontaneous genetic variability in somatic tissues in vivo.

Alternative Plant Breeding Methods

Because *A. graveolens* is readily induced to undergo callus formation and somatic embryogenesis and is also a significant crop species, suggested applications to plant improvement have been numerous. One of the most active areas has been herbicide resistance. Since vegetative portions of the plant are consumed, control of compounds applied as selective postemergence herbicides has been stringent. Moreover, these compounds are generally the most expensive to develop and selectivity often narrows the potential market volume. The logical alternative using in vitro culture is to select cells (and ultimately plants) resistant to a broader spectrum of compounds that are less toxic to human handling and consumption and/or exhibit faster biodegradability. Metcalf and Collin (1978) found that the triazine herbicide simazine inhibited both cell division and expansion in celery suspension cultures at 5.0-50.0 μM, and speculated that it was acting as a respiratory uncoupler. Although this suggests that in vitro selection for simazine-resistant plants is a theoretical possibility, no such mutants have yet been reported. The herbicide asulam was shown to have similar toxic effects in both differentiated and undifferentiated tissues (Watts and Collin, 1979). Stable spontaneous asulam-resistant (to 8×10^{-5} M) variants were selected from celery suspension cultures, but plants from selected clones have not been regenerated (Merrick and Collin, 1980). Mutagenesis with EMS greatly increased the efficiency with which glyphosate- (Roundup) resistant clones (to 0.1-1.0 mg/l) could be selected from celery suspension cultures. These variants are now showing some signs of regeneration (Orton and Lassner, unpublished).

Other schemes advanced which utilize in vitro mutant selection in *A. graveolens* breeding include disease resistance and flavor enhancement. Rappaport (1980, 1981) has provided compelling evidence that a toxic compound (or compounds) is involved in Fusarium yellows pathogenesis. Lyophilized ethanol extracts of disease lesions completely inhibited growth of suspension cultures of cv. Tall Utah 52-70 R (susceptible to the pathogen) at a concentration of 10%, whereas those of PI 169001 (resistant to the pathogen) were only partially inhibited. Moreover, the extract and pathogen have been shown to incite similar responses on excised petioles and germinating seeds, further suggesting that toxins in the extract are involved in the disease. Clones of the former cultivar resistant to 10% extract were selected and regenerated, but resulting plants showed no evidence of enhanced resistance or tolerance to the pathogen. Efforts to purify the toxic constituent(s) of these extracts, and to determine relationships with pathogenesis are in progress (Rappaport, personal communications).

Phthallide derivatives, the main flavor compounds of *A. graveolens*, were found to be present in "differentiated callus" (i.e., consisting of globular embryos) but not in undifferentiated callus (Al-Abta et al., 1979). Moreover, whole plants and differentiated callus were quantitatively and qualitatively similar with respect to phthallide composition. Thus it is possible that gas chromatography/mass spectrography could be used to screen somatic clones of differentiated callus for altered phthallide composition, ultimately to select genetic variants for enhanced flavor.

Very little work has been conducted toward developing genetic modification schemes utilizing protoplasts (e.g., somatic hybrids, cybrids, transformation via uptake). Successful isolation of *A. graveolens* protoplasts was first reported by Tseng et al. (1975), who used leaf mesophyll as source tissue. No attempts to foster further growth of these protoplasts were described. Recently, techniques have been successfully adapted from carrot for the isolation and culture of protoplasts from suspension cultures (Table 2) (Knuth, unpublished). Calcofluor fluorescence was used to demonstrate cell wall deposition on isolated protoplasts within 2-4 days after isolation. Dr. G. Pullman has modified this protocol by substituting 10% strength MS basal salts (no NH_4NO_3) and 0.3 M sorbitol + 0.075 M sucrose as the osmotica in the protoplast isolation medium. Regenerated plants were successfully obtained after colonies were plated on hormoneless medium.

Cloning for Field Production

As discussed previously, much of the world celery crop is grown from transplants. The possibility that field transplants could be grown from somatic embryos has led to speculation that in vitro cloning could be used to produce a crop superior to that attainable by existing techniques. For example, celery transplants grown from seed for the California grower market are produced as depicted in Fig. 4. The ultimate cost of transplants raised in this manner presently ranges from 0.015 to 0.020 U.S. dollars per plant. The main limitations of the scheme are (a) celery (and celeriac) seed germination is characteristically slow, variable, and incomplete, and (b) the genetic status of each unit with respect to field value is uncertain since open pollination among genetically variable populations is almost universally employed for seed production. In vitro cloning, in contrast, offers the possibility of overcoming both of these limitations. Because the embryogenic process is readily induced in callus and suspension cultures of *A. graveolens*, and resulting embryos exhibit differences in size and shape, it should be possible to engineer methods to produce uniform suspensions of developmentally synchronized somatic embryos from large batch cultures. Genetic limitations could be overcome by choosing a superior plant (or

Table 2. Modified Protocol of Knuth (unpublished) for the Isolation of Protoplasts from Celery Suspension Cultures and Subsequent Culture of Proliferating Tissue

---

Preparation and Culture of Celery Protoplasts

1. Subculture 3-4 days prior to protoplast isolation. The preferred medium is as follows: MS basal salts, 2.3 μM 2,4-D, .47 μM KIN, and 0.09 M w/v sucrose.
2. Pour 50 ml of suspension into a sterile centrifuge tube and let stand for 20 min.
3. Draw off supernatant and resuspend cells in enzyme solution (1:1 v/v). Add the following to the above medium: 2.0% Cellulysin (Calbiochem), 0.1% Pectolyase (Kikkoman), and 0.6 M mannitol.
4. Agitate on a gyrotary shaker (40 rpm) at room temperature under ambient light 3-4 hr.
5. Filter sequentially through sterile miracloth (Calbiochem) and 60 M nylon mesh.
6. Centrifuge at approximately 20 x g for 1 min.
7. Remove supernatant and replace with an equal volume of medium containing 0.6 M mannitol.
8. Repeat Steps 6 and 7.
9. Suspend pellet in equal volume of medium containing 0.6 M mannitol and pipet onto a sterile 7-cm Gelman millipore filter in a 100 x 15 plastic petri dish.
10. Culture at 25 C under light until signs of division are observed.
11. Transfer filter paper to the surface of medium containing 0.6 M mannitol and solidified with 0.5% w/v agarose.

---

plants) as explant donor, possibly an F_1 hybrid of superior inbred parents. Potential problems (e.g., vulnerability) imposed by complete genetic uniformity in production fields could be avoided by employing artificial mixtures of somatic embryos or cloned transplants. The in vitro production scheme pictured in Fig. 4 will no doubt necessitate higher transplant cost than that from seed. The ultimate economic question is whether genetic advantages of culture-derived transplants vs. those derived from seed can offset this additional cost. The main factors bearing on success will be: (a) genotype of explant donor, (b) batch maintenance costs, including genetic purity, (c) development of efficient, effective methods to synchronize somatic embryos with respect to uniformity at harvest, and (d) development of efficient, effective methods to process synchronized embryos into transplants capable of withstanding intense environmental and ecological pressures.

Another possible point at which in vitro cloning technology may find application is in hybrid celery seed production. Genetic male sterility is generally used to produce hybrid seed, but female parent lines are difficult to propagate as a consequence. Where self- or sib pollination can be used to propagate parental lines, such as with celery or celeriac, inbreeding depression becomes a problem. In the ideal scenario,

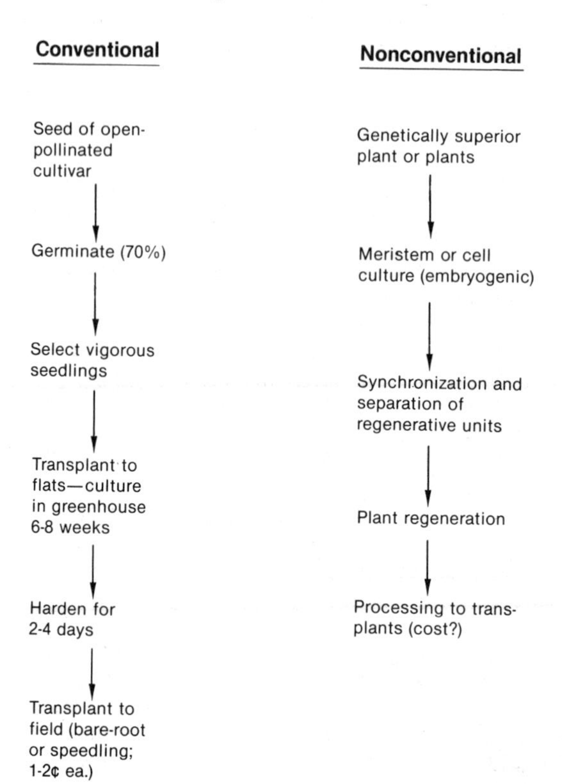

Figure 4. A comparison of the standard commercial procedure for producing field transplants of celery from seed vs. that envisioned from in vitro cultures.

stock tissues of known male sterile and fertile inbred lines, female and male parents, respectively, of a hybrid variety, would be kept frozen in liquid nitrogen to insure genetic purity. When a hybrid seed lot is desired, a small amount of stock tissue could be withdrawn from storage, amplified in vitro, regenerated, and transplanted to the field for seed production.

Most of the work conducted toward developing such a scheme has, up to now, been exploratory. Both Chen (1976) and Williams and Collin (1976a) determined the basic manipulations necessary to initiate, propagate, and regenerate celery in vitro. Chen (1976) demonstrated the possibility of producing hardened plants in this manner, but reported no data regarding synchrony or mortality. Williams and Collin (1976a) suggested some possible improvements of their crude production scheme, which required 14 weeks from explant to the first batch of somatic embryos. Although no effort was made to synchronize embryos, they noted that mortality was much lower than that encountered in similar work with carrot. Further work demonstrated that this scheme characteristically resulted in high genetic fidelity with respect to chromosome number and economic plant characteristics (Williams and Collin, 1976b). Despite initially promising results, these groups have not reported further work toward improving efficiency and scale-up. During their period of involvement with bio-agricultural research (1974-1980) the Union Carbide Corporation reportedly made great progress toward an economically feasible system for in vitro transplant production in celery, although the program was abandoned before it could reach fruition (R. Lawrence, personal communication).

Rappaport (1980) and Fujii (1982) have suggested some modifications of reported procedures to improve the efficiency of cloning green Utah celery. The preferred explant tissue was determined to be dormant axillary shoots of vegetatively mature plants, as opposed to young petiole sections used previously. Moreover, they found that higher 2,4-D levels (9.0 μM) and BA in place of KIN were preferrable for callus initiation, although this may have been specific to the explant of choice. They have conducted extensive greenhouse field evaluations of plants produced using their in vitro cloning procedures. Morphological anomalies observed among greenhouse populations were often lost after transplanting to the field. Additionally, variations often appeared during later stages of vegetative growth which were not observed in the greenhouse. Among three field plantings in different locations and dates along the California coast, 65-70% of the plants were classified as marketable at maturity. The remaining 30-35% exhibited undesirable features, such as premature flowering, dwarfing, off color, and multiple shoots. The genetic component of this variability is, where possible, being determined.

The apparent differences in fidelity between the methods reported by Williams and Collin (1976a,b) and Rappaport (1980) and Fujii (1982) may lie in the concentration of 2,4-D used to initiate callus. The former used 2.3 μM, which gave rise to a perpetual mixture of undifferentiated callus and globular embryos while the latter used 9.1 μM, which resulted in only undifferentiated callus, even after transfer and maintenance on 2.3 μM 2,4-D. Both Chen (1976) and Al-Abta and Collin (1978a) have observed that progressive culture of celery on 2,4-D concentrations in excess of 4.5 μM results in a corresonding loss of embryogenic potential. Since embryogenic potential is closely linked to karyological anomalies, the variation reported by Fujii (1982) may have been a consequence of variation arising early in the initiation process and persisting during maintenance. Another consequence of this may be that their cultures progressively lost embryogenic potential (Thompson, unpublished) when serially subcultured or in 2.3 μM 2,4-D.

Although early reports were highly optimistic about the potential role of in vitro cloning in field production of celery, the published literature shows little evidence of progress toward this goal beyond demonstrations of feasibility. Information generated up to now has been mostly academic. It will be necessary for horticulturists and agricultural engineers to cooperate toward further developing proposed production schemes before commercial in vitro propagation is realized.

FUTURE PROSPECTS

The prevailing attitude approximately 10 years ago when the full potential of in vitro culture was first realized, was that the ease of manipulation previously achieved with tobacco and carrot could be readily extended to other plant species by exhaustive experimentation with culture conditions. It is now clear that differences exist among phylogenetic entities with respect to the ease with which they are cultured. These genetic differences may, in some cases, be insurmountable by simple alterations of medium or environment. For example, tobacco, potato, and tomato are relatively closely related yet differ with respect to ability to regenerate from long-term cultures, especially those cultures derived from protoplasts. Differences are even evident in this respect within a species, such as those described in this chapter for *A. graveolens* (Browers, 1981). The preferred approach, when attempting to culture a new species at present, is to choose a few basic media which seem to work across a broad spectrum, and to utilize explant material from a broad genetic base. Evidence is also accumulating which suggests that callus "type" is important, and that tissue proliferation and ability to regenerate often hinge on callus type selection and maintenance (Vol. 3, Chap. 5).

A. graveolens cultures, under most circumstances, are easily regenerated, the only exceptions being old cultures or those exposed to high concentrations of 2,4-D, cultures of certain cultivars (e.g., PI 171500), and (thus far) those from protoplasts. Because most cultures of *A. graveolens* are relatively stable genetically and retain embryogenic potential over long periods, they hold excellent potential for studies of molecular events during embryogenesis. It would be interesting to see whether the phenomenology now well established regarding somatic embryogenesis in carrot can be extended to celery as well. By sorting out similarities in events during embryogenesis across a broad spectrum of genera, strides will be made toward understanding determination and subsequent development.

For all of the research and rhetoric of the past 10–15 years, in vitro culture currently enjoys surprisingly little application in agriculture. In celery, the conditions are ripe for the successful application of in vitro cloning for field transplant production. As was discussed earlier, the remaining details which need to be worked out in this system are synchronization of somatic embryos, and handling and hardening processes. As with so much of the tissue culture technology of potential application generated in academia, this system has reached a dead end at the interface between fundamental and applied biology. The risks and development costs are apparently greater than potential economic gains over existing methods. Hopefully, the gap between potential and realization will be bridged in celery.

KEY REFERENCES

Al-Abta, S. and Collin, H.A. 1978a. Control of embryoid development in tissue cultures of celery. Ann. Bot. 42:773–782

Merrick, M.M.A. and Collin, H.A. 1980. Selection for asulam resistance in tissue cultures of celery, *Apium graveolens* L. var. dulce cv. New Dwarf White. Plant Sci. Lett. 20:291-296.

Sims, W.L., Welch, J.E., and Rubatzky, V.E. 1977. Celery production in California. Leaflet 2673, Cooperative Extension, Division of Agricultural Sciences, University of California, Berkeley.

Williams, L. and Collin, H.A. 1976a. Embryogenesis and plantlet formation in tissue cultures of celery. Ann. Bot. 40:325-332.

REFERENCES

Al-Abta, S. and Collin, H.A. 1978b. Cell differentiation in embryoids and plantlets of celery tissue cultures. New Phytol. 80:517-522.

______ 1979. Endogenous auxin and cytokinin changes during embryoid development in celery tissue cultures. New Phytol. 82:29-36.

Al-Abta, S, Galpin, I.J., and Collin, H.A. 1979. Flavor compounds in tissue cultures of celery. Plant Sci. Lett. 16:129-134.

Arus, P. and Orton, T.J. 1983. Inheritance and linkage of seven isozyme loci in celery (*Apium graveolens* L.) J. Hered. (in press).

Bayliss, M.W. 1973. Origin of chromosome number variation in cultured plant cells. Nature 246:529-530.

______ 1975. The effects of growth in vitro on the chromosome complement of *Daucus carota* L. suspension cultures. Chromosoma 51: 401-411.

______ 1980. Chromosomal variability in plant tissues in culture. Int. Rev. Cytol. 11A:113-144.

Browers, M.A. 1981. Chromosomal Variability in Tissue and Cell Cultures of Celery (*Apium graveolens* L.). M.S. Dissertation, University of California, Davis.

______ and Orton, T.J. 1982a. A factorial study of chromosomal variability in callus cultures of celery (*Apium graveolens*). Plant Sci. Lett. 26:65-73.

______ and Orton, T.J. 1982b. Transmission of gross chromosomal variability from suspension cultures into regenerated plants of celery. J. Hered. 73:159-162.

Chen, C.H. 1976. Vegetative propagation of the celery plant by tissue culture. Proc. S. Dakota Acad. Sci. 55:44-48.

D'Amato, F. 1977. Cytogenetics of differentiation in tissue and cell cultures. In: Plant Cell, Tissue, and Organ Culture (J. Reinert and Y.P.S. Bajaj, eds.) pp. 343-357. Springer-Verlag, Berlin and New York.

Emsweller, S.L. 1933. An hereditary pithiness in celery. Amer. Soc. Hort. Sci. Proc. 29:480-485.

Endo, R.M. 1981. Etiology, biology, and control of the Fusarium yellows disease. In: California Celery Research Program (F. Pusateri, ed.) pp. 9-15. California Celery Research Advisory Board, Bakersfield, California.

Fujii, D.S. 1982. In Vitro Propagation of Celery (*Apium graveolens* L.). M.S. Dissertation, University of California, Davis.

Hart, L.P. and Endo, R.M. 1978. The reappearance of Fusarium yellows of celery in California. Plant Dis. Rep. 62:138-142.

Honma, S. and Lacy, M.L. 1980. Hybridization between pascal celery and parsley. Euphytica 29:801-805.

Kandeel, N., Kaufmann, F., Borner, R., and Benkenstein, I. 1981. Kallusentwicklung and sterilvermehrung bei sellerie (*Apium graveolens* L. var. rapaceum Mill. Gaud.). Archiv. fur Gartenbau 29:71-76.

Larkin, P.J. and Scowcroft, W.R. 1981. Somaclonal variation—A novel source of variability from cell cultures for plant improvement. Theor. Appl. Genet. 60:197-214.

Lorenz, O.A. and Maynard, D.N. 1980. Knott's Handbook for Vegetable Growers, 2d ed., Wiley-Interscience, New York.

Marks, G.E. 1978. The consequences of an unusual Robertsonian translocation in celery (*Apium graveolens* var. dulce). Chromosoma 69: 211-218.

McWilliam, A.A. and Street, H.E. 1974. The origin and development of embryoids in suspension cultures (*Daucus carota*). Ann. Bot. 38:243-250.

Metcalf, E.C. and Collin, H.A. 1978. The effect of simazine on the growth and respiration of a cell suspension culture of celery. New Phytol. 81:243-248.

Murata, M. 1982. An efficient air-drying technique for observation of plant chromosomes. Stain Technol. (in press).

______ and Orton, T.J. 1982. Chromosome structural changes in cultured celery cells. In Vitro 19:83-89.

Opgenorth, D.C. and Endo, R.M. 1979. Sources of resistance to Fusarium yellows of celery in California. Plant Dis. Rep. 63:165-169.

Orton, T.J. 1981. Breeding celery for disease resistance and improved quality. In: California Celery Research Program 1980-1981 Annual Report (F. Pusateri, ed.) pp. 41-63. California Celery Research Advisory Board, Bakersfield, California.

______ 1983. Spontaneous electrophoretic and chromosomal variability in callus cultures and regenerated plants of celery. Theor. Appl. Genet. (in press).

______ and Arus, P. 1982. Genetic studies of populations of wild celery in California. Ann. Bot. 49:461-468.

Rappaport, L. 1980. Cloning celery for plant propagation, inducing variation, and screening for disease resistance. In: California Celery Research Program—1979-1980 Annual Report (F. Pusateri, ed.) pp. 19-31. California Celery Research Advisory Board, Bakersfield, California.

______ 1981. Cloning celery for plant propagation, inducing variation, and screening for disease resistance. In: California Celery Research Program—1980-1981. Annual Report (F. Pusateri, ed.) pp. 29-36. California Celery Research Advisory Board, Bakersfield, California.

______ and Grogan, R.G. 1978. Selecting for resistance to *Septoria* leaf spot in celery. In: California Celery Research Program—1977-1978 Annual Report (E. Pusateri, ed.) pp. 27-30. California Celery Research Advisory Board, Bakersfield, California.

______, Fujii, D.S., and Thompson, R.H. 1980. From cells to celery (*Apium graveolens*): Callus formation and plant regeneration in tissue and liquid suspension cultures. HortScience 15:416.

Reinert, J., Backs, D., and Krosing, M. 1966. Faktoren der embryogenese in gewe bekulturen aus kulturformen von Umbelliferen. Planta 68:375-378.

Sharma, A.K. and Bhattacharya, N.K. 1959. Further investigations on several genera of Umbelliferae and their interrelationships. Genetica 30:1-62.

Simmonds, N.W. 1976. Evolution of Crop Plants. Longman, London.

Thompson, H.C. and Kelly, W.C. 1957. Vegetable Crops. McGraw-Hill, New York.

Tseng, T.-C., Liu, D.-F., and Shiuo, S.-Y. 1975. Isolation of protoplasts from crop plants. Bot. Bull. Acad. Sin. 16:55-60.

Watts, M.J. and Collin, H.A. 1979. The effect of asulam on the growth of tissue cultures of celery. Weed Res. 19:33-38.

Williams, L. and Collin, G.A. 1976b. Growth and cytology of celery plants derived from tissue cultures. Ann. Bot. 40:333-338.

Zee, S.-Y. and Wu, S.C. 1979. Embryogenesis in the petiole explants of Chinese celery (*Apium graveolens*). Z. Pflanzenphysiol. 93:325-336.

______ 1980. Somatic embryogenesis in the leaf explants of Chinese celery (*Apium graveolens* var. dulce). Aust. J. Bot. 28:429-436.

______, and Yue, S.B. 1979. Morphological and sodium dodecyl sulfate polyacrylamide gel electrophoretic studies of proembryoid formation in the petiole explants of Chinese celery. Z. Pflanzenphysiol. 95:397-404.

SECTION V
Root and Tuber Crops

CHAPTER 10
Cassava

W.M. Roca

Cassava (*Manihot esculenta* Crantz) is one of the most important staple food crops of the lowland tropics. It constitutes a major source of calories to nearly 500 million people in over 60 countries of Africa, Southeast Asia, the Far East, Latin America, and Oceania. Present world cassava output reaches 130 million tons, with most of the production taking place in Southeast Asia, Africa, and tropical Latin America (Table 1).

Cassava is also known as yuca (most Spanish speaking countries), mandioca (Brazil, Paraguay, Argentina), tapioca (Asia), and manioc (French-speaking Africa).

HISTORY OF THE CROP

Cassava originated in the Americas; it has been suggested that the ancestors of cassava were among the first plants used by man in Central and South America (Jennings, 1976). Since cassava is not known in the wild state, its evolution has probably been influenced by man. The north Amazonia was suggested as its place of domestication (Nassar, 1978), where selection was oriented toward large, starch-containing roots, more erect and less branched growth, and easy establishment by stem cuttings (Jennings, 1976). From South America, cassava spread rapidly during the post-Colombian era to Africa and from there to Southeast Asia and India (Leon, 1977).

The present distribution of cassava is worldwide between 30 degrees north and 30 degrees south of the equator, and its altitudinal limit is

Table 1. World Cassava Cultivation and Production[a]

REGION	AREA (ha x 10^{-5})	PRODUCTION (MT x 10^{-5})
Oceania[b]	0.2	2.0
Southeast Asia and the Far East[b]	46.0	537.0
Africa[b]	73.0	467.0
Latin America[c]	17.0	148.0
Tropical	14.0	128.0
Andean	2.0	14.0
Central	0.2	1.0
Caribbean	0.8	5.0
World Total	153.2	1302.0

[a]Sources: FAO (1981) and CIAT (1981b).

[b]Data corresponds to 1980.

[c]Data corresponds to 1972-1976.

approximately 2000 m, with annual precipitation from 500 mm to 8000 mm and soil pH between 4.5 and 8.0. Cassava tolerates practically any hot climate, but when the mean temperature is less than 20 C, growth and yield decrease rapidly (Irikura et al., 1979). The cassava ecological zone coincides with the less developed countries of the world (Lozano, 1977). Cassava is traditionally cultivated by small farmers under agricultural conditions in which climate and soils makes it difficult to grow other crops without very costly inputs (CIAT, 1981a).

UTILIZATION OF CASSAVA

In the majority of producing countries cassava is mainly utilized for human consumption and is available in the fresh market or processed as flour, the latter especially in the rural sectors of Latin America (Table 2).

Besides large quantities of carbohydrates, the cassava storage roots contain varying concentrations of cyanogenic glycoside, which upon autolysis releases HCN. Cassava processing rapidly removes the free cyanide content, but it is less effective on the bound cyanide fraction. Total cyanide levels of more than 50 mg/kg of fresh peeled root are generally considered toxic to humans. Traditional processing techniques most often involve boiling the fresh roots and eating them either hot or cold. Other processes lead to the production of cassava bread or cassave; a starch product called tapioca; or a dried and crisp product called "gari," the most important form of cassava consumption in West Africa and the Brazilian farinha (Lancaster et al., 1982). Despite its protein deficiency, cassava remains a valuable source of energy, cheaper in economic terms than many alternative foods. Furthermore, cas-

Table 2. Utilization of Cassava in Latin America[a]

REGION/ COUNTRY	HUMAN CONSUMPTION[b]			ANIMAL FEED[b]	TOTAL (MT x 10^{-5})
	Fresh	Processed	Starch		
Tropical					
Brazil	6	51	3	40	117.0
Paraguay	36	16	5	43	8.4
Venezuela	39				
Andean					
Colombia	61	-	-	39	7.7
Peru	56	-	-	44	3.0
Ecuador	92	-	-	8	1.3
Bolivia	37	-	-	63	1.9
Central					
Panama	75	-	10	15	0.4
Nicaragua	100	-	-	-	0.2
El Salvador	100	-	-	-	0.2
Costa Rica	100	-	-	-	0.1
Caribbean					
Dominican Republic	100	-	-	-	1.8
Cuba	100	-	-	-	1.5
Jamaica	100	-	-	-	0.1
Total					146.7

[a]Source: CIAT (1981b)

[b]Expressed as percent of total consumption in each region.

sava leaves, which are a good source of protein and vitamins, are commonly used as food in Africa (Lancaster et al., 1982).

Cassava, a carbohydrate source with low unit production cost, has the potential to enter alternative markets as a flour substitute, as a carbohydrate source in animal feed concentrates, and as a raw material in ethanol production (CIAT, 1980b). The animal feed uses of cassava have become important in recent years in Latin America (Table 2), and in Southeast Asia it constitutes an important export commodity in Thailand, Indonesia, and India (FAO, 1979). Cassava roots are considered a valuable source of starch for use in the baking, textile, paper, and adhesive industries; and as shown in Brazil (Barreto, 1980) cassava is also very efficient as raw material for the production of fuel alcohol.

CASSAVA PRODUCTION CONSTRAINTS

Despite the potential of cassava as an energy food for underdeveloped areas and as a raw material for industry, research on the crop has been limited. This can be attributed to the traditional cropping of cassava by low-income people in the tropics. Cassava is bulky, low in protein, has high energy requirements, and normally deteriorates quickly after harvest. World average yields range from 6 to 11 tons per ha per year; however, potential exists to double or even triple such yields with the use of improved technology.

In the last decade, comprehensive programs of research on cassava have been established in the Centro Internacional de Agricultura Tropical (CIAT) and the International Institute of Tropical Agriculture (IITA). A number of national research institutes, such as those in India, Malaysia, and Brazil have also initiated research efforts with this crop. Although these efforts are still limited in contrast to other tropical crops, they are expected to increase yields in the near future.

Cassava research strategies, such as those at CIAT, have been based on the use of improved germplasm. The production technology tends to minimize the inclusion of expensive inputs and thus exploits the comparative advantage of the crop for growing under marginal conditions. Agronomic practices, biological control of insect pests, phytosanitary control of diseases, and efficient techniques for fertilizer use must complement the genetically improved materials (CIAT, 1981a). The cassava world germplasm assembled at CIAT comprises nearly 3000 accessions collected throughout Latin America. Evaluations of some of this germplasm under minimal inputs (without fertilization and irrigation) in good climate and fertile soils gave yields of over 20 tons per ha per year, and under partial drought and acid savanna soils yielded from 8 to 12 tons per ha per year (Kawano et al., 1978a). New germplasm and hybrid lines have been identified, the former with the potential to yield 18 to 24 tons and the latter yielding up to 35 tons per ha per year (CIAT, 1982a).

Although the potential exists to increase cassava yields dramatically, average commercial yields are still low, averaging 9 tons per ha per year (FAO, 1981). This has been attributed to inadequate agronomic practices and the lack of high-yielding varieties tolerant to pests and diseases (Lozano et al., 1980). Main cassava disease constraints (Lozano and Booth, 1974) and economically important insect pests (Bellotti and Schoonhoven, 1978) have been identified.

Cyanide toxicity related to cassava consumption should be considered in improvement programs. Although no acyanogenic lines have been found in screening both cassava and wild *Manihot* germplasm (Jennings, 1976), material has been identified with low cyanide content (CIAT, 1976; Nassar, 1980).

General Botanical and Breeding Characteristics

Cassava is a perennial shrub which comprises cultivars with branched or non-branched stems. Leaves are palmate and the inflorescences develop at the site of apical branches; flowers may be pistillate or staminate. The fruit is a dry dehiscent capsule that contains three loculi. Storage roots develop at the base of the stem by secondary thickening of previously formed fibrous roots. The storage roots are normally cylindrical in form and consist of a periderm, a cortex, and a pith rich in starch.

The cassava crop is propagated vegetatively by means of stem cuttings. The cuttings, also known as stakes, estacas or cangres (Spanish), and manivas (Portuguese) are normally obtained from older previous crops or neighboring fields. For breeding purposes sexual propagation is important. The wild *Manihot* species generally flower and set seed readily; however, in cassava a wide range of flowering responses occur depending on the genotype and the growing conditions. Approximately 20% of CIAT's germplasm collection flowers only rarely or never (Hershey, personal communication) and in various Southeast Asian countries the use of new germplasm is limited due to lack of flowering. Seed set through controlled pollination in cassava tends to be low; an average of about 0.8 seeds per female flower has been reported (CIAT, 1980a).

All *Manihot* species so far studied have 36 chromosomes and it has been suggested that cassava is an allotetraploid (basic number x = 9) (Magoon et al., 1969).

Genetically, cassava is highly heterozygous; both cross- and self-pollination occurs naturally and, though variable among genotypes, selfing produces strong inbreeding depression. Natural selfing may be high in single clone plots; however, the fact that male and female flowers of an inflorescence mature at different times tends to reduce selfing during open pollination. Use of male sterile clones, which occur at a low frequency, also permits the use of open pollination without problems of selfing (Kawano et al., 1978b).

The amount of variation found in cassava, both among plants and within plants (as heterozygosity) is very large and is continously maintained by out-crossing within the species and with other *Manihot* species. Such variability is preserved through vegetative propagation since inbreds have little chance of survival (Martin, 1976).

Important characters such as harvest index (proportion of root yield to total plant weight) and root dry-matter content were found to be highly heritable and to follow an additive gene model (Kawano, 1978); furthermore, resistance to cassava bacterial blight can be transmitted with relative ease to the progeny (CIAT, 1976). Owing to vegetative propagation, once a desirable genotype has been identified it can be utilized and maintained indefinitely (Bellotti and Kawano, 1980).

A total of 98 wild cassva species have been identified, with greatest diversity in Brazil, Paraguay, and Mexico (see Jennings, 1976). Cassava can be readliy crossed with wild *Manihot* species (Nassar, 1980). *M. glassiovii*, *M. saxicola*, *M. dichotoma*, and *M. melanobasis* have been used as sources of disease resistance. However, except for resistance to the mosaic disease and to bacterial blight, no overall improvement of the crop was evident in interspecies hybrids (Hahn et al., 1973).

In cassava, an allotetraploid, when a given character is controlled by a recessive gene or if a character has tetrasomic inheritance, the breeding scheme becomes complicated (Bellotti and Kawano, 1980). It was found that cassava mosaic disease is under quantitative genetic control, the resistance appears to be recessive, and cyanide content seems to be regulated by a complex of recessive minor genes (Hahn et al., 1973).

Besides intervarietal and interspecific hybridization, other techniques such as polyploidy, mutation rectification and in vitro tissue culture methods have been suggested as potentially applicable to cavassa (Hrishi, 1978; Martin, 1976).

CASSAVA TISSUE CULTURE

Propagation by stem cuttings is the conventional means of planting cassava. This mode of propagation often exposes the crop to a wide range of pests and diseases, especially diseases caused by systemic organisms which can be transmitted with the stakes through successive generations. For instance, cavassa bacterial blight, African mosaic disease, superelongation disease, and frogskin disease can potentially produce up to 100% yield losses (Lozano and Booth, 1974). Thus the association of cassava pests and diseases with conventional vegetative propagation not only may affect the productivity of a variety in a locality, but also becomes an important constraint for the maintenance of germplasm collections and for the regional and international movement of cassava clones.

It is therefore justified that most of the cassava tissue culture work has been oriented toward the recovery of healthy clones from diseased varieties by meristem and shoot tip culture methods. More recently, however, cell, protoplast, and anther culture techniques have been investigated in cassava (Table 3).

MERISTEM AND SHOOT TIP CULTURE

General Responses

It wasn't until the last decade that meristem culture methods were used with cassava (see Kartha, 1981). The technique, as first estab-

Table 3. Summary of Cassava Tissue Culture Research

EXPLANT	OBJECTIVE[a]	RESULTS	REFERENCES
Shoot tips	Mosaic eradication	Symptom-free plants	Berbee et al., 1973 (see Kartha & Gamborg, 1975
Meristem tips	Plant regeneration	Over 90% plants	Kartha et al., 1975
Meristem tips	CMD eradication	Over 90% symptom-free plants	Kartha & Gamborg, 1975
Meristem tips	Cryopreservation	Tissue survival: 21%, plant regeneration: 13%	Bajaj, 1977
Shoot tips	Plant regeneration	40-50% plants	Rey & Mroginski, 1978
Shoot tips	Cryoprotectants effect	Phytotoxicity	Kartha & Gamborg, 1978
Meristem tips	CMD and CBSD eradication	30-100% symptom-free plants	Kaiser & Teemba, 1979
Meristem tips, internode sections	Hormonal effects	Best cytokinin: BA, callus and roots only	Nair et al., 1979
Shoot tips, single-node cuttings	Propagation, leaf mosaic	Multiple shoots, 85% symptom-free plants	CIAT, 1979
Shoot tips	Genotype effect, germplasm maintenance	Wide genotype technique, conservation technique	CIAT, 1980
Meristem tips	CMD eradication	Up to 100% symptom-free plants	Adejare & Coutts, 1981
Meristem tips, single-node cuttings	Frogskin disease and CBB eradication, germplasm exchange	Up to 100% symptom-free plants, international exchange technique	CIAT, 1982c
Meristem tips	Cryopreservation	Tissue survival: 90%, plant regeneration: up to 10%	Kartha et al., 1982
Stem sections	Callus growth	Callus growth, only rooting	Eskes et al., 1974
Storage root sections	Callus growth	Callus growth, only rooting	Prabhudesai & Narayanaswamy, 1975

Table 3. Cont.

EXPLANT	OBJECTIVE	RESULTS	REFERENCES
Stem, petiole, leaf sections	Callus growth, C and N sources	Callus growth, C and N sources established, callus greening, only rooting	Parke, 1978
Stem sections	Plant regeneration	Occasional leaf and shoot formation	Tilquin, 1979
Stem, petiole, leaf sections	Callus growth	Callus growth, only rooting	Rey et al., 1980
Seed cotyledons	Embryogenesis	Somatic embryos, whole plants	Stamp & Henshaw, 1982
Anthers	Organogenesis, haploid induction	Callus growth, only rooting, no haploidy	Liu & Chen, 1978
Anthers	Organogenesis, haploid induction	Callus growth, only rooting, some haploids	CIAT, 1982
Mesophyll protoplasts	Protoplast isolation, organogenesis	Colony formation, occasional shoot	Shahin & Shepard, 1980

[a]Abbreviations in this column are as follows: CMD = Cassava Mosaic Disease; CBSD = Cassava Brown Streak Disease; CBB = Cassava Bacterial Blight.

lished for a few cassava cultivars, employed the MS medium (Murashige and Skoog, 1962), containing vitamins as in the B5 medium (Gamborg et al., 1968), supplemented with NAA, BA, and GA at 1.0, 0.5, and 0.1 µM, respectively (Kartha et al., 1974). Later it was found that BA, amongst several cytokins, was best suited for plant regeneration in the presence of NAA and GA, and that GA has a stimulating effect on shoot growth from meristem cultures (Nair et al., 1979).

In fact, cassava apical meristems can be kept alive for only a short time in MS medium. The use of up to 0.5 µM BA promotes shoot initiation and growth. Further increase of BA concentration retards shoot growth, inhibits rooting, and stimulates callus formation. Addition of NAA to this medium further enhances callus growth and rooting in some varieties. On the other hand, the addition of GA to the BA-enriched medium may cause tissue deterioration (Roca, unpublished). It was also found that sucrose interacts with BA in meristem culture of cassava (CIAT, 1980a). At low BA concentrations, shoot elongation is practically doubled when sucrose is increased from 0.03-0.06 M. However, shoot elongation is almost negligible at higher sucrose concentrations, being retarded or inhibited at 0.12-0.15 M sucrose. On the other hand, increasing sucrose from 0.03-0.08 M at low BA concentrations, can result in the promotion of rooting. At higher BA concentrations, an increase in sucrose concentration hardly overcomes the inhibition of rooting due to BA, though callus formation increases. Further increase in sucrose concentration causes root browning and tissue deterioration probably as a consequence of osmotic stress.

The incubation of cultures at 22-25 C counteracts the inhibition of rooting caused by BA. Thus high sucrose concentration on the one hand, and low temperatures on the other, tend to produce similar effects on established shoot tip cultures. While root formation is promoted, shoot elongation is retarded, and leaf abscision as well as anthocyanic pigmentation are elicited (Roca, unpublished).

Recent findings on the effect of major elements indicate that low total nitrogen concentration (20-40 mM) promotes the growth of roots in length and girth at the expense of shoot growth (Roca and Reyes, unpublished).

BASIC PROTOCOL FOR MERISTEM AND SHOOT TIP CULTURE

Plant Material

1. Cut stakes 10-15 cm long with several dormant buds. Preferably obtain the stakes from the middle of the plant.
2. Disinfest the stakes by submersion in fungicide-insecticide mix; then let the stakes dry for a couple of hours.
3. Seal the upper ends of the stakes with melted paraffin.
4. Plant the stakes in pots containing sterilized soil, and place them in a greenhouse or growth chamber at a temperature of 25-30 C. At this stage, the potted stakes can be subjected to thermotherapy, if needed.
5. Irrigate the pots with Hoagland's nutrient solution at one-third strength or with a soluble fertilizer.

Preparation of Sterile Tissue

1. After about 2 weeks, the sprouts will have grown sufficiently to remove the terminal bud together with a short stem section from each. Rapidly growing vegetative but not flowering buds are most suitable as meristem donors. Collect the buds in a bag or beaker with moist paper towel, but do not place them in water. The outer appendages of the bud may carry contaminants; therefore, preventive disinfestation may be needed.
2. Disinfest the buds by rinsing them quickly in 70% ethanol, then soaking them 2-3 min in a 0.5% solution of sodium or calcium hypochlorite. Wash the buds under sterile conditions three to four times with distilled water. Finally, leave the buds in sterilized water.

Dissection of Buds and Explant Isolation

Basic tools needed include a stereo-microscope (10x-50x magnification) with incident illumination, two No. 11 scalpels, two pairs of tweezers, and two microscalpels made with 5-mm pieces of razor blades cemented to a wooden handle. Dissecting needles, 3-5 mm long, made from No. 25 hypodermic needles are also needed. The stereo-microscope should be equipped with zoom-type lenses and care should be taken to prevent overheating the tissues with the illuminator. Sterility can be maintained by the use of a laminar flow hood.

The dissecting tools must be sterilized, preferably in the autoclave, prior to use. However, they may become contaminated during work. It is convenient to work with several sets of the same tools, maintaining the asepsis by immersing them in 70% alcohol and then quickly flaming them. Excessive flaming causes loss of temper and oxidizes the metal; furthermore, the accumulation of burned organic material is difficult to remove.

1. With a pair of tweezers, support a disinfested bud with one hand on the microscope stage. Routinely clean the stage with 70% ethanol and allow it to dry by evaporation. At a 10x-15x magnification, remove the external appendages (leaves and stipules) of the bud using one scalpel.
2. When the internal structures of the bud have been reached, these appear as pale green and three to four leaf primordia remain. At this stage, the shoot tip is 0.8-1.5 mm long and can be cut easily along its morphological neck. The isolated tip adheres to the knife and can be transferred onto the culture medium.
3. The following steps must be done as quickly as possible, otherwise the succulent structures of the meristem dehydrate. Continue dissecting the leaf primordia inward until, at a magnification of about 40x, the youngest primordia can be seen partially encroaching the meristem tip.
4. Remove the young primordia using one curved needle, taking care not to damage the meristem, and cut away the remains of leaf bases as well as other adjacent tissue.
5. All that remains is the dome-shaped apical meristem accompanied by one or two of the youngest primordia. The meristem tip measures 0.4-0.6 mm (see Fig. 2a).
6. Slant the tip so it can be seen in profile and, using one microscalpel, make the final cut perpendicular to the vertical axis. An alternative to this would be to maintain the vertical tip and make oblique cuts into the opposite sides of the meristem so that they cross in the center.
7. The meristem tip should attach to the point of the microscalpel and be brought to the test tube and transferred onto the surface of the medium. It is desirable to place the meristem vertically on the agar.

Incubation of Cultures

Single temperature and light conditions have been used successfully for the incubation of cassava meristem cultures (Kartha et al., 1974; Nair et al., 1979). However, work (Rodriguez and Roca, unpublished) with a wide range of cassava genotypes has shown the requirements for optimal incubation conditions at four critical stages of meristem development (Fig. 1).

STAGE I: INITIATION. A shoot, 1-2 cm in length, often with a small basal callus and without roots, develops from the explant. Depending on the variety this stage may last 2-4 weeks.

STAGE II: ROOTING. The tip of the shoot, as formed at I, is removed and planted on a rooting medium. If further growth of the

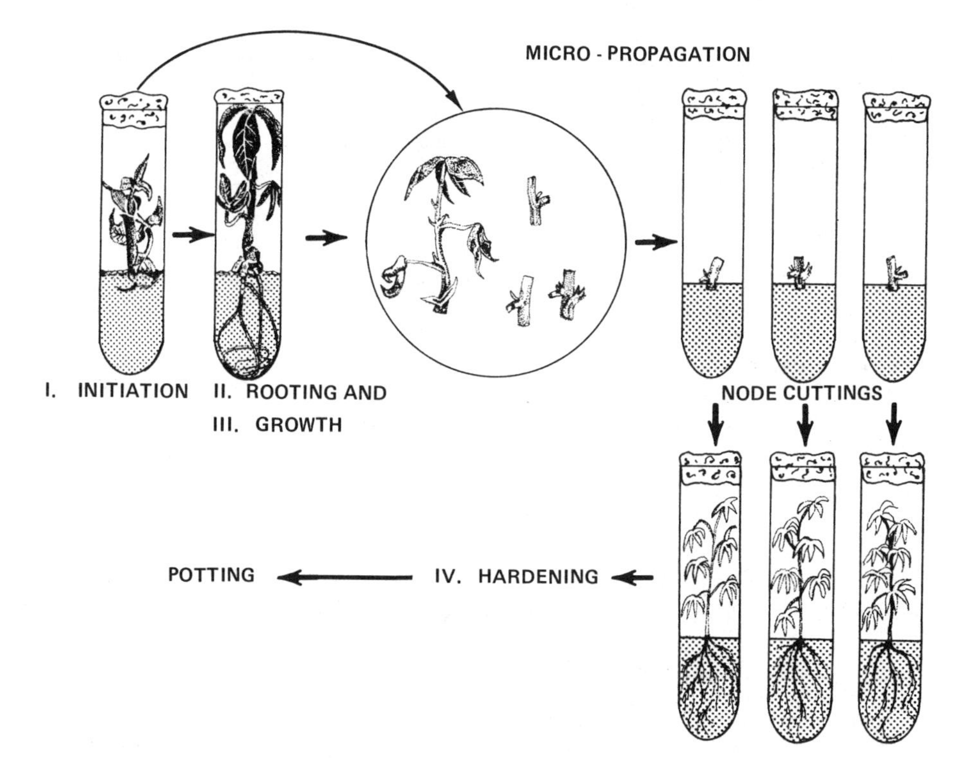

Figure 1. Technique for growing whole plants from cassava meristems. Regardless of rooting, a meristem gives rise to a small shoot (Stage I); rooting is induced on apical (Stage II) and single-node cuttings (Stage III). After a hardening treatment (Stage IV), plantlets are potted. Media composition and cultural conditions are described in the text.

shoot is allowed at Stage I, two to three single nodes can be cut from every shoot and each of these also rooted (Fig. 1). The approximate timing for this stage is 1 week.

STAGE III: FURTHER GROWTH. After root initiation has begun and the first leaf has unfolded, growth of roots and shoots occurs until a complete plantlet, 4-6 cm in length, is formed. Large test tubes allow faster and larger final growth than small tubes.

STAGE IV: HARDENING. For successful potting, the cultures require a conditioning treatment of about 1 week.

Conditions

Temperature: 27-28 C for Stages I, II, and III; 24-25 C for Stage IV.
Illumination: not higher than 2000 lux for Stages I and II; 4000-5000 lux for Stage III; 8000-10,000 lux for Stage IV.
Photoperiod: maintained at 14-16 hr throughout.
Light quality: day-light flourescent lamps throughout.

Culture Media

Stage I: MS medium with 1.0 μM thiamine-HCl, 0.6 mM m-inositol, and 0.058 M sucrose. NAA (0.1-0.2 μM), BA (0.05-0.1 μM), and GA (0.1 μM) are used as growth regulators. The medium is solidified with 0.6% Difco agar.

Stages II, III, and IV: MS (one-third strength) with 3.0 μM thiamine-HCl, 0.6 mM m-inositol, and 0.058 M sucrose. The only growth hormone added to this medium is 0.05-0.1 μM NAA. Agar, 0.8%.

Potting

The hardening period affords the cultures a chance to tolerate water stress and to quickly adapt to the environment after potting.

1. Move the test tubes containing the cultures to a clean bench in the greenhouse.
2. Use substrate comprised of one part soil and three parts fine sand, properly sterilized. "Jiffy" type, plastic, or clay pots can be utilized.
3. Uncap the tubes and remove the plantlets with the aid of forceps. Holding the plant with one hand, wash the roots thoroughly with clean running water.
4. Place the roots, plus one-third of the shoot, in a hole made in the center of the substrate. Then firm the substrate around the plant.

5. Immediately water the pots. The use of a soluble fertilizer rich in phosphorous greatly enhances the growth of the plantlets (CIAT, 1982c). A fertilizer with formula N-P-K (10-52-10) produced a 50- to 100-fold increase in fresh weight within 1 month, compared to watering with Hoagland's nutrient solution or tap water, respectively (Fig. 2f, inset).
6. Place the pots under high relative humidity. This is achieved by placing the pots on trays containing sterilized dampened soil and covering them with a plastic hood. The chamber must be kept away from the sun and strong winds (Fig. 2f). Conditions inside the potting chamber should be 30-40 C during the day and 18-25 C at night, with illumination around 20,000 lux, and relative humidity near saturation during the day and 70-80% at night. One chamber of 1 x 2 m in size can hold up to 200 "jiffy" type pots.
7. One week after potting, open the lid of the chamber gradually until the plants are completely exposed to the greenhouse environment. Continue watering with high phosphorus until the fourth week.

Transplanting to the Field

1. Plants 10-15 cm tall and comprised of 8-10 leaves are appropriate for transplanting to the field.
2. Choose a cloudy day. Otherwise, make the transplant late in the afternoon. The soil moisture should be at field capacity. Do not move the pots from the trays; carry them together to the field.
3. Remove all the large leaves from the plants, and place each one in a hole large enough to cover up to the middle of the shoot. Press the soil around the plant. If "jiffy" pots or plastic bags have been used for potting, there is no need to expose the roots completely; it is only necessary to remove the bottom of each pot to allow faster establishment.
4. Up to about 1 month from transplanting, maintain high soil humidity and watch for possible insect damage. At this point, meristem-derived plants can be treated conventionally (Fig. 2g).

Effect of Cassava Genotype on Meristem Culture

The successful regeneration of plants from meristem culture depends on the interaction of the genotype with the culture medium and with the physical conditions of culture. Thus it can be expected that different cassava varieties would react differently to meristem culturing.

Work conducted with over 300 cassava cultivars from Latin America indicated a general tendency among varieties to form shoots more readily than shoots and roots in the same medium. This tendency was related to increasing concentrations of BA. Attempts to overcome the lack of rooting with higher NAA level resulted in callus proliferation.

A two-step technique was devised for the regeneration of whole plants from a wide range of cassava genotypes (CIAT, 1980a). In the first step, meristem tips are cultured in a medium designed to enhance

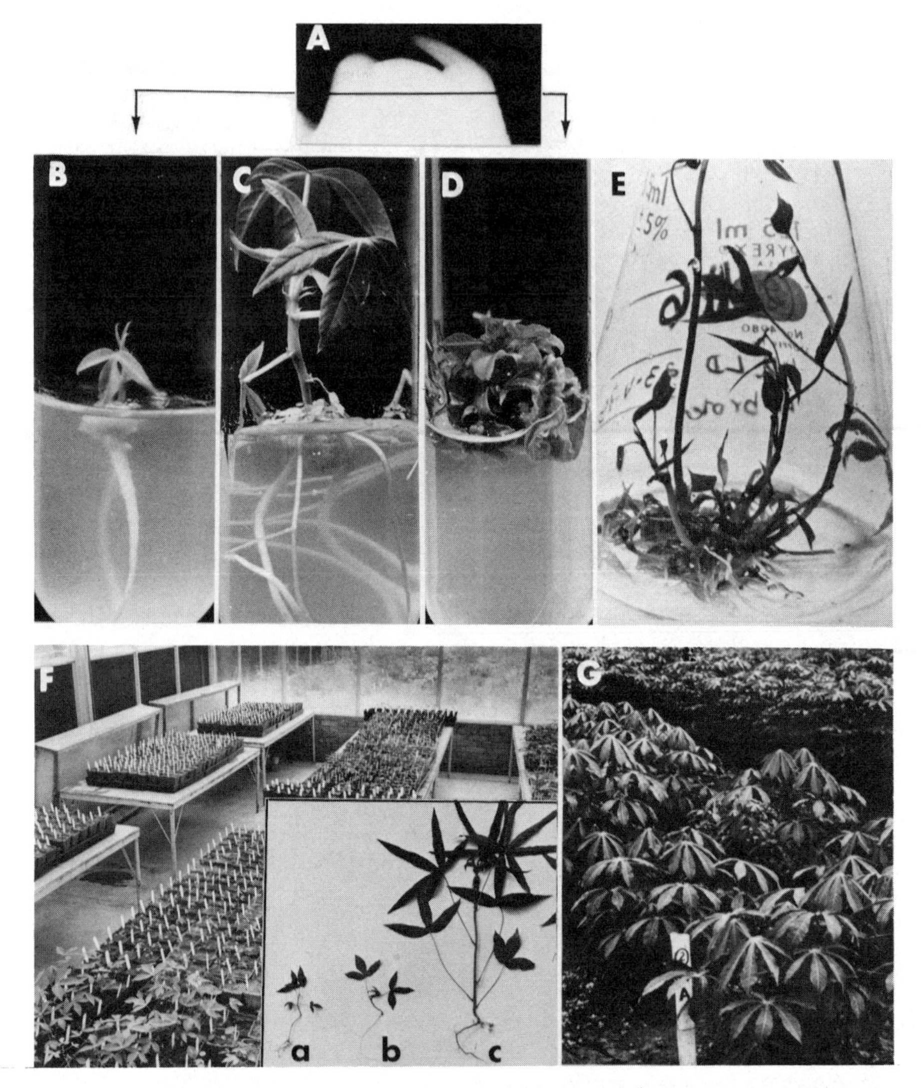

Figure 2. Propagation of cassava by means of meristem culture. A meristem tip (A) normally grows into a single plantlet (B and C), but can be induced to form rosette cultures (D) whose axillary buds grow into multiple shoots (E) in liquid rotated medium. Apical and single-node cuttings from multiple shoots are rooted to form whole plants for potting (F). Watering with a fertilizer rich in phosphorus (c), compared to tap water (a) and Hoagland's solution (b), enhances vigorous growth for successful field establishment (G).

shoot formation regardless of rooting; in the second step, apical segments are removed from each shoot and planted on a rooting medium (Fig. 1). Media composition for steps one and two are those described earlier for Stages I and II, respectively. This technique not only allows quick rooting irrespective of the variety, but also avoids or diminishes callus formation at the shoot-root transition zone, a desirable condition for successful potting.

Multiple Shoot Cultures

Previous work with cassava shoot tip culture (Table 3) failed to exploit the potential of the technique for rapid propagation.

The apical dominance, evident in the uppermost four to six primordial nodes of a cassava vegetative shoot apex, can be overcome by altering the composition of the culture medium (CIAT, 1979). Meristem tips cultured in MS medium with 0.05 μM NAA and increasing concentrations of BA, gradually developed into shortened shoots comprising many nodes; the number of nodes increased in proportion to BA concentrations of 2.5–5.0 μM, and then decreased above 5.0 μM BA. Conversely, shoot elongation gradually decreased with BA concentration (Fig. 3). At optimal BA concentrations, rosette cultures were formed which were comprised of 10–20 nodes each, depending on the variety (Fig. 2). Microscopic observations of the rosettes showed incipient growth of the axillary bud at each node. However, further growth of axillary buds occurred preferentially when the concentration of BA was reduced to

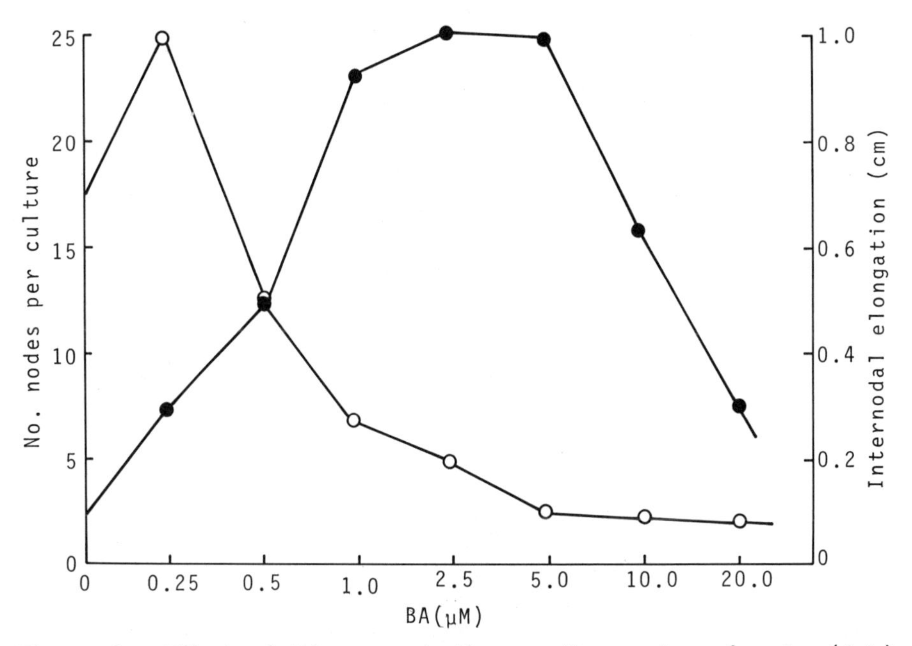

Figure 3. Effect of BA concentration on the number of nodes (dots) and the elongation of internodes (circles) of rosette cultures developed from cassava meristems. Data represent the average of four cultures. For other cultural conditions, see text.

0.25 μM in the presence of 0.1 μM GA and 0.1 μM NAA in rotated liquid MS medium. Compared to the well-known single shoot culture (Figs. 2b and c), the growth of axillary buds on rosette cultures gave rise to multiple shoot cultures (Fig. 2e). Apical and single node cuttings were "harvested" at weekly intervals and transferred to a rooting medium for recovery of plantlets (Fig. 1). Up to 20 apical and nodal cuttings could be harvested weekly from each multiple shoot culture; however, the rate of shoot formation declined with time (Fig. 4).

Rapid in vitro propagation techniques could be used profitably within schemes on maintenance and international exchange of cassava germplasm in which in vitro propagation would be used to produce healthy planting material.

APPLICATIONS OF MERISTEM CULTURE METHODS TO CASSAVA

Production of Healthy Clones

Despite the paucity of knowledge on the etiology and accurate diagnosis of cassava viral diseases, previous work has demonstrated that

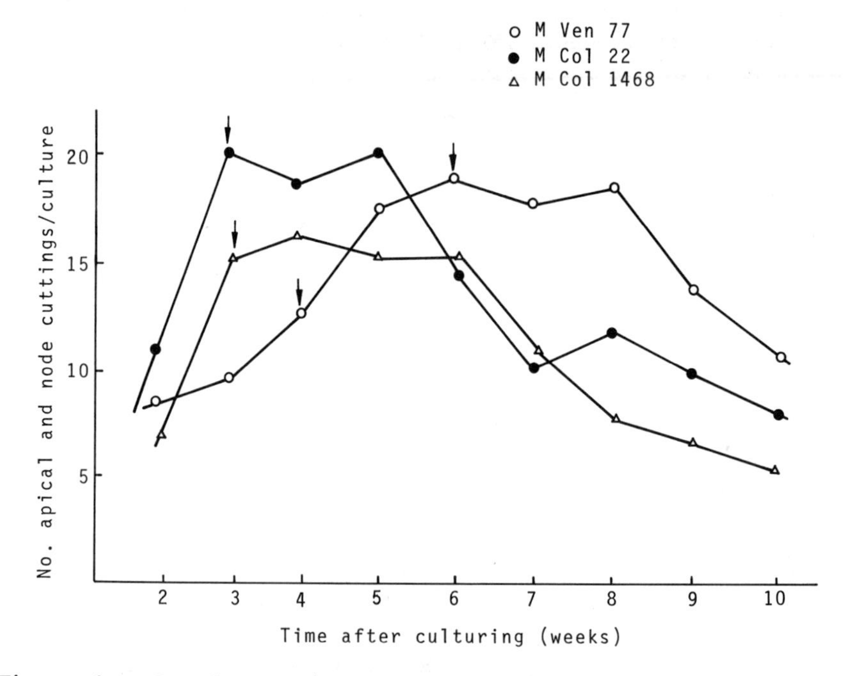

Figure 4. In vitro propagation of cassava through multiple shoot cultures. Production of apical and single-node cuttings "harvested" from multiple shoot cultures of three varieties in liquid rotated medium. Arrows indicate replenishing of fresh medium. See the text for cultural conditions.

healthy (symptom-free) cassava plants can be produced by a proper combination of thermotherapy and meristem culture (Table 3). Although mechanical transmission of the brown streak disease and the African mosaic disease has been reported (see Kaiser and Teemba, 1979; Adejare and Coutts, 1981) the exact etiological agents have not conclusively been identified. A similar situation currently exists for the frogskin disease and other mosaic diseases of possible viral etiology (CIAT, 1982b). Certain indexing techniques such as sap inoculation on indicator plants, grafting of resistant material onto sensitive varieties, and visual evaluation of symptoms provide relative diagnostic criteria. However, the meristem culture-derived plants can only be considered symptom-free until more sensitive indexing methods become available. Recent results have demonstrated the presence of electrophoretic unique proteins associated with a viral disease known as frogskin (CIAT, 1982). In addition, detection of a mosaic disease and of the so called Caribbean cassava disease have been possible through serological and grafting tests, respectively.

The number of disease-free plants produced by meristem culture depends on the virus, the variety, and the proper use of the technique. The general principle that the relative number of virus particles decreases acropetally toward the apical meristem seems applicable to cassava (see Kartha, 1981). The number of plants free of the African mosaic disease was very small when the meristem explant exceeded 0.4 mm in size; however, when heat therapy was applied to the infected stakes, nearly 100% symptom-free plants were produced even with meristem explants measuring up to 0.8 mm (Kartha and Gamborg, 1975). Work on the eradication of a mosaic disease from the Caribbean (CIAT, 1979), and frogskin disease (CIAT, 1982b) resulted in 85% and nearly 100% symptom-free plants, respectively. The culture of small meristem explants, after heat therapy of infected stakes, was essential. The frogskin symptom-free plants have been vegetatively propagated through consecutive cycles with the result that such condition still remains.

Ribavirin, a chemotherapeutic compound with broad antiviral activity against both DNA- and RNA-containing viruses (Lerch, 1977), has been used experimentally for controlling frogskin disease. Preliminary results showed high phytotoxicity when applied to sand-grown sprouts but not when it was applied to shoot tips in vitro (Roca and Roa, unpublished data).

Recovery of Varietal Yield and Vigor

A gradual decline or degeneration in yield and vigor of local cultivars may result from the accumulation of diseases that may become manifested through symptoms or remain latent.

Root yield increases have been obtained at CIAT due to cleaning cassava cultivars infected with viral-like pathogens. In the last 4 years, nearly 600 cassava cultivars have been processed through thermotherapy-meristem culture and handed to the breeders and agronomists for use in performance trials under various agroclimatic systems, along with local "uncleaned" cultivars and new breeding lines.

Yields of Secundina, a popular variety from the Colombian north coast, have been diminished due to a viral disease named Caribbean cassava mosaic (CCM) (CIAT, 1982b). The symptoms often pass unnoticed in the farmer's field, but show up through grafting or under greenhouse conditions. Clean material has been produced through thermotherapy followed by meristem culture, then propagated and sent back to the north coast for in-farm trials. Root yield increases of 70% in fresh weight and starch content were obtained in comparison to the use of traditional planting material of Secundina; no differences in yield were found between the meristem culture-derived Secundina plants and a breeder's hybrid (CM 342-170) selected for the same region (Lozano et al., 1983).

The cassava cultivar Llanera was quite preferred by farmers until a few years ago when it gradually began to diminish its yield in the Colombian Cauca Valley. The growth of meristem culture-derived plants has not only increased general vigor, but fresh root yield is 30-40% as compared to conventional plantations (Roca and Coral, unpublished results). Graft-indexing seems to suggest the existence of a latent viral type disease in the farmers' stocks (Jayashinghe, personal communication).

Recently another cassava virus-like disease called frogskin has been shown to reduce yields drastically in many cultivars. Thermotherapy, followed by meristem culture, has been very effective in producing healthy stocks and recovering yield and vigor. Nearly 400 varieties from the germplasm collection of CIAT have been cleaned from the frogskin disease (Roca and Roa, unpublished results) in the last 3 years.

These examples show that farmers' crop yields can be substantially increased through simple meristem culture methods. This approach can be especially useful in the case of traditional cultivars in the poor regions of developing countries, where short-term solutions to agricultural problems are needed, since the development of new varieties through breeding is a time-consuming task.

PROTOCOL FOR THE PRODUCTION OF HEALTHY CLONES

The overall meristem culture-mediated control of cassava diseases entails: (a) etiology of the disease, (b) application of thermotherapy followed by meristem culture, (c) indexing for freedom of pathogens, and (d) propagation of healthy clones under conditions to minimize reinfection. Steps (a) and (c) are beyond the scope of this chapter.

1. **Prepare the plant material as indicated in the basic meristem culture protocol.**
2. **Place potted stakes in a growth chamber at 35 C, with 6000 lux illumination and a 14-hr photoperiod for about 1 week.**
3. **Gradually increase the day temperature (1 C per day) up to 40 C; maintain the night temperature at 35 C.**
4. **After 3 weeks of thermotherapy (40 C during the day and 35 C at night), remove the apical bud from every sprout and proceed to surface-sterilize.**

5. Meristem isolation and culture, as well as potting, etc., should be conducted as specified in the basic protocol.
6. Carry out indexing with available techniques. If possible, indexing could begin at the test tube level prior to potting, then confirmed with potted plants.
7. Keep healthy plants under conditions that prevent or minimize recontamination, especially if the disease is transmitted by insect vectors or is highly transmissible through mechanical means, soil, or water.
8. Propagation of healthy plants follows. In vitro propagation is the most secure method to practically eliminate recontamination of healthy materials by means of insects, soil, water, and even through the air.

As shown before, cassava can be multiplied in vitro by a combination of multiple-shoot culture (Fig. 2d and e) and single node cuttings (Fig. 1). In practice, this propagation can provide enough basic material for use in more conventional rapid multiplication. Conventional propagation of cassava is very slow; only 10–20 stakes can be produced per year per mother plant. Two improved techniques have been adapted. One utilizes sprouts grown on 2-node stakes (Cock et al., 1976) and the other utilizes single leaf-bud cuttings obtained directly from the mother plants (Roca et al., 1980). The former can potentially yield up to 36,000 stakes and the latter, up to 300,000 stakes per year per mother plant.

Germplasm Conservation

The potential danger of genetic erosion of both cultivated and wild cassava germplasm resources may be attributed to the replacement of primitive cultivars by new varieties or hybrids and the incorporation of new land to agriculture in the areas of genetic diversity (Hershey, personal communication). Such danger, and the requirement of genetic variability for use in the improvement of the crop, justifies cassava germplasm conservation efforts.

Conventional maintenance of cassava germplasm collections is done by continuous vegetative field cultivation. New germplasm plantings often use freshly cut stakes from old fields. Besides the high costs, field maintenance often exposes the valuable germplasm to insect attack, disease infection, and soil or climatic problems. Freshly cut stakes can only be kept for a short time because of premature sprouting and insect or microbial attack. Chemically treated stakes have been maintained for up to 6 months, but vigor of the planting material, as well as yield, decreased (Sales-Andrade and Leihner, 1980). Furthermore, because of their bulkiness, stakes can potentially harbor systemic contaminants. Seeds also can be used to maintain cassava germplasm. Seeds stored at 5 C and 60% relative humidity have maintained their viability for several years (Hahn et al., 1973). Although cassava seedlings can be free of most diseases, adapted genotypes can not be preserved by seeds due to their high genetic segregation. But if sufficient number of seeds can be collected from random crosses, the nonfixed alleles of an accession could be preserved (IBPGR, 1982).

Meristem culture methods can be used for maintenance of cassava germplasm because of their freedom from microorganisms and their small space requirement, coupled with their potentially high propagation rates and high phenotypic stability.

Maintenance of cassava germplasm by means of meristem culture can be done in combination with cryogenic techniques, or through minimum-growth storage conditions.

CRYOGENESIS. Prior results from cassava freeze-preservation studies have shown low tissue survival (Henshaw and Stamp, personal communication) and low plant regeneration after retrieval of shoot tips from liquid nitrogen (Bajaj, 1977). In addition, it was found that cassava meristems were very sensitive to many cryoprotectants which could arrest or modify organogenesis (Kartha and Gamborg, 1978). Recent findings, however, have demonstrated the feasibility of cryogenesis with cassava. Meristem tips, 0.4 and 0.5 mm in size, were frozen in droplets of MS medium with dimethyl sulfoxide and sucrose as cryoprotectants. A terminal temperature of -25 C, prior to storage in liquid nitrogen, resulted in 90% tissue survival and up to 10% whole plant regeneration (Kartha et al., 1982).

MINIMUM GROWTH STORAGE. Recent research has provided a means to maintain cassava clones in vitro. The storage temperature, illumination, and variations in the composition of the medium (osmotic level, nutrient limitation, growth hormones, and other factors, such as the addition of activated charcoal to the medium and the use of large culture vessels for storage) all had an influence on the rate of growth and viability of single node cultures (CIAT, 1980). Throughout 18-24 months of storage at 20 C, the rate of shoot elongation decreased to about one-fifth that of cultures kept at 25-30 C (Figs. 5a,b). Storage temperatures lower than 18 C were detrimental to a number of cassava varieties if the illumination was kept high. However, culture viability could be maintained at even lower temperatures (10-15 C) as long as the illumination was also lowered to less than 500 lux. Increasing BA from 0.044 to 0.22 μM, on the one hand, and sucrose from 0.058 to 0.12 M, on the other, also slowed down shoot elongation, with over 90% viability. However, if low-temperature storage is combined with high BA and sucrose levels, the growth of cultures is arrested to the degree that most of them become deteriorated after 3 months. Recent findings (Roca et al., 1983) indicate that culture growth is decreased when the total nitrogen content of the medium is lowered to 20 mM at 27-28 C and to 40 mM at 20-22 C. Mannitol at 5-25 mM was found effective in arresting growth, but decreased tissue viability at the lower storage temperature. However, if mannitol is added to the medium, along with 0.088 and 0.18 M sucrose, culture viability significantly increases at both low and high storage temperatures.

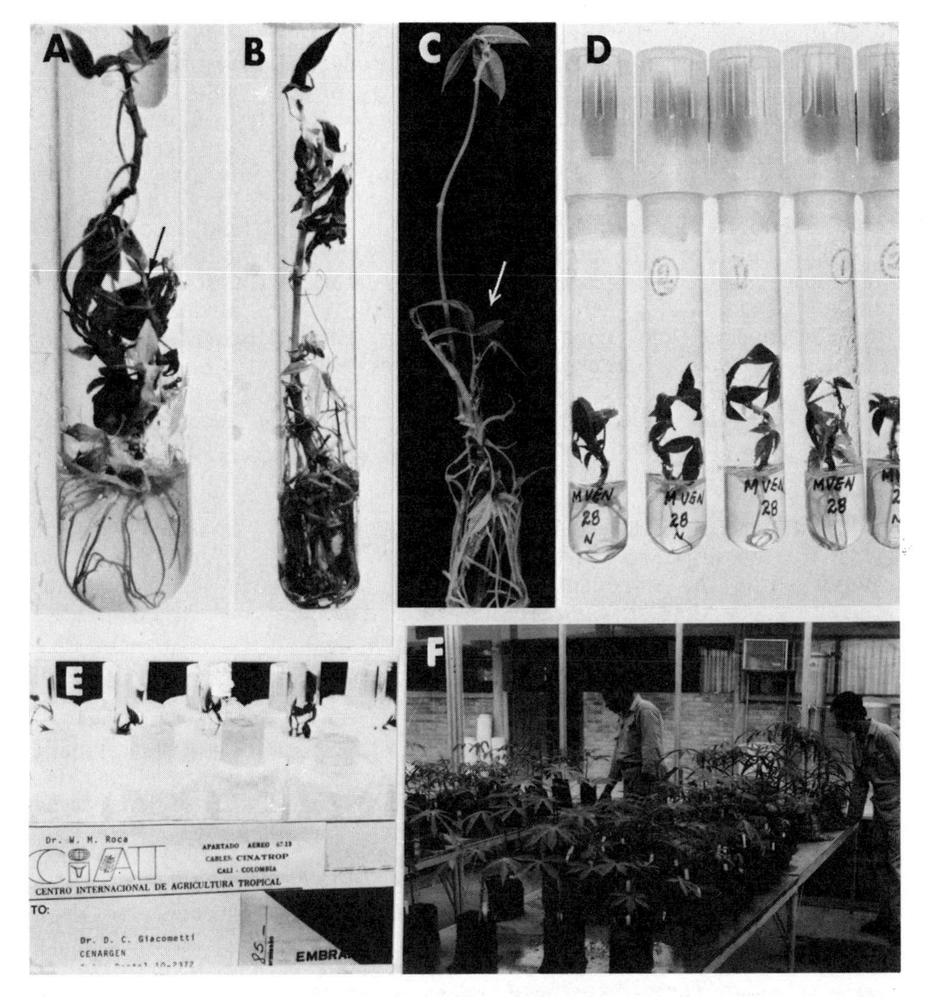

Figure 5. In vitro maintenance and international exchange of cassava germplasm. After 18-24 months of storage at 20-22 C (A) culture viability is maintained through axillary buds (arrow), while storage at 27-28 C (B) diminishes viability. Retrieved culture from 20-22 C after 1 year of storage (C); note axillary branching (arrow). Single rooted plantlets (D) as prepared for packing and international distribution (E). Potted plants under sanitary conditions after recovery from introductions as in vitro cultures (F).

More than 1500 cassava varieties from CIAT's germplasm collection have been transformed into in vitro cultures for storage. Depending on the variety, these cultures can be maintained for up to 54 months with periodic transfers to fresh media. Varieties differ in their relative tolerance to the low temperature and in their relative rate of growth. Furthermore, old cultures from certain varieties tend to deteriorate as a consequence of the oxidation of phenolic-type exudates from the roots. Throughout storage, the cultures produce axillary buds (Fig. 5c). The number of axillary buds per variety is directly related to culture viability, and hence, to plant regeneration upon retrieval from storage. At the end of each storage cycle, the axillary buds are transferred to a fresh medium in order to initiate a new cycle. The germplasm bank in vitro (a room 5 x 6 x 2 m in size) can potentially hold 5000 cassava accessions that otherwise require over 8 ha of land area.

Sample cultures are retrieved from storage once or twice per year, micropropagated, and grown in the field along with stake-propagated plants. Evaluations of phenotypic stability are under way using morphoagronomic and biochemical criteria, but, in general, the plants look true-to-type. Narrowing of leaf lobes has been observed in a few varieties following retrieval from storage and growth in the field, but reversion to the wider lobe type began after the second vegetative growth cycle (Roca and Coral, unpublished).

INTERNATIONAL EXCHANGE OF GERMPLASM. Quarantine regulations for cassava vary from country to country, from those which readily allow the introduction of stakes (some Latin American countries) to those that strictly prohibit their entrance (several countries in Southeast Asia, Africa, and Latin America).

As mentioned above, several cassava diseases are caused by systemic organisms such as virus, bacteria, and fungi. These pathogens can be disseminated, often without noticeable signs, within the stakes. Frequent introductions of pathogenic organisms to a given locality increases the probability of their establishment and prevalence (see Hewitt and Chiarappa, 1977). This can be especially critical if cassava is moved from its center of origin—which may also be a center for diverse pests and diseases—to other areas of the world. On the other hand, certain cassava diseases are geographically confined or their occurrence has not been reported elsewhere. For instance, the transfer of materials from Africa and India to America has been restricted due to the apparent absence of the African mosaic disease in the latter (Lazano, 1977). Similarly, every precaution needs to be taken in moving materials to various Southeast Asian countries which seem to remain "clean" of various important viruslike diseases.

A plantlet (derived from meristem culture of heat-treated stakes) maintained in vitro in a sterile nutritive artificial medium should be free of insects, mites, nematodes, fungi, and bacteria (Kahn, 1977). Should the latter be present, it could be detected readily because of media contamination. If a fastidious organism is present, a special media to support its growth would be needed. To test the absence of virus or virus-like organisms, available indexing techniques should be used. Thus, the use of in vitro cultures for the international exchange of cassava germplasm constitutes an additional safeguard for minimizing the risks of pest and disease dissemination.

International exchange of germplasm (CIAT, 1982c) involves the following steps:

1. The establishment of in vitro cultures. Single rooted plantlets (Fig. 5d) are the simplest form of in vitro cassava clones for shipping. The rooted plantlets are derived from apical and single-node cuttings. Except for the use of 1% agar, the cultures are prepared as described in the basic protocol. The materials distributed in vitro from CIAT include selected varieties, basic germplasm, and promising hybrid lines.
2. The packing and shipment of cultures. The cultures are packed in polystyrene boxes (Fig. 5e). Each package is properly labelled to expedite rapid clearance from the customs office to the Plant Health Services and from there to the institution from which the request originated. The proper phytosanitary documents should be included in the package, the list of materials with their protocol on the handling of the cultures at the receiving end. The cultures should be shipped by air, preferably as accompanied luggage; otherwise they can be air freighted or air mailed. Cassava is highly sensitive to protracted darkness. Shipments longer than 2 weeks cause etiolation, chlorosis, and finally tissue deterioration. Such detrimental effects can be partially prevented if short, vigorous, plantlets are prepared for shipment. This is accomplished through the exposure of the cultures prior to shipment to 8000 lux illumination at 24-25 C. Furthermore, the use of 2,4-D in lieu of NAA diminished etiolation, and when 2,4-D was added together with BA, both defoliation and chlorosis due to darkness were decreased.
3. The handling of cultures at the receiving end is the next part of the process. This is the most critical step in the international exchange of in vitro cassava. Successful handling of cultures after arrival depends on two factors: the time elapsed between shipment and arrival and the physical and personal facilities of the receiver. Actually, the in vitro system is only effective as a tool if managed by well-trained personnel. The overall task is to move the plantlets, after their arrival, from the test tube to the field. Through training and follow-up programs it has been possible to organize a network of collaborating institutes in various Latin American and Southeast Asian countries. Fairly good facilities for handling in vitro cultures now exist in Brazil, Mexico, Costa Rica, Cuba, Venezuela, the Phillipines, Thailand, Malaysia, and Indonesia. Several of the collaborating institutes are able to recover plants from the cultures through the in vitro node-cutting technique (Fig. 1) and carry them up to the field for further multiplication and testing. Minimum handling in other countries only involves direct potting of the cultures after a period of hardening.

The in vitro culture methodology has been accepted by various countries as one of the safest means of receiving vegetative cassava material. This is a recommended method for the transfer of genetic resources both from the collection site to the main germplasm centers and from these to the national programs (IBPGR, 1982).

Between 1979 and 1982, more than 360 cassava materials (varieties and hybrids) were distributed from CIAT to numerous countries in

America as in vitro cultures. Similarly, nearly 50 cultures were shipped to Southeast Asian countries and 15 clones to other countries.

The in vitro system also has been utilized to introduce to CIAT new cassava germplasm from various countries in Latin America. Between 1979 and 1982, nearly 800 accessions were introduced as meristem cultures. Following in vitro micropropagation, these materials were moved to the greenhouse for phytosanitary observation (Fig. 5f), then to the field for further multiplication and use in germplasm trials.

Summary

To summarize, meristem and shoot tip culture techniques have been utilized only in the last decade mainly as a means for ridding selected cassava varieties of viruses. More recently, the use of these techniques has been extended to the maintenance and international exchange of cassava germplasm. The future of cassava cryogenic storage is promising. Minimum-growth storage is now a viable method for maintaining large collections in small spaces free of pests and disease risks. International movement of in vitro cassava provides a valuable safeguard for minimizing the dangers of pest and disease dissemination. The various applications of meristem culture are presented diagramatically in Fig. 6.

OTHER TISSUE CULTURE METHODS IN CASSAVA

Compared to meristem and shoot tip culture, the development of cell, callus, protoplast, and anther culture in cassava is still in its infancy. More extensive work is needed on this subject.

Embryo Culture

There have not been any attempts to recover plants from the culture of immature embryos of cassava. Past experience in the interspecific hybridizations of *Manihot* (Nassar, 1980) shows that there are no substantial barriers to successful hybridization. Nevertheless, embryo rescue techniques may become a valuable tool in the recovery of certain crosses.

On the other hand, plants have been grown from cultured embryos dissected from mature seeds of wild cassava species. Using this technique some plants have been recovered that otherwise would not have survived because of the very poor germination of seeds in several *Manihot* species (Rodriguez and Roca, unpublished).

Cell and Callus Culture

CALLUS GROWTH. Callus has been induced from stem, petiole, leaf, and even root sections of cassava (Table 3). In general, stem sections seem best suited for callus induction.

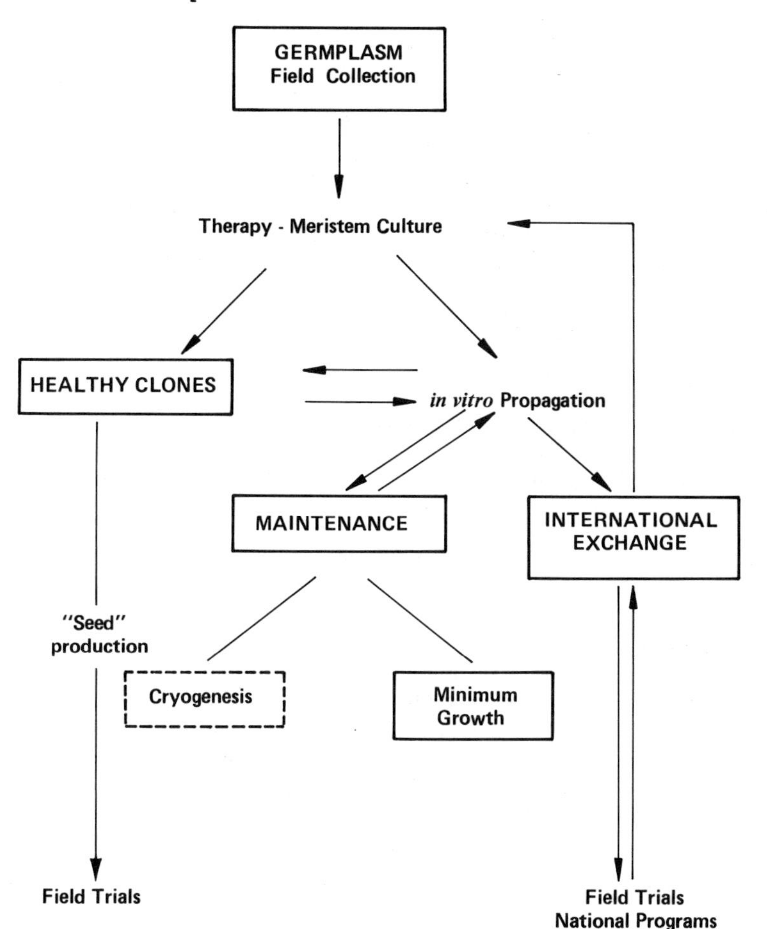

Figure 6. Flow diagram showing the utilization of cassava meristem culture for the production of healthy clones, and the maintenance (through minimum growth storage and cryogenesis) and international exchange of germplasm in clonal form.

The main requirements for callus initiation, growth, and maintenance have been established. The MS medium seems better than WH medium. Sucrose (0.03–0.087 M) is a good carbon source; although the dry weight of callus increases with up to 0.087 M sucrose, the callus turns brown at higher concentrations.

As auxins, 2,4-D and NAA seem better suited than IAA for callus initiation, growth and maintenance. More rapid growth is achieved with a combination of 2,4-D (5.0–13.0 μM) and a cytokinin. With KIN or BA (2.0–8.0 μM), callus growth is rapid, but greening may occur more readily with 2iP or ZEA (10 μM).

Common nitrogen sources such as ammonium nitrate (20 mM) and potassium nitrate (19 mM) support callus growth as well as the combination of ammonium chloride (20 mM) and potassium succinate (10 mM), though the latter may be better for greening. Organic additives such as CW (10-18%) can increase callus growth in the presence of 2,4-D, but browning also has occurred with CW in the medium.

ORGANOGENESIS. Root formation in callus cultures is readily obtained with the aforementioned media, especially when NAA is used as the auxin. Addition of BA up to a certain concentration seems to promote rooting, but at higher concentrations all cytokinins counteract the rooting effect of auxins. Thus, for the maintenance of undifferentiated callus, both auxins and cytokinins need to be present.

Tilquin (1979) claimed that leaves and shoots were occasionally regenerated from callus grown from stem sections on the same medium as that used for meristem culture (Kartha and Gamborg, 1975). However, attempts to reproduce those results have been unsuccessful (Rodriguez and Roca, unpublished).

SOMATIC EMBRYOGENESIS. Somatic embryos and whole plants have been regenerated using cotyledonary explants dissected from mature cassava seeds (Stamp and Henshaw, 1982). The highest frequency of embryo formation occurred in MS medium with high 2,4-D (20 µM); development of embryos was enhanced with lower 2,4-D (0.05 µM) and with the addition of BA (0.5 µM). Embryogenesis was always observed to occur from cotyledonary tissue and not from callus tissue. Embryogenesis obtained from genetically segregating tissues, such as the seed cotyledons, may help identify the requirements for somatic embryogenesis in callus of clonal origin in cassava.

Protoplast Culture

Protoplasts from leaf mesophyll cells have been isolated successfully and induced to regenerate cell walls and form callus. However, shoot formation was only observed occasionally (Bidney, personal communication; Shahin and Shepard, 1980).

Protoplasts can be isolated from plants grown in growth chambers, from shoot tip cultures, or from cell suspensions (Rodriguez and Roca, unpublished). Macerozyme and cellulase or macerase and driselase have been used for enzymatic digestion of cell walls. Colony formation and callus growth were induced following a series of passages on various media (Shahin and Shepard, 1980).

Anther Culture

Callus and root formation, but not shoot formation, were reported for anthers cultured in MS medium supplemented with BA and NAA (Liu and Chen, 1978). Green areas developed on callus with the addition of either GA or ABA. Activated charcoal arrested callus growth; however, cold pretreatment of anthers enhanced callusing. Examinations of preparations of squashed, stained callus cells indicated that the callus originated from somatic tissue.

In recent work with cassava anther culture (CIAT, 1982c), floral buds of 1.5-2.5 mm, corresponding to late tetrad through late uninucleate microspore, were utilized. Callus and then roots formed in MS medium supplemented with BA, NAA, and CW. Chromosome counts indicated haploidy in a few root tips. In some varieties nearly 100% of the anthers formed callus, while in others callus formation was variable both between plants and even within anthers of the same plant.

Conclusion

It can be concluded that the inability to regenerate whole plants from cell and callus cultures of cassava, using reproducible procedures, currently constitutes the main constraint to progress in research in this field.

The probability that genetically regulated factors block shoot differentiation in cassava cell and callus cultures can not be discarded. Given the fact that all attempts to induce whole plant differentiation have been done with very few genotypes, it would be worthwhile to screen large germplasm collections for regenerative ability in callus culture. Basic requirements would be determined with the selected genotypes and extended to other materials.

Inbreeding depression may already exert an effect at the cellular level in anther culture; thus genotypes with very low inbreeding depression should be selected for use in the development of the technique.

FUTURE PROSPECTS

Meristem and Shoot Tip Culture

Important problems in a vegetatively propagated crop such as cassava are associated with the production of disease-free stocks and germplasm storage. Meristem culture integrated with disease diagnostic techniques seems a logical approach for initiating the production of disease-free materials in national programs of individual countries. Large amounts of planting material could then be produced using clean donor stocks. The problem of varietal decline, due to some kind of degeneration and loss of resistance to diseases, may be approached through clean "seed" production.

The development of in vitro germplasm banks will shortly become an important support to conventional field maintenance of large collections and should be a valuable alternative in the conservation of cassava germplasm in the future. Cryogenic storage will make such an alternative even more attractive.

In vitro work with germplasm banks will expedite the exchange of clonal materials internationally with less risk of pest and disease transfer.

Cell, Protoplast, and Anther Culture

Cell culture methods should become an important aid to cassava germplasm improvement in the future. Breeding for characters controlled by recessive genes is difficult in cassava, and even more so if the character is tetrasomically inherited (Bellotti and Kawano, 1980). For example, no forms devoid of cyanogenic glycosides have been found in cassava or wild *Manihot* species (Jennings, 1976), and the incapability of producing the glycoside seems to be due to recessive genes (Hahn et al., 1973). Thus the homozygosity needed to express acyanogenesis would be difficult to achieve by conventional breeding methods. By using genotypes with low inbreeding depression, such as those which seem to exist in cassava (Kawano et al., 1978b), homozygous lines could be produced quickly through anther culture and chromosome doubling. Tetrasomic inheritance would be avoided in the doubled haploids, which could be used for genetic and physiological studies. The acyanogenic character would be maintained by hybridization with other anther culture-derived lines.

Hybrid seed may offer an alternative to the growth of cassava in the future. Because of inbreeding depression it is difficult to produce pure lines for use in hybrid seed production through successive inbreeding. Homozygous lines could be quickly produced from selected materials, thereby facilitating the maximization of heterosis in cassava.

Certain specific traits such as acyanogenesis or any other simply inherited trait (Hrishi, 1978) that may drastically affect an otherwise highly desirable cultivar, could be amenable to rectification through mutation techniques or induction of intraclonal variability. Vegetative cassava buds have been treated with mutagens (Nayar, 1975). Changes in stem and leaf morphology and cyanide content were observed after treatment with gamma radiation. Increases in ploidy as well as in protein content were produced with the use of colchicine. It may be possible to expose cell cultures—preferably haploid—or even meristem cultures to acute and chronic radiation as well as to chemical mutagens and select desirable phenotypes. If, for instance, cyanide freedom is due to a recessive mutant gene, such treatments could give rise to material with altered glycoside content.

The use of haploid cells in mutation work would facilitate mutant selection since both recessive and dominant mutants would appear immediately and could be stabilized through chromosome doubling. Haploid propagates could be even more amenable to mutagenesis than cell and meristem cultures since the latter do not consist of single cells and can not be subjected uniformly to mutagenic agents (Gamborg et al., 1974).

Haploids could also help breeding efforts if they are able to provide any phylogenetic clues on the nature of allopolyploidy in cassava. Furthermore, haploids could help in the determination of genetic ratios and gene action. Since dominance effects are absent in doubled haploids, the method could also serve to detect lethal genes which may have accumulated in heterozygous clones.

Vegetative propagation of cassava is an advantage in cell culture manipulations. Once valuable variability has been selected, its genotype can be maintained through vegetative multiplication. Even epigenetic variability, common in tissue culture and stable through mitosis (Carlson, 1979), could be maintained vegetatively. Because of vegetative propagation, chimeral plants would be regenerated from mixed callus of different genotypes (Carlson, 1977).

Finally, the future would witness important developments in molecular biology and genetics which would allow manipulation of cassava cells or protoplasts with physical stresses, specific pathogenic toxins, or metabolic analogs to select valuable phenotypes. The exchange of genetic information through cell fusion, organelle uptake, or DNA transformation processes would pave the way to the application of genetic engineering schemes to cassava.

ACKNOWLEDGMENTS

The research assistance of J. Rodriguez, J. Narvaez, J. Roa, G. Mafla, and J. Beltran is gratefully acknowledged. I thank Drs. D.R. Laing and C. Hershey for reviewing the manuscript.

KEY REFERENCES

Centro Internacional de Agricultura Tropical. 1979. Cassava program, cassava tissue culture. In: CIAT Annual Report 1979, pp. 82-88. CIAT, Cali, Colombia.

______ 1980a. Cassava program, cassava tissue culture. In: CIAT Annual Report, 1980, pp. 79-85. CIAT, Cali, Colombia.

Kartha, K.K. 1981. Meristem culture and cryopreservation. Methods and applications. In: Plant Tissue Cultre—Methods and Applications in Agriculture (T.A. Thorpe, ed.) pp. 181-211. Academic Press, New York.

______, Gamborg, D.L., Constabel, F., and Shyluk, J.P. 1974. Regeneration of cassava plants from shoot apical meristems. Plant Sci. Lett. 2:107-113.

Martin, F.W. 1976. Cytogenetics and plant breeding of cassava. A review. Plant Breed. Abst. 46:909-916.

REFERENCES

Adejare, G.O. and Coutts, R.H.A. 1981. Eradication of cassava mosaic disease from Nigeria cassava by meristem-tip culture. Plant Cell Tissue Organ Culture 1:25-32.

Bajaj, Y.P.S. 1977. Clonal multiplication and cryopreservation of cassava through tissue culture. Crop Improv. 4:198-204.

Barreto, T.J. 1980. Produccion de alcohol de yuca. El caso Brasilero. In: Simp, Colombiano sobre Alcohol Carburante (T. Brekelbaum, J.C. Toro, and V. Izquierdo, eds.) pp. 133-144. CIAT, Cali, Colombia.

Bellotti, A. and Kawano, K. 1980. Breeding approaches in cassava. In: Breeding Plants to Insects (F.G. Maxwell and P.R. Jennings, eds.) pp. 314-335. John Wiley and Sons, New York.

Bellotti, A. and Schoonhoven, A.V. 1978. Mite and insect pests of cassava. Ann. Rev. Entomol. 23:39-67.

Carlson, P.S. 1977. Novel cellular associations formed in vitro. In: Molecular Genetic Modification of Eucaryotes (I. Rubenstein, ed.) pp. 43-56. Academic Press, New York.

______ 1979. Peptides of normal and variant cells of tobacco. Develop. Genet. 1:3-12.

Centro Internacional de Agricultura Tropical. 1976. Cassava program, varietal improvement section. In: CIAT Annual Report 1975, pp. B36-B49. CIAT, Cali, Colombia.

______ 1980b. Cassava program, varietal improvement section. In: CIAT Annual Report 1980, pp. 33-37. CIAT, Cali, Colombia.

______ 1981a. CIAT in the 1980s. A Long-Range Plan for the Centro Internacional de Agricultura Tropical. Cali, Colombia.

______ 1981b. Latin American Agriculture: Trends in CIAT Commodities. Internal Document Economy 1.6. CIAT, Cali, Colombia.

______ 1982a. Cassava program, germplasm development section. In: CIAT Annual Report 1981, pp. 125-142. CIAT, Cali, Colombia.

______ 1982b. Cassava program, pathology section. In: CIAT Annual Report 1981, pp. 87-112. CIAT, Cali, Colombia.

______ 1982c. Cassava program, cassava tissue culture. In: CIAT Annual Report 1981, pp. 113-123. CIAT, Cali, Colombia.

Cock, J.H., Wholey, D., and Lozano, J.C. 1976. A rapid propagation system for cassava. CIAT Series EE-20. CIAT, Cali, Colombia.

Eskes, A.B., Vargas, A., Staritsky, G., and Bruinasma, J.J. 1974. Callus growth and rooting of cassava (*Manihot esculenta* Crantz) stem segments cultured in vitro. Acta Bot. Neerl. 23:315-320.

Food and Agriculture Organization of the United Nations. 1979. Agricultural Commodity Projections 1975-1985. FAO, Rome.

______ 1981. Production Yearbook 1980. Vol. 34. FAO, Rome.

Gamborg, O.L., Miller, R.A., and Ohyama, K. 1968. Nutrient requirements of suspension cultures of soybean root cells. Exp. Cell Res. 50:148-151.

Gamborg, O.L., Constabel, F., Fowke, L., Kao, K.N., Ohyama, K., Kartha, K., and Pelcher, L. 1974. Protoplast and cell culture methods in somatic hybridization in higher plants. Can. J. Genet. Cytol. 16:737-750.

Hahn, S.K., Howland, A.K., and Terry, E.R. 1973. Cassava breeding at IITA. Proceedings of the International Tropical Root Crop Symposium. Ibadan, Nigeria.

Hewitt, W.B. and Chiarappa, L., eds. 1977. Plant Health and Quarantine in the International Transfer of Genetic Resources. CRC Press, Cleveland, Ohio.

Hrishi, N. 1978. Breeding techniques in cassava. In: Cassava Production Technology (N. Hrishi and R.G. Nair, eds.) pp. 1-5. Central Tuber Crops Research Institute, Trivandrum, India.

International Board of Plant Genetic Resources. 1982. Report of working group on cassava genetic resources. IBPGR, Rome.

Irikura, Y., Cock, J.H., and Kawano, K. 1979. The physiological basis of genotype-temperature interaction in cassava. Field Crops Res. 2: 227-239.

Jennings, D.L. 1976. Cassava, *Manihot esculenta* (Eupharbiaceae). In: Evolution of Crop Plants (N.W. Simmonds, ed.) pp. 81-84. Longman, London.

Kahn, R.P. 1977. Plant quarantine: Principles, methodology and suggested approaches. In: Plant Health and Quarantine in International Transfer of Genetic Resources (W.B. Hewitt and L. Chiarappa, eds.) pp. 289-307. CRC Press, Cleveland, Ohio.

Kaiser, W.J. and Teemba, L.R. 1979. Use of tissue culture and thermotherapy to free East African cassava cultivars of African cassava mosaic and cassava brown streak diseases. Plant Disease Rep. 63: 780-784.

Kartha, K.K. and Gamborg, O.L. 1975. Elimination of cassava mosaic disease by meristem culture. Phytopathology 65:826-828.

______ 1978. Meristem culture techniques in the production of disease-free plants and freeze-preservation of germplasm of tropical tuber crops and grain legumes. In: Diseases of Tropical Food Crops, Proceedings of the International Symposium (H. Maraite and J.A. Meyer, eds.) pp. 267-283. U.C.L. Lovaine, Belgium.

Kartha, K.K., Leung, N.L., and Mroginski, L.A. 1982. In vitro growth responses and plant regeneration from cryopreserved meristems of cassava (*Manihot esculenta* Crantz). Z. Pflanzenphysiol. 107:133-140.

Kawano, K. 1978. Genetic improvement of cassava (*Mannihot esculenta* Crantz) for productivity. Japan Trop. Agric. Res. Series No. 11:9-21.

______, Amaya, A., Rios, M., and Goncalves, W.M.F. 1978a. Evaluation of cassava germplasm for productivity. Crop Sci. 18:377-386.

______, Amaya, A., Daza, P., and Rios, M. 1978b. Factors effecting efficiency of hybridization and selection in cassava. Crop Sci. 18: 373-376.

Lancaster, P.A., Ingram, J.S., Lim, M.Y., and Coursey, D.G. 1982. Traditional cassava-based foods: Survey of processing techniques. Econ. Bot. 36:12-45.

Leon, J. 1977. Origin, evolution and early dispersal of root and tuber crops. In: Proceedings of the Fourth Symposium of the International Society for Tropical Root Crops (J. Cock, R. McIntyre, and M. Graham, eds.) pp. 20-36. International Development Research Centre, Ottawa, Canada.

Lerch, B. 1977. Inhibition of the biosynthesis of potato virus X by ribavirin. Phytopathology 89:44-49.

Liu, M.Ch. and Chen, W.H. 1978. Organogenesis and chromosome number in callus derived from cassava anthers. Can. J. Bot. 56:1287-1290.

Lozano, J.C. 1977. Cassava (*Manihot esculenta* Crantz). In: Plant Health and Quarantine in International Transfer of Genetic Resources (W.B. Hewitt and L. Chiarappa, eds.) pp. 103-109. CRC Press, Cleveland, Ohio.

______ and Booth, R.H. 1974. Diseases of cassava (*Manihot esculenta* Crantz). PANS 20:30-54.

______, Byrne, D., and Bellotti, A. 1980. Cassava-ecosystem relationships and their influence on breeding strategy. Trop. Pest Manage. 26:180-187.

______, Pineda, B., and Jayasinghe, U. 1983. Effect of cutting quality on cassava (*Manihot esculenta* Crantz) performance. In: Proceedings 6th International Tropical Root Crop Society Symposium, Lima, Peru (in press).

Magoon, M.L., Krishnan, R., and Bai, K.V. 1969. Morhphology of the pachytene chromosomes and meiosis in *Manihot esculenta.* Cytologia 34:612-624.

Murashige, T. and Skoog, F. 1962. A revised medium for rapid growth and bioassays with tobacco tissue cultures. Physiol. Plant. 15:473-497.

Nair, N.G., Kartha, K.K., and Gamborg, O.L. 1979. Effect of growth regulators on plant regeneration from shoot apical meristems of cassava (*Manihot esculenta* Crantz) and on the culture of internodes in vitro. Z. Pflanzenphysiol. 95:51-56.

Nassar, N.M.A. 1978. Conservation of the genetic resources of cassava (*Manihot esculenta*): Determination of wild species localities with emphasis on probable origin. Econ. Bot. 32:311-320.

______ 1980. Attempts to hybridize wild *Manihot* species with cassava. Econ. Bot. 34:13-15.

Nayar, G.G. 1975. Improving tapioca by mutation breeding. J. Root Crops 1:55-58.

Parke, D. 1978. Tissue culture of cassava on chemically defined media. Physiol. Plant. 42:195-201.

Prabhudesai, V.R. and Narayanaswamy, S. 1975. A tissue culture from tapioca. Plant Sci. Lett. 4:237-241.

Rey, H.Y. and Mroginski, L.A. 1978. Cultivo in vitro de apices caulinares de mandioca (*Manihot esculenta* Crantz). Phyton. 36:171-176.

______, and Fernandez, A. 1980. Induccion in vitro de callos y raices de explantes de seis cultivars de mandioca (*Manihot esculenta* Crantz). Phyton. 39:161-170.

Roca, W.M., Rodriguez, A., Patena, L.F., Barba, R.C., and Toro, J.C. 1980. Improvement of a propagation technique for cassava using single leaf-bud cuttings: A preliminary report. Cassava Newsletter No. 8:4-5. CIAT, Cali, Colombia.

Roca, W.M., Reyes, R., and Beltrán, J. 1983. Effect of various factors on minimal growth of tissue culture storage of cassava germplasm. In: Proceedings 6th International Root Crop Society Symposium, Lima, Peru (in press).

Sales-Andrade, A.M. and Leihner, D.E. 1980. Influence of period and conditions of storage on growth and yields of cassava. In: Cassava Cultural Practices (E.J. Weber, J.C. Toro, and M. Graham, eds.) pp. 33-37. Workshop Proceedings, Salvador, Bahia, Brazil. International Development Research Centre, Ottawa, Canada.

Shahin, E.A. and Shepard, J.F. 1980. Cassava mesophyll protoplasts: Isolation, proliferation, and shoot formation. Plant Sci. Lett. 17:459-465.

Stamp, J.A. and Henshaw, G.G. 1982. Somatic embryogenesis in cassava. Z. Pflanzenphysiol. 105:183-187.

Tilquin, J.P. 1979. Plant regeneration from stem callus of cassava. Can. J. Bot. 57:1761-1763.

Walkey, D.G.A. and Cooper, V.C. 1975. Effect of temperature on virus eradication and growth of infected tissue cultures. Ann. Appl. Biol. 80:185-190.

CHAPTER 11
Sweet Potato

J.H.M. Henderson, B.R. Phills, and *B.T. Whatley*

The sweet potato, *Ipomoea batatas,* is a member of the morning glory or bindweed family, the Convolvulaceae. The latter contains about 50 genera and over 1200 species. *Ipomoea* is derived from a Greek combination meaning worm bindweed. *Batatas* is the aboriginal American word for potato. The sweet potato is the only member of the genus *Ipomoea* whose roots are edible.

HISTORY OF THE CROP

The origin of the sweet potato is obscure. It is not known in the wild state. There seems little doubt that it originated in tropical America, but exactly where and by what parents are not known with certainty. The roots of the plant are said to have been first mentioned by an author named Pigafetta (Nicholls, 1906) who visited Brazil in the year 1519 and found the potato in use among the Indians as an article of food. It is therefore speculated to be a native of South America. Two of the ancient civilizations of tropical America, the Mayan and the Peruvian, grew and cultivated sweet potatoes.

Long before Europeans arrived in this hemisphere, the Incas of South America and the Mayas of Central America grew several cultivars. They called the plant "cassiri" (Kas-se-re). One cultivar was grown for food and the other cultivars were used to supply their artists with colors for use in paints. Early Spanish explorers are believed to have taken the sweet potato to the Philippines and East Indies; from there it was carried to India, China, and Malaysia by Portuguese explorers. The sweet potato apparently was introduced into Japan from China, sometime around 1700 by way of the Ryuku Islands (Boswell and Bostelman, 1949). In Kyushu today, it is called Kara-imo (Ka-ra-e-mo), meaning Chinese potato.

Primitive cultures of the South Pacific islands gave a central role to the sweet potato in their celebrations, suggesting that sweet potatoes were cultivated for food since ancient times in two widely separated parts of the world. Whether the crop originated in the New World and was transported to the Polynesian Islands or vice versa remains open to speculation.

Although Columbus noted the use of sweet potatoes by West Indian natives on his fourth voyage, there is no record of pre-Columbus cultivation of sweet potatoes by Indians in the Continental United States. Sweet potatoes were grown in Virginia as early as 1648, most likely from roots obtained in the West Indies (Bouwkamp, 1977; Edmond and Ammerman, 1971).

It should be mentioned that there is often a case of mistaken identity between the fleshy root, the sweet potato (a member of the Convolvulaceae family and of the class Dicotyledoneae), and the yam, a large fleshy (stem) tuber of the much smaller family, Dioscoreaceae, and a member of the class Monocotyledoneae. One cultivated yam species is also *batatus* (*Dioscorea batatas*) (Ammirato, Vol. 3, Chap. 13).

GEOGRAPHIC DISTRIBUTION AND LIMITATIONS

Although sweet potatoes were first recorded as being cultivated in Virginia and Louisiana, they are now propagated extensively in all the warmer regions of the United States and the world. In many countries sweet potatoes form a large part of the food of the people, ranking seventh among food crops in annual production in the world in 1976 (Harlan, 1976).

The sweet potato is adaptable to a wide range of climatic and soil regimes. It is normally grown from 40 degrees N to 32 degrees S; from sea level to 9000 ft elevation. But major growing areas occur where the average temperature is 23.8 C or more with a well-distributed annual rainfall of 30-50 in and an abundance of sunshine. A frost-free growing period of 4-6 months is essential for maximum yield of marketable roots for most cultivars. The cultivars Ga. Jet and Travis are exceptions to this rule, usually producing a high percentage of marketable roots in 90-95 growing days. Because of the climatic limitations of sweet potatoes, major production areas occur in the South with North Carolina and Louisiana producing approximately 90% of the total crop grown in the United States. Most states in the southern part of the United States produce at least 2000 commercial acres and many home gardeners will grow at least one row of sweet potatoes for home use.

ECONOMIC IMPORTANCE

The growing of sweet potatoes is a billion-dollar industry. However, the use of sweet potatoes in the United States is relatively low compared to other staple crops and generally regionalized for both growth and consumption. Its economic importance and general consumption appear to be increasing with the advent of improved varieties as well as improved storage and processing facilities. A reshuffling of southerners to the North and vice versa has also caused a larger number of people to utilize sweet potatoes that previously had not done so.

Over the past 100 years the sweet potato industry in the United States has fluctuated from the North to the South and recenty Westward. In the middle nineteenth century they were grown from Louisiana, around the Gulf states, and up the Atlantic Coast as far as New Jersey and Delaware. By the early twentieth century commercial districts began to form in clusters in most of the southern states (Louisiana, Alabama, Georgia, North Carolina, and South Carolina) and the border states of Virginia, Delaware, and New Jersey, with some cultivation in Tennessee. By the middle 1930s some of these areas had enlarged considerably, especially the lower Mississippi Valley (Louisiana, Mississippi, Tennessee), and along the coastal states of South Carolina and North Carolina, and Delaware. After the depression of the 1930s many of the commercial districts did not recover their predepression markets. Louisiana became especially strong as well as southeastern Virginia and the southern areas of Delaware and New Jersey. Significant production also began to appear in both Texas and California (Seufferle et al., 1940).

Regional research designed to develop alternative uses for sweet potatoes has served as a stimulus for creating new ways for utilization of the entire plant (root and foliage). Although this is a relatively new regional project (established in spring of 1981), research has already begun to evaluate everything from alcohol to sweet potato greens (Magee, unpublished minutes, Regional S-161 Progress Report, 1981).

BREEDING AND CROP IMPROVEMENT

Early Breeding Work and the Sexual Revolution

Although sweet potato breeding through sexual means was underway as early as 1904 (Tioutine, 1935), it was not until 1937 when Miller (1937) induced the sweet potato to flower and set seed that significant improvements were made in the United States. Miller spent a large portion of his professional career breeding sweet potatoes and developing ways to improve the crop from both a breeding and a cultural standpoint. Because of his vast contributions to sweet potato research, he is often referred to as the father of sweet potato breeding in this country and one of the founders of the Sweet Potato Collaborators Group (Bowers et al., 1969). Since his early work, a vast amount of research has been carried out to improve overall sweet potato production, and the magnitude and ramifications of this research has been astronomical. More than 50 varieties have been developed since this group was formed. The major objectives of the Sweet Potato Collaborators Group are the improvements of sweet potatoes in terms of yield, storage ability, nutritional composition, disease and insect resistance, and improved culinary properties. Most of the varieties developed through this group possess two to three times the carotene content of the older varieties and all of them possess multiple disease resistance.

Despite the fact that tremendous improvements have been made with sweet potatoes, very little progress has been made in terms of yield when compared to other crops such as corn or soybeans. Much of this stagnation in yield improvement can be attributed to the limited genetic base that most breeders have utilized. Very limited explorations have been made with species or genera as a means of transferring desirable germplasm into *Ipomoea batatas.* Incompatibility has caused numerous problems in hybridizing desirable parents and this is probably an even greater problem in interspecific or intergeneric hybridization. Thus there is a need to develop techniques that would be useful in overcoming such problems. New advancements in tissue culture and somatic hybridization offer great promise in this area. But to date these areas have received only limited exploration. It appears that with the present techniques of breeding sweet potatoes, the greatest improvements will be made through disease and insect resistance and/or improved culture practices. Tissue culture will play a major role in the growth and development of suitable cultivars. Rapid multiplication of genetically pure plant material that is disease free will be of great value to future plant breeders. Tissue culture will be very valuable in this respect.

Disease and Insect Pest Resistance

When environmental abnormalities do not pose a problem, the most limiting restraint to good sweet potato production would be the presence of diseases and insects (mainly sweet potato weevil). Diseases are responsible for the destruction of thousands of acres each year. They cause severe losses not only in the field, but in storage and the plant bed as well. Field diseases account for approximately 70-80% of the disease damage each year.

Fusarium wilt, incited by the fungus, *Fusarium oxysporum f.* sp. *batatus*, and root-knot nematode, caused mainly by *Meloidogyne incognita* and several other species, are probably the two most serious diseases affecting sweet potatoes in the United States and perhaps the world. These diseases have the capacity to destroy 50-90% of the entire crop in one growing season when left unchecked. Plant breeders have made tremendous progress in developing varieties resistant to these two diseases (Collins, 1977; Harmon, 1976; Jatala and Russel, 1972; Jones, 1969). Most varieties grown in the United States today would have at least resistance to one of those diseases and many would have some level of resistance to both diseases (Harmon et al., 1969; Edmond and Ammerman, 1971).

Another major area of disease research that is receiving increased attraction is that of viruses. Compared to other economically important crops, sweet potatoes have received very little attention with respect to virus diseases. With the exception of internal cork, very little is known about the economic impact of viruses on sweet potatoes. During the past ten years, several viruses have been identified in sweet potatoes (Hollings et al., 1976; Center for Overseas Post Research, 1978; Martin, 1970; Moyer and Kennedy, 1978; Mukiibi, 1977; Overdeli and Elliot, 1971).

While virus detection is continuing to increase, limited success is being made at indexing it, determining its vector, and lastly, developing virus-free plants. Alconero et al. (1975) were able to use meristem tip culture to develop virus-free plants in a short period of time. The use of tissue culture, either meristematic or otherwise, is seen as the only real and effective means of eliminating viruses from existing germplasm. It is highly possible that the slow rate of yield increase in sweet potatoes is due to the high virus index in present varieties and breeding lines. Significant yield increases could possibly be realized if virus-free material were available. Tissue culture certainly makes it possible for such avenues to be pursued.

Much of the insect work has involved the determination of the nature of resistance and then incorporation of this resistance into new varieties. In addition, research has also involved the development and evaluation of chemicals for the control of insect pests. Success along these lines has been slow because of the high cost of gaining approval for such chemicals for use on a crop whose growth is highly regionalized.

The sweet potato weevil, *Cylas formicarius elegantus*, is by far the most damaging and dreaded insect pest of sweet potatoes in the world. The presence of this insect in areas within the United States has limited the movement of sweet potatoes. Sweet potatoes grown in infested areas cannot be marketed outside of the area unless they are fumigated prior to shipment. Once fumigated, the shelf-life is greatly reduced and the sweet potato must be consumed in a matter of weeks. Since the banning of DDT, chemical treatments have had very limited success in controlling this pest (Jayaramaiah, 1975; Sing, 1973; Waddill and Conover, 1978).

The second most damaging insect pests are probably larvae from *Diabrotica balteata* and *Systena blanda*. These insects do damage mainly to the storage roots, usually causing damage to the skin surface. Breeders and entomologists are actively screening and developing varieties resistant to many of these soil insects.

In additon to the heavy breeding emphasis being placed on yield, disease, and insect resistance, efforts are being made to improve the sweet potato from many other aspects as well. These include such things as nutritional quality and composition, processing, appearance, culinary properties, foliage consumption (greens), and industrial uses such as gasohol and animal feed. The potential for animal feed looks very promising, since sweet potatoes can compete effectively with corn in the South for total yield per unit area of land. Several breeders of the National Sweet Potato Collaborators Group are actively screening breeding lines for industrial use and have met with good success.

Improved Cultural Practices

Increasing yield and storage ability has long been a major objective of sweet potato researchers, probably beginning with Dr. George W.

Carver of Tuskegee Institute (Carver, 1898, 1906, 1910, 1918, 1922; Whatley, 1975). Dr. Carver spent much of his career working with sweet potato farmers in and around Macon County, Alabama, and throughout the South. In an attempt to get them to stop growing cotton, he looked at sweet potatoes as a substitute cash crop. He found that the greatest limitations to productivity in the South were soil type and fertility level. His research efforts revealed that through proper selection of soil types and good fertility levels, good, high-quality yields could be obtained on a consistent basis. Proper soil type has been shown to play an even greater role than fertility in increasing productivity. Sweet potatoes may be grown on a wide range of soils, but production of high-quality roots with desirable sizes, shapes, appearance, and yields is best on fertile, well-drained sand and loam soil types. Heavy, poorly drained soils often result in ill-shaped roots of poor quality and reduced storage life. Where soil type is not a major constraint, fertility is usually the chief factor limiting productivity. No one fertilizer formula will fit all types of sweet potato fields. Much depends on the soil texture and general fertility. Where the field is sandy in nature, more fertilizer is needed than where the field is loamy. Potassium is probably the most critical of all the nutrients, since it plays a major role in root shape, a primary constituent in marketability (Speights et al., 1967).

Soil moisture has also been shown to play a large role in root growth and development. Sweet potatoes are normally considered a drought-resistant crop, but under severe drought-stress conditions yields could be reduced drastically. For this reason, a source of irrigation should be available, especially where large acreages are grown. The amount and frequency of irrigation depends on the amount of rainfall, the texture of the soil, and the amount of evapotranspiration. In general, the soil must not be allowed to become so dry that plants become wilted for long periods of time. This is especially true just before and during harvesting, since root development is probably greatest 30-40 days prior to harvest (Edmond and Ammerman, 1971; Jones, 1961).

Prospects for Improvement of the Crop

Potential usefulness of tissue culture in breeding sweet potatoes is seen as very promising. As stated earlier, it could be very useful in developing virus-free plant material that could then be effectively evaluated for yield and other growth parameters. Secondly, it could be very useful in maintaining large populations of genetically pure materials in a relatively small area until needed. Under present circumstances, because of the vegetative nature of propagation, the germplasm must be renewed every year. Approximately 10% of the crop must be saved each year for replanting. This is not only time consuming but extremely expensive. Tissue culture could eliminate a large portion of this expense, both in storage and bedding, and at the same time ensure that a disease-free genetically pure plant population is being grown. This would be of great value to the breeder and grower alike. It could also lead to a cheaper product for the consumer.

LITERATURE REVIEW OF TISSUE CULTURE METHODS

Although specific in vitro culturing of *I. batatas* has not been as productive as that of other species (e.g., carrot, tobacco, crucifers, sugar cane, corn, etc.), some headway has been made in the in vitro culturing of sweet potato.

It appears from a search of the literature during the past 10–15 years that the sweet potato has not been a favorite subject of plant tissue culturists, especially in the United States. Probably the earliest references to in vitro culture of the sweet potato is in Japan by Yamaguchi and his colleagues in the 1960s. In the United States, one of the earliest publications which appeared in 1972 was by Gunckel et al. Currently, there are probably less than a dozen published papers by American plant tissue culturists on *Ipomoea batatas*. There is voluminous work by the Canadian group of Velicky, Zink, Rose, et al. on *Ipomoea* species, but it is almost exclusively on the morning glory, *Ipomoea purpurea* (Velicky et al., 1977; Zink and Velicky, 1977, 1979; Zink, 1980).

There is research at several locations in the southeastern United States, particularly under USDA and state agricultural experiment stations (notably in Tennessee, North Carolina, Georgia, Florida, and Louisiana), and our own work here at Tuskegee Institute in connection with breeding programs in the Department of Plant and Soil Science (Horticulture).

Most of the literature cited here is from the laboratories in Japan, China, and South America. This is because *I. batatas* is a food crop important to the economy of these countries to which it is or has become indigenous.

Formation and Growth of Callus Tissue

Most tissue culture work on the sweet potato starts with the establishment of callus tissue from either stem or the root—preferably the latter because of its fleshy nature, abundance, and ease of manipulation. Therefore, most of the following studies have been initiated by this means. (Only the sections titled Embryogenesis, Androgenetic Culture, and Shoot Tip Culture are primarily exclusive of this initial step.)

The formation of callus (undifferentiated tissue) is a natural response to wounding in many plants. In the natural condition such a callus persists for a brief period and soon becomes saturated with polyphenol substances which seal off the wounded organ or tissue from further infestation or invasion from outside influences. However, in tissue culture (an artificial, in vitro environment) the callus may continue to develop indefinitely. Despite the homogeneity of the initial tissue piece and the intent to establish and continue the true homogeneous nature of the original explant on well-defined medium, much variation can and does occur, especially on subsequent subculturing. It is probably due to the true totipotency of the species or cultivar that the tissue tends to vary from its original homogeneous nature, since it contains all of the inherent biochemical potential of the parent plant(s). Therefore, the callus grown in vitro may contain "mixed populations" of cells. To avoid this and to attempt to establish the true homogeneity of a cell population, single-cell isolates may be the only solution--and even these over time and in varying medium compositions may change from their original nature.

For callus formation several media have been used (White, Gamborg, MS), but predominant among these is the MS medium (Murashige and Skoog, 1962) or modifications of it.

Yamaguchi and Nakajima (1972) used a modified White's medium to produce callus and adventitious roots and buds. They state that the sweet potato was chosen because the root has no genetic difference of totipotency among individual root tubers of the same variety. They cut pieces 4 mm in diameter by 3 mm in height by using a sterile cork borer and razor blade, placed them in 20 ml of agar in glassware 3 cm in diameter and 10 cm in height, and cultured them at 30 C in the dark. They used medium consisting of White's nutrient minerals, 5 g/l Difco yeast extract, and 146 mM sucrose at pH 5.8-6.0. Callus cultured on medium containing 5.4 μM NAA produced adventitious roots after 6 weeks, provided that the explant came from tubers that had been stored for 6 months. In order to obtain adventitious roots from freshly harvested tubers, cytokinins (KIN and ZEA) were needed in the medium at varying concentrations (0.5-50 μM). Whereas cytokinins inhibited adventitious bud formation, abscisic acid (ABA) in combination with low concentrations of 2,4-D with some varieties stimulated adventitious bud formation in sweet potato tubers. They concluded that cytokinins play an important role in organ formation and that ABA has an antagonistic effect on endogenous cytokinins in sweet potato tissue as far as adventitious bud and root formation are concerned. Regenerated plantlets were grown from the adventitious buds.

Hill (1965) reported on the polarity of sweet potatoes and the affected organogenesis in explants grown in vitro. He used White's basal medium with the addition of several growth substances: 2,4-D, IAA, n(2,4-D) anthranilic acid, n(phenoxyacetic) anthranilic acid, and GA at concentrations varying from 0.01 mg/l to 1.0 mg/l. The explants were designated as to root ends and stem ends and so oriented in the culture tubes, some in their normal (polar) position and others inverted. They concluded that the polarity inherent in sweet potato roots can be altered by growth substances, especially 2,4-D and IAA. The former produced callus growth in the stem end of explants, regardless of orientation in the culture tubes, while the latter (IAA) produced callus on the root end of the cylinders. The other chemicals were less demonstrative and definite in their influence on callus growth and polarity. Xylary differentiation and cellular activity were also affected to some extent by these growth substances.

Probably the earliest and most thorough research in the United States was the investigation by Gunckel et al. (1972) on the influence of polarity in the sweet potato tuber. They also studied the influence of a number of organic nutrient variants on root and shoot initiation in relation to callus production from the explant. Three cultivars of *I. batatas* (Yellow Jersey, Jersey Orange, and Centennial) were cultured on White's basal salt medium in various combinations of the following organic supplements: IAA, NAA, 2,4-D, adenine sulfate, KIN, GA, and CW. The cultures were grown under 500 foot candles at 25 C with a 12-hour photoperiod in French square bottles containing 10 ml of medium. Explants were cut from sterilized tubers as small cylinders using a 6-mm stainless steel cork borer.

The number of roots formed correlated well with the total amount of callus formed. The amount of callus or number of roots formed depended upon tissue polarity, explant orientation, cultivar response, and culture medium.

Whereas callus and root formation were the predominant responses, there was some shoot (leaf and bud) initiation related to polarity, orientation, and medium variation. Shoots formed in two cultivars only (Centennial and Yellow Jersey).

Antoni and Folquer (1975) cultured three cultivars of sweet potato (Tucumana Lisa, Tucumana Morada, and Centennial) on both WH and MS media. They were placed in constant light at 200 lux and 23 C. Both 2,4-D and CW were used. However, they were unable to obtain differentiation or new plantlets, even with the Yamaguchi modification of White's medium with ABA. Callus tissue was formed abundantly and easily and subcultured with equal success. Their conclusion is that media variation may result in organ formation similar to that in other cultivars.

Hozyo (1973) also reported callus formation from four cultivars of sweet potato and the influence of medium supplements, especially 2,4-D and KIN. It is concluded that the success in grafting in stock varieties is correlated with the ability and ease of callus formation.

Tsai and Lin (1973a,b) experimented with callus produced from the anthers of several sweet potato cultivars. They used Blaydes medium. In one investigation (1973a) they showed the influence of IAA, 2,4-D, KIN, and several plant extracts, including CW. These experiments were conducted in the dark at 30 C. In most cultivars, 2,4-D was the most effective auxin in both induction and growth of callus, however, 2,4-D and IAA together gave the best growth of callus in subcultures. Among the plant extracts, pea extract was the most effective in growth of callus. However, the most vigorous callus growth was obtained from Blaydes medium supplemented with 2,4-D, CW, and pea extract.

In another investigation by Tsai and Lin (1973b) the culture medium composition and cultural conditions effected the growth of callus from sweet potato anthers. By the use of a three-salt triangular method, optimum concentrations of various inorganic salts and vitamins were determined. They found that the subculture duration was influenced by the pH of the medium; short (30-d) subcultures with a pH between 6.25 and 6.00 were optimum, but for 60-d subcultures, pH optima were between 5.1 and 5.3. The concentration of sucrose for optimum growth of callus was in the range of 117-146 mM. Whereas higher concentrations (204 mM) caused growth retardation, the dry weights were greater than those where the concentration of sugar was lower, the weight of the latter being caused by a higher percentage of water in the tissue.

Sehgal (1975) excised leaves of *I. batatas* that produced callus readily on MS plus 2,4-D (2.3 μM) and KIN (0.5 μM) from immature leaves. Injury to the leaves promoted callus formation. This callus had the capacity for unlimited growth. When transferred to MS plus IAA or NAA it produced roots, but when adenine (ADE) (7.4-14.8 μM) or KIN (0.5-2.3 μM) was added both roots and shoots were produced.

The origin of organs (roots and shoots) from specific (anatomical) sections of the root was investigated by Hwang (1981, personal communication). Meristematic budlike centers (MBLCs) are formed from which shoots and roots develop. Disks of portions of the root were grown on MS medium. Segments taken from lateral sections formed significantly more shoots than segments taken from central cylinders. The authors attribute the origin of organ formation to anomalous cambia (secondary and tertiary) within the inner section of the root. The disks from the lateral sections produced more organized structures than disks from the central cylinder.

Cell Suspension Cultures and Enzyme Studies

Cell suspensions, or more accurately, "cell aggregates," are usually established from a callus culture, although in some instances a tissue or organ of a plant, properly sterilized, may produce a viable growing cell suspension from the sloughing off of cells from the tissue placed in sterile medium. However, the usual procedure is to establish a cell suspension culture from a friable callus culture in a liquid medium that is physically agitated or rotated by some mechanical device. The success of establishing a cell suspension is dependent upon the nature and propensity of the species to produce cells or fragments which will multiply in a specific media. There are very successful examples of well-established and rapidly growing suspension cultures similar to bacterial and other microbial cultures: *Acer*, *Glycine*, *Haplopappus*, *Nicotiana*, *Daucus*, *Saccharum*, etc. Unfortunately, to date, *I. batatas* has not been placed in this category, although other species of *Ipomoea* have succeeded quite well, notably the morning glory species (Veliky et al., 1977).

Although there have been a number of reports of cell suspension cultures of sweet potato, very few have been published. Some of the earliest work using sweet potato suspension cultures was that of Yoshida et al. (1970) who compared the ion absorption of Ca^{2+} and Mg^{2+} by "cultured free cells" of both tobacco and sweet potato. The suspensions were isolated from callus tissue of cv. Kokei No. 14 and cultured on MS medium containing 2,4-D, NAA, KIN, and thiamine-HCl. The tobacco (chlorophyllous) was cultured under light (2000 lux) while sweet potato was grown in the dark, both at 25 C. Stimulatory effects of Ca^{2+} on K^+ uptake were found in tobacco cells, but not in sweet potato. In sweet potato cells, K^+ uptake was always higher than Ca^{2+} and Mg^{2+} in media with varying K^+/Ca^{2+} ratios, whereas the opposite was true for tobacco cells. Other results were reported for the uptake of these ions.

Sasaki et al. (1972) initiated cell suspensions from tissue of the same cultivar (Kokei, No. 1) grown on modified White's medium containing

2,4-D and incubated in darkness at 23 C. When placed in liquid medium to initiate cell suspension cultures they used Gamborg's PRL-4 medium with casein amino acids. The cultures were shaken in darkness at 24-25 C on a gyratory shaker at 80 rpm. When the cultures were 3 weeks old, the cells were placed on sterile agar medium. Enzymes extracted after 3 weeks of growth on agar were analyzed to determine methods of control of isozyme biosynthesis.

Nakamura et al. (1981) initiated suspension cultures from sweet potato explants grown in MS medium and dispersed into the same liquid medium on a gyrotary shaker (100 rpm) at 27 C. Subcultures were maintained by transferring 10 ml of culture into 100 ml of fresh medium on different days selected for different plants, usually between 14 and 19 days. Assays were described for determining the influence of degradation products of dehydroascorbic acid on the growth of cell suspensions of carrot, tobacco, soybean, and sweet potato. Several fractions of degradation products were tested on the above species. The results suggest that one of the products of degradation, 2,3 diketogulonic acid inhibits cell growth. This does not exclude the participation of some of the other products from functioning as inhibitors as well.

Tisch (1981) used cell suspensions of a mutant cultivar, Red Jewel (IPJR), for studies on sulfur metabolism. The enzyme under study was the ATP sulfurylase. Selenate inhibition of growth and sulfate uptake were also studied. The sulfur sources were sulfate, cysteine, glutathione, and kjenkolate. The medium used was modified MS in which chloride was substituted for the normal concentrations of sulfate. The cultures were shaken in the dark (250 rpm) for 20 days with a 12- to 16-day period of enzyme treatment. Growth varied between a 2- and 10-fold increase (0.5 gm initial culture).

Nutritional Studies

In vitro growth of plants lends itself eminently well to the basic study of plant nutrition and metabolism. Absence from the normal variable of environment, especially competition from other plants, microorganisms, and animals, makes in vitro culturing an ideal method for determining the basic needs for inorganic and organic constituents.

Two kinds of studies have been reported: (1) the nutritional requirements for the optimum growth of the cultures themselves, and (2) studies to determine what the basic requirements are for the plant species or cultivar as a whole plant or for any of its tissues and organs.

Much emphasis has been placed on the first type of investigation: the basic requirements for optimum growth of the specific culture, i.e., the best medium on which it grows. There are numerous studies of this type with other plant species. In addition many investigators modify the original medium and either change the name to their own (if the medium is sufficiently different) or call it by the original name and simply indicate change by the addition of *revised*, *modified*, etc. Since the first media used for the tissue culture of plants by White, Gautheret, and others, there have been no completely "new" media, only minor changes in which different inorganic salts have been added, e.g. K_2HPO_4

in place of KH_2PO_4 or NaH_2PO_4, or KNO_3 in place of or in addition to NH_4NO_3, etc. Probably the area in which there have been truly new changes has been that of the organic consituents of the various growth factors, e.g., vitamins, phytohormones and auxins, nucleic acids, cytokinins, etc.

The use of cell suspensions has probably been the most desirable and the most easily analyzed technique for cell growth studies. It is also the method requiring the least time for the studies, since there is a rapid turn-over or doubling time. This assumes, of course, that the particular species or cultivar of interest produces adequate cell suspensions.

Three examples of nutritional studies with sweet potato are those of Gunckel et al. (1972), Tsai and Lin (1973a,b), and Yoshida et al. (1970); the first two with callus explants and the last with cell suspensions.

Gunckel et al. (1972) used 15 different combinations of organic constituents along with the basal medium (White) to study callus growth. These included IAA, NAA, 2,4-D, KIN, CW, sucrose, adenine sulfate, and GA_3. They report morphogenetic responses and root and shoot induction as being influenced by both polarity and growth factors. Another conclusion is that root induction correlates positively with the amount of callus produced. However, the latter is basically related to the constitution and composition of the culture medium. They state that the chemical effect of a sophisticated medium may nullify inherent tissue polarity and it is the latter which is the basis of the morphogenetic responses. They further state that the quantitative differences in morphological responses may be correlated with the orientation-dependent mechansim, which alters either the qualitative or quantitative uptake of certain constituents contained in the medium.

Tsai and Lin (1973b) investigated the influence of both inorganic salts ($MgSO_4$, $Ca(NO_3)_2$, and NH_4HO_3) as well as vitamins (thiamine, nicotinic acid, and pyrodoxin), auxins (IAA, 2,4-D) and KIN on the growth of sweet potato callus derived from anther tissue. In general they found that the effects of inorganic salts and vitamins on the growth of callus were more significant than those of auxins.

Yoshida et al. (1970) used cell suspensions to make nutritional studies. They observed the uptake and yield of K^+, Ca^{2+}, and Mg^{2+} in relation to their various compositions in the external media. Additionally, these authors attempted to find out if there were differences between the cell absorption mechanism of Ca^{2+} and Mg^{2+}.

Isolation of Natural or Secondary Products and Their Biosynthesis

For centuries man has used plants for medicinal purposes, food preparation, and many other uses. Some secondary metabolites or natural products synthesized by plants, such as alkaloids, flavanoids, phenolics, steroids, terpenoids, aliphatics, glycosides, etc., have been isolated from various plant parts (leaf, fruit, bark, seed, wood, tuber) to be used in many ways by man. In some cases there is no practical use for these products, other than the identification of them as characteristics of the plant or species. Most of these products are not essential to the metabolism and growth of the plant itself, but are by-products or waste products in the biosynthesis of the plant. Thus, they may simply be degradation products of no harm to the plant or tissue itself. In some cases, these products may act as antibiotics and inhibit growth of the plant. In most instances, however, they are antibiotic to invading or competing plants or animal organisms and not to the plant producing the chemical.

Through the technique of plant tissue culture many of these products have been produced and isolated, either by harvesting and isolating from the cultures themselves (endogenous), or preferably, by the diffusion of the product into the medium (exogenous). Over the past 30 years very sophisticated and successful pieces of equipment have been developed for these products. These have been confined, however, to those cultures which are high yielding, especially those used for medicinal purposes (e.g., chlorogenic acid from *Haplopappus*, phenolics from *Acer*, serpentine from *Catharanthus*, and nicotine from *Nicotiana*).

The most useful techniques have employed cell suspensions exclusively, either by large-scale batch cultures or closed or open continuous culture systems.

Unfortunately, the sweet potato is not known to contain important secondary metabolic products. Although the tuber does contain several phenolics, flavonoids, etc., none of these products are among those needed in medicine or industry. According to Robinson (1975) one class of steroids, sterolins, was first discovered in the common morning glory, *I. purpurea* and named for this genus, ipuranol. Other *Ipomoea* species known to contain resins are *I. orizabensis* or *I. violacea* pearly gates, and *I. rubrocoerulea* Praecox, both of which contain ergot alkaloids which are used as hallucination-inducing narcotics and as a cathartic. Another species, *I. purga*, also known as *Exogonium purga* and *E. jalopa* yields a compound called jalapa, consisting of a resin, gum, and sugar. This species has been used as a "drastic cathartic" (Merck Index, 8th ed., 1968).

As has been pointed out earlier in this chapter, the majority of *Ipomoea* species (Convovulaceae) originated in Central and South America (the tropics) and have been introduced and naturalized in the United States.

However, there are a few reports of secondary products from *I. batatas*. Wernicke and Kohlenbach (1976) report that in the vacuoles and in vitro cultures of *I. batatas* the mesophyll tissues contain iso-chlorogenic acid in high quantities. The resin is said to be contained in the vacuoles of the cells in so-called chlorogonoplasts that result from a supersaturation of chlorogenic acid.

A more detailed study has been made by Uritani and his colleagues (Oba and Uritani, 1979) working with cell suspensions of sweet potato. Using Heller's medium containing vitamins, 2,4-D, and yeast extract, callus tissue was induced from sprout tissue. In the callus, furano-terpenes (F-t) were scarcely detected; however, when cell suspensions were developed F-ts were rapidly produced. F-ts were suppressed by the causative agent of the black root fungal disease of sweet potatoes, *Ceratocystis fimbriata*.

This is a typical case for the use of tissue culture in studying diseases of commercial food crops such as the black rot fungus disease in sweet potatoes. The disease is caused by *Ceratocystis fimbriata* which invades and infests the root tissues. The chemical associated with the disease is a furanoterpene which accumulates in the tissue. One of these is ipomeamarone, a conversion product of acetate and mevalonate. The enzyme involved in the formation of ipomeamarone is 3-hydroxy-3-methylglutaryl coenzyme A reductase (HMG-CoA reductase), in sweet potato roots infected with *Ceratocystis fimbriata*, and some of the properties of the enzyme, have been reported (Suzuki et al., 1975).

In previous research with this system whole roots or slices have been used. However, it is desirable to apply tissue culture techniques to conduct more uniform and systematic studies of this problem. Thus the work of Oba and Uritani is encouraging.

Embryogenesis, Organogenesis, and Plantlet Regeneration

Gunckel et al. (1972) were able to produce roots and shoots from sweet potato explants by a combination of auxins, KIN, adenine, and GA from which plantlets were recovered. Yamaguchi and Nakajuma (1972) were able to regenerate plantlets by varying cytokinins and ABA. Adventitious root formation was induced by the former while adventitious buds were induced only when the ABA replaced the cytokinins. In later investigations Yamaguchi (1973) concluded that ABA acts antagonistically toward endogenous cytokinins in the sweet potato disks and promotes adventitious bud formation, simultaneously suppressing the formation of adventitious roots.

In an earlier paper Sehgal (1975) was able to produce roots and shoots from leaf callus tissue (see above). Later Sehgal (1978) used anthers of sweet potato at various stages to produce profuse callus on MS supplemented with adenine and 2,4-D. However, only diploid tissue was developed. Plantlets were regenerated when the callus was transferred to regular MS medium. Androgenic haploids were not recovered using any combinations of growth regulators (adenine, KIN, ZEA, IAA, NAA, and 2,4-D).

Litz and Conover (1978) propagated two varieties of white-fleshed sweet potatoes (White Star and PI [Plant Introduction] 315343) by using explants from shoot tips and axillary buds grown on MS medium containing BA, IAA, KIN, and activated charcoal. The latter was used to inhibit excess callus formation in favor of regenerated shoots. Small rooted transferable plantlets were formed from apical explants 8 weeks after the initial culture. Optimum shoot regeneration from White Star explants was induced by BA (4.4 μM) and from PI 315343 using KIN (4.6 μM) and IAA (5.7 μM).

Tsai and Tseng (1979) investigated embryo formation and plant regeneration from anther callus of five cultivars of sweet potato. They experimented with two basal media, Blaydes and MS, supplemented with various concentrations of IAA, 2,4-D, KIN, ABA, 2iP, BA, and yeast extract. Cultures were grown at 27 C. Callus induction (in darkness) occurred in both media containing auxins and KIN, and was especially successful in the presence of 2,4-D and KIN. For embryo formation (16-hr photoperiod) the medium required ABA. They suggested that in the presence of ABA, as in the work of Yamaguchi, endogenous cytokinins are antagonized and regulated and this then permits the formation of embryos. It is felt that varietal differences are due to different endogenous concentrations of cytokinins.

Using explants from stem, leaf, and root of the cv. Jewel, Carswell (1981) performed a factorial treatment incorporating NAA (0.5–5.4 μM) and BA (0.4–4.4 μM). Roots and shoots both were formed best at 5.4 μM NAA and 0.4 μM BA. Both roots and shoots were regenerated in culture; roots grew best in liquid medium and shoots grew best in solid medium. There is no mention of whole plant regeneration. Light was shown to have a beneficial effect on regeneration.

Hwang et al. (1981; personal communication) used explants of the cvs. Centennial, Jewel, and Redmar in their investigation. These were grown on MS high mineral salt medium containing Staby vitamins, myo-inositol, ascorbic acid, BA, and NAA. To form callus the latter two were replaced by 2,4-D (2.3 μM) and KIN (0.5 μM). No plantlets were produced from the callus; however, plants were recovered from the parental (explant) tissue. Callus production was prolific, from which adventitious shoots or roots were formed, but no plants were regenerated from callus tissue alone.

Anther Cultures

The obvious advantage of producing haploid plants from callus tissue derived from the gametophytic number of chromosomes is well known: recessive alleles can be expressed. Thus, this is the best method for the identification of mutations, especially recessive ones. The many benefits and advantages of using haploids to create new breeding lines are well documented (see Vol. 1, Chap. 6).

The real problem is making sure that callus produced from gametic sources is truly haploid and not formed from diploid tissue such as the connective filament of the anther.

Thus, one must determine by actual chromosome count whether the callus and regenerated plants are truly haploid in chromosome number. This state is not easily maintained even after the original callus formation, since rapidly growing callus in culture may revert to diploidy, polyploidy, or even aneuploidy.

To date no true haploids have been reported for *I. batatas*, but there have been a number of experiments using anthers as the source for callus tissue from which plantlets were developed.

Callus tissue derived from anthers of the sweet potato flower has been recovered (Tsai and Lin, 1973a,b; Sehgal, 1978; Tsai and Tseng, 1979). In the work of Tsai and Lin, plantlets were not regenerated. Sehgal, however, did develop plantlets, but only diploid plantlets. Tsai and Tseng successfully produced plants from embryos of one of the five cultivars used, but the report does not state its ploidy.

Kobayashi and Shikata (1975) attempted to produce haploid plants by the use of several media and supplements (Miller, White, Bonner, and Murashige-Skoog). Supplements were 2,4-D, IAA, KIN, ABA, pyridoxine, glycine, thymidine, and myo-inositol. The number of chromosomes counted in the pollen mother cells of redifferentiated plants was 2n = 45, indicating that the induced plants were not haploid. Thus, to date, no case of true haploid plants of sweet potato has been reported.

Meristem and Shoot Tip Culture

Meristem tip culture has been used by many investigators to obtain virus-free clones of vegetatively propagated plants. Two purposes can be served by this technique: to obtain virus-free plant material (callus tissue, organs or whole plants) in which viruses may be introduced for studying the infection mechanism, and to obtain virus-free plants for production and commercial purposes.

The crux of the problem is the position at which one excises to obtain such virus-free tissue. As a rule of thumb any tissue cut near the uppermost apical point, i.e., as near to 10 μ as possible, will have obtained what is technically the apical meristem (Murashige, unpublished).

The whole area above 100 μ may technically be called the shoot tip and may not be void of virus.

Close to 50 species have produced virus-free plantlets or whole plants by the technique of apical meristem culture (see Vol. 1, Chap. 5). An alternative to this technique is heat treatment in which the whole plant or tissue is exposed from a few hours a day to a few weeks at temperatures ranging from 35 to as high as 50 C. Many species would not be able to survive these conditions and remain viable.

Ipomoea batatas is a prime candidate for this technique since it is well known that sweet potato, especially in the United States, is almost always virus-infected. The implications of this were discussed above.

Therefore, there has been research attempting to free vegetative clones of sweet potato from viruses by non-tissue-culture practices, especially by heat treatment of germplasm.

In vitro methods have been attempted during the past 10 years or more. Probably the best-known work on sweet potato meristem culture is that of Mori (1971). He has been able to establish virus-free meristem cultures of several species, including the sweet and white potato. He has been able to eliminate three of the most prevalent and damaging viruses from these plants: internal cork, rugouse mosaic, and feather mottle viruses.

Elliot (1969) excised meristem tips from the cv. Kumara. The length of the apical explant influenced the initiation of roots and subsequent development of the plantlets. MS medium supplemented with various growth regulators and auxins accelerated the growth and development of root initiation and eventually plantlet production.

Alconero et al. (1975) used shoot (meristematic) tips 0.4-0.8 mm long of axillary shoots of 10 cultivars of sweet potato. They were placed in sterile MS medium containing NAA and several combinations of IAA and KIN. They developed into complete plants in 20-50 days, but varied markedly among the various cultivars. When NAA (5.4 μM) was used it shortened the time necessary to produce complete plants from meristem tips. Of 150 plants tested, 47% did not cause virus symptoms when grafted onto *I. setosa*, the virus-sensitive test plant. The use of small tips avoided the need for preliminary heat therapy, and the appropriate combination of growth factors allowed the use of a single culture medium (MS) for production of complete plants within a short time.

Liao and Chung (1979) produced virus-free sweet potato plants by a combination of heat treatment and shoot tip culture. Keeping 50-mm-long terminal meristems in a growth chamber at 38-42 C continuously for 30-90 days with a daily light period of 16 hr at 3300 lux light intensity failed to eliminate a yellow spotting virus from sweet potato cv. Tainung 63. Neither was the virus eliminated by culturing shoot tips 5 mm in length in modified MS medium. However, elimination of the virus was achieved by culturing meristem tips 0.3-0.6 mm long taken from virus-infected Tainung 63 which had previously been exposed to 38-42 C for 28 days. The meristem cultures were grown initially in modified MS medium containing 17.7-35.5 μM BA and 5.7-11.0 μM IAA and were placed under 25 ± 1 C for 30-50 days at a light intensity of 150 lux or below. Then they were transferred to modified MS with 9.3 μM KIN. A photoperiod of 16 hr at 3300 lux light was used. The temperatures during the alternative periods were 28-32 C and 25 ± 1 C, respectively. The excised meristem tissues developed into plantlets within 65-120 days. Of 24 such plants, 16 did not cause virus symptoms on *Ipomoea nil* by mechanical transmission and/or grafting. Single-node cuttings of these virus-free test tube plantlets developed into complete plants within 10-30 days after subculturing on modified MS medium.

Frison (1981) cultured sweet potato clones in growth chambers at 32 C until 3 cm high, when lateral buds were excised. These were aseptically dissected and between 0.25-0.4 mm tips were excised and placed on MS medium for growth. They were grown under 5000 lux white fluorescent lamps at 29 C day and 24 C night temperatures on a 12-hr photoperiod. When the plantlets were 3-5 cm in height they were transferred to peat pellets, then to pots with sterilized soil in an insect-free isolation room. After 1-2 months the plants were approach-grafted to *I. setosa* seedlings to test for virus. Negatively indexed plants (those showing no symptoms after being grafted three times to *I. setosa* at 2-month intervals) were multiplied by single-node cuttings in vitro on MS medium. After distribution, cuttings were cultured in plastic tubes containing 10 ml of multiplication medium. After about 2 weeks of growth the plantlets were checked for absence of fungal or bacterial contamination and were then ready for international transportation.

J.W. Moyer (1982, personal communication, N.C. State Univ., Raleigh, N.C.) investigated meristem-tip culture as a method to provide a very high percentage of sweet potato (cv. Jewel) free of known viruses and to reduce the time required for regeneration. Heat therapy has not proved satisfactory. Excised apical meristems devoid of leaf primordia (to avoid virus carry-over) are placed in MS medium modified with BA and NAA and later transferred to liquid medium by the use of filter paper bridges. Plantlets were transferred to soil in 6-12 weeks. Plants were indexed for virus by grafting onto *Ipomoea setosa* seedlings. They were then multiplied and grown under field conditions to evaluate for varietal characteristics. This is important because of the color variation (root skin and flesh). This selection is critical for commercial acceptance. Variation in the initial cycle far exceeds that observed in commercial, vegetative propagation. Selections are stabilized for increase by certified seed growers. A dark red-skinned selection was made from meristem-Jewel clones and will be released as a new cultivar.

Protoplast Culture

The advent of protoplast isolation and culture techniques probably offers the most promising area for the introduction of new plant hybrids and varieties. This technique has obvious advantages in the production of new sweet potato hybrids and varieties. However, to date very little research has been reported on protoplast experimentation with sweet potato.

Wu and Ma (1979) reported successful isolation of protoplasts from stem callus of *Ipomoea batatas*. Cellulase (2.0%) with an osmoticum of 0.8 M mannitol was required for protoplast release within 5 hr. Protoplasts plated in agar media containing 0.47 μM KIN, 0.16 μM NAA, and 0.45 μM 2,4-D regenerated cell walls within 7 days followed by cell division 10 days from culturing.

Using the cell layer-reservoir system (Shepard, 1980), Bidney and Shepard (1980) cultured sweet potato petiole protoplasts. Protoplast culture medium containing 0.27 μM NAA, 2.2 μM BA, and 0.3 M sucrose supported repeated divisions to regenerate callus within 3 weeks. Important factors cited for the cell layer-reservoir method were low auxin concentration, gradual reduction of sucrose in the cell layer, and influx of mannitol from the reservoir. Root regeneration from protoplast-derived callus has been observed, but as yet no shoot morphogenesis or embryogenesis has occurred.

PROTOCOL FOR CULTURING SWEET POTATO

The protocol for producing callus tissue and cell suspensions of sweet potato follows (Fig. 1).

Select a smooth-surfaced, well-shaped sweet potato with no breaks in the skin; wash it thoroughly with tap water. Then rinse with 70% ethanol in paper towel (Stage 1). Let it dry for a few minutes to several hours. All of the following procedures are conducted in a sterile culture room and under a culture hood. Next, using a Parker

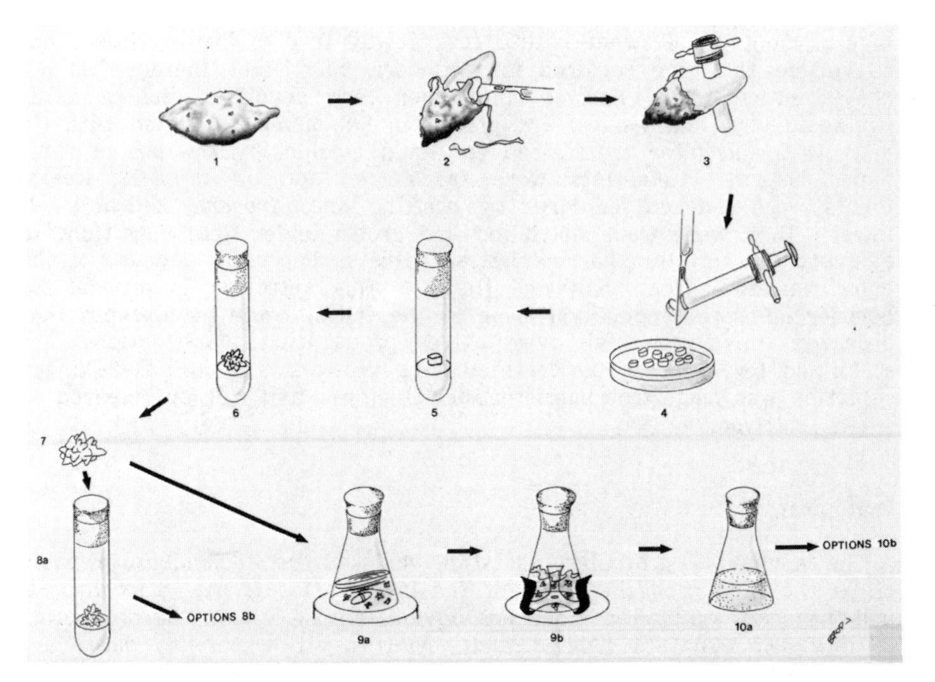

Figure 1. Protocol for in vitro culture of sweet potato (*Ipomoea batatas* Lam.).

Knife (no. 4 handle, no. 20 blade), cut back the outer skin (pericarp) to expose a clean, sterile surface (Stage 2). Then using a no. 7 cork borer (1-cm diameter), cut through the potato near the middle to obtain a cylinder of potato root tissue (Stage 3). Push the cylinder out slowly with a sterile cork extruder, and cut off cylindrical explants 3-5 mm thick, depositing them on a sterile (disposable) petri dish (Stage 4). Using sterile 8-inch forceps, place one of the pieces in a screw-capped test tube (25 x 150 mm) containing 5-15 ml of agar (0.8%) nutrient medium (MS or modified MS) and press firmly onto agar surface (Stage 5). Place in a test tube rack and let grow in darkness or light (16 hr light/8hr dark photoperiod) for 4-8 weeks (Stage 6). When the culture has grown to a size approaching the diameter of the test tube, transfer to the desired media (Stage 7). To continue on semisolid (agar) medium, cut callus into an appropriate number of subculture pieces (between 50-100 mg each) and place on new nutrient agar medium (Stage 8a). These may be used for several different growth or development options (Stage 8b). These options include cell suspensions, continuous subculturing of callus tissue, nutritional studies, embryogenesis and organogenesis, morphogenesis, plantlet regeneration, and habituation. The friable callus may be placed instead in a liquid medium in either a 125-ml flask (25-ml medium) or a 350-ml flask (50-ml medium) with a magnetic stirrer bar in the medium. This is placed on a magnetic stirrer and set to a brisk rate of speed (rpm) to break the callus into small fragments and cells (Stage 9a). The other alternative is to place the friable callus into the flask, breaking it up with a sterile spatula or knife, and place the flask on a shaker (New Brunswick) and allow it to rotate briskly (100–160 rpm) to break the callus into fragments and cells (Stage 10a). The resulting culture when established as a consistent "cell suspension" may be subcultured further for a variety of options (Stage 10b). These options include protoplast culture, mutagenesis, preservation of germplasm, isolation of secondary metabolic products, enzyme and nutritional studies, embryogenesis, plantlet regeneration, cytogenetic studies, and cell (growth) kinetics.

The medium most often used for the culturing of cell suspensions is a modified MS medium (Table 1) which contains abscisic acid.

Table 1. Constituents of Murashige-Skoog (MS) and Modified Murashige-Skoog (MS-Mod)

COMPONENT	MS	MODIFIED MS
Inorganic		
KNO_3	18.8 mM	18.8 mM
$MgSO_4 \cdot 7H_2O$	1.5 mM	1.5 mM
$CaCl_2 \cdot 2H_2O$	3.0 mM	3.0 mM
$FeSO_4$	0.18 mM	
Sequestrene 330Fe		28.0 mM
$MnSO_4 \cdot H_2O$	0.13 mM	0.13 mM
KH_2PO_4	1.24 mM	1.24 mM
H_3BO_4	0.1 mM	0.1 mM
$ZnSO_4 \cdot 7H_2O$	0.03 mM	0.03 mM
$Na_2MoO_4 \cdot 2H_2O$	1.0 μM	1.0 μM
$CuSO_4$	0.15 μM	0.15 μM
$CoCl_2 \cdot 6H_2O$	0.1 μM	0.1 μM
KI	5.0 μM	5.0 μM
NH_4NO_3	20.6 mM	20.6 mM
$Na_2 \cdot EDTA$	0.11 mM	0.11 mM
Organic		
Nicotinic acid		8.1 μM
Thiamine-HCl	3.0 μM	3.0 μM
Pyrodoxine-HCl		4.8 μM
Myo-inositol	0.55 mM	0.55 mM
Sucrose	0.58 M	0.58 M
2,4-D	13.5 μM	4.5 μM
2iP		1.2 μM
NAA		0.54 μM
Agar (optional)	8.0 g/l	8.0 g/l
ABA (abscisic acid)		7.6 μM

FUTURE PROSPECTS

With the exception of meristem-shoot tip culture, all other in vitro culturing begins with a fresh explant that produces a callus which eventually increases considerably in size. Such a callus may become more callus, cell suspensions, organ or embryo cultures, or plantlets. Plantlets may develop from several origins: callus, original explants, or cell suspensions (adventitious roots or stems). Secondary metabolites or biochemical by-products may come either from callus tissue or cell suspensions, usually the latter. Nutritional studies are done best with cell suspensions, but callus tissue may also be used to evaluate nutrients or organic constituents.

The production of embryos or plant organs and ultimately plantlets which can develop into mature plants must be the goal of this whole research effort on improvement of an important food crop such as sweet potato.

Several authors have concluded that the only way to overcome and combat the shortages of certain foods or to overcome the natural ravages of disease or illnesses in these plants is through the application of in vitro (tissue) culture.

The most productive of these areas would seem to be the following (in order of importance): meristem- and shoot tip culture, anther (haploid, androgenesis) culture, and protoplast culture.

True meristem culture is capable of ridding the germplasm of one of the most devestating, and certainly devious of the sweet potato problems. The several virus diseases, the worst of which is probably internal cork disease, have not been controlled to date by chemotherapeutic or horticultural means. These diseases cause tremendous losses to the grower. Therefore, the perfection and expansion of this technique would be of significant value in increasing the production and economics/marketing of this crop. The work thus far of such investigators as Mori (1971), Alconero (1975), Liao and Chung (1979), and more recently, Frison (1981) has gone a long way towards the production of virus-free clones.

The promise which haploid cultures give is based on segregation and selection against poor characteristics (disease susceptability, low or lack of certain nutrients, color or poor keeping quality—many recessive traits) in favor of capturing the good characteristics of the cultivar (high growth yields and nutrient levels, etc.).

Probably the most promising and potentially selective method is that of somatic hybridization with its many ramifications for gene transfer. To date this is the area of research least developed for sweet potato, but it could hold the key, especially in combination with other methods for the development in the future of a better sweet potato.

REFERENCES

Alconero, R., Santiago, A.G., Morales, F., and Rodriguez, F. 1975. Meristem tip culture and virus indexing of sweet potatoes. Phytopathology 65:769–773.

Antoni, H.J. and Folquer, F. 1975. In vitro tissue culture of sweet potatoes, *Ipomoea batatas* (L.) Lam., for the production of new cultivars. Rev. Agron. Noroeste Argent. 12:177-178.

Bidney, D.L. and Shepard, J.F. 1980. Colony development from sweet potato petiole protoplasts and mesophyll cells. Plant Sci. Lett. 18: 335-342.

Boswell, V.R. and Bostelman, E. 1949. Our vegetable travelers. Nat. Geog. Mag. 96:145-217.

Bouwkamp, John C. 1977. Sweet potatoes—buried treasure. In: Gardening for Food and Fun. Yearbook of Agriculture (Jack Hayes, ed.) pp. 212-216. USDA, Washington, D.C.

Bowers, J.L., Harmon, S.A., and Dempsey, A.H. 1969. History of group and early research. In: Thirty Years of Cooperative Sweet Potato Research 1939-1969. South. Coop. Ser. Bull. No. 159:1-7.

Carswell, G. 1981. Regeneration of sweet potato (*I. batatas* Lam.). M.S. Thesis, Univ. of Illinois, Urbana-Champaign.

Carver, G.W. 1898. Experiments with sweet potatoes. Bulletin Number 2. Tuskegee Institute Press, Tuskegee, Alabama.

______ 1906. Saving the sweet potato crop. Bulletin Number 10. Tuskegee Institute Press, Tuskegee, Alabama.

______ 1910. Possibilities of the sweet potato in Maco County Alabama. Bulletin Number 17, Tuskegee Institute Press, Tuskegee, Alabama.

______ 1918. How to make sweet potato flour, starch, and sugar. Bulletin Number 37, Tuskegee Institute Press, Tuskegee, Alabama.

______ 1922. How the farmer can save his sweet potatoes. Bulletin Number 38, Tuskegee Institute Press, Tuskegee, Alabama.

Centre for Overseas Post Research. 1978. Sweet potato diseases in C.O.P.R. Pest control in tropical root crops. London, C.O.P.R., M.O.D. PANS 4:57-95.

Collins, W.W. 1977. Diallel analysis of sweet potatoes for resistance to *Fusarium* wilt. J. Am. Soc. Hortic. Sci. 102:109-111.

Edmond, J.B. and Ammerman, G.R., eds. 1971. Sweet potatoes: Production, Processing, Marketing. AVI Pub., Westport, Connecticut.

Elliott, R.F. 1969. Growth of excised meristem-tips of kumara, *Ipomoea batatas* (linn.) Poir in axenic culture. N.Z. J. Bot. 7:158-166.

Frison, E.A. 1981. Tissue culture: A tool for improvement and international exchange of tropical root and tuber crops. ITTA Res. Briefs 2:1-4.

Gunkel, J.E., Sharp, W.R., Williams, B.W., West, W.C., and Drinkwater, W.U. 1972. Root and shoot initiation in sweet potato explants as related to polarity and nutrient media variations. Bot. Gaz. 133:254-262.

Hahn, S.K. 1979. Effects of viruses (SPVD) on growth and yield of sweet potato. Exp. Agric. 15:1-5.

Harlan, Jack R. 1976. The plants and animals that nourish man. Sci. Am. 235:89-97.

Harmon, S.A. 1976. Breeding and testing sweet potatoes for nematode resistance. HortScience 11:288.

______, Hammett, A.L., Hernandez, T., and Pope, D.T. 1969. Progress in the breeding and development of new varieties. In: Thirty Years of Cooperative Sweet Potato Research 1939-1969. South. Coop. Ser. Bull. No. 159:8-17.

Hill, R.A. 1965. Polarity studies in roots of *Ipomoea batatus* in vitro. M.S. Thesis, Howard Univ., Washington, D.C.

Hollings, Mo., Stone, O.M., and Bock, K.R. 1976. Purification and properties of sweet potato mild moddle—a whitefly-borne virus from sweet potato (*Ipomoea batatas*) in East Africa. Ann. App. Biol. 82: 511-528.

Hozyo, Y. 1973. The callus formation on tissue explant derived from tuberous roots of sweet potato plants, *Ipomoea batatas* Poiret. Tokyo Nat. Inst. Agr. Sci. Bull Ser. D 24:1-33.

Hwang, L.S. 1981. In vitro shoot formation from sweet potato (*Ipomoea batatas* Lam.). M.S. Thesis, Univ. of Illinois, Urbana-Champaign.

Jatala, P. and Russel, C. 1972. Nature of sweet potato resistance to *Meloidogyne incognita* and the effects of temperature on parasitism. J. Nematol. 4:1-7.

Jayaramaiah, M. 1975. Studies on the chemical control of the sweet potato weevil, *Cylas formicarius* (Fabricius) Olivier (Coleoptera: Curculionidae). Mysore J. Agric. Sci. 9:307-313.

Jones, A. 1969. Quantitative inheritance of *Fusarium* wilt resistance in sweet potatoes. J. Am. Soc. Hortic. Sci. 94:207-208.

Jones, S.T. 1961. Effect of irrigation at different levels of soil moisture on yield and evapotranspiration rate of sweet potatoes. Proc. Amer. Soc. Hort. Sci. 77:458-462.

Kobayashi, M. and Shikata, S. 1975. Anther culture and development of plantlets in sweet potato. Bull. Chugoku Agric. Exp. Stn., Ser. A 24:109-124.

Liao, C.H. and Chung, M.L. 1979. Shoot tip culture and virus indexing in sweet potato. J. Agric. Res. China 28:139-144.

Limasset, P. and Cornuet, P. 1949. Recherche de virus de la mosaique du tabac (Marmor Tabaci, Holmes) dans les meristemes des plantes infectees. In: Plant Tissue Culture: Methods and Applications in Agriculture (T.A. Thorpe, ed.) pp. 181-211. Academic Press, New York.

Litz, R.E. and Conover, R.A. 1978. In vitro propagation of sweet potato. HortScience 13:659-660.

Martin, W.J. 1970. Virus diseases (of sweet potato). In: Thirty Years of Cooperative Sweet Potato Research 1939-1969. South. Coop. Ser. Bull. No. 159:49-55.

Merck Index, 8th ed. 1968. Merck and Co., Rahway, New Jersey.

Miller, J.C. 1937. Inducing the sweet potato to bloom and set seed. J. Hered. 28:347-349.

Mori, K. 1971. Production of virus-free plant by means of meristem culture. Jn. Agric. Res. Q. 6:1-7.

Moyer, J.W. and Kennedy, G.G. 1978. Purification and properties of sweet potato feathery mottle virus. Phytopathology 68:998-1004.

Mukiibi, J. 1977. Effect of mosaic on the yield of sweet potatoes in Uganda. In: Proceedings of the Fourth Symposium of the International Society for Tropical Root Crops (J. Cock, R. MacIntyre, and M. Graham, eds.) pp. 169-170. CIAT, Cali, Columbia.

Murashige, T. and Skoog, F. 1962. A revised medium for rapid growth and bioassay with tobacco tissue cultures. Physiol. Plant. 15:473-476.

Nakamura, Y., Ikeda, Y., Ebihara, N., and Tabuchi, K. 1981. Inhibition of growth callus cells (from carrot, tobacco, soybean, sweet potatoes) by degradation products of dehydroascorbic acid. Agric. Biol. Chem. 45:759-760.

Nicholls, H.A.A. 4th ed. 1906. A Textbook of Tropical Agriculture. Macmillan. pp. 288-290.

Oba, K. and Uritani, I. 1979. Biosynthesis of furano-terpenes by sweet potato cell culture. Plant Cell Physiol. 20:819-826.

Overdeli, A.J. and Elliot, R.F. 1971. Virus infection in *Ipomoea batatas* and method for its elimination. N.Z. J. Agric. Res. 14:720-724.

Quak, F. 1970. Review of heat treatment and meristem tip culture as methods to obtain virus-free plants. Proc. 18th Int. Hortic. Cong., Jerusalem 3:12-25.

Robinson, T. 1975. The Organic Constituents of Higher Plants. Cordus Press, North Amherst, Massachusetts.

Sasaki, T., Tadokoro, K., and Suzuki, S. 1972. The distribution of glucose phosphate isomerase isoenzymes in sweet potato and its tissue culture. Biochem. J. 129:789-791.

Sehgal, C.B. 1975. Hormonal control of differentiation in leaf cultures of *Ipomoea batatas* Poir. Beitr. Biol. Pflanz. 51:47-52.

______ 1978. Regeneration of plants from anther cultures of sweet potato (*Ipomoea batatas* Poir.). Z. Pflanzenphysiol. 88:349-352.

Seufferle, C.H., Burdette, R.F., Hamilton, A.B. and Devault, S.H. 1940. Production and marketing of Maryland sweet potatoes. In: Sweet Potatoes: Production, Processing, Marketing (J.B. Edmond and G.R. Ammerman, eds.) pp. 4-8. AVI Pub., Westport, Connecticut.

Shepard, J.F. 1980. Abscisic acid-enhanced shoot initiation in protoplast-derived calli of potato. Plant Sci. Lett. 18:327-333.

Sing, S.R. 1973. Identification of resistance to and insecticidal control of sweet potato weevil, *Cylas puncticollis*. In: Third International Symposium on Tropical Root Crops, Vol. 3, pp. 1-8, International Institute of Tropical Agriculture, Ibadan.

Speights, W.E., Burns, E.E., Paterson, D.R., and Thames, W.H. 1967. Some vascular variations in the sweet potato root influenced by mineral nutrition. Proc. Am. Soc. Hort. Sci. 91:478-485.

Suzuki, H., Oba, K., and Uritani, I. 1975. The occurrence and some properties of 3-hydroxy-3-methylglutaryl coenzyme A reductase in sweet potato roots infected by *Ceratocystis fimbriata*. Physiol. Plant Pathol. 7:265-276.

Tioutine, M.G. 1935. Breeding and selection of sweet potato. J. Hered. 26:2-10.

Tisch, T. 1981. Regulation of sulfur metabolism in *Nicotiana tabacum* and *Ipomoea batatas* cultures. M.S. Thesis, North Carolina State Univ., Raleigh.

Tsai, H.S. and Lin, C.I. 1973a. The growth of callus induced from in vitro culture of sweet potato anthers. J. Agric. Assoc. China 81:12-19.

______ 1973b. Effects of the compositions of culture media and cultural conditions on growth of callus of sweet potato anther. J. Agric. Assoc. China 82:30-41.

Tsai, H.S. and Tseng, M.T. 1979. Embryoid formation and plantlet regeneration from anther callus of sweet potato. Bot. Bull. Acad. Sin. 20:117-122.

Vasil, I.K. and Vasil, V. 1980. Clonal propagation. In: Perspectives in Plant Cell and Tissue Culture (I.K. Vasil, ed.) pp. 145-173. Academic Press, New York.

Veliky, I.A., Rose, D., and Zink, M.W. 1977. Uptake of magnesium by suspension cultures of plant cells (*Ipomoea* sp.). Can. J. Bot. 55:1143-1147.

Waddill, V.H. and Conover, R.A. 1978. Resistance of white-fleshed sweet potato cultivars to the sweet potato weevil. HortScience 13: 476-477.

Wernicke, W. and Kohlenbach, H.W. 1976. Akkumulation von Chlorogensaure in isolierten Mesophyllzellen von *Ipomoea batatas* Poir; Accumulation of chlorogenic acid in isolated mesophyll cells of *Ipomoea batatas* Poir, sweet potatoes. Z. Pflanzenphysiol. 77:464-470.

Whatley, B.T. 1975. Historical Background of the World's Leading Horticultural Crops: The Sweet Potato. Skinner Printing Co., Montgomery, Alabama.

Wu, Y.W. and Ma, T.P. 1979. Isolation, culture and callus formation of *Ipomoea batatas* protoplasts. Acta Bot. Sin. 21:334-338.

Yamaguchi, I. 1978. Hormonal regulation of organ formation in cultured tissue derived from root tuber of sweet potato. Bull. Ser. B, Agric. Biol. 30:55-88.

Yamaguchi, T. and Nakajima, T. 1972. Effect of abscisic acid on adventitious bud formation from cultured tissue of sweet potato. Crop Sci. Soc. Jpn. Proc. 41:531-532.

______ 1973. Hormonal regulation of organ formation in cultured tissue derived from root tuber of sweet potato. In: Plant Growth Substances; Proceedings of the 8th International Conference (1974):1121-1127.

Yoshida, F., Kawaku, K., and Takaku, K. 1970. Functions of calcium and magnesium on ion absorption by cultured free cells of tobacco and sweet potato. Tamagawa Univ. Fac. Agric. Bull. 10:13-27.

Zink, M.W. 1980. Acid phosphatases of *Ipomoea* sp. cultured in vitro. 2. Influence of gibberellic acid on the formation of phosphatases. Can. J. Bot. 58:2171-2180.

______ and Veliky, I.A. 1977. Nitrogen assimilation and regulation of nitrate and nitrate reductases in cultured *Ipomoea* cells. Can. J. Bot. 55:1557-1568.

______ and Veliky, I.A. 1979. Acid phosphates of *Ipomoea* sp. cultured in vitro. 1. Influence of pH (hydrogen-ion concentration) and inorganic phosphate on the formation of phosphatases. Can. J. Bot. 57:739-753.

SECTION VI
Tropical and Subtropical Fruits

CHAPTER 12
Banana

A.D. Krikorian and *S.S. Cronauer*

HISTORY OF THE CROP

Bananas are one of the earliest crops cultivated by man and, for the most part, only clones derived from natural evolution are cultivated for food. The extensive work of Simmonds and Shepherd (1956) suggests that edible bananas evolved in the Old World by several mechanisms from two wild species—*Musa acuminata* Colla (*M. Cavendishii* Lamb ex Paxt.) and *M. Balbisiana* Colla. *M. acuminata*, which has a primary center of diversity in the Malaysia-Indonesia region, carries the so-called A genome. Plants occur as diploids in both wild (seeded) and parthenocarpic forms. The edible diploid forms (AA) probably arose by the establishment of varying degrees of parthenocarpy and female sterility. The edible (seedless) triploids (AAA) developed through occurrence of triploidy within *M. acuminata* itself. This presumably arose by fusion of a normal haploid male gamete with an egg cell containing a doubled or unreduced chromosome complement. Human selection and clonal (vegetative) multiplication ensured the persistence of the most desirable of these clones. Even so, somatic mutation could have occurred and plants with favored qualities again would have been selected and perpetuated.

M. Balbisiana, of the Indian subcontinent-Malayan region, carries the so-called B genome and occurs only as seeded diploids (BB). When *M. acuminata*, whether diploid (AA) or triploid (AAA), came into contact with *M. Balbisiana* there arose various new forms with the hybrid genomic constitution. The edible diploid forms of these chance hybridizations are seen as originating through the out-crossing of edible diploids of *M. acuminata* to seeded forms of *M. acuminata* and *M. Balbisiana*, followed by human selection among the edible (parthenocarpic) progeny of these hybridizations. The whole new range of edible triploids arose by haploid pollination of diploid egg cells, e.g., egg cells AA x male gamete B = AAB; egg cells AB x male gamete A = AAB; egg cells AB x male gamete B = ABB. (No edible diploid or triploid *M. Balbisiana* are known.)

Tetraploidy has also been instrumental in the evolution of edible forms of banana. This arose via polyploidy and hybridity. There are some natural edible tetraploids: AAAB, AABB, and ABBB (cf. Richardson et al., 1965); most have been produced artificially and by experimentation with colchicine treatment or pollination of triploids. Again, and as in the case of edible AA or AAA forms, desirable somatic mutations would have been selected and clonally maintained. Simmonds and Shepherd (1956) have pointed out that it is impossible to designate the precise place of origin of these "hybrid" edible bananas but they provide evidence that India was probably the main site of hybridization between *M. acuminata* and *M. Balbisiana*. Accordingly, the Malayan region is generally accepted as the primary center of origin and most of the diversity there is of *M. acuminata* origin. India is the secondary center and the majority of diversity involves the 'A' and 'B' hybrids.

There are, of course, hundreds of clones grown throughout the world. To be taxonomically accurate, the AA and AAA (and even the artificially produced AAAA) should be designated *M. acuminata* Colla. (These have been variously named *M. Cavendishii*, *M. chinensis*, *M. nana*, *M. zebrina*). The edible (seedless) hybrid forms (*M. acuminata* x *M. Balbisiana*) should be designated *Musa* x *paradisiaca* L. (*M.* x *sapientum* L.). (In the Genetics and Improvement section of this chapter, the "shorthand" system of designating clones is mentioned in greater detail). It is interesting to note that the 'B' genomes confer some degree of drought and disease tolerance to clones of hybrid origin. Also, increased starchiness and unpalatibility of the fruit when raw is associated with the 'B' genome or *Balbisiana* hybridity, but is not an invariable consequence of it. In some parts of the world these starchy cooking bananas are called plantains. All *M. acuminata* types and some hybrid clones have more or less sweet fruits and hence may be eaten uncooked. These are the so-called dessert bananas.

In its diffusion path westwards from Southeast Asia, bananas seem to have gone to Madagascar and from there to the African mainland. The Arab trade routes to the interior facilitated its movement throughout central Africa and ultimately to the west coast. From the Guinea coast, bananas were transported by the Portuguese to the Canary Islands, and from there to Santo Domingo (modern Dominican Republic) in 1516. While it is impossible to say precisely which bananas were introduced, it seems fairly certain that one of them was a cooking plantain. The so-called Canary banana (variously known as Chinese, Dwarf Chinese, Dwarf Cavendish, [Dwarf] Governor), originally from Cochin China (today known as South Vietnam), was not introduced to the Western Hemisphere until the 1820s. Dwarf Cavendish, one of a complex series of bud mutations called the "Cavendish group," is one of the most important and widely cultivated banana varieties in the world.

As one can appreciate, the nomenclatural history at the varietal level is extensive and somewhat confusing (cf., e.g., Simmonds, 1954a,b, 1966; DeLanghe, 1961).

It is impossible in this brief overview to do justice to the historical facts and readers are urged to supplement this necessarily superficial treatment by referring to Kervegant (1935), Reynolds (1951), and Simmonds (1962, 1966, 1976, 1979).

ECONOMIC IMPORTANCE

The importance of dessert and cooking bananas to tropical economies can hardly be exaggerated (Champion, 1963; Haarer, 1964; Simmonds, 1966; Purseglove, 1972). Dessert bananas frequently play a major role as a cash export and are inevitably a complementary food in local diets. Cooking bananas, sometimes called plantains, very often are a staple food and comprise a major part of the caloric intake of large numbers of people in the Caribbean, Central and South America, south central and Southeast Asia, and the tropical West, central, and East Africa (Baker and Simmonds, 1951, 1952; Massal and Barrau, 1956; Anonymous, 1962; Mukasa et al., 1970). Indeed, cooking bananas are increasingly being imported into large urban areas in temperate parts of the world with large ethnic populations who both appreciate and know how to prepare them. The food value of bananas—which has been equated to that of the potato except bananas are usually slightly lower in protein—has been appreciated for a very long time (Myers and Rose, 1917) and continuous efforts are being made to broaden and extend the form in which bananas are marketed. Notable in this regard are specialty items such as banana chips (Palmer, 1979). Some speculate that bananas will play an even greater role—at least where they are grown—as a source of carbohydrate for the production of alcohol (e.g., for the use as a gasoline extender as in gasohol). Needless to say, their importance in the manufacture of specialty beers and spirits (Masefield, 1938)—for instance Ugandan and Sudanese waragi—will be sustained and one occasionally encounters these too in Western markets.

Completely aside from its value as an economic plant, banana has special interest for its morphology, anatomy (see, for example, Barker and Steward, 1962a,b; Fahn, 1953; Fisher, 1978; Mohan Ram et al., 1962; Riopel and Steeves, 1964), physiology (Tai, 1977), and biochemistry (see Steward et al., 1960a,b; Palmer, 1971). From the perspective of growth and development, the presumed involvement of hormones in the flowering process and the subsequent tremendous increase in the pulp tissue of the fruit in parthenocarpic forms are still largely unexplored and challenging problems (see Simmonds, 1953; Steward and Simmonds, 1954).

GENETICS AND IMPROVEMENT

Naturally evolved seedless bananas are perhaps the most conspicuously sterile of all cultivated fruits. Dessert bananas seem to have derived from *M. acuminata* as edible diploids or triploids (Simmonds, 1962). The artifically produced edible bananas are derived from diploid (2n = 22) species *M. acuminata* and/or *M. Balbisiana*. Clones grown on a large scale include diploids, e.g., Pisang Lilan, Bande, Paka, and Ney Poovan; triploids, e.g., Valery, Grande Naine, Lacatan, and Robusta; and tetraploids. The latter are various hybrids predominantly bred by the Imperial College of Tropical Agriculture in Trinidad; the Banana Scheme of the Banana Board, Jamaica (cf. Purseglove, 1972); and more recently by the commercial dessert banana exporters (Purseglove, 1972; Menendez and Shepherd, 1974; Simmonds, 1976). Vegetatively propagated parthenocarpic triploids outnumber the others by far in terms of commercial cultivation and provide all export bananas and plantains.

Since the cultivated bananas are sterile hybrid forms, they cannot be given precise names and until relatively recently the nomenclature has been at best very cumbersome and confusing (cf. Cheesman, 1947, 1948; Moore, 1957, 1958; De Langhe, 1961). This situation was eased when Simmonds and Shepherd (1956) suggested use of the term *genome* or *haploid chromosome set* as a key to classification. Using some 15 characteristics these authors categorized the cultivars in terms of relative similarity of the ancestral seeded species. Although the genomic designations are known to be oversimplifications, most agree that they serve a very useful purpose in interpreting the origins and patterns of polyploidy. The genome with 11 chromosomes from *M. acuminata* is designated 'A.' The genome with 11 chromosomes from *M. Balbisiana* is designated 'B.' In this scheme, a triploid cultivar or clone of *M. acuminata* such as Valery, Grande Naine, Lacatan, or Gros Michel would be designated *Musa* (AAA group) Valery, Grande Naine, Lacatan, or Gros Michel, or for convenience and brevity AAA Valery, Grande Naine, etc. A triploid cooking plantain, which would frequently be a hybrid between *M. acuminata* and *M. Balbisiana* would be designated, for instance, AAB Laknau, Harton, Domenico, etc., or ABB Chato, Pelipita, Saba, etc. (Simmonds and Shepherd, 1956).

Selection by man for fruit has resulted in the most widely used edible cultivars. As a result of this selection, the ideal dessert fruit, especially in Western trade, is large, long, slender, and seedless, ripens to deep yellow, and is borne in bunches comprised of compact hands on a stalk. Ability to resist damage in transit and to ripen slowly and uniformly over a longish period of time have also played a major role in determining marketability, but by far the most important factors in edible banana production are, of course, parthenocarpy and sterility. A stimulus (or stimuli) to growth of the fruit pulp results in parthenocarpy. Simmonds (1953) and Steward and Simmonds (1954) suggested that auxins and cytokinins or cytokininlike substances are involved. It has been known for a very long time that pollination is not required for production of the edible banana fruit (d'Angremond, 1912). Sterility is the result of a complex of factors, the most important of which is triploidy and attendant meiotic anomalies. This results in seedlessness for, as mentioned above, edible diploids exist. Parenthetically, parthenocarpy apparently does not occur in *M. Balbisiana* since there are no edible (seedless) BB or BBB types (Dodds and Simmonds, 1948). It seems clear, therefore, that parthenocarpy and sterility have arisen via gene mutations in fertile diploids; selection and subsequent vegetative propagation ensured their success. A number of structural changes have accumulated in the chromosomal mechanism and this has further exaggerated the sterility (Simmonds, 1962).

De Langhe (1969) has presented at length both the possibilities and difficulties for improvement of banana. He emphasizes that "When all is said and done, the improvement of bananas amounts to a continual fight against fertility, a property, which is, however, the essential instrument of the improvement itself." Menendez (1973a) has provided a summary statement of the activities of the Jamaica Banana Board and again emphasizes the great difficulties. More recently, Rowe (1981, 1984) has provided a general summary of the Banana Breeding Program of the United Fruit Company in La Lima, Honduras. One can rest assured, moreover, that some breeding efforts are being made by a number of the major banana exporting countries but one cannot expect specific details to be generally available.

Because vegetative selection has, until recently, been said to give minimal improvement of dessert bananas, crosses using pollen from male-fertile (AA) cultivars of *M. acuminata* and flowers of male-sterile triploid (AAA) cultivars have been the prime strategy of breeders (Dodds, 1947; Menendez and Shepherd, 1975). Tetraploid progeny, resulting from the fusion of the triploid egg cell of the female and the haploid (A) gamete of the male parent, are produced in very small numbers because of the miniscule number of seeds produced. This, of course, restricts screening trials (Shepherd, 1960). Most breeding efforts have been directed towards the improvement of dessert bananas, the greatest consideration being fruit quality, dwarfness, and disease resistance. Resistance to important diseases such as Fusarium wilt caused by *F. oxysporum* f. sp. *cubense*, Sigatoka leaf spot caused by *Mycosphaerella musicola* (a more severe strain of *M. musicola* has not yet been found in the Western Hemisphere but has been known in the Pacific for many years—cf. Rowe, 1981), Black Sigatoka caused by *Mycosphaerella fijiensis* var. *difformis*, and resistance to the burrowing nematode *Radopholus similis* have been by far the main recent breeding objectives and this, in turn, has necessitated a continued search for germplasm and an interest in the origin of various cultivars and clones. There has been, and still is, considerable interest in assembling germplasm collections for conservation and breeding purposes. The great danger of the monoclonal culture of a disease-susceptible cultivar such as the tall-growing Gros Michel was made very clear in the middle part of this century with the near-decimation of dessert banana plantations in this hemisphere by the Panama or Fusarium wilt disease. Although use of the more dwarf, Fusarium wilt disease-resistant Cavendish AAA clones like Valery and Grande Naine has enabled the industry to survive and flourish, there are apparently some signs, especially in Taiwan, that the Fusarium pathogen can mutate and successfully attack the supposedly resistant Cavendish clones (cf. Rowe, 1981, and refs. there cited). All this emphasizes that efforts still must be made to not only expand the disease-resistant germplasm base but to integrate it into a product that combines all the desirable qualities. This is all the more urgent for it is significant that not a single new cultivar has been developed for commercial export use in the past 60 or so years (Rowe, 1981, 1984).

Modest programs have been initiated to explore the use of induced mutation breeding (Menendez, 1973a). As a purely clonal crop plant, it seems that bananas are admirably suited to this approach. It is quite clear that the effects of somatic mutation or bud sporting can be significant, since from a historical perspective in Jamaica alone, some six mutants of the once highly prized Gros Michel were detected in a 100-year period and it is surmised that many more have gone undetected (cf. Baker and Simmonds, 1951, 1952). Shepherd (1957) points out that somatic mutation has contributed much to the banana cultivars (clones) in East Africa. Champion (1963), De Langhe (1969), Broertjes and van Harten (1978), de Guzman et al. (1976), and Menendez and Loor (1979) all have emphasized that improvement by mutations induced by chemical mutagens or irradiation has great potential which needs further investigations. Since the theoretical and practical advantages afforded by aseptic culture techniques are particularly relevant here, discussion will be deferred to the next section.

REVIEW OF THE LITERATURE

Embryo Culture

The erratic and generally low germination levels of seeds from many *Musa* clones has rendered use of embryo culture very valuable. Thus this in vitro technique has been in use for a long time (Cox et al., 1960). Because the technique has been described in considerable detail, no attempt will be made to belabor the specifics. A modified Knudson's medium or Randolph and Cox (1943) medium containing 0.12 M sucrose (but without growth regulators) solidified with agar (0.5–0.7% w/v) can be used to rear young plantlets from the tiny embryos excised from banana seeds until they are large enough (that is, several cm tall) to be placed in the soil. The method of Cox et al. (1960) as published applies to *M. Balbisiana* and no references that we are aware of specifically deal with the growing of embryos of other species, but it is clear that this is more or less readily carried out (cf. Rowe and Richardson, 1975).

Excised Shoot Tip Culture

Conventional agricultural practices rely on planting of pieces of corm or rhizome which include at least one bud, small suckers which have only recently appeared above ground level (so-called peepers), sword suckers formed from buds low on the corm, large suckers which have reached the broad-leaved stage (so-called maiden suckers), portions of large corms, etc., etc. (Navarre, 1957; Haarer, 1964; Hamilton, 1965; Turner, 1968; Acland, 1971; Purseglove, 1972). In 1959 a system of maximum multiplication of the banana plant was described (Barker, 1959). The method capitalized upon the presence of "lateral" buds—

actually so-called leaf-opposed buds (Fisher, 1978)—that never achieve any major degree of development under field conditions. This morphological feature of the banana plant attracted Barker and he devised a means of forcing the buds, both those laterally exposed and those that never would be exposed. This involved repeated stripping and removal of the outer leaf sheaths to expose adventitious buds. (It is interesting to note that in the "ensat," sometimes called the Ethiopian banana but more frequently called the False Banana, *Ensete ventricosum* [Welw.] E.E. Cheesm., native farmers have learned to force buds by slicing corms into pieces or removing the central part of the pseudostem from the corm and burying them. New suckers are produced after several weeks. This remarkable practice emphasizes that even in a plant such as *Ensete* where suckering does not take place under field conditions, the potential to release dormant buds may be realized provided the right technique is used—cf. Taye Bezuneh and Asrat Felleke, 1966; Westphal, 1975; Duthie, 1977.)

Investigations have shown that, along the lines shown by Barker (1959) but including refinements made possible by aseptic culture methods and the use of exogenously added growth regulators, excised shoot tips of quite a few clones of dessert bananas are capable of yielding entire plants. Berg and Bustamante (1974) were the first to show that cucumber mosaic virus-free plants of the economically supreme Cavendish group AAA can be obtained from meristems from lateral buds of virally-infected plants by a combination of heat treatment and aseptic culture. In the technique as they describe it, apparently only one plant was obtained per excised shoot apex. The method involved use of Knudson's medium (1946) with the trace elements of Berthelot (1934), thiamine HCl, supplemented with CW 10% v/v and CH. Although the development of roots from such excised shoot tips was rather slow (even after 3 months they had not formed roots), additional work soon demonstrated that roots could be induced to form promptly enough (2 months) with the addition of 5.4 μM NAA to render the technique commercially useful. Studies from the Republic of China using a combination of semisolid media and liquid media were published around the same time (Ma and Shii, 1972, 1974). For a number of years and until recently, the late Dr. Emerita de Guzman and her colleagues at the University of the Philippines at Los Banõs had been making considerable progress in assessing the applicability of the excised shoot tip culture technique to a wide range of clones available to them through the extensive *Musa* germplasm bank at the University of the Philippines at Los Banõs. Special attention was given to abaca, *M. textilis*, the source of Manila hemp (de Guzman et al., 1976).

All this work has provided the necessary underpinning for the now extensive use of the shoot tip culture technique (Figs. 1 and 2). One of the major features of the technique as it is now used is that multiplication can be induced by releasing dormant buds at the leaf bases. Subculturing can be carried out from the proliferating mass of shoots which result. Protocormlike bodies can also form at a shoot base. Multiple shoots that originate from these can, in turn, be separated by cutting to produce individual plantlets (Figs. 2c and d and Tables 1-3).

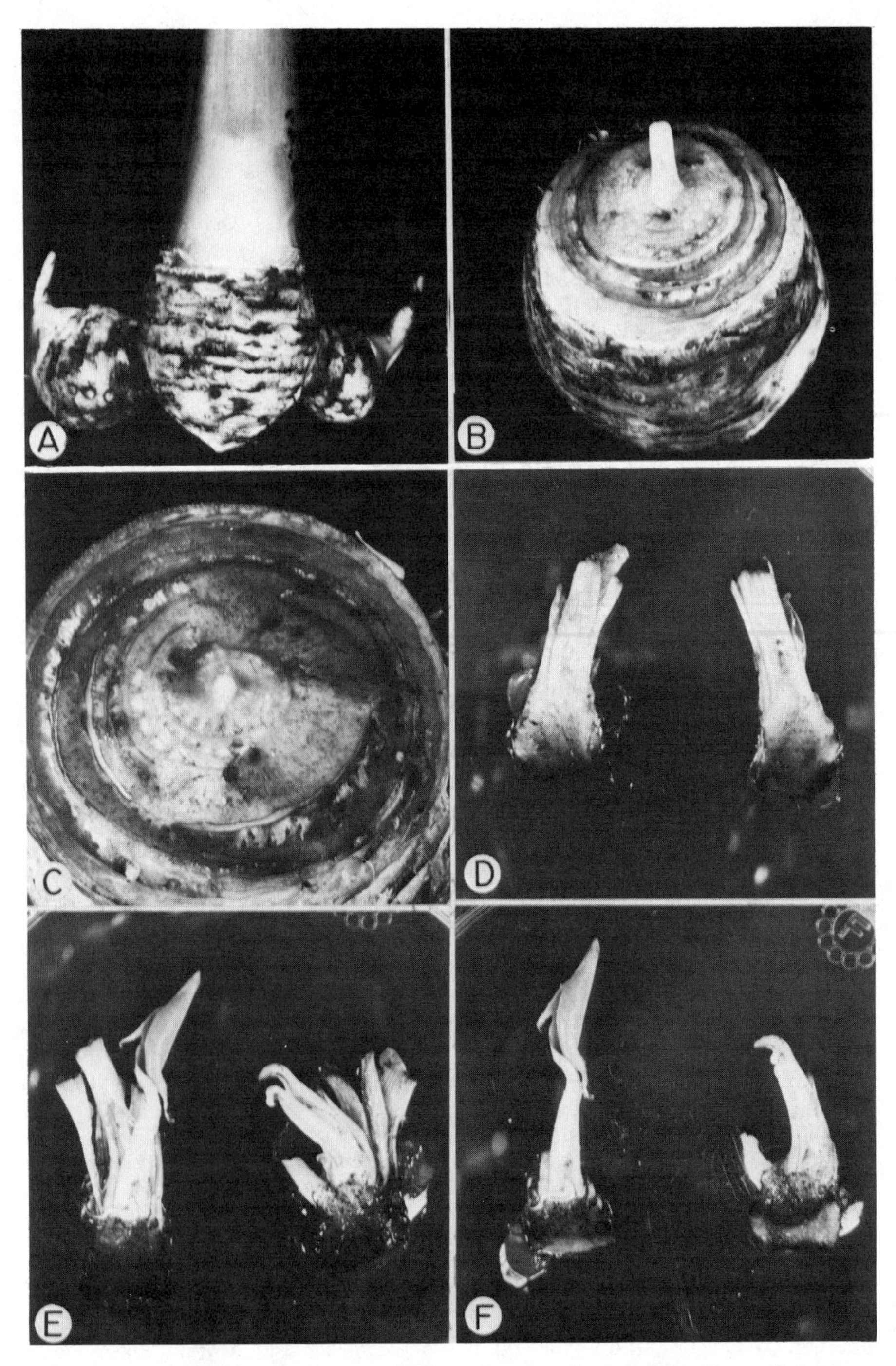

Figure 1. Stepwise sequence of the clonal micropropagation of banana via shoot tip culture. (a) Young plant (1.7 m tall) from greenhouse

with soil and roots removed to expose corm with two suckers. (b) Corm with side suckers and outer leaf bases of pseudostem removed. (c) Corm trimmed so as to expose the growing point. When explanted this would comprise the apical meristem with one or two leaf primordia. In cases where the material has been demonstrated to be virus-free, special precautions need not be taken to go down to the dome itself. In cases where virus-free plants are to be obtained, the heat therapy treatments according to Berg and Bustamante (1974) should be followed. (d) Cultured shoot split longitudinally through the apex. (e) Same shoot halves, 4 days later. (f) Shoot halves with outer leaves and blackened shoot bases trimmed. Note emergence of side shoots.

Functional roots can be initiated within a few days by using a combination of procedures. (We have observed spontaneous root formation on plants as small as 0.6 cm.) However, root induction can be achieved with far more control and predictability by using low levels of IBA (on the order of 0.01-1.23 μM). In addition, the rooting process can be greatly accelerated by the addition of activated charcoal (0.25% w/v) to a basal medium. Even very tiny shoots (smaller than 1 cm) will produce roots within 4-5 days (Fig. 2e). If an excised shoot tip is placed on a charcoal-containing basal medium, it will begin to produce roots shortly after a recognizable (to the naked eye) shoot has organized. Young plantlets placed on such a medium promptly produce whitish and cream-colored roots (Fig. 2e), and from these propagules plantlets can be established in the greenhouse or in growth chambers within 1 month provided precautions are taken to prevent drying out. In our hands and in the temperate climate of Long Island, New York, a "propagation bed" supplied with an automatic misting system (Mist-A-Matic) in the greenhouse has been very satisfactory.

Callus and Suspension Cultures from AAA *Musa* Clones

The earliest sustained efforts to establish tissue and cell suspension cultures of dessert banana were made in the late 1950s and in subsequent years at Cornell University (Mohan Ram and Steward, 1964). A few isolated attempts have been published since but they have not been impressive, and subculture of callus has not been reported in any papers that we are aware of (cf. Tongdee and Boon-Long, 1973; de Guzman, 1975). Since little of great insight has been published more recently, discussion of callus, suspension, and protoplast techniques will be deferred until later in this chapter.

Key Variables and Factors for Success

Not surprisingly, information is not available as to the full range of *Musa* clones, cultivars, or species that can be rendered pathogen-free or multiplied through the use of shoot tip culture. Since it is very likely that genomic differences in terms of sucker production (so-called stooling properties) may be expressed faithfully, exaggerated, or perhaps

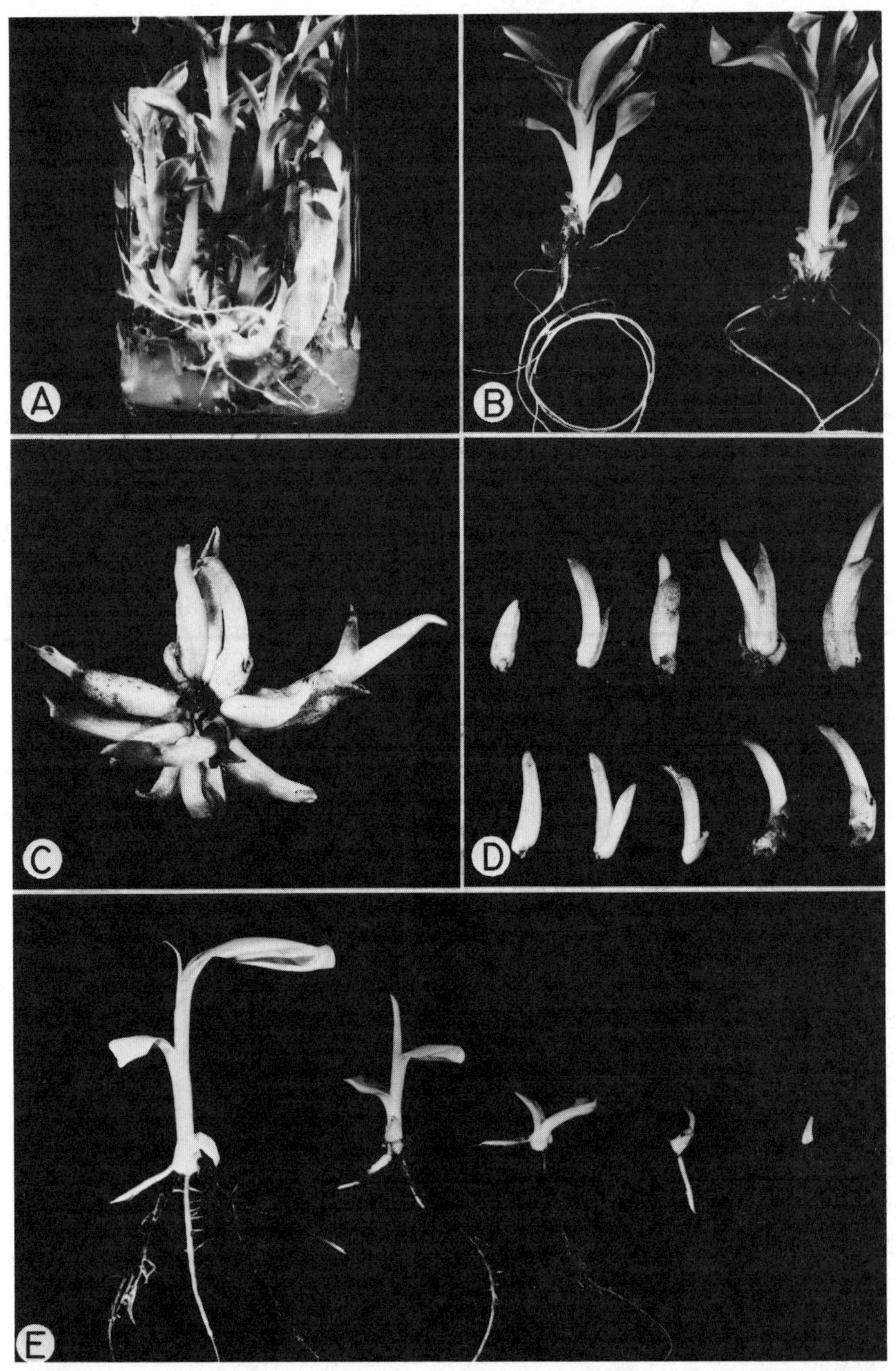

Figure 2. (a) Banana plantlets in culture jar (10 oz) several weeks after subculture. (b) Individual plants removed from the jar. Note

substantial root development. (c) Close-up showing multiple shoots formed on explant similar to that in Fig. 1f. This emphasizes the origin of shoots from otherwise dormant buds. (d) Individual shoots from the growth form at Fig. 2c separated by scalpel incisions. (e) Rooted banana plantlets. Note that all except the very smallest shoot tip explant has produced roots. Even the smallest can, however, be induced to produce roots (see text for details).

even overcome under conditions of aseptic culture, it will be very useful to have this information. No attempts that we are aware of have been made to ascertain why some produce more side branches than others. Since the plantlets produced in culture are very small (often times as small as a few cm), this could provide a very elegant assay system or means of exposing large numbers of growing points to test conditions. It is well known, of course that *M. Balbisiana* has very variable stooling properties but generally has a much lower tendency to produce suckers. *M. acuminata* subspecies show considerable variation, e.g., *M. acuminata* subsp. *malaccensis* and *siamea* stool quickly and *M. acuminata* subsp. *burmannica* and *banksii* stool slowly (cf. Anonymous, 1962; De Langhe, 1969). Improved stooling properties have been indicated as one of the major breeding aims in cooking bananas of special importance in Zaire (De Langhe, 1969, p. 66). Thus, even in a vegetatively propagated plant like banana one may be faced with the problem of an erratic, irregular or slow stooling response. Trials with cooking bananas (e.g., the clones Pelipita and Saba) suggest that the shoot tip culture methods are as successful as with dessert bananas but the multiplication rates are slower (see also Vessey and Rivera, 1981).

The shoot tip multiplication method obviously has great potential for producing specific pathogen-free planting materials in quantity, and provided precautions are taken, diseases could presumably be eliminated or at least reduced from nursery stock. While ensuring potential reinfection in the field would not be eliminated, use of the method could play a hygienic role in minimizing problems. This would apply especially to presumed virus-caused diseases such as Bunchy Top; leaf mosaic or infectious chlorosis; fungal pathogens such as the Panama Disease, Sigotoka leaf spot, Black Sigotoka, and Black leaf streak; and bacterial diseases such as Moko or bacterial wilt caused by *Pseudomonas solonacearum* (cf. Stover, 1972; Wardlaw, 1972).

From a commercial perspective, the need for rapid clonal multiplication in terms of introducing new clones seems, for the immediate future, to be of limited use because cultivars change exceedingly slowly and new industries are relatively scarce. Most helpful, however, would be the adaptation of the meristem or shoot tip culture technique to germplasm conservation or storage (cf. Withers, 1980, 1981). Ideally, one would like to see the excised shoot tips of as much and as diverse a range of germplasm as is available maintained in aseptic culture under conditions of media composition and/or environment that foster multiplication or branching at the absolute minimal level. Such a system would facilitate storage with a modest input of labor and upkeep. When needed, a culture would be removed from the relatively quiescent state and multiplied as needed by use of the appropriate stimulatory sequence of media or manipulation.

Table 1. Protocol for Banana Shoot Apex Isolation

1. Begin with a banana plant or sucker.
2. Cut off leaves and pseudostem about 30 cm above soil level.
3. Remove all soil and cut off roots to expose the corm.
4. Peel off outer sheathing leaf bases of the pseudostem one at a time until they become too small to remove carefully by hand.
5. Remove the last remaining leaves using a dissecting microscope and a scalpel.
6. Excise shoot apex with a scalpel.
7. Sterilize in 1% commercial bleach (0.0525% NaOCl) and Tween 20 for 5 min, swirling occasionally.
8. Wash 4 times with sterile distilled water.
9. Place in culture medium.[a]
   Liquid—20 ml in 50-ml Erlenmeyer flask
   Solid—0.7% agar in snap-lid petri dishes or screw-cap jars
10. After 3 weeks, transfer to culture medium minus coconut water, solidified with 0.7% agar in screw-cap jars.

[a]Culture medium:

MS mineral salts	BAP 22.0 μM	Inositol 5.55 mM
Thiamine HCl 2.97 μM	Sucrose 0.12 M	CW 15%
pH 5.8	Autoclave	

Table 2. Protocol for Stimulation of Multiple Shoot Formation

1. Begin with an aseptically cultured banana shoot or sucker 21 days after explanation.
2. Make a transverse cut to separate leaves. This yields a section of pseudostem approximately 3 cm long including an intact vegetative bud.
3. Trim the lower part of explant to remove darkened or necrotic tissue.
4. Cut pseudostem explant in half longitudinally through the apex.
5. Transfer halves to culture medium with 0.7% agar in screw-cap jar.
6. After 4-7 days, remove from culture jar. Trim off outermost leaves and blackened base of explant.
7. Transfer to fresh medium.
8. After 2-3 weeks, use a scalpel to cut the multiple shoots that have formed. Transfer shoots thus separated to fresh growth medium.
9. Maintain the multiple shoot cultures by transferring to fresh culture medium and separating the multiple shoots in the same way every 3-4 weeks.

Table 3. Protocol for Plantlet Development

1. Separate individual shoots from multiple-shoot cultures with scalpel incisions.
2. Transfer to culture medium supplemented with 0.25% (w/v) charcoal to induce root formation.
3. After 2-3 weeks, transplant into Pro-Mix-Vermiculite (1:1) in small plastic pots and place under mist system.
4. After 7-10 days move to normal greenhouse conditions.

Even now excised axenic shoot tip cultures facilitate germplasm exchange and passage of propagules through plant quarantine (cf. Kahn, 1979). We have used the method with success for banana but precautions must be taken to use a fairly firm agar medium (order of 1%), and a sterile cotton plug should be pushed carefully but firmly against the plantlet or the shoot tip being shipped. In that way any tendency for the plantlet to dislodge in transit is minimized.

Also, large numbers of clonal hybrids that might otherwise be limited to a relatively few specimens can be made available for continued breeding research and field testing. By extension, one can anticipate excising growing points from young seedlings produced through hybridization and inducing them to proliferate and multiply freely, saving much valuable time.

Whatever the rationale, justification, or use, the method must be profitable, for several laboratories are engaged in commercial production for export.

PROTOCOLS

Before presenting a precise protocol for isolation of shoot tips and their subsequent maintenance and growth in culture, it may be useful to restate that different investigators will, of course, utilize different procedures to achieve the same end. Some have recommended that no special effort be made to surface-sterilize the shoot tip since, by virtue of its special morphology, it will generally be free of surface contaminants and pathogens (Berg and Bustamante, 1974). Since the procedure given here is adaptable to obtaining a meristem or a growing point with one or two primordial leaves, we use a surface-sterilizing agent. In the protocol below we recommend the use of sodium hypochlorite so as to minimize the level of contamination. Far too much effort goes into the whole procedure to risk loss of valuable material and we urge that no short-cuts be taken without first establishing the sagacity of the move.

Similarly, some investigators recommend use of media with different components from the one we have outlined. For instance, de Guzman (1975 and personal communication) frequently used Nitsch and Nitsch vitamins (1969), citric acid, KIN (in addition to CW), and dextrose (instead of sucrose) as supplements to a basal medium. We examined this in some detail using the same cultivar, Philippine Lacatan, and have concluded that for our purposes the supplementation with KIN or citric acid is neither necessary nor helpful. Similarly, the complex vitamin mixture can be replaced with thiamine HCl (1 mg/l) alone and sucrose can be substituted for dextrose. The CW we use is not essential; it does help a bit. Therefore, investigators are urged to consider the protocol below as merely a guideline protocol and not a problem-free and certain method for all clones.

Shoot Apex Isolation (see Table 1)

1. Shoot apices are isolated from small plants or suckers. The leaves and upper portion of the pseudostem are removed so that a 30-cm stem piece remains. All roots are cut off at the corm and the soil is washed off (Fig. 1a). Any side suckers found on the corm are removed.
2. The outer leaf sheaths of the pseudostem are peeled off one at a time. This procedure is used until the inner leaf sheaths measure approximately 1 cm across at their bases and are 2 cm tall (Fig. 1b). At this point, they become too small to remove by hand.
3. Using a dissecting microscope and a scalpel, the remaining outer leaf bases are carefully removed until only one or two young leaf primordia remain (Fig. 1c).
4. The shoot apex is excised by making four incisions with a scalpel into the corm beneath the apex.
5. The excised apex is placed in 50 ml of a 1% Clorox solution with two drops of Tween 20 made in a sterile 125-ml Erlenmeyer flask. The apex is allowed to soak for 5 min and the solution is swirled occasionally.
6. Working under aseptic conditions, the Clorox solution is decanted and the apex is rinsed four times with sterile distilled water. When the fourth rinse water is decanted, it is poured through a sterile 864 μm (#20) sieve to catch the apex.
7. The sterilized apex is transferred from the sieve to the culture medium using sterile forceps. The culture medium used consists of MS mineral salts supplemented with 5.55 μM inositol, 22.0 μM BAP, 2.97 μM thiamine HCl, 0.12 M sucrose, and 15% CW. It is adjusted to pH 5.8 using KOH. This medium is used as a liquid. Twenty ml are placed in a 50-ml Erlenmeyer flask, which is then stoppered with a foam plug and autoclaved. Apex cultures are maintained at 30 C on a 16-hr light schedule. Liquid cultures are kept on a rotary shaker (100 rpm). The medium can also be solidified with 0.7% Difco agar and poured into 50 x 12 mm snap-tight petri dishes (Falcon #1006). The excised apex can be placed gently on top of the agar and maintained in stationary culture under the same conditions as the liquid cultures.
8. Within 10 days, the growing apex starts to turn green. After an additional 11 days, a small green shoot is clearly visible to the naked eye.

9. When the shoot becomes about 1 cm long and starts to produce new leaves, it is transferred to the same growth medium minus the CW. This is solidified with 0.7% agar. In this laboratory, the young shoots are often grown in 1-ounce "French square" screw-top bottles under the same growth conditions.
10. An excised apex usually grows into a single shoot or plantlet. Roots usually do not appear on these young shoots although excised shoots only 14 days old have in our hands produced roots longer than 2 cm. Excised shoots can also grow into multiple shoot forms. This type of growth can also be induced as described in the next step (Table 2).
11. A large (3 cm tall) single shoot can be forced into producing many smaller shoots simply by cutting it in half longitudinally through the apex. Using a sterile petri dish as a working surface, the procedure is begun by cutting across the pseudostem to remove the leaf blades, thus producing a stem piece about 3 cm long. If the shoot has produced any roots, they are trimmed off. The stem is cut longitudinally through the apex (Fig. 1d). Each half is placed upright in a culture jar containing culture medium solidified with 0.7% agar.
12. After 4-7 days, the 2 shoot halves are removed and the outer leaves and blackened shoot bases are trimmed off (Fig. 1e). New side shoots are clearly visible at this time (Fig. 1f).
13. After 2-3 weeks, the multiple shoots that have formed are separated by a scalpel incision. These new, smaller shoots should continue to multiply.
14. Our cultures are routinely transferred every 3-4 weeks. Multiple shoots are separated into smaller groups of shoots. This separation procedure induces further shoot multiplication. As the shoots become larger, the cultures are moved into larger jars (Figs. 2a and 2b).
15. Roots are routinely induced by transferring single shoots to culture medium supplemented with 0.25% (w/v) charcoal (Fig. 2e and Table 3). The addition of IBA at the level of 0.1 μM enhances rapid root formation but roots still form without it. In the presence of charcoal, white or cream-colored roots can be seen at the shoot base in 4-5 days. Plantlets can be potted in Pro-Mix-Vermiculite in small plastic pots in about 2-3 weeks. They can be moved to normal greenhouse conditions in 7-10 days.

Mutation Breeding and Mutagenesis

Champion (1963), Stotzky et al. (1964), De Langhe (1969), Menendez (1973a), de Guzman (1975), de Guzman et al. (1976), Broertjes and van Harten (1978), and others all have seen and emphasised the benefits that could accrue if the various methodologies for producing dessert banana and plantain mutants could be worked out, applied, and assessed. In most of the above cases, mutagenesis of suckers using γ-irradiation prior to excision and culturing of the growing point has been the point of departure (cf. Broertjes and van Harten, 1978, pp. 256-258), but Menendez (1973b) has also reported use of the chemical mutagen ethyl methane sulphonate on seeds of *M. acuminata*. The field is only now beginning to realize some of its potential (Menendez and Loor, 1979). De Guzman and her colleagues regularly used about 2.5 Krad to irradiate suckers prior to culture but they encountered much variation in responsiveness depending on the clone (de Guzman, personal communication). Some clones actually grow faster as a result of irradiation and in others the growth is attended by a dramatic slowing down. The gross morphology of the cultures is also affected, those being irradiated tending more towards a compact protocormlike habit. No attempt is made here to outline the procedures since general principles are available (cf. Broertjes and van Harten, 1978).

From our perspective the most challenging and rewarding work in this area is sure to derive from the eventual application of mutation breeding techniques to totipotent cell suspensions or protoplasts but this will not be easy to achieve (Krikorian and Cronauer, 1983). Even so, work in this laboratory with a recalcitrant perennial monocotyledon, the daylily (*Hemerocallis*), has shown that perseverance can lead to significant progress (cf. Krikorian and Kann, 1979, 1980, 1981; Krikorian et al., 1981; Fitter and Krikorian, 1981).

The production from banana of true callus and/or rapidly growing morphogenetically competent cell suspensions will be a major step towards generating improvements in banana as a crop. These techniques will greatly enhance the chances of selecting potentially useful mutations if plantlets can be caused to arise from single cells. Consequently, much interest has been expressed in the initiation of totipotent callus cultures and subsequent suspension cultures that have a capacity for production of plantlets either by initiation of adventitious shoot and root growing points on a cell-derived callus, or by the formation of adventive or somatic embryos (Steward et al., 1975; Steward and Krikorian, 1979; Krikorian, 1982; Krikorian and Cronauer, 1983).

An examination of the literature demonstrates how recalcitrant the banana has been in terms of the induction of totipotent cell suspensions or callus tissue. Mohan Ram and Steward (1964) used a variety of auxins to induce callus formation on sections of mature and immature fruits. The callus tissues and the cell suspensions they obtained when placed in liquid medium were slow-growing and showed no signs of organized development. Although later workers (Tongdee and Boon-Long, 1973; de Guzman, 1975) also reported the production of callus from banana fruits of several clones, they were equally unsuccessful in stimulating any organogenesis. Attempts in this laboratory to evoke callus which can be subcultured from immature and mature but pre-climacteric fruit sections have not been encouraging. Although proliferating tissue masses can be obtained, "growth" is predominantly by cell enlargement, and the potential for any morphogenesis or organogenesis from such fruit-derived tissues, given the state of the art, seems extremely limited at present. A major problem to be overcome involves the substantial tendency of fruit tissues to blacken due to oxidation of polyphenols. In our view it is much more likely that cultures derived from other somatic tissues will have a greater potential for producing totipotent cell suspensions and work is progressing in that direction.

Cell suspensions might also be obtained through protoplast culture. Such cultures could offer yet another means of inducing or even introducing variation into banana. We have been able to prepare protoplasts reproducibly from various parts of dessert bananas and are working towards establishing cell suspensions from these preparations (Cronauer and Krikorian, unpublished). Since no plants have yet been recovered from protoplasts, we hesitate to provide details here.

FUTURE PROSPECTS

The distinctive evolutionary origin of edible bananas and plantains and their high potential for clonal multiplicaton provides many opportunities to the tissue culture worker. Wild (seeded) and edible (seedless) *M. acuminata* exist as normal diploids as well as triploids. Only seeded diploid types of *M. Balbisiana* occur but even some of these are economically valuable. Parthenocarpy and sterility within triploids; outcrossing of edible diploids with seeded forms of *M. acuminata* and *M. Balbisiana*, followed by human selection; occurrence of triploidy within *M. acuminata*; and triploidy in crosses between *M. acuminata* and *M. Balbisiana* (giving rise to two genetically different kinds of triploids), have each played their part in the origin of the many kinds of edible bananas available today. Dwarf, medium, and tall growth habits with varying intergrading forms are known. Variation in habit, fruiting behavior, fruit color, and quality abound. Somatic mutations have further expanded the base of diversification.

Artificial means of inducing tetraploidy and higher levels of polyploidy by use of colchicine emphasized some time ago that man can further intervene in manipulating the genomic composition (cf. Vakili, 1967). The use of radiation and chemicals for mutation breeding has shown that a real potential exists for each. But as the kinds of techniques described in Volume 1 of this series are developed and perfected for application to *Musa* clones and species, a whole new level of potential will emerge. We have mentioned the use of totipotent cell suspensions for rapid clonal multiplication. Mutagenesis using free morphogenetically competent cells or protoplasts has also been mentioned. To this may be added use of controlled protoplast fusion to achieve a still greater combination of characters hitherto unobtainable (Krikorian and Cronauer, 1983). Androgenesis and pollen culture, if achievable, could play a significant role in all this. In short, the potential is there, one merely has to develop it. Unfortunately banana must be classified as a rather recalcitrant system at this time for all procedures other than embryo culture and shoot tip culture. Whether this state of affairs is a reflection of the lack of research emphasis until rather recently, or the fact that we lack a fundamental scientific base from which to operate, will soon become apparent. This is because several groups are now engaged in basic tissue, cell, and protoplast studies (cf. Withers, 1981).

ACKNOWLEDGMENTS

Investigations providing the background for this chapter were supported by various grants. Prominant among them has been support from the National Aeronautics and Space Administration (Grant NSG 7270, Plant Cells, Embryos and Morphogenesis in Space). Support from the National Science Foundation, Division of International Programs (2EA-794) permitted A.D.K. to visit the Philippines early in 1979. The late Professor Emerita de Guzman served as hostess and scientific counterpart during that visit. More recent travel to the banana- and plantain-growing regions of Colombia by one of us (A.D.K.) was also supported by NSF Division of International Programs. All of this help is gratefully acknowledged.

KEY REFERENCES

Berg, L.A. and Bustamante, M. 1974. Heat treatment and meristem culture for the production of virus-free bananas. Phytopathology 64: 320-322.

Cox, E.A., Stotzky, G., and Goos, R.D. 1960. In vitro culture of *Musa Balbisiana* Colla embryos. Nature 185:403-404.

de Guzman, E.V., Ubalde, E.M., and Del Rosario, A.G. 1976. Banana and coconut in vitro cultures for induced mutation studies. In: Improvement of Vegetatively Propagated Plants and Tree Crops through Induced Mutations, pp. 33-54. International Atomic Energy Agency, Technical Document 194, Vienna.

Mohan Ram, H.Y. and Steward, F.C. 1964. The induction of growth in explanted tissue of the banana fruit. Can. J. Bot. 42:1559-1579.

Vakili, N.G. 1967. The experimental formation of polyploidy and its effect on the genus *Musa*. Am. J. Bot. 54:24-36.

REFERENCES

Acland, J.D. 1971. East African Crops. FAO of the U.N. Longman, London.

Anonymous. 1962. *Musa*. In: The Wealth of India (B.N. Sastri, chief ed.) Vol. 6, pp. 448-480. Council of Scientific and Industrial Research, New Delhi.

Baker, R.E.D. and Simmonds, N.W. 1951. Bananas in East Africa. Pt. I. The botanical and agricultural status of the crop. Emp. J. Exp. Agric. 19:283-290.

______ 1952. Bananas in East Africa. Pt. II. Annotated list of varieties. Emp. J. Exp. Agric. 20:66-76.

Barker, W.G. 1959. A system of maximum multiplication of the banana plant. Trop. Agric. (Trinidad) 36:275-284.

______ and Steward, F.C. 1962a. Growth and development of the banana plant. I. The growing regions of the vegetative shoot. Ann. Bot. (London) 26:389-411.

______ and Steward, F.C. 1962b. Growth and development of the banana plant. II. The transition from the vegetative to the floral shoot in *Musa acuminata* cv. Gros Michel. Ann. Bot. (London) 26: 413-423.

Berg, L.A. and Bustamante, M. 1974. Heat treatment and meristem culture for the production of virus-free bananas. Phytopathology 64: 320-322.

Berthelot, A. 1934. Nouvelles remarques d'ordre chimique sur le choix des milieux de culture naturels et sur la maniere de formuler les milieux synthetiques. Bull. Soc. Chim. Biol. 16:1553-1557.

Broertjes, C. and van Harten, A.M. 1978. Application of Mutation Breeding in the Improvement of Vegetatively Propagated Crops. Elsevier Sci. Pub. Co., Amsterdam.

Champion, J. 1963. Le Bananier. G.P. Maisonneuve & Larose, Paris.

Cheesman, E.E. 1947. Classification of the bananas. Kew Bull. 57:97-117.

______ 1948. Classification of the bananas. Kew Bull. 58:11-28; 145-157; 323-328.

d'Angremond, A. 1912. Parthenokarpie und Samenbildung bei Bananen. Ber. Dtsch. Bot. Ges. 30:686-691.

de Guzman, E.V. 1975. Project on production of mutants by irradiation of in vitro cultured tissues of coconut and bananas and their mass propagation by the tissue culture technique. In: Improvement of Vegetatively Propagated Plants through Induced Mutations, pp. 53-76. Internatonal Atomic Energy Agency, Technical Document 173, Vienna.

De Langhe, E. 1961. La taxonomie du bananier plantain en Afrique equatorial. J. Agric. Trop. Bot. Appl. 8:417-449.

______ 1969. Bananas, *Musa* spp. In: Outlines of Perennial Crop Breeding in the Tropics (F.P. Ferwerda and F. Wit, eds.) pp. 53-78. Miscellaneous Papers 4. Landbouwhogeschool Wageningen, the Netherlands.

Dodds, K.S. 1947. The genetic system of banana varieties in relation to banana breeding. Emp. J. Exp. Agric. 11:89-98.

______ and Simmonds, N.W. 1948. Sterility and parthenocarpy in diploid hybrids of *Musa*. Heredity 2:101-116.

Duthie, D. 1977. Propagation of *Ensete ventricosum (Musa ensete)* purple form. Int. Plant Prop. Soc. Comb. Proc. 27:329-330.

Fahn, A. 1953. The origin of the banana inflorescence. Kew Bull. 1953 (No. 3):299-306.

Fisher, J. 1978. Leaf-opposed buds in *Musa*: Their development and a comparison with allied monocotyledons. Am. J. Bot. 65:784-794.

Fitter, M.S. and Krikorian, A.D. 1981. Recovery of totipotent cells and plantlet production from daylily protoplasts. Ann. Bot. 48:591-597.

Haarer, A.E. 1964. Modern Banana Production. L. Hill, London.

Hamilton, K.S. 1965. Reproduction of banana from adventitious buds. Trop. Agric. (Trinidad) 42:69-73.

Kahn, R.P. 1979. Tissue culture applications for plant quarantine. In: Practical Tissue Culture Applications (K. Maramorosch and H. Hirumi, eds.) pp. 185-201. Academic Press, New York.

Kervegant, D. 1935. Le Bananier et son Exploitation. Société d'Éditions Géographiques, Maritimes et Coloniales, Paris.

Knudson, L. 1946. A new nutrient solution for the germination of orchid seed. Bull. Am. Orchid Soc. 15:214-217.

Krikorian, A.D. 1982. Cloning higher plants from aseptically cultured tissues and cells. Bio. Rev. 57:151-218.

_____ and Cronauer, S. 1983. Tecnicas de cultivo aséptico por el mejoramiento del banano y plátano. Informe mensual (Union de Paises Exportadores des Banano, Panama) Año 7/No. 55, 42-47.

_____ and Kann, R.P. 1979. Micropropagation of daylilies through aseptic culture techniques: Its basis, status, problems and prospects. Hemerocallis J. 33(No. 1):44-61.

_____ and Kann, R.P. 1980. Mass blooming of a daylily clone reared from cultured tissues. Hemerocallis J. 34(No. 1):35-38.

_____ and Kann, R.P. 1981. Plantlet production from morphogenetically competent cell suspensions of daylily. Ann. Bot. 47:679-686.

_____, Staicu, S., and Kann, R. 1981. Karyotype analyses of a daylily clone reared from aseptically cultured tissues. Ann. Bot. 47:121-131.

Ma, S.-S., and Shii, C.-T. 1972. In vitro formation of adventitious buds in banana shoot apex following decapitation. Chung-Kuo Yuan I Hsueh Hui (China Horticulture) 18:135-142. (In Chinese with English summary.)

_____ 1974. Growing banana plantlets from adventitious buds. Chung-Kuo Yuan I Hsueh Hui (China Horticulture) 20:6-12. (In Chinese with English summary.)

Masefield, G.B. 1938. The production of native beer in Uganda. East Afr. Agric. J. 3:362-364.

Massal, E. and Barrau, J. 1956. Food Plants of the South Sea Islands. South Pacific Commission Technical Paper No. 94. South Pacific Commission, Noumea, New Caledonia.

Menendez, T. 1973a. Application of mutation methods to banana breeding. In: Induced Mutations in Vegetatively Propagated Plants (proceedings of a panel on mutation breeding) pp. 75-83. International Atomic Energy Agency, Vienna.

_____ 1973b. A note on the effect of ethyl methane sulphonate on *Musa acuminata* seeds. In: Induced Mutations in Vegetatively Propagated Plants (proceedings of a panel on mutation breeding) pp. 85-90. International Atomic Energy Agency, Vienna.

_____ and Loor, F.H. 1979. Recent advances in vegetative propagation and their application to banana breeding. In: Proceedings of the Fourth Conferences of ACORBAT (Association for Cooperation in Banana Research in the Caribbean and Tropical America). UPEB (Union de Paises Exportadores de Banano), Panama 5, Panama.

_____ and Shepherd, K. 1975. Breeding new bananas. World Crops 27:104-112.

Mohan Ram, H.Y., Ram, M., and Steward, F.C. 1962. Growth and development of the banana plant. III.A. The origin of the inflorescence and the development of the flowers. B. The structure and development of the fruit. Ann. Bot. 26:657-673.

Moore, H.E. 1957. *Musa* and *Ensete*. The cultivated bananas. Baileya 5(No. 4):166-194.

_____ 1958. *Musa* and *Ensete*. Additions and corrections. Baileya 6(No. 1):69-70.

Mukasa, S.K., Thomas, D.G., Ingram, W.R., and Leakey, C.L.A. 1970. Bananas (*Musa* spp.). In: Agriculture in Uganda, 2d ed. (J.D. Jameson, ed.) pp. 139-152. Oxford Univ. Press, Oxford, England.

Myers, V.C. and Rose, A.R. 1917. The nutritional value of the banana. JAMA 68:1022-1024.

Navarre, E. 1957. Multiplication des *Musa* a feuilles rouges. Rev. Hort. 129:712.

Nitsch, J.P. and Nitsch, C. 1969. Haploid plants from pollen grains. Science 163:85-87.

Palmer, J.K. 1971. The banana. In: The Biochemistry of Fruits and Their Products (A.C. Hulme, ed.) pp. 65-105. Academic Press, New York.

______ 1979. Banana products. In: Tropical Foods: Chemistry and Nutrition (G.E. Inglett and G. Charalambous, eds.) pp. 625-635. Academic Press, New York.

Purseglove, J.W. 1972. Musaceae. In: Tropical Crops, Monocotyledonous, pp. 343-384. John Wiley & Sons, New York.

Randolph, L.F. and Cox, L.G. 1943. Factors influencing the germination of iris seed and the relation of inhibiting substances to embryo dormancy. Proc. Am. Soc. Hort. Sci. 43:284-300.

Reynolds, P.K. 1951. Earliest evidence of banana culture. J. Am. Oriental Soc. 71 (Suppl.):v, 28pp., and 8 plates.

Richardson, D.L., Hamilton, K.S., and Hutchinson, D.J. 1965. Notes on banana. I. Natural edible tetraploids. Trop. Agric. (Trinidad) 42: 125-137.

Riopel, J.L. and Steeves, T.A. 1964. Studies on the roots of *Musa acuminata* cv. Gros Michel. 1. The anatomy and development of main roots. Ann. Bot. 28:475-490.

Rowe, P. 1981. Breeding an 'intractable' crop—Bananas. In: Genetic Engineering for Crop Improvement (K. Rachie and J. Lyman, eds.) pp. 66-83. Rockefeller Foundation, New York.

______ 1984. Breeding bananas and plantains. Plant Breed. Rev. 2:(in press).

______ and Richardson, D.C. 1975. Breeding Bananas for Disease Resistance, Fruit Quality, and Yield. Bulletin No. 2. Tropical Agriculture Research Services (SIATSA). La Lima, Honduras.

Shepherd, K. 1957. Banana cultivars in East Africa. Trop. Agric. (Trinidad) 34:277-286.

______ 1960. Seed fertility of edible bananas. J. Hort. Sci. 35:6-20.

Simmonds, N.W. 1953. The development of the banana fruit. J. Exp. Bot. 4:87-105.

______ 1954a. A survey of the Cavendish group of bananas. Trop. Agric. (Trinidad)31:126-130.

______ 1954b. Varietal identification in the Cavendish group of bananas. J. Hort. Sci. 29:81-88.

______ 1962. The Evolution of Bananas. Longman, London.

______ 1966. Bananas, 2nd ed. Longman, London.

______ 1976. Bananas *Musa* (Musaceae). In: Evolution of Crop Plants (N.W. Simmonds, ed.) pp. 211-215. Longman, London and New York.

______ 1979. Principles of Crop Improvement. Longman, London.

______ and Shepherd, K. 1956. The taxonomy and origins of cultivated bananas. J. Linn. Soc. Lon. Bot. 55:302-312.

Steward, F.C. and Krikorian, A.D. 1979. Problems and potentialities of cultured plant cells in retrospect and prospect. In: Plant Cell and

Tissue Culture: Principles and Applications (W.R. Sharp, P.O. Larsen, E.F. Paddock, and V. Raghavan, eds.) pp. 221-262. Ohio State Univ. Press, Columbus.

Steward, F.C. and Simmonds, N.W. 1954. Growth-promoting substances in the ovary and immature fruit of the banana. Nature 173:1083-1084.

Steward, F.C., Hulme, A.C., Freiburg, S.R., Hegarty, M.P., Pollard, J.K., Rabson, R., and Barr, R.A. 1960a. Physiological investigations on the banana plant. I. Biochemical constituents detected in the banana plant. Ann. Bot. 24:83-116.

Steward, F.C., Freiburg, S.R., Hulme, A.C., Hegarty, M.P., Barr, R.A., and Rabson, R. 1960b. Physiological investigations on the banana plant. II. Factors which affect the nitrogen compounds of the fruit. Ann. Bot. 24:117-146.

Steward, F.C., Israel, H.W., Mott, R.L., Wilson, H.J., and Krikorian, A.D. 1975. Observations on growth and morphogenesis in cultured cells of carrot (*Daucus carota* L.). Philos. Trans. R. Soc. London, Ser. B, 273:33-53.

Stotzky, G., Cox, E.A., Goos, R.D., Wornick, R.C., and Badger, A.M. 1964. Some effects of gamma irradiation on seeds and rhizomes of *Musa*. Am. J. Bot. 51:724-729.

Stover, R.H. 1972. Banana, Plantain and Abaca Diseases. Commonwealth Mycological Institute, Kew, Surrey.

Tai, E.A. 1977. Banana. In: Ecophysiology of Tropical Crops (P. de T. Alvim and T.T. Kozlowski, eds.) pp. 441-460. Academic Press, New York.

Taye, B. and Asrat, F. 1966. The production and utilization of the genus *Ensete* in Ethiopia. Econ. Bot. 20:65-70.

Tongdee, S.C. and Boon-Long, S. 1973. Proliferation of banana fruit tissues grown in vitro. Thai. J. Agric. Sci. 6:29-33.

Turner, D.W. 1968. Micropropagation of bananas. Agric. Gaz. N.S.W. 79:235-236.

Vessey, J.C. and Rivera, J.A. 1981. Meristem culture of bananas. Turrialba 31:162-163.

Waldlaw, C.W. 1972. Banana Diseases, 2d ed. Humanities Press, Atlantic Highland, New Jersey.

Westphal, E. 1975. Agricultural Systems in Ethiopia. Center for Agricultural Publishing and Documentation, Wageningen, Netherlands.

Withers, L.A. 1980. International Board for Plant Genetic Resources: IBPGR Technical Report /80/8. Tissue Culture Storage for Genetic Conservation, IBPGR Secretariat, Rome.

______ 1981. International Board for Plant Genetic Resources. IBPGR Consultant Report /81/30. Institutes Working on Tissue Culture for Genetic Conservation, IBPGR Secretariat, Rome.

CHAPTER 13
Papaya

R.E. Litz

HISTORY OF THE CROP

The papaya or pawpaw (*Carica papaya* L.) is a member of the small dicotyledonous family Caricaceae. There are only four genera within this family: *Carica, Cylicomorpha, Jacaratia,* and *Jarilla.* With the exception of *Cylicomorpha,* which is indigenous to Africa, all are native to the tropics of North and South America. There are approximately 22 species within the genus *Carica.* The papaya (2n = 18) is believed to have originated in Central America through natural hybridization between *C. peltata* Hook. and Arn. and another wild species (Purseglove, 1968). It is the only *Carica* species that is widely cultivated, although *C. goudotiana* Solms-Lauback, *C. quercifolia* Benth. and Hook., *C. monoica* Deof., *C. cauliflora* Jacq., *C. pentagona* Heilborn, and *C. erythrocarpa* Heilborn are also edible.

The papaya is a rapidly growing arborescent herb with a single straight hollow stem and a crown of palmately lobed leaves (Fig. 1). There are three primary sex types: pistillate, staminate, and hermaphroditic. Plants of indeterminate sex type also occur. Flowers are borne on cymose inflorescences that originate in the leaf axils. Laticifers occur in all organs.

Plants can be grown efficiently from seed and this has probably contributed to the current widespread distribution of papayas. Following the discovery and exploration of Central America by the Spanish, the papaya was spread rapidly by colonizers throughout the Caribbean region. It was reported in the Philippines and elsewhere in Southeast Asia by the middle of the sixteenth century (Popenoe, 1920; Storey,

Figure 1. The papaya (*Carica papaya* L.). A pistillate plant approximately 8 months old.

Figure 2. The effect of papaya ringspot virus (PRV) on a 5-month-old planting of papaya. This is the most widespread and troubling problem with large-scale papaya production in the tropics and subtropics.

1969) and in India by 1596 (Knight, 1980). Papaya-growing had been established in East Africa by the late eighteenth century. Currently papayas are grown throughout the tropics and subtropics.

ECONOMIC IMPORTANCE

The papaya plant is frost-sensitive and can only be grown between latitudes 32 degrees N and S (Table 1). The leading papaya-producing countries are Zaire, Mexico, Brazil, India, and Indonesia (FAO, 1980).

Propagation has been by seed as there have not been efficient methods for large-scale propagation of papaya by grafting or cuttings. Allan (1964) has reported a procedure for the rooting of papaya crowns. The papaya is a rapidly growing plant with a short juvenile stage of 3-4 months. Ripe fruit are ready for harvest 10-12 months after germination of the seed. Healthy trees bear fruit continuously thereafter for the lifetime of the plant, i.e., 20-25 years in the absence of disease pressure. Papaya trees can yield as much as 50,000 kg of fresh fruit per ha during the first year of production. Individual fruit can weigh 0.5-9.1 kg. Because of the extraordinary productivity of this plant and its attractiveness not only as a fruit crop but also as a source of biochemicals, it is one of the principle horticultural crops of

Table 1. World Production of Papayas in 1000 MT, 1969–1980[a]

	1969–1971	1978	1979	1980
Africa	201	225	228	229
N. America	197	420	422	426
S. America	300	569	593	614
Asia	471	609	645	631
Oceania	17	16	17	17
Total	1186	1839	1905	1917

[a]FAO Yearbook, 1980.

the tropics and subtropics. The annual worldwide production of papaya fruit is approximately equal to that of strawberries, apricots, or avocados (Knight, 1980).

Papayas are usually grown as a breakfast or dessert fruit, and closely resemble a melon in flavor and texture. The fruit may be spherical, pyriform, or cylindrical. Flesh color varies from pale yellow to red. The edible portion of the flesh consists primarily of water (86.0%) and carbohydrate (12.8%), and has been reported to be a good source of provitamin A and ascorbic acid (Chan and Tang, 1979). The sugar composition of ripe papayas is 48.3% sucrose, 29.8% glucose, and 21.9% fructose (Chan and Kwok, 1975). Extracts from plants have some medicinal value as a purgative and vermifuge. Chan and Tang (1979) reported that the alkaloid carpaine occurs in papaya and has been used as a heart depressant, amoebicide, and diuretic.

Large papaya plantations exist in Zaire and in the Lake Manyara area of northern Tanzania where production is primarily involved with the recovery of the proteolytic enzyme papain. This product is derived from sun-dried latex tapped from immature papaya fruit. It is used as a meat tenderizer, for clearing beer, for treatment of gangrenous wounds and bed sores, as a tool for detecting stomach cancer and as a cosmetic. The principal importing country is the United States.

Generally, dioecious papayas have been preferred for large-scale production and for dooryard plants in the Caribbean area. The highly inbred hermaphroditic Solo cultivars which were developed in Hawaii, are now grown extensively in Southeast Asia, Oceania, Africa, and Brazil. There are currently several Solo cultivars (Table 2), each of which has unique horticultural characteristics or is adapted to specific environmental conditions. The South African cultivars Hortus Gold and Honey Gold are dioecious papayas whose integrity has been maintained through carefully controlled pollination. In addition, dioecious papaya lines have been identified that yield fruit with high levels of papain, e.g., Coimbatore-2 and Peradeniya (Singh and Sirohi, 1977), or that are tolerant to infection by papaya ringspot virus (Conover and Litz, 1978). Rigid control over seed production of these dioecious lines is necessary; otherwise they would be lost like the former Betty cultivar of Florida.

Table 2. Some Currently Grown Papaya Cultivars

CULTIVAR	COUNTRY OF ORIGIN	SEX TYPE
Kapoho Solo	Hawaii, United States	Hermaphrodite
Masumoto Solo	Hawaii, United States	Hermaphrodite
Sunrise Solo	Hawaii, United States	Hermaphrodite
Waimanalo Solo	Hawaii, United States	Hermaphrodite
Higgins Solo	Hawaii, United States	Hermaphrodite
Wilder Solo	Hawaii, United States	Hermaphrodite
Kariya Solo	Hawaii, United States	Hermaphrodite
Blue Stem Solo	Hawaii, United States	Hermaphrodite
Coorg Honey Dew	India	Hermaphrodite
Hortus Gold	South Africa	Dioecious
Honey Gold	South Africa	Dioecious
Coimbatore-1	India	Dioecious
Coimbatore-2	India	Dioecious
Peradenija	India	Dioecious

BREEDING AND CROP IMPROVEMENT

Disease Problems

VIRUS AND VIRUS-LIKE DISEASES. Papaya ringspot virus (Purcifull, 1972) occurs in most of the papaya-producing regions of the world, e.g., North America (Conover, 1964), Africa (Kulkarni, 1970), the Caribbean (Acuna and Zayas, 1939), Hawaii (Holmes et al., 1948), and India (Capoor and Varma, 1958). There is no resistance to infection within the species, although Conover (1976) demonstrated that a dioecious papaya accession from Colombia was tolerant to papaya ringspot virus. Tolerance has been transferred to horticulturally accepted papaya types and appears to be conferred by a complex of genes (Conover and Litz, 1978). This virus is probably the greatest single threat to papaya production in the world (Fig. 2). Plantings in affected areas become almost 100% infected before they are a year old (Conover, 1964). Fruit from infected trees is marred by greasy ringspot patterns, has a low sugar content, and is often unmarketable. Conover (1976) has observed that the popular but highly inbred Solo papayas develop very severe symptoms following controlled and natural infection with papaya ringspot virus.

Papaya mosaic virus (Purcifull and Hiebert, 1971) causes a minor disease of papayas in Florida and Hawaii (Conover, 1962; Yee et al., 1970), in Venezuela (Cook and Zettler, 1970), and possibly India (Capoor and Varma, 1958) and East Africa (Kulkarni and Sheffield, 1968). There have been numerous reports of mosaic symptoms on papaya; however, the etiology of these diseases is uncertain. Singh and Misra (1969) have noted that papaya mosaic virus-infected plants yield lower quantities of papain. Conover (1964) observed that mixed infections of papaya mosaic virus with papaya ringspot virus cause a disease that is more severe than infection by either virus.

Papaya apical necrosis virus was first observed in Venezuela by Lastra and Quintero (1981). Papaya production was devastated in parts of that country because of recent and rapid spread of this severe disease. A similar disease has also been reported in Florida by Wan and Conover (1981). This disease is potentially as threatening to papaya production as papaya ringspot virus. Plantings with nearly 100% infection have been observed with nearly total loss of production.

Leaf curl virus is apparently restricted to India and the Philippines, but causes significant reductions in yield and quality wherever it has been found (Reyes et al., 1959; Sen et al., 1946). A disease caused by spotted wilt virus has been reported in the state of Hawaii on the islands of Kauai, Hawaii, and Oahu (Trujillo and Gonsalves, 1967). Because of the severity of this disease, papaya production has been abandoned in the affected islands. Existing cultivars or seed lines apparently have no tolerance to infection. Yellow crinkle virus causes considerable losses in Australia--as much as 30% reduction in yield (Jensen, 1949). Infected plants do not bear fruit. Tobacco ringspot virus, although apparently restricted to papayas in the Rio Grande Valley, can cause a very serious disease (Lambe, 1963). Infected plants die after developing terminal necrosis.

Papaya bunchy top has been a major problem in papaya-growing regions in the Caribbean (Cook, 1931; Acuna and Zayas, 1939). This disease has been associated with a MLO by Story et al. (1968). Infected papayas are often killed. Control measures such as tetracycline therapy are uneconomical in large-scale plantings.

A disease of unknown etiology referred to as "dieback" has been reported in Puerto Rico (Adsuar, 1946), Trinidad (Baker, 1936), and Australia (Simmonds, 1965), where it is probably the primary factor limiting papaya production in the state of Queensland. During severe epidemics, more than half of the papaya acerage has been affected.

FUNGAL DISEASES. Anthracnose, caused by *Colletotrichum gloeosporioides*, is a disease of papaya fruit and older petioles. It has been considered to be the most important disease of papayas in Hawaii (Yee et al., 1970). Allan (1976) has indicated that the dioecious South African cultivar Honey Gold has some resistance to anthracnose.

Black spot of papaya, caused by *Cercospora papayae*, is a serious fruit and leaf spot disease. Although little damage is sustained on fruit, infection of leaves results in considerable defoliation and a significant reduction in yield.

Phytophthora parasitica causes a serious stem and fruit blight of papaya, particularly in wet weather. In addition to causing damping-off of seedlings, stems of bearing plants may be girdled, thereby killing the top of the plant. Fruit of any age may also be affected. *Pythium* causes a similar disease.

The major disease threats to papaya production are for the most part caused by viruses or by pathogens that produce viruslike symptoms. Many papaya-producing areas have relied on the genetically uniform Solo cultivars which are highly susceptible to diseases such as ringspot. The major challenge to papaya breeders must be the development of breeding lines that possess resistance or tolerance to the most important virus or viruslike diseases.

Horticultural Problems

The papaya fruit, like many other tropical fruits, often possesses a musky flavor and odor which is perhaps the reason this and other tropical fruits are not readily accepted by people unaccustomed to them. Papayas might have greater appeal if types with better taste and odor could be developed. Fruit from virus-infected trees is of poor quality and often has an unpleasant taste. In areas with cool winter months, the quality of papaya fruit is often inferior, with low levels of sugar. Extending the geographic range of papayas by selection for cold tolerance or early maturity might improve papaya quality in many areas.

Increase in latex content or the percentage of papain in the latex could result in greater efficiency in papain production. There does not appear to be a significant difference in latex production among most cultivars, although the Indian cultivar Coimbatore-2 apparently yields higher quantities of papain. Virus infections also cause reduced yields of latex, either through a change in physiology or by reducing the life span of plantings.

Harvesting of papaya fruit from mature trees is difficult because of tree height (6-7.5 m) and fruit size (0.5-9 kg). Trees with low bearing habit and with short internodes would lower the cost of harvesting. Development of cultivars with long peduncles would make harvesting easier and would prevent the occurrence of misshapen fruit caused by overcrowding of fruit around the main stem.

Papaya fruit are very perishable, easy to bruise, and subject to anthracnose during ripening and while in transit. This limits their marketability. Development of firmer fruit would reduce this problem considerably.

With the exception of Hawaii, there has been little comprehensive research on production practices, breeding for quality, disease control, etc. Because the relatively homogeneous Solo types of papaya have often been developed for use in narrowly defined geographical limits, they do not grow well in other countries with different growing conditions.

The efficient regeneration of haploid plants from cultured anthers and recovery of homozygous diploid plants from these haploids would be of enormous benefit for papaya breeding, as this would eliminate much of the variability in crosses involving dioecious types of papayas.

REVIEW OF THE LITERATURE (Table 3)

Callus Induction

Several workers have described callus formation from different types of papaya explants including seedling petioles (DeBruijne et al., 1974),

Table 3. Primary Contributions for Developing In Vitro Systems for *Carica papaya* L.

TYPE OF RESPONSE	EXPLANT SOURCE	REFERENCE
Organogenesis	Seedling stem segment	Yie & Liaw, 1977; Arora & Singh, 1978c
Shoot tip culture	Seedling	Yie & Liaw, 1977
	Mature plant	Litz & Conover, 1978a
Somatic embryogenesis	Seedling petiole segment	DeBruijne et al., 1974
	Seedling stem segment	Yie & Liaw, 1977
	Ovule	Litz & Conover, 1981a, 1982
Embryo culture	*C. papaya* x *C. cauliflora*	Khuspe et al., 1980
Haploid induction	Anthers	Litz & Conover, 1978b

seedling stem segments (Arora and Singh, 1978a,b,c; Medora et al., 1973, 1979; Yie and Liaw, 1977), cotyledons (Litz et al., 1983), and ovules (Litz and Conover, 1981a). In addition, induction of callus from peduncle segments of a related species, *C. stipulata* Badillo, has been described (Litz and Conover, 1980). Most workers have utilized the MS basal medium although WH basal medium has also been used. According to Medora et al. (1979) the optimum growth regulator formulation for callus induction and growth from seedling stem pieces is 1.8 μM 2,4-D. They also demonstrated that ammonium nitrate appeared to be essential for callus growth, and suggested that the cytokinin BA was inhibitory. On the other hand, DeBruijne et al. (1974), Arora and Singh (1978a,b,c), and Yie and Liaw (1977) achieved good callus growth with NAA; Arora and Singh (1978a) determined that both IAA and 2,4-D inhibited callus formation. Furthermore, Arora and Singh (1978a,c) and Yie and Liaw (1977) supplemented their media with KIN (0.5–2.2 μM). DeBruijne et al. (1974) achieved best callus yields when the medium was supplemented with 10 μM 2iP. *Carica stipulata* peduncle explants also responded to 2.2 μM BA and 0.54 μM NAA by forming good callus cultures (Litz and Conover, 1980). It is apparent that optimum conditions for induction and growth of papaya callus are provided by NAA together with a cytokinin, KIN, 2iP, or BA.

Papaya callus is generally light brown or white in color and spongy in texture. Papaya explants appear to be unable to produce a typical friable callus culture. Callus is first evident 6–7 days after explanting and grows relatively vigorously. Litz et al. (1983) were able to distinguish three distinct in vitro callus induction responses on papaya cotyledons on media containing different concentrations of NAA and BA. Optimum conditions for callus induction from midrib explants and vein tissue occurred with 1.3-8.8 μM BA and 2.7-16.2 μM NAA, whereas the optimum response for callus induction from lamina explants required much higher concentrations of both NAA and BA: 6.5-27.0 μM NAA and 2.6-13.2 μM BA. A hard, dark-green callus was formed from both lamina and midrib explants when relatively low concentrations of NAA (0-1.1 μM) and moderate concentrations of BA (0.2-4.4 μM) were used. The growth regulator requirement for callus induction is clearly dependent on the source or nature of the explant (Table 4).

Somatic Embryogenesis

DeBruijne et al. (1974) described the induction of somatic embryos from papaya callus derived from cultured seedling petiole segments according to a three-stage procedure. Callus was first induced from the explants in the dark on MS medium with 1.0 μM NAA and 10.0 μM 2iP. Embryogenic competency was established under these conditions and globular embryos were regenerated upon transfer of the cultures into the light and by subculturing the callus on WH medium containing 0.01 μM BA, 0.1 μM NAA, and 29.0 mM sucrose, or on WH medium without vitamins, but containing 0.001 μM ZEA, 0.01 μM BA, and 0.58 M sucrose. Mature embryos were only obtained after transfer of globular embryos to 0.1-strength MS medium without vitamins and containing 0.01 μM BA and 0.10 μM NAA. They suggested that embryogenesis seemed to be dependent on a relatively low sucrose concentration (29.0 mM), although Litz and Conover (1981a) have demonstrated high-frequency regeneration of mature somatic embryos on WH medium containing much higher sucrose concentrations (175 mM) and 20% CW (Fig. 3).

Yie and Liaw (1977) demonstrated a two-stage procedure for induction of somatic embryogenesis in callus of seedling stem origin. Following the induction of callus on MS medium with 5.4 μM NAA and 0.5 μM KIN, somatic embryos could be regenerated from this callus on a second growth medium containing 0.3 μM IAA and 4.7-9.4 μM KIN.

Litz and Conover (1981a) described the induction of polyembryony in cultured papaya ovules (Fig. 4) and the control of somatic embryogenesis from ovular callus (Litz and Conover, 1982). Embryogenic competency was induced in ovular callus on WH medium with different growth regulator formulations: no growth regulators; 0.4, 0.9, 2.2, and 4.4 μM BA; and 20% CW. Upon subculture of callus from this medium to solid or liquid WH medium with 175 mM sucrose and 2.7 mM glutamine, a very high frequency of somatic embryogenesis resulted. This was strikingly more efficient in liquid culture. Mature embryos have been induced to germinate upon transfer to solid WH medium containing 87.6 mM sucrose, 2.7 mM glutamine, 0.5-10.8 μM NAA, and 0.2-0.9 μM BA (Litz and Conover, 1982), and in liquid medium (Fig. 5). It is not clear from the results of DeBruijne et al. (1974) or Yie and Liaw (1977) if they were able to regenerate plantlets from somatic embryos.

Table 4. Comparison of Explant Type with Exogenous Growth-Regulator Requirements for Callus Induction and Growth

EXPLANT TYPE	AUXIN	AUXIN CONC. (μM)	CYTOKININ	CYTOKININ CONC. (μM)	REFERENCE
Petiole segment (seedling)	NAA	1.0	2iP	10.0	DeBruijne et al., 1974
Stem segment (seedling)	NAA	5.4	KIN	0.47	Yie & Liaw, 1977
	2,4-D	1.8	-	-	Medora et al., 1979
	NAA	10.8	KIN	2.35	Arora & Singh, 1978c
Peduncle segment	NAA	1.0	BA	2.2	Litz & Conover, 1980
Cotyledon mid-rib	NAA	2.7-16.2	BA	1.3-8.8	Litz et al., 1983
Cotyledon lamina	NAA	6.5-27.0	BA	2.6-13.2	Litz et al., 1983

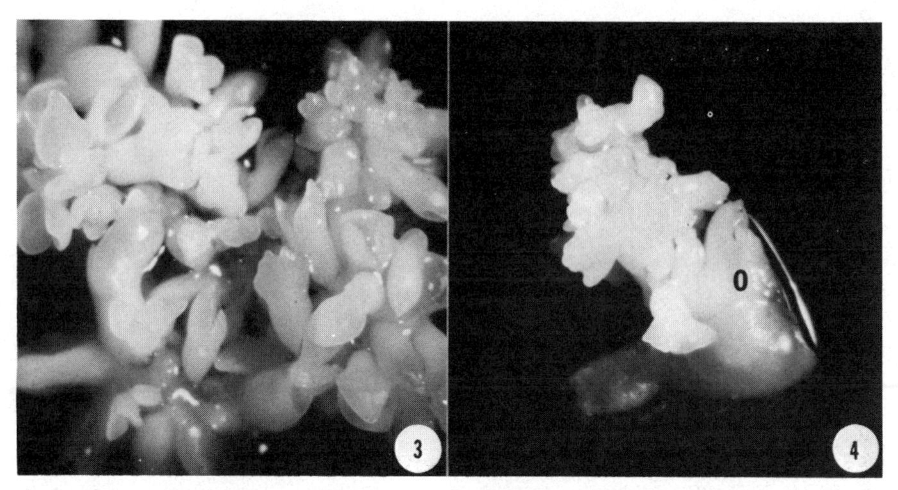

Figure 3. Proliferating culture of *Carica* somatic embryos on WH medium with 175 mM sucrose and 2.7 mM glutamine.

Figure 4. In vitro polyembryony in a papaya ovule placed into culture 35 days after pollination with *C. cauliflora*. Ovule has ruptured and released highly embryogenic callus and somatic embryos (ovule = O).

Somatic embryogenesis was observed in *C. stipulata* suspension cultures derived from peduncle explants in liquid MS medium (Litz and Conover, 1980); however this only occurred in treatments consisting of 2.2 μM BA, 0.5 μM NAA, and 10 g/l activated charcoal. This procedure has not been attempted with papaya callus cultures.

Organogenesis

Yie and Liaw (1977) not only were able to induce somatic embryos from papaya seedling stem callus, but also observed adventitious shoots from callus on the same culture medium (MS with 0.3 μM IAA and 4.7 μM KIN or with 9.4 μM KIN alone). Arora and Singh (1978c) found that it was necessary to transfer papaya callus from a callus induction medium (LS with 10.8 μM NAA, 2.3 μM KIN, 0.74 mM adenine sulfate, and other additives) to a shoot induction medium (LS with 9.2 μM KIN and 1.1 μM NAA), whereas root induction from callus was caused by 0.9 μM KIN with 2.7 μM NAA.

Litz et al. (1983) have demonstrated that adventitious roots were formed from midrib callus from papaya cotyledons on MS medium with 0.5–81.0 μM NAA and 0–2.2 μM BA. Adventitious meristems were formed from both lamina and midrib callus on culture media containing 0–1.1 μM NAA and 0.2–4.4 μM BA.

Shoot Tip Culture

Mehdi and Hogan (1976) established papaya seedling shoot tips on MS medium containing 4.7 μM KIN, and rooted them on medium with 4.7

Figure 5. Precocious germination of somatic *Carica* embryos in liquid WH medium containing 175 mM sucrose, 2.7 mM glutamine, and 5% CW.

Figure 6. Shoot tip propagation of papaya. Proliferating culture derived from mature, field-grown papaya. Medium consists of MS with 87.6 mM sucrose, 2.2 μM BA, and 0.54 μM NAA.

Figure 7. Plant regeneration from papaya shoot tip culture. MS medium with 87.6 mM sucrose and 10 μM NAA.

Figure 8. Protoplasts isolated from papaya cotyledons in modified MS medium (see text) in 0.3 M mannitol. Protoplasts shown here two days after isolation.

μM KIN and 24.5 μM IBA. They did not observe multiple shoot formation. Yie and Liaw (1977) established papaya seedling shoot tips in vitro and obtained proliferative growth on MS medium containing 0.3 μM IAA and either 23 μM KIN or 2.2–4.4 μM BA.

Because of the inherent variability in commercially important dioecious papaya types, Litz and Conover (1977, 1978a) developed a procedure for establishment and culture of excised shoot tips from field-grown, mature papayas. Shoot tips (2-3 mm) were cultured on establishment medium consisting of MS basal medium with 87.6 mM sucrose, 47 μM KIN, and 10.8 μM NAA. Enlarged explants were subcultured after 2-3 months onto shoot proliferation medium: MS medium with 2.2 μM BA and 0.54 μM NAA (Fig. 6). Establishment time and rate of proliferation were both dependent on age of stock plants, time of year, sex types, and the presence of bacterial contaminants and/or a rhabdovirus of unknown etiology (Litz and Conover, 1981b). Staminate plants responded more rapidly than pistillate plants and had a greater proliferation. Explants responded best when taken from stock plants during periods of rapid growth. The rate of increase during proliferation varied among clones, but generally was seven- to eightfold between subcultures. The ability to maintain proliferating cultures was lost after 8-13 subcultures. Apical dominance and rooting could be induced by subculture of lateral buds on MS medium with 5.4-16 μM NAA (Fig. 7). Plantlets have been successfully established in the field.

Embryo Culture

One of the primary goals of papaya improvement schemes has been to hybridize the papaya with the sexually incompatible *Carica* species, particularly *C. cauliflora* Jacq., which is resistant to infection by papaya ringspot virus. Incompatibility between papaya and *C. cauliflora* is caused by failure of endosperm formation. Phadnis et al. (1970) successfully cultured mature papaya embryos on WH medium containing 0.24 μM KIN alone or 0.24 μM KIN, 0.3 μM GA, and 0.6 μM IAA. DeBruijne et al. (1974) evidently were also successful in culturing mature papaya embryos on MS medium without meso-inositol, casein hydrolysate, or vitamins, containing instead 58.4 mM sucrose. Khuspe et al. (1980) demonstrated the practicality of using embryo culture to rescue immature embryos that resulted from *C. papaya* x *C. cauliflora* crosses. Solid WH medium without vitamins but supplemented with 58.4 mM sucrose, 2.0 μM cupric chloride, and 1.2 μM sodium molybdate was used to germinate excised immature hybrid embryos in the dark. Subsequently, the young plantlets were subcultured in liquid WH medium with vitamins and glycine, where they were able to grow to transplantable size.

Anther Culture

The use of anther culture as a means for recovery of haploid papaya plants was reported by Litz and Conover (1978a). Cymose staminate inflorescences were removed from screenhouse-grown plants and the flower buds were segregated according to size. Tetrad formation was generally observed in anthers from 7- to 8-mm flower buds; the first mitotic division occurred in 16- to 20-mm flower buds. Anthers were removed from flower buds reflecting every stage of development and were placed into MS culture medium. Optimum results were obtained by culturing anthers from 16- to 20-mm flower buds in liquid MS medium supplemented with 87.6 mM sucrose, 8.8 μM BA, 2.7 μM NAA, and 10 g/l activated charcoal following a pretreatment of 4 C for 3-4 days. Only a small number of haploid plantlets have been recovered using this technique, i.e., 1 plantlet for each 1000 cultured anthers. Chromosome counts have verified that the regenerated plants have the haploid number ($n = 9$).

Protoplast Isolation and Culture

Although Litz and Conover (1979) have reported the large-scale isolation of papaya protoplasts (Fig. 8), this procedure has been substantially modified as the optimum preconditioning and isolation steps have been improved. For a more detailed description of this procedure, please refer to the section on Protoplast Isolation and Culture in the following protocols. Limited callus induction, but with a low plating efficiency, has been achieved (Litz, Hendrix, and Conover, unpublished data).

PROTOCOLS

Organogenesis (Yie and Liaw, 1977; Arora and Singh, 1978c)

1. Germinate and grow 6-12 papaya seedlings in peat:vermiculite (1:1). Use 0.5 to 1.0 cm petiole or stem segments for explants. Surface-sterilize segments in 5% Clorox (v/v) in distilled water with 1-2 drops of Tween 20 for approximately 10 min. Wash thoroughly with 3 rinses of sterile distilled water. Culture in sterile, nutrient broth with 3-5% sucrose for 2-3 days. Discard contaminated cultures.
2. Transfer explants to solid WH, LS, or MS medium supplemented accordingly: 5-6 μM NAA and 0.4-0.5 μM KIN (Yie and Liaw, 1977), or 10-11 μM NAA, 2.0-3.0 μM KIN, and 3.0 μM GA with 13.3 μM glycine, 1.0 g/l casein hydrolysate, 500 mg/l malt extract, 740 μM adenine sulfate, 0.04 mM nicotinic acid, 3.0 μM thiamine HCl, and 4.9 μM pyridoxine HCl (Arora and Singh, 1978c). All media should include 0.087 M sucrose and 8 g/l Difco Bacto agar. Adjust pH to 5.7 with 0.1 N HCl or NaOH and autoclave for 15 min at 1.1 kg/cm^2 and 120 C. Maintain cultures in a growth chamber at 25 C with a photoperiod of 16 hr (1000 lux).
3. Subculture callus regularly at 2-4 week intervals onto the same medium.
4. To induce shoot differentiation, transfer callus to sterile medium supplemented accordingly: 0.3 μM IAA with 5-10 μM KIN (Yie and Liaw, 1977) or 1.0 μM NAA with 10 μM KIN (Arora and Singh, 1978c).
5. Re-establish apical dominance by transfer of adventitious buds to cytokinin-free medium. Rooting can be induced simultaneously on media with 5-15 μM NAA or IBA.

6. Transfer rooted plantlets to soil-free potting mixture and establish under intermittent mist in a protected area for approximately 2 weeks.

Shoot Tip Culture (Litz and Conover, 1978b)

1. Remove shoot tips (2-3 cm) from mature, field-grown papaya plants. (Papaya plants generally do not branch; however, they can be made to do so by decapitation, and at least 6 side shoots can thereby be induced.) Remove large petioles and inflorescences until the shoot tip is 3-5 mm.
2. Surface-sterilize shoot tip in 5-10% Clorox (v/v) in distilled water with 1-2 drops of Tween 20 for approximately 10 min. Wash thoroughly with 3 rinses of sterile distilled water. Culture in sterile, nutrient broth with 87.6-146 mM sucrose for 2-3 days. Discard contaminated cultures.
3. Transfer clean shoot tips to solid MS, WH, or LS medium supplemented accordingly: 50 μM KIN with 11 μM NAA, 87.6 mM sucrose and 8 g/l Difco Bacto agar. (Adjust pH of medium to 5.7 with 0.1 N HCl or NaOH prior to autoclaving for 15 min at 1.1 kg/cm^2 and 120 C.)
4. Maintain cultures in a growth chamber at 25 C with a 16-hr photoperiod (1000 lux).
5. After 1-1/2 to 2-1/2 months on this medium, shoot tips become enlarged; however, apical dominance is suppressed. Transfer the enlarged primary explant onto proliferation medium, i.e., MS medium containing 2.2. μM BA with 0.5 μM NAA, 87.6 mM sucrose, and 8 g/l agar.
6. Induction of lateral bud growth occurs after 2 weeks on proliferation medium (Fig. 6). Subculture by excision of lateral buds and transfer to fresh proliferation medium following 3-week intervals.
7. Re-establish apical dominance and plantlet formation by subculture of lateral buds onto MS medium with 5-15 μM NAA or IBA. Rooting occurs on this medium (Fig. 7).
8. Remove plantlets from culture vessels. Remove agar from roots and transfer to soil-free potting mixture under intermittent mist for 2 weeks in a protected area.

Efficient Somatic Embryogenesis from Papaya Ovular Callus (Litz and Conover, 1982)

1. It has not been possible to induce polyembryony in unfertilized papaya ovules, and low-frequency polyembryony occurs in ovules following papaya x papaya crosses. For best results, pollinate pistillate papayas having cylindrical fruit with *C. cauliflora* pollen. Bag the fruit and leave it for 30-120 days before harvesting.
2. Surface-sterilize harvested fruit by immersion in 20% (v/v) Clorox for 20 min. Rinse thoroughly with sterile distilled water.

3. Bisect fruit and remove the ovules. Place into sterile solid culture medium in 100 x 15 mm petri dishes. Culture medium should consist of half-strength MS major salts and Fe•EDTA, MS minor salts and organics, 2.7 mM glutamine, 175 mM sucrose, 20% (v/v) filter-sterilized CW, and 8 g/l Difco Bacto agar. (Adjust pH of medium to 5.7 with 0.1 N HCl or NaOH prior to autoclaving for 15 min at 1.1 kg/cm and 120 C.)
4. Maintain cultures in a growth chamber at 25 C with a 16-hr photoperiod (1000 lux) for 1.5 months for youngest ovules to 3 months for the oldest ovules.
5. Youngest ovules generally rupture and release somatic embryos and callus 1.5 months after culturing (Fig. 3). It is necessary to dissect the older ovules.
6. Remove callus from ovules. Callus is hard and honey-colored. Subculture callus on fresh sterile medium as described above. Limited somatic embryogenesis will occur on this medium.
7. Subculture callus after 1-2 months on solid or in liquid medium but without 20% CW. Maintain liquid cultures at 100 rpm on a rotary shaker. Somatic embryogenesis occurs. It is possible to recover fully mature embryos from this medium.

Protoplast Isolation and Culture

1. Papaya seeds are germinated in a soil-free potting mixture in a growth chamber with a 16-hr photoperiod at 25 C. Trays are watered and fertilized regularly.
2. Cotyledons are harvested 4-8 days after germination and are floated in a solution of 1.0 mM NH_4NO_3 and 1.0 mM $CaCl_2 \cdot 2H_2O$ with 22 μM BA and 5.4 μM NAA in the dark for 1.5-2.5 days.
3. Following sterilization for 5 min in 5% (v/v) Clorox with 1-2 drops Tween 20, the midribs are discarded and 1-mm lamina strips are soaked in 0.25-strength MS with 3.0 mM $CaCl_2 \cdot 2H_2O$, 0.3 M mannitol, and 500 mg/l casein for approximately 1 hr under subdued light.
4. The soak solution is replaced by filter-sterilized enzyme mixture consisting of: 0.25 strength MS with 3 mM $CaCl_2 \cdot 2H_2O$, 500 mg/l casein, 200 mg/l Garramycin, 22.5 mM tetracycline, 1.1 mM ampicillin, 0.3 M mannitol, 0.5% macerozyme, and 1.5% cellulase (both enzymes from Kinki Yakult. Mfg. Co.). Leaf tissue is vacuum-infiltrated with the enzyme mixture and incubated in the dark on a rotary shaker for 4 hr at 40 rpm.
5. Protoplasts are then isolated by passage through 63 μM filtration fabric and centrifugation at 70-80 x g for 5 min. Supernatant is removed and replaced with an identical solution used for soaking. This is repeated 2-3 times.
6. Protoplasts (Fig. 8) are resuspended in culture medium—0.25-strength MS major salts less NH_4NO_3, 3 mM $CaCl_2 \cdot 2H_2O$, 500 mg/l casein, 0.3 M mannitol, MS minor salts, 0.25-strength MS Fe•EDTA, MS organics, 2.7 mM glutamine, 10% CW, 58.4 mM sucrose, and 5

mM NH_4NO_3—at 1-2 x 10^5 protoplasts/ml, and plated in thin liquid layers in petri dishes at densities of 5 x 10^4-1 x 10^5 protoplasts/ml.

FUTURE PROSPECTS

Some of the wild *Carica* species have useful characters—e.g., disease resistance, cold tolerance, small fruit size, and dwarf growth habit—that would be desirable in papaya. The nature of incompatability between many of these species and papaya has not been demonstrated; however, through the application of embryo or ovule culture, in vitro fertilization, and even controlled protoplast fusion, it should be possible to overcome at least some of these incompatability barriers.

Suspension cultures from which somatic embryos or adventitious plantlets can be regenerated efficiently would be of utmost importance for developing in vitro selection systems for papaya. Papaya plants are extremely susceptible to injury by pesticides and herbicides, and have little defense against a number of fungal pathogens, e.g., *Phytophthora* sp., *Pythium* sp., etc. In vitro selection for resistance to chemical injury and for toxin resistance would substantially reduce production losses. The currently applied control measures for many fungal diseases involve the abandoning of severely affected fields or the elimination altogether of papaya production in certain areas.

The manipulation of papaya cell cultures for efficient recovery of biochemicals such as carpain or papain has hardly been examined. Medora et al. (1973) have reported that proteolytic enzyme activity occurs in papaya callus cultures derived from seedling stems, and have characterized the substrate specificity of some of these enzymes (Mell et al., 1979). Proteolytic activity was not observed in papaya fruit callus (Krikorian and Steward, 1965); however, there is very little papain in ripe papaya fruit. Papain is a sulfhydryl enzyme and the sulfhydryl group is necessary for enzyme activity. In most papain preparations, only a small proportion of the enzyme is active due to the blockage of the sulfhydryl group. Blockage can be overcome by the addition of cysteine to enzyme preparations (Drenth et al., 1970). It may be possible to recover papain from cell cultures through incorporation in the medium of an appropriate precursor to papain. Alternatively, by unblocking sulfhydryl groups by the addition of cysteine, it may be possible to activate existing proteolytic enzymes in vitro.

Because of the inbred nature of hermaphroditic Solo papaya types, these cultivars are apparently more susceptible to environmental stress and disease pressure than dioecious papaya cultivars. Somatic variability in plants derived clonally from tissue culture may be useful for overcoming crop-threatening situations involving papaya cultivars. Some interesting variability has been observed in plantlets derived from *Carica* somatic embryos, including a high proportion of plants with 3 or 4 cotyledons, differences in petiole pigmentation, and alteration of internode length (Litz and Conover, unpublished data). It has also been observed that a low proportion of hermaphroditic plants derived from cultured shoot tips will revert to the female or pistillate sex type (personal communication from Grace Mee, Hawaiian Sugar Planter's Assoc.). Papaya plants, whether derived from leaf protoplasts by organogenesis, from proliferating shoot tip cultures, or from somatic embryos, may possess sufficient useful variability to be utilized in a papaya improvement program.

There is an urgent need for a procedure that would efficiently yield haploid plants from cultured anthers. Haploid plants that survive disease and environmental stress under field condtions would be useful for producing homozygous diploid stock by conventional chromosome doubling. Homozygous, staminate papaya plants used in the production of hybrid cultivars would eliminate much of the variability among dioecious papaya types.

The development of a complete protocol for regeneration of plants from papaya protoplasts will be of interest to plant breeders as genetic engineering techniques, such as the use of recombinant DNA or organelle uptake, become directly applicable to papaya improvement.

The use of meristem cultures for producing virus- or disease-indexed papaya at this time is of little interest, because of the widespread distribution of alternate hosts and of vectors for most papaya viruses. Shortly after (virus-free) seedlings are transferred to the field they become infected with PRV. Until the spread or papaya viruses can be effectively controlled under field conditions, virus-indexed stock will have little importance.

ACKNOWLEDGMENTS

I am grateful for the technical assistance provided by Bunny Hendrix and Sara Walker, and for helpful advice and gentle criticism from my good friend and former colleague Robert A. Conover (deceased). Much of this work has been supported by the Rockefeller Foundation and by USDA Cooperative Agreement No. 58-7B30-9-116.

Florida Agricultural Experiment Stations Journal Series No. 3649.

KEY REFERENCES

Arora, I.K. and Singh, R.N. 1978a. Growth hormones and in vitro callus formation of papaya. Sci. Hortic. 8:357-361.

DeBruijne, E., DeLanghe, E., and van Rijck, R. 1974. Action of hormones and embryoid formation in callus cultures of *Carica papaya*. Int. Symp. Fytofarm. Fytiat. 26:637-645.

Litz, R.E. and Conover, R.A. 1978a. In vitro propagation of papaya. HortScience 13:241-242.

______ 1981a. In vitro polyembryony in *Carica papaya* L. ovules. Z. Pflanzenphysiol. 104:285-288.

Yie, S. and Liaw, S.I. 1977. Plant regeneration from shoot tips and callus of papaya. In Vitro 13:564-567.

REFERENCES

Acuna, J. and Zayas, F. 1939. Fruta bomba o papaya. Rev. Agric. (Cuba) 23:40-43.

Adsuar, J. 1946. Studies on virus diseases of papaya (*Carica papaya*) in Puerto Rico. Tech. Paper Nos. 1-4. Univ. of Puerto Rico Agric. Exp. Sta., Rio Piedras, Puerto Rico.

Allan, P. 1964. Pawpaws grown from cuttings. Farming S.A. 101:1-6.

______ 1976. Out-of-season production of pawpaws (*Carica papaya* L.) in cool subtropical areas. Acta Hortic. 57:97-103.

Arora, I.K. and Singh, R.N. 1978b. Callus initiation in the propagation of papaya (*Carica papaya* L.) in vitro. J. Hortic. Sci. 53:151.

______ 1978c. In vitro plant regeneration in papaya. Curr. Sci. 47: 867-868.

Baker, R.E.D. 1939. Pawpaw mosaic disease. Trop. Agric. (Trinidad) 16:159-163.

Capoor, S.P. and Varma, P.M. 1958. A mosaic disease of papaya in Bombay. Indian J. Agric. Sci. 28:225-233.

Chan, H.T. and Kwok, S.C.M. 1975. Importance of enzyme inactivation prior to extraction of sugars from papaya. J. Food Sci. 40:770-771.

Chan, H.T. and Tang, C.S. 1979. The chemistry and biochemistry of papaya. In: Tropical Foods (G.E. Inglett and G. Charolambous, eds.) Vol. I, pp. 33-53. Academic Press, New York.

Conover, R.A. 1962. Virus diseases of the papaya in Florida. Phytopathology 52:6.

______ 1964. Mild mosaic and faint mottle ringspot, two papaya virus diseases of minor importance to Florida. Fla. State Hort. Soc. 77: 444-448.

______ 1976. A program for development of papaya tolerant to the distortion ringspot virus. Proc. Fla. State Hort. Soc. 89:229-231.

______ and Litz, R.E. 1978. Progress in breeding papayas with tolerance to papaya ringspot virus. Proc. Fla. State Hort. Soc. 91:182-184.

Cook, A.A. and Zettler, F.W. 1970. Susceptibility of papaya cultivars to papaya ringspot and papaya mosaic virus. Plant Dis. Rep. 54:893-895.

Cook, M.T. 1931. New virus diseases of plants in Puerto Rico. J. Puerto Rico Dept. Agric. 15:193-195.

Drenth, J., Jansonius, J.N., Koekock, R., Sluyterman, L.A.A., and Wolthers, G. 1970. IV. Cysteine Proteinases. The structure of the papain molecule. Philos. Trans. R. Soc. London Ser. B 257:231-236.

Food and Agricultural Organization of the U.N. 1980. 1979 FAO Production Yearbook. FAO, U.N., Rome.

Holmes, F.O., Hendrix, J.W., Ireka, W., Jensen, D.D., Linder, R.C., and Storey, W.B. 1948. Ringspot of papaya (*Carica papaya*) in the Hawaii islands. Phytopathology 38:310-312.

Jensen, D.D. 1949. Papaya ringspot virus and its insect vector relationship. Phytopathology 39:212-220.

Khuspe, S.S., Hendre, R.R., Mascarenhas, A.F., and Jagannathan, V. 1980. Utilization of tissue culture to isolate interspecific hybrids in *Carica* L. In: Plant Tissue Culture, Genetic Manipulation, and Somatic Hybridization of Plant Cells (P.S. Rao, M.R. Heble, and M.S. Chadha, eds.) pp. 198-205. BARC, Bombay, India.

Knight, R.J. 1980. Origin and world importance of tropical and subtropical fruit crops. In: Tropical and Subtropical Fruits: Composition, Properties, and Uses (S. Nagy and P.E. Shaw, eds.) pp. 1-120. AVI Publishing, Westport, Connecticut.

Krikorian, A.D. and Steward, F.C. 1965. Biochemical differentiation: The biosynthetic potentialities of growing and quiescent tissue. In: Plant Physiology (F.C. Steward, ed.) Vol. 5B, pp. 227-326. Academic Press, New York.

Kulkarni, H.Y. 1970. Decline viruses of pawpaw (*Carica papaya* L.) in East Africa. Ann. Appl. Biol. 66:1-9.

______ and Sheffield, F.M.I. 1968. Interim report on virus diseases of pawpaw in East Africa. E. Afr. Agric. For. J. 33:323-324.

Lambe, R.C. 1963. Terminal necrosis and wilt of papayas. J. Rio Grande Val. Hortic. Soc. 17:128-129.

Lastra, R. and Quintero, E. 1981. Papaya apical necrosis, a new disease associated with a rhabdovirus. Plant Dis. 65:439-440.

Litz, R.E. and Conover, R.A. 1977. Tissue culture propagation of papaya. Proc. Fla. State Hort. Soc. 90:245-246.

______ 1978b. Recent advances in papaya tissue culture. Proc. Fla. State Hort. Soc. 91:182-184.

______ 1979. Development of systems for obtaining parasexual *Carica* hybrids. Proc. Fla. State Hort. Soc. 92:180-182.

______ 1980. Somatic embryogenesis in cell cultures of *Carica stipulata*. HortScience 15:733-735.

______ 1981b. Effect of sex type, season, and other factors on in vitro establishment and culture of *Carica papaya* L. explants. J. Am. Soc. Hortic. Sci. 106:792-794.

______ 1982. In vitro somatic embryogenesis and plant regeneration from *Carica papaya* L. ovular callus. Plant Sci. Lett. 26:153-158.

Litz, R.E., O'Hair, S.K., and Conover, R.A. 1983. In vitro growth of *Carica papaya* L. cotyledons. Sci. Hortic. 19:287-293.

Medora, R.S., Campbell, J.M., and Mell, G.P. 1973. Proteolytic enzymes in papaya tissue cultures. Lloydia 36:214-216.

Medora, R.S., Bilderback, D.E., and Mell, G.P. 1979. Effect of media on growth of papaya callus cultures. Z. Pflanzenphysiol. 91:79-82.

Mehdi, A.A. and Hogan, L. 1976. Tissue culture of *Carica papaya*. HortScience 11:311 (abstr.).

Mell, G.P., Medora, R.S., and Bilderback, D.E. 1979. Substrate specificity of enzymes from papaya callus cultures. Z. Pflanzenphysiol. 91: 279-282.

Phadnis, N.A., Budrukkar, N.D., and Kaulgud, S.N. 1970. Embryo culture technique in papaya (*Carica papaya* L.). Poona Agric. Coll. Mag. 60:101-104.

Popenoe, W. 1920. The papaya and its relatives. In: Manual of Tropical and Subtropical Fruits, pp. 225-249. Hafner Press, New York.

Purcifull, D.E. 1972. Papaya ringspot virus. In: Descriptions of Plant Viruses (A.J. Gibbs, B.D. Harrison, and A.F. Murant, eds.) No. 84. Commonwealth Mycological Institute and Association of Applied Biologists, Kew, England.

______ and Hiebert, E. 1971. Papaya mosaic virus. In: Descriptions of Plant Viruses (A.J. Gibbs, B.D. Harrison, and A.F. Murant, eds.) No. 56. Commonwealth Mycological Institute and Association of Applied Biologists, Kew, England.

Purseglove, J.W. 1968. In: Tropical Crops. Vol. I: Dicotyledons. John Wiley and Sons, New York.

Reyes, G.M., Martinez, A.L., and Chinte, P.T. 1959. Three virus diseases of plants new to the Philippines. FAO Plant Prot. Bull. 7:141-143.

Sen, P.K., Ganguli, B.D., and Malik, P.C. 1946. A note on a leaf curl disease of papaya (*Carica papaya* L.). Indian J. Hortic. 3:38-40.

Simmonds, J.H. 1965. Pawpaw diseases. Queensland Agric. J. 76:666-677.

Singh, B.P. and Misra, A.K. 1969. Studies on papain from virus infected and noninfected fruits of *Carica papaya.* Indian Phytopathol. 22:221-224.

Singh, I.D. and Sirohi, S.C. 1977. Breeding new varieties of papaya in India. In: Fruit Breeding in India: Symposium on Fruit Crop Improvement (G.S. Nijjar, ed.) pp. 186-196. Oxford and IBH Pub., New Delhi, India.

Storey, W.B. 1969. Papaya (*Carica papaya* L.). In: Outlines of Perennial Crop Breeding in the Tropics (F.P. Ferwerde and F. Wit, eds.) pp. 389-407. H. Veenman en Zonen B.V., Wageningen.

Story, E., Halliwell, R.S., and Smith, L.R. 1968. Investigacion de los virus de papaya (*Carica papaya* L.) en la Republica Dominica, con apuntes especiales sobre la asociacion de un organismo del tipo mycoplasma con la enfermedad "Bunchy Top" Boletin 14. Instituto Superior de Agricultura, Division de Investigaciones Agricolas, Santiago de las caballeros, Dominican Republic.

Trujillo, E.E. and Gonsalves, P. 1967. Tomato spotted wilt in papaya. Phytopathology 57:9.

Wan, S.H. and Conover, R.A. 1981. A rhabdovirus associated with a new disease of Florida papayas. Proc. Fla. State Hort. Soc. 94:318-321.

Yee, W., Akamine, E.K., Aoki, G.M., Haramoto, F.H., Hine, R.B., Holtzmann, O.V., Hamilton, R.A., Ishida, J.T., Keeler, J.T., and Nakasone, H. 1970. Papayas in Hawaii. University of Hawaii Cooperative Extension Service Circular 436.

SECTION VII
Temperate Fruits

CHAPTER 14
Apple

R.H. Zimmerman

HISTORY OF THE CROP

The apple is probably the most widely grown fruit in the world, being cultivated throughout the temperate zones of both the Northern and Southern hemispheres. Its history, origin, and early development have been reviewed recently by Brown (1975). The cultivated apple is the result of intercrossing among various species of *Malus*. As a result, some disagreement exists on the appropriate botanical name for it with *M. pumila* Mill., *M. sylvestris* Mill., and *M. domestica* Borkh. being most commonly used.

The native habitat of the species ancestral to the modern apple is southeastern Europe and southwestern Asia. Selections from the wild have been made for several thousand years, since apples were cultivated by the Greeks and Romans. Named cultivars are known from the thirteenth century and clonal propagation by budding or grafting goes back at least 2000 years (Brown, 1975). The cultivated apple was introduced to North America by the early settlers and was often grown from seed. As desirable types adapted to new conditions were selected, they were vegetatively propagated.

ECONOMIC IMPORTANCE

Apple-growing regions are located throughout the temperate zones of the world. Factors which restrict apple-growing in these regions include soil type, depth and fertility, amount and timeliness of precipitation, amount of insolation, minimum winter temperatures, length of the growing season, and average day and night temperatures during the growing season. A fuller discussion of these factors is provided by Childers (1976).

The leading countries in apple production are the U.S.S.R, the United States, China, France, Italy, and the Federal Republic of Germany (Table 1), with world production totaling at least 35 million metric tons. Production in the United States has been increasing (Table 2) and similar increases are occurring in some other countries. Apples are produced for fresh market, for processing, and, in some countries, for cider. The leading U.S. cultivars, Delicious (including the many red strains) and Golden Delicious (Table 2), are also the leading cultivars worldwide. Other cultivars are popular in certain countries or regions in which they are well adapted, e.g., Cox's Orange Pippin in the United Kingdom, but are not grown elsewhere because they do not adapt well to different environmental and/or cultural conditions. Information on cultivar characteristics and countries in which they are grown has been published (Childers, 1976).

BREEDING AND CROP IMPROVEMENT

The chief objective of modern apple breeding programs is to produce high-quality fruit at the lowest possible cost (Brown, 1975). Because of modern marketing methods, this objective means emphasizing fruit with multiple uses having flavor and appearance that will appeal to the largest number of consumers. High yields are essential and to achieve these, trees must have a high photosynthetic efficiency, a strong branching structure to bear the weight of the fruit, and must produce annual fruit crops. Precocious flowering is desirable since it provides an early return on investment to the grower.

Resistance to diseases and insect pests is important since both reduce yield. Furthermore, disease- and pest-control measures are expensive and use of only the absolute minimum amount of pesticides is desirable. The program for breeding resistance to apple scab (incited by *Venturia inaequalis* (Cke) Wint.) is a model on which other large-scale breeding programs can be patterned (Brown, 1975). Several cultivars from this program have now been introduced and are being evaluated for commercial potential.

The trend in apple production has been toward higher density plantings which give greater yield per unit area. This trend was stimulated first by the availability of clonal, size-controlling rootstocks which permitted denser plantings. More recently, as slower growing, more compact, spur-type strains of important cultivars became available, the use of seedlings as rootstocks for these spur-types has increased since they provide a strong root system and can still be used in high-density planting. The increased demand for trees for these plantings, combined with the current general inflation in the economy, has led to sharp increases in the nursery price of trees. As a result, renewed interest has been directed toward producing trees with shoots and roots of the same genotype, both to meet increased demand and to propagate trees at lower cost. Tissue culture propagation stands out as the logical

Table 1. Leading Apple-Producing Countries of the World, 3-Year Average Annual Production, 1978-1980[a]

COUNTRY	METRIC TONS (thousands)	COUNTRY	METRIC TONS (thousands)
Argentina	913	Japan	885
Australia	298	Korea (DPR)	443
Austria	282	Korea (Rep)	427
Belgium-Luxemburg	294	Mexico	256
Bulgaria	299	Netherlands	477
Canada	462	Poland	894
China (People's Rep.)	2841	Romania	473
France	2964	South Africa	364
Germany (Dem. Rep.)	563	Spain	1048
Germany (Fed. Rep.)	1871	Switzerland	380
Greece	246	Turkey	1183
Hungary	848	United Kingdom	360
India	763	United States	3696
Iran	453	USSR	6892
Italy	1936	Yugoslavia	421

[a]1980 FAO Production Yearbook, Vol. 34, FAO Statistics Series No. 34, Food and Agriculture Organization of the United Nations, Rome. 1981.

Table 2. Leading Apple Cultivars and Current Apple Production in the United States

	METRIC TONS (thousands)	
CULTIVAR	1980	1981
Delicious	1589	1476
Golden Delicious	673	685
McIntosh	360	258
Rome Beauty	290	252
Jonathan	198	176
York Imperial	159	143
All others	732	694
Total	4013	3672

method to use for own-rooted trees because it has the potential for large-scale increases in plants and because conventional cuttings of apple cultivars are so difficult to root.

REVIEW OF THE LITERATURE

Development of Micropropagation Methods

In a continuation of studies of cytokinins in apple xylem sap, Jones (1967) cultured shoot tips from the rootstocks M.26 and M.7 in vitro and found that BA stimulated growth of the tips; he reported no effect on shoot proliferation. Pieniazek (1968) extended these results to shoot tips taken from 5-week-old apple seedlings and to dormant buds from older seedlings. She also found that BA stimulated growth of lateral and collateral buds on the shoot tips from young seedlings. Later, Elliot (1972) reported that shoot apices taken from adult phase trees of Granny Smith in the field grew into leafy shoots in vitro when supplied with BA or ZEA but not when supplied with KIN. No shoot proliferation was noted. No rooting occured on shoot tips in the presence of cytokinin in any of these studies. However, both Jones (1967) and Pieniazek (1968) reported that shoot tips on control medium sometimes rooted and these root tips also grew, although not as well as those supplied with BA. Shoots were also produced in vitro from dormant axillary buds of apple seedlings, with multiple shoot formation sometimes occurring when BA was included in the medium (Dutcher and Powell, 1972). Subsequently, apple plantlets were regenerated from meristem-tips dissected from seedlings (Walkey, 1972) and M.26 rootstock (Quorin, 1974) without any proliferation of shoots prior to rooting. The addition of polyvinylpyrrolidone (PVP) in liquid medium was nessary for the first 4-8 weeks of culture to eliminate the inhibitory effect on growth of the meristem-tips by oxidized phenolic compounds (Walkey, 1972). However, more recent research by other workers indicates that PVP is not necessary for successful initiation of cultures.

Proliferation of apple shoots in vitro, followed by rooting of these shoots, was accomplished by Abbott and Whiteley (1976). Starting from meristem-tips taken from seedling- and adult-phase greenhouse-grown trees of Cox's Orange Pippin, they used the Murashige and Skoog revised medium (MS) (1962) with simplified organic constituents, no auxin, and 0.5–4.6 μM KIN. Shoot masses formed from which individual shoots could be recultured repeatedly. Adult-phase explants proliferated only through growth of axillary shoots but seedling explants produced both axillary shoots and adventitious shoots from basal callus on the explants. Levels of KIN below 0.5 μM and above 23.0 μM did not induce shoot proliferation. Rooting of shoots from both seedling- and adult-phase sources was obtained on media containing IBA but the results were not consistent. Most subsequent work with apples has used BA as the cytokinin. Soon after the above report was published, Jones (1976) reported the striking effect of phloridzin (PZ) and phloroglucinol (PG) on proliferation of M.7 and M.26 rootstock shoots. Shoot proliferation occurred on a modified MS meduim containing BA and IAA; addition of PZ or PG increased shoot number, length, and weight, whereas elimination of IAA stopped proliferation. Shoots from these cultures rooted readily when transferred to cytokinin-free medium.

Further rooting studies (Jones and Hatfield, 1976) showed that PG and phloretic acid, in combination with auxin, more than doubled the number of M.7 shoots that rooted, whereas pyrogallol, caffeic acid, and catechol had no effect. The most effective auxin treatment was 5 µM IBA. The above results led to the description by Jones et al. (1977) of the first complete system for apple micropropagation using M.26 rootstock; this same procedure also worked for M.27 rootstock.

Subsequent research extended the technique to numerous other rootstock and scion cultivars (Table 3). Huth (1978) found that 8.9 µM BA and 1.1 µM NAA was best for the establishment of very small (0.2 mm) Jonathan meristem-tips but that the NAA had to be reduced to 0.3 µM for optimal shoot growth. Up to 70% rooting was achieved using 5.4 or 54.0 µM auxin with improved results when 0.3 µM GA_3 was added to the medium. No rooting occurred at auxin concentrations below 5.4 µM. However, better than 90% rooting of Jonathan has been achieved consistently with 0.5–1.5 µM of IBA or 0.5–1.6 µM NAA (Zimmerman, unpublished results). Difficulties in acclimatizing the rooted cuttings led Huth (1978) to develop a successful technique for grafting the tissue-culture-produced shoots onto apple seedlings in the greenhouse. Lane (1978) established the cultivar Macspur in culture from meristem-tips taken from buds on actively growing shoots as well as from dormant buds. Both NAA and GA_3 in the culture medium reduced shoot proliferation as did BA concentrations above or below 5 µM. Root development was enhanced when the shoots were transferred to a medium without growth regulators after an initial period on a medium with NAA. Temperatures below 28 C also reduced rooting (Lane, 1978). Better rooting of Spartan, Ozark Gold, and MM 106 was obtained by using a liquid half-strength MS medium with a perlite-vermiculite (1:1) support than by using an agar medium (Zimmerman, 1978). This technique produced roots as well without IBA as with it.

Phloroglucinol was not tested by Huth (1978) or Lane (1978) and was not effective in Zimmerman's (1978) tests. Nevertheless, PG was found to stimulate rooting of five additional apple cultivars (Jones et al., 1979). Whether in this case the PG acted directly on shoot proliferation and rooting or whether it acted indirectly by controlling bacterial contaminants in the cultures could not be determined. Phloroglucinol also enhanced the rooting of the rootstock M.9 (James and Thurbon, 1979b, 1981a,b). Shoot proliferation was not improved but rooting was better if the cultures were grown on PG during the proliferation stage. The best rooting was obtained by leaving the cuttings on medium containing IBA and PG for 4 days, then transferring them to medium without growth regulators. Optimum concentration of IBA was 14.8 µM and of PG was 10 mM. Rooting differed between two lines of M.9 and the differences persisted over several subcultures (James and Thurbon, 1981a).

The effectiveness of different cytokinins in stimulating shoot proliferation of Golden Delicious cultures was investigated by Lundergan and Janick (1980). The most effective was BA, the least effective was 2iP, and KIN was intermediate; however, the effect on shoot elongation was in the reverse order. Shoots produced at the most effective concentra-

Table 3. Apple Rootstocks and Cultivars That Have Been Micropropagated

ROOTSTOCK/CULTIVAR	REFERENCE
Rootstock	
A 2	Welander & Huntrieser, 1981
Antonovka KA 313	Cheng, 1978
M.4	Liu et al., 1978
M.7 (EMLA-7)	Jones, 1976; Cheng, 1978; Werner & Boe, 1980
M.9 (EMLA-9)	James & Thurbon, 1979, 1981a,b; Cheng, 1978
M.25	Jones, 1979b
M.26	Jones et al., 1977; Quoirin et al., 1977
M.27	Jones et al., 1977; Quoirin et al., 1977
MAC 9	Cheng, 1978
MM 104	Snir & Erez, 1980
MM 106	Quoirin et al., 1977; Snir & Erez, 1980
MM 109	Snir & Erez, 1980
MM 111	Quoirin et al., 1977; Snir & Erez, 1980
P.16	Liu et al., 1978
Cultivar	
Annurca	Ancora et al., 1981
Bramley's Seedling	Oehl, 1980
Cox's Orange Pippin	Abbott & Whiteley, 1976; Quoirin et al., 1977; Jones et al., 1979
Delicious	Sriskandarajah et al., 1982; Zimmerman, 1983a,b
Earlistripe Red Delicious	Zimmerman, 1981
Redspur Delicious	Zimmerman, 1983b
Skyspur	Zimmerman, 1983b
Starkspur Red Delicious	Nemeth, 1981
Supreme Red Delicious	Anderson, 1981
Wellspur Delicious	Anderson, 1981
Gala	Oehl, 1980; Zimmerman, 1981
Golden Delicious	Quoirin et al., 1977; Jones et al., 1979; Zimmerman, 1981
Perleberg 3	De Paoli, 1979
Granny Smith	Sriskandarajah & Mullins, 1981
James Grieve	Jones et al., 1979
Jonagold	Quoirin et al., 1977

Table 3. Cont.

ROOTSTOCK/CULTIVAR	REFERENCE
Cultivar	
Jonathan	Huth, 1978; Zimmerman, 1981; Sriskandarajah et al., 1982
Malling Greensleeves	Jones et al., 1979
Malling Kent	Jones et al., 1979
Malling Suntan	Jones et al., 1979
MacSpur (McIntosh)	Lane, 1978
Mutsu (Crispin)	Quoirin et al., 1977; Oehl, 1980; Zimmerman, 1981
Northern Spy	Zimmerman, 1981; Zimmerman & Broome, 1981
Nugget	Zimmerman, 1981
Ozark Gold	Zimmerman, 1981
Rome Beauty	Zimmerman, 1981
Spartan	Zimmerman, 1981; Zimmerman & Broome, 1981
Spuree Rome	Zimmerman, 1981; Zimmerman & Broome, 1981
Stayman	Zimmerman, 1981; Zimmerman & Broome, 1981
Summer Rambo	Zimmerman, 1981; Zimmerman & Broome, 1981
York Imperial	Zimmerman, 1981; Zimmerman & Broome, 1981

tions of BA for proliferation (13.2-22.2 μM) were stunted. Normal growth could be obtained by transferring the cultures to medium containing 2iP or 4.4 μM BA plus 4.9 μM IBA.

Techniques have been reported for in vitro propagation of the rootstocks M.7 (Werner and Boe, 1980), MM 104, MM 106, and MM 109 (Snir and Erez, 1980). For M.7, explant establishment and shoot proliferation were readily accomplished on half-strength MS medium containing only 2.2 μM BA. Shoots rooted readily on a soft agar (0.27%) medium with one-third strength MS medium supplemented by 4.9-14.8 μM IBA.

In contrast, Snir and Erez (1980) used the medium of Jones et al. (1977) without the PG, since they found no response to PG in their cultures. They did find an increase in shoot proliferation by culturing shoot sections in liquid medium for 4 days on an orbital shaker. By transferring shoots from rooting medium to one without IBA but containing activated charcoal, two ancillary effects of the IBA, callus production at the base of the rooted cuttings and inhibition of the shoot growth, were eliminated. Wounding the shoot cuttings at the base using a scalpel seemed to increase the number of roots formed per cutting. Use of a liquid medium with continuous agitation of the cultures on a reciprocating shaker was the only method in which adequate rooting (80% vs. 0-4%) could be obtained with the Granny Smith cultivar (Sriskandarajah and Mullins, 1981).

The influence of PG on rooting a number of apple scion cultivars in vitro was evaluated by Zimmerman and Broome (1981). The data in Table 4 show that only cuttings of the cultivar Spartan gave a positive response to PG at all the tested concentrations of IBA, whereas Nugget and Stayman responded only at a single concentration. Similar results were obtained when the number of roots per cutting were evaluated (Zimmerman and Broome, 1981). Furthermore, earlier tests with Spartan from a different explant source did not show any advantage to using PG. Autoclaving did not alter the influence of PG on rooting. However, PG did reduce the formation of callus at the base of cuttings grown for 4 weeks on medium containing 4.9 μM IBA. Recently, rooting of Delicious and Redspur Delicious shoots was found to be enhanced by 1 mM PG (162 mg/l) in the rooting medium while rooting of Gala was stimulated somewhat by 1 mM PG in the proliferation medium prior to rooting in medium without PG (Zimmerman, unpublished results).

Table 4. Effect of Phloroglucinol (PG) and IBA on In Vitro Rooting Percentage after 4 Weeks of Tissue Culture-Produced Shoots of 6 Apple Cultivars[a]

	IBA (μM)					
	0		0.5		4.9	
CULTIVAR	-PG[b]	+PG	-PG	+PG	-PG	+PG
Northern Spy	80	50	100	85	100	100
Nugget	0	0	35	35	26	61
Spartan	2	20	17	47	25	85
Spuree Rome	34	20	55	59	77	75
Stayman	18	38	87	74	98	100
Summer Rambo	70	76	100	98	98	100

[a] Zimmerman and Broome, 1981.

[b] -PG = 0 mg/liter; +PG = 162 mg/liter (10^{-3} M).

Further indication of the varying effects that PG can have on rooting was shown with the rootstock A2 (Welander and Huntrieser, 1981). For shoots derived from an adult-phase plant, 1 mM PG inhibited and 0.1 mM PG stimulated rooting in the presence of 5 μM IBA; for shoots derived from a juvenile-phase plant, both concentrations of PG improved rooting. Callus formation on the cuttings was reduced by 1 mM PG at all IBA concentrations tested.

Cuttings made from proliferating cultures of Jonathan and Delicious grown with continuous illumination of 90-100 $\mu E\ m^{-2}\ s^{-1}$ (Osram MCFE

40W cool white fluorescent tubes) at 26 C rooted better than those from cultures grown on 16-hr photoperiods of 20-30 $\mu E\ m^{-2}\ s^{-1}$ at 23 C (Sriskandarajah et al., 1982). Rooting percentages and number of roots increased with increasing number of subcultures so that 32% of Jonathan cuttings rooted after the first subculture and 95% after the ninth. Rooting of Delicious increased from 15% after the first subculture to more than 70% after 31 or more subcultures. Concurrent with the increasing number of subcultures, the leaves on the proliferating shoots changed in shape and the shoot diameter was reduced, possibly indicating a rejuvenation of the tissue. These changes correlated with the increased rooting of cuttings from later subcultures (Sriskandarajah et al., 1982).

Various auxins and other compounds have been evaluated for their potential for inducing roots on apple shoots in vitro. Nemeth (1981) tested several synthetic auxins and reported that 2-chloro-3-(2,3 dichlorophenyl) proprionitrile was more effective than IBA in stimulating rooting of M.26, M.27, MM104, and Stark Spur (probably Starkspur Red Delicious). The best concentration was 5 µM. However, the rooting percentages achieved with IBA seem low in comparison with other reports using the same rootstock cultivars (Snir and Erez, 1980; Jones et al., 1977).

Little research on effects of environmental factors on in vitro culture of apples has been reported and when it has, more concern (correctly or not) has been given to light intensity than to light quality. Standardi (1979) compared different photoperiods, light intensities, and light sources on proliferation and rooting of M.26. Both proliferation and rooting were better in 16-hr rather than 12-hr photoperiods. The best results were obtained with 3100 lux from 40W Sylvania Lifeline Gro-lux tubes and 7600 lux from Phillips (TL40W/33) tubes, although the latter was less effective when roots per cutting were counted. The number of roots produced per cutting was inversely related to light intensity but root length was directly related, possibly because there were fewer roots formed. Increasing light intensity from ca. 50 $\mu E\ m^{-2}\ s^{-1}$ (Sylvania Lifeline F40WWX) up to 420 $\mu E\ m^{-2}\ s^{-1}$ (General Electric F48T12-CW-1500) cut the rooting percentage of Jonathan cuttings by more than half with almost as great a reduction in the number of roots per cutting (Zimmerman, unpublished data).

A number of rootstocks as well as seedlings were established in culture by incubating shoot tips in the dark on MS medium (Liu et al., 1978). After the cultures were initiated, they were returned to low light (500 lux) for 1 week to allow the etiolated shoots to regreen. These cultures were then returned to darkness for several months; the etiolated shoots were cut into 1- to 2-bud segments, transferred to light (3000-5000 lux), and allowed to proliferate. The shoots were used for rooting. Alternating culture in dark and light increased the rate of proliferation and the uniformity of the shoots produced (Liu et al., 1978).

Etiolation of the proliferating cultures for 2 weeks, followed by several days in a low light intensity of 1000 lux (to permit the cultures to regreen), dramatically increased rooting percentages of Supreme Red Delicious from 2% to 53% and of Wellspur Delicious from

10% to 82% (Anderson, 1981). Etiolation without regreening resulted in formation of callus proximal to the shoot apex with subsequent abscission of the apex in many cases. With increasing time of regreening up to 8 days, the abscission of the shoot apex was eliminated. Although cuttings on which the apex abscised sometimes rooted and were acclimatized, they did not grow well in the greenhouse. In general, etiolated cuttings were weak and difficult to acclimatize. Similar but less dramatic improvements have also been found in rooting cuttings of Gala, Mutsu (Crispin), Delicious, and Redspur Delicious following etiolation and regreening of the proliferating cultures (Zimmerman, unpublished data). The problem of abscission of the shot apex was also severe on Delicious and Redspur Delicious, but not on Gala and Mutsu.

Since 1977, apple rootstocks, and limited quantities of a few apple scion cultivars, have been produced in commercial tissue culture laboratories in steadily increasing numbers. In many cases, this work has involved the adaptation and modification of the published methods, but these changes and the reasons for making them are generally not available in the published literature. This fact does emphasize, however, that a considerable amount of work on apple micropropagation methods has been done in addition to that cited above.

Adventitious Shoots

The production of such shoots from basal callus on explants of Cox's Orange Pippin seedlings was noted by Abbott and Whiteley (1976). More recently, subcultured shoots of M.26 rootstock were reported to produce adventitious shoots from basal callus (Nasir and Miles, 1981). Adventitious shoots have also been found to arise directly from leaves of M.27 rootstock in proliferating cultures (Constantine et al., 1980a).

Organogenesis from Callus

Endosperm isolated from young, developing apple seeds was induced to form callus on MS medium containing casein hydrolysate, plus 4.6 µM KIN and 2.3 µM 2,4-D (Mu et al., 1977). When some of the callus was transferred to a similar medium containing BA and NAA, one plantlet developed within 60 days. Roots and leaves were also differentiated from the callus (Mu et al., 1977). Plants of M.9 rootstock have been regenerated from callus in vitro (Chen et al., 1979). Callus was induced from stem segments on MS medium containing 2.2 µM BA, 10.7 µM NAA, and 100 mg/l casein hydrolysate (CH). It was then transferred to a similar medium without NAA and shoots differentiated from nodules in the callus. These shoots were then rooted in vitro on a medium containing IBA. Callus cultures derived from various parts of seedlings of Golden Delicious have regenerated leaves, shoots, and roots (Mehra and Sachveda, 1979; Liu et al., 1981). The shoots produced from callus could then be rooted in vitro to obtain plants (Liu et al., 1981).

Somatic Embryogenesis

Nucellar tissue was isolated from developing seeds of Golden Delicious, separated into micropylar and chalazal halves, and cultured on growth-regulator-free MS medium in the dark (Eichholtz et al., 1979). Several of the micropylar halves of the nucellus produced a few adventive embryos. When the embryos were recultured in the dark, additional embryolike structures formed on the cotyledons. None of the embryos grew to maturity, however.

PROTOCOLS FOR MICROPROPAGATION

Several different protocols have been published for each stage of apple micropropagation. In the following discussion, the one used in my laboratory will be described first, followed by a description of one or more different or modified protocols used elsewhere.

Explant Establishment from Actively Growing Shoots

Three- to 5-cm tips from actively growing shoots (Fig. 1a) are collected in the greenhouse or field and kept in a plastic bag to prevent wilting until they can be processed in the laboratory. The leaves are snapped off the stem, taking care to remove the stipules and not to damage the axillary buds or to strip the epidermis from the stem. As many as possible of the furled leaves surrounding the apex are removed. The prepared shoots are accumulated in a container of distilled water with up to 40 shoots per 100 ml to which 2 drops of liquid detergent are added. The shoots are stirred gently for 5 min on a magnetic stirrer after which the water is poured off. The shoots are then rinsed 3 times with distilled water to remove debris. A saturated stock of calcium hypochlorite is prepared by adding 60 g to a liter of distilled water, stirring 15 min, and then filtering using a vacuum after the residue has settled. Freshly prepared stock solution, diluted to one-half strength with distilled water and containing 0.01% Tween 20, is added to the container and the shoots are gently stirred for 20 min. In a laminar flow hood, the calcium hypochlorite is drained off and the shoots are rinsed briefly in sterile distilled water followed by a second 5-min rinse with stirring. The shoots are then held in fresh sterile distilled water until final trimming before placing them in culture medium. For this, shoots are removed to a sterile dish or surface, cut below a node to a final length of 1-2 cm, and placed in 15 ml of modified MS liquid medium (see details of medium below) contained in a 125-ml Erlenmeyer flask capped with aluminum foil (Fig. 1b). The flasks are clamped horizontally to disks which are rotated continuously at 1 rpm for 2-4 days in a culture room maintained at 26 C. Light is provided 16 hr per day at 50-70 $\mu E\ m^{-2}\ s^{-1}$ (General Electric F72PG17 Power Groove fluorescent lamps). This method speeds detection of bacterial and fungal contaminants, rinses phenolic compounds from the

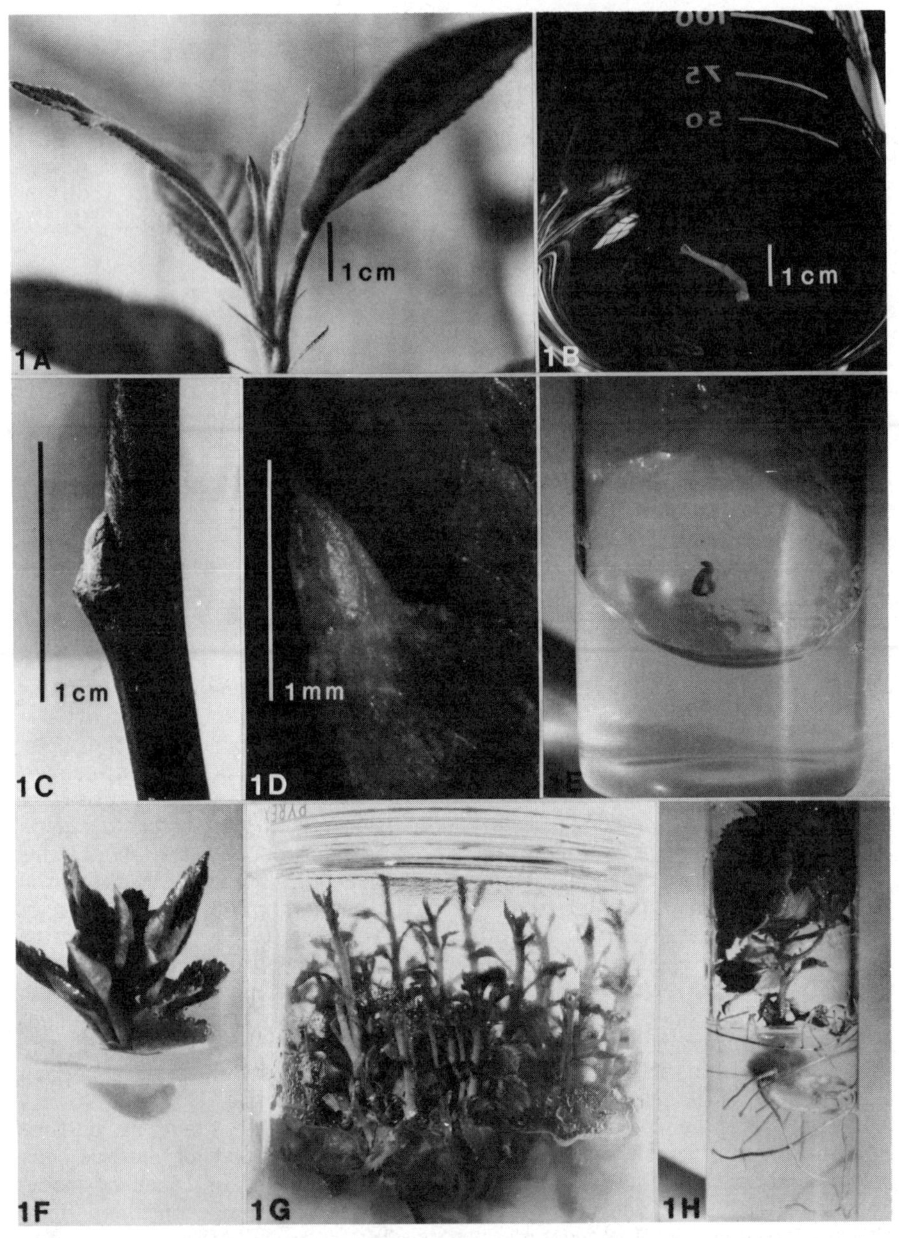

Figure 1. Establishment, proliferation, and rooting of apple tissue cultures. (A) Actively growing shoot tips of apple ready to be taken for culture establishment. (B) Shoot tip in liquid MS medium for explant establishment. (C) Dormant bud prior to dissection. (D) Dormant bud dissected to show meristem-tip with several leaf primordia prior to excision from stem. (E) Small shoot produced from meristem-tip after

several weeks in culture. (F) Shoot with rosette of leaves after 6-8 weeks of growth from meristem-tip. (G) Proliferating culture with vigorous shoots satisfactory for use as cuttings. (H) Rooted cutting after 4 weeks on rooting medium.

explants, and provides maximum contact between the explant and the medium. The explants are then transferred to solid medium of the same composition. The resulting growth of the explants is superior to that achieved when the explants are placed on solid medium initially.

Growth of the terminal and lateral buds usually begins in 7-10 days. By 3 weeks, the leaf primordia present at the time of explanting have developed into long unfurling leaves, sometimes pushing the explant free of the medium. These leaves are trimmed off and the explant is transferred to fresh medium. Subsequent transfers are made every 3-4 weeks, dividing and separating the newly produced shoots as necessary.

Several alternative procedures have been developed by Jones et al. (1977, 1979). The terminal 2-4 cm of the shoots are collected in distilled water, then dipped briefly in 0.01% wetting agent before placing in sodium hypochlorite solution (0.14% w/v available Cl) for 1 min and washing 3 times in sterile distilled water (Jones et al., 1977). The shoot tips are then placed on a medium containing MS salts, sugar, and agar overnight after which they are again dipped briefly in the wetting agent followed by 40 min in sodium hypochlorite solution (0.42% w/v available Cl). After 3 rinses in distilled water, the terminal 1- to 2-cm section is dissected from the shoot tip and transferred to MS medium. This procedure is normally used for shoot tips taken shortly after bud-break from a previously cold-stored plant (Jones, 1979a). Washing in running tap water for 1 hr prior to a 40-min immersion in sodium hypochlorite was effective for tips collected from field-grown trees (Jones et al., 1979).

Other alternatives have been to dip the shoot tip in 70% ethanol for 5 seconds (Lane, 1978) or 30 seconds (Welander and Huntrieser, 1981), then dissect a meristem tip of 0.2-0.5 mm with 1-4 leaf primordia from both terminal and axillary buds.

Explant Establishment from Buds

Dormant shoots are collected in the field after the rest period has been satisfied or are collected earlier and cold-stored for several months until needed. The shoots are then cut into sections 2-4 cm long with the bud near the distal end (Fig. 1c). These sections are dropped into 95% ethanol containing 1 drop of Tween 20 per 100 ml and agitated on a magnetic stirrer for 10 min. The alcohol is poured off and replaced with saturated calcium hypochlorite solution (60 g/l, prepared as described above) containing 0.01% Tween 20. After constant stirring for 20 min, the stem sections are rinsed with sterile distilled water for 5 min, again with constant stirring. The segments are then kept in a fresh change of sterile distilled water until dissected.

The buds are dissected by sequentially removing the bud scales and then the leaf primordia to expose the growing point. The sterile scalpels and forceps used for dissecting the meristem-tip are changed several times during the procedure to reduce the chance of contaminating the tip. The growing point together with several leaf primordia (Fig. 1d) is excised and explanted with the base resting on the surface of the medium, which is that of Lepoivre (Quoirin et al., 1977) slightly modified (Table 5), containing 0.4 μM BA and 0.05 μM IBA. The vessel used is a screw cap vial with the agar solidified at a slant to facilitate removal of the explant from the tip of the scalpel blade. If dissection and explanting of the meristem tip is done in a laminar flow hood, the tissue desiccates rapidly. Therefore, this step is done in the laminar flow hood with the blower turned off or outside the hood in a quiet corner of the laboratory relatively free of air currents.

Meristem-tip explants grow into small shoots in 3-4 weeks (Fig. 1e) and can have a well-developed rosette of leaves on a short shoot in 6-8 weeks (Fig. 1f). At this point, the explant is transferred to the MS proliferation medium and handled as described above for cultures established from actively growing shoot tips.

A similar technique differing in some important details is used at a commercial laboratory in British Columbia (D.I. Dunstan, personal communication). Dormant budwood is collected in early autumn and the shoots are cut into 1-node pieces about 2-4 cm long. The bark is cut away from the stem all around the bud with a scalpel and the outer bud scales are removed. The buds together with about 5 mm of subtending wood are cut out from the stem section and accumulated in distilled water. When about 30 buds have been accumulated, the water is replaced with sodium hypochlorite (0.6-1.2% active Cl) for 10-15 min. The concentration and length of treatment vary with the material being dissected. The buds are then rinsed in sterile distilled water and transferred to a sterile petri dish. Four buds are removed and placed in separate quadrants of a second sterile petri dish in which they are then dissected in sequence. Dissection and explanting take about 2 min per bud. The tissue is clasped at the woody base with forceps and held upright. A scalpel is then used to flick away green bud scales and leaves following the phyllotaxic arrangement until only the meristem-tip and 2-4 leaf primordia remain. The wood is then trimmed on the sides, shoulders, and base to leave the meristem-tip attached to a small stub which is then inserted into MS medium with 2.2-4.4 μM BA.

A technique for establishing explants from buds without dissecting the buds has been described recently (Sriskandarajah et al., 1982). Shoots are collected when about 40 cm long and 5 mm in diameter from greenhouse-grown plants, the leaves are removed, and the shoots are cut into segments 8-10 cm long. These segments are surface-sterilized by shaking for 15 min in sodium hypochlorite solution (1% available Cl) containing 0.05% Tween 20. After several washes with sterile distilled water, nodal segments are explanted onto 15 ml of medium in 25-mm diameter test tubes. The MS medium contains 8 g agar, 87.6 mM sucrose, and 10 μM BA. When the cultures are grown at 26 C under continuous illumination from cool white fluorescent tubes (90-100 μE m^{-2} s^{-1}), bud burst begins in about 1 week and extension growth of the shoot occurs in about 2 weeks.

Table 5. Composition of Media Used for Establishment, Proliferation, and Rooting of Apple Cultivars as Modified for Use in the USDA Fruit Laboratory[a]

COMPONENT	MS[b,c] (mM)	Lepoivre[d] (mM)	WPM[e] (mM)
NH_4NO_3	20.6	5.0	5.0
$Ca(NO_3)_2 \cdot 4H_2O$	—	5.1	2.3
KNO_3	18.8	17.8	—
$CaCl_2 \cdot 2H_2O$	3.0	—	0.9
$MgSO_4 \cdot 7H_2O$	1.5	1.5	1.5
KH_2PO_4	1.2	2.0	1.2
K_2SO_4	—	—	5.7
$FeSO_4 \cdot 7H_2O$	0.1	0.1	0.1
Na_2EDTA	0.1	0.1	0.1
$MnSO_4 \cdot H_2O$	0.1	0.1	0.1
$ZnSO_4 \cdot 7H_2O$	0.03	0.03	0.03
H_3BO_3	0.1	0.1	0.1
KI	5 μM	5 μM	—
$Na_2MoO_4 \cdot 2H_2O$	1 μM	1 μM	1 μM
$CoCl_2 \cdot 6H_2O$	0.1 μM	0.1 μM	—
$CuSO_4 \cdot 5H_2O$	0.1 μM	0.1 μM	1 μM
Myo-Inositol	0.56	0.56	0.56
Nicotinic acid	—	4.1 μM	4.1 μM
Pyridoxine HCl	—	2.4 μM	2.4 μM
Thiamine HCl	1.2 μM	1.2 μM	3.0 μM
Glycine	—	—	27 μM
Sucrose	87.6	58.4	58.4
Agar	7 g/l	7 g/l	7 g/l

[a]See protocols in text for concentrations of BA, IBA, and GA_3.

[b]Murashige and Skoog, 1962.

[c]Half-strength salts and sucrose used for rooting.

[d]Quoirin et al., 1977.

[e]Lloyd and McCown, 1980.

Proliferation

Once the cultures have been established, shoots are proliferated on modified MS medium (Table 5) containing 4.4 μM BA, 0.5–4.9 μM IBA, and 1.5 μM GA_3. Newly established tips, initially from actively growing shoots, are place horizontally on the medium and embedded slightly in the agar. Alternatively, the Woody Plant Medium (WPM) (Lloyd and McCown, 1980) has been used. This has produced vigorous growth at the beginning of the proliferation phase and marginally better proliferation rates than the MS medium, but tissue vitrification has been a greater problem using WPM.

Proliferation rates vary with cultivar but normally are in range of 3:1 to 6:1 in 4 weeks. When the proliferation rate slows or the shoots produced lack vigor, the culture can be renewed using the tips of the in-vitro-produced shoots only, discarding the basal portion that forms the clump of shoots. The proliferation rate can be improved, in some cases, by removing the leaves from the shoots and inverting the shoots in agar.

Growing conditions for the proliferating cultures are a constant temperature of 24-26 C, 16-hr photoperiods, and a light intensity of 50-70 $\mu E\ m^{-2}\ s^{-1}$ (400-700 nm) provided by Sylvania Lifeline 40W warm white or cool white fluorescent lamps.

The main variation in technique for shoot proliferation is the inclusion of 1 mM PG in the modified MS medium used by Jones (1976) and used for proliferating several different apple rootstocks and cultivars (Jones, 1979a; Jones et al., 1977, 1979; James and Thurbon, 1979b, 1981a,b).

Snir and Erez (1980) proliferate shoots on solid MS medium, using the longer shoots for rooting and the shorter ones for further proliferation. These short shoots are cultured in liquid MS medium (as Jones et al., 1977, but without PG) on an orbital shaker for 4 days on 16-hr photoperiods. The growth rate is faster in liquid medium and intense proliferation occurs when the shoots are placed back on solid medium. However, the 5-fold multiplication rate per month reported is comparable to results reported by others using solid medium (Jones et al., 1977; Lane, 1978; Zimmerman, unpublished results).

Rooting

Vigorously growing shoots from proliferating cultures (Fig. 1g) are used to make cuttings about 2-3 cm long. These cuttings are then rooted on half-strength MS medium (Table 5) containing IBA at a concentration that varies depending upon the cultivar and containing no BA or GA_3. For most cultivars, 0.5-1.5 µM IBA gives optimum rooting but 4.9 µM has been better for cultivars that are very difficult to root, e.g., Delicious. Concentrations higher than 4.9 µM IBA inhibit rooting and stimulate callus production; even 4.9 µM IBA induces considerable callus formation on most cultivars. Cuttings are well rooted in 4 weeks (Fig. 1h) and can then be removed from the culture vessel. Growth-room conditions for rooting are the same as those described above for proliferation. Inclusion of PG in the medium has not been useful except in the case of Spartan (Zimmerman and Broome, 1981), Delicious, and Redspur Delicious (Zimmerman, unpublished results).

In liquid medium with cuttings supported by vermiculite, perlite, or sand, cuttings rooted as well with 0.05 µM IBA, or even no IBA with some cultivars, as with higher concentrations up to 49.2 μM (Zimmerman and Broome, 1980). The root systems seem to develop somewhat better than in agar medium but more labor is required to prepare the culture tubes using this technique.

To counteract the problem of excess callus production at the base of the stem and to improve root elongation following initiation, several workers have transferred the cuttings from a medium containing auxin (IBA or NAA) to one with no growth regulators after 4 days to 4 weeks (James and Thurbon, 1979b, 1981a,b; Snir and Erez, 1980; Lane, 1978). All used MS medium: Lane (1978) at full strength with 10 μM NAA, James and Thurbon (1979b, 1981a,b) at full strength with 1 mM PG and 4.9-24.6 μM IBA or 28.5 μM IAA, and Snir and Erez (1980) at half strength with 4.9 μM IBA. Following the initial root-initiation phase on these media, the cuttings were transferred to the same medium without growth regulators after 4 weeks (Lane, 1978), to half-strength MS without growth regulators after 4-6 days (James and Thurbon, 1981a,b), and to half-strength MS without growth regulators and with 0.25% activated charcoal after 6-8 days (Snir and Erez, 1980). However, Jonathan cuttings rooted best when on half-strength MS medium with 1.5 μM IBA for 4 weeks as opposed to cuttings transferred to medium without IBA after 4, 8, or 12 days (Zimmerman, unpublished results). Thus it appears that the best auxin concentration is the minimum necessary to give satisfactory root initiation, since this concentration will allow the greatest root development and will also stimulate less callus growth.

Rooting of M.7 rootstock on one-third strength MS medium containing only 0.27% agar and 4.9-14.7 μM IBA was reported by Werner and Boe (1980). Cuttings 5-10 mm long floated on this medium and rooted well in 4 weeks. Use of liquid half-strength MS medium continuously shaken at 70 strokes/min on a reciprocating shaker resulted in 80% rooting of Granny Smith (Sriskandarajah and Mullins, 1981). The medium contained 29.2 mM sucrose and 10 μM IBA. Culture conditions were continuous light of only 10-15 μE m^{-2} s^{-1} and a temperature of 26 C.

Acclimatization

Rooted cuttings are removed from the culture vessel 4-6 weeks after being placed on the rooting medium, the agar is washed off the roots and the plant is planted in a soil mix (peat-lite mix:vermiculite:perlite, 1:1:1). Extreme care must be taken to prevent the leaves from drying at this stage since they desiccate within a few minutes. The potted plants are moved to the greenhouse and placed under a mist system capable of maintaining a thin film of water on the leaves without overwatering the soil mixture (Fig. 2a). This is done by running the mist for only 1 or 2 seconds every 3-6 min. Once active growth of the shoot tip has begun, the mist can be discontinued. Better results can be attained if the potted plants are put under higher light intensity (125-150 μE m^{-2} s^{-1} from Sylvania cool white VHO fluorescent tubes) at high humidity (90-100%) for several weeks before transferring to mist. If the rooted cuttings are taken from agar and the roots

Figure 2. Tissue-cultured apples being acclimatized under mist in the greenhouse and growing in the field. (A) Apple plants from in vitro culture 1 month after removal from the culture container in which they were rooted. (B) Tissue-cultured trees of the cultivars Northern Spy (foreground) and Golden Delicious (second row) after three growing seasons in the orchard, and of Rome Beauty (background) after two growing seasons. Trees were 40-60 cm tall when planted in early June.

placed directly in an aerated nutrient solution at this higher light intensity and humidity, both the shoots and roots grow very rapidly. However, the resulting plants are more difficult to pot and are extremely susceptible to wilting.

Preliminary results suggest that cuttings rooted in 25-mm diameter vials or test tubes can be acclimatized by removing the caps 5 or 6 days before taking the plants from the tubes. Some minor desiccation of leaf margins occurs but the plants do not wilt. After the plants are transplanted to a soil mix, they can be exposed to ambient conditions in the laboratory and shoot growth continues with almost no interruption. Rooted plants growing in containers of large diameter (70-100 mm) wilt rapidly and suffer severe desiccation injury when treated in a similar manner.

Jones et al. (1979a) reported that after 3 weeks, small, well-rooted plants were removed from the rooting medium and placed aseptically on filter paper wicks in test tubes of liguid medium of similar composition but having half-strength salts, 58.4 mM sucrose, and no IBA or PG. These tubes were put in the greenhouse and, after 10 days, the plants were potted in compost and set in humid propagating cases in a shaded greenhouse. After 14 days, the plants could be grown under normal greenhouse conditions.

DISCUSSION OF PROBLEMS IN APPLE MICROPROPAGATION

Bacterial Contamination of Cultures

A common problem in the culture of apples from shoot or meristem-tips is the appearance of bacterial contaminants after several months of culture. The sudden appearance of these organisms from tissue apparently free of contaminants and their apparent emergence only from freshly cut surfaces leads one to suspect that the organisms may have existed within or on the tissue of the original explant and were not destroyed by the original disinfection procedure. The most common organism found in our cultures has been *Acinetobacter calcoaceticus*, which is present in soil and water, is often isolated from animals and man, but is not known to be pathogenic to apple. This organism has also been isolated from apple cultures at a commercial tissue culture laboratory in Oregon (G. Suttle, personal communication). Chlortetracycline has been the most effective antibiotic for controlling the growth of *A. calcoaceticus* in tests in our laboratory; however, it has not been possible to eliminate it from cultures and this antibiotic usually inhibits growth of the culture as much as it does the organism. Rinsing affected shoots in dilute sodium hypochlorite seems to be the most effective means of keeping the organism under control. Rooting of shoots does not appear to be affected by the presence of *A. calcoaceticus* and it has not been possible to determine conclusively the effect it might have on shoot proliferation.

Bacillus pumilis has been tentatively identified as one of the contaminants appearing in cultures in the same manner as *A. calcoaceticus* (Constantine et al., 1980b). However, *B. pumilis* could survive on flamed forceps so that it may have been introduced to the cultures inadvertently at some time after explant extablishment. Some cultures were freed of infection by incubation with 0.03 mM vancomycin for 72 hr.

Similar results found at the East Malling Research Station (James and Thurbon, 1978; Jones et al., 1978) indicated that at least 5 types of bacteria survived surface-sterilization (James and Thurbon, 1979a). Even when tips screened monthly on yeast peptone agar showed no contaminants for 5 successive times, 25% of the cultures were found to be infected when screened a sixth time (James and Thurbon, 1979a). Thus some latent bacteria are apparently not completely eliminated by the usual surface-sterilization procedures, but whether these are external or internal has not yet been established.

Tissue Vitrification

Apple shoots grown on media containing BA often produce succulent leaves and stems variously described as vitreous, waxy, water-soaked, or translucent. The leaves are malformed, strap-like, and curled in many cases, and stem elongation is inhibited. Werner and Boe (1980) refer to BA toxicity appearing in some cultures after several months without any description of the symptoms. They reduce or eliminate this toxicity by transferring the cultures regularly to medium containing 24.6 μM 2iP. Since transferring vitreous cultures from medium containing BA to that containing 2iP has been shown to correct this problem (E. Strahlheim, personal communication; Zimmerman, unpublished data), presumably the BA toxicity of Werner and Boe is the same as tissue vitrification. However, Quoirin et al. (1977) stated that the leaf succulence observed in their cultures was due to the salt formulation used rather than to the BA concentration used. The MS salts produced more vitrification than did those of Lepoivre. In another study, MS salts produced less vitrification than the WPM salts, whereas the Lepoivre salts seemed to be no better than MS (Zimmerman, unpublished data).

Vitrification seems to occur more rapidly on liquid than on solid medium and sometimes varies according to the type of agar used. Since the type of agar reducing vitrification in one laboratory increased it in another, the role of agar source may be important.

Acclimatization

Apple plants regenerated in vitro are not as easy to adapt to greenhouse conditions as plants of other fruit crops, e.g., strawberry, because the leaves desiccate very rapidly when the plants are removed from the culture container. This has been thought to result from the leaves having an incomplete cuticle (Zimmerman and Broome, 1980) when coming from the humid environment of the culture container. A similar phenomenon has been observed in carnation (Sutter and Langhans, 1979). That the problem is more complex is evidenced by the fact that the stomata of plants cultured in vitro do not respond to stress by closing (Brainerd and Fuchigami, 1982a), with the result that the leaves desiccate rapidly. However, the apple plants can be acclimatized successfully by opening the jars for 4-5 days at 30-40% relative humidity (RH) (Brainerd and Fuchigami, 1981) or for 2 days at 65% RH (Brainerd and Fuchigami, 1982). Stomatal anatomy did not seem to change during the acclimatization procedure (K.E. Brainerd, personal communication).

We have found that once new shoot growth occurs on plants being acclimatized under mist, the mist can be discontinued (Zimmerman and Broome, 1980). Desiccation of leaves developed in the culture containers may still occur but this does not seem to affect survival of the transplants. At this stage, good root growth appears to be more important to ultimate survival of the transplants. Therefore the composition of the potting mix and water application to it must be carefully controlled. Too much water seems to cause as many problems as too little water.

Variability of Response Among Cultivars

Different cultivars of apple do not respond in the same way during establishment, proliferation, and rooting in vitro. Furthermore, the same cultivar does not always respond similarly to apparently similar conditions in different laboratories. Some of these differences are doubtless due to unrecognized, and thus uncontrolled, variables in the procedures used. Differential sensitivity to the particular cytokinin and auxin, to the concentration of each, and to the ratio between them are important factors, especially in the establishment and proliferation of cultures. Probably all apple cultivars are difficult to root when first established in culture. Once adapted to culture conditions, however, some cultivars become consistently easy to root while others remain difficult (Zimmerman and Broome, 1981; Sriskandarajah et al., 1982; Zimmerman, unpublished results). Even the latter may be rooted at acceptable rates when appropriate treatments are developed, e.g., M.9 (James and Thurbon, 1981a,b) and Delicious (Sriskandarajah et al., 1982; Anderson, 1981; Zimmerman, 1983a).

Phenotypic Stability of Regenerated Plants

Micropropagation will be a useful technique only if the plants produced are phenotypically identical to the original cultivar, whether scion or rootstock, and are as stable genetically as plants propagated by other vegetative methods. Confirming the identity and stability of apple cultivars entails long-term field testing (Fig. 2b) because information must be obtained both on flowering and fruiting characteristics of the micropropagated trees on their own roots or on micropropagated rootstocks, and on their field performance. Only when such testing has been done can growers be fully confident in planting trees produced through micropropagation.

Juvenility

Apples, as well as other crops, in tissue culture show a reversion to juvenile leaf forms (Zimmerman, 1981; Sriskandarajah et al., 1982; Lane, personal communication). The increased ease of rooting demonstrated by many cultivars after an increasing time in culture may also reflect a reversion to juvenility for this characteristic (Zimmerman, 1981; Sriskandarajah et al., 1982). However, preliminary data would indicate that this apparent reversion to juvenile form does not extend to flowering. Thus a number of cultivars have flowered after 1 or 2 years in the orchard, some in as little as 14 months after removal from the culture container (Zimmerman, 1981; Gayner et al., 1981; Lane, personal communication). Further experimentation is required to determine how precocious in flowering these own-rooted cultivars are in comparison to conventionally propagated trees of these same cultivars, particularly where dwarfing rootstocks are used for the latter.

Cost of Production

Assuming that no unforeseen problems arise, the ultimate commercial acceptance of apple micropropagation will depend upon the cost of producing a plant in this way compared to the cost by using conventional methods. For clonal rootstocks, the price competition will be very strong and profitable use of micropropagation will depend upon improving proliferation rates, rooting percentages and survival during acclimatization as well as developing efficient methods for growing rootstocks to a usable size. It should be possible to produce own-rooted scion cultivars more cheaply by micropropagation than by conventional methods since the cost per conventional tree is much higher than for rootstocks. However, this use will depend upon such trees performing satisfactorily in the orchard. A higher cost per tree is acceptable when micropropagation can facilitate rapid increase of new selections or virus-indexed trees for testing or development of commercial varieties.

FREEING PLANTS OF VIRUSES

Thermotherapy is the technique normally used for freeing apples of viruses (Fridlund, 1980) but a few viruses, e.g., apple stem-grooving, are rarely eliminated by this method. Accordingly, meristem-tip culture, as already described, followed by appropriate indexing techniques can be used for virus elimination.

An alternate method has been to graft a meristem tip in vitro onto a 15-day-old seedling, germinated in vitro, and then decapitated either 15 mm above the cotyledons (Alskieff and Villemur, 1978) or immediately below the cotyledons (Huang and Millikan, 1980). Successful graft unions have been obtained on 23% of the grafts in the former case and 65% in the latter. Grafting success varied with time of year; those done in April through June were most successful (Huang and Millikan, 1980). This method was used with 2 clones to produce plants free of apple stem-grooving virus (Huang and Millikan, 1980).

GERMPLASM PRESERVATION

Little work has been done to date on low-temperature storage of apple as a means of maintaining desirable germplasm. Although dormant apple buds have survived storage in liquid nitrogen for 23 months (Sakai and Nishiyama, 1978), no successful storage of tissue cultured apple tissue or organs at such low temperatures has been reported. Single shoots of Golden Delicious were successfully stored in vitro for 1 year at 1 and 4 C and proliferated new shoots when returned to standard culture conditions at 26 C (Lundergan and Janick, 1979). Proliferating cultures of a number of apple cultivars have been stored at 2-4 C for periods up to 1 year and grow well when transferred to fresh medium and returned to standard culture conditions although frequent transfers may be necessary at first (Zimmerman, unpublished data). Some cultivars will even root while stored in vitro at 2-4 C, but in vitro rooted cuttings placed into low temperature do not store as well as the proliferating cultures.

FUTURE PROSPECTS

Although micropropagation of apples is rapidly becoming an established commercial technique, certain important problems remain to be solved. One of these is the extension of the technique to a number of important cultivars, especially Delicious and its many different strains. Progress has been made with these cultivars as indicated in various papers cited earlier but the technology for micropropagating them economically on a large scale does not yet exist. Similarly, progress has been made on developing an efficient technique for acclimatizing cuttings rooted in vitro, but losses and delays in growth at this stage are still too great to be commercially acceptable.

The production of plants by somatic embryogenesis could become a very useful technique for large-scale propagation. This would depend, of course, on the somatic embryos producing genetically stable plants that would be faithful copies of the source clone. Even if this were not the case, variation among the plants could be used as a tool for genetic improvement by screening the seedlings for resistance to diseases and insects and for other important horticultural characteristics.

Since apple has a long breeding cycle, techniques to shorten or circumvent it would be very useful. If methods can be developed to regenerate apple plants from callus, cells, or protoplasts, either by somatic embryogenesis or by organogenesis, then it is conceivable to do screening in vitro for resistance to diseases, insects, and various environmental stresses. Further, mutagenic treatments could be applied in vitro to cells, plant parts, or young plants to induce mutations which might prove useful in breeding or in production of trees and fruit. The ability to combine mutagenic treatments with in vitro screening would provide apple breeders with a very powerful tool for crop improvement.

Other methodologies such as production of haploids for breeding purposes, protoplast fusion, gene transfer, and uptake of organelles by protoplasts are the subject of current research on certain crops. However, the application of any of these methodologies to apple improvement lies far in the future.

ACKNOWLEDGMENT

I wish to thank Olivia C. Broome for preparing the figures, assisting in the research reported from this laboratory, and many helpful discussions. Mention of a trademark, proprietary product, or vendor does not constitute a guarantee or warranty by the U.S. Department of Agriculture and does not imply its approval to the exculsion of other products or vendors that may also be suitable.

KEY REFERENCES

Lane, W.D. 1982. Tissue culture and in vitro propagation of deciduous fruit and nut species. In: Application of Plant Cell and Tissue Culture to Agriculture and Industry (D.T. Tomes, B.E. Ellis, P.M. Harney, K.J. Kasha, and R.L. Peterson, eds.) pp. 163-186. Plant Cell Culture Centre, University of Guelph, Guelph, Ontario.

U.S. Department of Agriculture. 1980. Proceedings on the Conference on Nursery Production of Fruit Plants Through Tissue Culture—Applications and Feasibility. Science and Education Administration, Agricultural Research Results, ARR-NE-11. Beltsville, Maryland.

Zimmerman, R.H. 1983. Tissue culture. In: Methods in Fruit Breeding (J.N. Moore and J. Janick, eds.) pp. 124-135. Purdue University Press, West Lafayette, Indiana.

REFERENCES

Abbott, A.J. and Whiteley, E. 1976. Culture of *Malus* tissues in vitro. I. Multiplication of apple plants from isolated shoot apices. Sci. Hortic. 4:183-189.

Alskieff, J. and Villemur, P. 1978. Greffage in vitro d'apex sur des plantules decapitees de pommier (*Malus pumila* Mill.). C. R. Acad. Sci. Paris, Ser. D. 287:1115-1118.

Ancora, G., Belli-Donini, M.L., and Cuozzo, L. 1981. Propagazione di fruttiferi mediante la coltura in vitro di apici vegetativi. L'Ottenimento di piante autoradicate nella cv. di melo "Annurca." Riv. Ortoflorofruttic. Ital. 65:59-66.

Anderson, W.C. 1981. Etiolation as an aid to rooting. Proc. Int. Plant Prop. Soc. 31:138-141.

Brainerd, K.E. and Fuchigami, L.H. 1981. Acclimatization of aseptically cultured apple plants to low relative humidity. J. Am. Soc. Hortic. Sci. 106:515-518.

______ 1982. Stomatal functioning of in vitro and greenhouse apple leaves in darkness, mannitol, ABA, and CO_2. J. Exp. Bot. 33:388-392.

Brown, A.G. 1975. Apples. In: Advances in Fruit Breeding (J. Janick and J.N. Moore, eds.) pp. 3-37. Purdue Univ. Press, West Lafayette, Indiana.

Chen, W.-L., Yang, S.-Y., Wang, H.-X., and Cui, C. 1979. Organogenesis of *Malus* (rootstock M.9) callus in vitro (in Chinese). Acta Bot. Sin. 21:191-194.

Cheng, T.-Y. 1978. Clonal propagation of woody plant species through tissue culture techniques. Proc. Int. Plant Prop. Soc. 28:139-155.

Childers, N.F. 1976. Modern Fruit Science, 7th ed. Horticultural Publications, New Brunswick, New Jersey.

Constantine, D.R., Abbott, A.J., and Wiltshire, S. 1980a. Microvegetative propagation of fruit and woody ornamental plants. In: Long Ashton Research Station Report, 1979, pp. 72-73.

Constantine, D.R., Wiltshire, S., and Beddows, C. 1980b. Contamination of cultures. In: Long Ashton Research Station Report, 1979, p. 74.

DePaoli, G. 1979. Efficienza di alcuni mezzi di coltura nella moltiplicazione e radicazione "in vitro" dell' "M.26" e della varieta' "Perleberg 3." In: Tecniche di Colture "In Vitro" per la propagazione su Vasta Scala Delle Specie Ortoflorofrutticole, pp. 127-133. Pistoia, Italy.

Dutcher, R.E. and Powell, L.E. 1972. Culture of apple shoots from buds in vitro. J. Am. Soc. Hortic. Sci. 97:511-514.

Eichholtz, D.A., Robitaille, H.A., and Hasegawa, P.M. 1979. Adventive embryony in apple. HortScience 14:699-700.

Elliott, R.F. 1972. Axenic culture of shoot apices of apple. N. Z. J. Bot. 10:254-258.

Fridlund, P.R. 1980. Maintenance and distribution of virus-free fruit trees. In: Proceedings of the Conference on Nursery Production of Fruit Plants through Tissue Culture—Applications and Feasibility, pp. 86-92. USDA, Science and Education Administration, Agricultural Research Results, ARR-NE-11, Beltsville, Maryland.

Gayner, J.A., Hopgood, M.E., Jones, O.P., and Watkins, R. 1981. Propagation in vitro of fruit plants. In: Report of the East Malling Research Station for 1980, p. 145.

Huang, S.C. and Millikan, D.F. 1980. In vitro micrografting of apple shoot tips. HortScience 15:741-743.

Huth, W. 1978. Kultur von Apfelpflanzen aus apikalen Meristemen (in German, English summary). Gartenbauwissenschaft 43:163-166.

James, D.J. and Thurbon, I.J. 1978. Culture in vitro of M.9 apple. In: Report of the East Malling Research Station for 1977, pp. 176-177.

______ 1979a. Culture in vitro of M.9 apple. In: Report of the East Malling Research Station for 1978. pp. 179-180.

______ 1979b. Rapid in vitro rooting of the apple rootstock M.9. J. Hortic. Sci. 54:309-311.

______ 1981a. Shoot and root initiation in vitro in the apple rootstock M.9 and the promotive effects of phloroglucinol. J. Hortic. Sci. 56:15-20.

______ 1981b. Phenolic compounds and other factors controlling rhizogenesis in vitro in the apple rootstocks M.9 and M.26. Z. Pflanzenphysiol. 105:1-10.

Jones, O.P. 1967. Effect of benzyl adenine on isolated apple shoots. Nature (London) 215:1514-1515.

______ 1976. Effect of phloridzin and phloroglucinol on apple shoots. Nature (London) 262:392-393.

______ 1979a. Propagation in vitro of apple trees and other woody fruit plants: Methods and applications. Scientific Hortic. 30:44-48.

______ 1979b. Metodi ed applicazioni della propagazione in vitro delle piante da frutto. In: Tecniche di Colture In Vitro per la Propagazione su Vasta Scala delle Specie Ortoflorofrutticole, pp. 95-106. Pistoia, Italy.

______ and Hatfield, S.G.S. 1976. Root initiation in apple shoots cultured in vitro with auxins and phenolic compounds. J. Hortic. Sci. 51:495-499.

______, Hopgood, M.E., and O'Farrell, D. 1977. Propagation in vitro of M.26 apple rootstocks. J. Hortic. Sci. 52:235-238.

______, Billing, E., and Crosse, J.E. 1978. Bacterial contamination. In: Report of the East Malling Research Station for 1977, p. 177.

______, Pontikis, C.A., and Hopgood, M.E. 1979. Propagation in vitro of five apple scion cultivars. J. Hortic. Sci. 54:155-158.

Lane, W.D. 1978. Regeneration of apple plants from shoot meristem-tips. Plant Sci. Lett. 13:281-285.

Liu, J.R., Sink, K.C., and Dennis, F.G., Jr. 1981. Regeneration of *Malus domestica* cv. Golden Delicious callus, cotyledon, hypocotyl, and leaf explants. HortScience 16:460 (abstr).

Liu, S.F., Chen, W.-L., Wang, H.-H., and Yang, S.-Y. 1978. Shoot tip culture of apple rootstocks and apple seedlings in vitro. In: Proceedings of Symposium on Plant Tissue Culture, May 25-30, 1978, Peking. pp. 485-489, Pl. I-IV. Science Press, Peking.

Lloyd, G. and McCown, B. 1980. Commercially feasible micropropagation of mountain laurel, *Kalmia latifolia*, by use of shoot-tip culture. Proc. Int. Plant Prop. Soc. 30:421-427.

Lundergan, C. and Janick, J. 1979. Low temperature storage of in vitro apple shoots. HortScience 14:514.

______ 1980. Regulation of apple shoot proliferation and growth in vitro. Hortic. Res. 20:19-24.

Mehra, P.N. and Sachveda, S. 1979. Callus cultures and organogenesis in apple. Phytomorphology 29:310-324.

Mu, S.-K., Liu, S.-Q., Zhou, Y.-K., Qian, N.-F., Zhang, P., Xie, H.-S., Zhang, F.-S., and Yan, Z.-L. 1977. Induction of callus from apple endosperm and differentiation of the endosperm plantlet. Sci. Sin. 20:370-376.

Murashige, T. and Skoog, F. 1962. A revised medium for rapid growth and bioassays with tobacco tissue cultures. Physiol. Plant. 15:473-479.

Nasir, F.R. and Miles, N.W. 1981. Historical origin of EMLA 26 apple shoots generated during micropropagation. HortScience 16:417 (abstr.)

Nemeth, G. 1981. Adventitous root induction by substituted 2-chloro-3-phenyl-propionitriles in apple rootstocks cultured in vitro. Scientia Hortic. 14:253-259.

Oehl, V.H. 1980. Shoot tip culture of scion varieties and rootstocks. In: Report of the East Malling Research Station for 1979, p. 51.

Pieniazek, J. 1968. The growth in vitro of isolated apple-shoot tips from young seedlings on media containing growth regulators. Bull. Acad. Pol. Sci. Ser. Sci. Biol. 16:179-183.

Quoirin, M. 1974. Premiers relustats obtenus dans la culture in vitro du meristeme apical de sujets porte-greffe de pommier. Bull. Rech. Agron. Gembloux 9:189-192.

_____, Lepoivre, P., and Boxus, P. 1977. Un premier bilan de 10 annees de recherches sur les cultures de meristemes et la multiplication in vitro de fruitiers ligneux. In: C.R. Rech. 1976-1977 et Rapports de Synthese, Stat. Cult. Fruit. et Maraich. Gembloux, pp. 93-117.

Sakai, A. and Nishiyama, Y. 1978. Cryopreservation of winter vegetative buds of hardy fruit trees in liquid nitrogen. HortScience 13:225-227.

Snir, I. and Erez, A. 1980. In vitro propagation of Malling Merton apple rootstocks. HortScience 15:597-598.

Sriskandarajah, S. and Mullins, M.G. 1981. Micropropagation of Granny Smith apple: Factors affecting root formation in vitro. J. Hortic. Sci. 15:71-76.

_____, and Nair, Y. 1982. Induction of adventitious rooting in vitro in difficult-to-propagate cultivars of apple. Plant Sci. Lett. 24:1-9.

Standardi, A. 1979. Indagine preliminare sull 'influenza della luce nella micropropagazione di alcune specie legnose. In: Tecniche di Colture In Vitro per la Propagazione su Vasta Scala delle Specie Ortoflorofrutticole, pp. 107-118. Pistoia, Italy.

Sutter, E., and Langhans, R.W. 1979. Epicuticular wax formation on carnation plantlets regenerated from shoot tip culture. J. Am. Soc. Hortic. Sci. 104:493-496.

Walkey, D.G. 1972. Production of apple plantlets from axillary-bud meristems. Can. J. Plant Sci. 52:1085-1087.

Welander, M. and Huntrieser, I. 1981. The rooting ability of shoots raised 'in vitro' from the apple rootstock A2 in juvenile and adult growth phase. Physiol. Plant. 53:301-306.

Werner, E.M. and Boe, A.A. 1980. In vitro propagation of Malling 7 apple rootstock. HortScience 15:509-510.

Zimmerman, R.H. 1978. Tissue culture of fruit trees and other fruit plants. Proc. Int. Plant Prop. Soc. 28:539-545.

_____ 1981. Micropropagation of fruit plants. Acta Hortic. 120:217-222.

_____ 1983a. Rooting apple cultivars in vitro: Interactions among light, temperature and phloroglucinol. In Vitro 19:264 (abstr.).

_____ 1983b. Factors affecting in vitro propagation of apple cultivars. Acta Hortic. 131:171-178.

_____ and Broome, O.C. 1980. Apple cultivar micropropagation. In: Proceedings of the Conference on Nursery Production of Fruit Plants through Tissue Culture—Applications and Feasibility, pp. 54-58. USDA Science and Education Administration, Agricultural Research Results, ARR-NE-11, Beltsville, Maryland.

_____ and Broome, O.C. 1981. Phloroglucinol and in vitro rooting of apple cultivar cuttings. J. Am. Soc. Hortic. Sci. 106:648-652.

CHAPTER 15

Grapes

W.R. Krul and *G.H. Mowbray*

HISTORY OF THE CROP

Vitis is the only genus of the family Vitaceae (formerly Ampelidaceae) that is of commercial interest. According to fossil records, it has been on this planet much longer than man and it is likely to have served as a food source for our early, foraging antecedants. Just when and where man first learned to cultivate *Vitis* sp. probably will never be known for certain; however, some have surmised, perhaps not entirely in jest, that man only ended his roaming and foraging ways and settled down to agriculture so that he could be assured a dependable supply of grapes with which to make his wine. Whatever the reason and wherever the place, it happened a long time ago.

Scholars generally concede that cultivation of the grape began in Asia Minor from whence it spread both east and west. It seems likely, from what is known of diffusion patterns and the movement of people, that grape culture was practiced by the sixth millenium B.C. in Caucasia and in Mesopotamia. By 400 B.C. it had spread west into Egypt and other parts of North Africa and is known to have reached China by 2000 B.C. (Lichine, 1973). About the same time, the Greeks were colonizing the eastern Mediterranean shores and islands and used grapes and wine. Over the next thousand years they extended their influence westward to Sicily, Italy, and to the shores of southern France around Marseilles, taking vines with them. The Romans later populated the entire Italian peninsula with vines and then, as part of their colonizing efforts, spread them north along the valley of the Rhone. By the advent of the Dark Ages, much of what we know today as modern Europe practiced viticulture (Dion, 1959).

The single species of *Vitis vinifera* with its thousands of variants, known by at least twice as many names as there are varieties, is for all practical purposes, the only important commercial species. Explorers, colonizers, and settlers have spread it to all of the hospitable areas of both the Northern and Southern Hemispheres where it was previously unknown. In some instances, as for example in North America, accidental interspecific crosses of *V. vinifera* with native wild species have produced some useful cultivars. The Concord grape is one such variety. Later, in response to perceived needs for new varieties with greater disease and cold resistance, deliberate interspecific crosses were made in both Europe and America. Some of these hybrids are now gaining importance in the wine industry of the eastern United States.

GEOGRAPHICAL DISTRIBUTION OF GRAPE PRODUCTION

Vitis is a temperate climate plant. It cannot withstand the extreme cold of high latitudes and high elevations and, for optimum maturity of its fruit, sufficient summer heat and light are required. For all practical purposes, then, the vine does not succeed beyond the fiftieth parallel in the Northern Hemisphere and the fortieth parellel in the Southern Hemisphere (see Fig. 1). Within those limits there exist numerous climatic regions in which the grape can be commercially exploited with varying degrees of success.

Temperate climates offer the widest scope for success since this is the plant's native habitat, yet in temperate regions there exist regional variations caused by high mountains, large land masses, and oceanic currents.

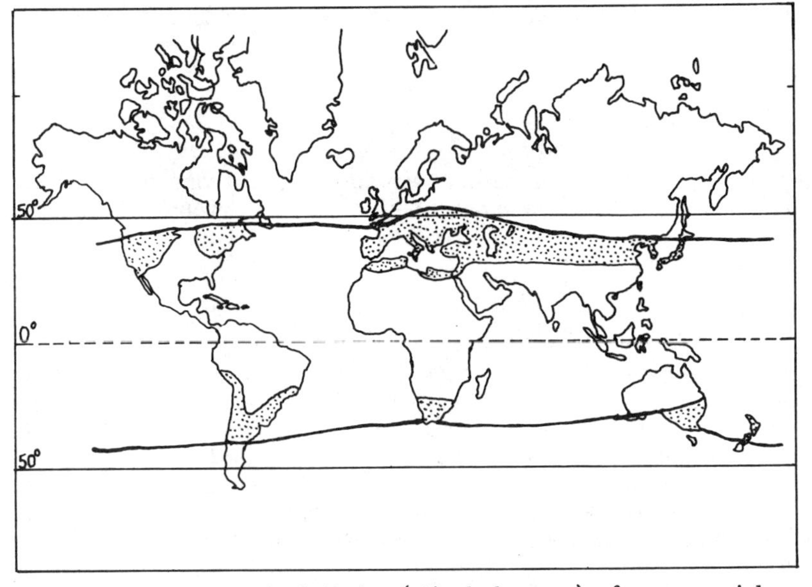

Figure 1. The geographical limits (stippled areas) of commercial grape production. Redrawn from Pongracz, 1978.

In north and south temperate zones, elevation quickly becomes a restricting factor in the successful production of grapes. It has been estimated that, on the average, every 30 m of altitude retards vegetative growth by one day (Galet, 1973). This can be a distinct advantage at lower latitudes where less rather than more heat is desirable, and in some countries successful vineyards are located between 1500 and 2500 m altitude. In France, however, vineyards of the highest quality do not exceed 300 meters of elevation while more ordinary ones can attain nearly 1000 m.

The highest vineyards in the world are in Bolivia at around 2500-3000 m, in Afganistan at 1500-2500 m, in Peru at 1400 m, in Andalusia up to 1300 m, and in Italy on the slopes of Mt. Etna and Vesuvius around 1200 m and in the Val d'Aosta, at 1180 m. As altitude increases, so does the necessity for rational selection of the vineyard site to take advantage of angle to the sun and local precipitation patterns.

With several restrictions, grapes can be cultivated in several climates around the world. Ocean climates have the advantage of heavy rainfall and moderate temperatures. Continental climates where air temperature does not exibit great extremes can also be good for grape cultivation.

The vine flourishes in Mediterranean climates and it is here that most of the world's grapes are grown. Mediterranean-type climates are found in Europe and North Africa from the western shore of Portugal and Morocco all the way to Greece, the south of the Crimea and the border of Asia Minor, as well as in all of the Mediterranen isles. The coasts of South Africa, southwest Australia, California, and Chile also have a Mediterranean climate. Grapes can also be grown in desert climates if the region is supplemented with irrigation.

Finally, with good management, grapes can be cultivated in tropical climates, and with careful pruning may even produce two crops per year.

PRODUCTS OF THE VINE

Depending on the preponderant use made of any one of its products, the vine can be classified among any of several groups of agricultural plants.

Table Grapes

The fresh grape is a fruit that like peaches, apples, and pears, is consumed for food. In the Northern Hemisphere, the appearance of fresh grapes on the market occurs toward the end of July and extends through the summer and into the autumn until mid-November. Extension of the season can be accomplished by importation from North Africa and the Middle East and even prolonged throughout the year by importation from the Southern Hemisphere—South Africa and Australia, in particular.

It is difficult to state precisely the worldwide production of table grapes. Certain producing countries do not provide statistics and others are hampered in their efforts because of indifferent reporting from individual producers. Many table varieties are used for wine in years of surplus yields and producers of wine grapes often consume a portion of their production at the table or sell them for consumption on the open market. An estimate for 1971 puts worldwide production at 6,300,000 MT, or close to 7 million short tons. Based on world population of 3.6 billion, that represents a per capita consumption of 1.7 kg (3.74 lb). This is certainly a conservative figure in view of the fact that many producing countries annually consume more than 10 kg (22.2 lb) of table grapes per capita.

Fresh grapes have important nutritional qualities. They provide about 700 cal/kg (2.2 lb) and in addition contain calcium, phosphorous, and vitamins.

Raisins

The dried grape constitutes yet another possibility for extending the consumption of a seasonal product throughout the whole year and offers a method to conserve what is essentially a perishable food product. The technique of raisin production has been known and used since very early times, particularly in the grape-growing regions of the Mediterranean basin, where summer climates are cooperative, and in the central valley of California. Raisins require high temperature and low humidity during the drying period.

The raisin is a high-energy food, yielding 3340 cal/kg, making it comparable to dried almonds. It is very rich in sugars which make up approximately 60% of its total weight.

The consumption of raisins is high among the Moslem peoples of the old Ottoman Empire, and the inhabitants of the Middle East and central Asia. In Europe, the principal consuming countries are Great Britian, Greece, Germany, the Low Countries, and Russia, while in the Western Hemisphere they are the United States and Canada.

Worldwide production of raisins varies from 650,000 to 750,000 MT, and since 100 kg of fresh grapes are required to give 20-30 kg of raisins, that corresponds to a production of 2.6-3.0 million MT of grapes from about 1,400,000 acres of specialized grapes.

Grape Juice

The fresh juice of pressed grapes is either bottled for consumption as a beverage or is converted to jellied products. Again accurate worldwide figures are unavailable, but it has been suggested that from 7,200,000 to 8,400,000 acres of grapes are devoted to this use. The prinicipal producers are summarized in Table 1.

Concentrated Grape Juice

Various evaporative processes are used to extract water from grape juice and to concentrate the sugars and other soluble solids for the

Table 1. Principal Grape-Juice-Producing Countries of the World and Their Estimated Annual Production in Millions of Gallons

COUNTRY	EST. ANNUAL PRODUCTION	COUNTRY	EST. ANNUAL PRODUCTION
United States	275.00	Morocco	1.24
USSR	20.50	Turkey	1.06
France	18.50	Austria	0.79
Syria	5.94	Bulgaria	0.79
Italy	5.40	Spain	0.66
Algeria	2.64	Germany	0.29
Switzerland	2.06		

production of edible condiments. These techniques have been known and exploited for years in areas where viticulture is predominant. During the difficult years of 1940-1944 in France, for example, great quantities of grapes were used in this fashion to supplement the diets imposed by occupation forces. In Moslem countries more than a third of the grapes grown are converted to concentrated musts that are used in numerous ways to produce high-energy foods that are nonperishable and easy to transport.

Within the last 10 years a surging increase in interest in wine, coupled with a back-to-the-earth, do-it-yourself movement, has generated a large market for concentrated grape juice. Home wine-makers in urban areas and regions where fresh grapes cannot be obtained have learned that extremely potable wine can be made from reconstituted concentrates. The exact size of this market is unknown but it is appreciable and growing.

Wine

By far the largest percentage of the world's grapes are used for the production of wine. The 15 largest wine-producing countries of the world with their estimated production and grape acreage are listed in Table 2.

In a world where 26 million acres or more are devoted to grape production, the countries listed in Table 2 represent over 70% of the total. Although they produce wine, these countries don't necessarily drink it all. There are several densely populated, relatively affluent nations that cannot produce wine and are forced to import. The consumption per capita of the 10 major wine-consuming nations of the world along with their production per capita are listed in Table 3.

Of the ten countries listed in Table 3, only three--Chile, Switzerland, and Austria--consume more wine than they produce, and so are importers as well as producers. It is also interesting to note that many large industrial countries have a relatively low per capita consumption of wine, as for instance, West Germany (4.8 gal), South

Table 2. The Fifteen Largest Wine-Producing Countries of the World in 1973[a]

COUNTRY	AREA PLANTED (acres)	WINE PRODUCTION (millions of gallons)
Italy	2,927,631	1,696.75
France	3,265,750	1,619.14
Russia	2,727,374	758.29
Spain	3,914,826	625.02
Argentina	740,012	582.20
United States	580,450	314.40
Portugal	864,749	237.02
Algeria	716,300	217.80
Rumania	867,659	203.28
W. Germany	227,398	159.11
Yugoslavia	621,498	144.04
South Africa	250,285	146.07
Greece	525,575	125.95
Hungary	548,340	117.72
Chile	314,134	105.60

[a]Galet, 1973.

Table 3. Annual Consumption Per Capita of the 10 Largest Wine-Consuming Countries of the World and Their Wine Production Per Capita

COUNTRY	ANNUAL CONSUMPTION PER CAPITA (gal)	ANNUAL WINE PRODUCTION PER CAPITA (gal)
Italy	29.3	30.1
France	28.3	30.2
Portugal	24.1	25.0
Argentina	22.5	24.3
Spain	15.9	19.0
Chile	11.6	10.6
Switzerland	10.8	3.7
Greece	10.6	14.3
Austria	10.5	6.5
Hungary	8.5	11.4

Africa (3.0 gal), the United States (2.0 gal), Sweden (1.7 gal), England (1.0 gal), and Canada (0.8 gal). Within the last few years the per capita consumption of France and Italy has been declining, but that of the more affluent industrial countries just listed has been rising. All projections indicate that the rise will continue well into the next century.

MISCELLANEOUS USES. Wine is distilled in many countries for the production of several different items. Some is used for high-proof drinks e.g., cognac, armagnac, brandy, marc, and grappa. Some is used to fortify standard wines such as sherry, port, and dessert wines. Still more distilled wine goes into the manufacture of flavored aperitifs such as Vermouth, Arak, Anis, and others. There is a significant proportion that is used in the form of ethanol to extend motor fuel and for pharmaceuticals, perfumes, and aromatic powders. Finally the residues of the distillation of wine yield tartrates used in the production of tartaric acid and cream of tartar, as well as glycerine; ultimately, they are useful as fertilizer.

Wine vinegar is another widely used product to which large quantities of wine are devoted. In France alone about 7,920,000 gal are converted to vinegar annually. This is about as much wine as Israel, for instance, produces each year.

The vegetative parts of vines also have their uses. The leaves themselves are used in many parts of the world as a food. Grape leaves stuffed with meat and rice are common dishes in many grape-growing regions. The seeds contain oils that are useful in the preparation of soap, painting media, and edible unsaturated oils. The hardened, mature canes of producing vines and the trunks of old vines serve as fuel for cooking and heating, for the production of charcoal, and when shredded and composted, as agents for restoring organic matter to agricultural land. Finally, as decortive motifs, vine leaves, vine canes, and grape bunches add much to the esthetic pleasures of living.

GRAPE PROBLEMS AND PROSPECTS FOR IMPROVEMENT

The single grape species, *Vitis vinifera*, with its many variants, is today the predominant species for the production of high-quality wine and table fruit. It evolved in climates that were mild and relatively devoid of harmful pathogens. Exchange among peoples has changed that. The transplantation of *V. vinifera* to regions in which it does not naturally occur has uncovered its weaknesses with respect to winter hardness and its lack of resistance to cryptogamic diseases and soil-inhabiting insects. On the other hand, commerce between nations has introduced those same diseases and insects into regions where the vine had evolved without them, with resultant economical turmoil. In 1845, a malady probably originating in North America was discovered in Europe. This is variously known as oidium, or powdery mildew, and its taxonomic designation is *Uncinula necator*. By 1851, all of the vineyards in Europe were infested (Ribereau-Gayon and Peynaud, 1971). In short order followed invasions of the root louse, *Phylloxera vastatrix* (1865), downy mildew, *Plasmopora viticola* (1878), and finally black rot *Guignardia bidwellii* (1885), all introduced from North America. Nearly all of the grape-growing regions of the world have since become infected.

With ingenuity, hard work, and great expense, solutions to all of these problems and more, have been found. The airborne diseases have succumbed to modern spray materials, but the latter are expensive to produce, and costly and time-consuming to apply. The soil-derived problems, specifically those related to phylloxera and nematodes, have been ameliorated after years of research that led to the development of resistant rootstock varieties, requiring that sensitive cultivars be grafted. Selective interspecific breeding programs, dating from the late nineteenth century and still on-going, have produced new cultivars with varying degrees of resistance to low temperatures, diseases, and insect predation. Breeding programs have made possible the commercial planting of vines in areas previously considered hostile. With a very few exceptions, these hybrid crosses only approach but do not equal in quality the fruit produced by *V. vinifera.*

Modern research methods have helped grape specialists identify a number of viruses to which the vine is susceptable. The work of locating and controlling vectors, of testing for the presence of infestations, and of developing reliable methods of producing virus-free stock is continuing. Accomplishments to date have been of immense value to the grape-growing industry in terms of increased yields of high-quality fruit, but progress is slow. The highly technical nature of the task combined with the necessarily high costs of such research are to blame.

There seems to be great promise for the future in a new line of attack on all of the problems associated with grape-breeding. Tissue-culturing of grape material combined with concerted efforts to delineate the detailed genetics of *Vitis* could make it possible to produce cultivars with outstanding taste characteristics, yet with resistance to low winter temperatures, cryptogamic diseases, and predation by insects, whether soil-inhabiting or airborne. This, admittedly, is a large order, but it is certainly a goal worth pursuing.

LITERATURE REVIEW

The first reports of successful grape culture came from the laboratory of Morel (1944a). He determined the optimum concentration of auxin for cell proliferation (Morel, 1944b), and developed hormone-independent cell lines, and cell lines with lowered auxin requirements (Morel, 1947). He also described the origin of callus from stem explants (Morel, 1945). Shortly thereafter, Kulescha (1949) demonstrated that habituated and crown gall cells (also habituated) produced greater quantities of auxins than auxin-requiring cells.

Most of the in vitro research during the 1950s focused on the anatomical (Gautheret, 1955), chemical, and enzymatic differences (Czosnowski, 1952a), and vitamin requirements (Czosnowski, 1952b) of cultured grape crown gall cells. The regulation of anthocyanin production in habituated cells by light, sugars, and osmotic stress was reported (Slabecka-Szweykowska, 1952) and the attempt to establish cell cultures of several species and cultivars was described (Fallot, 1955). The enhancement of cell growth by CW and its interactions with auxin were explored (Duhamet, 1951a,b) and single-cell clones of grape phylloxera galls were established (Hildebrandt, 1958). Graftage of auxin-requiring, auxin-independent, and crown gall cells to young vines, demonstrated that growth of auxin-requiring and habituated cells was controlled by the host, but tumor cells were not (Braun and Morel, 1950).

Many of the publications of the 1960s were confined to the systematic determination of growth-factor requirements for single-cell clones isolated from phylloxera galls and normal *V. riparia* tissues (Pelet et al., 1960; Arya et al., 1962a,b, 1963; Hildebrandt, 1963, Arya, 1965a,b). The physiological and biochemical differences between insect gall and normal clones (Arya, 1964, 1965c) and their differential sensitivities to gamma radiation were investigated (Arya and Hildebrandt, 1969a,b). Single-cell clones of crown gall cells were established on defined medium (Reinert, 1963) (clones obtained previously were grown on media supplemented with CW or other complex organics) and intact berries were cultured, some of which produced callus (Radler, 1964).

The demonstration that photoperiodic treatment of parent vines could influence morphogenic potential of explants (Alleweldt and Radler, 1962) has recently been exploited for optimization of micropropagation methods (Chee, 1980; Chee and Pool, 1982).

Perhaps the most practical work in the 1960s concerned the development of methods for the production of virus-free clones by thermotherapy treatment and solutions to some of the attendant problems of growth and organ regeneration of heat-treated shoot explants (Galzy, 1961, 1962, 1964a,b, 1969a; Galzy and Compan, 1968; Gifford and Hewitt, 1961; Hoeffer and Gifford, 1964).

The diversity and quantity of in vitro grape research increased dramatically during the 1970's. Some of the earlier physiological research was abandoned and many new and exciting efforts initiated.

The research on phylloxera gall and normal cells was apparently terminated after the physiological correlates between cell and nuclear size were mathematically analyzed (Goyal and Goyal, 1970, 1971a,b). Virus-elimination research increased, and major efforts were made to optimize the chemical and physical parameters of the media for growth and development of heat-treated shoot tips (Galzy, 1971, 1972a, 1977; Bini, 1976; Grenan, 1979a,b; Legin et al., 1979; Mur, 1979; Valat et al., 1979). Methods for grafting apical domes from heat-treated plants onto seedling or herbaceous root stocks were also described (Ayuso and Pena-Inglesias, 1978; Englebrecht and Schwerdtfeger, 1979).

The production of cell cultures from various explants on the vine such as berries (Hawker et al., 1973) and endosperm (Mu et al., 1977), and from tetraploid and diploid clones of *V. vinifera* cultivars (Staudt et al., 1972) were explored. The conditions and media formulations for successful long-term culture of cells (which is a problem for many *vinifera* cultivars) were investigated (Jona and Webb, 1978).

The first reports of protoplast production (Benbadis and Baumann, 1973; Skene, 1974, 1975; Burgess and Linstead, 1976) indicated that cell division and regeneration of walls may be difficult in grape. However, the production of walls and callus by pericarp callus protoplasts suggests that success is not impossible (Skene, 1975).

Anther cultures produced haploid callus which subsequently perished (Gresshoff and Doy, 1974), plantlets and dormant embryolike structures which also did not survive (Hirabayashi et al., 1976), and callus which produced viable somatic embryos with diploid genomes (Rajasekaran and Mullins, 1979). Research directed toward control of plant regeneration has been particularly successful in grapes. Somatic embryos have been obtained from unfertilized ovules of *V. vinifera* and another *Vitis* species (Mullins and Srinivasan, 1976, 1978); from immature stems, flower clusters, and young leaves of a *Vitis* spp. hybrid (Krul and Worley, 1977); and from petiole and leaf interveinal explants of *Vitis* spp. (Favre, 1977). The first commercial planting of vines from somatic embryos (*Vitis* sp. cv. Seyval) was established at Montbray Vineyards, Silver Run, Maryland, during the spring of 1977.

Three potential methods for rapid clonal multiplication of grape vines were developed during the 1970s. Some of the requirements for propagation via internode cuttings were reported (Pool and Powell, 1975; Ecevit, 1979), and the requirements for rapid multiplication of shoot tips and virus elimination were reported (Jona and Webb, 1978: Harris and Stevenson, 1979). A novel method for rapid multiplication involving the induction of adventive buds from fragmented apical meristem explants was also described (Barlass and Skene, 1978).

The developmental anatomy (Bernard and Mur, 1979) and regulation of morphogenesis of isolated grape shoots has been extensively studied. It was demonstrated that tendril formation, flower, and fruit production could occur on a limited scale in vitro (Srinivasan and Mullins, 1978; Mur, 1978; Favre and Grenan, 1979). The demonstration that anlagen (a group of uncommitted meristematic cells) could produce tendril or flower primordia under an appropriate hormone regime was an important contribution to the understanding of the floral stimulus for grape (Srinivasan and Mullins, 1978). The production of vines with juvenile characteristics was reported for somatic embryos, adventive buds, or repeatedly subcultured shoot cuttings (Mullins and Srinivasan, 1976; Barlass and Skene, 1978; Mullins and Nair, 1979).

The problem of long-term cultures of isolated grape cells continues to plague researchers and methods for plantlet regeneration from a more diverse range of cultivars and species is still required. To help solve these problems, the effect of age and morphological position of explants (Brezeanu et al., 1980), and the effects of hydrogen ion concentration and light quality and intensity (Lai and I, 1980) on growth and differentiation of grape cells have been investigated.

The effect of in vitro culture and thermotherapy on subsequent morphological changes and field performance of "cured" root- and scion stocks have been reported (Grenan, 1980; Valat et al., 1981).

The prospects and strategies for grape improvement via tissue culture have been discussed (Jona, 1980; Meredith, 1981) and several exciting opportunities exist for application of tissue culture to crop improvement.

The propagation of grape using fragmented shoot apices or shoot multiplication methods, may be commercialized during the 1980s. The origin of adventive buds (Barlass and Skene, 1980a, 1981a) and the chemical and physical parameters for shoot production continue to be revised and optimized for several cultivars and species (Barlass and Skene, 1980b; Jona and Vallania, 1980; Chee and Pool, 1982). Shoots from fragmented apices have been used for selection and adaptation to salt stress (Barlass and Skene, 1981a).

Efforts to develop ovule culture systems to produce seeds in normally seedless grape cultivars (Cain, 1980) will facilitate breeding programs and contribute to the understanding of embryo development.

The production of somatic embryos by ovule culture has recently been extended to several other cultivars and species (Srinivasan and Mullins, 1980) and the physical and chemical factors which regulate the production of somatic embryos from other somatic embryos and the relationship of this process to embryo dormancy and development were described (Krul and Myerson, 1980; Myerson, 1981).

We hope to see a dramatic increase in research directed toward the selection of cells and excised shoot cultures resistant to stresses such as salt, boron, Pierce's disease toxin, and low temperature. There will be a resurgence of research examining the co-culture of host and pathogen. The objectives of these efforts are to select clones tolerant to fungal pathogens (mildews) and nematodes, and to determine biochemical, physiological, and genetic nature of host-pathogen specificity. The refinement of thermotherapy methods and propagation of grape by adventive bud production or shoot tip multiplication may be combined to provide a commercial process for rapid multiplication and storage of virus-indexed plants. The nature of the transition from the adult to the juvenile state, via in vitro culture, should also be pursued. The use of juvenile tissue as a source of totipotent cells is also likely to be explored.

PROTOCOLS

Development of Cell Culture Systems

The use of grape cell cultures for the selection of clones tolerant to applied chemical or physical stress is one of the options available for grape improvement (Meredith, 1981). However, two major problems confront investigators attempting to use these strategies: (1) cell cultures from grape explants of some species and cultivars are difficult to establish and maintain (Hawker et al., 1973; Krul and Worley, 1977; Jona and Webb, 1978), and (2) when established, cultures rarely produce adventive shoots or embryos (Brezeanu et al., 1980; Hawker et al., 1973; Jona and Webb, 1978; Lai and I, 1980; Krul and Worley, 1977). The difficulty of long-term grape culture appears to have a genetic as well as a physiological-biochemical basis. Long-term cultures have been obtained from *Vitis* species (Pelet et al., 1960) and from habituated or crown gall tumor cells of *V. vinifera* cultivars (Morel, 1947; Kulescha, 1949). The hormone-independent lines appear to be extremely stable with respect to longevity in culture. In recent years long-term cultures of *V. vinifera* and other *Vitis* species have been obtained

(Hawker et al., 1973; Skene, 1975; Brezeanu et al., 1980). The parameters for successful long-term culture are discussed below.

EXPLANT SOURCE. The formation of organs, tissues, and cells in plants and animals requires the differential expression of gene programs (Brown, 1981). The stability of gene expression for a given isolate can be maintained or lost in culture with the result that cells from the same plants but from different organs can display different requirements for in vitro growth (Pelet et al., 1960; Arya, 1965a,c; Arya and Hildebrandt, 1969a) and often produce cells with different phenotypes (Brezeanu et al., 1980).

Relatively stable and fast-growing cell cultures have been isolated from young herbaceous stem explants, petioles, tendrils, and buds, and slower growing isolates were obtained from mesophyll, inflorescence, and fruit cells (Brezeanu et al., 1980). Stable isolates were also obtained from berries (Hawker et al., 1973) and regenerated pericarp protoplasts (Skene, 1975), and our group has obtained a hormone-independent cell line from somatic embryos of *Vitis* sp. cv. Seyval. There are no reports of cultures produced from root explants. The longevity of a cell culture does not appear to be a function of morphological position of the explant; however, the genotype of the isolates appears to be a major determinant.

Hawker et al. (1973) examined the production of callus from berries of several *V. vinifera* cultivars. All but two of the cultivars produced callus (Wortley Hall and Muscat Gordo Blanco) and the remaining eight produced moderate to rapidly growing callus. Staudt et al. (1972) showed that tetraploid clones of *V. vinifera* cvs. Riesling and Portugieser produced more callus than diploid clones of the same cultivars. The increase in growth of tetraploid clones was attributed to a greater cell volume (mostly water) as the dry weights for each ploidy level were similar.

SEASONAL INFLUENCE. Many cultivars of grape are photoperiodically sensitive and display a reduction in growth rate in response to decreasing day length (Alleweldt and Radler, 1962). Isolates produced successful cultures throughout the season but differences in growth rate and morphogenic potential have been observed (Alleweldt and Radler, 1962; Brezeanu et al., 1980). For instance, stem isolates from Kober 5BB, grown under shorter days, produced mainly roots and small amounts of callus, whereas similar explants from long-day plants produced callus. The seasonal effects on rapidity of callus initiation, growth, and morphological change has been determined for several grape cultivars (Brezeanu et al., 1980). The following observations were made: February cultures showed a long lag phase for callus establishment and slow growth for the first few passages but more rapid growth thereafter; April and May cultures showed rapid callus production and growth; in June and July, callus cultures produced roots after a few passages, and in October, callus readily formed roots.

LIGHT. As demonstrated above, the duration of light received by the parent plant influences the rate of callus initiation and the subsequent growth rate. The effect of light on established cell cultures is quite variable. Hawker et al. (1973) suggest that light did not benefit growth of berry callus cells of several *V. vinifera* cultivars. In contrast, Lai and I (1980) demonstrated a linear increase in growth of callus to increasing intensities of white light of *V. vinifera* cv. Sentenial. Intensities above 3 Klux suppressed growth. Brezeanu et al. (1980) report that growth of callus from *vinifera* cultivars was optimal with light intensities of 5-6 Klux.

The quality of light may also influence the growth of grape cells. Lai and I (1980) demonstrated that blue light enhanced growth of callus relative to red, green, or yellow light, or dark. It is not known whether cells received equal amounts of energy from each spectral region.

Callus cultures of grape appear to be insensitive to photoperiodicity. Although most investigators find that 16-hr light periods produce adequate growth (Brezeanu et al., 1980), others (Jona and Webb, 1978; Staudt et al., 1972) found relatively short photoperiods of 12-hr adequate.

TEMPERATURE. Grape cells display dramatic growth and development responses to temperature. Both dormant buds and seeds require a period of chilling prior to resumption of growth. These observations suggest that *Vitis* contains temperature-sensitive genes which regulate growth and development.

Nearly all reports concerning in vitro cell growth indicate that temperatures of 25-27 C provide adequate growth. However, the demonstration that cells from *Phylloxera* galls have a lower optimal temperature than normal cells (36-38 C) suggests that additional information on temperature requirements may be desirable.

H^+ CONCENTRATION. The growth of cells from *V. riparia Phylloxera* galls was optimum at a pH of 4.5 whereas cells from normal clones grew best at a pH of 5.0 (Pelet et al., 1960). However, when normal cells were exposed to gamma irradiation they grew best at a pH somewhat lower than irradiated gall cells (Arya and Hildebrandt, 1969b). The optimum range of pH was between 5.2 and 5.6 for growth of *V. vinifera* cv. Sentenial cells (Lai and I, 1980). This latter range of H^+ concentration appears to be useful for most grape cell culture systems reported to date. However, the data which show that *V. riperia* grows optimally at lower pH than *V. vinifera* suggest that H^+ ion concentration of the medium should be considered, in the initiation of cultures of Ca_2^+-tolerant and susceptible cultivars.

NUTRIENT SALTS. To our knowledge, manipulations of mineral salt compositions to achieve maximal growth and development of grape cells have not been reported. Grape cell cultures have been established on a wide array of mineral salt formulations and there are no reports which indicate that one formulation is superior to another.

CARBON SOURCE. Adequate growth of grape cell cultures was obtained with sucrose concentrations of 58.4-87.6 mM. Grape cells will also grow on glucose (Reinert, 1963) and hydrolyzed starch (Slabecka-Szweykowska, 1952). The starch hydrolysate was superior to sucrose for the growth of grape tumor cells.

Clones of *Phylloxera* leaf gall cells and normal cells displayed several interesting contrasts in ability to grow on various carbon sources. For instance, gall cells grew more rapidly on mannose than did normal cells and normal cells grew faster on fructose than did gall cells. Both cell types displayed different optimal concentrations for individual sugars and one clone of gall cells grew as well on galactose as sucrose (Arya, 1965a). The latter observation is of interest because galactose is considered toxic to many plant tissues (Burstrom, 1948; Ferguson et al., 1958) and has been shown to regulate morphogenic activity (Moore et al., 1969; Colclasure and Yopp, 1976; Kochba et al., 1978).

VITAMINS. Grape tumor cells appear to require thiamine for adequate growth (Reinert, 1963), whereas other clones of tumor cells did not require vitamins for growth (Czosnowski, 1952b). Calcium pantothenate increased growth of normal but not *Phylloxera* gall cells, and pyridoxine and choline chloride did not influence growth of either clone (Arya, 1965c). Because a systematic determination of vitamin requirements has not been made for normal cells of *V. vinifera* cultures, it may be useful to determine whether reported lack of success of long-term cultures may be related to a vitamin deficiency.

MISCELLANEOUS ORGANIC SUBSTANCES. Although not required for growth of grape cell cultures, certain complex organic addenda have provided increased growth of isolated cells. CW (10% v/v) (Dahammet, 1951a,b; Hildebrandt, 1958), yeast extract (10% w/v) (Pelet et al., 1960), adenine sulfate plus casein hydrolysate (40 mg/l and 3-10 g/l, respectively) (Pelet et al., 1960), casein hydrolysate (2 g/l) (Skene, 1975), and yeast RNA (10-41 mg/l) (Arya, 1963) have all been reported to enhance the growth rate of isolated cells of grape.

GROWTH REGULATORS. Many of the early workers utilized either IAA, NAA, or IBA with or without CW for induction and continuous growth of isolated grape cells (Morel, 1944b; Duhamet, 1951b; Hildebrandt, 1958). Subsequent investigators found that CW could be deleted from the medium if a cytokinin was added.

Phenoxyacetates (2,4-D, PCPA, or NOA) are useful for callus initiation (Krul and Worley, 1977; Jona and Webb, 1978; Srinivasan and Mullins, 1980; Brezeanu et al., 1980) and may be essential for somatic embryo induction (Srinivasan and Mullins, 1980), but maintenance of long-term growth, in their presence, is not always possible.

Several cytokinins have been utilized for the initiation and long-term growth of isolated grape cells. However, grape cells do not demonstrate a requirement for cytokinin or a preference for a particular molecular configuration. BA is the most widely used cytokinin, although KIN (Hawker et al., 1973) or ZEA (Jona and Webb, 1978) are also effective.

The effect of GA_3, ABA, or ethylene on the initiation and growth of grape cells has not been reported.

CONCLUSIONS. At this writing it is not possible to maintain long-term callus or cell suspensions of many important grape cultivars. Because of this deficiency, the use of grape cells for newly emerging crop improvement strategies (mutant and stress selection, protoplast fusion, etc.) is severely limited. It is possible that some cultivars are sensitive to media components such as boron or calcium (e.g., rhododendron cultures are extremely sensitive to "normal" K levels in MS medium) or require an as yet untested vitamin or growth regulator. Systematic exploration of media components, although tedious, may provide information for maintenance of long-term cultures.

Protoplast Culture

The use of protoplasts for the introduction of foreign germplasm, somatic hybridization, and for studies of hormone action, membrane structure and function, and cell wall synthesis, are well documented for model plant systems (Wagner et al., 1978). The use of grape protoplasts as gene acceptors will not develop at the same rate as the tobacco model system, because regeneration of grape protoplasts has not been demonstrated and because the economic and human resources have not been mobilized for the task (Meredith, 1981). The production of grape hybrids by somatic cell fusion may not be realized for three reasons. As mentioned above, grape protoplasts have not been regenerated, tetraploid vines have not been productive (Olmo, 1976), and haploid plants have not been produced.

Protoplasts of grape have been prepared from pericarp callus (Skene, 1974, 1975), stem callus (Burgess and Linstead, 1976), and crown gall cells (Benbadis and Baumann, 1973). Protoplasts from the above sources produced cell walls and divided one or two times. To date, only protoplasts from the pericarp have produced callus capable of continued cell division (Skene, 1975).

There are no reports of protoplast isolation from differentiated leaf or other tissues and organs of grape plants. Callus or liquid cell suspension cultures have been used for protoplast isolation for all published reports. Protoplast production was more efficient from callus cells than from suspensions (Skene, 1975).

Callus was incubated for 16 hr at 26 C in a mixture of 2% cellulase (Onozuka SS) and 1% pectinase (macerozyme). Optimal osmotic conditions were maintained by a KCl-$CaCl_2$ solution (0.1 and 0.14 M, respectively) and the pH was adjusted to 5.5. Isolation was completed in the dark. Wall digestion could be speeded up and protoplasts isolated in less than 2 hr when cells were digested with a sequence of enzymes at 35 C, but protoplasts isolated in this fashion formed a peculiar wall and did not divide (Burgess and Linstead, 1976).

PROTOPLAST PURIFICATION. The mixture of cells and enzymes are passed through a series of stainless steel sieves (150 and 60 μM) and protoplasts are centrifuged at 100 x g for 3 min. Protoplasts are then washed 3 times with KCl-$CaCl_2$ and pelleted after each wash by centrifugation (50 x g for 3 min).

The protoplast pellet was suspended in an appropriate culture medium and adjusted to a cell density of 10^5 protoplasts/ml. Three-ml aliquots of protoplasts were transferred to 6-cm petri dishes which were then sealed with Parafilm. Cells were incubated in the dark at 26 C. Survival percentages were greater in the dark and generally improved by using agar medium (6 g/l) relative to liquid suspension.

The medium which supported wall regrowth and continued cell division consisted of the nutrient salts of Ohyama and Nitsch (1972) and the vitamins of Gamborg and Eveleigh (1968). These formulations were supplemented with the following: casein hydrolysate (2 g/l), glutamine (13.7 mM), sucrose (29.2 mM), ribose (1.67 mM), glucose (1.39 mM), xylose (1.0 mM), arabinose (1.0 mM), glycine (0.05 mM), inositol (5.5 mM), serine (0.95 mM), and BA (4.4 μM).

Cell multiplication was faster with NAA (5.4 μM) and sorbitol (0.24 M), as opposed to 2,4-D (0.45 μM) and sucrose (0.23 M) in wall regeneration medium. Cell development did not occur if the osmolality of the medium exceeded 0.5 M. Cell colonies were transferred from wall regeneration medium to Hawker et al. (1973) callus medium where they continued to divide for several passages.

Anther Culture

Grape breeding would be facilitated by the availability of homozygous inbred lines. The long sexual generation time and the complexity of available genotypes suggests that inbreeding [which results in sterility after two to three generations (Olmo, 1976)] and selection are difficult for grape. "Instant" inbreds from colchicine-treated pollen embryos have been incorporated into the breeding program of several crop plants (Collins and Legg, 1979). In order to obtain homozygous lines of grape it may be necessary to use doubled haploids because natural haploid plants have not been observed in the genus *Vitis* and inbreeding is limited to only a few cycles (Olmo, 1976).

Haploid callus from anthers was obtained from only 3 of the 27 cultivars examined by Gresshoff and Doy (1974). None of the calli produced plantlets, but a few formed green cells which contained vascular tissue. All cultures died after 3 months. Plantlets and structures resembling dormant embryos were produced from anther callus of *Vitis thumbergii* (Hirabayashi et al., 1976). Plantlets did not survive and their chromosome number was not reported. Somatic embryos and viable plants were obtained from anther callus of Glory vine, a *V. vinifera* x *V. rupestris* hybrid (Rajasekaran and Mullins, 1979), using the following procedure.

EXPLANT PRETREATMENT. Anthers from flowers 2-3 mm long, with pollen grains predominantly in the uninucleate stage, were chilled for 72 hr at 4 C prior to culture. All filaments were removed and special care was taken not to injure the anther wall during excision.

INDUCTION MEDIUM. The salts of Nitsch and Nitsch (1969) supplemented with 0.02 M $FeSO_4 \cdot 7H_2O$, 1 μM BA, and 5 μM 2,4-D proved to be superior for the production of callus which subsequently gave rise to embryos. Media supplemented with NOA and/or higher BA concentrations produced callus which did not survive or ceased growing after subculture. Anthers were cultured in 100 ml Erlenmeyer flasks with 25 ml of liquid medium for 25 anthers. Cultures were maintained on a gyratory incubator (80 oscillations/min) at 27 C. Agar-based medium was not satisfactory for the induction process. Both prechilling of anthers and maintenance of cultures in darkness were essential for embryo induction. Callus from ruptured anthers was visible 10-20 days after culture initiation.

EXPRESSION MEDIUM. Callus and aggregates grown on basal medium or basal medium supplemented with GA_3 produced numerous somatic embryos when cultured in the light (16 hr illumination and 2.5 W/m^2). Mature embryos formed 45 days after initiation.

EMBRYO DEVELOPMENT. Development of embryos occurred slowly on agar-based basal medium and only 2% of the embryos formed plants. GA_3 (1 μM) increased the frequency of normal plant development (47%) and chilling the embryos for 2-8 weeks at 4 C increased the frequency even further (83-94%).

The chilling treatment was effective for all stages of embryogeny. Plants which developed without chilling produced the first foliage leaf in 30 days whereas those which received chilling treatment formed them within 7 days after transfer to 27 C.

CONCLUSIONS. Embryos were produced only from vines that were male or had strong tendencies toward maleness. The hermaphrodite, Cabernet Sauvignon, produced callus that eventually browned and died.

All plants regenerated from callus had the normal diploid complement of chromosomes (2n = 38). It was not clear whether embryos were dihaploid or derived from somatic tissue. Many dwarf and abnormal vines were produced. There were variations in viability and vigor as only 25% of the plants survived transplantation to soil.

Somatic Embryogenesis

One of the prerequisites for success in the genetic modification of grape by current genetic engineering strategies or the selection of stress-tolerant clones from large cell populations, is the technology to produce plants from isolated cells. Development of this technology might also be used for the mass propagation of new plant introductions and/or the maintenance and production of virus-indexed plants. The recent development of a stable system for continuous embryo production from epidermal cells and the isolation of hormone-independent grape cells with embryogenic competence (Krul and Myerson, 1980) may provide systems for the study of gene expression during embryo induction and development.

The process of embryo production from isolated parenchyma cells involves three identifiable stages that are based on cell morphology and response to growth regulators. These stages are: (a) induction, or the conversion of somatic parenchyma cell to a proembryonic mass (PEM), (b) expression, or the conversion of the PEM into an embryo, and (c) germination of the embryo and development of the plant.

INDUCTION. The conversion of a highly vacuolate parenchyma cell into a cluster of small, thick-walled, densely cytoplasmic cells with prominent nuclei and nucleoli, by internal partitioning is defined here as the process of induction (also see Kohlenbach, 1978). The precise location of inducible cells in grape remains unknown. However, they appear to arise from the nucellar callote (Srinivasan and Mullins, 1980; Mullins and Srinivasan, 1976), the superficial cells of isolated callus (Krul and Worley, 1977) single epidermal cells of the junction of the root and shoot of somatic embryos (Krul and Myerson, 1980), and suspensors (Srinivasan and Mullins, 1980). Tissues or organs that contain induced or inducible cells are nucellar callote (Mullins and Srinivasan, 1976), anther (Hirabayashi et al., 1976; Rajasekaran and Mullins, 1979), leaf petiole and interveinal area (Favre, 1977), immature flower clusters and green stems (Krul and Worley, 1977), embryo suspensors (Srinivasan and Mullins, 1980), and epidermal cells at the junction of the hypocotyl and root of somatic embryos (Krul and Myerson, 1980) (see Fig. 2).

The genetic composition of the parent plant plays a significant role in the inductive process. The induction of somatic embryos in *V. vinifera* cultivars has been confined to nucellar explants, whereas wild *Vitis* species and interspecific hybrids contain a more diverse range of inducible tissues or organs (Favre, 1977; Hirabayashi et al., 1976; Krul and Worley, 1977; Rajasekaran and Mullins, 1979).

The precise chemical and physical requirements for the inductive process remain to be defined. It was postulated that an auxin is required for induction (Kohlenbach, 1978), but experiments to confirm this hypothesis have not been performed. It will be difficult to demonstrate this requirement because auxin is required for the initiation of callus. If the initial callus contains competent cells, then a case for the role of auxin might be argued. But if competence arises later, other substances should be considered.

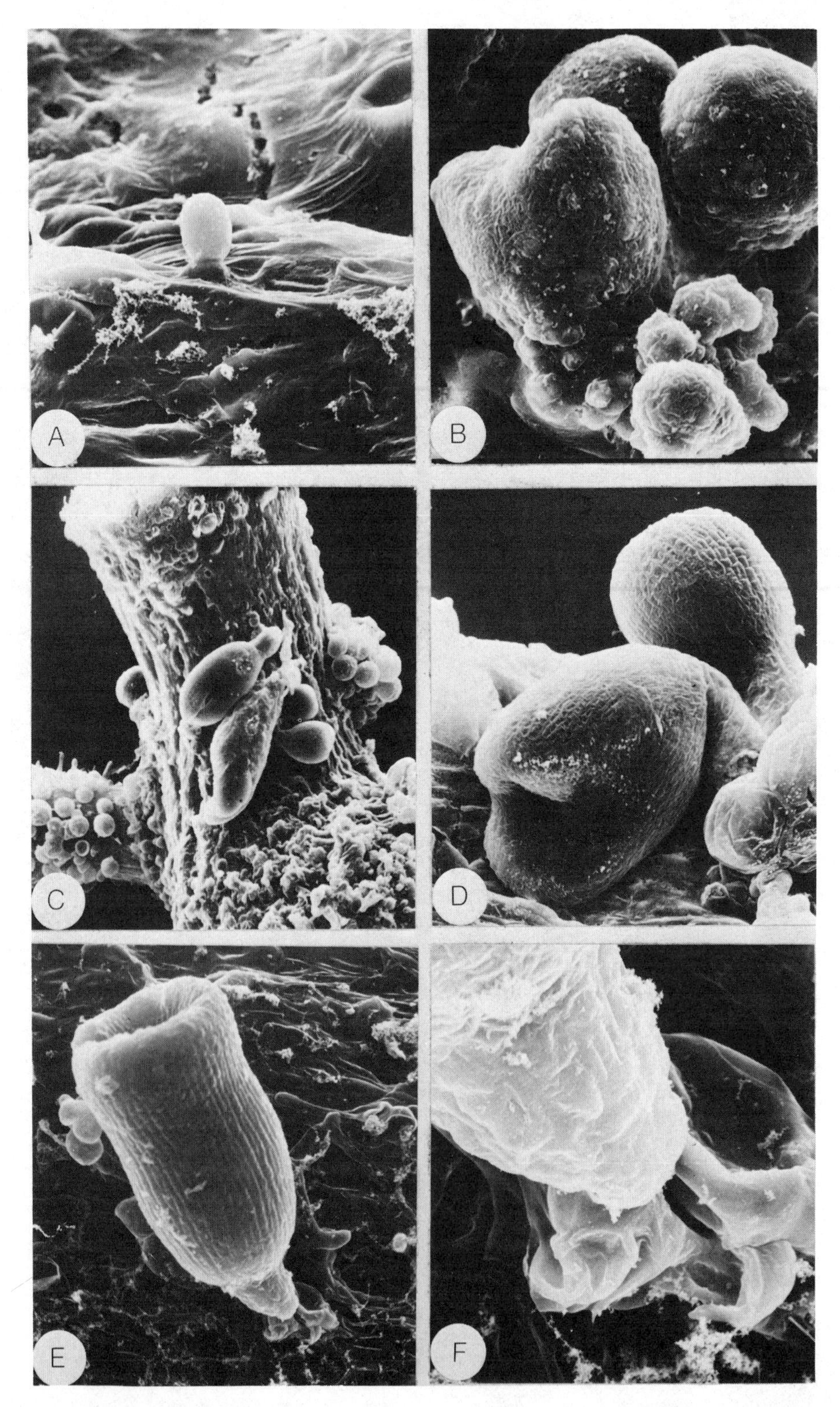
A
B
C
D
E
F

Figure 2. Origin and development of somatic embryos of *Vitis* sp. Seyval. (a) Putative single celled embryo originating from an epidermal cell. (b) Proembryonic masses (foreground), early cotyledonary stage embryo (midground), and globular embryos (background). (c) Colony of epidermal embryos at various stages of development on the transition zone between the hypocotyl and root and on a secondary root emerging from the transition zone. Hypocotyl is rough textured area in foreground. (d) Cotyledonary stage embryo and collapsed suspensor. (e) Early torpedo stage embryo and suspensor. (f) Attachment site of suspensor to swollen epidermal cell. Same embryo as in (e).

Embryo induction from nucellar tissue is more efficient when the auxin contains an oxyacetic acid side-chain (Srinivasan and Mullins, 1980). Auxins with acetate side-chains also induced embryolike structures and plantlets, but frequencies were low and plantlet development was impaired (Hirabayashi et al., 1976; Favre, 1977). Cytokinin may also be required for induction because a sequential treatment, consisting of ovule culture on basal medium with BA transferred to basal medium with both BA and NOA, was the most efficient way to produce nucellar somatic embryos (Srinivasan and Mullins, 1978). The cytokinin requirements of non-nucellar explants and for non-*vinifera* cultivars remain to be determined.

The production of embryos from the epidermis of *Vitis* sp. cv. Seyval occurs on a hormone-free medium (Krul and Myerson, 1980). The inductive substances are not known, but IAA (0.1 μM), GA (0.001-0.1 μM), and sucrose (0.2-0.4 M) enhance embryo production (Krul and Myerson, 1980; Myerson, 1981). Induction of embryos in this system was reduced by activated charcoal (0.01-1.0%), ethanol (0.001-1.0%, v/v), abscisic acid (0.01-100 μM), benzyladenine (0.01-100 μM), or low temperature (Myerson, 1981).

The effect of nutrient salts on the inductive process for *Vitis* has not been systematically explored. The production of somatic embryos has occurred on the basal salts of Nitsch and Nitsch (1969) (Mullins and Srinivasan, 1976; Hirabayashi et al., 1976; Rajasekaran and Mullins, 1979) or the basal salts of Murashige and Skoog (1962) (Favre, 1977; Krul and Worley, 1977). In one case the development of plantlets and embryolike structures occurred on the Nitsch and Nitsch salts but not on those of Murashige and Skoog (Hirabayashi et al., 1976).

The physical state of the induction medium appears to be a critical determinant for some explants. Liquid medium was superior to an agar-based medium for production of embryos from ovular explants (Srinivasan and Mullins, 1980) and anthers (Rajasekaran and Mullins, 1979). Production of embryos from somatic embryos was not consistently influenced by the physical state of the medium (Myerson, 1981).

Light inhibits the production of embryos from anthers (Rajasekaran and Mullins, 1979) and reduces the number of bipolar embryolike structures produced by leaf callus (Favre, 1977). The development of embryos from very young somatic embryos was greater under 16-hr photoperiods than under 8-hr photoperiods (Myerson, 1981). The effects of light quality and intensity on embryo induction in callus or from somatic embryos remain to be explored. The effect of light duration on regeneration of roots (Allweldt and Radler, 1962) and shoots (Chee and Pool, 1982) indicate that these parameters deserve further consideration.

The production of embryos from anthers was enhanced by prechilling at 4 C for 72 hr (Rajasekaran and Mullins, 1979). In contrast the production of embryos was repressed when parent embryos were chilled at 4 C for 30 days (Krul and Myerson, 1980). This latter effect has been ascribed to a change in correlative control mechanisms. The optimum temperature range for embryo production from somatic embryos lies between 25 and 35 C. The higher temperature range, however, influenced the quality and viability of induced embryos in an adverse manner because many were brown and developed calluslike outgrowths. Embryo production from somatic embryos was repressed by temperatures below 15 C (Myerson, 1981).

EXPRESSION OF COMPETENCE. In the carrot model system the conversion of PEM cells to embryos occurs when auxin is depleted from the medium. The conversion is enhanced by mineral ions (NH_4^+ and K^+) and repressed by growth regulators other than auxin. Competence is lost when cultures are dominated by non-PEM cells (large, highly vacuolate cells). Whether the loss in competence is due to overgrowth of cell cultures by non-PEM cells, or conversion of PEM to non-PEM cells, or both, remains to be determined (Wetherell, 1978; Kohlenbach, 1978).

The requirements for the conversion of anther callus cells to recognizable embryo precursors are the same as those for carrot, e.g., culture on a hormone-free medium. In contrast, callus from ovules (Mullins and Srinivasan, 1976) and vegetative tissues (Krul and Worley, 1977) appears to require a transition from a more active auxin to a less active auxin. For ovules the transition was from 2,4-D to NOA to hormone-free medium and for callus from vegetative tissues, the transition was from 2,4-D to NAA to hormone-free medium that yielded viable embryos. The expression of competence of somatic embryos derived from the epidermis of somatic embryos appears to be spontaneous. This conclusion is based on the observation that no embryo precursors have been observed on the surfaces of embryos treated with cytokinin or cold (Krul, Crusburg, Lowe, El-Fiki, and Dellaporta, personal communication).

The physical and chemical factors that modulate expression of competence in grape cultures need further exploration. Other areas requiring attention are detection of chemical markers to differentiate between induced and noninduced cells, and between induced cells that are capable and those incapable of expression under standard conditions. Some progress has been made in this direction with cultures of carrot but the information obtained was ambiguous (Sung and Okimoto, 1981).

The use of grape PEM clusters as a source of protoplasts may have advantages over other tissues or cells because the latter (at least in *V. vinifera*) have not displayed organogenic competence (Srinivasan and Mullins, 1980).

EMBRYO DEVELOPMENT. Grape seeds require a period of low temperature (5 C for 60-90 days) for germination. Embryos produced from anthers (Rajasekaran and Mullins, 1979), ovules (Mullins and Srinivasan, 1976), leaf petioles (Favre, 1977), and somatic embryos (Krul and Myerson, 1980) develop slowly and often do not produce a normal plant. The embryos from anthers develop green cotyledons but fail to form a shoot (Rajasekaran and Mullins, 1979) and those derived from somatic embryos fail to develop separated, expanded, green cotyledons and shoots (Krul and Myerson, 1980). Embryos from anthers develop normally if chilled for 2-8 weeks at 4 C, whereas those from ovules develop plants that are normal in appearance when cultured on a cytokinin-gibberellin-enriched medium followed by a hormone-free medium with reduced salt content (Srinivasan and Mullins, 1980). Several physical and chemical factors have been examined in an attempt to produce normal plantlets from secondary embryos, but to date no single factor has produced the desired result (Krul and Myerson, 1980; Myerson, 1981). Cold, cytokinin, gibberellin, combinations of cytokinin and gibberellin, and deletion of sucrose folowed by refeeding, have produced progress toward normal plant development (greening and separation of cotyledons), but as yet none of these factors has increased the frequency of normal plant production (5-30%).

The initiation of somatic embryos from the epidermis of somatic embryos appears to be under a form of correlational control. Embryos that develop normally rarely produce somatic embryos. In contrast embryos with dormant cotyledons and shoot meristems produce embryos with a frequency of nearly 80%. A seasonal variation in the frequency of normal plant production and embryos from secondary embryos was observed. The frequency of normal plant development during the winter months may reach 30% and the frequency of embryo production becomes almost zero. In contrast, an inverse relationship exists during the summer months. Physical and chemical factors that promote development towards the normal state repress somatic embryogenesis. This inverse relationship between plant development and somatic embryo formation may provide an interesting tool for study of both dormancy and embryogenic mechanisms and perhaps will illustrate a new set of correlative control mechanisms for plant development.

GENETIC STABILITY AND PERFORMANCE OF REGENERATED PLANTS. Plantlets regenerated from callus of *Vitis* sp. cv. Seyval displayed several characteristics which differed from the parent vines. For instance, the leaves were darker blue-green, the young canes had a definite reddish cast in contrast to the usual green, and the fruit clusters were cylindrical rather than conical. Interestingly the description of the regenerated vines fits the description of the original hybrid (Mowbray, 1981). The regenerated vines have produced an equivalent of 5 tons of fruit per acre at 20 Brix when 3 years old. Cuttings from regenerated vines rooted earlier and displayed more uniform and vigorous top growth relative to cuttings from the parents. It is speculated that return to the original phenotype and increased vigor may result from the elimination of latent viruses that have slowly accumulated and modified growth and development of the parent vines.

Micropropagation

The propagation of grape by the rooting of forced axillary or adventitious buds, in vitro, may provide a means to rapidly increase new plant introductions, maintain and propagate virus-indexed foundation plants, propagate nongrafted hybrids, and preserve grape germplasm in clonal repositories.

Two methods have been developed for the shoot tip multiplication of grape. One is a traditional method based on the pioneering work of Murashige (1974), in which a small explant from an apical or axillary bud is forced to produce multiple side shoots which are subsequently rooted (Harris and Stevenson, 1979; Chee and Pool, 1982; Jona and Vallania, 1980). The other is a rather novel method in which adventitious shoots are produced from small leaves obtained from a fragmented shot apex (Barlass and Skene, 1978, 1980a,b, 1981a; Skene and Barlass, 1981).

EXPLANT SOURCE. Apical explants (1-2 cm) are generally obtained from rapidly expanding shoots obtained from cold-treated, woody canes (Barlass and Skene, 1978; Harris and Stevenson, 1979) or collected from the field during a period of rapid growth (Chee and Pool, 1982).

STERILIZATION. Explants were washed with sodium or calcium hypochlorite (1.2 or 5% w/v respectively) with Tween-20 (0.01-0.1% v/v) for 10-20 min and then rinsed with sterile water (three times at 5 min each) or soaked in 70% ethanol for 1 min followed by sterile water rinses. Segments of the apical bud (0.5-4 mm) were removed from sterile explants or small apical domes were dissected into approximately 20 fragments (Barlass and Skene, 1978) and placed on shoot induction medium. Fragments containing fewer than two leaf primordia did not produce adventive shoot tips (Barlass and Skene, 1980a).

SHOOT INDUCTION MEDIUM. The salt formulations of Murashige and Skoog (1962) or Jona and Webb (Jona and Vallania, 1980) were adequate for shoot induction. The vitamins and organic supplements, excluding growth regulators, are, with one exception (Chee and Pool, 1982), the same as those in the original MS formulation. In this case the vitamin formulation of Galzy (1972a) minus calcium pantothenate and biotin was used.

A cytokinin concentration between 5 and 20 μM and an auxin concentration of 0-0.5 μM provided adequate shoot development for all investigators. The most effective cytokinin was BA. Generally an auxin was not required for bud induction, but Chee and Pool (1982) found that NAA (0.5 μM) was required for optimum bud formation in cultures of *Vitis* sp. cv. Rougeon. Shoot proliferation was enhanced by the addition of ADE (0.3 mM) and monobasic sodium phosphate (1.4 mM) to the induction medium (Harris and Stevenson, 1979).

The induction medium included agar (6-8 g/l) or, in the case of fragmented shoot apices where a liquid medium was superior, agar was deleted (Barlass and Skene, 1978). In the latter case, one fragmented apex was cultured per 5 ml of medium.

A temperature of 25-27 C was optimal for shoot induction. Elevation of temperature to 35 C during induction hastened the appearance of adventive primordia (18 vs. 35 days), but the new primordia consisted of branched tendrils that contained inflorescencelike structures. Reduction of the culture temperature to 27 C inhibited the development and production of tendril-inflorescence structures (Barlass and Skene, 1981b).

Light is required for production of shoots or adventitious buds from fragmented shoot apices. No shoots or greenish callus clusters were formed when tissues were cultured without light (Skene and Barlass, 1981; Chee and Pool, 1982). The optimum light intensity for most cultivars was 2000-3000 lux with a photoperiod of 16 hr (Harris and Stevenson, 1979; Skene and Barlass, 1981). However, shoot multiplication of *Vitis* sp. cv. Rougeon apices required a photoperiod of 10 hr. No shoot development was observed with 15-hr photoperiods or continuous light (Chee and Pool, 1982).

SHOOT MULTIPLICATION. The continuous multiplication of shoots requires the presence of cytokinin and the concentration used for induction was generally sufficient to permit continued growth (Harris and Stevenson, 1979; Skene and Barlass, 1981). Shoot proliferation on fragmented apices was greater when leaf fragments were transferred to semisolid medium after 2-4 weeks on induction medium (Skene and Barlass, 1981). Concentrations of cytokinin above 20 μM resulted in repression of shoot length and subsequent difficulty in rooting of explants (Harris and Stevenson, 1979).

ROOT INITIATION. Shoot tips of grape were not difficult to root. Shoots 1.5-3 cm long were inserted into gelled medium (Barlass and Skene, 1978) or placed on filter paper support in liquid medium (Harris and Stevenson, 1979). Some cultivars do not require an auxin for root initiation (Skene and Barlass, 1981; Harris and Stevenson, 1979). However, the addition of 0.1-2 μM of IAA, NAA, or IBA to the culture medium was required for those cultivars that are slightly more difficult to root (Skene and Barlass, 1981; Chee and Pool, 1982). Root initiation was not influenced by salt concentration but root growth was enhanced when the salt content of the rooting medium was reduced (Harris and Stevenson, 1979; Skene and Barlass, 1981). Root growth was also enhanced if cuttings were removed from root initiation medium after the first root primordia were visible (Skene and Barlass, 1981).

The current method of acclimatization to greenhouse and field conditions requires transfer of rooted cuttings to Jiffy 7 peat pellets, and growth for 9 days in an enclosed glass tank. The lid was removed from the tank and after 2 weeks the plants were moved to pots with peat-sand (1:1 v/v) mix and transferred to a glass-house (Skene and

Barlass, 1981). A substantial reduction in the cost of production could be realized if cuttings were rooted directly in pots or propagation beds with mist or other means to maintain high relative humidity.

GENETIC STABILITY AND PERFORMANCE OF REGENERATED PLANTS. Both the conventional shoot tip and apical fragmentation methods can produce 20–24,000 plants from a single apex in a year (Barlass and Skene, 1978; Harris and Stevenson, 1979).

Plants regenerated from fragmented shoot apices contained the normal chromosome complement of 38 (Barlass and Skene, 1978). However, they display juvenile characteristics (entire leaves, lack of tendrils, and spiral phyllotaxy) that disappear when plants are removed from in vitro conditions (Barlass and Skene, 1980b). Because shoots from fragmented apices arise adventitiously (Barlass and Skene, 1981), the probability of genetic change may be increased relative to the standard method of forcing development of existing buds.

Harris and Stevenson (1979) established shoot tip cultures of 14 cultivars which consisted of *V. vinifera* and *Vitis* sp. hybrids. Of the 14, the root stocks 5BB, 5C, and SO_4, and the *viniferas*, Pinot Noir, Okanagan Riesling, and Rotberger were difficult to establish. Barlass and Skene (1980b) examined 11 cultivars and found that *V. rupestris* (Rupestris St. George) and *V. rupestris* x *V. berlandieri* (R-99) did not proliferate as rapidly as five *V. vinifera* cultivars, two *V. champini* cultivars and the complex hybrid rootstock Harmony.

The production of plants by shoot tip multiplication or induction of adventive shoots on leaf fragments appears to have commercial potential. However, data concerning field performance of in vitro and conventionally propagted plants (Skene and Barlass, 1981) are required prior to commercial application. A second need would be to establish a cost accounting of conventional and shoot tip methods to determine which method might be more profitable for the propagator.

Pathogen Elimination

Nearly all grape cultivars are propagated by asexual methods. Some clones date back to Roman and perhaps earlier times (Penning-Rowsell, 1971). The continuous clonal multiplication of ancient cultivars, without intervening sexual cycles, favors the infection, multiplication, and dissemination of virus, mycoplasmalike organisms, and fungal and bacterial pathogens.

Methods for pathogen elimination include heat therapy of whole plants or excised shoot tips (Glazy, 1961; Nyland and Goheen, 1969; Valet et al., 1979; Grenan, 1980), nucellar embryogeny (Mullins and Srinivasan, 1976, 1978; Srinivasan and Mullins, 1980), and micrografting of heat-treated apical domes (Ayuso and Pena-Iglesias, 1978; Englebrecht and Schwerdtfeger, 1979). Each of the above methods has one or more problems which prevent adaptation to all species and cultivars of grape.

The heat treatment of excised shoot tips or cultivars of shoot tips from heat-treated plants for virus elimination has two problems which prevent immediate application. The most important problem is that not all viruses can be eliminated by heat treatment (Nyland and Goheen, 1969; Valat et al., 1979), and the second involves difficulty in regulation of the growth and development of the excised, nonvascularized, shoot apical meristems (Galzy, 1971, 1977; Grenan, 1979a; Mur, 1979).

SHOOT APEX MICROGRAFTS. The most recent strategy for dealing with the above problem involves a micrograft between an excised apical dome from a heat-treated plant onto an appropriate root stock. Shoot meristems 0.1-0.5 mm in size (with 2-5 leaf primordia) were placed on the vascular ring of a decapitated seedling grape (Engelbrecht and Schwerdtfeger, 1979) or inserted into a cubic depression cut from a sterile, green, rootstock cutting (Ayuso and Peña-Iglesias, 1978). Shoots developed 3 weeks after graft initiation.

The success of the graft union was dependent on scion size. The larger the apical dome the greater the probability of a successful union. When dome size was 0.5 mm or larger the frequency of successful union was more than 50%. Unfortunately, with increasing dome size there is a greater probability of pathogen propagation.

The use of seedling rootstocks was discouraged because of high variability in graft union survival. The use of somatic embryos from rootstock species could provide a more genetically homozygous acceptor for grafted domes.

Elimination of a heat-stable virus by this method was suggested, but no quantitative data were presented (Engelbrecht and Schwerdtfeger, 1979).

SHOOT APEX CULTURE. The production of scion or root stocks from isolated, heat-treated, apical domes, by in vitro culture and subsequent grafting, offers a technologically simpler means to produce virus-indexed plants. However, this method also has some difficulties. As mentioned above, the growth of apical explants is proportional to their size. Small domes, 0.2 mm or less, do not elongate readily and once elongated may be difficult to root. Much of the research, to date, has focused on the physical and chemical factors which modify the rooting response of apical domes (2-4 leaf primordia) and to a lesser extent the growth of the domes in vitro (Galzy, 1971, 1977). The focus on root initiation rather than shoot extension may be related to the realization that once rooted, the small explants would be supplied with shoot growth factors from the roots. Shoots without roots elongate slowly or not at all (Galzy, 1977; Pool and Powell, 1975). However, the demonstration that axillary buds (Pool and Powell, 1975) and small apical explants (Chee and Pool, 1982) grow without roots in the presence of cytokinin suggests that a new strategy for thermotherapy may evolve.

ROOT INITIATION AND GROWTH. The initiation of roots on small apical explants is a function of the concentration (Galzy, 1972) and type (Grenan, 1979a) of auxin and the grape genotype (Galzy, 1964a; Galzy and Compan, 1968). Rooting percentages of *V. vinifera* explants were acceptable with IAA or IBA (0.1 μM) and less acceptable with NAA (Grenan, 1979a). In contrast explants of *V. rupesrtis* displayed greater rooting percentages with NAA (0.1 μM) than with the indole auxins (Grenan, 1979a; Galzy, 1972). Concentrations of auxin above 0.1 μM caused basal callus formation and reduced shoot extension (Galzy, 1972). The duration of auxin treatment and concentration influences the initiation and subsequent rate of root elongation. Root elongation was more rapid on hormone-free medium (Galzy, 1972a; Grenan, 1979a).

Gibberellic acid inhibited root formation but stimulated root elongation (Galzy, 1971). The effect of other growth regulators on root initiation and growth were not reported.

The addition of K^+ and NH_4^+ together, or NO_3^- alone, to Knops salts (which lack NH_4^+) enhanced root percentages of *V. rupestris* explants (Galzy, 1977). A high K^+:NH_4^+ ratio favored both initiation and subsequent growth of roots, but once initiated, long-term root growth was favored by reduced concentrations of NO_3^- or NH_4^+. The rooting response to auxin or elevated K^+:NH_4^+ concentration was about equal, and combinations of the two did not further enhance rooting (Galzy, 1971). Root growth was repressed when explants were cultured in media enriched with both gibberellin and K^+/NH_4^+, even though individually each treatment promoted root elongation (Galzy, 1971).

Root initiation was not dramatically influenced by sucrose or combinations of sucrose with six vitamins. However, growth of roots over a long period of time was enhanced by elevated concentrations of sucrose and vitamins (Galzy, 1969a).

SHOOT GROWTH. Apical explants from vines displaying symptoms of fan leaf virus displayed a low percentage of rooting. Heat treatment of excised shoots (35 C for 90 days) improved rooting percentages and permitted recovery of virus-symptomless plants. Exposure of virus-infected apical explants to 35 C for as little as 5 days increased rooting percentage to that of noninfected cuttings but no mention was made of removal of virus symptoms (Galzy, 1962, 1964b, 1969a,b).

Initial development of shoots on small apical explants was rapid if roots were initiated, but in the absence of roots, shoot growth is slow or ceases (Pool and Powell, 1975). Therefore, many of the factors that enhanced the initiation and growth of roots, as discussed above, also facilitated shoot development. For instance, shoot development was enhanced by auxins and elevated levels of K^+, NH_4^+, NO_3^-, sucrose, vitamins, and heat treatment. Factors that retard root development also retard shoot development, e.g., the continued presence of auxin or elevated mineral salts (Galzy, 1969a, 1971, 1977).

Heat treatment and/or prolonged in vitro culture can result in phenotypic change in regenerated grapes. Some of the observed changes are: altered leaf morphology, repression of flower cluster number and size, production of hairs on normally glaborous leaves, enhanced red pigmentation of stems, and development of the symptoms of the juvenile state (lack of tendrils, altered phyllotaxy, and leaf shape). Some of the changes are transient and are manifest only when plants are maintained in vitro, while others appear to be quite stable (Galzy, 1971; Grenan, 1979b; Mullins and Nair, 1979; Mur, 1979; Valat et al., 1981).

CONCLUSIONS. The success of thermotherapy methods depends in part on the survival and development of small apical fragments. The survival of even smaller explants, desirable for elimination of thermostable viruses, might result from the inclusion of a cytokinin (Chee and Pool, 1982) into the medium previously optimized for salts and organic components (Galzy, 1971, 1977).

The possibility that nucellar embryogeny results in virus-free clones (Srinivasan and Mullins, 1980) should be pursued by the production of nucellar embryos from virus-infected and noninfected vines. The dramatic changes in phenotype (induced by in vitro thermotherapy methods) may provide a novel approach with which to understand the regulation of gene expression associated with phase change.

Embryo and Ovule Culture

Several cultivars of seedless grape develop embryos that abort shortly after anthesis. The rescue of embryos from seedless cultivars would permit sexual transmission of the seedless characteristic and permit genetic analysis of the phenomenon.

Some progress has been made in this direction. A recent abstract (Cain, 1980) indicates that ovule culture of some seedless cultivars resulted in seed production. The seeds germinated after cold stratification. Embryos excised from cold-stratified seed which did not germinate, produced plants when cultured on a salt medium containing cytokinin.

Galzy and Galzy (1964) showed that embryos obtained from self-pollinated *V. vinifera* cv. Aramon seed required CW (10% v/v) for germination. Embryos did not require CW after root length exceeded 1 cm. Embryos from fan leaf virus-infected Anamon seed were less viable and many seeds were embryoless or contained dead embryos.

Apparently 25% of the seeds contained temperature-sensitive genes that regulate chloroplast development. When "mutant" seedlings were grown at 20 C the leaves were variegated. However, when leaves of variegated plants developed at 35 C, they were completely green. Variegation or greenness of successive leaves could be manipulated by shifting temperature during leaf formation.

Regulation of Metabolite Production

Grape is not noted for the production, in commercial quantities, of biologically active or useful compounds. Investigations of metabolite production in grape have been used to answer questions concerning the partitioning of organic acids (Skene and Hale, 1971) and to examine the biogenesis of anthocyanin pigments (Slabecka-Szweykowska, 1952).

The question of whether immature grape berries accumulate or synthesize malic and tartaric acids was determined by in vitro culture of excised flower clusters. Berries, which developed slowly, contained malate but not tartrate, indicating that the latter may be imported from other sources.

Anthocyanin production by hormone-independent grape callus was enhanced by elevation of sucrose concentration, reduction in NO_3^- concentration, and by various combinations of sucrose and light. Anthocyanin production was inversely related to the growth of cells. The enhancement of pigment by sucrose was due to an increase in the osmotic potential of the medium because production of pigment could be enhanced at low-sucrose concentration by the addition of osmotically active concentrations of NaCl.

Stress Physiology

The effect of NaCl on the induction of adventive shoots from fragmented shoot apices and the growth and survival of established shoot tip cultures was determined for several *V. vinifera* and hybrid cultivars (Barlass and Skene, 1981a). Multiplication of shoots was enhanced by 5 mM NaCl and repressed by 50 mM NaCl. A differential sensitivity of cultivars to NaCl was demonstrated by prolonged growth of shoot tips (11 passages) on NaCl-enriched (0-100 mM) medium. The *V. vinifera* cultivars grew more vigorously than hybrids on NaCl-enriched media, but the difference could not be ascribed entirely to NaCl because hybrids were less vigorous in the absence of salt. Shoot cultures of Cabernet Sauvignon and Ramsey displayed a recovery of growth rate after several passages on NaCl-enriched medium. The hybrid cultivar (Ramsey) displayed recovery at concentrations up to 50 mM and Cabernet at concentrations up to 20 mM NaCl. Comparisons of NaCl tolerance between shoot tip cultures and rooted cuttings suggested good agreement between the most- and least-tolerant cultivars. The use of shoot cultures as a preliminary screen for the growth of root stocks under saline conditions appears feasible. The possibility that cultures were acclimated to high concentrations of NaCl remain to be explored. Adaptation of shoot tip cultures of root stocks to applied stress may be a valuable tool for grape improvement because such adaptations may remain stable as long as the plant was asexually propagated.

In Vitro Vine Development

The juvenile state of the grape vine is recognized by an altered, phyllotaxy, leaf shape, and by the lack of tendrils or flowers (Mullins and Nair, 1979). The juvenile state can be induced by repeated subculture of shoot explants (Mullins and Nair, 1979; Mur, 1979), induction of adventitious buds (Barlass and Skene 1981a), or somatic embryos (Mullins and Srinivasan, 1976; Krul and Worley, 1977). Cytokinin feeding may hasten the formation of juvenile buds by enhancing production of adventitious shoots or the development of incipient axillary meristems (Mullins and Nair, 1979). The transition to the juvenile state of in vitro shoot cuttings can be delayed by supplementing the medium with vitamins (Favre and Grenan, 1979) or CH (Mur, 1978). In the presence of vitamins several grape cultivars produced tendrils, and some produced tendrils, flowers, and fruit (Favre and Grenan, 1979). A rapid transition to the adult form occurs when in vitro grown plants are transferred to ambient conditions (Skene and Barlass, 1981b).

One of the interesting characteristics of the juvenile state which might be exploited for grape improvement is the ease of organ regeneration of juvenile growth. Juvenile forms of many woody perennials root readily, whereas adult forms do not. Callus or tissue explants from juvenile tissues of woody plants, such as seeds or cotyledons, regenerate adventive shoots or embryos, whereas callus initiated from adult tissue may remain incompetent (Mott, 1978). Therefore, the induction of juvenility by repeated subculture of apical explants of grape or other woody species may provide a useful starting point for the production of cells with embryonic competence or for the production of protoplasts with greater potential for regeneration. The obvious advantage of this method is that the clonal genotype of the juvenile tissue would be preserved.

Seedling grapes require 3-5 years growth (with some exceptions, Skene and Barlass, 1981) prior to the transition to the adult or flowering stage. Shortening the length of the juvenile period would facilitate progress in grape breeding.

Tendrils and flowers of grape are homologous organs which originate from a group of uncommitted meristematic cells known as the anlagen (Srinivasan and Mullins, 1981). Perception of different environmental and/or chemical cues determine which organ will be produced by the anlagen (Srinivasan and Mullins, 1978; Srinivasan and Mullins, 1981). Highly branched anlagen induced by cytokinin result in flower primordia, whereas weakly branched anlagen induced by GA_3 yield tendrils (Srinivasan and Mullins, 1981). Cytokinin applications to tendrils resulted in flower and fruit formation on 6-month-old vines (Srinivasan and Mullins, 1978). Tendrils cultured in vitro with various cytokinins form rudimentary flowers which lack male and female sporophytes (Srinivasan and Mullins, 1978). The factors which regulate meiosis in mitotic cells of in vitro cultured flowers remain to be indentified.

Cytokinin application to tendrils of intact plants produces fertile flowers capable of setting fruit, while tendrils cultured in vitro produce flowers lacking male and female sporophytes. The lack of meiotic divisions in flowers from tendrils cultured in vitro may be due to some factor translocated in the xylem or phloem. Exudates from each tissue system may provide factors needed for complete flower development. Identity of "meiotic" regulators would facilitate in vitro flower production and might also provide a means of producing haploid cells from diploids.

The in vitro conversion of an uncommited meristem to the floral state in response to a balance of growth regulators, provides a valuable tool for probing the mechanism of gene regulation of floral induction and its associated metabolism.

Host-Parasite Interaction

The continuous production of *Plasmophora viticola* Berl. and Toni (downy mildew) on isolated cells of grape was demonstrated by Morel (1948). The coexistence of host and pathogen provide an opportunity to explore such questions as: (1) Could cell culture be used to screen for mildew-tolerant clones? (2) Is the interaction between host and parasite the same in vitro as in vivo? (3) Are specific metabolites of the host required for growth of the parasite? Systems to probe these and other questions concerning host-parasite interaction are in the developmental stage (Aldwinckle and Buturac, 1980a,b, 1981).

Agrobacterium tumefaciens, a gram-negative bacterium, incites tumor formation on many dicotyledonous genera. Tumor formation is elicited by transfer of a portion of bacterial plasmid DNA to the host (Chilton et al., 1977). Host range of the bacterium is determined by genes of the plasmid (Thomashow et al., 1980). Two basic groups of plasmids have been identified; those which infect a large number of genera and contain wide host range (WHR) plasmids, and those which infect primarily grape and occasionally raspberry (Panagopoulos et al., 1978) and are said to contain limited host range (LHR) plasmids. The tumor-inducing DNA (T-DNA) of the WHR plasmid contains only a few sequences in common with the DNA from LHR plasmids (Thomashow et al., 1981), indicating that genes for tumor initiation are different.

Agrobacterium containing WHR plasmids do not produce tumors on *V. vinifera* (Thomashow et al., 1981), but we have obtained (in vitro) tumors on the *vitis* sp. cv. Seyval, elicited by both WHR and LHR strains of the bacterium (Lowe and Krul, unpublished). Further, we have demonstrated that DNA of both wide and limited host range strains is incorporated into Seyval DNA (Dellaporta, Lowe, and Krul, unpublished).

The identity of the DNA sequences of WHR and LHR plasmids and their relation to host range in *Vitis* species need to be defined before plasmids can be used as shuttle vectors for the genetic modification of grape. The production of tumors in *Vitis* by plasmids with nonhomologous T-DNA sequences may provide interesting insights into the genetic basis of tumor induction.

FUTURE PROSPECTS

The potential for genetic improvement of grape by development of gene vectors or selection of variants against imposed chemical or physical stress has the same probability for success as that of most other crop genera, provided adequate resources (material and human) are provided (Meredith, 1981).

The barriers that remain do not appear to be insurmountable. There is a need for protoplast systems with high potential for plant regeneration. Possible sources of organogenically competent protoplasts might be somatic embryos, adventitious shoots, or shoot tips which display juvenile characteristics after repeated subculture. Similarly, the lack of in vitro organogenic competence of most grape cultivars might be remedied by the use of explants from induced juvenile shoots.

Efforts to isolate and identify the T-DNA of limited host range plasmids of *Agrobacterium tumefaciens* and further clarification of host-plasmid interaction should facilitate the development of gene vectors for grape. Once the above barriers are surmounted there remains the decision as to which single gene or small group of genes would enhance grape productivity. The latter is not a small problem because the genetics of *Vitis* is still primitive relative to other major crop plants. Hopefully this deficiency can be reduced by implementation of newly developed methods to reduce the juvenile period of the vine (Srinivasan and Mullins, 1978).

ACKNOWLEDGMENTS

We wish to express our gratitude to scientists who made our task less difficult by providing reprints of their research. We also thank those investigators who shared descriptions of their current and anticipated research efforts. This spirit of open communication will, no doubt, enhance progress toward our goal of more efficient grape production.

Our special thanks to Jean Krul and Phyllis Mowbray for their efforts in typing and editing this manuscript.

This is Rhode Island Agricultural Experiment Station Contribution Number 2073.

REFERENCES

Aldwinckle, H.S. and Buturac, I. 1980a. Culture of grape cultivars from apical meristems. In: Seventh International Council for the Study of Viruses and Vines Diseases of Grape Vine (A.J. McGinnis, ed.) pp. 339-341. Agriculture Canada Research Branch, Ottawa, Canada.

______ 1980b. In vitro techniques for studying obligate pathogens of *Vitis*. Third International Symposium on Grape Breeding (H.P. Olmo, ed.) pp. 87-91. Dept. of Pomology and Viticulture, Univ. of California, Davis.

______ 1981. In vitro culture of grapevine for study of obligate pathogens. Environ. Exp. Bot. 21:439.

Alleweldt, G. and Radler, F. 1962. Interrelationship between photoperiodic behavior of grapes and growth of plant tissue cultures. Plant Physiol. 37:376-379.

Arya, H.C. 1963. In vitro growth of phylloxera gall and grape stem single cell clones with inositol, naphthalene acetic acid, and nucleic acids. Indian J. Exp. Biol. 1:148-153.

______ 1964. Changes induced by nucleic acid contents of phylloxera gall and grape stem single cell clones in tissue culture. Indian J. Exp. Biol. 2:44-48.

______ 1965a. Cultural behavior of insect gall and normal plant single cell clones. In: Tissue Culture (C.V. Ramakrishnan, ed.) pp. 293-304. W. Junk Pub., The Hague, Netherlands.

_____ 1965b. Effect of naphthalene acetic acid on the nitrogen and nucleic acid contents of phylloxera gall and normal grape stem single cell clones in tissue culture. In: Tissue Culture (C.V. Ramakrishnan, ed.) pp. 382-387. W. Junk Pub., The Hague, Netherlands.

_____ 1965c. Vitamins as growth factors for phylloxera gall and grape stem single cell clones in tissue culture. Indian J. Biol. Sci. 3:126-129.

_____ and Hildebrandt, A.C. 1969a. Differential sensitivities to gamma radiation of phylloxera gall and normal grape stem cells in tissue culture. Can. J. Bot. 47:1623-1628.

_____ and Hildebrandt, A.C. 1969b. Effect of gamma-radiation on callus growth of phylloxera gall and normal grape stem tissues in culture. Indian J. Exp. Biol. 7:158-162.

_____, Hildebrandt, A.C., and Riker, A.J. 1962a. Clonal variation of grape stem and phylloxera gall callus growing in vitro in different concentrations of sugars. Am. J. Bot. 49:368-372.

_____, Hildebrandt, A.C., and Riker, A.J. 1962b. Growth in tissue culture of single cell clones from grape stem and phylloxera gall. Plant Physiol. 37:387-392.

Ayuso, P. and Peña-Iglesias, A. 1978. Shoot apex (meristem) grafting: A novel and promising technique for regeneration of virus-infected grape vines. Proceedings of the Sixth Conference on Virus and Virus Diseases of the Grape Vine. Colleccion Monografias 18, Instituto Nacional Investigaciones Agrarias, Madrid, Spain.

Barlass, M. and Skene, K.G.M. 1978. In vitro propagation of grapevine (*Vitis vinifera* L.) from fragmented shoot apices. Vitis 17:335-340.

_____ 1980a. Studies of the fragmented apex of grapevine. I. The regenerative capacity of leaf primordial fragments in vitro. J. Exp. Bot. 31:483-488.

_____ 1980b. Studies on the fragmented shoot apex of grapevine. II. Factors affecting growth and differentiation in vitro. J. Exp. Bot. 31:489-495.

_____ 1981a. Relative NaCl tolerances of grape vine cultivars and hybrids in vitro. Z. Pflanzenphysiol. 102:147-156.

_____ 1981b. Studies on the fragmented shoot apex of grapevine. III. A scanning electron microscope study of adventitious bud formation in vitro. J. Exp. Bot. 32:1079-1083.

Benbadis, A. and Baumann, F. 1973. Etude comparative de protoplastes obtenus par traitement enzymatique a partir de divers tissue: Etude untrastructurale et culture in vitro. In: Colloques Internationaux du Centre National de la Rechsci Protoplastes et Fusion de Cell Somatiques Vegetales (B. Ephrussi, G. Morel, and J. Tempe, eds.) pp. 189-205. Center National de la Recherche Scientifique, Versailles, France.

Bernard, A.C. and Mur, G. 1979. Observations sur l'organogenese des bourgeons de plants de *Vinis vinifera* cultives in vitro. Ann. Amelior. Plant. 29:311-323.

Bini, G. 1976. Prove di coltura in vitro di meritemis apicali di *Vitis vinifera* L. Riv. Ortoflorofruttic. Ital. 60:289-296.

Brezeanu, A., Iordan, M., and Rosu, A. 1980. The micropropagation of callus from tissue of somatic origin. Rev. Roum. Biol. Ser. Biol. Vegetale 25:135-142.

Braun, A. and Morel, G. 1950. A comparison of normal, habituated, and crown gall tumor tissue implants in the European grape. Am. J. Bot. 37:499-501.

Brown, D.D. 1981. Gene expression in eukaryotes. Science 211:667-674.

Burgess, J. and Linstead, P.J. 1976. Ultrastructural studies of the binding of concanavalin A to the plasmalemma of higher plant protoplasts. Planta 130:73-79.

Burstrom, H. 1948. Observations on the influence of galactose on wheat roots. Physiol. Plant. 1:209-215.

Cain, D.W. 1980. In vitro culture of seeded and abortive grape ovules. HortScience 15:415.

Cassells, A.C. and Barlass, M. 1978. A method for the isolation of stable mesophyll protoplasts from tomato leaves throughout the year under standard conditions. Physiol. Plant. 42:236-242.

Chee, R. 1980. The Effects of Growth Substances and Photoperiod on Shoot Apices of *Vitis* Cultured In Vitro and Their Effects on Subcultured Shoot Tips. Masters Thesis, Cornell University, Ithaca, New York.

_____ and Pool, R.M. 1982. The effects of growth substances and photoperiod on the development of shoot apices of *Vitis* cultures in vitro. Sci. Hortic. 16:17-27.

Chilton, M.-D., Drummond, M.H., Merlo, D.J., Sciaky, D., Montoya, A.L., Gordon, M.P., and Nester, E.W. 1977. Stable incorporation of plasmid DNA into higher plant-cells: The molecular basis of crown gall tumorigenesis. Cell 11:263-271.

Colclasure, G.C. and Yopp, J.H. 1976. Galactose-induced ethylene evolution in mung bean hypocotyls: A possible mechanism for galactose retardation of plant growth. Physiol. Plant. 37:298-302.

Collins, G.B. and Legg, P.D. 1980. Recent advances in the genetic applications of haploidy in *Nicotiana*. In: The Plant Genome, Proceedings of the Fourth John Innes Symposium, 1979 and Second International Haploid Conference (D.A. Davies and D.A. Hopwood, eds.) pp. 197-213. The John Innes Charity, John Innes Institute, Norwich, England.

Czosnowski, J. 1952a. Charakterystyka fizjologiczna trzech typow tkanex *Vitis vinifera*: Normalnej, tumora bakteryjnego (crown gall): Tumora cheicznego hodowanych in vitro. Prace Kom. Biol. Pozn. Tow. Przyj. 12:189-208.

_____ 1952b. Badania nad gospodarka witaminowa tkawzrostowymi typu auksyny. Prace Kom. Biol. Pozn. Tow. Przyj. 12:209-221.

Dion, R. 1959. Histoire de la Vigne de Du Vin en France des Origines au XIXe Siecle. Flammarion, Paris.

Duhamet, L. 1951a. Action du lait de coco sur la croissance des cultures des tissus de crown-gall de Vigne, de tabac, de topinambour, et de scorsonere. C.R. Acad. Seances Soc. Biol. Paris 145:1781.

Ecevit, F.M. 1979. Asamin steril ortamada (in vitro) mineral beslenmesi. Ege University Zirat Fak. Derg. 16:95-108.

Englebrecht, D.J. and Schwerdtfeger, U. 1979. In vitro grafting of grape vine shoot apices as an aid to the recovery of virus-free clones. Phytophylactica 11:183-185.

Fallot, J. 1955. Culture de tissue de quelques especes et varietes du genve *Vitis* et leur compartment. Bull. Soc. Hist. Nat. Toulouse 90: 163-172.

Favre, J.M. 1977. Premiers resultats concernant l'obtention in vitro de neoformations caulinaires chez la vigne. Ann. Amelior. Plant 27:151-169.

_____ and Grenan, S. 1979. Sur la production de vrilles, de fleurs et de baies, chez la vigne cultivee in vitro. Ann. Amelior. Plant 29: 247-252.

Ferguson, J.D., Street, H.E., and David, S.B. 1958. The carbohydrate nutrition of tomato roots. VI. The inhibition of excised root growth by galactose and mannose and its reversal by dextrose and xylose. Ann. Bot. 22:525-538.

Galet, P. 1973. Precis de Viticulture. Paul Dehan, Montpellier, France.

Galzy, R. 1961. Confirmation de la nature virale du court-noue de la vigne par des essais de Thermathorapie sur des cultures in vitro. C.R. Acad. Seances Soc. Biol. Paris 253:706-708.

_____ 1962. Essais de Thermotherapie du court-noue de la vigne sur des cultures in vitro. 3rd conference internationale pour l'etudes des maladies a virus de la vigne. Lisbon. 1962. Bull. O.I.V. 36:41-44.

_____ 1964a. Technique de thermotherapie des viroses de la vigne. Epiphyties 15:245-256.

_____ 1964b. Premieres observations sur la distribution de l'infection chez des souches de vigne atteintes de court-noue. C.R. Acad. Seances Soc. Biol. Paris 259:1761-1763.

_____ 1969a. Remarques sur la croissance de vitis rupestris cultivee in vitro sur differents milleux nutritifs. Vitis 8:191-205.

_____ 1969b. Recherches sur la croissance de vitis rupestris schlee sain et court noue cultive in vitro a differences temperatures. Ann. Phytopathol. 1:149-166.

_____ 1971. Recherches sur la Croissance de la Vigne Saine et Court-nouee Cultivee In Vitro. Printed by G. Taris, 20, Rue Condillac, Bordeaux, France.

_____ 1972a. La culture in vitro des apex de vitis rupestris. C.R. Acad. Seances Soc. Biol. Paris 275:210-213.

_____ 1972b. Remarques sur la nutrition minerale des apex vitis rupestris. C.R. Acad. Seances Soc. Biol. Paris 275:561-564.

_____ 1977. Study of a mineral medium for the in vitro culture of vitis rupestris apices including three foliar outlines. In: La Culture des Tissues et des Cellules des Vegataux (R.J. Gautheret, ed.) pp. 134-146. Masson, Paris.

_____ and Compan, H. 1968. Thermotherapie de quelques varieties de vigne presentent des symptomes de virose. Vignes Vins 168:1-8.

_____ and Galzy, P. 1964. Action de la temperature sur la panachure blanche de la vigne. Prog. Agric. Viti. 162:1-9.

Gamborg, O.L. and Eveleigh, D.E. 1968. Culture methods and detection of glucanases in suspension cultures of wheat and barley. Can. J. Biochem. 46:417-21.

Gautheret, R.J. 1955. Sur la variabilite des properietes physiologiques des cultures de tissus vegataux. Rev. Gen. Bot. 62:5-112.

Gifford, E.M., Jr. and Hewitt, W.M.B. 1961. The use of heat therapy and in vitro shoot tip culture to eliminate fan leaf vines from the grape vine. Am. J. Enol. Vitic. 12:129-135.

Goyal, S.P. and Goyal, A.N. 1970. Note on the cumulative cell, nucleus, nucleoli counts in growth patterns of phylloxera gall and normal grape stem single cell clones in tissue culture. Nat. Inst. Sci. India Proc. 36:80-85.

______ 1971a. On the growth patterns of phylloxera vastatrix gall and normal grape stem single cell clones in tissue culture. Nat. Inst. Sci. India Proc. 36:305-310.

______ 1971b. Correlations between phylloxera vastatrix gall and normal grape stem single cell clones in tissue culture. Marcellia 37: 219-224.

Grenan, S. 1979a. Rhizogenesis de bourgeons apicaux de boutures de vigne cultives in vitro. Conaiss. Vigne Vin. 13:125-136.

______ 1979b. Possibilities d'elimination des modifications foliaires, apparues sur la variete Granache N. apres un passage prolonge en culture in vitro. Prog. Agric. Vitic. 96:152-157.

______ 1980. La thermotherapie un traitement contre les virus. France Agric. 21:3-6.

Gresshoff, P.M. and Doy, C.H. 1974. Derivation of a haploid cell line from *Vitis vinifera* and the importance of the stage of meiotic development of anthers for haploid culture of this and other genera. Z. Pflanzenphysiol. 73:132-141.

Harris, R.E. and Stevenson, J.H. 1979. Virus elimination and rapid propagation of grapes in vitro. Proc. Int. Plant Prop. Soc. 29:95-108.

Hawker, J.S., Downton, W.J.S., and Mullins. 1973. Callus and cell culture from grape berries. HortScience 8:398-399.

Hildebrandt, A.C. 1958. Stimulation or inhibition of virus infected and insect-gall tissues and single cell clones. Proc. Natl. Acad. Sci. U.S.A. 44:354-363.

______ 1963. Growth of single cell clones of diseased and normal tissue origins. In: Plant Tissue Culture and Morphogenesis (J.C. O'Kelly, ed.) pp. 67-72. Scholars Library, New York.

Hirabayashi, T., Kozaki, I., and Akihama, T. 1976. In vivo differentiation of shoots from anther callus in vitis grape research. Hortic. Sci. 11:511-512.

Hoeffer, L.L. and Gifford, E.J., Jr. 1964. Growth in vitro of excised stem tips of *Vitis vinifera*. Am. J. Bot. 51:677 (abstr.).

Jona, R. 1980. La Cultura di tessuti NEL miglioramento genetico. Situazione attuale e prospettive per la vite. Accad. Ital. Vite Vino, Siena, Atti 22:1-10.

______ and Vallania, R. 1980. Stem elongation and root initiation in proliferating shoot of *Vitis vinifera*. In: Plant Cell Cultures: Results and Perspectives (F. Sala, B. Parisi, R. Cella, and O. Cieferri, eds.) pp. 313-315. Elsevier/North-Holland Biomedical Press, Amsterdam.

Kochba, J., Spiegel-Roy, P., and Saad, S. 1978. Tissue culture studies with citrus: (1) The effect of sugars on embryogenesis and (2) Application of citrus tissue cultures for selection of mutants. In: Production of Natural Compounds by Cell Culture Methods (A.W. Alfermann and E. Reinhard, eds.) pp. 223-232. Gesellschaft Fur Strahlen- und Umweltforshung MBH, Munich.

Kohlenbach, H.W. 1978. Comparitive somatic embryogenesis. In: Frontiers of Plant Tissue Culture 1978. International Association for Plant Tissue Culture Congress Proceedings, Calgary, Alberta, Canada (T.A. Thorpe, ed.) pp. 59-66. IAPTC, Calgary, Alberta.

Krul, W.R. and Myerson, J. 1980. In vitro propagation of grape. In: Proceedings of the Conference on Nursery Production of Fruit Plants through Tissue Culture: Applications and Feasibility (R. Zimmerman, ed.) pp. 35-43. USDA Science and Education Administration, Agricultural Research Results, ARR-NE-11. Beltsville, Maryland.

Krul, W.R. and Worley, J.F. 1977. Formation of adventitious embryos in callus cultures of 'Seyval,' a French hybrid grape. J. Am. Soc. Hortic. Sci. 102(3):360-363.

Kulescha, Z. 1949. Recherches sur l'e'laboration de substances de croissance par les cultures de tissue de vigne (*Vitis vinifera* L.). C.R. Acad. Seances Soc. Biol. Paris 143:1499-1500.

Lai, P.C. and I, H.T. 1980. Studies on tissue culture of grape (*Vitis vinifera* L.): Influence of environmental factors on the growth and differentiation of grape callus. J. Agric. Res. China 29(2):157-165.

Legin, R., Bass, P., and Vuittenez, A. 1979. Premiers resultats de guerison par thermotherapie et culture in vitro d'une maladie de type cannelure (legno riccio) produite par le greffage du cultuvar Servant de *Vitis vinifera* sur le porte-greffe *Vitis riparia* x *V. berlandieri* Kober 5 BB. Comparison avec diverses viroses de la vigne. Phytopathol. Mediterr. 18:207-210.

Lichine, A. 1973. Encyclopedia of Wines and Spirits. Alfred A. Knopf, New York.

Meredith, C.P. 1981. Genetic engineering—The outlook for grapes. Wine Investor 5:1-4.

Moore, D.J., Rau, B., Viera, R., Stancy, T., and Dion, T. 1969. Environmental regulation of experimental leaflet abscission. Proc. Indiana Acad. Sci. 78:146-159.

Morel, G. 1944a. Sur le development de tissus de vigne cultives in vitro. C.R. Acad. Seances Soc. Biol. Paris 138:62.

______ 1944b. Action de l'acide indole-B-acetique sur la croissance des tissus de Vigne. C.R. Acad. Seances Soc. Biol. Paris 138:93.

______ 1945. Caracteres anatomiques des tissus de vigne cultives in vitro. C.R. Acad Seances Soc. Biol. Paris. 139:674-676.

______ 1947. Transformations des cultures des tissus de vigne produites par l'hetero-auxine. C.R. Acad. Seances Soc. Biol. Paris 141: 280-282.

______ 1948. Recherches sur la culture associee de parasites obligitoires et de tissues vegetaux. Ann. Epiphyt. 14:1-112.

Mott, R.L. 1978. Tissue culture propagation of conifers. In: Propagation of Higher Plants through Tissue Culture (K.W. Hughes, R. Henke, and M. Constantin, eds.) pp. 125-131. University of Tennessee Symposium Proceedings. National Technical Information Service, U.S. Dept. of Commerce, Springfield, Virginia.

Mowbray, G.H. 1981. Clone me around again, Willie. Virginia Farm Vine-to-Wine Letter 1:4-5.

Mu, S.C., Kwei, Y.L., Liu, S.C., Chang, F.C., Lo, F.M., Yang, M.Y., and Wang, F.H. 1977. Induction of callus in *Vitis* grape endosperm cultured in vitro. Chih Wu Hsueh Pao. 19:93-94.

Mullins, M.G. and Nair, Y. 1979. Rejuvenation in vitro: Induction of juvenile characters in an adult clone of *Vitis vinifera* L. (Grape). Ann. Bot. 44:623-627.

Mullins, M.G. and Srinivasan, C. 1976. Somatic embryos and plantlets from an ancient clone of the grapevine (cv. Cabernet-Sauvignon) by apomixis in vitro. J. Exp. Bot. 27(100):1022-1030.

______ 1978. Plantlets of Cabernet-Sauvignon grapes by nucellar embryony in vitro. Genetique et Amelioration de la Vigne C.R. 2eme Symposium International Amelioration, Vigne Bordeaux (R. Pouget and J.P. Doazan, eds.) pp. 12-15. INRA, Paris.

Mur, F. 1978. Obtention de vrilles et de baies chez *Vinis vinifera* en culture in vitro. Prog. Agric. Vitic. 95(21):609-611.

Mur, G. 1979. Thermotherapie de varietes de *Vitis vinifera* par la methode de culture in vitro. Quelques observations, quelques resultats. Prog. Agric. Vitic. 96(7):148-151.

Murashige, T. 1974. Plant propagation through tissue cultures. Ann. Rev. Plant Physiol. 25:135-166.

______ and Skoog, F. 1962. A revised medium for rapid growth and bioassays with tobacco tissue cultures. Physiol. Plant. 15:473-497.

Myerson, J. 1981. Adventitious embryogenesis and plantlet development in cultures of *Vitis* species 'Seyval.' M.S. Thesis, Univ. of Rhode Island, Kingston.

Nitsch, J.P. and Nitsch, C. 1969. Haploid plants from pollen grains. Science 163:85.

Nyland, G. and Goheen, A.C. 1969. Heat therapy of virus diseases of perennial plants. Ann. Rev. Phytopathol. 7:331-354.

Ohyama, K. and Nitsch, J.P. 1972. Flowering haploid plants obtained from protoplasts of tobacco leaves. Plant Cell Physiol. 13:229-236.

Olmo, H.P. 1976. Grapes. In: Evolution of Crop Plants (N.W. Simmonds, ed.) pp. 294-298. Academic Press, New York.

Panagopoulos, C.G., Psallidas, P.G., and Alivizatos, A.S. 1978. Studies on biotype 3 of *Agrobacterium radiobacter* var. tumefaciens. Proceedings of the 4th International Conference on Plant Pathology and Bacteria, INRA, Angers, France. pp. 221-228.

Pelet, F., Hildebrandt, A.C., Riker, A.J., and Skoog, F. 1960. Growth in vitro of tissues isolated from normal stems and insect galls. Am. J. Bot. 47:186-195.

Penning-Rowsell, E. 1971. The Wines of Bordeaux. International Wine and Food Pub., London.

Pongracz, D.P. 1978. Practical Viticulture. David Philip, Cape Town, South Africa.

Pool, R.M. and Powell, L.E. 1975. Influence of cytokinins on in vitro shoot development of Concord grape. J. Am. Hortic. Soc. 100:200-202.

Radler, F. 1964. Versuche zur kultur isolierter Beeren der Rebe. Vitis 4:365-367.

Rajasekaran, K. and Mullins, M.G. 1979. Embryos and plantlets from cultured anthers of hybrid grape vines. J. Exp. Bot. 30:399-407.

Reinert, J. 1963. Growth of single cells from higher plants on synthetic media. Nature (London) 200:90-91.

Ribereau-Gayon, J. and Peynaud, E. 1971. Sciences et Techniques de la Vigne. 2 vols. Dunod, Paris.

Skene, K.G.M. 1974. Culture of protoplasts from grapevine pericarp callus. Aust. J. Plant Physiol. 1(3):371-376.

_____ 1975. Production of callus from protoplasts of cultured grape pericarp. Vitis 14(3):177-180.

_____ and Barlass, M. 1982. Micropropagation of grapevine. Int. Plant Prop. Soc. Proc. 32 (in press).

_____ and Hale, C.R. 1971. Organic acid synthesis by grape berries cultured in vitro. Phytochemistry 10:1779-1781.

Slabecka-Szweykowska, A.E. 1952. Warunki tworzenia sie antocjanu w tkance *Vitis vinifera* hodowanej in vitro. Acta Soc. Bot. Pol. 21:537-576.

Srinivasan, C. and Mullins, M.G. 1978. Control of flowering in the grapevine (*Vitis vinifera* L.): Formation of inflorescences in vitro by isolated tendrils. Plant Physiol. 61(1):127-130.

_____ 1980. High frequency somatic embryo production from unfertilized ovules of grape. Sci. Hortic. 13:345-352.

_____ 1981. Physiology of flowering in the grapevine—A review. Am. J. Enol. Viti. 32:47-63.

Staudt, G., Borner, H.G., and Becker, H. 1972. Untersuchungen uber die Kallusbildung von di- and tetraploiden Reben in vitro. Vitis 11(1):1-9.

Sung, Z.R. and Okimoto, R. 1981. Embryonic proteins in somatic embryos of carrot. Proc. Natl. Acad. Sci. U.S.A. 73:3683-3687.

Thomashow, M.F., Panagopoulos, C.G., Gordon, M.P., and Nester, F.W. 1980. Host range of *Agrobacterium tumefaciens* is determined by the Ti plasmid. Nature (London) 283:794-796.

Thomashow, M.F., Knauf, V.C., and Nester, E.W. 1981. Relationship between the limited and wide host range octopine-type Ti plasmids of *Agrobacterium tumefaciens*. J. Bacter. 146:484-493.

Valat, C., Grenan, S., Auran, G., and Bonnet, A. 1979. Guerison de quelques maladies a virus de la vigne par thermotherapie de plantules cultivees in vitro. Vignes Vins, 289:19-22.

Valat, C., Grenan, S., and Auran, G. 1981. Thermotherapie in vitro: Premiers observations sue des aptitudes de quelques varieties de porte-greffes et de *Vitis vinifera* traitess. Vignes Vins 289:17-23.

Wagner, G.J., Butcher, H.C., and Siegelman, H.W. 1978. The plant protoplast: A useful tool for plant research and student instruction. BioScience 28:95-101.

Wetherell, D.F. 1978. In vitro embryoid formation in cells derived from somatic tissues. In: Propagation of Higher Plants through Tissue Culture (K.W. Hughes, R. Henke, and M. Constantin, eds.) pp. 102-124. U.S. Dept. of Energy, Technical Information Center.

SECTION VIII
Fiber and Wood

CHAPTER 16
Conifers

T.A. Thorpe and *S. Biondi*

It has been predicted that world demand for forest products will rise sharply over the next few decades. This prediction can be confirmed easily if one considers the ever-increasing value given to wood by the pulp, paper, timber, and furniture industries, as well as by the ecological needs of reforestation, and the possibilities for mitigating increasing energy demand by the production of gasohol from plant products (Keays, 1974; Karnosky, 1981). In addition, the rapid and disastrous effects of disease, pests, and fires today may jeopardize the very existence of certain tree species. In all cases there is an urgent requirement for large numbers of improved, fast-growing trees with shortened rotation. Present-day tree improvement and propagation methods offer only limited possibilities to achieve this goal.

Historically, investment in forestry-related research has not been considered profitable. This was true as long as wood was a plentiful and cheap end-product. Wild stands of forest trees, however, are no longer able to supply the future needs of humanity for forest products. Today it is clear that a considerable effort in time and money is necessary to solve the problems mentioned above, just as in the past, a similar effort in breeding and propagating nonforest species, especially fruit, brought productivity of these species to the levels present today.

Today, trees of poor or heterogeneous quality and slow or unreliable propagation methods are no longer acceptable. There is an urgent need to make a broad and concerted effort in domesticating and breeding trees with superior wood quality, optimal stem form and uniformity, rapid growth rates, short rotations and a high production index (stem: total tree biomass), resistance to disease and pests, the ability to adapt to new climates and extreme environmental variables (including pollution), the ability to respond well to silvicultural practices (e.g., fertilization) and, ultimately, as artificial fertilizers become increasingly uneconomical, the ability to fix nitrogen directly.

To date, breeding in forestry has lagged behind breeding in agriculture. Even disease resistance is poorly understood (Mott, 1981a). While some species are genetically uniform and offer limited potential for genetic improvement, others (e.g., Douglas fir and white spruce) are more variable and show promise for considerable genetic improvement (Durzan, 1980). Long life cycles, usually 15-20 years from seed to flowering, are the main cause for this lag, since it is difficult to breed enough successive generations to develop both uniformity and superiority. In addition, it has been suggested that one-half to three-quarters of the knowledge accumulated by generations of foresters has been lost through loss of seed sources and lack of program continuity, owing again to the long time spans occurring between the initiation of a program, and evaluation and implementation of the results. Tissue culture has an important role to play in solving these various problems (Durzan, 1980; Karnosky, 1981; Mott, 1981a; Sommer and Brown, 1979; Winton and Huhtinen, 1976).

IMPORTANCE OF VEGETATIVE PROPAGATION

Since even small increases in yield due to tree improvement result in considerable increases in the commercial value of planted forests, these programs are a good investment (Carlisle and Teich, 1971). For example, small gains in wood quality and stem uniformity which facilitate mechanical harvesting greatly affect daily profits made by pulp mills (Davis, 1969). Carlisle and Teich (1971) conclude that there is good evidence that the costs of producing genetically superior seed are more than offset by small increases in yield of 2-5%.

Kleinschmit (1974) has indicated that at least a 10% increase in gain can be expected from planting selected clonal propagules rather than selected seed families. Rediske's (1979) work with western hemlock substantiates this claim. Thus, vegetative propagation is an important part of the domestication and breeding process. The reasons for this are that (a) it has the potential for increased gains from any one generation of a tree improvement program, since multiplication of a superior selection is possible at any stage; (b) selection and multiplication of both additive and nonadditive gene effects are possible, thus increasing gains; (c) selected clones can be maintained in a "gene bank" represented by a clonal orchard; then genes can be recombined via controlled pollination; (d) genotype-environment interactions—genetic and environmental covariances between characteristics—can be better studied. (The ratio of variability due to the environmental component to total variability can be well estimated in good clonal studies, and this describes the plasticity of a given trait. It is a useful indicator of the success that silvicultural manipulations such as fertilizing and thinning may have.); (e) artificial clonal selection of genotypes can be used for production plantations; (f) competition between adjacent trees, host-pathogen relations, and mycorrhizal symbiosis can be better understood by using intraclonal studies; (g) new hybrid cultivars arising from breeding programs or chance mutation can be increased quickly for progeny testing as well as perpetuated to ensure program continuity; and (h) genetically uniform parent plants can be produced in large numbers for large-scale hybrid seed production.

The early practice of foresters was to select the best trees available in natural stands, propagating these by grafting (owing to the difficulty of rooting cuttings from mature trees) to establish clonal orchards from which, by breeding and progeny testing, improved seed was produced.

However, it is well known that with seed propagation there is little guarantee that the progeny will resemble the elite parent, especially with forest trees which, as outbreeders, are highly heterogenous and self-incompatible. On the other hand, the impact that cloning of the most desirable genotypes can have on forest productivity is easy to foresee. For example, by cloning several hundred vigorous selections followed by performance tests on a variety of sites, one could quickly assay those clones best adapted to biomass production under short-rotation regimes, thus producing rapid genetic gains (Sommer and Brown, 1979). It is important to remember, however, that factors other than genetic constitution can influence the behavior of a tree in the field.

The production of interspecific hybrids followed by selection and cloning of the best offspring from each cross could likewise result in rapid genetic gain. Current methods for hybridization of pine, for instance, are costly and relatively inefficient except under special circumstances (Sommer and Brown, 1979). Sommer and Brown (1979) quote the case of a reforestation program in S. Korea which uses the *Pinus taeda* x *P. rigida* F_1 hybrid. Only the exceedingly low labor costs existing there have allowed enough controlled crosses to be made to produce enough seed for mass production of the hybrid. Similarly, New Zealand has an annual requirement for 75 million *Pinus radiata* seedlings for its afforestation program (Aitken et al., 1981). This would require 500,000 controlled crosses—clearly not a very viable proposition. A relatively small amount of superior seed can be greatly increased with cell culture methods.

Clonal propagation of planting stock saves time. It now takes at least 16 years after selection for either a Douglas fir or loblolly pine seed orchard to attain full production and thus for the gains of the improvement effort to be fully realized. Although this period can be reduced by treatments (e.g., gibberellins, Ross et al., 1983) in some cases, Rediske (1979) anticipated that this 16-year period could be shortened to 4 years or less with vegetative propagation. Furthermore, he estimated a significant reduction in capital and operating costs. If 400 cuttings/m are obtainable annually for Douglas fir or loblolly pine, this is approximately equivalent to 2 million cuttings/ha per year, while a seed orchard will produce about 800,000 seedlings/ha annually. Assuming 80% rooting success with the cuttings, the clonal orchard need only cover half the land area of the seed orchard.

The traditional methods for vegetative propagation are rooted cuttings or rooted needle fascicles (also known as brachyblasts, short shoots, dwarf spurs) for the pine species, and grafting. As far as forest tree species are concerned, *Cryptomeria japonica* has been propagated by cuttings for centuries in Japan. More recently, Norway spruce (*Picea abies*) has been propagated in a similar manner on a large scale (2.5 million cuttings annually in Sweden). The breeding program carried out in Sweden by the Hilleshog Company is based on this method of propagation. The gain is at least 15 years. However, for the majority of conifers, propagation by rooted cuttings is in various stages of development and is often characterized by a rapid loss of rooting capacity of the ramet (cutting) with increasing age of the ortet (parent plant). On the other hand, one of the main aims of vegetative propagation is to multiply trees old enough to have demonstrated their superior characteristics. Percent rooting, speed of rooting, root length and number, and survival and growth in and after the year of rooting, all decline, particularly when the parent plant is more than 10 years old (Girouard, 1974). Furthermore, in many species, rooted branch cuttings tend to continue to grow with a horizontal orientation and bilateral symmetry (plagiotropy) until, after varying periods of time, the terminal meristem changes to radial symmetry and vertical growth (orthotropy). This reversal to normal growth frequently displays intra- and interclonal variations and thus is erratic so that the evaluation of genotypes in a selection experiment becomes almost impossible (Libby, 1974). While plagiotropic growth does not appear to be a problem for Norway spruce and radiata pine (Aitken, personal communication), it certainly is for Douglas fir (Abo El-Nil, personal communication).

Finally, vegatative propagules display slower height and diameter growth as the age of the ortet increases (Libby, 1974). Sweet (1973) has estimated a loss of about 40% in the lower stem volume growth of ramets from mature radiata pine compared to juvenile ramets or seedlings.

Maturation is thus one of the key obstacles to successful vegetative propagation. Numerous methods have been proposed which may retard or possibly even reverse the maturation process and will be dealt with in a later section. These "rejuvenation" techniques (grafting, hedging, etc.) are as important for rooting cuttings as for micropropagation but are not, as yet, of widespread use since—as will be seen later in this chapter—most conifer cell culture systems to date are initiated with young material. Assuming then that many of the problems associated with propagating mature trees by cuttings can be solved by rejuvenation of the parent plant and that morphogenetic capacity in vitro is enhanced by a similar procedure, then one must begin to evaluate the pros and cons of using one or the other technique.

The main advantage of using cell culture as a tool in breeding programs and mass production is its potential for enormous (potentially unlimited) multiplication rates. Cell culture provides a much greater control over plant selection and can force expression of totipotency. Thus, while a rooted cutting can produce a single plant from which, several years later, further cuttings are available, even the most limited cell culture systems—that of resting buds—with today's techniques can produce several axillary as well as adventitious shoots. Both these types of shoots can be in turn induced to form axillary additional and/or adventitious shoots, often within a matter of weeks or, at most, months.

Cell culture generally opens up a score of other possibilities which justify its continued use and the worldwide interest displayed by scientists. These include: (a) production of homozygous lines (via haploid cultures), (b) production of incompatible hybrids (via embryo cultures), (c) production of disease-free clones (via meristem cultures), (d) production of somatic hybrids (via protoplast fusion), (e) production, screening, and multiplication of mutants (via suspension cultures), (f) study of host-pathogen relations, (g) study of mycorrhizal associations, (h) physiological studies, including the prediction of phenotypic expression (early-testing), discovery of juvenility-maturation parameters, etc., (i) germplasm storage, and (j) natural secondary products formation.

HISTORICAL BACKGROUND

Since the 1940s considerable progress has been made in applying cell culture techniques in breeding programs of horticultural and agricultural crops (Nickell and Torrey, 1969). The same cannot be said for forest species yet trees were among the first plants to be cultured in vitro. Thus as early as 1934, Gautheret cultured cambial tissue of woody species including *Pinus pinaster* and *Abies alba*. In 1936, La Rue grew embryos of *Pinus resinosa, Thuja occidentalis, Picea canadensis, Tsuga canadensis*, and *Pseudotsuga menziesii* in culture and produced normal-looking seedlings. While in the 1940s several reports were published regarding organogenesis in callus cultures of some broad-leaf species, it was in 1950 that Ball was able to report on the differentiation of buds in callus cultures of a conifer, *Sequoia sempervirens*. Such buds did not grow into shoots (Skoog, 1954) and bud-forming capacity declined rapidly after the fourth subculture (Gautheret, 1957), yet it was the first continuous culture of a coniferous species. The culture was initiated from young adventitious shoots growing on burls.

In the 1960s, conifer cell culture benefited from research which established in detail the nutritional requirements for optimal growth of normal and tumor tissues of *Picea glauca* (White and Risser, 1964; Risser and White, 1964). Soon after Brown and Lawrence (1968) adapted the Murashige Skoog (1962) medium to significantly improve callus growth of several pine species.

At the time of the IAPTC Congress in Leceister in 1974, 26 papers had been published or were in press reporting regeneration of angiosperm trees in vitro (Winton, 1978); and complete conifer plantlets had yet to be produced. Nonetheless adventitious bud and shoot formation had been achieved with a number of gymnosperms (mostly conifers) of which about half belonged to the genus *Pinus* (Brown and Sommer, 1977). In addition roots had been induced on callus of *Pseudotsuga* (Bethel, 1972) and *Pinus gerardiana* (Konar, 1963), and cotyledons of *Thuja (Biota) orientalis* had formed "embryoids" which, when cultured alone, formed shoots but not roots (Konar and Oberoi, 1965). This was the first indication that buds from gymnosperm cell culture could form shoots.

The first complete conifer plantlet obtained in vitro was one of *Pinus palustris* (longleaf pine) produced in Brown's laboratory in 1974 (Sommer et al., 1975). The plantlets originated from adventitious buds formed along the cotyledons of embryos in culture. From 1975 onward the number of reports have increased rapidly. Plantlets have been regenerated from *Picea glauca* (Durzan and Campbell, 1976), *Pinus radiata* (Reilly and Washer, 1977), *Thuja plicata* (Coleman and Thorpe, 1977), *Pseudotsuga menziesii* (Cheng and Voqui, 1977; Boulay, 1979b), *Tsuga heterophylla* (Cheng, 1976), *Pinus taeda* (Mehra-Palta et al., 1978), *Sequoia sempervirens* (Boulay, 1979a), *Pinus pinaster* (David et al., 1978), and others (see Table 1). Table 1 contains only those species for which it is clear that plantlets and not separated shoots and roots were formed in vitro.

The first case of cytodifferentiation (of vascular elements) in a cell suspension culture was reported in 1971 for white spruce (Durzan and Chalupa, 1971), and for protoplast isolation followed by cell division, the first case was reported in 1974 with Norway spruce (Chalupa, 1974). Again with Norway spruce, Huhtinen obtained that same year root formation from haploid megagametophyte-derived callus; later shoot regeneration was also reported (Bonga, 1977a; Steinhauer and Huhtinen, 1978). We remain somewhat unconvinced that somatic embryogenesis from conifer cell suspension cultures has as yet been achieved, despite some statements to that effect in conference- and review-type articles.

CELL CULTURE METHODS

The early success with cambial explants from which it was relatively easy to induce and maintain continuous callus cultures was probably due to their high content of endogenous cytokinins (CK) (Bonga, 1977b). In this way, even though the role of CKs as cell division regulators was as yet unknown, it was sufficient to supply an auxin as the only growth hormone (Nitsch, 1963). Morphogenesis, on the other hand, is much more difficult to achieve and requires a precise auxin-CK balance (Skoog and Miller, 1957). Where endogenous CK levels are suboptimal, an exogenous supply is necessary. Originally KIN was most commonly used, but now it is very often replaced by BA. Occasionally, ZEA and 2iP are used, alone or in combination with other CKs (Mehra-Palta et al., 1978; Cheng, 1975; von Arnold and Eriksson, 1978).

Most cell culture systems also require an exogenous supply of auxin, but several exceptions exist such as embryo or cotyledon cultures of *Pinus radiata* (Reilly and Washer, 1977; Aitken et al., 1981) and hypocotyl cultures of *Picea glauca* (Campbell and Durzan, 1976). These differences in hormonal requirements probably reflect differences in endogenous hormone concentrations present within the explant and indicate that the auxin-CK balance mentioned earlier must take into account the physiological state of the explant (Thorpe, 1980; Thorpe and Biondi, 1981).

However, due in part to our poor knowledge of endogenous hormone contents, the mechanisms of hormonal action, and the interactions at tissue, cellular, and molecular levels, varying the auxin-CK ration does not always lead to morphogenesis. To complicate the matter further, there is little doubt that many chemical factors other than auxins and CKs play a role in morphogenetic control, not to mention the as yet poorly studied physical factors of light, temperature, osmotic potential,

Table 1. Plantlet Regeneration in Gymnosperms

FAMILY	GENUS AND SPECIES	REFERENCE
Auraucariaceae		
	Araucaria angustifolia	Winton, 1978
	A. cunninghamii	Haines & de Fossard, 1977
Cupressaceae		
	Thuja plicata	Coleman & Thorpe, 1977
Pinaceae		
	Picea glauca	Campbell & Durzan, 1976; Rumary & Thorpe, 1983
	P. mariana	Rumary & Thorpe, 1983
	P. sitchensis	Webb & Street, 1978
	Pinus contorta	Patel & Thorpe, unpublished
	P. radiata	Reilly & Washer, 1977
	P. pinaster	David et al., 1978
	P. taeda	Mehra-Palta et al., 1978
	P. palustris	Sommer et al., 1975
	P. resinosa	Bonga, 1977b
	P. rigida	Brown & Sommer, 1977
	P. sabiniana	Brown & Sommer, 1977
	P. virginiana	Brown & Sommer, 1977
	P. elliotti	Brown & Sommer, 1977
	Pseudotsuga menziesii	Cheng & Voqui, 1977; Winton & Verhagen, 1977; Boulay & Franclet, 1977; Boulay, 1979b
	Tsuga heterophylla	Cheng, 1976
Taxodiaceae		
	Sequoia sempervirens	Ball, 1978; Boulay, 1979a

etc. One report claims that bud initiation in Douglas fir cultures is enhanced by red light at 660 nm if applied during the third, fourth, or fifth week of culture, i.e., during bud primordium development (Kadkade and Jopson, 1978).

At least one way to cope with the myriad of unanswered questions which the cell culturist faces is to carefully select the initial explant. Most workers in this field recognize the importance of this factor and its intimate relationship with the condition of the source plant, yet it remains probably the least understood and least researched aspect of cell culture (Murashige, 1974; Sommer and Caldas, 1981).

In general, the more juvenile the tissue, in a state of active growth, the better the response to the process of de novo organ differentiation. Cells in culture that are induced to become meristematic (i.e., high nucleus:cytoplasm ratio, small vacuoles, dense cytoplasm) must, in some way, also show a considerable degree of plasticity in order to divert their normal developmental pathway.

Aitken et al. (1981) noted the importance which explant selection had on adventitious bud formation in radiata pine cultures. By using excised cotyledons from embryos they doubled the shoot-forming capacity of whole embryos, and by using cotyledons excised from newly germinated seeds, shoot-forming capacity was increased 12.5-fold over that of cotyledons from embryos. Mott et al. (1977) also used this approach to increase bud production in loblolly pine cultures. Furthermore, the radiata pine system has shown that even a difference of a few days (e.g., 5-10 days of germination) greatly affects the morphogenetic capacity of cotyledons.

Often external factors, which ultimately influence the physiological state of the explant, can affect responses in culture and are easier to control and standardize. Where the explant source is of seed origin, storage and imbibition time and temperature, degree of stratification, and time and conditions of germination can be important. Thus while longleaf pine seeds have no dormancy and can be put directly into culture, cotyledon cultures of loblolly pine produce more buds if the seeds are partially stratified (Sommer and Caldas, 1981).

A similar criterion can be applied to the choice of the parent plant since its physiological state can play a decisive role on the explant's behavior in culture. One example of this is the aspect which involves use of rejuvenating treatments. Another example was presented by von Arnold (1982) who found that bud proliferation on resting buds of Norway spruce varied according to the growth conditions of the source plant: The response improved considerably when going from trees growing in a natural stand, to a greenhouse or a phytotron.

Another consideration which frequently arises is that of the variability of response from explant to explant or seed to seed (in this case, these reflect genotypic variations). The problem is particularly noticeable with tree species because they are highly heterogeneous (Sommer and Caldas, 1981). With embryo cultures of radiata pine, a variation of 1-200 buds/embryo has been observed (Reilly and Washer, 1977). Most reports on conifer cell culture do not provide much quantitative data but a similar degree of variability both in the number of explants forming buds, as well as the number of buds per explant, is likely to occur with other species.

In theory, this variability should be drastically reduced under optimal conditions. Certainly media manipulations and explant selections can reduce it significantly (Campbell and Durzan, 1976; Mott, 1979; Aitken, personal communication).

Durzan and Campbell (1974) subdivide the methods of micropropagation into two broad categories: The first consists of organized partial systems, which include bud cultures, stem segments (including cuttings), needle fascicles (brachyblasts or short shoots) for pine species, and shoot or seedling apices. These systems are characterized by the fact that bud proliferation occurs as a result of the release of inhibited axillary buds; i.e. meristems are preformed. Nonetheless it should be emphasized that such cultures are often capable of forming both axillary and adventitious buds, or sometimes only the latter, as has been shown for Douglas fir (Mapes et al., 1981; Thompson and Zaerr, 1981).

The second category for micropropagation consists of unorganized systems, in which adventitious buds and/or embryoids are formed de novo. Callus or cell suspension cultures and embryonic tissues (where adventitious buds are formed directly without intermediate callus phase) are examples of this.

Theoretically, the developmental path with organized systems should be shorter and this could be an advantage. Experience has shown, however, that a good unorganized system can also produce shoots within a few weeks. The use of preformed meristems implies that the number of progeny is finite. However, if a rapid "generation time" can be established (say 5-6 weeks from the formation of the first axillary shoots to the formation of axillary shoots from the initial explant) then this method can be useful. With unorganized systems, cultures theoretically can be initiated from any part of the plant and the number of progeny is theoretically infinite, especially with suspension cultures. Many scientists are apprehensive of the fact that propagules from callus in particular and, perhaps to a lesser extent, from certain organ explants may show genetic deviations from the original plant (Larkin and Scowcroft, 1981). It is essential that the process of propagating plants by cell culture should not in itself significantly increase the mutation rate, otherwise the time and energy spent in discarding unwanted plantlets (especially if the mutations only become apparent after planting in the field) could easily offset the advantages which the method offers. How widespread or serious the phenomenon of genetic instability is in tree cultures is not yet known (Tominaga and Oka, 1970; de Torok and White, 1960; Partanen, 1963; Bonga, 1974a). Furthermore, cytoplasmic epigenetic factors play a significant role in cell growth, rootability, graftability, juvenility, and aging in forest trees (Stanley, 1970).

It is generally assumed that the shoot meristem offers great genetic stability and that the probability of producing abnormal plants via adventitious shoots on organ explants is intermediate between that of axillary and callus-derived shoots (Hussey, 1978).

Table 2 is a brief survey of some of the types of explants used to date which have shown bud, shoot, and/or plantlet regeneration. The most common explant involves the use of embryos or seedling parts. Such embryonic material is generally very responsive. Older material is generally more recalcitrant, but some success has been obtained with buds of mature trees.

Although callus cultures initiated from a variety of explants have been established for many conifers (for a review see Brown and Sommer, 1975 or Winton, 1972a), there is little evidence that subcultured callus will generally give rise to shoots or plantlets, exceptions being Douglas fir (Cheng, 1975; Winton and Verhagen, 1977), *Pinus wallichiana* (Konar and Singh, 1980), and *Sequoia sempervirens* (Ball, 1950).

THE STAGES IN PLANTLET FORMATION

Plantlet formation in conifers can be conveniently divided into the following four stages (Thorpe, 1977).

Table 2. Examples of Parts and Ages of Conifer Species Used as Explants for the Induction of Primordia

SPECIES	EXPLANT[a]	AGE	REFERENCE
Picea abies	Embryos	-	von Arnold & Eriksson, 1978
P. glauca	Hypocots	6-12 d	Campbell & Durzan, 1975, 1976
	Epicots	25-28 d	Rumary & Thorpe, 1979
P. mariana	Epicots	25-28 d	Rumary & Thorpe, 1979
Pinus palustris	Embryos	-	Sommer et al., 1975
P. radiata	Cots	5-7 d	Aitken et al., 1981
	Embryos	-	Reilly & Washer, 1977
P. taeda	Cots	7 d	Mehra-Palta et al., 1978; Mott, 1978
Psuedotsuga menziesii	Cots	2-4 w	Cheng, 1975, 1976, 1979
Tsuga heterophylla	Cots	2-5 w	Cheng, 1976
Thuja plicata	Cots	10-14 d	Coleman & Thorpe, 1977
Abies balsamea	Dormant buds	15-20 yr	Bonga, 1977a, 1981
Picea abies	Dormant buds	2-50 yr	von Arnold, 1982
P. sitchensis	Shoot apices	2 yr	Webb & Street, 1978
Pinus pinaster	Needle fascicles	2-3 yr	David & David, 1977
	Needle fascicles	3-11 yr	David et al., 1979
P. radiata	Sec needles	7.5 mo	Reilly & Brown, 1976
Pseudotsuga menziesii	Dormant buds	20-50 yr	Thompson & Zaerr, 1982
	Dormant buds	2-40 yr	Boulay, 1979b
Sequoia sempervirens	Stem segments with buds	20-100 yr	Boulay, 1979a
	Basal shoot tips		Ball, 1978
Thuja plicata	Lateral shoot tips	4-10 yr	Coleman & Thorpe, 1977

[a] Cots = cotyledons; epicots = swollen epicotyls.

Initiation of Shoot Buds

Typically the selected explant is placed in a high-salt medium such as Murashige-Skoog (1962) or Schenk and Hildebrandt (1972) in the presence of a cytokinin (0.5–30 μM). In some cases a low level of auxin (0.005–0.5 μM) may be useful. For other species, e.g., radiata pine, the latter is unnecessary and auxin may even enhance callus formation rather than primordium initiation (Biondi and Thorpe, 1982). In some cases, double cytokinins and auxins are more effective than single phytohormones, e.g., Douglas fir responds well to BA plus 2iP and IAA plus IBA (Cheng, 1975).

During adventitious bud formation on embryonic explants, cell division occurs at the periphery of the explant to yield small cells with large nuclei and dense cytoplasm (meristematic cells) (Mott, 1979; Aitken et al., 1981; Cheng, 1979; Sommer et al., 1975). Epidermal and hypodermal cells are directly involved (Cheng, 1979; Yeung et al., 1981). The proliferation of these cells results in swelling of the explant and the formation of a nodular tissue called "meristematic bud centers" by Mott (1981a) and "meristematic tissue" by Reilly and Brown (1976). Within 3-6 weeks in culture, increasingly definite "dome-like swellings" (Coleman and Thorpe, 1977), "meristemoids" (Yeung et al., 1981), or "tube-like protrusions" and "scale-like organs" (von Arnold and Eriksson, 1978) appear.

At these early stages, the subdivision of meristematic aggregations into smaller pieces is possible and results in increased bud production. This method has been reported for *Pinus radiata* (Reilly and Washer, 1977). For *Pinus radiata* the explant is usually subdivided at about 9 weeks in culture (Aitken et al., 1981), when the large number of shoots growing on the explant may mutually inhibit growth and development. Similarly, of the 10–30 buds formed per cultured embryo of Norway spruce, only 2 or 3 will elongate. If buds which have reached 3–5 mm in height are isolated, new buds will start to elongate (von Arnold and Eriksson, 1978). Experiments at the Forest Research Institute of Rotorua (N. Zealand) indicate that by adding a liquid culture step, a furthur 10–25-fold multiplication is possible (Horgan and Aitken, 1981).

A new experimental approach to induce shoot proliferation in resting buds of adult trees, which goes beyond exogenous auxin-cytokinin variations, is that used by Bonga (1977a, 1981). In his early studies, Bonga found that if embryonic shoots (dormant buds minus bud scales) of 15–20 year-old *Abies balsamea* were soaked either for 24 hr in distilled water or for 15 min in a solution of 100 mg/l caffeic acid, organogenesis occurred upon transfer to a nutrient medium with the formation of adventitious buds. He could not establish which of the various soak treatments was most effective, but since it is assumed that soaking leaches inhibitory substances from the explant (Cresswell and Nitsch, 1975), he presumed that the high-growth regulator content in the soaking solutions acted simply by accelerating the leaching effect which occurred anyway, though at a slower rate, with the water alone. As a follow-up to this work, Bonga (1981) recently reported that the inclusion of a large number of auxins, cytokinins, growth inhibitors, herbicides, antioxidants, and other chemicals in the soaking solutions had little effect except for malonic acid (9.6 mM for 15 min) following a 24 hr soak in the water. Some time after transfer to Romberger, Varnell, and Tabor's medium as modified by Bonga (1977a), clusters of needle organs and small shoots grew at the base of the needles. Malonic acid may act by inhibiting the TCA cycle and thus cause an accumulation of succinate, other TCA-cycle acids, glutamate, and glutamine, or by switching part of the oxidative metabolism to the pentose phosphate pathway which, as has already been suggested (Thorpe and Laishley, 1973; Brown and Thorpe; 1980), may be an important aspect of shoot primordia formation.

Knowledge in the area of the physiology and regulation of bud induction is still limited. Although many aspects have been investigated using a number of herbaceous species (for recent reviews, see Thorpe, 1980, and Thorpe and Biondi, 1981), few such studies have been performed on conifer tissue cultures.

For example, plant organs or tissues used in studies of in vitro regeneration are seldom investigated for their growth-regulator content. De Yoe and Zaerr (1976) and Caruso et al. (1978) have demonstrated the presence of IAA in 15-year-old trees, and in shoot tips of trees and seedlings of Douglas fir. In both studies, concentrations of IAA found in growing shoots was over an order of magnitude greater than in the vegetative tissues of herbaceous species such as *Pisum* or *Phaseolus*. The work of Caruso and co-workers also showed that seedlings at about the same developmental stage as those which were found to be regenerative in culture have a surprisingly high level of endogenous IAA. Von Arnold (1979) noted that while a low-auxin concentration slightly increased the percentage of Norway spruce needles (from 5-year-old plants) forming adventitious buds in culture, a similar stimulation by auxin was not found when embryos were cultured instead of needles. The status of endogenous CK and the mode of action of exogenous CK in conifer tissue cultures remain to be elucidated. However, work with other species has demonstrated that BA stimulates protein synthesis (Kulaeva and Romanko, 1967), polyribosome formation (Short et al., 1974), and qualitative changes in the spectrum of newly synthesized proteins (Fosket et al., 1977).

A study by Hasegawa and co-workers (1979) has shown that differences in the electrophoretic profiles of newly synthesized proteins in bud-forming versus nonbud-forming Douglas fir cotyledon explants were detectable as early as Day 2 in culture, thus preceding even the earliest visible histological changes which only started after Day 4 (Cheah and Cheng, 1978). Work in our laboratory with radiata pine cotyledon cultures has shown that, at the time of excision from the seedling (soon after radicle protrusion), the explants are in a metabolically active state: There was active incorporation of ^{3}H-amino acids; the reserve substances present at Day 0 in culture (lipids, free sugars, starch, proteins) were rapidly depleted; and protein-N, after an initial drop (breakdown of storage proteins) rose again. These observations indicated that radiata pine cotyledons from germinated seeds are rich in reserve substance at the time of excision and already metabolically very active in terms of DNA, RNA, and protein synthesis rather than being quiescent or dormant like many other types of explants. The respiratory peak occurring between Days 0 and 3 in culture ensured that the major storage substances could be mobilized rapidly in order to provide carbon, nitrogen, and energy for subsequent use in the high-energy-requiring process of adventitious bud formation, while a steady protein content was maintained presumably by continued protein synthesis (Biondi, 1980; Biondi and Thorpe, 1982). Much of the energy presumably comes from the rapid breakdown of lipid, as the decrease of lipid is much greater than is needed for the production of membrane lipids, etc. (Douglas et al., 1982).

Development of Buds into Shoots

In most cases the formation of tree shoot apices and juvenile leaf primordia requires transfer onto a medium with altered nutritional and/or hormonal levels. In other cases, once buds are formed, a change of medium is necessary to allow for stem elongation. This is particularly important because it makes separation of individual shoots possible, and because the subsequent step—rooting—is facilitated if the shoot has a recognizable stem portion and is in active growth (Mehra-Palta et al., 1978).

Thus, for example, radiata pine cultures are transferred for 3-4 weeks on Schenk and Hildebrandt's salts followed by a further 3-4 weeks on Gresshoff and Doy's salts prior to rooting onto basal medium lacking hormones and with reduced sucrose levels (1-2% instead of 3%) (Horgan and Aitken, 1981). The removal of the growth regulators (both auxins and cytokinins) present in the initial culture medium is most common (Aitken et al., 1981; Cheng, 1975, 1976; Mehra-Palta et al., 1978; von Arnold and Eriksson, 1978). In some cases, the macro- and micronutrients are diluted by half (Cheng, 1975; Mehra-Palta et al., 1978). The inclusion of activated charcoal (0.1-1%) is sometimes found to exert beneficial effects on shoot elongation (Mehra-Palta et al., 1978; Boulay, 1979b; David et al., 1979; Rumary and Thorpe, 1983) but its role is not yet clear. Activated charcoal probably absorbs excess hormones or inhibitory substances (Fridborg et al., 1978). It possibly plays some additional role as activated charcoal of different origins may have different degrees of effectiveness. Conifer-derived charcoal was found to be best for shoot development in white and black spruce (Rumary and Thorpe, 1983).

Other factors that may enhance shoot elongation are vitamin D (5 μM), far-red light, and a high light intensity (Bornman and Jansson, 1982). Theoretically, gibberellic acid (GA) ought to favor internode elongation, but this phytohormone has not been shown to be effective in conifer cultures.

Since it would appear that the transfer of buds to a CK-free medium after a prolonged culture period on relatively high CK is not always sufficient to reduce endogenous CK to concentrations that permit elongation and rooting ("carry-over effect"), Mott (personal communication) has suggested that hormonal "pulsing" may be desirable. This means exposing the explant to high concentrations of the hormone for a very short period of time (such as a few days) and transferring to a CK-free medium before neoformed organs are visible. This approach was used by Cheng (1975) who pipetted high concentrations of CK (0.5-1.0 mM BA) around Douglas fir tissue which was later transferred to phytohormone-free medium. However, Biondi and Thorpe (1982) have found that CK is most effective in inducing shoot formation in radiata pine if it is present during the first 21 days in culture. Shorter exposures reduce meristematic tissue formation.

Rooting of Shoots

Rooting of angiosperm trees in culture has been relatively easy while gymnosperms have presented more problems (Sommer and Caldas, 1981). These problems have been overcome to some extent in *Pinus taeda*, *Tsuga heterophylla*, *Pseudotsuga menziesii*, *Thuja plicata*, *Pinus radiata*, *Sequoia sempervirens* and *Pinus pinaster*. Some of the culture conditions commonly employed to induce rooting of shoots obtained in vitro include such factors as reducing the mineral content of the medium (for example, using half-strength mineral salts), reducing the sucrose content of the medium (0.5-1%), reducing the temperature regime to approximately 20 C, and placing the cultures in total darkness for the first 10 days of the rooting treatment and subsequently back in the light. As far as hormonal treatments are concerned, auxins represent the critical factor. IBA alone (0.5-49 μM), a combination of two auxins (IBA + IAA or IBA + NAA), or tumbling overnight in a concentrated solution (590-685 μM) of IAA or IBA have all been tried with success. Sometimes a low concentration of CK (0.04-0.4 μM) can be beneficial (Mehra-Palta et al., 1978), but again, once root primordia are formed, transfer to a hormone-free medium is necessary to allow root elongation to take place.

Increasingly it is being found that rooting in nonsterile conditions is equivalent or superior to rooting aseptically. Furthermore, it is also easier to plant a shoot without roots than a rooted shoot in soil. The transfer of plantlets to nursery beds or the field is sometimes a problem because the vascular connection between shoot and root is incomplete. This is probably less likely to occur with nonsterile rooting. Agar-grown roots sometimes lack root hairs and this too is an undesirable characteristic which can be avoided with nonsterile rooting.

Thus, loblolly pine shoots (5 mm long) can be rooted in vermiculite-sand or peat moss-sand (1.5:1) mixtures under mist (10 sec at 6 min intervals). Bottom heat (26 C) and a high light intensity (10,000 lux) were applied (Mehra-Plata et al., 1978). Mehra-Palta et al. claimed that roots formed in soil were more vigorous than those formed in agar.

Pinus radiata shoots (1-2 cm long) are given a 5-day pretreatment under nonsterile conditions on a water-agar medium containing 11.0 μM IBA and 2.7 μM NAA. They are then transferred to a peat-pumice (1:1) rooting mix and kept under mist during the first week (Horgan and Aitken, 1981). Trimming of a single elongated root encourages growth of laterals.

Cheng (1976) rooted western hemlock shoots in a "Mica-Peat" soil mix to which a solution of NAA (15 µM, pH 6.2) was added at 2-day intervals for 2 weeks. A plastic cover maintained a relatively high humidity around the shoots.

Evidence has been accumulating indicating that the poorly understood phenomenon of mycorrhizal symbiosis improves rooting and plantlet growth. A few reports have been published recently on this subject (e.g., Powell, 1980; Ruehle, 1980). One such report (Gay, 1981) indicated that the ectomycorrhizal fungus *Hebeloma hiemale* increased the rooting percentage of *Pinus halepensis* hypocotyls, particularly of cultures under reduced auxin regimes where the morphogenetic effect of the hormonal treatment was low.

Preparation of Plantlets for Field Planting

A few general principles have been enumerated by Sommer and Caldas (1981) for the transfer to soil. The essential points are as follows. First, there should be a reasonable balance between roots and shoot. Either should be capable of supporting the other and neither should predominate. An unbalanced root-shoot system can lead to improper growth forms which can be mistaken for, among other things, plagiotropism. Second, the plantlet must undergo a gradual transition (over 2-3 weeks) from a constant high humidity regime to one of varying and low humidity in the nursery or field. This is accomplished by misting in a greenhouse or humidity tent (plastic cover), in a tray covered with a glass sheet (air bubbled through water passes through holes in the glass), or under a glass fabric cover saturated with water (AFOCEL, 1979). Third, care must be taken to remove all traces of agar as they provide a substrate for pathogen growth. Treatment with a fungicide (125 mg/l benlate, Boulay, 1979a) may be good practice. Finally, some shading is necessary to prevent leaf burn during the plantlet's transition from laboratory conditions to full sunlight.

Care must also be taken in selecting the containers in which plantlets are rooted and grown. Pots which do not allow root growth through the sides and bottom cause root malformation and spiralling. Paper pots, Jiffy pots, peat pots, trays, or specially developed rooting supports are recommended. For example, AFOCEL (France) routinely uses a rectangular bag of unwoven fabric containing a rooting mix and a slow-release fertilizer pellet, which is wrapped around the shoot stem (Motte Melfert[R]). Rooted plantlets can probably be planted directly into nursery beds.

POTENTIAL FOR LARGE-SCALE PROPAGATION

A survey of the conifer cell culture systems established to date indicates that the use of in vitro methods for large-scale propagation is as yet impossible, problematic, or at best, uneconomical. Yet the existence of even a few highly productive systems is encouraging, particularly in view of the fact that we are dealing generally with very recalcitrant and heterogeneous species and with an area of research that only yielded its first true success less than 10 years ago.

At AFOCEL for instance, cell culture is the method used to establish clonal orchards from 10- to 500-year-old elite trees of *Sequoia sempervirens*. AFOCEL does not operate on an industrial scale, yet their in vitro techniques are regularly applied for the production of 30,000 plantlets from approximately 200 selected clones (Poissonnier et al., 1980).

Similarly, a small production laboratory has been built at the Forest Research Institute of Rotorua (N. Zealand) where micropropagation techniques can produce 100,000 radiata pine plantlets annually (FRI, 1980). Once these pioneer efforts have shown that the field performance of these plantlets is comparable to that of seedlings, and so far it appears to be so (Franclet, 1979; Aitken, personal communication), and once techniques have been further refined so that cost/benefit comparisons are possible, then mass propagation at industrial levels can be undertaken. At present, micropropagated plantlets of radiata pine at FRI are about 10 times more expensive than seedling planting stock (30-40£ vs. 3-4£, Smith et al., 1981). However, planting stock costs are only one component in the total cost of producing a mature stand of trees. If indeed the cell culture derived plantlets perform better than seedlings of even selected seed, and if planting of the former can be done at or close to final plantation density, then the savings obtained from pruning and final thinning of unwanted trees, etc., will be great enough that the initial higher cost of the propagule would be insignificant. We are hopeful that conifers will follow in the footsteps of other forest tree species such as *Eucalyptus* where planting of 10,000 acres with 7.5 million test-tube trees for methanol production is currently envisaged in Florida (Biomass Energy Systems Inc., Lakeland, FL, 1981).

Presently, however, researchers with experience in this field concur that, because of excessive costs, micropropagation alone is an uneconomical way of producing vast quantities of planting stock directly for reforestation. Instead it can be used as an intermediate phase, a way of rapidly establishing a clonal orchard from selected trees or control-pollinated seed, either full or half sib, from which further propagules will be produced by rooted cuttings. With redwood, the cell culture stage offers the added advantage of rejuvenating very old material which otherwise could not be propagated (Poissonnier et al., 1981).

As with all species, the real breakthrough in mass propagation will come about when somatic embryogenesis in cell suspension cultures becomes a reality. This technique has been tried with many woody species, but only a few angiosperms have responded so far. A significant advantage offered by cell suspension cultures is that instead of going through several stages to obtain plantlets with the high cost that this involves in terms of labor, time, and equipment, the formation of embryos which could lead directly to plantlets would be achieved. Secondly, the multiplication rate of cell suspensions is enormous. Durzan (1976) has calculated that if 80% of the cells form plants, 100 liter of culture medium could produce enough seedlings to plant 100,000 acres at a 12 x 12 ft spacing.

Embryolike structures have appeared occasionally in conifer cell cultures. As mentioned earlier, Konar and Oberoi (1965) observed such structures on cotyledons of *Thuja* but these did not form complete plantlets. Bonga (1977a) also observed similar structures along the lower needles of elongated balsam fir dormant buds. Chalupa and Durzan (1973) cultured cells and aggregates of Norway spruce callus in suspension cultures and after repeated transfers some cells formed small embryolike structures of 5-10 cells. Suspension cultures of Douglas fir were established by Winton (1972b) which, in a simple medium containing only inorganic salts, iron, and sucrose, produced small embryolike structures. From callus of white spruce (Durzan and Steward, 1968; Durzan et al., 1973) and jack pine (Durzan and Steward, 1968), suspension cultures composed of cells and cell aggregates were established in which small embryolike structures appeared. These structures have also been observed in Douglas fir and loblolly pine cell suspension cultures (Durzan, 1980). However, to date no somatic embryos leading to plantlets have been obtained with any conifer, as far as we are aware. Unfortunately, under these conditions, our understanding of the regulation of embryogenesis is as hazy as that of adventitious shoot formation (see Thorpe, 1982).

Webb (1981) has established cell suspension cultures of *Pinus contorta* in which cytodifferentiation was observed as tracheary element formation. This developmental pathway could be altered by removing boron from the nutrient medium in which case phloem-type elements are formed instead.

One of the problems which frequently arises in cell suspension cultures is that of a build-up in phenolic compounds in the medium which is lethal to the pro-embryos (Parham and Kaustinen, 1977). Some control of tannin production has been achieved by Durzan et al. (1973), and recently Litvay et al. (1981) have developed a conifer suspension culture medium using analytical data from developing seeds of Douglas fir. This medium (Table 3) is reported to allow tissues of juvenile and mature trees of Douglas fir and loblolly pine to be grown and maintained as fine cell suspensions for prolonged periods. Phenolic production is strongly reduced with this medium (Verma, personal communication).

TREE IMPROVEMENT

The primary goal of plant cell culture research is crop improvement (Evans et al., 1981). However, it has been recognized since 1929 that for trees a special physiological technique must be elaborated for crop improvement investigation (Bailey and Spoehr, 1929) to circumvent the problems which forest tree species pose due to their complexity, heterogeneity, and longevity. Cell suspension culture is such a technique since it allows the application of parasexual (somatic) hybridization, mutagenesis, or DNA uptake. Regeneration from haploid tissues is also important since the availability of homozygous lines offers a significant contribution to breeding programs (Melchers, 1972). Durzan (1980) has listed the advantages of using cell suspensions. These

Table 3. Components of Conifer Cell Suspension Culture Medium[a]

COMPOUND	LEVELS	COMPOUND	LEVELS
NH_4NO_3	0.02 M	$CuSO_4 \cdot 5H_2O$	2.0 μM
KNO_3	0.02 M	$CoCl_2 \cdot 5H_2O$	0.6 μM
$MgSO_4 \cdot 7H_2O$	7.5 mM	$FeSO_4 \cdot 7H_2O$	0.05 mM
KH_2PO	2.5 mM	Na_2EDTA	0.11 mM
$CaCl_2 \cdot 2H_2O$	0.15 mM	Myoinositol	0.56 mM
KI	0.02 mM	Nicotinic acid	4.1 μM
H_3BO_3	0.5 mM	Pyridoxine·HCl	0.5 μM
$MnSO_4 \cdot H_2O$	0.12 mM	Thiamine·HCl	0.3 μM
$ZnSO_4 \cdot 7H_2O$	0.15 mM	Sucrose	0.09 M
$Na_2MoO_4 \cdot 2H_2O$	5.2 μM		

[a]Litvay et al., 1981.

include (a) few cells are required as starting material, (b) cells can be grown like bacteria and are potentially totipotent, (c) space utilization and system of propagation is economical and controlled, (d) mutations can be introduced and identified, since large populations of cells can be screened and selected for mutant traits, (e) time and cost of regenerating plantlets is greatly reduced (cell to nursery in 18 months is possible), (f) with haploids, mutant traits are not masked, thus diploidization of haploid cells allows for genetic analysis and improvement (hybrid vigor), and (g) cell walls can be removed to produce protoplasts for genetic manipulation (somatic hybrids by fusion or DNA uptake, etc.).

Spontaneous Mutation

Phenotypic (or somaclonal) variation has been observed in plants regenerated in vitro. This may be useful when applied to existing breeding programs. Variations of chromosome numbers of plants in long-term cultures has been well documented (Kao et al., 1970) and reviewed extensively (e.g., D'Amato, 1978; Krikorian et al., Vol. 1, Chap. 16). However, although angiosperm triploids (for example, triploid aspen) have demonstrated a superior performance compared to diploids (an exceptionally fast growth rate), in pine species both natural and induced polyploidy seem to result in an overall reduction of growth and the formation of morphologically abnormal seedlings (Mergen, 1958, 1959).

Adding Traits

Prospects for producing genetic combinations which cannot be obtained by sexual crossing are good. The basic requirements to achieve this are that the tree material can be grown as undifferentiated cells and that regeneration of plantlets from such cells is possible.

When a new trait is introduced artificially it will usually occur in a very small fraction of the treated population. Carlson et al. (1972) found 33 altered cells from a population of 2 x 10^7 cells. Thus the system must be capable of handling enormous quantities of cells in the way bacteria are handled; cell suspension cultures offer this kind of system. Secondly, a method for screening and selecting altered cells must be developed. The possibility of plating cells onto a medium on which only the altered cells will grow would be a fast and easy method. Alternately, since most traits in trees are pleiotropic (Wright, 1968), one could envisage using one trait (not the determining one) to select cells altered in the desired trait.

Most mutations, whether induced or spontaneous, are recessive and thus can only be expressed and possibly recovered by using haploid cultures. At the same time natural haploidy is very rare among forest trees. Furthermore, inbreeding in long-lived tree species has not been practical because of the length of time required to produce successive inbred generations. Thus the culture of haploid pollen grains or female gametophytes followed by the regeneration of haploid plantlets offers a new and important approach for conifer improvement programs. Of value to such programs would also be the doubling of the chromosome number before or after plantlet regeneration so as to obtain homozygous diploids.

The pioneering work of La Rue (1954) and Tulecke (Tulecke, 1953, 1957; Tulecke and Sehgal, 1963) showed that callus could be formed in culture from gymnosperm pollen. Much later, when haploid plantlet regeneration from anther cultures was shown possible for tobacco (Bourgin and Nitsch, 1967), Bonga and Fowler (1970) began to cultivate in vitro microsporangia, microsporophylls, and mature pollen of *Pinus resinosa*. Some haploid callus was obtained, but mature pollen did not form callus. Sommer and Brown (1979) have attempted sporophyll cultures of loblolly, longleaf, and Virginia pines. Some callus was formed with all three species but haploid callus was obtained only in Virginia pine.

In general, pollen-derived callus is composed of a mixoploid population of cells, and only limited organogenesis has been reported (Isikawa, 1972; Bonga, 1974b). One of the complications which arises when using anther/microsporangium cultures is that the pollen grains all have different genomes. Thus unless plantlets are regenerated from a single cell of the callus, or embryos are formed from a single pollen grain, chimeras result (Bonga, 1977b). Another complication is represented by the multinucleate cells which characterize such cultures (Bonga, 1981). Finally, one should be aware of the fact that many tree species carry a load of deleterious/lethal recessive genes and that when homozygous diploids are produced, they may display growth depression or even death.

Gymnosperms offer a unique opportunity for establishing haploid cultures, since the megagametophyte is a large haploid structure composed of many cells derived from a single megaspore, and thus are all genotypically identical. In recent years, Huhtinen (1972, 1976) has done extensive work on female gametophyte cultures of Norway spruce. Success in establishing such cultures was low (less than 5%) on a modified MS medium supplemented with glutamine, KIN, and 2,4-D. Ninety percent of the callus were haploid and remained haploid even after subculturing. Some cultures in the presence of 4.6-23.0 μM KIN formed small embryolike shoots that did not root. Later, Steinhauer and Huhtinen (1978) were able to achieve shoot regeneration from such cultures.

Bonga and Fowler (1970) and Bonga (1974b) cultured the megagametophyte of various pine species. Callusing was not uncommon and, among the callus examined, nearly all showed haploid metaphases. Only Mugo pine callus showed some cytodifferentiation (circular growth centers that sometimes contained tracheids). Most of Sommer and Brown's work has been with longleaf pine (Sommer and Brown, 1979). They have obtained sporadic callusing and cells are predominantly haploid.

Bonga (1977b) published a photograph of a plantlet arising from a female gametophyte callus of Norway spruce cultured on Brown and Lawrence's (1968) medium supplemented with IAA, and Ball (1981) has obtained plantlet regeneration from *Sequoia sempervirens*. Unfortunately none of the plantlets were haploid. This is one of the difficulties which currently faces researchers in this area. In addition, Huhtinen (1976) has mentioned two others: the risk of contaminating the gametophyte with diploid cells from the embryo during the excision procedure, and a low rate of callus formation. While a solution to the latter problem has yet to be proposed, for the former, Huhtinen reports that vernalizing the seeds in the cones under high humidity conditions makes embryo removal easier.

It is possible to blend characters and produce new genetic combinations by asexual means in order to overcome crossability barriers via somatic cell hybridization and genetic engineering. Walls of plant cells in culture can be removed and the resulting protoplasts induced to fuse. From there on, the following sequence can be induced: cell wall regeneration, cell division, formation of cell clumps (colonies), callus proliferation, and finally, organogenesis.

There exist relatively few publications concerning the production of protoplasts from tree tissues. For conifers, Winton et al. (1975), Kirby and Cheng (1979), Kirby (1980), Huhtinen and Winton (1973), Chalupa (1974), and David and David (1979) have reported protoplast isolation and in some cases callus formation from Douglas fir, Norway spruce, maritime pine, and other species.

Fusion of protoplasts from different species has been attempted with limited success at the Institute of Paper Chemistry in Appleton, Wisconsin (Durzan, 1980). Many fundamental problems still need to be overcome such as the identification of specific markers for each trait or species, effective screening methods for the hybrids produced, and last but certainly not least, further development of the techniques which will allow reliable plantlet regeneration from conifer callus.

Genes can also be introduced into cells by genetic engineering using cells and protoplasts. By "transcession," plant cells incubated with bacterial suspensions may acquire bacterial genes; by "transduction," specific genes can be introduced (e.g., phage-carrying bacterial genes for galactose utilization); and during "transformation" or "transgenosis," cells are exposed to foreign DNA isolated from another source. The successful introduction of foreign DNA into plant cells requires the uptake and integration of the DNA into the host genome and then its replication, expression, and transmission to daughter cells. Years of basic research will be necessary to understand how this type of genetics may apply to trees (Durzan, 1980). Yet even before fully comprehending the details of this process it may be possible to obtain potentially useful products. The most frequently cited example is that of transforming higher plant cells, including conifers, by conferring on them the ability to fix inorganic nitrogen from the atmosphere (Giles, 1979). This has been attempted with many species and a recent report (Davey et al., 1980) indicated, for example, that the fusion of legume root nodule protoplasts with nonlegume protoplasts (*Petunia*, carrot) was possible and that the *Rhizobium* bacteroids initially present in the legume root nodule cells retained their structural integrity and nitrate-reducing ability even in the cultured heterokaryons.

JUVENILITY AND MATURATION

The most severe difficulties with the use of vegetative propagules in forestry are associated with phase change and the maturation of meristems (Libby, 1974). Since maturation is believed to occur in meristems (Schaffalitsky de Muckadell, 1959), clonal propagules from meristems of mature plants are physiologically older than those from juvenile plants and will often display such undesired characteristics as plagiotropic growth, reduced growth rate, etc. (Bonga, 1980). This means that we are confronted with the considerable problem of having to evaluate and identify superior clones when they mature, by which time they are hard to propagate or, if propagated, may not grow with the form and vigor of the selected ortet. Nevertheless, the rapid and reliable gains which one can expect to make by proceeding in this manner has meant that interest in this problem, which is of fundamental relevance to both basic research and commercial application, has been growing rapidly in recent years.

The basic approaches utilized today to pretreat ("rejuvenate") adult trees so that they may respond to vegetative propagation procedures are based on traditional horticultural/silvicultural practices. One of these is repeated pruning of all new growth above a specified height (hedging). It is used for radiata pine (Libby et al., 1972) and seems to maintain, with a fair degree of accuracy, the maturation state at the time of pruning (Garner and Hatcher, 1962). This may be a general feature for woody perennials.

Maturation can also be slowed down by serial propagation. In other words, if a ramet taken from the original ortet is rooted and ramets taken from that are rooted, and so forth, rootability may increase with successive generations. Thus Pawsey (1971) found that ramets from fourth-generation scions grew in a more juvenile manner than ramets from preceding grafts. Grafting onto young seedlings is also utilized at

AFOCEL to "rejuvenate" Douglas fir. With successive generations, the grafted material assumes an increasingly juvenile (seedlinglike) morphology and is more responsive to in vitro proliferation. Radiata pine buds which begin to flush after hedging also appear more amenable to proliferation in culture (Aitken, personal communication). This approach is also possible with *Pinus pinea*, *P. pinaster*, *P. nigra*, *P. rigida*, and *P. halepensis* (Franclet, 1979).

"Rejuvenated" material can also be obtained by cutting down trees to release stump sprouts. Alternatively, species which form coppice, basal sprouts, or suckers spontaneously are a good source of "juvenile" material. However, such phenomena are rare among conifers, with a few exceptions (such as *Sequoia sempervirens* and *Pinus rigida*) (Franclet, 1979).

Another procedure recently introduced by AFOCEL is that of spraying the ortet with BA. This seems to release the inhibition of inactive meristems (by inducing flushing of buds).

An in vitro phase may, in some cases, reverse the maturation process per se. Thus with material taken from a 100-year-old redwood, Franclet (1979) reported that with each subculture, percentage survival increased (from 2/35 in the initial culture to 12/12 in the fourth), shoots were mostly orthotropic by the fifth subculture, and one shoot rooted spontaneously.

The term rejuvenation must be used with caution. Bonga (1980) has pointed out that the shoots obtained from hedging or cutting down trees are not really rejuvenated because they arise from buds formed early in the life of the tree that have remained dormant. In radiata pine, however, the buds used are formed less than 6 months previously, when the short shoot (needle fascicle) was being laid down (Smith, personal communication). Natural rejuvenation only occurs when buds arise adventitiously from relatively unorganized tissues such as wound tissue or sphaeroblasts. However such adventitious bud formation is rare or unknown in most conifer species. The need arises now among cell culturists and foresters to study some very basic anatomical questions and to provide a consensus governing terminology, so that the confusion which now reigns in the use of such terms as trace buds, latent buds, epicormic buds, etc., is cleared up.

While much remains to be done in understanding the chemical-physiological differences between juvenile and mature tissues (presence of inhibitors, enzyme activities, redox state, etc.), in practice, considerable success has been obtained by the application of the earlier-mentioned procedures. Their refinement and widespread use can provide us with some very important information as well as significant aid in the propagation of mature specimens (see Bonga, 1980).

PROTOCOLS

Pinus radiata: Embryo and Cotyledon Cultures (Reilly and Washer, 1977; Aitken et al., 1981; F.R.I. Publication, 1980; Horgan and Aitken, 1981)

1. Surface sterilize seeds with 50% (v/v) commercial bleach for 15 min. Wash overnight under running water.

2. Soak seeds in 6% H_2O_2 for 10 min and rinse 3 times in sterile water.
3. Store at 5 C for 48 hr.
4. For embryo culture, aseptically remove seed coat and endosperm and dissect out embryo.
5. For cotyledon culture, place seeds on sterile vermiculite soaked in water or onto filter paper bridges and incubate in continuous darkness for 5–7 days.
6. Seedlings are ready when radicles have emerged. Radicle length varies from 1–30 mm; cotyledon length will vary accordingly (optimal cotyledon length is 3–5 mm). Remove seed coat and endosperm and cut cotyledons at the node under aseptic conditions. There will be 5–10 cotyledons per seed.
7. For culture, place individual cotyledon segments or embryos horizontally on solid medium based on Schenk and Hildebrandt (1972) as modified by Reilly and Washer (1977): KNO_3 (0.025 M), $MgSO_4 \cdot 7H_2O$ (1.62 mM), $NH_4H_2PO_4$ (2.6 mM), $CaCl_2 \cdot 2H_2O$ (1.36 mM), $FeSO_4 \cdot 7H_2O$ (0.03 mM), Na_2EDTA (0.06 mM), $MnSO_4 \cdot 4H_2O$ (0.09 mM), H_3BO_3 (0.08 mM), $ZnSO_4 \cdot 7H_2O$ (3.4 μM), KI (6 μM), $CuSO_4 \cdot 5H_2O$ (0.8 μM), $Na_2MoO_4 \cdot 2H_2O$ (0.8 μM), $CoCl_2 \cdot 6H_2O$ (0.8 μM), supplemented with myo-inositol (5.5 mM), thiamine HCl (14.8 μM), nicotinic acid (0.04 mM), pyridoxine HCl (2.4 μM), sucrose (87.6 mM), BA (22 μM) and solidified with agar (Difco Bacto, 8 g/l), pH 5.6–5.8.
8. Medium (20–25 ml) is contained in 100 x 15 mm petri dishes and cultures are incubated in a 16 hr photoperiod with light intensity of 80 $\mu Em^{-2}s^{-1}$ and a day/night temperature regime of 28 ± 1 C/ 24 ± 1 C.
9. After 3 weeks, remove developing embryos or cotyledon segments to the same medium but with sucrose at 58.4 mM and lacking BA.
10. Tissue is transferred every third or fourth week. After 9 weeks, use Pyrex jar.
11. When shoots are 1–2 cm long, remove from sterile conditions and place on water-agar mixture containing 4.9 μM IBA and 2.7 μM NAA for 5 days to promote rooting.
12. Transfer to peat-pumice (1:1) rooting mix under mist for the first week. The average time for bud development (when shoots can be separated) is 9–12 weeks. The mean number of rootable shoots per shoot-forming embryos is 9–15 (variation: 1–200). The mean number of rootable shoots per seed (using excised cotyledons) is 180. Approximately 87% of the embryos produce shoots; 100% of the excised cotyledons do so.

Picea albies: Embryo and Needle Cultures (von Arnold and Eriksson, 1977, 1978, 1979)

This method is also applicable to embryo cultures of *P. sylvestris* and *P. contorta* and, according to preliminary evidence (von Arnold, 1979) also to needles of *P. sylvestris*.

1. Newly flushed buds are taken from 5-year-old greenhouse-grown plants. They should be cut off close to the stem.

2. For surface sterilization, use 7.5% calcium hypochlorite solution, soaking seeds 20–30 min and flushed buds 10 min. Transfer to 70% ethanol for 2 min then rinse in sterile water, once for seeds, three changes for flushed buds.
3. Using aseptic procedures, dissect embryos from the seeds. Cut 3-mm-tips from the upper half of the needles. There are typically 30 needles per bud.
4. Place the explants on 5 ml of a modified LP medium (von Arnold and Eriksson, 1977) that is contained in 5-cm petri dishes. The basal medium is a mixture of two, one for pea mesophyll protoplasts (Eriksson, 1965 as modified by Wallin et al., 1974) and the other is an embryo culture medium (Norstog, 1973). The LP medium is modified by omitting sorbitol, 2,4-D, and 2iP and reducing the Ca^{2+} concentration from 12 to 1.2 mM. The medium is supplemented with sucrose (100 mM), 2iP (5 μM) for embryos or BA (10 μM) for needles, and gelled with agar (5–7 g/l). Low auxin concentrations (IBA or NAA at 0.5–10 nM) slightly increase the percentage of needles forming buds, but not of embryos.
5. Embryo cultures are incubated for 5 weeks, needle cultures for 6 weeks. For further development of buds, transfer to cytokinin-free LP medium, using half-strength basal medium. Transfer all cultures to fresh media every third or fourth week.
6. Cultures are grown at 10–16 C for the first 4–14 days after excision, then at 20 C. Incubate under continuous light from a Philips white fluorescent tube (400–700 nm) 20 W m^{-2}.
7. The average time in culture for bud development is 7 weeks for embryo cultures, 10 weeks for needles. For embryo cultures, on the average 60% of the explants have adventitious primordia after 5 weeks in culture. For needles, on the average there are 10 adventitious buds per explant after 12 weeks, 43% of the explants have primordia after 6 weeks, 22% of the explants have adventitious buds after 12 weeks.
8. There are two types of adventitious bud primordia: (a) "scale-like organs," which are predominant at lower cytokinin concentrations on embryo cultures, resemble bud primordia and eventually form buds with an apical dome and needle primordia; and (b) "tube-like protrusions," which are predominant at higher cytokinin concentrations on embryo cultures, are composed of small cells rich in cytoplasm surrounding central xylem elements. These eventually form buds enclosed in green "bud scales." The type (a) primordia is found almost exclusively on needles.

Pseudotsuga menziesii: Cotyledon Cultures (Cheng, 1975, 1977, 1979; Cheng and Voqui, 1977)

1. Remove seed coats and surface sterilize seeds in 5.25% sodium hypochlorite (commercial bleach) for 8 min and rinse with sterile water.
2. Germinate seeds on half-strength MS medium with 0.06 mM $FeSO_4 \cdot 7H_2O$, 0.11 mM Na_2EDTA, 87.6 mM sucrose and supplemented with myo-inositol (1.39–2.78 mM), thiamine (7.4–14.8 mM),

and gelled with agar (6-8 g/l), pH 5.5. Incubate at 25 C with an 18 hr daylength.

3. One alternate method (Cheng and Voqui, 1977) is to sow seeds directly in a soil mixture (sorghum peat moss and vermiculite, 60:40) and place in a growth chamber, 18 hr daylength, 25 C/19 C day/night temperature. After 2-4 weeks, the cotyledons are sterilized in 6% commercial bleach in agitation for 8 min and then rinsed three times in sterile water.
4. A second alternative (Abo El-Nil, personal communication) is to store seeds at 0 C, sterilize in 6% commercial bleach and, after rinsing with sterile water, place the seeds on moist filter paper and store at 4 C for 3 weeks. The seeds are then germinated in soil mix (peat-perlite-sand).
5. Two to four weeks after germination, the cotyledons are cut into 3-mm-long pieces and placed either on 50 ml of agar-solidified medium contained in 100 x 200 mm plastic petri dishes or on fabric tissue support (100% polyesther fleece, 3 mm thick, Pellon Corp., Lowell, Mass.) contained in plastic petri dishes. This type of support requires liquid medium and, for change of medium, the old one can be syphoned off.
6. The medium is supplemented with BA (5 μM) and NAA (0.5-5 μM) or BA (5 μM), IBA (2.5 μM), and IAA (2.5 μM) for bud formation. Cultures are grown at 25 C with 18 hr daylength. About 97% of the explants usually form buds.
7. After 4-6 weeks, transfer cultures to hormone-free basal medium for bud elongation.
8. When shoots are about 2 cm long, they are excised and cultured individually in culture tubes with 20 ml agar-solidified medium at 19 C with reduced sucrose (5 g/l) and NAA (0.25 μM). This treatment results in about 80% rooting.
9. For root elongation, remove to media without NAA.
10. Upon transfer to soil, there is approximately 90% survival.

Thuja plicata: Mature Tissues (Coleman and Thorpe, 1977)

1. Remove lateral vegetative branches from 4-10-year-old trees. Seal cut ends with molten wax.
2. Surface sterilize with 20% sodium hypochlorite (commercial bleach) for 25 min and rinse four times with sterile water.
3. Cut off 8-10-mm-long shoot tips (with a visible pair of lateral scale-like leaves).
4. For shoot formation, place on MS mineral salts supplemented with myo-inositol (555 μM), asparagine (757 μM), thimaine-HCl (1 μM), pyridoxine-HCl (1 μM), nicotinic acid (1 μM), BA (50 μM), NAA (0.1 μM), sucrose (87.6 mM), and agar (8 g/l), pH adjusted to 5.2 ± 0.2.
5. Cultures are incubated at 25 ± 2 C under continuous light of 2000 lux for 6-8 weeks.
6. For rooting, transfer developed shoots to half-strength MS salt solution with vitamins (above) at 1 μM, sucrose (87.6 mM), IBA (50 μM), and agar (8 g/l). Incubate in a 16-hr photoperiod.
7. After 4-6 weeks, transfer plantlets to soil mix (peat-vermiculite, 1:1) under mist in greenhouse with a similar 16-hr photoperiod.

FUTURE PROSPECTS

There are many areas in which various aspects of in vitro technology could profitably be employed in the improvement of conifers. While cell culture techniques will not replace traditional tree-breeding efforts, they can play a major role as a complementary approach to the problems of tree-breeding, genotype evaluation, etc. To fulfill this role, a concerted research effort is required in the following areas:

1. The use of haploids as a means of testing structural linkage among genetic loci responsible for specific traits. Trees often show tight genetic linkage of traits and are thus recalcitrant to showing classical genetic segregation (Durzan, 1980).
2. Screening for disease resistance. In forest trees, resistant stock is often the only way to combat disease problems (Mott, 1981a).
3. Studying mycorrhizal symbiosis. Early studies showed that isolated roots could be used to study mycorrhizal infection (Fortin, 1966). Both endo- and ectomycorrhizal associations need to be examined.
4. Juvenility/maturation relations must be studied. Biochemical markers such as those that exist in animal cells must be sought. Rejuvenation may be the key to success with mature tissues.
5. The regulation of organogenesis, embryogenesis, and rooting. Without success in regenerating plantlets from callus or cell suspension cultures of conifers, it will not be possible to utilize any advances made in cell modification by parasexual methods, etc.
6. Understanding the correlation between the composition of nitrogenous compounds in in vivo and in vitro tissues with growth rate. This is important because N is often a limiting factor in forest soils (Durzan, 1980).
7. Correlation between one or more traits of cultured tissues with the future performance of the progeny. For example, is there a correlation between the growth rate of callus and plantlet growth rate (Mathes and Eisnpahr, 1965)? Methods for early testing of at least some traits are urgently needed and any possibility for evaluation during the earliest phase (i.e., the cell culture stage) would merit serious study (Mott, 1979). This would significantly accelerate breeding programs and permit a rapid evaluation of genetic gains.
8. Formation of natural compounds, including fibers and wood. There is a need to study which factors regulate this, and whether in vitro control is possible. Unfortunately no good xylogenic cell culture system exists in woody plants (Thorpe, 1982b), and this could delay required studies.

We agree with Sommer and Brown (1979), however, that the most urgently needed role for cell culture in forest tree improvement lies in the regeneration of plantlets from somatic and gametic tissue, followed by field testing. This means obtaining plantlets not only from juvenile material, but also from mature sources. In addition, these should come from large enough selections, families, and provenances to avoid any risk of monoculture. The regenerated plantlets must duplicate parent phenotypes and genotypes and be available for genetic testing.

The culture of cells, tissues, and organs in vitro offers unparalleled opportunities for forest tree improvement (Karnosky, 1981). Forest tree cell culture research is in its infancy, and research must be expanded to realize the full potential available from cell culture. Thus massive infusions of manpower and capital are required to address the many problems outlined in this chapter. Without such an infusion, continued slow progress is inevitable, and the potential opportunities to gear up for the predicted shortage of this renewable resource by the end of this century (Keays, 1974) will be lost, perhaps even irrecoverably.

KEY REFERENCES

AFOCEL. 1979. Micropropagation d'Abres Forestiers. No. 12-6/79, Association Foret-Cellulose, Nangis, France.

Bonga, J.L. and Durzan, D.J. (eds.). 1982. Tissue culture in forestry. Martinus Nijhoff/Dr. W. Junk, The Netherlands.

Durzan, D.J. 1980. Progress and promise in forest genetics. Proceedings 50th Anniversary Conference, Paper Science and Technology—The Cutting Edge, Institute of Paper Chemistry, Appleton, Wisconsin, May 8-10, 1979, pp. 31-60.

Karnosky, D.F. 1981. Potential for forest tree improvement via tissue culture. BioScience 31:114-120.

Mott, R.L. 1981a. Trees. In: Cloning Agricultural Plants via In Vitro Techniques (B.V. Conger, ed.) pp. 217-254. CRC Press, Boca Raton, Florida.

Sommer, H.E. and Brown, C.L. 1979. Application of tissue culture to forest tree improvement. In: Plant Cell and Tissue Culture: Principles and Applications (W.R. Sharp, P.O. Larsen, E.F. Paddock, and V. Raghavan, eds.) pp. 461-491. Ohio State Univ. Press, Columbus.

Sommer, H.E. and Caldas, L.S. 1981. In vitro methods applied to forest trees. In: Plant Tissue Culture: Methods and Applications in Agriculture (T.A. Thorpe, ed.) pp. 349-358. Academic Press, New York.

REFERENCES

Aitken, J., Horgan, K., and Thorpe, T.A. 1981. Influence of explant selection on the shoot-forming capacity of juvenile tissue of radiata pine. Can. J. For. Res. 11:112-117.

Aitken, J., Smith, D.R., and Horgan, K.J. 1982. Micropropagation of radiata pine. In: Colloque International sur la Culture in vitro des Essences Forestieres, IUFRO International Workshop, Fontainebleau, France, August 31 to September 4, 1981. pp. 191-196. AFOCEL, Nangis, France.

Bailey, I.W. and Spoehr, H.A. 1929. The Role of Research in the Development of Forestry in North America. Macmillan, New York.

Ball, E.A. 1950. Differentiation in a callus culture of *Sequoia sempervirens*. Growth 14:295-325.

______ 1978. Cloning in vitro of *Sequoia sempervirens*. Abstract No. 1726, 4th International Congress of Plant Tissue and Cell Culture, Aug. 20-25, 1978, Calgary, Canada.

Bethel, J.S. 1972. Influence of environment on quality of wood of Douglas fir. In: Forest Research Progress in 1971, pp. 22-23. Cooperative Extension Research Service, USDA, Washington, D.C.

Biondi, S. 1980. Some Aspects of Plant Organogenesis in Cultured Explants. M.Sc. Thesis, Univ. of Calgary, Calgary, Canada.

______ and Thorpe, T.A. 1982. Growth regulator effects, metabolite changes, and respiration during shoot initiation in cultured cotyledon explants of *Pinus radiata*. Bot. Gaz. 143:20-25.

Bonga, J.M. 1974a. Vegetative propagation: Tissue and organ culture as an alternative to rooting cuttings. N.Z. J. For. Sci. 4:253-260.

______ 1974b. In vitro cultures of microsporophylls and megagametophyte tissue of *Pinus*. In Vitro 9:270-277.

______ 1977a. Organogenesis in in vitro cultures of embryonic shoots of *Albies balsamea* (balsam fir). In Vitro 13:41-48.

______ 1977b. Application of tissue culture in forestry. In: Plant Cell, Tissue, and Organ Culture (J. Reinert and Y.P.S. Bajaj, eds.) pp. 93-108. Springer-Verlag, Berlin, Heidelberg, and New York.

______ 1980. Plant propagation through tissue culture emphasizing woody species. In: Plant Cell Cultures: Results and Perspectives (F. Sala, B. Parisi, R. Cella, and O. Ciferri, eds.) pp. 253-264. Elsevier/North Holland Biomedical Press, Amsterdam.

______ 1981. Organogenesis in vitro of tissues from mature conifers. In Vitro 17:511-518.

______ 1982. Vegetative propagation in relation to juvenility, maturity, and rejuvenation. In: Tissue Culture in Forestry (J.M. Bonga and D.J. Durzan, eds.) pp. 387-412. Martinus Nijhoff/Dr. W. Junk, The Hague, Netherlands.

______ and Fowler, D.P. 1970. Growth and differentiation in gametophytes of *Pinus resinosa* cultured in vitro. Can. J. Bot. 48:2205-2207.

Bornman, C.H. and Jansson, E. 1982. Regeneration of plants from the conifer leaf with special reference to *Picea abies* and *Pinus sylvestris*. In: Colloque International sur le Culture in vitro de Essences Forestries, IUFRO International Workshop, Fountainbleau, France, Aug. 31-Sept. 4, 1981, pp. 41-54. AFOCEL, Nangis, France.

Boulay, M. 1979a. Multiplication et clonage rapide du *Sequoia sempervirens* par la culture in vitro. In: Micropropagation d'Arbres Forestiers, Annales AFOCEL, No. 12, 6/79, pp. 49-56. AFOCEL, Nangis, France.

______ 1979b. Propagation in vitro du Douglas [*Pseudotsuga menziesii* (Mirb.) Franco] par micropropagation de germination aseptiques et culture de bourgeons dormants. In: Micropropagation d'Arbres Forestiers, Annales AFOCEL, No. 12, 6/79, pp. 67-75. AFOCEL, Nangis, France.

______ and Franclet, A. 1977. Recherches sur la propagation du Douglas: *Pseudotsuga menziesii* (Mirb.) Franco. C.R. Acad. Sci. (Paris) 284:1405-1407.

Bourgin, J.P. and Nitsch, J.P. 1967. Obtention de *Nicotiana* haploides a partir d'etamines cultivees in vitro. Ann. Physiol. Veg. 9:377-382.
Brown, C.L. and Lawrence, R.H. 1968. Culture of pine callus on a defined medium. For. Sci. 14:62-64.
Brown, C.L. and Sommer, H.E. 1975. An Atlas of Gymnosperms Cultured In Vitro: 1924-1974. Georgia Forest Research Council, Macon, Georgia.
______ 1977. Bud and root differentiation in conifer cultures. Tappi 60:72-73.
Brown, D.C.W. and Thorpe, T.A. 1980. Adenosine phosphate and nicotinamide adenine dinucleotide pool sizes during shoot initiation in tobacco callus. Plant Physiol. 65:587-590.
Campbell, R.A. and Durzan, D.J. 1975. Induction of multiple buds and needles in tissue cultures of *Picea glauca.* Can. J. Bot. 53:1652-1657.
______ 1976. Vegetative propagation of *Picea glauca* by tissue culture. Can. J. For. Res. 6:240-243.
Carlisle, A. and Teich, A.H. 1971. The Costs and Benefits of Tree Improvement Programs. Department of the Environment, Canadian Forest Service Publication No. 1302, Ottawa, Ontario.
Carlson, P., Smith, H., and Dearing, R. 1972. Parasexual interspecific plant hybridization. Proc. Natl. Acad. Sci. U.S.A. 69:2292-2294.
Caruso, J.L., Smith, R.G., Smith, L.M., Cheng, T.-Y., and Doyle Davis, G. 1978. Determination of indole-3-acetic acid in Douglas fir using a deuterated analog and selected ion monitoring. Plant Physiol. 63:841-845.
Chalupa, V. 1974. Isolation and division of protoplasts of some forest tree species. Third International Congress of Plant Tissue and Cell Culture, abstr. no. 174, Leicester, England.
______ and Durzan, D.J. 1973. Growth of Norway spruce [*Picea abies* (L.) Karst] tissue and cell cultures. Commun. Inst. For. Cech. 8:111-125.
Cheah, K.-T. and Cheng, T.-Y. 1978. Histological analysis of adventitious bud formation in cultured Douglas fir cotyledons. Am. J. Bot. 65:845-849.
Cheng, T.-Y. 1975. Adventitious bud formation in culture of Douglas fir [*Pseudotsuga menziesii* (Mirb.) Franco]. Plant Sci. Lett. 2:97-102.
______ 1976. Factors affecting adventitious bud formation of cotyledon culture of Douglas fir. Plant Sci. Lett. 9:179-187.
______ 1979. Recent advances in development of in vitro techniques for Douglas fir. In: Plant Cell and Tissue Culture: Principles and Applications (W.R. Sharp, P.O. Larsen, E.F. Paddock, and V. Raghavan, eds.) pp. 493-508. Ohio State Univ. Press, Columbus.
______ and Voqui, T.H. 1977. Regeneration of Douglas fir plantlets through tissue culture. Science 198:306-307.
Coleman, W.C. and Thorpe, T.A. 1977. In vitro culture of western red cedar (*Thuja plicata*). I. Plantlet formation. Bot. Gaz. 138:298-304.
Cresswell, R. and Nitsch, C. 1975. Organ culture of *Eucalyptus grandis.* Planta 125:87-90.
D'Amato, F. 1978. Chromosome number variation in cultured cells and regenerated plants. In: Frontiers of Plant Tissue Culture 1978 (T.A. Thorpe, ed.) pp. 287-295. Univ. of Calgary Press, Canada.

Davey, M.R., Pearce, N., and Cocking, E.C. 1980. Fusion of legume root nodule protoplasts with non-legume protoplasts: Ultrastructural evidence for the functional activity of *Rhizobium* bacteroids in a heterokaryotic cytoplasm. Z. Pflanzenphysiol. 99:435-447.

David, A. 1982. In vitro propagation of gymnosperms. In: Tissue Culture in Forestry (J.M. Bonga and D.J. Durzan, eds.) pp. 72-108. Martinus Nijhoff/Dr. W. Junk, The Hague, Netherlands.

______ and David, H. 1977. Manifestations de diverses potentialites organogenes d'organes ou de fragments d'organes de Pin maritime (*Pinus pinaster* Sol.) cultives in vitro. C.R. Acad. Sci. (Paris) 284: 627-630.

______ and David, H. 1979. Isolation and callus formation from cotyledon protoplasts of pine (*Pinus pinaster*). Z. Pflanzenphysiol. 94:173-177.

______, Faye, M., and Isemukali, K. 1979. Culture in vitro et micropropagation du Pin maritime (*Pinus pinaster* Sol.). In: Micropropagation d'Arbres Forestiers, Annales AFOCEL, No. 12, 6/79, pp. 33-40. AFOCEL, Nangis, France.

David, H. Isemukali, K., and David, A. 1978. Obtention de plants de Pin maritime (*Pinus pinaster* Sol.) a partir de brachyblastes ou d'apex caulinaire de tres jeunes sujets cultives in vitro. C.R. Acad. Sci. (Paris) 287:245-248.

Davis, L.S. 1969. Economic models for program evaluation. Proceedings from the Second World Consultation on Forest Tree Breeding, Washington, D.C., FO-FTB-69-13/2, p. 14.

de Torok, D. and White, P.R. 1960. Cytological instability in tumors of *Picea glauca*. Science 131:730.

de Yoe, D.R. and Zaerr, J.B. 1976. Indole-3-acetic acid in Douglas fir. Plant Physiol. 58:299-303.

Douglas, T.J., Villalobos, V.M., Thompson, M.R., and Thorpe, T.A. 1982. Lipid and pigment changes during shoot initiation in cultured explants of *Pinus radiata*. Physiol. Plant. 55:470-477.

Durzan, D.J. 1976. Biochemical changes during gymnosperm development. Symposium on juvenility in woody perennials. Acta Hortic. 56:183-194.

______ and Campbell, R.A. 1974. Prospects for the mass production of improved stock of forest trees by cell and tissue culture. Can. J. For. Res. 4:151-174.

______ and Campbell, R.A. 1976. Vegetative propagation of *Picea glauca* by tissue culture. Can. J. For. Res. 6:240-243.

______ and Chalupa, V. 1971. Cell and organ culture of white spruce and jack pine. Proc. Can. Soc. Plant Physiol. 11:33-34.

______ and Steward, F.C. 1968. Cell and tissue culture of white spruce and jack pine. Bi-Mon. Res. Notes Can. For. Serv. 24:30.

______, Chafe, S.C., and Lopushanski, S.M. 1973. Effects of environmental change on sugars, tannins, and organized growth in cell suspension cultures of white spruce. Planta 113:241-249.

Eriksson, T. 1965. Studies on the growth requirements and growth measurements of cell cultures of *Haplopappus gracilis*. Physiol. Plant. 18:976-993.

Evans, D.A., Sharp, W.R., and Flick, C.E. 1981. Growth and behavior of cell cultures: Embryogenesis and organogenesis. In: Plant Tissue Culture: Methods and Applications in Agriculture (T.A. Thorpe, ed.) pp. 45-114. Academic Press, New York.

Fortin, J.A. 1966. Synthesis of mycorrhizae on explants of root hypocotyl of *Pinus sylvestris* L. Can. J. Bot. 44:1087-1092.

Fosket, D.E., Volk, M.J., and Goldsmith, J. 1977. Polyribosome formation in relation to cytokinin-induced cell division in suspension cultures of *Glycine max*. Plant Physiol. 60:554-562.

Franclet, A. 1979. Rejeunissement des arbres adultes en vue de leur propagation vegetative. In: Micropropagation des Arbres Forestiers, Annales AFOCEL No. 12, 6/79, pp. 3-18. AFOCEL, Nangis, France.

F.R.I. 1980. Micropropagation—A new aid in tree improvement? "What's New in Forest Research" No. 87, F.R.I., Rotorua, New Zealand.

Fridborg, G., Pedersen, M., Landstrom, L.E., and Eriksson, T. 1978. The effect of activated charcoals on tissue cultures: Adsorption of metabolites inhibiting morphogenesis. Physiol. Plant. 43:104-106.

Garner, R.J. and Hatcher, E.S. 1962. Regeneration in relation to vegetative vigor and flowering. Proc. XVI Int. Hortic. Cong. Brussels 3: 105-111.

Gautheret, R.J. 1934. Culture du tissu cambial. C.R. Acad. Sci. (Paris) 198:2195-2196.

______ 1957. Histogenesis in plant tissue cultures. J. Nat. Cancer Inst. 19:555-590.

Gay, G. 1982. Influence d'un champignon ectomycorhyzien sur l'enracinement d'hypocotyles excises de *Pinus halepensis* Mill. In: Colloque International sur la Culture in vitro des Essences Forestieres, IUFRO International Workshop, Fontainebleau, France August 31 to September 4, 1981. pp. 269-276. AFOCEL, Nangis, France.

Giles, K.L. 1979. The transfer of nitrogen-fixing ability to nonleguminous plants. In: Plant Cell and Tissue Culture: Principles and Applications (W.R. Sharp, P.O. Larsen, E.F. Paddock, and V. Raghavan, eds.) pp. 807-829. Ohio State Univ. Press, Columbus.

Gresshoff, P.M. and Doy, C.H. 1972. Development and differentiation of haploid *Lycopersicon esculentum* (tomato). Planta 107:161-170.

Girouard, R.M. 1974. Propagation of spruce by stem cuttings. N.Z. J. For. Sci. 4:140-149.

Haines, R.J. and de Fossard, R.A. 1977. Propagation of hoop pine (*Araucaria cunninghamii* Ait.). Acta Hortic. 78:297-302.

Hasegawa, P.M., Yasuda, T., and Cheng, T.-Y. 1979. Effect of auxin and cytokinin on newly-synthesized proteins of cultured Douglas fir cotyledons. Physiol. Plant. 46:211-217.

Horgan, K. and Aitken, J. 1981. Reliable plantlet formation from embryos and seedling shoot tips of radiata pine. Physiol. Plant. 53: 170-175.

Huhtinen, O. 1972. Production and use of haploids in breeding conifers. IUFRO Genetics SABRAO Joint Symposium, Tokyo, D-3 (I):1-8.

______ 1976. In vitro cultures of haploid tissues of trees. In: XVI IUFRO World Congress Proc. Div. II, pp. 28-30. Norwegian IUFRO Congress Comm. OS-NLH, Norway.

_____ and Winton, L.L. 1973. Cell and tissue culture for production of haploid, polyploid, and clonal propagation. In: IUFRO Workshop for Methods in Biochemical Genetics of Forest Trees, Gottingen, Germany, Section 2, pp. 1-12.

Hussey, G. 1978. The application of tissue culture to the vegetative propagation of plants. Sci. Prog. (London) 65:185-208.

Isikawa, H. 1972. Culture of cells and tissues and differentiation of organs in forest trees. IUFRO Genetics SABRAO Joint Symposium, Tokyo, D-2 (I):1-13.

Kadkade, P.G. and Jopson, H. 1978. Influence of light quality on organogenesis from the embryo-derived callus of Douglas fir (*Psuedotsuga menziesii*). Plant Sci. Lett. 13:67-73.

Kao, K.N., Miller, R.A., Gamborg, O.L., and Harvey, B.L. 1970. Variation in chromosome number and structure in plant cells grown in suspension cultures. Can. J. Genet. Cytol. 12:297-301.

Keays, J.L. 1974. Full-tree and complete-tree utilization for pulp and paper. For. Prod. J. 24:13-16.

Kirby, E.G. 1980. Factors affecting proliferation of protoplasts and cell cultures of Douglas fir. In: Plant Cell Cultures: Results and Perspectives (F. Sala, B. Parisi, R. Cella, and O. Ciferri, eds.) pp. 289-293. Elsevier/North Holland Biomedical Press, Amsterdam.

_____ and Cheng, T.-Y. 1979. Colony formation from protoplasts derived from Douglas fir. Plant Sci. Lett. 14:145-154.

Kleinschmit, J. 1974. A programme for large-scale cutting propagation of Norway spruce. N.Z. J. For. Sci. 4:359-366.

Konar, R.N. 1963. Studies on submerged callus culture of *Pinus gerardiana* Wall. Phytomorphology 13:165-169.

_____ and Oberoi, Y. 1965. In vitro development of embryoids on cotyledons of *Biota orientalis*. Phytomorphology 15:137-140.

_____ and Singh, M.N. 1980. Induction of shoot buds from tissue cultures of *Pinus wallichiana*. Z. Pflanzenphysiol. 99:173-177.

Krikorian, A.D. 1983. Chromosome number variation and karyotype stability in cultures and culture-derived plants. In: Handbook of Plant Cell Culture, Vol. 1 (D.A. Evans, W.R. Sharp, P.V. Ammirato, and Y. Yamada, eds.) pp. 541-581. Macmillan Press, New York.

Kulaeva, O.N. and Romanko, E.G. 1967. Effect of 6-benzylaminopurine on isolated chloroplasts. Dokl. Akad. Nauk SSSR Ser. Biol. 177:464-467.

Larkin, P.J. and Scowcroft, W.R. 1981. Somaclonal variation—A novel source of variability for cell cultures for plant improvement. Int. J. Breeding Res. & Cell Genetics 60:197-214.

La Rue, C.D. 1936. The growth of plant embryos in culture. Bull. Torrey Bot. Club 63:365-382.

_____ 1954. Studies on growth and regeneration in gametophytes and sporophytes of gymnosperms. Brookhaven Symp. Biol. 6:187-207.

Libby, W.J. 1974. The use of vegetative propagules in forest genetics and tree improvement. N.Z. J. For. Sci. 4:440-447.

_____, Brown, A.G., and Fielding, J.M. 1972. Effects of hedging radiata pine on production, rooting and early growth of cuttings. N.Z. J. For. Sci. 2:263-283.

Litvay, J.D., Johnson, M.A., Verma, D., Einspahr, D., and Weyrauch, K. 1981. Conifer suspension culture medium development using analytical data from developing seeds. IPC Technical Paper No. 115, Nov. 1981. pp. 17, Institute of Paper Chemistry, Appleton, Wisconsin.

Mapes, M.O., Young, P.M., and Zaerr, J.B. 1982. Multiplication in vitro du Douglas (*Pseudotsuga menziesii*) par induction précoce d'un bourgeonnement adventif et axillaire. In: Colloque International sur la Culture "in vitro" des Essences Forestieres, IUFRO International Workshop Fontainebleau, France, August 31-September 4, 1981, pp. 109-114. AFOCEL, Nangis, France.

Mathes, M. and Einspahr, D. 1965. Comparison of tree growth and callus production in aspen. For. Sci. 11:360-363.

Mehra-Palta, A., Smeltzer, H., and Mott, R.L. 1978. Hormonal control of induced organogenesis. Experiments with excised plant parts of loblolly pine. Tappi 61:37-40.

Melchers, G. 1972. Haploid higher plants for plant breeding. Z. Pflanzenzuecht. 67:19-32.

Mergen, F. 1958. Natural polyploidy in slash pine. For. Sci. 4:283-293.

______ 1959. Colchicine-induced polyploidy in pines. J. For. 57:180-190.

Mott, R.L. 1978. Interaction of seed pretreatments and growth regulators on bud regeneration from excised pine cotyledons. Fourth International Congress on Plant Tissue and Cell Culture, Aug. 20-25, 1978, Calgary, Alberta, Canada, abstr. no. 107.

______ 1979. Tissue culture propagation of conifers. In: Propagation of Higher Plants through Tissue Culture: A Bridge Between Research and Application (K.W. Hughes, R. Henke, and M. Constantin, eds.) pp. 125-133. Conf. 7804111. Nat. Tech. Inf. Serv., Springfield, Virginia.

______ 1981b. Shoots from callus of *Pinus taeda*. IUFRO International Workshop on In Vitro Cultivation of Forest Tree Species, Aug. 31-Sept. 4, 1981, Fontainbleau, France.

______, Smeltzer, R.H., Mehra-Palta, A., and Zobel, B.J. 1977. Production of forest trees by tissue culture. Tappi 60:62-64.

Murashige, T. 1974. Plant propagation through tissue cultures. Ann. Rev. Plant Physiol. 25:135-166.

______ and Skoog, F. 1962. A revised medium for rapid growth and bioassays with tobacco cultures. Physiol. Plant. 15:473-499.

Nickell, L.G. and Torrey, J.G. 1969. Crop improvement through plant cell and tissue culture. Science 166:1068-1069.

Nitsch, J.P. 1963. Naturally-occurring growth substances in relation to plant tissue culture. In: Plant Tissue and Organ Culture—A Symposium (P. Maheshwari and N.S. Rangaswany, eds.) pp. 198-214. Int. Soc. Plant Morphologists, Delhi, India.

Norstog, K. 1973. New synthetic medium for the culture of premature barley embryos. In Vitro 8:307-308.

Parham, R.A. and Kaustinen, H.M. 1977. On the site of tannin synthesis in plant cells. Bot. Gaz. 138:465-467.

Partanen, C.R. 1963. Plant tissue culture in relation to developmental cytology. Int. Rev. Cytol. 15:215-244.

Pawsey, C.K. 1971. Development of grafts of *Pinus radiata* in relation to age of scion source. Aust. For. Res. 5:15-18.

Poissonnier, M., Franclet, A., Dumant, M.J., and Gautry, J.Y. 1981. Enracinement de tigelles in vitro de *Sequoia sempervirens*. Annales de Recherches Sylvicoles 1980, pp. 231-253. AFOCEL, Nangis, France.

Powell, C.L. 1980. Phosphate response curves of mycorrhizal and non-mycorrhizal plants. I. Response to super-phosphate. N.Z. J. Agric. Res. 23:225-231.

Rediske, J.H. 1979. Vegetative propagation in forestry. In: Propagation of Higher Plants through Tissue Culture: A Bridge between Research and Application (K.W. Hughes, R. Henke, and M. Constantin, eds.) pp. 35-43. Conf. 7804111, Nat. Tech. Inf. Serv., Springfield, Virginia.

Reilly, K. and Brown, C.L. 1976. In vitro studies of bud and shoot formation in *Pinus radiata* and *Pseudotsuga menziesii*. G. For. Res. Pap. 86:1-9.

Reilly, K. and Washer, J. 1977. Vegetative propagation of radiata pine by tissue culture. Plantlet formation from embryonic tissue. N.Z. J. For. Sci. 7:199-206.

Risser, P. and White, P.R. 1964. Nutritional requirements of spruce tumor cells in vitro. Physiol. Plant. 17:620-635.

Romberger, J.A., Varnell, R.J., and Tabor, C.A. 1970. Culture of apical meristems and embryonic shoots of *Picea abies*: Approach and techniques. USDA For. Serv. Tech. Bull. No. 1409.

Ross, S.D., Pharis, R.P., and Binder, W.D. 1983. Growth regulators and conifers: Their physiology and potential use in forestry. In: Plant Growth Regulating Chemicals (L.G. Nickell, ed.). CRC Press, Boca Raton, Florida (in press).

Ruehle, J.L. 1980. Growth of containerized loblolly pine with specific ectomycorrhizae after 2 years on an amended burrow pit. Reclam. Rev. 3:95-101.

Rumary, C. and Thorpe, T.A. 1979. A comparison of bud formation in tissues of black (*Picea mariana* Mill Brit) and white spruce (*P. glauca* Moench Voss) grown in vitro. In Vitro 15:177-178.

______ 1983. Plantlet formation in black and white spruce. I. In vitro techniques. Can. J. For. Res. (in press).

Schaffalitsky de Muckadell, M. 1959. Investigations of aging of apical meristems in woody plants and its importance in silviculture. Forstl. Forsøgsv. i Denmark. 25:307-455.

Schenk, R.U. and Hildebrandt, A.C. 1972. Medium and techniques for induction and growth of monocotyledonous and dicotyledonous plant cell cultures. Can. J. Bot. 50:199-204.

Short, K.C., Tepfer, D.A., and Fosket, D.E. 1974. Regulation of polyribosome formation and cell division in cultured soybean cells by cytokinin. J. Cell Sci. 15:75-87.

Skoog, F. 1954. Substances involved in normal growth and differentiation of plants. Brookhaven Symp. Biol. 6:1-21.

______ and Miller, C.O. 1957. Chemical regulation of growth and organ formation in plant tissue cultured in vitro. Symp. Soc. Exp. Biol. 11:118-131.

Smith, D.R., Aitken, J., and Sweet, G.B. 1981. Vegetation amplification: An aid to optimizing the attainment of genetic gains from *Pinus radiata*. In: Proc. Symp. on Flowering Physiology, XVII IUFRO World Congress (S.L. Krugman and M. Katsuta, eds.) pp. 117-123, Japan.

Sommer, H.E., Brown, C.L., and Kormanik, P.P. 1975. Differentiation of plantlets in longleaf pine (*Pinus palustris* Mill.) tissue cultured in vitro. Bot. Gaz. 136:196-200.

Stanley, R.G. 1970. Biochemical approaches to forest genetics. In: International Review of Forest Research (J.A. Romberger and P. Mikola, eds.) Vol. 3, pp. 253-309. Academic Press, New York.

Steinhauer, A. and Huhtinen, O. 1978. Haploid tissue culture in forest trees: Callus and shoot regeneration from megagametophyte cultures of *Picea abies*. Fourth International Congress on Plant Tissue and Cell Culture, Aug. 20-25, 1978, Calgary, Alberta, Canada.

Sweet, G.B. 1973. Effect of maturation on growth and form of vegetative propagules of radiata pine. N.Z. J. For. Sci. 3:191-210.

Thompson, D.G. and Zaerr, J.B. 1982. Induction of adventitious buds on cultured shoot tips of Douglas fir (*Pseudotsuga menziesii* (Mirb.) Franco). In: Colloque International sur la Culture "in vitro" des Essences Forestieres, IUFRO International Workshop, Fontainebleau, France, Aug. 31 to Sept. 4, 1981. pp. 167-174. AFOCEL, Nangis, France.

Thorpe, T.A. 1977. Plantlet formation of conifers in vitro. In: Vegetative Propagation of Forest Trees: Physiology and Practice, pp. 27-33. Gotab, Stockholm, Sweden.

______ 1980. Organogenesis in vitro: Structural, physiological, and biochemical aspects. Int. Rev. Cytol., Suppl. 11A:71-111.

______ 1982a. Callus organization and de novo formation of shoots, roots, and embryos in vitro. In: Techniques and Applications of Plant Cell and Tissue Culture in Agriculture and Industry (D. Tomes, B.E. Ellis, P.M. Harney, K.J. Kasha, and R.L. Peterson, eds.) pp. 115-138. Univ. of Guelph Press, Canada.

______ 1982b. Carbohydrate utilization and metabolism. In: Tissue Culture in Forestry (J.M. Bonga and D.J. Durzan, eds.) pp. 325-368. Martinus Nijhoff/Dr. W. Junk, The Netherlands.

______ and Biondi, S. 1981. Regulation of plant organogenesis. In: Adv. Cell Culture 1:213-239.

______ and Laishley, E.J. 1973. Glucose oxidation during shoot initiation in tobacco callus cultures. J. Exp. Bot. 24:1082-1089.

Tominaga, Y. and Oka, K. 1970. Cytological studies on the calli of *Pinus densiflora* in vitro. Hiroshima Jpn. Agric. Coll. Bull. 4:8-10.

Tulecke, W. 1953. A tissue derived from the pollen of *Gingko biloba*. Science 117:599-600.

______ 1957. The pollen of *Gingko biloba*. In vitro cultures and tissue formation. Am. J. Bot. 44:602-608.

______ and Sehgal, N. 1963. Cell proliferation from the pollen of *Torreya nucifera*. Contrib. Boyce Thompson Inst. 22:153-163.

von Arnold, S. 1979. Induction and development of adventitious bud primordia on embryos, buds, and needles of Norway spruce (*Picea abies* (L.) Karst.) grown in vitro. Doctoral Thesis, Univ. of Uppsala, Uppsala, Sweden.

______ 1982. In vitro propagation of Norway spruce. In: Colloque International sur la Culture "in vitro" des Essences Forestieres, IUFRO International Workshop, Fountainebleau, France, Aug. 31 to Sept. 4, 1981, pp. 87-100. AFOCEL, Nangis, France.

______ and Eriksson, T. 1977. A revised medium for growth of pea mesophyll protoplasts. Physiol. Plant. 39:257-260.

______ and Eriksson, T. 1979. Bud induction on isolated needles of Norway spruce (*Picea abies* (L.) Karst.) grown in vitro. Plant Sci. Lett. 15:363-372.

Wallin, A., Glimelius, K., and Eriksson, T. 1974. The induction of aggregation and fusion of *Daucus carota* protoplasts by polyethylene glycol. Z. Pflanzenphysiol. 74:64-80.

Webb, K.J. 1982. Growth and cytodifferentiation in cell suspension culture of *Pinus contorta*. In: Colloque International sur la Culture "in vitro" des Essences Forestierse, IUFRO International Workshop, Fountainebleau, France, Aug. 31 to Sept. 4, 1981, pp. 217-226. AFOCEL, Nangis, France.

______ and Street, H.E. 1978. Factors affecting morphogenesis of *Pinus contorta* and *Picea sitchensis* tissue cultures. In: Frontiers of Plant Tissue Culture 1978 (T.A. Thorpe, ed.). Univ. of Calgary Press, Canada.

White, P.R. and Risser, P. 1964. Some basic parameters in the cultivation of spruce tissues. Physiol. Plant. 17:600-619.

Winton, L.L. 1972a. Annotated bibliography of somatic conifer callus cultures. Genet. Physiol. Notes, No. 16, Inst. Paper Chem., Appleton, Wisconsin.

______ 1972b. Callus and cell cultures of Douglas fir. For. Sci. 18: 151-154.

______ 1978. Morphogenesis in clonal propagation of woody plants. In: Frontiers of Plant Tissue Culture 1978 (T.A. Thorpe, ed.) pp. 419-426. Univ. of Calgary Press, Canada.

______ and Huhtinen, O. 1976. Tissue culture of trees. In: Modern Methods in Forest Genetics (J.P. Miksche, ed.) pp. 243-264. Springer-Verlag, Berlin.

______ and Verhagen, S.A. 1977. Shoots from Douglas fir cultures. Can. J. Bot. 55:1246-1250.

______, Parham, R., and Kaustinen, H. 1975. Isolation of conifer protoplasts. Genet. Physiol. Notes No. 20, Inst. Paper Chem., Appleton, Wisconsin.

Wright, S. 1968. Evolution and the Genetics of Populations, Vol. I. Univ. of Chicago Press, Illinois.

Yeung, E.C., Aitken, J., Biondi, S., and Thorpe, T.A. 1981. Shoot histogenesis in cotyledon explants of radiata pine. Bot. Gaz. 142: 494-501.

CHAPTER 17

Special Problems: Adult vs. Juvenile Explants

D.J. Durzan

Explant source is one of the first practical considerations in the application of cell and tissue culture technology to crop plants. With woody perennial crops, a major problem has been the residual memory that the explant seems to have when isolated aseptically and forced to grow in vitro.

The memory, or "mnene," is a common trait of biological systems (Russell, 1930; Whyte, 1949). It reflects the summation of the past developmental history of cells in the whole plant body. Through proper nutrition and synthetic plant growth regulators (Matsubara, 1980; Thimann, 1977), the viability of explants can be maintained and organogenesis expressed (Skoog and Miller, 1957) in many but not yet in all species (Murashige, 1978; Durzan, 1980). Where explant growth and development cannot be predicted or controlled, the tendency has been to consider explant sources largely on the basis of size, position, growth pattern, age, microenvironment, and totipotency of cells (Steward et al., 1975). These variables tend to translate into the most suitable "physiological state" given the genetics of the cells in terms of ploidy, life form, and vegetative reproduction (Gustafsson, 1948).

Every explant will have a finite number of general properties, such as genome ploidy and heterozygosity, age, number of components (daughter-cell divisions), connectivity among components, nutrient inputs, reponse to growth regulators, redox states, and blocks in cell cycles. Any combination of factors is referred to as the "state" of the explant. In this chapter, special emphasis is placed on juvenility and maturity as physiological states contributing to the success or failure of vegetative propagation.

The hope remains that through some simple explant choice, substrate provision, nutrient formulation or growth regulator supplement, morphogenesis can be controlled so as to release or induce the fullest developmental capacity of the explant. Until now, the problem has been not so much the production of callus over long and indefinite periods of culture (Gautheret, 1955) but the establishment of the conditions for "true-to-type" morphogenesis, i.e., the step-by-step recapitulation of normal hereditary mechanics of cells, tissues, organs, and plants.

The focus of this chapter largely will be on explants from the more recalcitrant species and on perennial diploid sporophytes, especially those with long life cycles, strong internal correlations, and a complex physiology well-attuned to environmental fluctuations. From the propagator's viewpoint, it is desirable to rejuvenate or invigorate proven individuals and capture a trueness-to-type or the genetic gains on a massive scale (Libby, 1974; Rediske, 1974, 1977). The propagator may also want to introduce new traits in explants by using methods of agricultural microbiology (Brill, 1981).

SELECTION FACTORS

Juvenility and Cycles of Determination

Interest in the juvenile phase of the life cycle of a crop extends to several practical problems in vegetative propagation: First, how can the rooting potential be increased for the capture of genetic gains? Second, how can the juvenile period be controlled to produce early or late flowering? Early flowering would speed up breeding programs and seed or fruit production, whereas late flowering prolongs vegetative growth and leads to greater wood production. Third, how can the variability in growth performance and morphology of propagules from juvenile and adult phases be exploited in agroforestry and for horticultural varieties?

Other plant cycles often determine juvenility and the solution to the above problems (Fig. 1). One of these, the nuclear-cytoplasmic cycle of cell determination is an interpretation of Brink's views (1962) on phase change. Another, although not specifically illustrated, emerges from the nuclear-cytoplasmic cycle, namely the cell cycle (cf. Brown and Dyer, 1972; King, 1980).

A significant challenge for the plant propagator is to alter or avoid the phasic development in mature whole plants, so that individual cells are transformed into a physiological state suitable for vegetative propagation (Kester, 1976) or flowering (Scorza, 1982).

Although the control of maturation of juvenile cells may be considered a function of the development of the growing point (Kester, 1976), additional facts are emerging. In cottonwood, the position and number of leaf primordia in the apex are a function of vascular development at nodes well below the shoot apex (Larson, 1975). The duration of the juvenile phase seems to be determined genetically by the number and length of nodes rather than by chronological time (Hansche, 1982; Kester, 1984). Continuous growth of the donor plant will allow

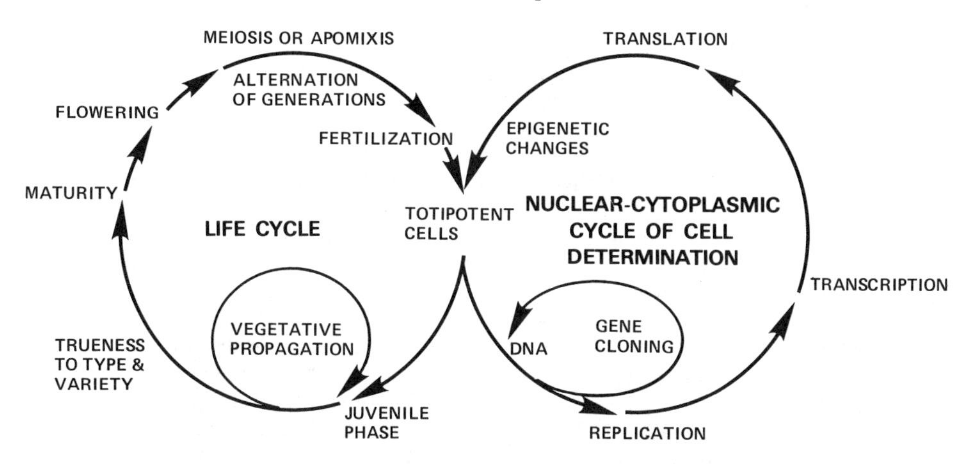

Figure 1. Life cycle from which juvenile tissues and totipotent cells are sought for vegetative propagation and apomixis largely to capture genetic gains. Stability of the trueness to type depends on epigenetic factors in the nuclear-cytoplasmic cycle of cell determination. The two interlocking cycles are, in fact, two faces of the same coin. Vegetative propagation and gene cloning are seen as subcycles serving the specific needs of the propagator.

growing points to transverse the juvenile phase in the shortest time and will result in early flowering. Juvenile growth has a strong genetic component and can be selected against to promote precocious flowering of dwarf peaches (Hansche and Beres, 1980).

The phasic development of the plant as a whole and over the life cycle needs to be considered in the determination of the fate of any given explant. In some woody perennial species the unusual patterns of vegetative and reproductive growth lead to alternate bearing, as in the pistachio (Crane and Iwakiri, 1981). Explants of female inflorescences taken during the "on" or "off" year may confuse results in vitro. This example illustrates the diversity of factors at the biological level that complicates explant choice.

Many plant physiologists prefer to avoid the use of the term aging because it conjures up analogies with animals that may be inappropriate. Hence the terms maturation, senescence, and senility are encountered more frequently in the plant literature when life cycles are described.

The control of developmental expressions in explants from all phases of the life cycles is believed to be generally "permissive" or "inductive" (Street, 1979). Sharp et al. (1980) distinguished two further aspects of explant development leading to propagule regeneration, viz. "direct" and "indirect." Whatever concepts these terms imply, sooner or later a better perspective on causality, chance, and the laws of nature (deterministic and nondeterministic) will have to be considered. Bunge (1961) has provided a philosophical basis for these concerns.

With woody perennials, different parts of the same plant tend to remain juvenile and others remain adult (Kester, 1976; Romberger, 1976). The position from which an explant is taken becomes important.

The position effect, called topophysis, is based on the observation that explants remember their position and continue to express the potential for that position. Another response called cyclophysis is based on performance related to the developmental age of the explant. A third, called periphysis, is caused by microenvironmental differences (e.g., shade, drought) arising in explants from similar positions and developmental ages.

The importance of tissue source is recognized by foresters in the use of the terms ortet for the original seedling source and ramet for the vegetative progeny (Wright, 1976; Greenwood, 1978; Franclet, 1977). With continuous cycles of vegetative reproduction, the expressions of the adult phase tend to become stabilized: selected adult characteristics of plant cultivars, such as dwarfing and fruiting, become more or less fixed. By contrast, in the body of the tree and during the normal life cycle, juvenile and adult tissues may occur at the same time as a function of developmental and physiological gradients (Bonga, 1982).

Some explants may contain an array of nonsexually transmitted traits or "epigenetic niches" (Hackett, 1980; Brink, 1962; Wright, 1968; Sharp et al., 1980; Sinnott, 1960). Once an epigenetic change occurs, the effect persists throughout the cloning process even after the initial stimulus for cloning is removed.

Epigenetic shifts in development may be reversed in cells that proceed through meiosis. Tissues, which exhibit strong developmental tendencies, and propagules retaining these traits, when passed through meiosis, may become highly rejuvenated, bearing juvenile plant characteristics. Rejuvenation or invigoration of older cells can be accomplished by a number of other methods described later in this chapter. At the molecular level (Fig. 1), we can postulate that editing mechanisms for the genetic code (e.g., Fersht, 1981) may somehow be involved in determining the extent of rejuvenation.

Continued vegetative propagation produces a group of plants referred to as a clone and which could be named a cultivar (Kester, 1984). Clones could start as either juvenile or adult, depending on the selection of the explant. Hence, the terms topoclone, cycloclone, and periclone are sometimes used. Mericlone is used to indicate the origin of the clone from meristematic tissues. According to Kester (1976), most fruit tree growers probably have never seen a juvenile tree, since all grafted cultivars are in the adult stage.

Variations in apomictic phenomena are important to recognize (Gustafsson, 1948). Apomixis is the name given to those cyclic processes in which seeds are formed without meiosis and fertilization (Fig. 2). In rare instances a single meristemoid may grow inside a cluster of cells and break through this mass much like a spore (Fig. 3). One of the major developments of recent plant propagation of crop plants has been the ability to duplicate apomixis in cell and tissue culture systems and produce somatic embryos in vitro.

Another major development is the progress with protoplast technology and expressions of protoplast totipotency (Fowke and Gamborg, 1980; Galun, 1981). Indeed the egg is itself a protoplast (Steward and Krikorian, 1979). With protoplasts the effect of newly introduced haploid nuclei on the development of the resulting hybrids may someday be determined with the aid of monoclonal antibodies and immunoassays (Daie and Wyse, 1982). This is especially interesting in species with conservative karyotypes that go through a multinuclear embryonic stage.

In conifers, for example, haploid maternal cells are readily available in seeds. Through protoplast fusions, hybrid protoplasts containing haploid nuclei could be constructed that have a range of homozygous to heterozygous nuclei (Durzan, 1982a). Fixing the degree of genetic homozygosity or heterozygosity may affect how well these multinuclear hybrid protoplasts become invigorated and recapitulate somatic embryogenesis.

General Considerations: Explant Source

The prime concerns about the origin of the explant are genetic identity, stability, and performance. Genetic features should be monitored by (a) establishment of genetic identity of the source plant, i.e., proof of the heritable nature and mode of transmission of traits under consideration; (b) use of vegetative progeny tests for horticultural and genetic verification; (c) maintenance of source identity so as to trace the initiation of any variant to one of the phases of propagation; and (d) identification and quantification of abnormal metabolites, storage substances and cellular states or threshholds that characterize the genetic stability or instability of development.

Given a specific genome the problems of explant selection are not unlike those in foliar analysis for the diagnosis of the nutritional state of the plants. Here elemental levels are sought (e.g., Muratov and Mitrofanov, 1973) to reflect nutrient status and to prescribe amendments to the soil to correct a problem (Tamm, 1981). In general the following factors are considered: age of the sampled tissue; season and time of day; position in the body of the plant (shoot, crown, root); effects on shaded tissues, especially on very shade-intolerant species; fruiting or vegetative state; weather conditions such as rain and drought; presence of mychorrhizae, disease, or insects; and exposure to pollution or atypical situations (acid rain, radiation, mine tailings, alkaline soils, etc.).

As with elemental analyses, statistical correlations are the only results obtainable as long as we know nothing about the genetic and physiological meaning of the nutrient levels and growth correlations found. Furthermore, when the explant is severed, whole plant-cell interactions are removed, metabolism is altered, and new compounds often appear and accumulate as levels of protein nitrogen decline (Durzan and Chalupa, 1976b). We need specific diagnostic tests for selecting explants with defined genetic gains that are responsive to in vitro technologies. Monoclonal antibodies may provide the required specificity to detect these gains (Green, 1981).

Until such diagnostic tests are developed, the laboratory test is the most direct and critical test of the value of an explant for the following reasons: (a) it evaluates the effect of a particular nutrient medium upon growth and development; (b) it optimizes nutrients, growth regula-

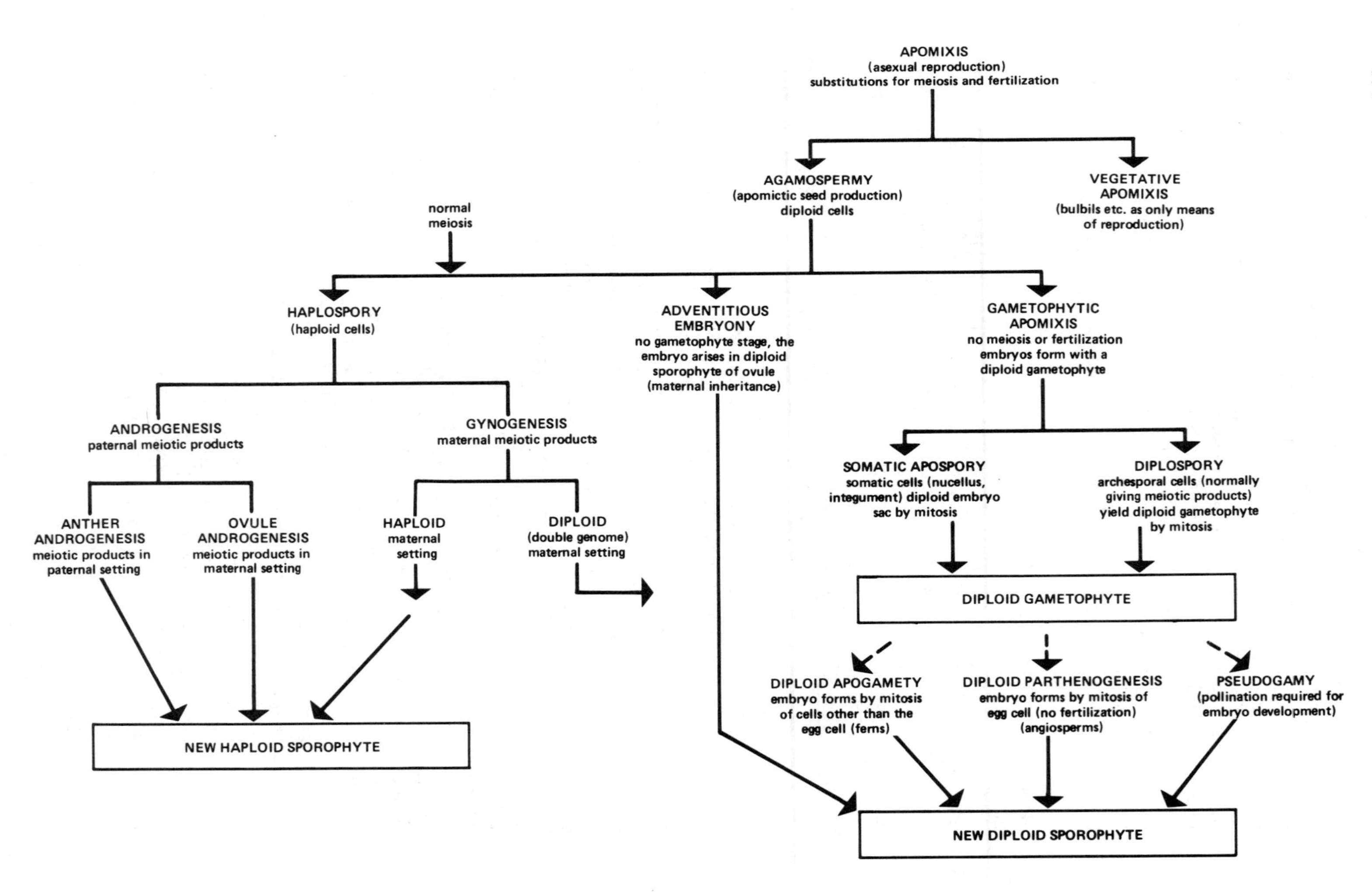
APOMIXIS
(asexual reproduction)
substitutions for meiosis and fertilization
AGAMOSPERMY
(apomictic seed production)
diploid cells
VEGETATIVE
APOMIXIS
(bulbils etc. as only means
of reproduction)
normal
meiosis
HAPLOSPORY
(haploid cells)
ADVENTITIOUS
EMBRYONY
no gametophyte stage, the
embryo arises in diploid
sporophyte of ovule
(maternal inheritance)
GAMETOPHYTIC
APOMIXIS
no meiosis or fertilization
embryos form with a
diploid gametophyte
ANDROGENESIS
paternal meiotic products
GYNOGENESIS
maternal meiotic products
SOMATIC APOSPORY
somatic cells (nucellus,
integument) diploid embryo
sac by mitosis
DIPLOSPORY
archesporal cells (normally
giving meiotic products)
yield diploid gametophyte
by mitosis
ANTHER
ANDROGENESIS
meiotic products in
paternal setting
OVULE
ANDROGENESIS
meiotic products in
maternal setting
HAPLOID
maternal
setting
DIPLOID
(double genome)
maternal setting
DIPLOID GAMETOPHYTE
DIPLOID APOGAMETY
embryo forms by mitosis
of cells other than the
egg cell (ferns)
DIPLOID PARTHENOGENESIS
embryo forms by mitosis of
egg cell (no fertilization)
(angiosperms)
PSEUDOGAMY
(pollination required for
embryo development)
NEW HAPLOID SPOROPHYTE
NEW DIPLOID SPOROPHYTE

Figure 2. Vegetative propagation or asexual reproduction (apomixis) can evolve into many alternatives when viewed as a substitution for meiosis and fertilization (Swanson, 1957; Heslop-Harrison, 1972). Parthenogenesis, which is defined as the development of a new individual from an egg without fertilization, is but one phase of apomictic behavior. In the normal life cycle, reduction in chromosome number by meiosis results in the formation of haploid gametes and is compensated for by fertilization so as to establish an alternation of generations. Hence, a functional apomictic or vegetative propagation cycle should have suitable substitutes for meiosis and fertilization with haploid or diploid cells. Apomixis is of two principle types: vegetative reproduction and agamospermy. Apomictic seed formation or agamospermy occurs by adventitious embryony (e.g., *Citrus*, *Euphorbia*) or gametophytic apomixis when a gametophyte is formed. The latter develops from somatic or archesporal cells, which normally would undergo meiosis and a reduction to form a haploid embryo sac. If somatic cells are involved (apospory) (e.g., *Malus*, *Poa* spp.), these are usually nucellar or integumental, and a diploid embryo sac is formed. If cells were of archesporal origin (diplospory) (e.g., *Hieracium*, *Taraxacum*), the meiotic processes are missing or abortive and a diploid gametophyte is obtained. Once the embryo sac is formed, the embryo may develop from the egg (parthenogenesis) or from the remaining cells (apogamety), e.g., ferns. Apomicts may be pseudogamous. Here development of the embryo does not proceed unless pollination has taken place. Once haploid cells are formed by meiosis, haploid plants may develop via androgenesis or gynogenesis. The meiotic products of pollen develop in the anther (anther androgenesis) or ovule (ovule androgenesis). The meiotic products in the ovule can develop within the ovule (gynogenesis). Should the genome be doubled, a diploid sporophyte could emerge.

tors, and their combinations above or below threshhold values; (c) where plant regeneration is obtained, it generates data for economic analysis; (d) it allows for the evaluation of contamination, genetics, and growth regulators in regard to nutritional and environmental variables; and (e) it confirms hypotheses concerning choice of explant.

Multitest Criteria

Because explants usually contain specific traits to be captured by cloning, the genetic gains and cultural traits for the individual must be defined carefully. The choice of explant should involve an ambitious test of the importance of cell composition, physiological status, size, position, developmental patterns, maturity and totipotency in terms of genetics, nutrition, environment, and specific factors such as osmolality of the medium and levels of auxins and cytokinins.

Even so, the expression of growth and development under individual and combinations of variables may be difficult to achieve and to interpret. The expected response may not appear immediately. Synergisms among factors may alter other parameters, such as growth rates and morphogenesis, in unexpected ways. Even the point of final responsive-

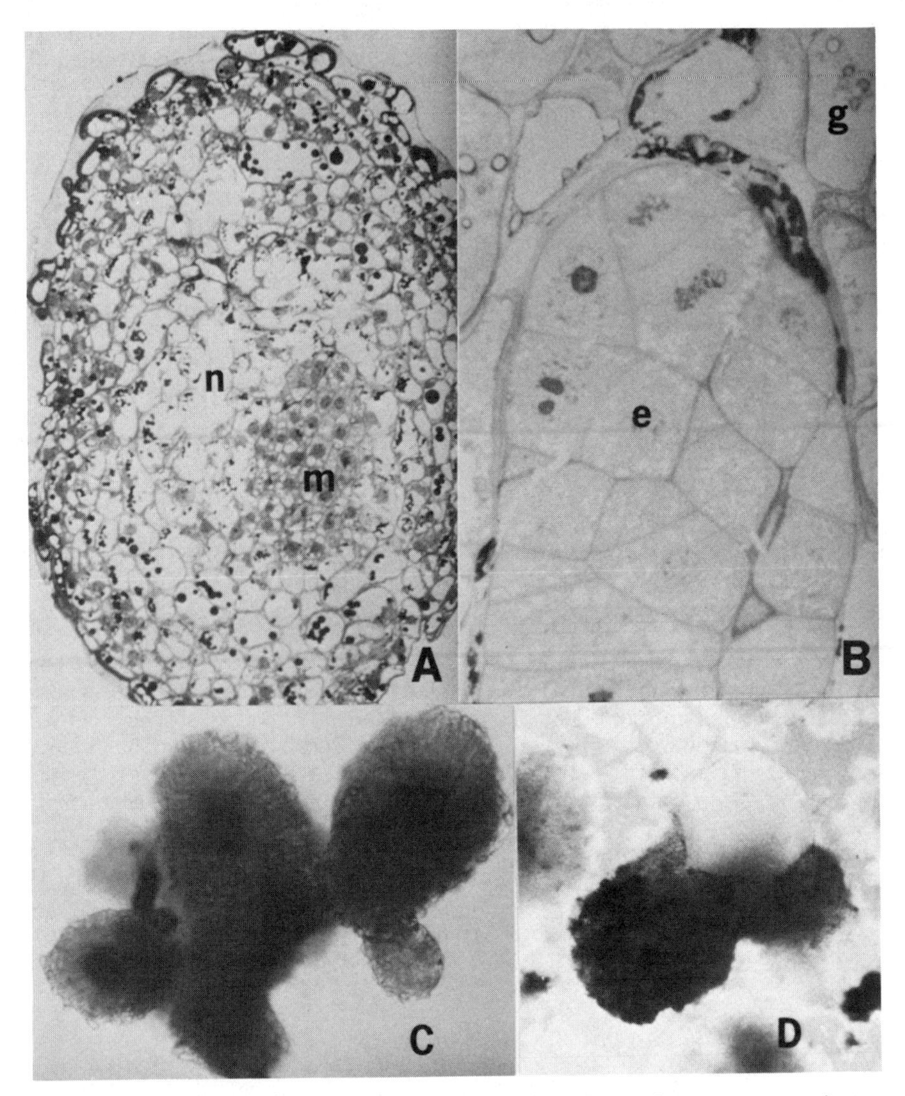

Figure 3. The development of cell suspensions (60–290 μ dia) into structures with internal meristemoids. Cell suspensions were derived from a callus of a lateral bud of a 4-year-old Douglas-fir sapling. Meristemoids eventually grow through the initial structure much like a germinating seed (unpublished data of S. Hwo and D.J. Durzan). Buds from more mature trees do not give the same results. (A) Cross section of the clump showing the meristemoid (m) with cytoplasmically dense cells having large nuclei. (B) Cross section of a developing seed showing the young embryo (e) embedded in the nutritive haploid female gametophyte (g) is provided for comparison with A. (C) After 6 weeks, the centers of the spherical clumps contain dark green centers, which represent internal meristemoids. The number of meristemoids per clump

is usually one, but this can vary as a function of exposure to auxins and cytokinins. (D) After 3 months, the internal meristemoid grows and develops at the expense of surrounding tissues and eventually ruptures the surface cells to emerge as a white, compact mass of cells much like a germinating spore. The cell source was derived from cotyledon callus of Douglas fir.

ness is not always known, and, more seriously, a morphogenetic response may be latent while a buildup of activity occurs at the biochemical level, so the evaluation of choice of explant may remain improperly diagnosed.

Serious attention has not always been directed to the nature or principles or the factors controlling or limiting the life cycle under field conditions. For medium development, often the testing of particular nutrient elements or combinations has been random or as a "broad-spectrum" experiment rather than based upon solid nutritional, genetic, developmental and ecological principles (Clarkson and Hanson, 1980; Durzan, 1974; Steward and Krikorian, 1979).

At least four stages will be needed to evaluate the responsiveness of explants: (a) recognition of a widespread effect of past developmental history; (b) selection of a short list of causal factors; (c) demonstration of a correlation between one of the factors and the effect; and (d) proof of a causal connection by further experiments.

Criteria of merit should be sought from the subcellular level to the whole donor plant. We should keep in mind that the activity of the whole plant cannot be fully explained in terms of the activities of a few isolated explants. It can be less explained, the more abstract are the parts distinguished. In the selection of explants, we must strive, at least initially, to correct abstract concepts by distinguishing factors that are concrete and biological, not just physico-chemical. We must carry out as complete a reconstitution or reintegration of factors as possible: no explant or any single process of any plant can be fully or properly understood in isolation from the structure and activities of the organism or its past environment as a whole.

The donor plant itself, at any one moment of its history, must be regarded as at merely a phase of a life cycle. It is the whole cycle that is the life of the individual, and this cycle is indissolubly linked with previous life cycles. The activities of the donor organism and its explant at any stage of manipulation can be understood only if they are reintegrated in the individual and the evolutionary life cycles (alternation of generations and the seed habit). This enables us to envisage the possibility of giving some rational account of the stereotyped repetition of thc propagation process and of the developmental rhythm which shows itself as heredity in the broadest sense. This includes stages of morphogenesis reminiscent of ancestral stages that earlier botanists recognized as laws of recapitulation. It also draws attention to the yet largely untapped potential of haploid sources, particularly in gymnosperms (LaRue, 1954; Tulecke, 1965).

Concerning the selection activities of explants, whether unicellular or multicellular, the following should be considered. First, genetic gains to be captured (e.g., rapid growth rate, resistance to disease) and the associated or underlying metabolic basis for these traits. Metabolism will deal with assimilation, respiration, biosynthesis, conditioning of the medium, and the maintenance and restoration of form both from the single cell and the arrangement of differentiated cells to reconstitute organs and organisms according to some set pattern. Juvenile tissues placed in specific environments tend to have a greater capacity for this restoration. Mature sources tend to revert quickly to more mature expressions of development and secondary traits. Differences in metabolism should be linked to the expression of these morphogenetic tendencies.

A second activity is the behavior of the explant. This should be viewed in relation to the external environment (Langridge, 1963; Went, 1956), to its own surface-volume dimensions (Richards, 1969) and to polarity (Greenwood and Goldsmith, 1970; Went, 1974). This includes stimulus perception (e.g., nutrition, geotropism, aeration) in culture systems that can extend the range of this set of factors for testing.

Third, explants should be sought with cells having special functions that can be invigorated and have the power of self-regulation, especially in the formation of new meristems (Campbell and Durzan, 1975, 1976). In these explants a close relation often exists between past development and anticipatory action. Earlier events or states may help explain those occurring in the future, e.g., the propensity for somatic embryogenesis may be easier to express in cells that are already embryonic.

In summary, the explant is regarded as a "bud" with all of the developmental tendencies of the parent. In the propagator's hands, this ancestral connection accounts for the orderly succession of development and the need for multitest criteria. Sequential developmental tendencies involve heredity because the propagules regenerated through this process resemble those in the parents. Failures in development underscore that the explant is only a shortened epitome of the developmental history of the parent. In some cases, the development of the explant will take shortcuts and amplify or telescope developmental traits (refer to Fig. 6). These responses, however, explain neither the driving force behind development of the explant nor the physiological rates for adaptation to the unusual conditions of in vitro existence.

DIAGNOSTIC SPECIFICITY

Physiological State

The remarkable feature of plant cells is that they retain the potential for a physiological state that is totipotent: given special methods and under appropriate conditions, cells will regenerate plants that can complete a normal life cycle or variations thereof. During this recapitulation of ontogeny, a homeostasis emerges regardless of most external perturbations. For organs, historically this degree of determination has been a function of nutrition and correlations based on growth regulators.

During maturity the turnover of macromolecules and the eventual loss of cellular organization lead to a climacteric burst of respiratory activity (Blackmann and Parija, 1928) and to a loss of totipotency. However, no amount of semantics and rationalization can conceal the fact that genetics and molecular biology per se have not yet been demonstrably successful in furnishing causal explanations of how complex organization works and how it relates to growth and development (Steward and Krikorian, 1979).

Although most explants are taken from actively growing plants, this does not mean that resting cells lack useful physiological properties. Some dry seeds show a 24-hr rhythm of gas exchange that persists under constant conditions and seems independent of temperature. Since DNA replication, transcription, and translation do not occur in this quiescent state, the basic circadian oscillation in gas exchange does not seem to be derived directly from these processes (Bryant, 1972). In totipotent cells it is the information in the nucleus and the ability to eventually translate it in the cytoplasm that is believed to remain unimpaired throughout development.

We now have a wealth of observations to draw upon to characterize the physiological states of cells: genes for shape (Sinnott, 1960); polysomatic control over morphogenetic potential (Murashige and Nakano, 1967; Skirvin, 1978); physiological preconditioning (Rowe, 1964); apical dominance (Gregory and Veale, 1957); internal correlations (Sinnott, 1960); mutual incompatibility of loci (Wardlaw, 1968); contact pressure (Adler, 1974); organization resistance (Blackmann and Parija, 1928); binding sites for growth regulators (Murphy, 1980); calcium effects mediated through calmodulin and proton pumps (Kubowicz et al., 1982); reaction kinetics (Huennekens, 1959); microtubules and meromeres (Gunning and Hardham, 1979); explant maturity, position, and environment (Franclet, 1977; Nicholls et al., 1974); juvenility, maturity and senescence (Bonga, 1982; Shaffalitzky de Muckadell, 1959; Franclet, 1977); developmental fields (Romberger and Gregory, 1974; Waddington, 1966); retardation of senescence by cytokinins (Richmond and Lang, 1957); and "pre-embryonic determined" cells and "indirect embryogenesis" through the redetermination of differentiated cells (Sharp et al., 1980).

Factors are operational at physical (Bikerman, 1958; Green, 1980), chemical (Huennekens, 1959), nutritional (Murashige, 1978), or biological levels (Jacquiot, 1964; Steward and Krikorian, 1979; Sharp et al., 1980). Complexity is inevitable where factors become highly interdependent and autopoietic (Varella et al., 1974; Zeleny, 1977). Confusion becomes prevalent when one or more factors lead to the same end result. Indeed, plant genes are dynamic and often pleiotropic (having more than one phenotypic expression).

The detection and evaluation of the physiological state proceeds through five levels of sophistication:

1. Description of the genotype and phenotype, their presentation, history and pathological features. Irregular and abnormal cytology usually occurs if the plants have been selectively bred and are unusual hybrids.

2. Proof of the heritable nature and mode of transmission of the traits sought and their pathology.
3. Identification and quantitation of any specialized metabolism including critical levels of primary and secondary metabolites (indoles, purines, amino acids, etc.), as intermediates or conjugates (Hangarter et al., 1980), or the unusual appearance of abnormal intermediates and storage compounds (water, arginine, ureides, starch) that indicate that blocks occur in metabolism. Developmental tendencies that reflect certain evolutionary groups, such as monocotyledons and conifers, should be distinguished.
4. Demonstration of enzymatic competence or a defective or altered enzyme, structural protein, or nonhistone chromosomal protein.
5. Identification of the causal event itself.

An equivocal response of explants to treatment implies a potential failure of cells to discriminate among the different input treatments. Habituation, malfunction or failure of the physiological state in cells of the explant may ensue.

The input for every unit process in the physiological state of cells, such as assimilation, respiration, and photosynthesis, will come from several sources (medium, atmosphere, other cells, etc.); the inputs are no less than compound. Furthermore, most traits are affected by several genetic loci. Even key enzymes may be multifunctional. In jack pine explants, initial parameters such as adenylate energy charge, oxygen uptake and water, amino acid and nucleotide content in the donor have correlated positively with increased callus production (Durzan et al., 1976).

A chemical basis for the mnemic and maturation processes in juvenile and mature physiological states should be sought. Initially this was conceived to be a function of the property of colloidal systems (MacDougall, 1920) and later of the unraveling of gene-enzyme-forming (Brink, 1962; Steward and Durzan, 1965) or general systems (Zeleny, 1977).

A survey of the chemical nature of nitrogenous compounds of plants (Durzan and Steward, 1983) has revealed a wide range of post-translational modifications of proteins and multiple steps to the processing of DNA and transcribed nucleic acid messages that contribute a new level of structural and possibly functional diversity (Fig. 1). This structural polymorphism, based on the some 140 post-translational modifications of protein amino acid and the processing of messenger RNA (base methylation, splicing, polyadenylation, and transport) endows a degree of determinism in structure and function. Hence the enzymes that process and schedule macromolecular synthesis may allow the cell to quantify during development the effect of multiple nutritional environmental factors that impinge on daughter cells in the explant. This biochemical determinism is reinforced when enzymes are formed at specific locations (e.g., membranes, vacuoles, organelles) in the cell and convert substrates into products through assimilation, redox reactions, ligases, hydrolases, and other processes. This further "locks in" the fate of cells, especially if products are hormonal or inhibitory. Hence enzymatic reactions that contribute to the formation of homologues, analogues, cyclization, ring-closure, conjugation, and methylation among others become important (Bidwell and Durzan, 1965; Durzan and Steward, 1983; Durzan, 1982b). These processes occur progressively and sequentially over time as cells pass from totipotent to mature states in the life cycle (Fig. 1).

The importance of the structural organization and metabolic turnover of macromolecules in a totipotent cell is often ignored. Even in prebiological systems there are simple autopoietic rules with primary, secondary, tertiary and even quaternary levels of organization that determine the function and specificity of change (Zeleny, 1977). This leads to the premise that, as a cell undergoes successive stages of differentiation, changing patterns of structural organization in molecules are displayed in membranes and at the cell surfaces. This display is postulated as arising from DNA reshuffling during cell and life cycles.

Biological specificity arises from the interaction of complementary molecular structures (Anderson, 1976; Neumann, 1973; Huennekens, 1959; Pauling, 1974) and involves stereospecificity (Kuhn, 1940). Here hydrogen bonds are among the most important of the weak intermolecular forces (ionic and nonionic) between the interacting molecules (Pauling, 1974). The relationship between proteins and DNA in chromatin should be explored in juvenile and mature cells to characterize such intermolecular forces.

Concerning the role of chromatin in specific gene action, the nonhistone chromosomal proteins show considerable promise as controlling factors for phase changes (Pitel and Durzan, 1978, 1980). Physical-chemical aspects of development have been extended by Green (1980), Tanford (1978), and Anderson (1976) but have not yet been applied directly to problems of explant choice.

Size of Explant

Where very little baseline information is available, model cellular systems for somatic embryogenesis (wild carrot), organogenesis (tobacco), and sporulation (bacteria, e.g., Szulmajster, 1979) may be used for comparison. If a cell in an explant with desired traits is totipotent and connected with other cells such that the totipotent cell is affected by these associations, then the response to a given treatment will be a function of the number of cellular associations.

Disproportionate growth will inevitably ensue (Brown, 1964; Gordon, 1966; Richards, 1969). Changes in surface and volume of the explant can easily lead to differences in overall growth rate, e.g., monomolecular, Gompertz, autocatalytic, near-exponential (cf. Chalupa et al., 1976; Durzan and Chalupa, 1976a).

From a procedural viewpoint the size and complexity of donor (Romberger, 1976) and explant, from whatever source, should be interpreted in terms of cytogenetics, cell numbers, size, surface, and volume. In one example, nuclei and cells in explants or suspension cultures of conifers (Douglas fir, loblolly pine) are at least tenfold larger than those of the wild carrot (Price et al., 1973). Explants and screened populations of cells (e.g., 60-200 microns) carry different numbers of cells and this creates differences in inoculation densities sometimes with unexpected effects (e.g., Cass, 1972). Therefore, adjustments of scale must be made for comparative purposes. The interpretation of growth and size of tissues as a function of surface, volume, and organized growth has a historical precedent, and this has been described with the help of mathematics (Netter, 1969; Richards, 1969; Thompson, 1963).

Position of Explant

The perennial proliferation of new meristems in trees underscores the importance of amplified subsets of polarized activities along physiological gradients in specialized tissue systems. Tissue systems may contribute to a topological redundancy in the pattern of generation of new meristems or meristemoids. Explants from the same topological subset could contain cells with highly specialized and amplified functions (e.g., epidermal, transfer, and cambial cells) so that some degree of dedifferentiation of cellular components and the physiological state may be required.

Additional problems of interpretation of position arise because the plant body arises usually from a single cell (fertilized egg) or from primary and secondary meristems that develop later. The extent of derivation of an explant as a function of past developmental gradients becomes difficult to determine unless the meristems themselves are excised and studied by growth analysis (Romberger and Gregory, 1974). The extent and rate of derivation of daughter cells in meromeres within tissues (Gunning and Hardham, 1979; Brown, 1964) are often unknown and difficult to determine.

It is important to recognize potentially organizing patterns of cellular growth that may be carried into the explant or produced in a cell suspension culture (Fig. 4). Cells comprising these patterns in the original explant can serve as secondary sources of explants once cells become independent and cultures are established. Where cell suspensions are grown from callus the source of cells within the callus should be determined, such as in the parenchyma, cambium, subepidermis, and epidermis (Tran Than Van, 1981). The origin of cells relative to the original explant may lead to quite specific responses of plant growth regulators.

Even the position in which the explant is placed upon the agar is important, especially for the polarity of auxin transport and directed (vectorial) supply of nutrition (Sinnott, 1960). These factors have parallels in natural embryonic development and should be tested in new situations.

Recognition of Developmental Patterns

One advantage of uniform cell suspensions is that early developmental patterns can be monitored with live cells (Fig. 5). Studies with woody species indicate that many patterns of daughter cell association could be derived from the single cells (Detorok, 1968; Durzan and Steward, 1971; Durzan and Chalupa, 1976a) and that the evaluation of these developmental patterns needs a quantitative and genetic basis.

When cells organize into embryos or shoots, the early developmental patterns appear to have few alternatives at least initially. By contrast, the alternative and numerous "poor" patterns, tend to generate more callus or abortive structures. Abortive structures usually have chromosomal instabilities (e.g., Patel and Berlyn, 1982). The terminal point of the pattern may therefore reflect a defective biochemical sequence based on faulty genes, proteins and abnormal substrates. To show that certain daughter cell growth patterns develop into embryos a way will have to be found to produce the patterns consistently. The observer will need to know how many different developmental patterns can be discriminated in the end product before the problem of pattern regulation can be addressed at the biochemical level.

Mathematical models based on catastrophe theory (Thom, 1975) and stochastic (Gordon, 1966), autopoietic (Varella et al., 1974), and morphological processes (Dogra, 1978) are now available. Generally such models should eventually relate the biochemistry of growth regulators (Mitchison, 1981) to changes in chromatin, such as modification of the roles of histones and nonhistone chromosomal proteins (Pitel and Durzan, 1980). In the cytoplasm such models should consider the pattern of subcellular organization, e.g., microtubules, cell membranes, and walls (Green, 1980), and relate to sequential events in the cell cycle (Brink, 1962; Braun, 1980; Brown and Dyer, 1972).

From the total set of cellular patterns meaningful subsets may be distinguished so that polarity and symmetry become features of cell clusters especially for those showing internal division and a mother cell wall. The important relation to be stressed is that poor patterns are unorganized and have many alternatives and good patterns show symmetry and polarity and have few alternatives. It follows that the best patterns may be unique or at least highly organized in a sequential series of developmental events. The latter may be detected visually at an early stage through initial tactical displacments of daughter cells into a basic pattern.

The observer will not be able to tell one pattern from another unless the starting point of the sequence is known and the basic pattern recognizable. This is why natural embryonic development in the crop plant should be understood. The de novo occurence of a general symmetry of cellular associations may signify that perhaps sooner or later the temporal processes may, with appropriate stimuli, produce organized patterns regardless of how unorganized, and direct or indirect, the growth pattern was originally.

Our methods must discriminate among divisions of labor, segregation of functions, and the differentiation of cells within some overall basic pattern (Haccius, 1978). In the development of pine cones on mature trees, the location and growth of seeds can be described by a few initial relations involving the origin of the seed habit from an axil cell in the shoot (Thomson, 1945). The metabolism in the developing cone leading to this pattern would require a complex set of equations to describe the dynamics of this type of patterned development.

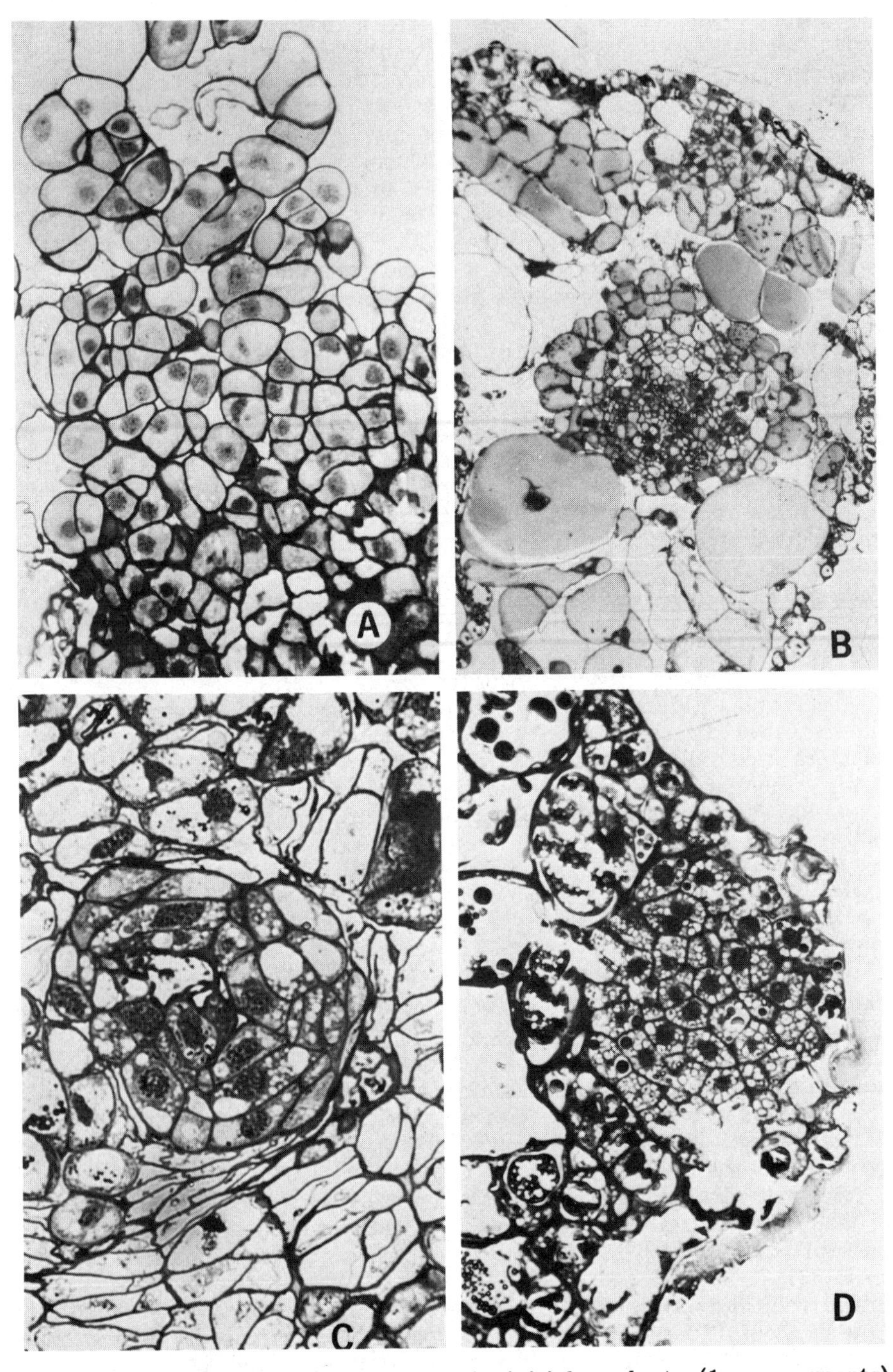

Figure 4. Growth patterns of cells in initial explants (1-cm segments) from juvenile sources of Loblolly pine and Douglas fir. (A) In 3-month-

old stems of Loblolly pine, daughter cells are produced from the surface (epidermal and subepidermal cells) under the influence of 25 µM NOAA and 0.44 µM BAP on a modified MS medium. The cross section shows a callus-like proliferation of cells with large prominent nuclei and with the potential to produce sphaeroblasts in suspension. (B) With the same explant source in A but under the influence of 11.0 µM NAA and 0.91 µM of zeatin and 2,4-D, daughter cells tend to form from the parenchyma and become highly vacuolated. Cells from this source show little potential for organized growth, although callus production can be great. (C) In 2-week-old Douglas-fir cotyledons on a modified MS medium, a sphaeroblast forms in the parenchyma under the influence of 25 µM NOAA and 0.4 µM BAP. With later development, the sphaeroblasts will distinguish themselves with a concentric ring of cambial-like cells and precocious vascular development in the center (cf. Fig. 6). (D) Again with Douglas-fir cotyledons under the same treatment as C, but with another individual explant a meristemoid forms just beneath the epidermis. The meristemoid grows into an adventitious shoot, and the cambial-like development and precocious vascularization is not seen as early as with sphaeroblasts.

With pattern recognition, it is important to be able to distinguish among apomictic and sexual reproductive process. One type of apomictic development relating to rejuvenescence particularly in trees is the formation of sphaeroblasts (Fig. 6). Stoutemeyer (1937) suggested that adventitious shoots from sphaeroblasts could be useful in the vegetative propagation of older fruit trees. Later Dermen (1948, 1953) induced sphaeroblasts by the drastic removal of buds.

According to Baldini and Mosse (1956) sphaeroblasts arise in stems after cambial activity has been inhibited by disbudding. Localized groups of cells in the phloem and cortex grow into globular masses of initially undifferentiated cells with meristematic potential. Sphaeroblasts develop a circumferential cambium and gradually become woody spheroids buried in the bark. Alternatively they can organize new apical meristems and generate adventitious and juvenile shoots and roots.

This phenomenon is not restricted to fruit trees. Figure 6 shows the development of sphaeroblasts from cells of suspension cultures from juvenile sources. In several cases, especially with white spruce sphaeroblasts, a low frequency of rooting (less than 3%) has been achieved (Durzan, unpublished; Durzan et al., 1973).

Organs and plants regenerated from cells and tissues may show considerable phenotypic variation (Went, 1956; Evans, 1980). This problem is discussed elsewhere in this series (see Reisch, Vol. 1, Chapter 25).

Maturity and Selection

Maturity of the "explant source" is a function of the life cycle and is explored in genetic, temporal, organizational and physiological terms. Most forest trees have zones that retain a degree of juvenility longer than other areas (Bonga, 1982; Romberger, 1976). However, in some

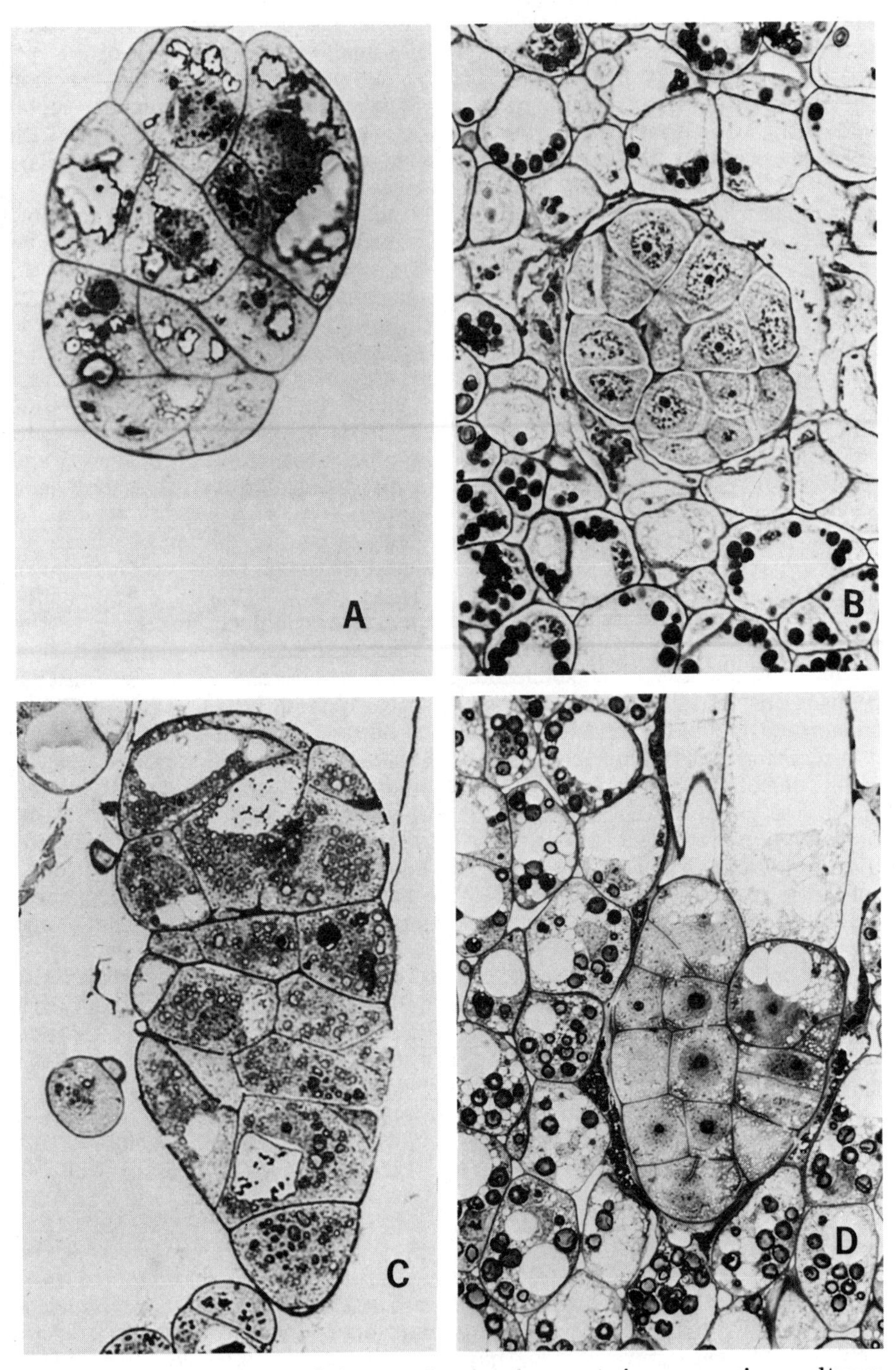

Figure 5. Comparison of the early development in suspension cultures of somatic cells of conifers (A,B) with natural embryogenesis (C,D). (A) Cross sections and B longitudinal sections of Loblolly pine cell clusters,

derived from cotyledonary callus and grown for 4 weeks in a modified MS medium under the influence of a 1/2 MS medium with 25 μM NOAA and 0.44 μM BAP. (B) Early stages of matural embryogenesis in seeds taken from a tree showing (C) cross sections and (D) longitudinal sections of the embryo imbedded in the nutritive (haploid) female gametophyte.

very old trees, senility of meristems can occur (Schaffalitzky de Muckadell, 1959). In others, hundreds of years old, tissues can be rejuvenated or at least invigorated successfully by grafting (Goublay de Nantois, 1980) and by in vitro procedures using auxins, cytokinins and gibberellins or a sequence of treatments (Bonga, 1981).

In this chapter, the correlation that seems most important is that once a genetic gain or trait has been proven in a mature or better yet, a juvenile specimen, how can the gain be selected, captured and cloned? The explant source should reflect this question. In certain cases, indirect selection methods may be required and become useful, especially if the questions raised by Romberger (1976), Visser (1976), and others can be addressed.

Indirect selection is defined as a procedure by which improvement of a desired trait is achieved by selecting the explant source on the basis of another trait which is generally correlated with the desired trait. Here environmental influences are a source of error. Genetic correlation between two traits occurs when their genes are closely linked (cf. Medina-Filho, Vol. 1, Chapter 33), or when one gene is pleiotropic and affects both traits (Flick, Vol. 1, Chapter 2), to give multiple phenotypic effects. Linkage and pleitropy may occur simultaneously.

If trueness to type were dependent on genetic linkages inherent in the explant and assuming no epistasis or interactions among genes, the genetic correlations among traits should be decreased by future breeding processes and by in vitro passages that encourage genetic instability. By contrast, if relationships among genes were due to pleiotropy, the genetic correlations would tend to remain unchanged over several cycles of breeding and selection. Therefore, the most useful markers evident in mature donors are based mainly on pleiotropy (Weissenberg, 1976). In some instances, environmental and physico-chemical factors can contribute to genetic instability and aging (Kuhn, 1940; Langridge, 1963; MacDougall, 1920).

With mature explants, the key is to create through the invigoration of a few cells new centers of physiological activity with the redirected migration of substances (e.g., Mothes and Engelbrecht, 1961; Miller, 1978, 1980). Once new correlations are established, totipotency may come into play. According to Hansche (1982), we now have solid genetic evidence that by applying selection we can effect dramatic reductions in the juvenile period. To do so with woody perennial species in a reasonable time we must apply intense selection pressure on this trait.

Totipotency

Concerning the search for totipotency in cells (cf. Steward and Krikorian, 1979), a simple initial laboratory test has already been

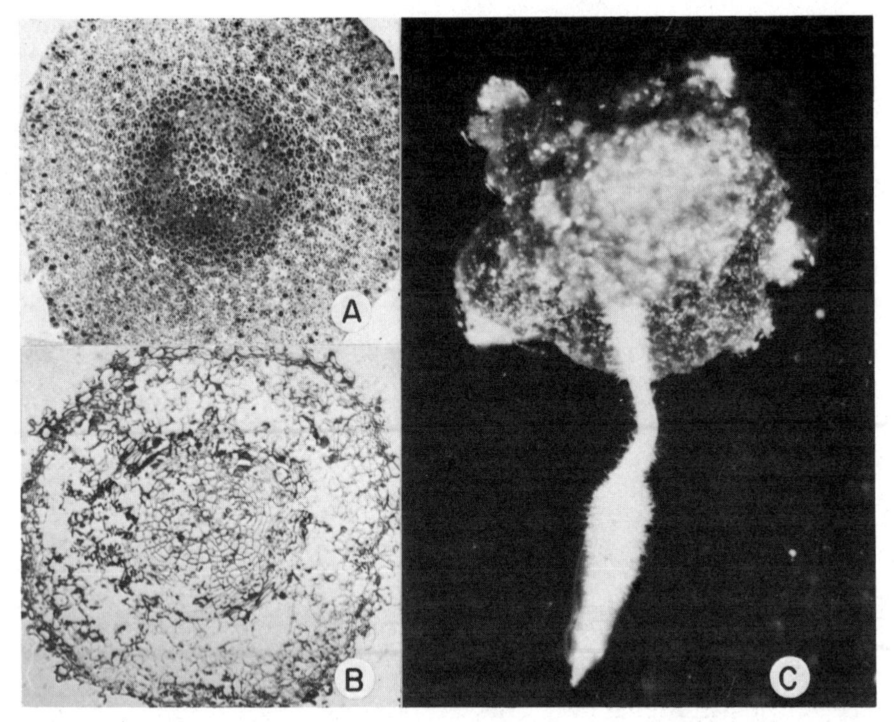

Figure 6. Development in cell suspension cultures of sphaeroblast-like structures from various explants of conifers. (A) Cross section of a true Douglas fir embryo excised from a dry seed for comparative purposes. (B) Cross section of a sphaeroblast of Douglas fir derived from cells taken from a callus of the apical dome from a lateral branch of a 2-year-old tree. Callus was initiated on a MS agar medium with 25 µM NOAA and 0.44 µM BAP. After the seventh monthly subculture, callus was transferred to a liquid culture of a simplified medium (0.25 MS) and lacking nitrate but with reduced nitrogen in the form of an L-amino acid supplement containing the same growth regulators as on agar. After 4 weekly subcultures the cells in suspension were screened to collect the 63-279 µ dia fraction and resuspended in the same medium lacking the synthetic growth regulators. The new medium was supplemented with 4.5 ppm Ca glycerophosphate and a 0.5% v/v aqueous extract from Douglas fir seeds. Two months later the globular sphaeroblasts were transfered to a fresh medium (0.5 MS) containing 0.1 M glycerol and 14.6 mM sucrose. Sections were prepared 5 weeks later. (C) Sphaeroblasts of white spruce derived from callus cells taken from a 1-week-old hypocotyl lacking apical meristems. A root (ca. 1.5 mm) emerged after 3 months of suspension culture in the medium of Durzan et al. (1973). The early development of cells and the effect of fluctuating environmental conditions is described by Durzan et al. (1973) and Chafe and Durzan (1973).

discussed to evaluate the potential of explants in terms of the life cycle (sporophyte, gametophyte), genetics, function, position, and maturity. This test is most useful for simple systems where totipotency is released by exposure of cells to auxins and cytokinins and by specific simple factors.

With recalcitrant species, however, many other problems require attention (cf. Steward and Krikorian, 1979). Levels of applied auxins, cytokinins (Matsubara, 1980), and gibberellins (Engelke et al., 1973; Pharis et al., 1976) used to evoke morphogenesis often tend to be several orders of magnitude higher than endogenous levels of corresponding categories of natural substances. Hence barriers exist (explant size, position, shape, amplified metabolic functions, conjugation, temperature) to diffusion and uptake, especially on agar plates. Although this can be overcome somewhat by using liquid cultures, the mechanism and specificity of action and binding sites for the change agents remain unknown.

In more recalcitrant situations, clearly auxins, cytokinins, and other combinations of growth regulators are not the entire story. In certain species there is more than one way to release totipotency. Apart from a change in individual genetic variation, these may include shifts in nutrients (Durzan, 1982b; Berlyn and Beck, 1980), growth regulators, aeration, cold and dark treatments (Pierik, 1967), and synergistic effects (Steward et al., 1975). Furthermore, while the propagator may not be able to deal directly with the molecular aspects of specificity, additional control over the explant response can be gained through macro element or trace element requirement suited best to the juvenile or adult phase. Controlled specificity over events in the explant can be extended to the enzymatic level for (a) enzymes in which a specific metal has been shown to be an integral component (e.g., nickel ions in amino and amide transferase reactions), or for highly specific proteins that bind elememts such as calcium (calmodulin) or iron (cytochromes) (Dixon et al., 1976; Ibers and Holm, 1980); and (b) enzymes that may have one or more metals as an activator. Categories of enzymes fall into the activities of hydrolases, oxido-reductases, isomerases, and synthetases (Ibers and Holm, 1980).

One category of specific change agent that may regulate totipotency comprises inhibitors of various types. These may include inhibitors of:

1. Enzymes, e.g., α-methylornithine to block ornithine decarboxylase in morphogenetic systems under the influence of 2,4-D (Montague et al., 1979), or phenolics to block IAA oxidase (Stafford, 1974). With proteases or nucleases, trayslol or natural inhibitors (Ryan, 1978) minimize the release of arginine from storage proteins during wounding and hence the production of polyamines (e.g., Montague et al., 1979) via arginine (Durzan, unpublished) or of conjugates of related intermediary metabolites (Montague et al., 1981).
2. Ion intake, such as tetrodotoxin that blocks Na^+ channels to study effect of sodium or fusicoccin to study effect on calcium and proton pumps (Kubowitz et al., 1982).

3. Respiration and growth, such as cytokinins (Miller, 1980; Tuli et al., 1964), a wide range of monosubstituted guanidines (Durzan and Chalupa, 1976c; Durzan and Steward, 1982), and secondary products of glutamine metabolism related to substituted succinic acid (Bonga, 1982; Durzan, 1973, 1976).
4. Polyvinylpyrrolidone (Siegel and Enns, 1979) or charcoal (Carré et al., 1979) may be used to remove excess phenolic products (e.g., Stafford and Cheng, 1980; Miller, 1978) that bind with proteins especially during the error-prone repair of damaged cells. Even in juvenile sources, such as conifer seeds, coumarin is an inhibitor of germination that should be washed out, removed or bound by a heavy metal (Hatano, 1967). The role of complex carbohydrates in controlling morphogenesis at the cell wall remains unknown.
5. Cell cycle activities, such as the purinones (Braun, 1980), or signaling at cell membranes (Raff, 1979).

Morphogenesis in explants may involve the production of inhibiting agents (Richards, 1969). Study of this mechanism calls for certain assumptions. It is often supposed that one region of the explant will develop fastest and will develop into some specific structure. In doing so the structure produces some substance, which can diffuse into the surroundings. This substance, when high enough in concentration, can bring the original diffusion process to an end and impede the tendency in neighboring cells to develop or differentiate in the same direction. The limitation of this viewpoint alone is that most evidence suggests that embryonic tissues tend to induce the differentiation of their like (Lang, 1973) rather than to suppress it.

Another approach is to supply cofactors to explants that are no longer supplied by the donor plant (e.g., Virtanen, 1949). These may include auxins and cytokinins, including their conjugates, gibberellins, vitamins, photosynthate, amino acids, seed extracts, and other factors. Even gaseous atmospheres enriched with carbon dioxide or other gases may be used to advantage to control development (Tisserat and Murashige, 1977; Côme, 1971). Here, the key is to provide a controlled rate of supply of these factors under appropriate environmental conditions (Whyte, 1949). Only recently have mechanisms and technologies become available to feed cells slowly (Urquhart, 1981) and to understand their nutrition in continuous feed systems (Ingestad, 1977).

Where the specificity and direction of metabolic events in the explant are not known, the use of isotopic tracers and stable isotopes can assist in the development of baseline information.

PRECONDITIONING AND PREDICTION

With difficult-to-propagate and mature species, the physiological state of the explant may be improved by pretreating the donor plant by: (a) severe removal of buds, pruning, or hedging, (b) multiple grafting onto juvenile root stocks, (c) use of stump sprouts at bole of tree or root suckers, (d) selection of shoot tips after successive cycles of shoot production, (e) acclimatization of plants to in vitro conditions before explants are taken, (f) application of sprays, pastes, or injections of synthetic plant growth regulators, (g) alteration of the balance of vegetative and reproductive growth through girdling, light treatment (cf. Durzan et al., 1979), root pruning, and addition of nitrogenous fertilizers (Barnes and Bengston, 1968) or growth retardants, (h) removal of constraints that involve heat treatment, chilling, photoperiod, and darkness and the removal of inhibitors with hydrogen peroxide, heavy metals, and auxin protectors (cf. Stafford, 1974), and (i) complete change or shift in nutritional status from a simple autotrophic medium to a heterotrophic medium rich in substrates, cofactors, and reduced nitrogen and vice-versa. This may include an initial starvation signal of short duration or the control of enzyme (protease, amylase, and nuclease) activity.

The excision of an explant complicates the interpretations of pretreatment by effects due to changes arising from hydration (MacDougall, 1920), wounding (Mullick, 1977), surface sterilization, and oxidation (Yemm, 1937); and new environmental (Puckaki et al., 1980) and nutritional stimuli (Ames, 1982). Excision of tissues when endogenous levels of growth regulators are high may improve the performance of explants in the test tube. Nevertheless, this shock may evoke an immediate physical and chemical impact, especially on calcium distribution, proton pumps, ion intake, and nucleotide biosynthesis (Kubowicz et al., 1982). In cells of the explant, the initial wounding leaves an enduring physiological trace lasting for several hours that may or may not be transmitted to daughter cells. In some donor trees, a new abscission layer in female bud inflorescences may form when an explant is taken. This is true in the pistachio (Crane and Iwakiri, 1981).

When the physiological state of cells exceeds certain thresholds, the new state may lead to the launching of morphogenesis, but very little is known about such a process. No doubt a close analogy exists between these initial events and phase changes in somatic cells at least in terms of specificity, orderliness, spontaneity, and automatic responses and in relative independence from environmental variables. Success or failure is difficult to predict.

Predictions and concepts of juvenility and maturity deal with time. Although it may seem anomalous to leave out the temporal variable curves of metabolite levels in whole trees and explants can be developed without the time scale. This is done by defining the trajectories of the system (the solution of the two equations describing the concentrations or factor over time) and by producing a phase plane (cf. Minorsky, 1962).

These grapho-analytical methods deal with complex nonlinear variables often associated with sequential developmental events and emphasize relationships between the variables. For example, a phase plane for light and temperature over the course of a year shows hysteresis in the elliptical relationship between the two variables. There is a lag in the rise of temperature of the air or soil based on the quantity of light in that environment. Similarly, in a spruce tree under field conditions, the main storage compound, arginine, and its contribution to the total soluble N of the bud may be correlated strongly with one of the environmental variables (Durzan and Steward, 1983). This curve of hysteresis and metabolic pursuit can be related to phenology and the biochemical stability of the tree (cf. Cunningham, 1963).

Hysteresis in biological systems has a cybernetic significance for the interpretation of mnemic aspects of development and endogenous rhythms (Neumann, 1973). In many cases, the relationship between two variables seen in this fashion reflects the analogy of a simple pendulum moved by a sinusoidal driving force that may be environmental or metabolic. How these relationships and phase planes behave in juvenile and mature trees and in explants from strategic locations still remains to be worked out.

Poincaré commented that most problems are never entirely solved but only "more or less solved." In practice, the understanding of the influence of explant source will not be reached until the problem and its functions, numerous as they may be, can be reduced to tabular form. This will enable the propagator to decide on a specified order of approximations that can be established over a prescribed range of choices and treatments.

CONCLUSION

The selection of an explant is a fundamental and causal step (deterministic and nondeterministic) in clonal propagation and in the genetic engineering of crop plants, particularly for woody perennial species that display stages of juvenility and maturity.

ACKNOWLEDGMENT

I wish to thank The Institute of Paper Chemistry, Appleton, Wisconsin, for permission to use Figures 2, 3, 4, 5, and 6.

KEY REFERENCES

Bonga, J.M. 1982. Phase change and sexual reproduction in relation to vegetative propagation. In: Tissue Culture in Forestry (J.M. Bonga and D.J. Durzan, eds.) pp. 387-412. Martinus Nijhoff/Dr. W. Junk, Amsterdam.

Brink, R.A. 1962. Phase change in higher plants and somatic cell heredity. Q. Rev. Biol. 37:1-22.

Franclet, A. 1981. Rajeunissement et micropropagation des ligneux. Colloque international sur la culture in vitro des essences Forestières. IUFRO Section S2015, Faontainebleau, France, 55-64. Association Forêt-Cellulose 77370, Nangis, France.

Romberger, J.A. 1976. An appraisal of prospects for research on juvenility in woody plants. Acta Hortic. 56:301-317.

REFERENCES

Adler, I. 1974. A model of contact pressure in phyllotaxis. J. Theor. Biol. 45:1-79.

Ames, B. 1982. Mutagens, carcinogens and anticarcinogens. In: Genetic Toxicology. Basic Life Science, Vol. 21 (R.A. Fleck and A. Hollander, eds.) pp. 489-508. Plenum, New York.

Anderson, N.G. 1976. Interactive molecular sites. I. Basic theory. J. Theor. Biol. 60:401-412.

Baldini, E. and Mosse, B. 1956. Observations on the origin and development of sphaeroblasts in the apple. J. Hortic. Sci. 31:156-162.

Barnes, R.L. and Bengston, G.W. 1968. Effect of fertilization, irrigation and cover cropping on flowering and on nitrogen and soluble sugar composition in slash pine. For. Sci. 14:172-180.

Berlyn, G.P. and Beck, R.C. 1980. Tissue culture as a technique for studying meristematic activity. In: Control of Shoot Growth in Trees (C.H.A. Little, ed.) Proceedings IUFRO Workshop, Xylem and Shoot Physiology, pp. 305-324. Maritimes Forest Research Centre, Fredericton, New Brunswick.

Bidwell, R.G.S. and Durzan, D.J. 1975. Some recent aspects of nitrogen metabolism. In: Historical and Current Aspects of Plant Physiology. A Symposium Honoring F.C. Steward (P. Davies, ed.) pp. 162-227. New York State College of Agricultural Science, Cornell Univ., Ithaca.

Blackmann, F.F. and Parija, P. 1928. Analytical studies in plant respiration. I. Respiration of a population of senescent ripening apples. Proc. R. Soc. London Ser. B 103:412-445.

Bonga, J.M. 1981. Organogenesis in vitro of tissues from mature conifers. In Vitro 17:511-518.

Braun, A.C. 1980. Genetic and biochemical studies on the suppression and of recovery from the tumorous state in higher plants. In Vitro 16:38-48.

Brill, W.R. 1981. Agriculture microbiology. Sci. Am. 245:199-215.

Brown, R. 1964. Protein synthesis during cell growth and differentiation. In: Meristems and Differentiation. Brookhaven Symp. Biol. 16: 157-169.

______ and Dyer, A.F. 1972. Cell division in higher plants. In: Plant Physiology: An Advanced Treatise (F.C. Steward, ed.) pp. 49-90, Volume 6C, Academic Press, New York.

Bryant, T.R. 1972. Gas exchange in dry seeds: Circadian rhythmicity in the absence of DNA replication, transcription and translation. Science 178:634-636.

Bunge, M. 1961. Causality, chance and law. Am. Sci. 49:432-448.

Campbell, R.A. and Durzan, D.J. 1975. Induction of multiple buds and needles in tissue cultures of *Picea glauca* (Moench) Voss. Can. J. Bot. 53:1652-1657.

______ 1976. The potential for cloning white spruce via tissue culture. In: Proceedings 12th Lake States Forest Tree Improvement Conf., USDA Forest Service Genetics Technical Report NC-26, pp. 158-166, Washington, D.C.

Carré, M., Martin-Tanguy, J., Mussilon, P., and Martin, C. 1979. La culture de méristèmes et la multiplication végétative "in vitro" au service de la pépinière. Institut National de la Recherche Agronomique, INVUFLEC (France) Bulletin Petit Fruits No. 14.

Cass, C.E. 1972. Density-dependent fluctuations in membrane potential in logarithmically growing cells. Exp. Cell. Res. 73:140-144.

Chafe, S.M. and Durzan, D.J. 1973. Tannin inclusions in cell suspension cultures of white spruce. Planta 113:251-262.

Chalupa, V., Durzan, D.J., and Vithayasai, C. 1976. Growth and metabolism of cells and tissue of jack pine (*Pinus banksiana* Lamb.). 2. The quantitative analysis of the growth of callus from hypocotyls and radicles. Can. J. Bot. 54:446-455.

Clarkson, D.T. and Hanson, J.B. 1980. The mineral nutrition of higher plants. Annu. Rev. Plant Physiol. 31:239-298.

Côme, D. 1971. Dégazage des enveloppes séminales lors de leur imbibition. II. Cas des graines de pommier. Physiol. Vég. 9:447-452.

Crane, J.C. and Iwakiri, B. 1981. Morphology and reproduction of pistachio. Hortic. Rev. 3:376-393.

Cunningham, W.J. 1963. The concept of stability. Am. Sci. 51:425-436.

Daie, J. and Wyse, R.E. 1982. Immunoassays—An alternative for the detection and quantification of small molecules in plants. Hortic. Sci. 17:307-310.

Dermen, H. 1948. Chimeral apple sports and their propagation through adventitious buds. J. Hered. 39:235-242.

______ 1953. Periclinal cytochimeras and origin of tissues in stem and leaf of peach. Am. J. Bot. 40:154-158.

Dixon, N.E., Gazzola, C., Blakely, R.L., and Zerner, B. 1976. Metal ions in enzymes using ammonia or amides. Science 191:1144-1150.

Detorok, D. 1968. The cytological and growth characteristics of tumor and normal clones of *Picea glauca*. Cancer Res. 28:608-614.

Dogra, P.D. 1978. Morphology, development and nomenclature of conifer embryo. Phytomorphology 28:307-322.

Durzan, D.J. 1973. Nitrogen metabolism of *Picea glauca*. V. Metabolism of uniformly labelled ^{14}C-L-proline and ^{14}C-L-glutamine by dormant buds in late fall. Can. J. Bot. 51:359-369.

______ 1974. Nutrition and water relations of forest trees: A biochemical approach. In: Proceedings 3rd North American Forest Biology Workshop, Sept. 9-12 (G.P.P. Reid and G.H. Fechner, eds.) pp. 15-63, Colorado State Univ. Press, Fort Collins.

______ 1976. Biochemical changes during development. Acta Hortic. 56:183-194.

______ 1980. Progress and promise in forest genetics. In: Proceedings 50th Anniversary Conference Paper Science and Technology: The Cutting Edge (May 8-10, 1978) pp. 31-60. Institute of Paper Chemistry, Appeton, Wisconsin.

______ 1982a. Cell and tissue culture in forest industry. In: Tissue Culture of Forest Trees (J.M. Bonga and D.J. Durzan, eds.) pp. 36-71. Martinus Nijhoff/Dr. W. Junk, Amsterdam.

______ 1982b. Nitrogen metabolism and vegetative propagation of forest trees. In: Tissue Culture in Forestry (J.M. Bonga and D.J. Durzan, eds.) pp. 256-324. Martinus Nijhoff/Dr. W. Junk, Amsterdam.

_____ and Chalupa, V. 1976a. Growth and metabolism of cells and tissues of jack pine (*Pinus banksiana*). 3. Growth of cells in liquid suspension cultures in light and darkness. Can. J. Bot. 53:456-467.

_____ and Chalupa, V. 1976b. Growth and metabolism of cells and tissues of jack pine (*Pinus banksiana*). 4. Changes in amino acids in callus and in seedlings of similar genetic origin. Can. J. Bot. 54: 468-482.

_____ and Chalupa, V. 1976c. Growth and metabolism of cells and tissues of jack pine (*Pinus banksiana*). 5. Changes in free arginine and Sakaguchi-reactive compounds during callus growth and in germinating seedlings of similar genetic origin. Can. J. Bot. 54:483-495.

_____ and Steward, F.C. 1971. Morphogenesis in cell culture of gymnosperms: Some growth patterns. International Union Forest Research Organizations Section 22 Workshop (May 28-June 5, Helsinki, Finland) Abstr. Int. For. Fenn. 74:16.

_____ and Steward, F.C. 1983. Metabolism of organic nitrogenous compounds. In: Plant Physiology: An Advanced Treatise (F.C. Steward and R.G.S. Bidwell, eds.) Vol. 8, pp. 55-265. Academic Press, New York.

_____, Chafe, S.M., and Lopushanski, S.M. 1973. Effects of environmental changes on sugars, tannins and organized growth in cell suspension cultures of white spruce. Planta 113:241-249.

_____, Chalupa, V., and Mia, A.J. 1976. Growth and metabolism of cells and tissues of jack pine (*Pinus banksiana*). 1. The establishment and some characteristics of a proliferated callus from jack pine seedlings. Can. J. Bot. 54:437-445.

_____, Campbell, R.A., and Wilson, A. 1979. Inhibition of female cone production in white spruce by red light treatment during night under field conditions. J. Exp. Environ. Bot. 19:133-144.

Engelke, A.L., Hamzi, H.Q., and Skoog, F. 1973. Cytokinin-gibberellin regulation of shoot development and leaf form in tobacco plantlets. Am. J. Bot. 60:491-495.

Evans, D.A. 1980. Genetic variability of somatic hybrid plants. Int. Assoc. Plant Tissue Culture Newsl. 33:6-9.

Fersht, A.R. 1981. Enzymatic editing mechanisms and the genetic code. Proc. R. Soc. London Ser. B. 212:351-379.

Fowke, L.C. and Gamborg, O.L. 1980. Applications of protoplasts to the study of plant cells. Int. Rev. Cytol. 68:9-51.

Franclet, A. 1977. Phases du développement et propagation végétative des conifères. Proceedings Conference lors d'une Session de Formation Continué à Versailles, May 11, 1977 AFOCEL. Association Ioret Cellulose, Nangis, France.

Galun, E. 1981. Plant protoplasts as physiological tools. Annu. Rev. Plant Physiol. 32:237-266.

Gautheret, R.J. 1955. Sur la variabilité des propiétés physiologiques des cultures de tissus végétaux. Rev. Gen. Bot. 62:5-112.

Gordon, R. 1966. On stochastic growth and form. Proc. Natl. Acad. Sci. USA 56:1497-1504.

Goublay de Nantois, T. de la 1980. Rajeunissement chez le Douglas (*Pseudotsuga menziesii*) en vue de la propagation végétative. In: Memoire l' Université Pierre et Marie Curie, Paris VI, p. 44.

Green, A.M. 1981. Monoclonal antibodies: Obvious and obscure applications. In: Biotechnology: Present Status and Future Prospects, Proceedings International Conference (June 1-2, 1981, Tarrytown) Chap. 13. R.L. First, Inc., White Plains, New York.

Green, P.B. 1980. Organogenesis: A biophysical view. Annu. Rev. Plant Physiol. 31:51-82.

Greenwood, M.D. 1978. Flowering induced on young Loblolly pine grafts by out-of-phase dormancy. Science 201:443-444.

Greenwood, M.S. and Goldsmith, M.H.M. 1970. Polar transport and accumulation of indole-3-acetic acid during root regeneration by *Pinus lambertiana* embryos. Planta 95:297-313.

Gregory, F.G. and Veale, J.E. 1957. A reassessment of the problems of apical dominance. Symp. Soc. Exp. Biol. 11:1-20.

Gunning, B.E.S. and Hardham, A.R. 1979. Microtubules and morphogenesis in plants. Endeavor 3:112-117.

Gustafsson, A. 1948. Polyploidy, life-form and vegetative reproduction. Hereditas 34:1-22.

Haccius, B. 1978. Questions of unicellular origin of nonzygotic embryos in callus cultures. Phytomorphology 28:74-81.

Hackett, W.P. 1980. Control of phase change in woody plants. In: Control of Shoot Growth Trees (C.H.A. Little, ed.), Proceedings International Union Forest Research Organizations Workshop, Xylem and Shoot Growth Physiology, pp. 275-272. Maritimes Forest Research Centre, Fredericton, New Brunswick.

Hangarter, R.P., Peterson, M.D., and Good, N.E. 1980. Biological activities of indoleacetylamino acids and their use as auxins in tissue culture. Plant Physiol. 65:761-767.

Hansche, P.E. 1982. Response to selection. In: Methods for Fruit Breeding (J. Janick and R. Moore, eds.) pp. 154-171. Purdue Univ. Press, West Lafayette, Indiana.

______ and Beres, W. 1980. Genetic remodeling of fruit and nut trees to facilitate cultivar improvement. HortScience 15:710-715.

Hatano, K. 1967. Detection of coumarin and o-coumaric acid in seed coats of *Pinus densiflora*. J. Jpn. For. Soc. 49:205-208.

Heslop-Harrison, J. 1972. Sexuality of angiosperms. In: Plant Physiology: An Advanced Treatise (F.C. Steward, ed.) Vol. 6C, pp. 133-289.

Heunnekens, F.M. 1959. Biological reactions. In: Technique of Organic Chemistry (A. Weissberger, ed.) Volume 8, pp. 535-627. Interscience, New York.

Ibers, J.A. and Holm, R.H. 1980. Modeling coordination sites in metallobiomolecules. Science 209:223-235.

Ingestad, T. 1977. Nitrogen and plant growth: Maximum efficiency of nitrogen fertilizers. Ambio 6:146:151.

Jacquiot, C. 1964. Application de al technique de culture des tissue végétaux à l'étude de quelques problèmes de la physiologie de l'arbre. Ann. Sci. For. (Paris) XXI:465 p.

Kester, D.E. 1976. The relationship of juvenility to plant propagation. Int. Plant Prop. Soc. 26:71-84.

______ 1984. The clone in horticulture. Hortic. Sci. (in press).

King, P.J. 1980. Plant tissue culture and the cell cycle. Adv. Biochem. Eng. 18:1-38.

Kubowicz, B.D., Vanderhoef, L.N., and Hanson, J.B. 1982. ATP-dependent calcium transport in plasmalemma preparations from soybean hypocotyls. Plant Physiol. 69:187-191.

Kuhn, W. 1940. Theory of ageing. Adv. Enzymol. 20:432-448.

Lang, A. 1973. Inductive phenomena in plant development. In: Basic Mechanisms in Plant Morphogenesis, Brookhaven National Laboratory Symposium No. 25 (June 4-6) pp. 129-144. Office of Technical Services, Dept. of Commerce, Washington, D.C.

Langridge, J. 1963. The genetic basis of climatic response. In: Environmental Control of Plant Growth (L. Evans, ed.) Symposium Paper 20, Academic Press, New York.

Larson, P.R. 1975. Development and organization of the primary vascular system in *Populus deltoides* according to phyllotaxy. Am. J. Bot. 62:1084-1099.

LaRue, C.D. 1954. Studies on growth and regeneration in gametophytes and sporophytes of gymnosperms. Brookhaven Symp. Biol. 6:187-208.

Libby, W.J. 1974. The use of vegetative propagules in forest genetics and tree improvement. N. Z. J. For. Sci. 4:440-453.

MacDougall, D.T. 1920. Hydration and growth. Carnegie Inst. Washington Publ. 297.

Matsubara, S. 1980. Structure-activity relationships of cytokinins. Phytochemistry 19:2239-2253.

Miller, C.O. 1978. Cytokinin modification of metabolism of p-coumaric acid by a cell suspension of soybean (*Glycine Max* L. Merrill). Planta 140:193-199.

______ 1980. Cytokinin inhibition of respiration in mitochondria from six plant species. Proc. Natl. Acad. Sci. USA 77:4731-4735.

Minorsky, N. 1962. Nonlinear Oscillations. Van Nostrand, Princeton, New Jersey.

Mitchison, G.J. 1981. The polar transport of auxin and vein patterns in plants. Philos. Trans. R. Soc. London Ser. B 295:461-471.

Montague, M.J., Armstrong, T.A., and Jaworski, E.G. 1979. Polymine metabolism in embryonic cells of *Daucus carota*. Plant Physiol. 63: 341-345.

Montague, M.J., Enns, R.K., Siegel, N.R., and Jaworski, E.G. 1981. Inhibition of 2,4-dichlorophenoxyacetic acid conjugation to amino acids by treatment of cultures soybean cells with cytokinins. Plant Physiol. 67:701-704.

Mothes, K. and Engelbrecht, L. 1961. Kinetin-induced directed transport of substances in excised leaves in the dark. Phytochemistry 1: 58-62.

Mullick, D.B. 1977. The nonspecific nature of defense in bark and wood during wounding, insect and pathogen attack. Recent Adv. Phytochem. 11:395-411.

Murashige, T. 1978. The impact of plant tissue culture on agriculture In: Proceedings 4th International Congress Plant Cell and Tissue Culture, August 20-25, 1978 (T.A. Thorpe, ed.) pp. 15-26. Univ. Calgary, Calgary, Alberta.

______ and Nakano, R. 1967. Chromosomal complement as a determinant of the morphogenetic potential of tobacco cells. Am. J. Bot. 54:963-970.

Muratov, Y.M. and Mitrofanov, D.P. 1973. Distribution of cations in differently oriented young pine shoots. Fiziol. Rast. 20:1062-1064.

Murphy, G.J.P. 1980. A reassessment of the binding of naphthalene-acetic acid by membrane preparations from maize. Planta 149:417-426.

Netter, H. 1969. Theoretical Biochemistry, pp. 602-604. Wiley-Interscience, New York.

Neumann, E. 1973. Molecular hysteresis and its cybernetic significance. Agew. Chem. (Int. Ed.) 12:356-369.

Nicholls, J.W.P., Brown, A.G., and Pedrick, L.A. 1974. Wood characteristics of sexually and vegetatively reproduced *Pinus radiata*. Aust. J. Bot. 22:19-27.

Patel, K.R. and Berlyn, G.P. 1982. Genetic instability of multiple buds of *Pinus coulteri* regenerated from tissue culture. Can. J. For. Res. 12:93-101.

Pauling, L. 1974. Molecular basis for biological specificity. Nature (London) 248:769-771.

Pharis, R.P., Ross, S.D., Wample, R.W., and Owens, J.N. 1976. Promotion of flowering in conifers of the *Pinaceae* by certain of the gibberellins. Acta Hortic. 56:155-162.

Pierik, R.L.M. 1967. Regeneration, vernalization and flowering in *Lunaria annua* L. in vivo and in vitro. H. Veenman and Zonen N.V., Wageningen.

Pitel, J.A. and Durzan, D.J. 1978. Chromosomal proteins of conifers. 2. Tissue-specificity of the chromosomal proteins of jack pine (*Pinus banksiana* Lamb.). Can. J. bot. 56:1928-1931.

______ 1980. Chromosomal proteins of conifers. 3. Metabolism of histones and nonhistone chromosomal proteins in jack pine (*Pinus banksiana* Lamb.) during germination. Physiol. Plant. 50:137-194.

Price, H.J., Sparrow, A.H., and Nauman, A.F. 1973. Evolutionary and developmental considerations of the variability of nuclear parameters in higher plants. I. Genome volume, interphase chromosome volume, and estimated DNA content of 236 gymnosperms. In: Basic Mechanisms in Plant Morphogenesis, Brookhaven National Laboratory Symposium No. 25 (June 4-6) pp. 390-421. Office of Technical Services, Dept. of Commerce, Washington, D.C.

Puckaki, P., Giertych, M., and Chalupka, W. 1980. Light filtering function of bud scales in woody plants. Planta 150:132-133.

Raff, M.C. 1979. Cell membranes and cell signalling. Interdiscip. Sci. Rev. 4:140-148.

Rediske, J.H. 1974. The objectives and potential for tree improvement. Yale University School Forestry Environmental Studies Bulletin No. 85, pp. 3-18. New Haven, Connecticut.

______ 1977. Tissue culture in forestry. In: Proceedings 4th North American Tree Biology Workshop (H.E. Wilcox and A.F. Hamer, eds.) pp. 165-171. State University College Environmental Science and Forestry, Syracuse, New York.

Richards, F.J. 1969. The quantitative analysis of growth. In: Plant Physiology: An Advanced Treatise (F.C. Steward, ed.) pp. 3-76. Academic Press, New York.

Richmond, A.E. and Lang, A. 1957. Effect of kinetin on protein content and survival of detached *Xanthium* leaves. Science 125:612-651.

Romberger, J.A. and Gregory, R.A. 1974. Analytical morphogenesis and the physiology of flowering in trees. In: Proceedings Third North American Forest Biology Workshop (C.P.P. Reid and G.H. Fechner, eds.) pp. 132-147. Colorado State Univ., Ft. Collins.

Rowe, J.S. 1964. Environmental preconditioning with special reference to forestry. Ecology 45:399-403.

Russell, E.S. 1930. The interpretations of development and heredity. Oxford Univ. Press, Clarendon and London.

Ryan, C.A. 1978. Proteinase inhibitors in plant leaves. A biochemical model for pest-induced natural plant protection. Trends Biochem. Sci. July:148-150.

Schaffalitzky de Muckadell, M. 1959. Investigations on ageing of apical meristems in woody plants and its importance in silviculture. Forstl. Forspg. Dan. 25:310-455.

Scorza, R. 1982. In vitro flowering. Hortic. Rev. 4:106-127.

Sharp, W.R., Sondahl, M.R., Caldas, L.S., and Maraffa, S.B. 1980. The physiology of in vitro asexual embryogenesis. Hortic. Rev. 2:268-310.

Siegel, N. and Enns, R.K. 1979. Soluble polyvinylpyrrolidine and bovine serum albumin adsorb polyphenols from soyabean suspension cultures. Plant Physiol. 63:206-208.

Sinnott, E.W. 1960. Plant Morphogenesis. McGraw-Hill, New York.

Skirvin, R.M. 1978. Natural and induced variation in tissue culture. Euphytica 27:241-266.

Skoog, F. and Miller, C.O. 1957. Chemical regulation of growth and organ formation in plant tissues cultured in vitro. Soc. Exp. Biol. Symp. 11:118-131.

Stafford, H.A. 1974. Metabolism of aromatic compounds. Annu. Rev. Plant Physiol. 25:459-480.

______ and Cheng, T.-Y. 1980. The procyanadins of Douglas-fir seedlings, callus and cell suspension cultures derived from cotyledons. Phytochemistry 19:131-135.

Steward, F.C. and Durzan, D.J. 1965. The metabolism of organic nitrogenous compounds. In: Plant Physiology: An Advanced Treatise (F.C. Steward, ed.) pp. 379-685. Academic Press, New York.

Steward, F.C. and Krikorian, A.D. 1979. Problems and potentialities of cultured cells in retrospect and prospect. In: Plant Cell and Tissue Culture: Principles and Applications (W.R. Sharp, P.O. Larsen, E.F. Paddock, and V. Raghavan, eds.) pp. 221-262. Ohio State Univ. Press, Columbus.

Steward, F.C., Israel, H.W., Mott, R.L., Wilson, H.J., and Krikorian, A.D. 1975. Observations on the growth and morphogenesis in cultured cells of carrot (*Daucus carota* L.). Proc. R. Soc. London Ser. B 273:33-53.

Stoutemeyer, V.T. 1937. Regeneration in various types of apple wood. Iowa Agric. Home Econ. Exp. Stn. Res. Bull. 220:307-352.

Street, H.E. 1979. Embryogenesis and chemically induced organogenesis. In: Plant Cell and Tissue Culture: Principles and Applications (W.R. Sharp, P.O. Larsen, E.F. Paddock, and V. Raghavan, eds.) pp. 123-153. Ohio State Univ. Press, Columbus.

Swanson, C.P. 1957. Cytology and Cytogenetics. Prentice-Hall, New Jersey.

Szulmajster, J. 1979. Is sporulation a simple model for studying differentiation? Trends Biochem. Sci. January:18-22.

Tamm, C.O. 1982. Nitrogen cycling in undisturbed and manipulated boreal forests in the nitrogen cycle. Proc. R. Soc. London Ser. B. 296:419-425.

Tanford, C. 1978. The hydrophobic effect and the organization of living matter. Science 200:1012-1018.

Thimann, K.V. 1967. Hormone Action in the Whole Life of Plants. Univ. Massachusetts Press, Amherst.

Thom, R. 1975. Structural stability and morphogenesis: An outline of a general theory of models. W.A. Benjamin.

Thompson, D.W. 1963. On Growth and Form. Cambridge Univ. Press.

Thomson, R.B. 1945. Polyembryony, sexual and asexual embryo initiation and food supply. Trans. R. Soc. Can. Sect. 5 39:143-169.

Tisserat, B. and Murashige, T. 1977. Probable identity of substances in *Citrus* that repress asexual embryogenesis. In Vitro 13:785-789.

Tran Tanh Van, K.M. 1981. Control of morphogenesis in in vitro cultures. Annu. Rev. Plant Physiol. 32:291-311.

Tulecke, W. 1965. Haploidy versus diploidy in the reproduction of cell type. Symp. Soc. Dev. Biol. 24:217-241.

Tuli, V., Dilley, D.R., and Wittwer, S.H. 1964. N^6-Benzyladenine: Inhibitor of respiratory kinases. Science 146:1477-1479.

Urquhart, J. 1981. Drug delivery systems. In: Biotechnology: Present Status and Future Prospects. Proceedings of Conference (June 1-2, Tarrytown) pp. 4-1 to 4-12. R.L. First, Inc., White Plains, New York.

Varella, F.G., Maturana, H.R., and Uribe, R.B. 1974. Autopoiesis: The organization of living systems, its characterization and a model. Biosystems 5:187-196.

Virtanen, A.L. 1949. On the role of substances present in the seeds and arising in them during germination in the growth of plants. Experientia 8:313-317.

Visser, J. 1976. A comparison of apple and pear seedlings with reference to the juvenile period. II. Mode of inheritance. Euphytica 25: 339-342.

Waddington, C.H. 1966. Fields and gradients. In: Major Problems in Developmental Biology (M. Locke, ed.) pp. 105-124. Academic Press, New York.

Wardlaw, C.W. 1968. Morphogenesis in plants. Metheun and Co., London.

Weissenberg, K.V. 1976. Indirect selection for improvement of desired traits. In: Modern Methods in Forest Genetics (J.P. Miksche, ed.) pp. 49-77. Springer-Verlag, New York.

Went, F.W. 1956. The role of environment in plant growth. Ann. Sci. 44:378-398.

_____ 1974. Reflections and speculations. Annu. Rev. Plant Physiol. 25:1-26.

Whyte, R.O. 1949. Crop production and environment. Faber and Faber, London.

Wright, J.W. 1976. Introduction to Forest Genetics. Academic Press, New York.

Wright, S. 1968. Evolution and the Genetics of Populations. I. Genetic and Biometric Foundations. Univ. Chicago Press, Chicago.

Yemm, E.W. 1937. Respiration in barley plants. III. Protein catabolism in starving leaves. Proc. R. Soc. London Ser. B 123:243-273.

Zeleny, M. 1977. Self-organization of living systems: A formal model of autopoiesis. Int. J. Gen. Syst. 4:13-28.

SECTION IX
Extractable Products

CHAPTER 18
Date Palm

B. Tisserat

Palms belong to the Arecaceae (Palmae) family which contains over 200 genera and over 2500 species (Corner, 1966; Tomlinson, 1961). Several palm members, such as date (*Phoenix dactylifera* L.), coconut (*Cocos nucifera* L.), and oil palm (*Elaesis quineensis* Jacq.) are widely cultivated for their fruit crop products (Table 1). In addition, numerous lesser known palm crops important to tropical subsistence agricultural communities are utilized (Blatter, 1926; Corner, 1966; Tomlinson, 1961).

The palm family contains several diverse groups that occupy numerous tropical habitats and are described on the basis of distinct anatomical and morphological features (McCurrach, 1960; Moore, 1973; Tomlinson, 1961). Palms are woody perennial monocots. They are characterized as having trunks comprised of cortical layers with dispersed numerous vascular bundles and fibers. Some palms may grow as tall as 100 m while others are subterranean in their growth habitat (Corner, 1966; Tomlinson, 1961). Generally, the palm stem terminates into a single shoot tip that gives rise to alternating leaves with encircling sheath tissue (Moore, 1973). Palms are usually pictured as being solitary stemmed (e.g., coconut). However, some species are branched and others exhibit a cluster appearance through production of additional shoots by suckering (e.g., date). Flowering in palms may be hermaphroditic, monoecious, or dioecious. The inflorescence is usually solitary, emanating from the axil of the leaf. The vast majority of palms are polycarpic, flowering annually for many years; however, some palms are monocarpic, flowering once and dying after fruiting.

Table 1. Major Palm Crop Producing Countries[a]

CROP	COUNTRIES	WORLD PRODUCTION (1000 MT)	WORLD EXPORTS (1000 MT)
Dates	Algeria, Chad, Egypt, India, Iran, Iraq, Israel, Lebanon, Libya, Morocco, Pakistan, Qatar, Saudi Arabia, Spain, Sudan, Syria, Tunisia, United Arab Emirates, United States, Yemen	2622	293
Coconuts	Burma, Dominican Republic, Ecuador, Ghana, India, Indonesia, Ivory Coast, Kenya, Malaysia, Mexico, Mozambique, Nigeria, Papua New Guinea, Philippines, Sri Lanka, Tanzania, Thailand, Vanuatu, Viet Nam	4513 (copra)	430 (copra) 1007 (copra cake) 134 (desiccate fruit) 1136 (coconut oil)
Palm oil	Angola, Bangladesh, Benin, Brazil, Cameroon, China, Columbia, Costa Rica, Ecuador, Ghana, Guinea, Honduras, Indonesia, Ivory Coast, Liberia, Malaysia, Nigeria, Sierra Leone, Surinam, Togo, Zaire	5036	2294
Palm kernels	Angola, Bangladesh, Benin, Brazil, Cameroon, China, Columbia, Ghana, Guinea, Indonesia, Ivory Coast, Liberia, Malaysia, Mexico, Nigeria, Paraguay, Sierra Leone, Surinam, Togo, Zaire	1744	504 (palm kernel cake) 162 (palm nut kernels) 340 (palm kernel oil)
Fiber	Brazil, Columbia, India, Indonesia, Malaysia, Paraguay, Philippines	290 (cori)	—
Palmito	Brazil, Paraguay	—	9.5
Arecanuts	India	140	—
Wax	Brazil	14	—

[a]Statistics obtained from FAO, 1980, 1981; Kitzke and Johnson, 1975; Quast and Bernhardt, 1978.

HISTORY OF THE CROP

Historically, the introduction and utilization of palm crops as important agricultural commodities for western countries has been slow. At the death of Linnaeus in 1778, only 15 palm species were formally recognized. By 1849, 430 palm species were formally described (Blatter, 1926). Since then, palm tree crops such as oil, arecanut, coconuts, and date palms were established on plantations and have increased in agricultural importance to become major agricultural export commodities in several countries (FAO, 1980).

Palms are ubiquitous throughout the tropics and occur to a lesser extent in some subtropical regions (e.g., date palm). Notable palms occurring in nontropical habitats are: *Chamaerops humilis*, which is native to Mediterranean Europe, *Nanophos*, a genus indigenous to the Himalayas, and date palm, a zerophytic species cultivated in the arid subtropical regions of North Africa and the Middle East (Blatter, 1926; Moore, 1973). Palms are native to all continents except Antarctica, and range across the Atlantic, the Indian and Pacific ocean islands, and the Northern and Southern Hemispheres. The majority of palm genera (about 126) are native to the Old World; fewer genera are endemic in the New World (about 79); and several dozen genera exist in both the Old and New Worlds (Moore, 1973). Furthermore, man has introduced a number of palms successfully to new continents and regions (e.g., the introduction of date palm to California from the Middle East and North Africa; and the introduction of the oil palm from Africa to Malaysia and Indonesia) (Moore, 1973). South America has the highest concentration of naturally occurring palm species—1200, of which 500 are native to Brazil (Markley, 1956).

The purpose of this review is twofold: first, to familiarize the reader with the economic importance of palms to world agriculture, mentioning problems involved in their cultivation; and second, to introduce plant cell culture techniques as practical tools to aid in studying and propagating palms.

ECONOMIC IMPORTANCE

After the grass family, the Arecaceae is the second most important and useful plant family to man (McCurrach, 1960). Palms have had a myriad of economic and agricultural uses throughout man's history (Blatter, 1926; Goor, 1967; McCurrach, 1960; Zohary and Spiegel-Roy, 1975). Date palm has the distinction of being one of the oldest cultivated trees (Zohary and Spiegel-Roy, 1975). Because of their indispensible utilization in the economy and domestic life of inhabitants of palm-growing countries, certain of the palms have been called the "tree of life"—e.g., coconut (Martin, 1978), wax (Johnson, 1972), and date (Popenoe, 1913). The entire tree of several palms, such as the coconut and date palm, is employed to provide food, shelter, fuel, fiber, clothing, furniture, and other miscellaneous implements and products.

The advent of global western civilization, and with it the mass introduction of new crops to all parts of the world since the 1800s, has severely reduced the importance of native crops (such as palms) to tropical subsistence farming. However, new importance is now placed on palm fruit and foliage crops in place of the multiple-utilization palm tree. Emphasis is placed on improvement of specific palm crops that provide commodities useful for domestic and international exploitation. The oil palm, particularly, has evolved into a plant with a bright future in the world's vegetable oil scene (Eber, 1978; Khera, 1976; Packard, 1974).

Unlike other dicot fruit trees in which the fruit are individually derived from separate flowers and occur scattered over the foliage of the tree, palm inflorescences occur in compact bunches terminating at the ends of inflorescence buds and separated from the vegetative leaves. Palm fruits are either berries or drupes. Palm flowers always occur in multiple branched inflorescences with either simple or compound spikes. Most palms produce fruit annually. The inflorescence arrangement in palms aids in its ease in harvesting.

Table 1 summarizes the countries that grow and export various palm products. Currently, palms are being utilized as a source of food, fiber, fuel, timber, building materials, ornamentals, wax, medicine, paper, sugar, wine, dyes, and vegetable oils (Alston, 1973; Deppe and Hoffmann, 1977; De Ramecourt, 1976; Eber, 1978; Hawkes and Robinson, 1979; Hodge, 1975; Johnson, 1972; Kitzke and Johnson, 1975; Lyman, 1972; Markley, 1956; Martin, 1978; McCurrach, 1960; Quast and Bernhardt, 1978). In terms of important international trade commodities, coconuts, dates, palm kernels, and vegetable oil are the major exports from palm-growing countries, and these crops also provide staple food. Palm crops giving rise to waxes, fibers, palmito (palm hearts), and arecanuts are minor export commodities, provided by a number of tropical countries. Production and domestic consumption of many palm crop products by native populations within tropical regions cannot be accurately estimated, owing to the intimate relationship between palms and their growers.

In terms of acreage planted and production, the three largest palm crops are oil palm, coconuts, and date palm. In 1979, about 1,878,000 metric tons (MT) of dates were produced, of which 293,266 MT were exported with a U.S. dollar value of $126 million (Anonymous, 1980). Also at this time, 4,410,000 MT of copra was produced from which a total of 1,567,334 MT of copra and coconut oil was derived, worth $1,298 million. World exports of palm oil amounted to 2,294,672 MT valued at $1,452 million in 1979, from a total production of 4,563,300 MT. Currently, the leading three producers of dates, coconuts, and palm oil are as follows: Saudi Arabia, Egypt, and Iraq; Philippines, India, and Indonesia; and Malaysia, Zaire, and Indonesia, respectively (FAO, 1980, 1981).

The palm oil industry has made remarkable strides within the last few decades to become one of the world's major producers of vegetable oils (Khera, 1976; Packard, 1974). In 1971, world production of fats and oils amounted to 41 million MT, and palm oil contributed only 4% of this total; by 1980 its contribution was 8%, accounting for 20% of the increase in the world's fats and oils (Khera, 1976). The increase in palm oil production is due to utilization of plantation systems, massive planting of superior yielding hybrids, and recent breakthroughs in manufacturing methods (De Ramecourt, 1976; Packard, 1974). However, palm oil production centers have shifted since the 1960s. African countries like Nigeria and Zaire, once the major sources of palm oil exports, have produced and exported less palm oil within the last two decades owing to civil strife and increased domestic consumption. In contrast, Malaysia has become the leading palm oil exporter, producing almost 60% of the world's total (FAO, 1980). Malaysia has about 2,619,000 acres of oil palm under cultivation currently (Denney, 1973). The trend toward increased production of this type of vegetable oil will continue (Eber, 1978).

Some of the major reasons for oil palm success is its high crop yield per acre and low maintenance and harvesting requirements compared to other oil-producing crops. Oil palm is the most efficient oil-producing crop in the world (Eber, 1978). Mature oil palms annually produce 3000 lb of oil per acre, while soybeans, peanuts and sunflowers, and cottonseed only produce one-tenth, one-fourth, and one-thirteenth of this amount, respectively.

GENETICS AND CROP IMPROVEMENT

The genetics, morphology, morphogenesis, and physiology of palms is somewhat less understood than other fruit-tree crops. Palms have been difficult to study because they are native to tropical regions, have long life cycles, and have diverse and unique growth habits compared to other fruit-producing trees. Only within the last few decades has serious consistent scientific research and methodology been applied to palm agriculture. Problem areas in palm culture deserving the most attention include the study of palm diseases coupled with the breeding of resistant cultivars, development of a rapid propagation method to mass produce superior palm cultivars, and development of new high-yielding hybrids through breeding programs.

Palm breeding is a long-term endeavor (Carpenter, 1979). Selection of high fruit-producing seedlings must await flowering. Most palms do not flower until 3-7 years after the germination of the seed. Further, no viable means exists for identifying male and female progeny in dioecious palm species. Palm breeding may be impaired by seed germination problems too. Interspecific and intergeneric hybridizations in palms often result in nonviable seed due to a defective endosperm (Hodel, 1977). Some palms, such as *Arenga engler* Becc., *Attalea* spp., *Orbignya* spp., and oil palm are noted as having seed that is notoriously slow or difficult to germinate owing to physical or physiological factors (Hodel, 1977; Rabéchault, 1962; Rabéchault et al., 1968; Rabéchault and Ahée, 1966).

Plantation cultivation of palms is most efficient when genetically uniform high-yielding plants are grown (e.g., the date palm industry in California). One of the most outstanding problems in palm cultivation that prevents rapid crop improvement is the lack of an adequate method of asexual reproduction. No means is known to promote adventitious budding or branching in palms on demand. Also, the natural means of vegetative propagation occurs at only a limited scale in palms. In the date palm, for example, offshoots are only produced in the juvenile life cycle, usually during the first 3-7 years of growth; thereafter lateral buds give rise to fruit bunches. Furthermore, only a few offshoots are produced in the lifetime of a palm (e.g., 1-20 offshoots in date palm depending on the cultivar).

Many palms are devoid of any natural means of vegetative propagation and can only be reproduced by seed (e.g., oil and coconut palms) (Davis, 1969; Kiem, 1958). Production of palms by seed is not a satisfactory means of preserving the characteristics of a desired clone. Seedling palms are quite heterozygous and the resulting seedling progeny is a hybrid. In addition, fruit quality produced from seedling progeny is usually inferior for cultivating clones (Carpenter, 1979; Carpenter and Ream, 1976).

Application of growth regulators and attempts to simulate environmental changes on test palms have been unproductive in understanding the physiological mechanisms involved in the control of the juvenile and adult life cycles in palms (Fisher, 1976; Oppenheimer and Reuveni, 1972). Also, the role of growth regulators in anthesis and fruit growth and development is poorly understood in palms.

Palms are plagued with numerous diseases such as lethal yellowing in coconut (a mycoplasma-like organism) and Bayoud, caused by *Fusarium oxysporum* Schlect. var. albedinis (Killian and Maire), Malencon in date palm. Lethal yellowing has seriously disrupted coconut production in the Caribbean region since the 1900s. Since its identification in the Miami area of Florida in 1971 it has killed more than 80% of the coconut palms (McCoy et al., 1976). Bayoud disease first noted in the late 1800s has killed about 10,000,000 of the 15,000,000 date palms in Morocco and western Algeria (Carpenter and Ream, 1976). Replanting devastated areas with resistant varieties of palms is restricted owing to the lack of an adequate means of rapid vegetative propagation. Repopulation of just the existing date acreage in the United States (about 4000 acres) with a superior date clone would require several decades using offshoots.

PALM CROP IMPROVEMENT THROUGH TISSUE CULTURE TECHNIQUES

Propagation of herbaceous ornamentals by cell culture techniques has been described as having the potential to rapidly increase desired clones (De Fossard, 1976; Holdgate, 1977). Similar methodology is sought for the development of a commercial system to clone tree crops in vitro (Bonga, 1977; Button and Kochba, 1977; Zimmerman, 1980). Micropropagation of palms through cell culture techniques may enable the mass production of elite high-yielding or disease-resistant clones in large numbers necessary for plantation conditions (Reuveni and Lilien-Kipnis, 1974).

Palm breeding and genetic studies should benefit from the application of embryo culture to accelerate embryo germination from poor- or slow-sprouting seeds so that seedling progeny can be more quickly evaluated (Abraham and Thomas, 1962; Hodel, 1977; Rabéchault, 1967). Embryo culture may also be employed to germinate unique interspecific or intergeneric hybrids that do not survive in nature (Hodel, 1977).

Tissue culture techniques may be used to study host-pathogen relationships for palm diseases within axenic conditions (Fisher and Tsai, 1979). They could also serve as a model to study palm morphogenesis and physiology (De Mason and Tisserat, 1980; Tisserat, 1982). Potentially, ideal bioassay plant populations can be developed from clonal plantlets to test the effects of growth regulators and environmental stimuli on palm growth. By altering the nutrient media composition and physical environment, the mechanism of flowering and suckering in palms might be elucidated. Transferring this understanding to plants grown in the soil would then be attempted. Thus far, application of chemicals to palms in the field has been performed mainly on seedlings, which are genetically diverse, and without respect to environmental growth conditions (Oppenheimer and Reuveni, 1972).

LITERATURE REVIEW

The earliest palm tissue culture study involved the growth of excised coconut embryos on a modified White's medium containing CW (Cutter and Wilson, 1954). Since then, embryo culture has been explored extensively in the coconut (Abraham and Thomas, 1962; Balaga and De Guzman, 1970; De Guzman and Del Rosario, 1964; Fisher and Tsai, 1978, 1979; Ventura et al., 1966), oil palm (Martin et al., 1972a; Rabéchault, 1962, 1967; Rabéchault and Ahée, 1966; Rabéchault et al., 1968, 1969, 1970b, 1972a,b, 1976; Rabéchault and Cas, 1974), and date palm (Reuveni and Lilien-Kipnis, 1974; Schroeder, 1970; Tisserat, 1979a).

Table 2 summarizes past palm tissue culture studies. The majority of these studies have been performed since 1970, with the bulk of the research conducted on coconut, date, and oil palms. Production of callus from explant tissue has now been reported in 40 species representing 29 genera. Initiation of asexual embryos and embryo-like structures have been preliminarily identified in 11 species. The subsequent formation of free-living palms from asexual embryos has been only documented in 2 species: oil (Corley, 1977; Corley et al., 1976, 1979, 1981; Lioret, 1981) and date palm (Rhiss et al., 1979; Tisserat, 1979a; Tisserat et al., 1981). Production of free-living seedlings from the culture of excised palm embryos has only been reported in four species (Hodel, 1977; Rabéchault et al., 1973; Schroeder, 1970). Production of complete plantlets from cultured shoot tips through formation of adventitious roots has only been reported in oil (Staritsky, 1970) and date palms (Poulain et al., 1979; Reuveni and Lilien-Kipnis, 1974; Rhiss et al., 1979; Tisserat, 1979a). Table 2 depicts the current state of the art for the tissue culture of palms. The list should continue to expand within the next few years as more economically important palms are studied in vitro.

Callus has been produced from a variety of explant sources in palms (Table 2). However, generally only the more meristematic explants

Table 2. Tissue Culture of Arborescent Monocotyledonous Species

PLANT	TISSUE SOURCE	RESPONSE	REFERENCE
Aiphanes caryotifolia (H.B.K.) H. Wendl.	Zygotic embryo	Callus	Author's data[a]
Arecastrum romanzoffianum (Cham.) Becc.	Zygotic embryo	Germination	Author's data
Arenga mindorensis Becc.	Zygotic embryo	Callus	Author's data
Brahea armata (Mart.) Wats (*Erythea armata*)	Zygotic embryo	Callus/asexual embryos	Author's data
B. dulcis (H.B.K.) Mart.	Zygotic embryo	Callus or germination	Author's data
Butia capitata (Mart.) L.	Zygotic embryo	Callus or germination	Author's data
Caryota urens L.	Zygotic embryo	Germination	Wang & Huang, 1976
Chamaedorea costaricana Oerst.	Zygotic embryo	Callus/asexual embryos	Reynolds, 1979; Reynolds & Murashige, 1979
C. humilis (Liebm.) Burret	Zygotic embryo	Callus or germination	Author's data
C. radicalis Mart.	Zygotic embryo	Callus or germination	Author's data
Chelycarpus thindera Wend.	Zygotic embryo	Callus or germination	Author's data
Cocos nucifera L.	Inflorescence	Callus	Eeuwens, 1976; Eeuwens & Blake, 1977
	Leaf	Callus	Eeuwens, 1976, 1978
	Root	Lateral root initiation	Fulford et al., 1976

PLANT	TISSUE SOURCE	RESPONSE	REFERENCE
Cocos nucifera L.	Stem	Callus/roots	Apavatjrut & Blake, 1977; Eeuwens, 1976; Eeuwens & Blake, 1977
	Zygotic embryo	Callus	Fisher and Tsai, 1978; De Guzman et al., 1979
	Zygotic embryo	Germination	Abraham & Thomas, 1962; Balaga & De Guzman, 1970; Cutter & Wilson, 1954; De Guzman & Del Rosario, 1964, 1972; De Guzman et al., 1971; Fisher and Tsai, 1978, 1979; Sajise & De Guzman, 1972; Ventura et al., 1966
Corypha elata L.	Zygotic embryo	Callus or germination	Author's data
Erythea edulis S. Wats.	Zygotic embryo	Callus/asexual embryos or germination	Author's data
Elaeis guineensis Jacq.	Apical tip	Callus/asexual embryos	Corley et al., 1976; Jones, 1974a,b; Rabéchault et al., 1972; Smith & Thomas, 1973
	Apical tip	Leaf/root initiation	Staristky, 1970
	Apical tip	Leaf initiation	Ong, 1977; Rabéchalt et al., 1972
	Inflorescence	Flower development	CELOS, 1971; Ong, 1977

Table 2. Cont.

PLANT	TISSUE SOURCE	RESPONSE	REFERENCE
E. guineensis Jacq.	Leaf	Callus/asexual embryos	Lioret, 1981; Rabéchault & Martin, 1976
	Root	Root elongation/callus	Ong, 1977; Martin et al., 1972a,b
	Root	Callus/asexual embryos	Corley et al., 1976; Jones, 1974a,b; Smith & Thomas, 1973
	Zygotic embryo	Callus/asexual embryos	Jones, 1974b; Rabéchault et al., 1970a; Smith & Jones, 1970; Smith and Thomas, 1973
	Zygotic embryo	Germination	Bouvinet and Rabéchault, 1965; Ong, 1977; Martin et al., 1972b; Rabéchault, 1962, 1967; Rabéchault & Ahée, 1966; Rabéchault et al., 1968; 1969, 1970b, 1972, 1973, 1976; Rabéchault & Cas, 1974
Heterospathe elata Scheff.	Zygotic embryo	Callus or germination	Author's data
Howeia forsteriana Becc.	Zygotic embryo	Callus/asexual embryos	Reynolds, 1979; Reynolds & Murashige, 1979
Livistona decipiens Becc.	Zygotic embryo	Callus/asexual embryos or germination	Author's data

PLANT	TISSUE SOURCE	RESPONSE	REFERENCE
L. merrillii Becc.	Zygotic embryo	Callus	Author's data
L. saribus Merr. ex. A. Cheval	Zygotic embryo	Germination	Author's data
Mascarena lagenicaulis L.	Zygotic embryo	Germination	Wang & Huang, 1976
M. vershaffeltii L.	Zygotic embryo	Germination	Wang & Huang, 1976
Opsiandra maya Cook.	Zygotic embryo	Callus or germination	Author's data
Phoenix canariensis Hort. ex Chabaud.	Zygotic embryo	Callus or germination	Author's data
P. dactylifera L.	Apical tips, lateral buds	Leaf differentiation	El-Hennawy & Walley, 1978; Oppenheimer & Reuveni, 1972; Reuveni et al., 1972; Reuveni & Lilien-Kipnis, 1974; Schroeder, 1970; Tisserat, 1979a,b, 1981a,b
	Apical tips, lateral buds	Leaf differentiation, root production	Oppenheimer & Reuveni, 1972 Poulain et al., 1979; Reuveni et al., 1972; Reuveni & Lilien-Kipnis, 1974; Rhiss et al., 1979; Tisserat, 1979a, 1981a; Tisserat et al., 1979

Table 2. Cont.

PLANT	TISSUE SOURCE	RESPONSE	REFERENCE
P. dactylifera L.	Apical tips, lateral buds	Callus	Oppenheimer & Reuveni, 1972; Reuveni, 1969; Reuveni et al., 1972; Reuveni & Lilien-Kipnis, 1974
	Apical tips, lateral buds	Callus/asexual embryos	Tisserat, 1979a,b, 1981b, 1982; Tisserat & De Mason, 1980; Tisserat et al., 1979, 1981
	Flower buds, inflorescence	Anthesis/flower differentiation	De Mason & Tisserat, 1980; Tisserat, 1981a; Tisserat et al., 1979
	Fruit mesocarp	Callus	Sharma et al., 1980
	Inflorescence	Callus/asexual embryos	Reynolds, 1979; Reynolds & Murashige, 1979; Tisserat, 1979a; Tisserat et al., 1979
	Meristele	Callus/asexual embryos	Tisserat, 1979a; Tisserat et al., 1979
	Petiole	Callus/roots	Eeuwens & Blake, 1977
	Petiole	Callus	Sharma et al., 1980
	Polyembryonic embryos	Callus/asexual embryos	Reuveni, 1979
	Root	Plant initiation	Smith, 1975
	Seed	Callus/asexual embryos	Ammar & Benbadis, 1977

PLANT	TISSUE SOURCE	RESPONSE	REFERENCE
P. dactylifera L.	Zygotic embryo	Callus/asexual embryos	Ammar & Benbadis, 1977; Reynolds, 1979; Reynolds & Murashige, 1979; Tisserat, 1979a,b, 1981a; Tisserat et al., 1979
	Zygotic embryo	Germination	Oppenheimer & Reuveni, 1972; Reuveni & Lilien-Kipnis, 1974; Reuveni, 1972; Sharma et al., 1980; Tisserat, 1979a, 1981a; Tisserat & De Mason, 1980
P. hamfarra formasana	Zygotic embryo	Callus or germination	Author's data
P. pusilla Gaertn.	Zygotic embryo	Callus/asexual embryos	Author's data
P. reclinata Jacq.	Zygotic embryo	Callus	Author's data
P. sylvestris (L.) Roxb.	Zygotic embryo	Callus/roots or germination	Author's data
Pinanaga copelandii Becc.	Zygotic embryo	Germination	Author's data
Prestoea sp.	Zygotic embryo	Callus/asexual embryos or germination	Author's data
Pritchardia kaalae Rock	Zygotic embryo	Germination	Hodel, 1977

Table 2. Cont.

PLANT	TISSUE SOURCE	RESPONSE	REFERENCE
Rhopalostylis sapida H. Wendl. and Druce	Zygotic embryo	Callus or germination	Author's data
Sabal domingenesis Becc.	Zygotic embryo	Callus or germination	Author's data
S. minor (Jacq.) Pers.	Zygotic embryo	Callus/asexual embryos or germination	Author's data
Thriniax radiata Lodd. ex. Desf.	Zygotic embryo	Callus or germination	Author's data
Trachycarpus fortunei Wendl.	Zygotic embryo	Callus or germination	Author's data
Veitchia joannis H.	Zygotic embryo	Germination	Hodel, 1977
Washingtonia filifera Wendl.	Zygotic embryo	Callus/roots or germination	Author's data
W. robusta Wendl.	Zygotic embryo	Germination	Author's data

[a]Author's data is obtained by the culture of excised zygotic embryos on MS medium with 0.3% activated charcoal for a germination response; or on MS medium with 100 mg/l 2,4-D, 3 mg/l 2iP, and 0.3% activated charcoal to produce callus and sometimes asexual embryos.

give rise to calli that are capable of giving rise to asexual embryos (Tisserat, 1979a). Embryogenic palm callus has been derived from zygotic embryos, lateral buds, shoot tips, and immature inflorescences (Table 2). Embryogenetic palm callus has been shown to be highly heterogenous in nature, composed of nonsynchronously developing embryos surrounded by nondividing accessory tissue (Tisserat and De Mason, 1980) (Fig. 1). Early oil palm somatic embryos were distinguished at the proembryo stage as pearly white globular structures with a well-defined outline (Jones, 1974a). These asexual embryos paralleled zygotic embryos by containing storage proteins and lipid bodies. Through elongation and division these structures acquired polarity and germinated. In date palms, meristematic clusters probably derived from single cells were embedded in the callus, and divided to form meristematic loci that also acquired polarity when transferred to a low-auxin media (Tisserat and De Mason, 1980) (Fig. 2).

Interestingly, the haustorium of asexual embryos is vestigal and doesn't contribute to the absorption of nutrients from the media for the developing embryoid. Zygotic and asexual embryogenic precursors were

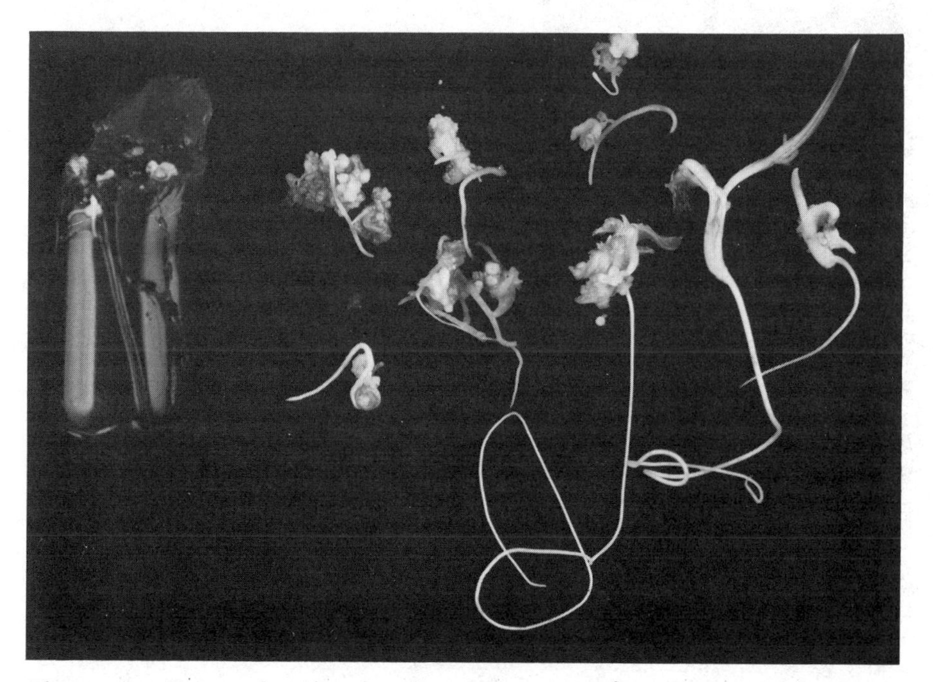

Figure 1. Date palm tissue culture contents after 8 weeks in culture. Note the nonsynchronous growth condition with asexual embryos and plantlets occurring in a variety of developmental states.

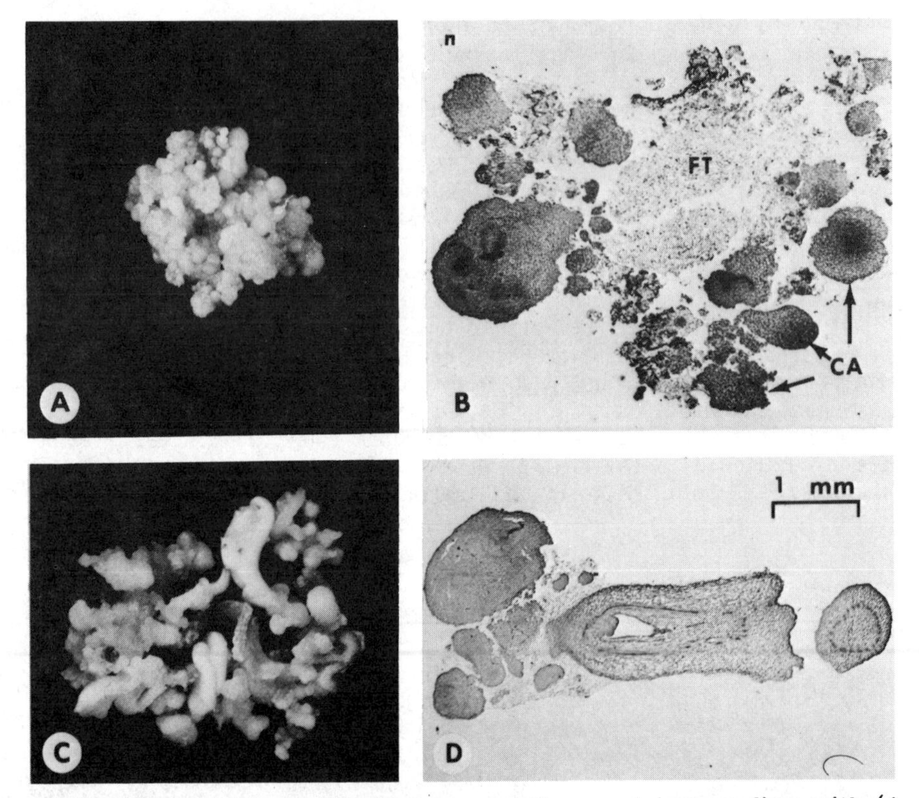

Figure 2. Comparison of date palm growth on nutrient medium with (A and B) and without (C and D) 0.45 mM 2,4-D and 0.0147 mM 2iP included in the nutrient media. (A) Example of date palm culture exhibiting early embryo initiation in vitro after 4 weeks in culture. Culture is composed of nodules that are embryogenetic compact aggregates and loose friable matrix tissue. (B) Transverse section of callus composed of compact aggregates (CA) that are dispersed among friable tissue (FT). (C) Example of callus giving rise to asexual embryos after 4 weeks in culture. Nodular bodies have acquired polarity and are undergoing germination. (D) Transverse section of callus culture and an asexual embryo. Note that the root pole region of the asexual embryo is embedded within the callus.

found to exhibit parallel development to form a seedling (Tisserat and De Mason, 1980) (Fig. 3). Coconut callus derived from somatic tissues, unlike that of oil and date palm, was not embryogenic and was found to consist of an orderly arrangement of cell ranks terminating in loose chains (Apavatjrut and Blake, 1977). However, recently a case of embryogenic coconut callus has been reported from excised embryos (De Guzman et al., 1979).

In palm embryo culture, germination for most cultured species is initiated within the first week. The embryo enlarges, and production

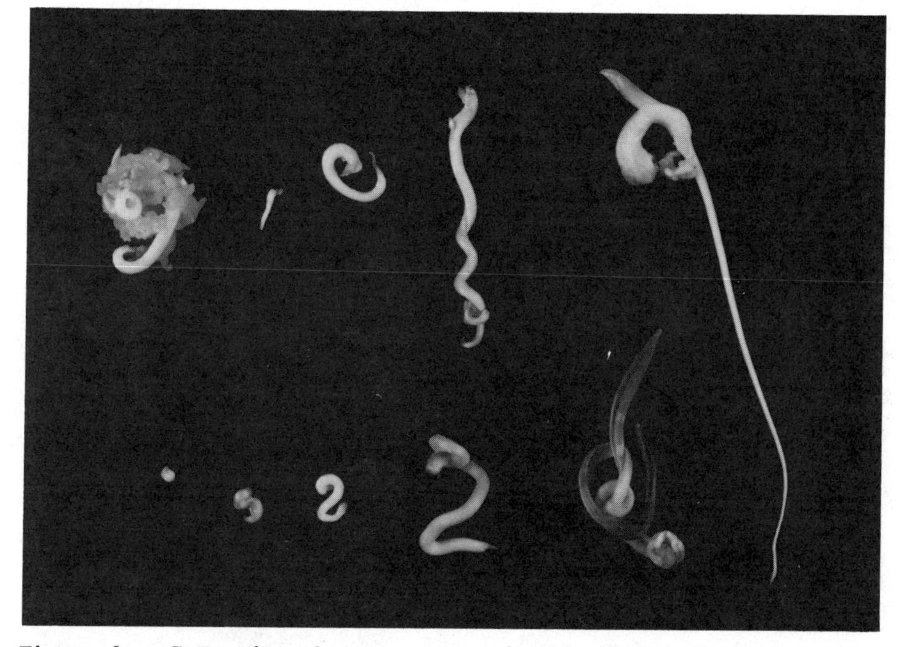

Figure 3. Comparison between asexual and zygotic embryogenesis in vitro. Asexual embryogenesis sequence of development, top, from left to right: Callus mass giving rise to asexual embryos. Early asexual embryo exhibiting cotyledonary elongation. Later cotyledonary elongation stage. Germination of the primary root. Asexual plantlet with first foliar leaves and primary root system. Zygotic embryogenesis sequence of development, bottom, from left to right: Isolated embryo, embryo undergoing cotyledonary elongation. Further cotyledonary elongation stages. Zygotic plantlet with two foliar leaves but with the primary root development suppressed. Note in both sequences of development the cotyledonary haustorium is much reduced in size.

of a primary root system and first foliage leaf is achieved after 4-8 weeks in culture (Hodel, 1977; Tisserat, 1979a; Tisserat and De Mason, 1980; Wang and Huang, 1976) (Fig. 3). Cultured oil palm embryos may require a longer period of time to germinate (Rabéchault, 1962, 1967). Like asexual embryos the cotyledonary haustorium organ is vestigal in excised zygotic embryos in the absence of the endosperm (Figs. 3 and 4). The culture of isolated embryo segments has been performed to study the development and early organogenesis of isolated embryo parts and the interaction of different organs involved in oil palm germination (Rabéchault and Cas, 1974).

Isolated date palm shoot tips and apical meristems have been established in vitro (Fig. 5). Production of adventitious roots was promoted by reculturing these shoots to nutrient media containing 0.54 μM NAA (unpublished data) (Fig. 6). Some additional axillary shoots have been initiated from cultured date palm shoots, but the mechanism of control remains to be elucidated (Fig. 5). Cultured tips exhibit enlargement

Figure 4. Idealized representation of production of asexual plantlets from callus in date palm. From left to right: Callus mass giving rise to embryos. Enlargement of embryos and formation of foliar leaves and primary roots. Isolated asexual embryo. Enlarged single asexual plantlet ready to be transferred to soil conditions.

and show initiation of additional leaves within 2 weeks after planting. After 8 weeks in culture numerous green photosynthetic leaves are prominent (Fig. 5). Rooted date palm tips have been successfully transferred to soil conditions (unpublished data).

Whole immature inflorescence buds and mature and immature individual flower buds have been established in vitro in dates and oil palms (CELOS, 1971; De Mason and Tisserat, 1980; Ong, 1977; Tisserat, 1979a). In the presence of nutrient media containing NAA, date inflorescences and single-flower buds exhibit anthesis; and in some cases male flower buds exhibited outgrowths of prominent vestigial pseudocarpels (De Mason and Tisserat, 1980). Under in vitro conditions cultured flower buds exhibit greening, enlargement, and morphogenesis, but still retain their transitory life, rarely surviving more than a few weeks or months in culture.

CRITICAL FACTORS

In palm tissue culture studies the following requirements must be fulfilled by the nutrient media: first, provide the basic nutrients necessary for explant survival and growth, and second, minimize the browning of explants and media.

Figure 5. Propagation of date palm through shoot tip culture. From left to right: Shoot tip about 2 weeks old undergoing enlargement and leaf initiation. Shoot tip culture giving rise to additional shoots through axillary branch outgrowths. Rooted single shoot ready to be transplanted to soil medium.

Media Components

BASAL MEDIA FORMULATIONS. Development of a characteristic nutrient media unique to a particular plant cultured in vitro is a common practice for plant micropropagation (e.g., orchid, gerbera, and carnation media; Reinert and Bajaj, 1977). A single optimum nutrient medium does not appear to exist for culturing palms. Comparisons between various nutrient media and their effects on particular palm tissues have not been performed. Relatively few critical palm tissue culture studies have been conducted. Eeuwens (1976) has attempted to determine the optimum inorganic media components necessary for coconut explant proliferation. He tested several common mineral formulations including WH, Heller's, MS, and his own Y3 medium on petiole, stem, and inflorescence tissue while keeping the levels of carbohydrate,

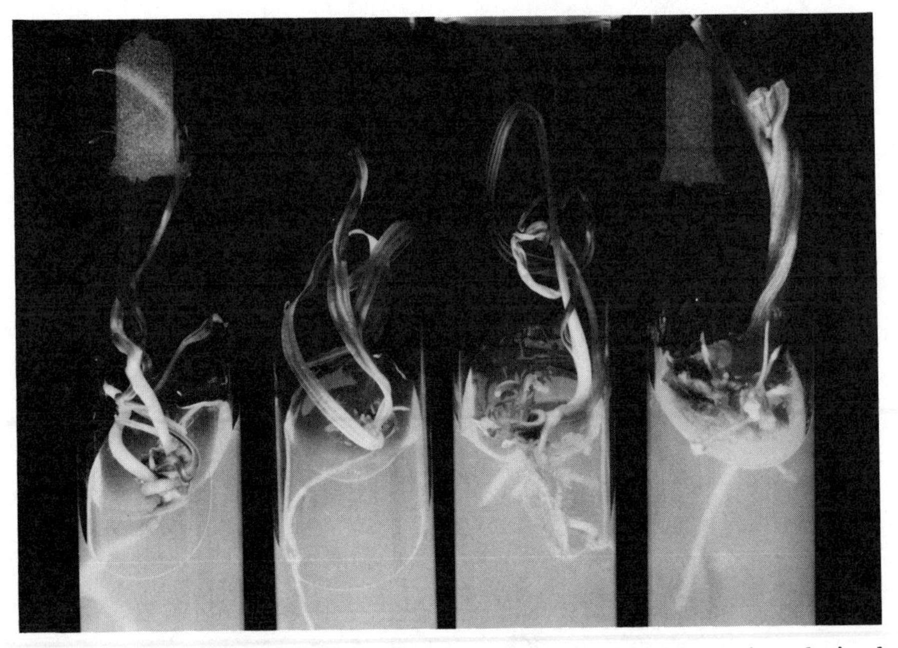

Figure 6. Plantlets produced from single isolated shoot tips derived from seedlings. Plantlets are cultured on a root-producing medium containing 0.5 μM NAA. Note the formation of adventitious roots.

organic, and hormonal additives constant. Best callus growth was obtained from MS and Y3 inorganic formulations, both of which were high in nitrogen. Conversely, media low in nitrate or without ammonium (WH and Heller's formulations) produced substantially less callus growth from cultured explants. Inclusion of iodine enhanced callus production from somatic coconut explants (Eeuwens, 1976). Similarly, Miniano and De Guzman (1978) found that coconut embryos grew better in liquid WH medium when chlorine was added through $MgCl_2$ or KCl additional supplements.

Several investigators have employed a modified WH mineral medium supplemented with CW, auxins, and cytokinins to germinate coconut embryos in vitro (Abraham and Thomas, 1962; De Guzman and Del Rosario, 1964; Ventura et al., 1966). Schroeder (1970) employed a modified WH medium supplemented with IAA and KIN to culture various date palm tissues. Ammar and Benbadis (1977) employed Knob's medium to germinate date palm seeds in vitro.

To obtain callus from somatic date explant tissues a modified MS medium supplemented with an auxin (either 2,4-D or NAA) is suitable (Ammar and Benbadis, 1977; Reuveni and Lilien-Kipnis, 1974). Heller's medium supplemented with CW, 2,4-D, and KIN produced embryonic callus from oil palm embryos (Rabéchault et al., 1970a). Inclusion of activated charcoal to a modified MS medium and increasing the auxin levels to 0.05-0.5 mM allowed various date palm explants to produce prolific callus (Reynolds and Murashige, 1979; Tisserat, 1979a,b).

Reuveni (1979) advocated reducing the inorganic salts of MS medium by half and adding CH, NAA, KIN, IBA, and 2,4-D with activated charcoal to obtain embryogenic callus from excised polyembryonic embryos.

The importance of the carbohydrate source to the survival of excised date palm shoot tips has been documented (Tisserat, 1979a). Date palm explants were able to survive and grow on nutrient media containing only 3% sucrose, regardless of the presence or absence of inorganic salts, vitamins, and hormones; however, the latter ingredients were necessary for complete differentiation in vitro.

COMPLEX UNDEFINED ADDITIVES. Inclusion of various complex addenda to nutrient media to induce growth of palm explants has been reported by several investigators (Abraham and Thomas, 1962; Ammar and Benbadis, 1977; Apavatjrut and Blake, 1977; Balaga and De Guzman, 1970; Cutter and Wilson, 1954; De Guzman and Del Rosario, 1964; Eeuwens, 1976; Fisher and Tsai, 1978; Oppenheimer and Reuveni, 1972; Reuveni and Lilien-Kipnis, 1974; Smith and Thomas, 1973; Wang and Huang, 1976). Germination of coconut embryos has been found to be stimulated by the addition of 5-20% CW (Abraham and Thomas, 1962; Balaga and De Guzman, 1970; Cutter and Wilson, 1954; De Guzman and Del Rosario, 1964; Fisher and Tsai, 1978; Sajise and De Guzman, 1972; Ventura et al.,1966). Wang and Huang (1976) employed a nutrient media containing 15% CW, 0.5% ME and 0.05 or 0.3% activated charcoal to germinate excised embryos of *Mascarena lagenicaulis* L., *M. vershaffletii* L., and *Caryota urens* L. in vitro. However, addition of CW to nutrient media for the germination of date palm embryos is not necessary (Reuveni and Lilien-Kipnis, 1969; Tisserat, 1979a). Further, the author has been able to germinate excised embryos from several palm species on a modified MS medium containing 0.3% charcoal, without hormones or CW. Also, Hodel (1977) germinated excised embryos of *Pritchardia kaalae* Rock and *Veitchia joannis* H. Wendl. on a simple defined media. Coconut- and oil-palm-isolated embryos require special media and treatments for adequate germination. Fisher and Tsai (1978) found that coconut embryos germinate best on a liquid modified MS medium comtaining 57.0 μM IAA, 98.0 μM IBA, 22.0 μM BA, 20% CW, and 12.5-25.0 μM 2iP; transfer to agar medium stimulated rooting. Rabéchault and his colleagues in France critically studied the germination and culture of oil palm embryos in vitro. Aside from the nutrient-media composition, embryo germination was found to be greatly influenced by seed-moisture content, previous duration of seed storage, and the relationship between seed dormancy and water content (Rabéchault, 1967; Rabéchault et al., 1968, 1969).

Production of callus from somatic explants has been achieved on nutrient medium containing CW (Ammar and Benbadis, 1977; Apavatjrut and Blake, 1977; Eeuwens, 1976; Reuveni and Lilien-Kipnis, 1974; Smith and Thomas, 1973). However, prolific callus production from palm explant tissues has also been obtained on various chemically defined media without inclusion of any complex addenda (Tisserat, 1979a,b; Tisserat and De Mason, 1980).

OTHER MEDIA ADDITIVES. From the available information, the influence of vitamins on production of callus or growth from palm tissues is uncertain. However, inclusion of 0.56 mM i-inositol and 1.2 μM thiamine•HCl to media has enabled the growth of various palm species in vitro (Reynolds and Murashige, 1979; Tisserat, 1979a; Table 1). Date palm shoot tips have also been grown on a more complex vitamin formulation containing calcium pantothenate (2.0 μM), nicotinic acid (8.1 μM), pyridoxine•HCl (4.9 μM), biotin (0.04 μM), and i-inositol (0.56mM) (Poulain et al., 1979; Rhiss et al., 1979). Addition of various amino acids to nutrient media to stimulate growth of coconut explants has been of little benefit (Apavatjrut and Blake, 1977).

Unquestionably, the most critical nutrient media component for the production of callus is auxin (Reuveni and Lilien-Kipnis, 1974; Tisserat, 1979a). Both the type and concentration of auxin is critical to the induction of callus from somatic palm tissues (Tisserat, 1979a). When activated charcoal is employed in the nutrient medium, the auxin concentration must be raised to extremely high levels such as 0.15-0.5 mM to induce callus formation (Tisserat, 1979a). Auxins and cytokinins are not necessary for the germination of most palm embryos (Rabéchault et al., 1976; Table 1).

ABSORBENTS. Nutrient additives that combat explant and media browning are necessary in the culture of most palm explants in vitro (Reuveni and Lilien-Kipnis, 1974; Tisserat, 1979a). Date tissues are especially susceptible to lethal browning substances discharged from the explant (Al-Mehdi and Hogan, 1979; Oppenheimer and Reuveni, 1972; Reuveni and Lilien-Kipnis, 1974; Tisserat, 1979a). Prominent browning problems have also been noted in the culture of coconuts (Apavatjrut and Blake, 1977; Fisher and Tsai, 1978), oil (Ong, 1977; Smith and Thomas, 1973), and other palms (Wang and Huang, 1976). To combat this browning, activated charcoal has been included in the medium with beneficial results (Fisher and Tsai, 1978, 1979; Oppenheimer and Reuveni, 1972; Reuveni and Lilien-Kipnis, 1974; Tisserat, 1979a,b; Tisserat et al., 1979; Wang and Huang, 1976). To minimize browning from date palm shoots a mixture of substances has been employed in the media, including polyvinylpyrrolidone, adenine, glutamine, and ammonium citrate (Poulain et al., 1979; Rhiss, et al., 1979). Reuveni and Lilien-Kipnis (1974) investigated the antibrowning effects of a number of substances such as ascorbic acid, dihydroxynaphthalene, dimethyl sulfoxide, and polyvinylpyrrolidone but they were all found to be ineffective. Smith and Thomas (1973) advocated excision of browning tissues during culture to prevent this problem. Apavatjrut and Blake (1977) suggested that browning could be eliminated by employing a nutritionally balanced medium.

Explant Sources

A critical variable in the establishment of any tissue culture in vitro is selection of the explant material. Best callus initiation and growth is derived from meristematic regions of the date palm, which includes zygotic embryos, shoot tips, and leafy lateral buds (Tisserat, 1979a,b, 1981a). Meristele, rachillae, petiole, and root and leaf explants, although easily procured, have been of little value in the successful establishment of palm callus with embryogenic potential (Apavatjrut and Blake, 1977; Eeuwens, 1976, 1978; Eeuwens and Blake, 1977; Reuveni and Lilien-Kipnis, 1974; Reuveni et al., 1972; Tisserat, 1979a). Excised zygotic embryos have been the source of embryogenic callus for several palm species including coconut (De Guzman et al., 19, 1979), date (Ammar and Benbadis, 1977; Reynolds and Murashige, 1979; Tisserat, 1979a,b) and oil palms (Rabéchault et al., 1970a; Smith and Jones, 1970; Smith and Thomas, 1973). Both immature (Reynolds and Murashige, 1979) and mature (Tisserat, 1979a,b) date embryos have served as sources of embryogenic calli.

Following germination of the embryo, the totipotency of palm explants derived from seedlings and trees have shown a diminished capacity to produce callus and undergo asexual embryogenesis (Ammar and Benbadis, 1977; Reuveni and Lilien-Kipnis, 1974; Schroeder, 1970). Embryogenic callus has been obtained from the cotyledonary sheath of germinated date palm embryos (Ammar and Benbadis, 1977). Callus has been obtained from stem and leaf tissue of date (Eeuwens, 1978; Eeuwens and Blake, 1977; Sharma et al., 1980) and stem and inflorescence of coconut (Apavatjrut and Blake, 1977; Eeuwens, 1976). However, these calli did not give rise to asexual embryogenic structures. Often explants derived from such specialized organs give rise to a callus that will produce roots only (Eeuwens, 1978; Tisserat, 1979a). Rabéchault and Martin (1976) noted that callus production was obtained from younger leafs. Inflorescences of palms have been cultured in vitro and have continued to exhibit flower development (CELOS, 1971; Tisserat, 1981a; De Mason and Tisserat, 1980). Meristematically active lateral buds with preformed leaves, as well as shoot tips of date palm, have been noted to continue development in vitro through leaf initiation and their subsequent enlargement (El Hennawy and Wally, 1978; Tisserat, 1979a). In contrast, nonmeristematic buds characterized by the absence of leaves rarely survive more than a few weeks in culture (Tisserat, 1979a; Fig. 7).

Contamination Problems

Palm cultures suffer from habitual contamination problems. Contaminants usually occur after planting of explants following surface sterilization. Problems in obtaining adequately surface-sterilized explants have been noted by several palm investigators (Fisher and Tsai, 1978; Reuveni et al., 1972; Smith and Thomas, 1973; Tisserat, 1979a). Several chemical sterilants and treatments have been applied to obtain clean cultures, including alcohol soak and flame treatments (Oppenheimer and Reuveni, 1972; Reuveni and Lilien-Kipnis, 1974; Schroeder, 1970; Smith and Thomas, 1973), soaking with chloramine-T, 8-hydroxyquinoline, peracetic acid, and mercury chloride (Smith and Thomas, 1973), and soaking with sodium hypochlorite solutions (Eeuwens, 1976; Reuveni and

Figure 7. Examples of growth responses obtained from various date palm explant buds, after 8 weeks in culture. From left to right: Undifferentiated bud, exhibiting browning only. Shoot tip culture greening and initiating leaves. Leafy lateral bud greening and initiating leaves. Inflorescence flower bud exhibiting continued flower development and enlargement.

Lilien-Kipnis, 1974; Reynolds and Murashige, 1979; Tisserat, 1979a,b). The most common and effective means to surface sterilize palm explants involves submerging the explant in a sodium hypochlorite solution (0.26–2.6% containing a few drops of emulsifier) for 15–30 min. The explants are then rinsed in sterile water a few times before planting to remove residual disinfectant. Sometimes, a subsequent dip for 5–10 sec in a sodium hypochlorite solution before planting has been found to be advantageous for shoot-tip and lateral-bud cultures (Tisserat, 1979a). Sterile inflorescence tissues can be obtained from unopen spathes without the necessity of any surface-sterilization treatments (Eeuwens, 1976, 1978).

The occurrence of contamination days, weeks, or even months after culture establishment on nutrient media has been reported (Cutter and Wilson, 1954; Fisher and Tsai, 1978; Smith and Thomas, 1973). The source of these contaminants is internal, and occurs despite the method of surface sterilization. Fisher and Tsai (1978) and Smith and Thomas (1973) suggested that antibiotics could be included in the medium to combat such contaminants. However, addition of antibiotics were ineffective against internal contaminants found in date tissues (Reuveni and Lilien-Kipnis, 1974).

Production of Free-living Palms

Free-living palms have been obtained from various palm species through embryo culture, plantlets produced from embryogenic callus, and rooting of shoot tips (Corley, 1977; Corley et al., 1976, 1979, 1981; Hodel, 1977; Jones, 1974b; Lioret, 1981; Poulain et al., 1979; Rhiss et al., 1979; Tisserat, 1979a, 1981a,b, 1982; Figs. 8 and 9).

Corley et al. (1976) reported that poor root systems were common on clonal plantlets derived from oil palm callus. However, enhanced rooting of oil palm plantlets could be achieved by altering the lighting conditions during culture (Corley, 1977). Similarly, plantlets derived from date palm callus were noted to have poor root systems due to the lack of adventitious roots (Tisserat, 1981a,b, 1982). Culture of date palm plantlets on an agar nutrient medium devoid of charcoal containing 0.005 μM NAA stimulated adventitious root formation. The initial primary root system is unnecessary for further plantlet growth in date and other palms, and should be trimmed to 1-2 cm in length to facilitate easier transfer in vitro. The size and number of leaves is another critical factor in the successful transfer of plantlets to soil (Tisserat, 1981a,b). Date plantlets with 2 or 3 foliar leaves with a shoot length greater than 10 cm, and with a well-developed adventitious root system of 5-10 cm in length, will have nearly a 100% survival rate when potted into free-living conditions. Corley et al. (1976) noted that the

Figure 8. Example of asexual plantlets that have been under free-living conditions for almost 3 months. These palms have the juvenile leaves similar to seedlings.

Figure 9. Asexual palms propagated from callus cultures that are almost 16 months old. Palms are at the adult leaf stage and spines have been produced. These palms may be transferred into the field.

use of antitranspirant on leaves and partial leaf defoliation aids in the successful establishment of oil palm plantlets into soil.

Quality Control Tests

A special note should be made regarding the clonal nature of these asexual plantlets derived from callus. Genetic variation among plantlets propagated from callus has been noted in several other families and should not be overlooked in the palms (D'Amato, 1965; Dulieu, 1972; Ibrahim, 1969; Liu and Chen, 1976; Saalbach and Koblitz, 1977; Shimamoto and Hayward, 1975; Thakur et al., 1976). Variation commonly noted in tissue culture plantlets includes loss or addition of chromosomes, gene mutation, or a combination of these factors. The best means to evaluate the clonal nature of plants derived from tissue culture is to grow them to maturity and compare their vegetative and fruiting characteristics with the parental clone. Already, researchers at the Unilever Corporation (Bedford, United Kingdom) and the Bakawit Clonal Oil Palm Research Unit (Banting, Selangor, Malaysia) have begun to grow experimental oil palm plantlets in the field to determine their clonal nature (Corley et al., 1981). Thus far, fruiting yield variation is less among tissue cultured palms derived from a single clone than from palms of a seedling population. Similarly, the oil palm group located at the Plant Physiology Laboratory, Orston's Central Scientific Departments (Bondy, France) has succeeded in obtaining fruiting oil palms from tissue cultured plantlets, following more than 10 years of study (Rabéchault et al., 1970a; Rabéchault and Martin, 1976; Lioret, 1981). Unfortunately, palms require 3-7 years from the time of seedling germination to fruit production. Once planted in soil, asexual plantlets from oil palm callus require about 2-3 years to fruit. Generally 6-12 months in culture is required to produce oil palm plantlets in vitro from initial somatic explants (Lioret, 1981). This is confirmed in date palm tissue culture research (Tisserat, 1981).

Preliminary chromosome analysis studies in oil palm (2n = 32) has confirmed that asexual plantlets have the same number of chromosomes as the parent clone (Corley et al., 1976; Lioret, 1981). Gene mutations are not apparent in such studies. Tisserat (1981a) has found that isozyme variation among 2 tissue cultures and their parental clones for 7 gene-enzyme systems were negligible. Selection of palms on the basis of vegetative characteristics and gross morphological and biochemical abnormalities may serve to eliminate genetically aberrant palms from growing to maturity.

PROTOCOLS

Palm tissue culture studies can be divided into three separate categories, each with distinct objectives: (a) rapid propagation, (b) embryo culture, and (c) physiological studies of growth and development. Rapid propagation of palms can be achieved either through asexual embryogenesis, i.e., plantlet formation by the initiation and germination of an asexual embryo; or through organogenesis, i.e., sequential formation of roots and shoots from cultured tissues or organs. Rapid propagation of palms is sought to obtain large numbers of genetically uniform trees. Embryo culture is the excision and germination of an isolated embryo, and may be employed to preserve a rare cross or facilitate seedling development which would not occur naturally. Tissue culture techniques provide promising avenues to develop an understanding of heretofore, little known aspects of palm growth, morphogenesis and physiological processes.

Although the Arecaceae family is large and diverse, a general similarity appears to exist regarding the response of palm tissues to in vitro conditions (Table 2). The author has been able to obtain callus and/or germination from excised embryos of numerous palm species by employing techniques and nutrient media developed for the date palm (Figs. 10 and 11). The following protocols should not be taken as absolute, but rather as a workable guide that may be modified through later research efforts for individual palms.

Germination of Palm Embryos

1. Soak seeds for 24-48 hr in tap water to hydrate the embryo and facilitate opening of the seed.

Figure 10. Growth of palm seedlings in vitro after 24 weeks in culture. Seedlings are cultured on nutrient media containing 0.5 µM NAA. Top cultures from left to right: *Arecastrum romanzoffianum*, *Butia capitata*, *Erythea edulis*, *Livistona decipiens*, *Trachycarpus fortunei*, *Brahea armata*, *Phoenix pusilla*, *Prestoea* sp., *Sabal minor*, *P. dactylifera* cv. Sayer. Bottom cultures from left to right: *Opsiandra maya*, *Thirnax radiata*, *L. saribus*, *Corypha elata*, *P. dactylifera* cv. Deglet Nour, *Washingtonia robusta*, *Heterospathe elata*, *W. filifera*.

2. Immerse seeds in a 2.6% sodium hypochlorite solution (containing 1 drop of Tween-20 per 100-ml solution) for about 15-20 min; drain off disinfectant solution and transfer seeds to a 15 x 150 Petri dish. Perform all subsequent operations in a laminar air-flow cabinet using aseptic technique.
3. Using a 7.5-inch-long bayonet forceps select individual seeds from Petri dish for extraction of the embryo. Position seed between thumb and index finger and using an anvil hand cutter (presoaked

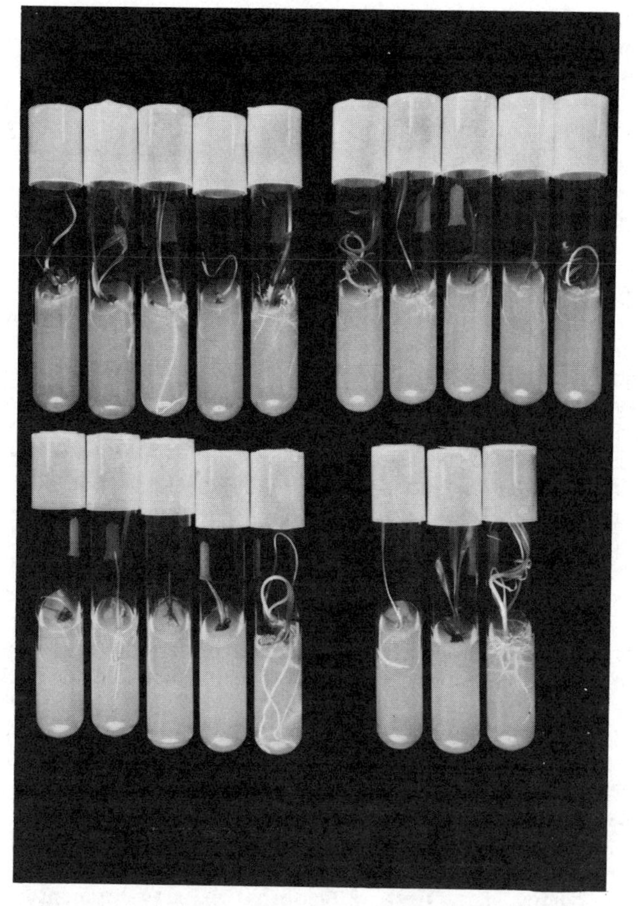

Figure 11. Growth of palm embryos on nutrient media containing .45 mM 2,4-D after 24 weeks in culture. Top from left to right: *Erythea edulis*, *Livistona decipiens*, *Phoenix dactylifera* cv. Sayer, *Sabal minor*, *Prestoea* sp., *P. pusilla*, *Brahea armata*, *P. dactylifera* cv. Delget Nour, *Corypha elata*, *Thrinzx radiata*. Bottom from left to right: *Opsiandra maya*, *Rhopalostylis sapida*, *P. reclinata*, *Washingtonia filifera*, *Heterospathe elata*, *L. merrillii*, *P. sylvestris*, *Butia capitata*.

in 95% ethanol and flame treated) administer a 2 cm longitudinal incision at opposite ends of the furrow. The halved seed will fracture apart allowing the embryo to become exposed without its damage.

4. Using a surgeon's scalpel fitted with a No. 11 surgical blade, remove the exposed embryo from the halved seed either by (a) piercing the haustorium end and the embryo (i.e., the end most embedded within the seed), or (b) lifting the embryo out of the endosperm-embryo cavity by applying pressure, using the blunt side of the scalpel blade.

5. Gently place the excised embryo on the surface of the agar medium. Avoid any further embryo damage and immersion of the embryo into the agar medium.
6. Incubate the embryos on the nutrient medium described in Table 3 (embryo germination medium) under 50 fc light intensity provided by Gro-lux fluorescent lights at 28 C in an environmental chamber.
7. Embryos will begin to enlarge and exhibit germination within 1-2 weeks after planting. After 8 weeks in culture, plantlets will be 2-6 cm in length with a primary root system and first foliar leaf. It may be necessary to reposition the embryo during this first culture transfer so that the root portion is embedded in the agar medium and the leaves grow upwards.
8. To enhance adventitious root formation, reculture seedling with primary root trimmed to 1-2 cm in length to adventitiuos rooting medium (Table 3). Continue reculturing procedure every 8 weeks for 2 or 3 culture passages until seedlings reach a length of 10 cm, with 2-3 leaves and a prolific adventitious root system.
9. To transfer plantlets to free-living conditions, plantlets are carefully removed from agar medium without damage to root system and are soaked in distilled water for 15 min to avoid dehydration and to remove excess adhering media. Plantlets are rinsed 3 times with distilled water, sprayed with 0.5% benolate fungicide solution (DuPont, Wilmington, Delaware) and transferred to soil medium. Soil medium consists of sterile peat moss and vermiculite in a 1:1 v/v ratio. Plantlets are planted in either 3-inch-diameter plastic pots or jiffy peat pots and enclosed within a transparent tent composed of two interlocking clear polystyrene tumblers.
10. Administer weekly applications of 0.5% benolate to the foliage to minimize fungal growth. Water pots every other day with distilled water and once a week with quarter-strength Hoagland's solution during the first 2 months of development. Incubate plants initially in an environmentally controlled chamber under 800 fc light intensity, 16-hr photoperiod at 28 C for 2 weeks. Transfer to a shaded greenhouse. Gradually acclimate plantlets to the greenhouse humidity conditions by punching holes in the plastic cover. After 2 months, covers may be removed and the plant treated as a palm seedling.

Plantlets from Embryo Callus

1. Handle excision and planting of zygotic embryos as described in steps 1-5 of protocol for embryo germination.
2. Incubate embryos on callus-production medium (Table 3) at 28 C in the dark.
3. Reculture explants at 8-week intervals. Callus initiation is evident after 2-3 culture passages.
4. When friable, white, nodular callus is prominent subculture 1-cm^2 pieces to embryo-germination medium (Table 3), and incubate cul-

Table 3. Composition of Various Palm Culture Media

COMPONENTS	MEDIA TYPES			
	Embryo Germination	Callus Production	Shoot Tip	Adventitious Rooting
INORGANIC SALTS				
MS formulation	+	+	+	+
CARBOHYDRATE SOURCE				
Sucrose (M)	0.087	0.087	0.087	0.087
VITAMIN SOURCES				
Meso-Inositol dihydrate (mM)	0.46	0.46	0.46	0.46
Thiamine·HCl (μM)	1.2	1.2	1.2	1.2
COMPLEX ADDENDA				
Phytagar (%)	0.8	0.8	0.8	0.8
Charcoal, activated, neutralized (Sigma) (%)	0.3	0.3	0.3	—
PHYTOHORMONES				
2,4-D (μM)	—	440	44.0	—
2iP (μM)	—	14.5	—	—
NAA (μM)	—	—	—	0.54

tures at 28 C under a 16-hr photoperiod of 50 fc intensity. Asexual embryos and green plantlets usually will become apparent within 2-4 weeks in culture.
5. Follow steps 7-10 of protocol for embryo germination to obtain free-living plantlets.

Plantlets from Shoot Tip and Lateral Bud Callus

1. Dissect offshoots or trees using a hachet and serrated knife. Remove leaves acropetally, exposing lateral buds at the axil of each leaf. Shoot tips are removed from the shoot terminal after all mature leaves are peeled away. Store buds and tips in cold antioxidant solution (0.73 mM citric acid and 0.57 mM absorbic acid). Keep explants in refrigerator at 0 C until surface-sterilization procedure.
2. Trim outermost leaves of buds and tips to obtain explants that are 0.5 cm^2.
3. Sterilize explants by wrapping in cheesecloth to prevent loss in handling procedures and position in 25 x 150 mm culture tube. Sterilize in 2.6% sodium hypochlorite solution (containing 1 drop of Tween-20 per 100 ml solution) for 15 min. Dislodge air bubbles from tissues by periodic agitation of the tube. Pour off bleach solution and rinse 3 times with sterile water. Remove explants and transfer aseptically to the sterile Petri dish (15 x 150 mm diameter).
4. Remove additional leaves from shoot tip and bud explants to obtain a culture that is 1-3 mm^2. A 10-sec dip of this explant into bleach solution may reduce contaminants prior to planting.
5. Plant explant on the surface of callus-production agar medium defined in Table 3.
6. Follow steps 3-5 in the protocol to obtain plantlets from embryo callus to procure free-living palms via callus.

Propagation of Palms via Rooting of Shoot Tips

1. Repeat protocol for excision and planting of shoot tips and lateral bud explants as described in steps 1-4 in protocol to obtain plantlets from shoot tips and lateral bud callus.
2. Plant explants on the surface of shoot tip medium described in Table 3.
3. Incubate cultures under 50 fc intensity of a 16-hr photoperiod at 28 C in an environmental chamber.
4. Cultures will initiate leaves and enlarge considerably in size within the next 4-6 weeks in culture. Reculture explant to fresh media at the end of the 8-week culture passage.
5. Follow steps 8-10 of protocol for germinating palm embryos to root buds and tips and obtain free-living palms.

FUTURE PROSPECTS

Several palms have great agricultural benefit but have not yet been commercially exploited (Table 4). Currently, a number of underexploited palms that are harvested in the wild could provide a substantial crop and source of revenue if they were cultivated under plantation conditions (Balick, 1979; Kitzke and Johnson, 1975). Balick (1979) identified 26 different palm species located within the Amazonia region alone that could be exploited to produce palm oil.

Table 4. Unexploited Palm Species with High Potential Agricultural Benefits

PALM	HABITAT	PRODUCTS	REFERENCE
Astrocaryum tucuma Mart.	Amazonia	Fiber, vegetable oil	Schultes, 1977
Bactris gasipaes H.B.K.	Tropical Central America	Edible food via fruit; vegetable oil	Kitzke & Johnson, 1975
Elaeis oelifera (H.B.K.) Cortes	Amazonia	Vegetable oil	Balick, 1979
Hyphaene natalensis	Subtropical South Africa	Fiber source; food source via sap	Moll, 1971
Manicaria saccifera Gaert.	Amazonia	Fiber, vegetable oil	Balick, 1979
Mauritia vinifera Mart.	Amazonia	Timber, vegetable oil	Balick, 1979
Syagrus coronata (Mart.) Becc.	Amazonia	Wax source via leaf	Kitzke & Johnson, 1975

Exploitation of palms has been hampered by the lack of an available mass-cloning technique. Tissue culture techniques could provide an answer to this problem. It is the author's opinion that palm tree crops are much neglected in terms of understanding their growth, physiology, and biochemistry. Techniques conducted under in vitro conditions would allow these tree crops to be critically studied in order to utilize them more efficiently for man's benefit. Currently, the only tissue-cultured palm close to being propagated commercially is the oil palm (Corley, 1977; Corley et al., 1981; Lioret, 1981). Research is necessary to improve culture techniques for commercial application and to develop a suitable quality control test to verify the clonal nature of callus-derived plantlets. Palm tissue culture research in the future should focus on (a) mass production of palm from callus using an assembly-line system and (b) defining the mechanism of division and rooting of lateral buds and shoot tips in vitro.

Palm tissue culture techniques may be used by other scientific disciplines as a tool. Micropropagation may be coupled with cryogenic storage to preserve rare or endangered palm genotypes for future breeding and genetic studies (Tisserat et al., 1981). Germplasm conservation programs with palms are expensive, labor intensive, and often impractical due to the lack of an adequate method of vegetative propagation. Preliminary studies have indicated that date palm callus may be stored at low temperatures with minimum cost and maintenance requirements (Tisserat et al., 1981). Genetic engineering studies of palms using tissue culture techniques such as protoplast fusion and anther culture may rapidly advance palm breeding studies (Eeuwens and Blake, 1977).

ACKNOWLEDGMENTS

Much of this study is the culmination of research endeavors covering the period between June, 1977 and May, 1980 while I was located at the U.S. Date and Citrus Station, Indio, California. I wish to express my deep appreciation for the technical help provided by the staff of the U.S. Date and Citrus Station during this study including J.D. Dennis, V.M. Enriquez, G. Foster, I.R. Huerta, and M.M. McQueen. Special appreciation should be mentioned to J.B. Carpenter, Station Superintendent. Grateful acknowledgment is made to collaborators D.A. De Mason and J.I. Stillman, University of California, Riverside, for histological studies, and B.J. Finkle and J.M. Ulrich, Western Regional Research Center, Albany, California. Thanks are also given to M.D. Nelson and A. Zaid for technical contributions while I was at the Fruit and Vegetable Chemistry Laboratory, Pasadena, covering the period from July, 1980 to June, 1981.

KEY REFERENCES

De Guzman, E.V., del Rosario, A.G., and Ubalde, E.M. 1979. Proliferative growths and organogenesis in coconut embryo and tissue cultures. Philipp. J. Coconut Stud. 7:1-10.

Eeuwens, C.J. and Blake, J. 1977. Culture of coconut and date palm tissue with a view to revegetative propagation. Acta Hortic. 78: 277-286.

Rabéchault, H., Ahée, J., and Guénin, G. 1970a. Colonies cellulaires et formes embryoids obtenues in vitro a partir de cultures d'embryons de palmier a huile (*Elaeis guineensis* Jacq. var. dura Becc.). C. R. Acad. Sci. 270:233-237.

Reuveni, O. and Lilien-Kipnis, H. 1974. Studies of the in vitro culture of date palm (*Phoenix dactylifera* L.) tissues and organs. Volcani Inst. Agric. Res. Div. Sci. Publ. Pam. 145, Bet Dagan, Isreal.

Reynolds, J.F. and Murashige, T. 1979. Asexual embryogenesis in callus cultures of palms. In Vitro 15:383-387.

Tisserat, B. 1979a. Propagation of date palm (*Phoenix dactylifera* L.) in vitro. J. Exp. Bot. 30:1275-1283.

REFERENCES

Abraham, A. and Thomas, K.H. 1962. A note on the in vitro culture of excised coconut embryos. Indian Coconut J. 15:84-88.

Al-Mehdi, A.A. and Hogan, L. 1979. In vitro growth and development of papaya (*Carica papaya* L.) and date palm (*Phoenix dactylifera* L.). Plant Physiol. (Suppl.) 63:100 (Abstr.).

Alston, A.S. 1973. Coconut palm timber: Interim report on properties and potential uses. Suva No. 60:1-9.

Ammar, S. and Benbadis, 1977. Multiplication vegetative du Palmier-dattier (*Phoenix dactylifera* L.) par la culture de tissus de jeunes plantes issues de semis. C.R. Acad. Sci. 284:1789-1792.

Apavatjrut, P. and Blake J., 1977. Tissue culture of stem explants of coconut (*Cocos nucifera* L.). Oleagineux 32:267-271.

Balaga, H.Y. and De Guzman, E.V. 1970. The growth and development of coconut "Makapuno" embryos in vitro. II. Increased root incidence and growth in response to media composition and to sequential culture from liquid to solid medium. Philipp. Agric. 53:551-565.

Balick, M.J. 1979. Amazonian oil palms of promise: A survey. Econ. Bot. 33:11-28.

Blatter, E. 1926. The Palms of British India and Ceylon. Oxford Univ. Press, London, New York, Bombay, Calcutta, Madras.

Bonga, J.M. 1977 Applications of tissue culture in forestry. In: Plant Cell, Tissue and Organ Culture (J. Reinert and Y.P.S. Bajaj, eds.) pp. 93-108. Springer-Verlag, Berlin, Heidelberg, New York.

Bouvinet, J. and Rabéchault, H. 1965. Effets de l'acide gibbèrellique sur les embryons de Palmier a huile (*Elaeis guineensis* Jacq. var. Dura) en culture in vitro. C. R. Acad. Sci. 260:5336-5338.

Button, J. and Kochba, J. 1977. Tissue culture in the citrus industry. In: Plant Cell Tissue and Organ Culture (J. Reinert and Y.P.S. Bajaj, eds.) pp. 70-92. Springer-Verlag, Berlin, Heidelberg, New York.

Carpenter, J.B. 1979. Breeding date palms in California. Date Grow. Inst. Rep. 54:13-16.

______ and Ream, C.L. 1976. Date palm breeding: A review. Date Grow. Inst. Rep. 53:25-33.

Centrum voor Landbonw Kundig onderzuck in Surinam (CELOS) 1971. Tissue culture of the oil palm (*Elaies guineensis* Jacq.). CELOS Bull. 13:10-11.

Corley, R.H.V. 1977. First clonal oil palms planted in the field. Planter (Kuala Lumpur) 53:331-332.

______, Barrett, J.H. and Jones, L.H. 1976. Vegetative propagation of oil palm via tissue culture. In: Malasian International Agriculture Oil Palm Conference, Incorporated Society of Planters, Kuala Lumpur, 1976, pp. 1-7.

______, Wooi, K.C., and Wang, C.Y. 1979 Progress with vegetative propagation of oil palm. Planter, Kuala Lumpur. 55:377-380.

______, Wong, C.Y., Wooi, K.C., and Jones, L.H. 1981. Early results from the first oil palm clone trials. In: Malaysian International Agriculture Oil Palm Conference, Incorporated Society of Planters, Kuala Lumpur, Report No. A-14, pp. 1-27.
from the first oil palm clone trails.

Corner, D.J.H. 1966. Natural History of Palms. Univ. California Press, Berkeley, Los Angeles.

Cutter, V.M. and Wilson, K.S. 1954. Effect of coconut endosperm and other growth stimulants upon the development in vitro of embryos of *Cocos nucifera*. Bot. Gaz. 115:234-240.

D'Amato, F. 1965. Endoploidy as a factor in plant development. In: Proceedings International Conference Plant Tissue Culture (P.R.White and A.R. Grove, eds.) pp. 449-462. McCutchan Press, Berkeley.

Davis, T.A. 1969. Clonal propagation of the coconut. World Crops 21: 253-255.

De Fossard, R.A. 1976. Tissue Culture for Plant Propagators. Department of Continuing Education, University of New England, Armidale, Australia.

De Guzman, E.V. and Del Rosario, D.A. 1964. The growth and development of *Cocos nucifera* L. 'Makapuno' embryos in vitro. Philipp. Agric. 48:82-94.

______ 1972. Tissue culture: A tool in coconut propagation and varietal improvement. Kalikasan Philipp. J. Biol. 1:241-242 (Abstr.).

______, and Eusebio, E.C. 1971. The growth and development of coconut 'Makapuno' embryo in vitro. III. Resumption of root growth in high sugar media. Philipp. Agric. 53:566-579.

De Mason, D.A. and Tisserat, B. 1980. The occurrence and structure of apparently bisexual flowers in the date palm, *Phoenix dactylifera* L. (Arecaceae). Bot. J. Linn. Soc. 81:283-292.

Denny, E.W. 1973. Malaysia's palm oil output and trade continue dramatic climb. Foreign Agric. 11:6-7.

Deppe, H.J. and Hoffman, A. 1977. Zur verwendung von catole-palme fur die herstel-ung von spanplatten. Holz Roh Werkst. 35:91-94.

De Ramecourt, B. 1976. Continuous processing of palm fruit. J. Am. Oil Chem. Soc. 53:256-258.

Dulieu, H. 1972. The combination of cell and tissue culture with mutagenesis for the induction and isolation of morphological developmental mutants. Phytomorphology 22:283-296.

Eber, F. 1978. Palm oil and shortenings as utilized by the baking industry. Proc. Annu. Meet. Am. Soc. Baking Eng. 54:78-84.

Eeuwens, C.J. 1976. Mineral requirements for growth and callus initiation of tissue explants excised from mature coconut palms (*Cocos nucifera*) and cultured in vitro. Physiol. Plant. 36:23-28.

______ 1978. Effects of organic nutrients and hormones on growth and development of tissue explants from coconut (*Cocos nucifera*) and date (*Phoenix dactylifera*) palms cultured in vitro. Physiol. Plant. 42:173-178.

El-Hennawy, H.M. and Wally, Y.A. 1978. Date palm "*Phoenix dactylifera*" bud differentiation in vitro. Egypt. J. Hortic. 5:81-82.

Food and Agriculture Organization. 1980. FAO Monthly Bulletin of Statistics 3(12):1-68.

______ 1981. FAO Monthly Bulletin of Statistics 4(4):32.

Fisher, J.B. 1976. Induction of juvenile leaf form in a palm (*Caryota mitis*) by gibberellin. Bull. Torrey Bot. Club 103:153-157.

______ and Tsai, J.H. 1978. In vitro growth of embryos and callus of coconut palm. In Vitro 14:307-311.

______ and Tsai, J.H. 1979. A branched coconut seedling in tissue culture. Principes 23:128-131.

Fulford, R.M., Justin, S.H.F.W., and Passey, A.J. 1976. Vegetative propagation of coconuts. In: Report of the East Malling Research Station, 1975, Abstr. No. 06009.

Goor, A. 1967. The history of the date through the ages in the holy land. Econ. Bot. 21:320-340.

Hawkes, A.J. and Robinson, A.P. 1979. The utilization of coconut palm timber as an aggregate with cement. Philipp. J. Coconut Stuc. 4:14-26.

Hodel, D. 1977. Notes on embryo culture of palms. Principes 21:103-108.

Hodge, W.H. 1975. Oil-producing palms of the world: A review. Principes 19:119-136.

Holdgate, D.P. 1977. Propagation of ornamentals by tissue culture. In: Plant Cell, Tissue and Organ Culture (J. Reinert and Y.P.S. Bajaj, eds.) pp. 18-43. Springer-Verlag, Berlin, Heidelberg, New York.

Ibrahim, R.K. 1969. Normal and abnormal plants from carrot root tissue cultures. Can. J. Bot. 47:825-826.

Johnson, D. 1972. The Carnauba Wax Palm (*Copernicia prunifera*). III. Exploitation and plantation growth. Principes 16:111-114.

Jones, L.H. 1974a. Plant cell culture and biochemistry: Studies for improved vegetable oil production. In: Industrial Aspects of Biochemistry (B. Spencer, ed.) Vol. 30, Part II, pp. 813-833. North-Holland, Amsterdam, London.

______ 1974b. Propagation of clonal oil palms by tissue culture. Oil Palm News 17:1-9.

Khera, H.S. 1976. Commodity control schemes for palm oil: Is there a need? J. Agric. Econ. Dev. 6:76-86.

Kiem, S.C. 1958. Propagation of palms. Principes 2:133-138, 142.

Kitzke, E.D. and Johnson, D. 1975. Commercial palm products other than oils. Principes 19:3-26.

Lioret, C. 1981. Vegetative propagation of oil palm by somatic embryogenesis. In: Malasian International Agriculture Oil Palm Conference, Incorporated Society of Planters, Kuala Lumpur, Report No. A-13, pp. 1-9.

Liu, M.-C. and Chen, W.-H. 1976. Tissue and cell culture as aids to sugarcane breeding. I. Creation of genetic variation through callus cultures. Euphytica 25:393-403.

Lyman, J.K. 1972. Palm oil production in West Africa: Its role in world palm oil trade. Foreign Agric. 10:8-10.

Markley, K.S. 1956. Mbocayá or Paraguay cocopalm: An important source of oil. Econ. Bot. 10:3-32.

Martin, J.F. 1978. New insights in the preservation of coconut timber: The Tutu insertion process. Aust. For. Res. 8:227-237.

Martin, J.P., Cas, S., and Rabéchault, H. 1972a. Cultures aseptiques de racines de Palmier à huile (*Elaeis guineensis* Jacq. var. dura Becc. et son hybride dura x var. pisifera Becc.). C.R. Acad. Sci. 274:2171-2174.

_____ 1972b. Note prêliminaire sur la culture in vitro de racines de palmier a huile (*Elaeis guineensis* Jacq. var. dura Becc. et son hybride dura x var. pisifera Becc.). Oléagineux 27:303-305.

McCoy, R.E., Thomas, D.L., and Tsai, J.H. 1976. Lethal yellowing: A potential danger to date production. Date Grow. Inst. Rep. 53:4-8.

McCurrach, J.C. 1960. Palms of the World. Harper & Brothers, New York.

Miniano, A.P. and De Guzman, E.V. 1978. Responses of non-makapuno embryos in vitro to chloride supplementation. Philipp. J. Coconut Stud. 3:27-44.

Moll, E.J. 1972. The distribution, abundance and utilization of the Lala palm, *Hyphaene natalensis*, in Tongaland, Natal. Bothalia 10: 627-636.

Moore, H.E., Jr. 1973. The major groups of palms and their distribution. Gentes Herbarum 11:27-141.

Murashige, T. and Skoog, F. 1962. A revised medium for rapid growth and bioassays with tobacco tissue cultures. Physiol. Plant. 15:473-497.

Ong, H.T. 1977. Studies into tissue culture of oil palm. Malasian International Agriculture Oil Palm Conference, Incorporated Society of Planters, Kuala Lumpur, 1976, pp. 9-14.

Oppenheimer, C. and Reuveni, O. 1972. Development of a Method for Quick Propagation of New and Superior Date Varieties. Agricultural Research Organization, The Volcani Center, Bet Daga, Israel.

Packard, R.L. 1974. Drop in Zaire's palm oil output may lead to oil imports. Foreign Agric. 12:7.

Popenoe, P.B. 1913. Date Growing in the Old World and New. West India Gardens, Altadena, California.

Poulain, C., Rhiss, A., and Beauchesne, G. 1979. Multiplication vègètative en culture in vitro du palmier-dattier (*Phoenix dactylifera* L.). C.R. Seances Acad. Agric. Fr. 11:1151-1154.

Quast, D.G. and Bernhardt, L.W. 1978. Progress in Palmito (heart-of-palm) processing research. J. Food Protect. 41:667-674.

Rabéchault, H. 1962. Recherches sur la culture in vitro des embryons de palmier a huile *Elaeis guineensis* Jacq. I. Effets de l'acide β-indole-acetique. Oléagineux 17:757-764.

_____ 1967. Relations entre le comportment des embryons de Palmier a huile (*Elaeis guineensis* Jacq.) en culture in vitro et la teneur en eau des graines. C.R. Acad. Sci. 264:276-279.

_____ and Ahée, J. 1966. Recherches sur la culture in vitro des embryons de palmier a huile (*Elaeis guineensis* Jacq.). III. Effets de lay grosseur et de l'age des graines. Oléagineux 21:729-734.

______ and Cas, S. 1974. Recherches sur la culture in vitro des embryons de palmier á huile (*Elaeis guineensis* Jacq. var. dura Becc.). X. Culture de segments d'embryons. Oléagineux 29:73-78.

______ and Martin, J.P. 1976. Multiplication végétative du palmier á huile (*Elaeis guineensis* Jacq.) á l'aide de cultures de tissus foliaries. C.R. Acad. Sci. 283:1735-1737.

______, Ahée, J., and Guénin, G. 1968. Recherches sur la culture in vitro des embryons de palmier a huile (*Elaeis guineensis* Jacq.). IV. Effets de la teneur en eau des noix et de la durèe de leur stockage. Oléagineux 23:233-237.

______, Guénin, G., and Ahée, J. 1969a. Recherches sur la culture in vitro des embryons de palmier a huile (*Elaeis guineensis* Jacq. var. dura Becc.). VI. Effets de la déshydratation naturelle et d'une réhydration de noix dormantes et non dormantes. Oléagineux 24:263-268.

______, Guénin, G., and Ahée, J. 1970b. Recherches sur la culture in vitro des embryons de palmier à huile (*Elaeis guineensis* Jacq. var. dura Becc.). VII. Comparison de divers milieux minéraux. Oléagineux 25:519-524.

______, Ahée, J., and Guénin, G. 1972a. Recherches sur la culture in vitro des embryons de palmier à huile (*Elaeis guineensis* Jacq. var. dura Becc.). VIII. Action du lait de coco autoclavé en presence ou non de gêlose et de lumiére et en raison de l'âge des graines. Oléagineux 27:249-254.

______, Martin, J.P., and Cas, S. 1972b. Recherches sur la culture des tissus de palmier à huile (*Elaeis guineensis* Jacq.). Oléagineux 27:531-534.

______, Guénin, G., and Ahée, J. 1973. Recherches sur la culture in vitro des embryons de palmier à huile (*Elaeis guineensis* Jacq. var. dura Becc.). IX. Activation de la sensibilite au lait de coco par une réhydratation des graines. Oléagineux 28:333-336.

______, Ahée, J., and Guénin, G. 1976. Recherches sur la culture in vitro des embryons de palmier à huile (*Elaeis guineensis* Jacq.). XII. Effets de substances de croissance a des doses supraoptimales. Relation avec le brunissement des tissus. Oléagineux 31:159-163.

Reinert, J. and Bajaj, Y.P.S. 1977. Applied and Fundamental Aspects of Plant Cell, Tissue, and Organ Culture. Springer-Verlag, Berlin, Heidelberg, New York.

Reuveni, O. 1979. Embryogenesis and plantlets growth of date palm (*Phoenix dactylifera* L.) derived from callus tissues. Plant Physiol. (Suppl.) 63:138 (Abstr.).

______ and Lilien-Kipnis, H. 1969. Date palm. Volcani Inst. Agric. Res. Div. Sci. Publ. Pam. (1960-69), pp. 143-180. Bet Dagan, Israel.

______, Adato, Y., and Lilien-Kipnis, H. 1972. A study of new and rapid methods for the vegetative propagation of date palms. Date Grow. Inst. Rep. 49:17-24.

Reynolds, J.F. 1979. Morphogenesis of palms in vitro. In Vitro 15:210 (Abstr.).

Rhiss, A., Poulain, C., and Beauchesne, G. 1979. La culture in vitro appliquee a la multiplication vegetative du palmier-dattier (*Phoenix dactylifera* L.). Fruits 34:551-554.

Saalbach, G. and Koblitz, H. 1977. Karyological instabilities in callus cultures from haploid barley plants. Biochem Physiol. Pflanz. 171: 469-473.

Sajise, J.U. and De Guzman, E.V. 1972. Formation of adventitious roots in coconut macapuno seedlings grown in medium supplemented with naphthalene acetic acid. Kalikasan Philipp. J. Biol. 1:197-206.

Schroeder, C.A. 1970. Tissue culture of date shoots and seedlings. Date Grow. Inst. Rep. 47:25-27.

Schultes, R.E. 1977. Promising structural fiber palms of the Colombian Amazon. Principes 21:72-82.

Sharma, D.R., Kumari, R., and Chowdhury, J.B. 1980. In vitro culture of female date palm (*Phoenix dactylifera* L.) tissues. Euphytica 29: 169-174.

Shimamto, Y. and Hayward, M.D. 1975. Somatic variation in *Lolium perenne*. Heredity 34:225-230.

Smith, S.N. 1975. Vegetative propagation of the date palm by root tip culture. Bull. Agron. Saharienne 1:67.

Smith, W.K. and Jones, L.H. 1970. Plant propagation through cell culture. Chem. Ind. N.Y. 44:1399-1401.

Smith, W.K. and Thomas, J.A. 1973. The isolation and in vitro cultivation of cells of *Elaeis guineensis*. Oléagineaux 28:123-127.

Staritsky, G. 1970. Tissue culture of the oil palm (*Elaeis guineensis* Jacq.) as tool for its vegetative propagation. Euphytica 19:288-292.

Thakur, S., Ganapathy, P.S., and Jori, B.M. 1976. Differentiation of abnormal plantlets in *Bacopa monnieri*. Phytomorphology 26:422-424.

Tisserat, B. 1979b. Tissue culture of the date palm. J. Hered. 70: 221-222.

_____ 1981a. Date Palm Tissue Culture. USDA/ARS Advances in Agricultural Technology, Western Series, No. 17. Agricultural Res. Ser., Oakland, California.

_____ 1981b. Production of free-living palms through tissue culture. Date Palm J. 1:43-54.

_____ 1982. Factors involved in the production of plantlets from date palm callus cultures. Euphytica 31:201-214.

_____ and De Mason, D.A. 1980. A histological study of the development of adventitive embryos in organ cultures of *Phoenix dactylifera* L. Ann. Bot. 46:465-472.

_____, Foster, G., and De Mason, D. 1979. Plantlet production in vitro from *Phoenix dactylifera* L. Date Grow. Inst. Rep. 54:19-23.

_____, Ulrich, J.M., and Finkle, B.J. 1981. Cryogenic preservation and regeneration of date palm tissue. HortScience 16:47-48.

Tomlinson, P.B. 1961. Anatomy of the monocotyledons. II. Palmae. Clarendon Press, Oxford.

Ventura, P.F., Zuniga, L.C., Figueroa, J.E., and Lazo, F.D. 1966. A progress report on the development of coconut embryo in artificial media. Philipp. J. Plant Ind. 31:81-87.

Wang, P.-J. and Huang, L.-C. 1976. Beneficial effects of activated charcoal on plant tissue and organ cultures. In Vitro 12:260-262.

Zimmerman, R.H. (ed.) 1980. Proceedings of the Conference on Nursery Production of Fruit Plants through Tissue Culture: Applications and Feasibility (April 21-22, 1980, Beltsville), USDA/ARS Agricultural

Research Results, Northeastern Region, ARR-NE-11. Beltsville, Maryland.
Zohary, D. and Spiegel-Roy, P. 1975. Beginnings of fruit growing in the Old World. Science 187:319-327.

CHAPTER 19
Rubber *(Hevea)*

Chen Zhenghua

The Brazilian rubber tree (*Hevea brasiliensis* Muell.-Arg.) belongs to the family Euphorbiaceae. It originates from the tropical rain forests of South America in the Amazon river basin and has been cultivated for only 100 years. Its culture on a large scale was started only 80 years ago.

H. brasiliensis Muell.-Arg. is an interspecific hybrid which is amphidiploid with a basic chromosome number of 9 (x = 9), and with 18 chromosomes in its gametes (n = 18) (Perry, 1943; Ong, 1975).

In 1876 H.A. Wickham collected several basketfuls of seed of *H. brasiliensis* in Brazil at a site along the Tapajoz River. Wickham succeeded in transporting these seeds across the dense forest to Santarem and down the Amazon River 640 km to Belem. From there they were shipped across the Atlantic to the Kew Botanical Gardens in England. Some seeds were still viable upon arrival at Kew, and approximately 2700 seedlings were obtained. The 2000 best seedlings were shipped from Kew to botanical gardens in Ceylon, Malaya, and Java. Some reports state that the bulk of them went to Ceylon and that only 22 seedlings out of the total shipment reached Malaya. Hence, the rubber trees in Southeast Asia—covering millions of hectares—were derived from a very few plants of Wickham's stock that originated on the banks of the Tapajoz (Imle, 1978).

There are nine species in the genus *Hevea* (College of Tropical Crops in South China, 1980):

1. *H. brasiliensis* Muell.-Arg.—the main cultivated species, high-yielding and adaptable.

2. *H. guianensis* Abul.—vigorous growth, but low yielding
   var. marginata Ducke and Schult
   var. lutea Ducke and Schult
3. *H. benthamiana* Muell.-Arg.—resistant to South American leaf blight and *Phytophthora* leaf fall, but superior quality, cold sensitive and low-yielding
4. *H. pauciflora* Muell.-Arg.—middle-sized crown, highly disease resistant.
   var. coriacea Ducke
5. *H. rididifolia* Muell.-Arg.—thin and sparse crown, low-yielding
6. *H. spruceana* Muell.-Arg.—low-yielding, vigorously growing, and high crossability to *H. brasiliensis*
7. *H. nitida* Muell.-Arg.—drought-tolerant, able to grow on sandy soil
   var. toxicodendroides Schult—Frutex, lower than 3m
8. *H. microphylla* Ule—very low-yielding, medium growth vigor
9. *H. camporum* Ducke—dwarf form, tolerant to drought; can be used as parent in breeding for dwarf form.

ECONOMIC IMPORTANCE AND GEOGRAPHICAL DISTRIBUTION

Rubber is one of the main raw materials used in transportation, industry, and national defense. More than 50,000 articles in the world are made purely from rubber. About 600 kg of rubber is needed for an airplane and 68,000 kg for a 35,000 ton warship. It is a good insulator and can be easily manipulated. Rubber is also an airtight and watertight material, and is very useful in making different kinds of waterproof apparatus. Innumerable machine fittings are made from rubber.

The polymerization of isoprene produces polymers having the same overall chemical composition as natural rubber, but physical properties of this synthetic rubber are inferior in every way. The growing understanding of the structural requirements for an elastomer has shown how well natural rubber is designed for this purpose. No method is available to polymerize isoprene to a material identical to natural rubber. It is necessary to mix the synthetic polymers of isoprene with 30-50% natural rubber before use. The complete replacement of natural rubber by synthetic materials has been impossible to date.

Brazilian *Hevea* is a megaphanerophyte in the rain forest, with an average stem girth of about 250 cm and a height of over 30 m. Its native habitats are regions with high temperature (mean annual temperature about 26-27 C) and high humidity (annual rainfall about 2500 mm, relative humidity over 90%). Since being cultivated, *Hevea* has spread widely to Asia and Africa. *Hevea* acreage has reached 8,000,000 ha (Table 1). The annual output of natural rubber is over 3,700,000 tons.

The major natural rubber producing countries in the world are Malaysia, Indonesia, Sri Lanka, China, Thailand, Viet Nam, Kampuchea, India, Burma, Bangladesh, Philippines, Papua New Guinea, Brazil, Mexico, Zaire, Liberia, Nigeria, Cameroon, Central Africa, The Ivory Coast, Ghana, and Singapore.

BREEDING AND CROP IMPROVEMENT PROBLEMS

The tremendous requirement for natural rubber is increasing every year. At the same time, a number of problems have been identified for the improvment of the rubber tree crop.

Yield is the main problem addressed by breeding programs. As a result of selective breeding, the yield of Brazilian *Hevea* has been enhanced from 300 kg/ha of original seedling-plants up to 1000 kg/ha, with a high report of 3000 kg/ha. However, high-yielding strains are not only required for fertile soil, but also for dry land.

Growth vigor has been shown to be a heritable character. Rubber trees with greater growth vigor can be harvested earlier, as they have shorter periods of immaturity and thereby tree turnover, resulting in an overall increase in yield.

In regions with frequent typhoons it is very important to produce wind-resistant strains with small leaves, thin crown, dwarfism, and great trunk and branch strength.

Disease resistance, particularly resistance to brown bark disease, is extremely important. Brown bark is a physiological disease, differing in susceptibility that has been detected between strains. Trees susceptible to brown bark disease have reduced yield.

The primary root diseases are: white root rot, *Rigidoporus lignosus* (Klitzsch) Imazeki; red root rot, *Ganoderma psuedoferreum* (Wakefield) van Over & Stein; and brown root rot, *Phellinus noxius* (Corner) G.H. Cunn. Important stem diseases are pink disease, *Corticium salmonicolor* Berk. & Br.; moldy rot, *Ceratocystis fimbariata* Ell. & Holst.; and black stripe or patch canker, *Phytophthora palmivora* (Butl.) Butl. Finally, the main leaf diseases may be listed as: South American leaf blight, *Microcyclus ulei* (P. Henn) v. Arx; powdery mildew *Oidium heveae* Steinm.; gloeosporium leaf disease, *Glomevella cingulata* (Stonem.) Spauld & Schrenk; bird's eye spot, *Drechslera heveae* (Petch) M.B. Ellis; and so-called *Phytophthora* leaf fall (Wastie, 1975).

Among the above-mentioned diseases breeders pay special attention to leaf diseases, particularly, the South American leaf blight.

POSSIBLE USES OF TISSUE CULTURE TECHNIQUES

Anther Culture

DIRECT UTILIZATION IN BREEDING NEW VARIETIES. The available varieties of the rubber tree are all highly heterozygous in nature. Hence, there is extensive segregation among microspores. It is hoped that pollen plantlets with superior genotypes can be recovered. The homozygous diploid pollen plantlets could thus be utilized directly in crop improvement. The possession of haploid and double haploid plants, on which both dominant and recessive characters would be expressed, considerably shortens the process of selection for desirable characters. This advantage is much more evident in perennial trees such as the rubber tree because of their longer generation. In addition to the advantage of producing various pure lines for use in plant breeding, anther culture offers the possibility of developing aneuploid and heteroploid pollen plants, as well as aneuploid and heteroploid sprouts on the pollen plants, which would serve as extremely valuable tools for improving crops (Fig. 1).

UTILIZATION OF PURE LINES FOR HETEROSIS. It is well known that the heterosis of good combinations of pure lines is much higher than that of crosses between two different varieties. However, 5-7 years are required for *Hevea* from sowing to blooming. In addition, the incidence of inbreeding during seed set is usually only a few out of 10,000. Therefore, using conventional breeding, it is impossible to

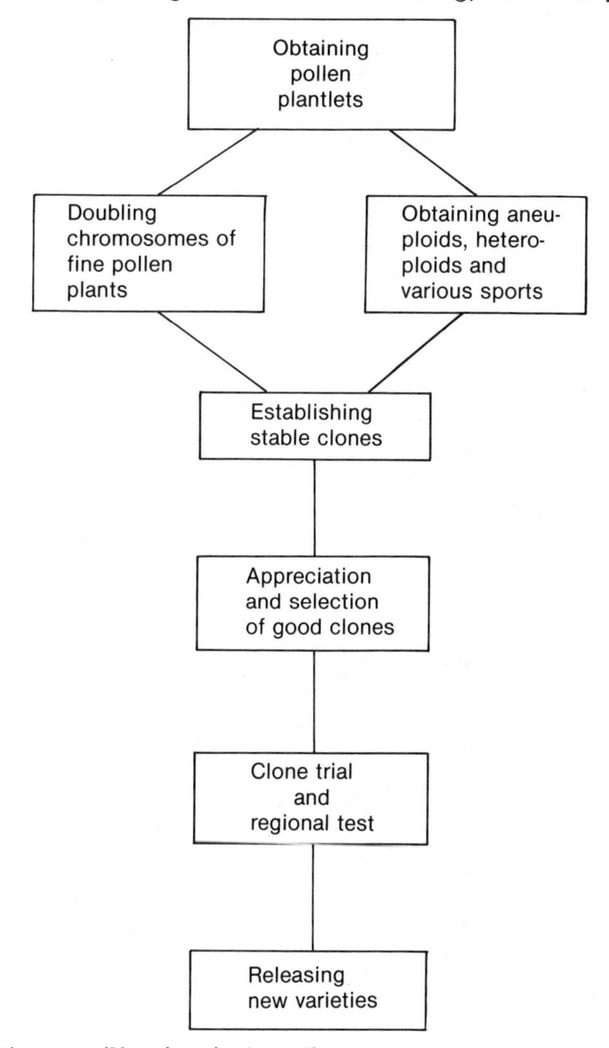

Figure 1. Direct utilization in breeding new varieties.

obtain pure lines by means of successive inbreeding. The establishment of anther culture techniques of *Hevea* enable us to obtain pure lines of different genotypes in a short time and thus pave the way to produce high-yielding, resistant, and uniform sexual lines, i.e., hybrid *Hevea*. With a lot of pure lines on hand, we can estimate the combining ability among different lines and select the superior combinations. Thus, hybrid seeds of good combinations will be produced. This will result in the replacement of budding with sexual reproduction. It can be expected that there will be a significant enhancement of yield and resistance, since the seedling plant has superiority over the bud-grafted stock, and the heterosis of pure lines with good combining ability will be much higher than that of available heterozygous varieties (Fig. 2).

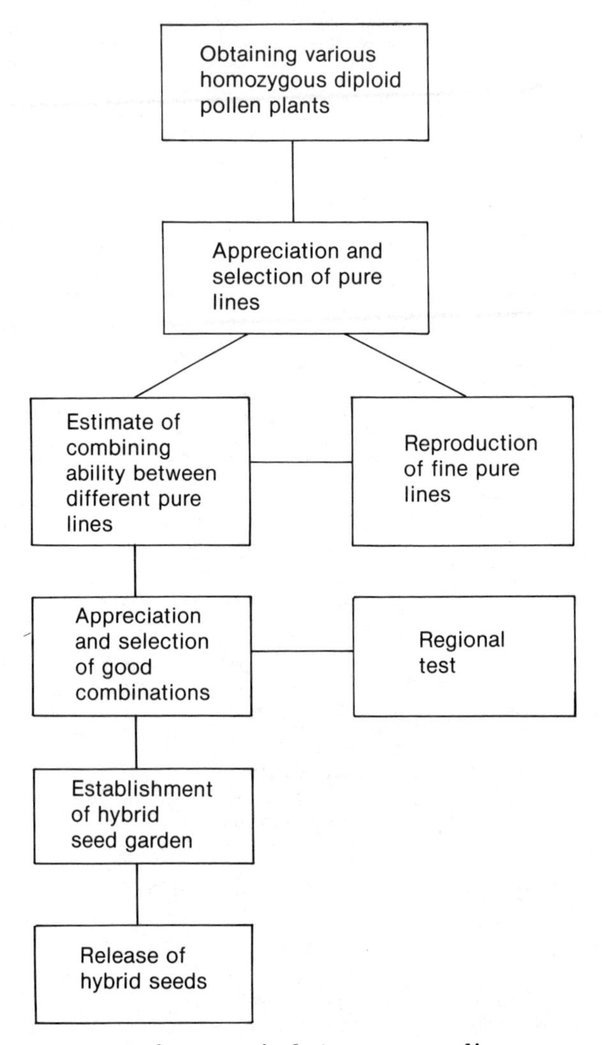

Figure 2. Utilization of heterosis between pure lines.

UTILIZATION OF HAPLOID CELL LINES. The haploid cell lines have only one set of chromosomes, so they are superior to diploids for induction and selection of mutations, to hybridize cells, and to begin work on genetic engineering. Although what has been done in this field at present falls far short of practical use, the course of its present development indicates that it has a very great future (Fig. 3).

Tissue Culture

INDUCTION AND SCREENING OF MUTATIONS. In recent years great attention has been paid to the induction of polyploids and heteroploids. The combination of mutation induction with tissue culture would greatly reduce the phenomenon of chromosome mosaicism, since

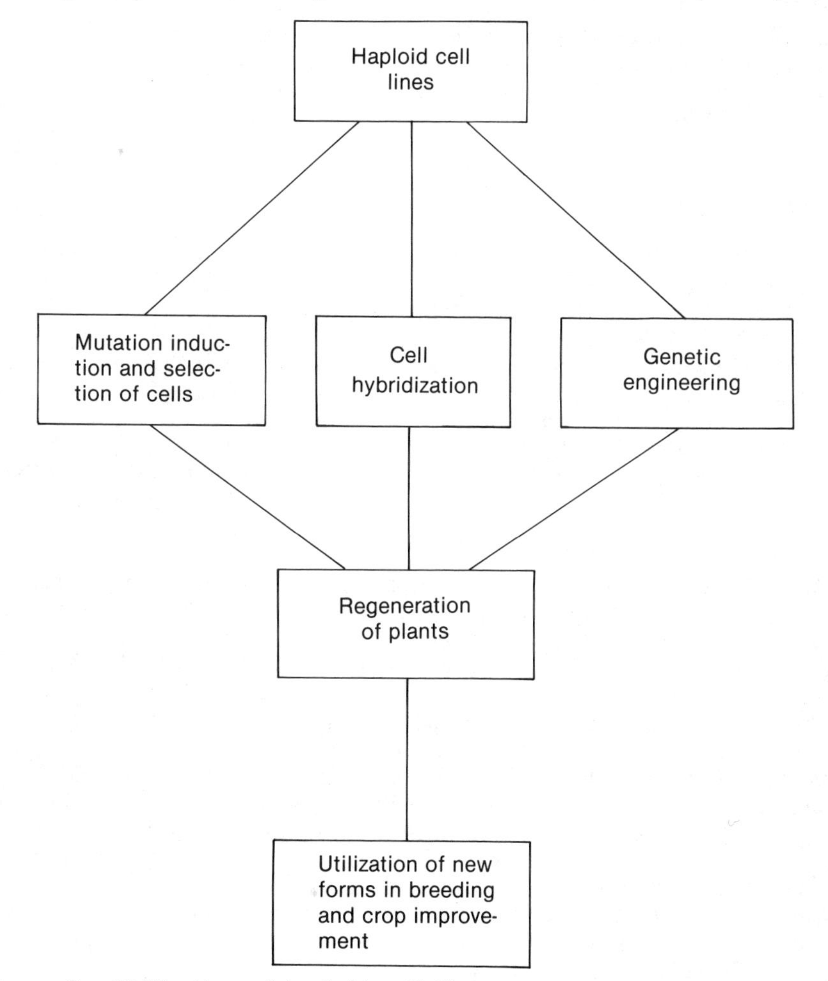

Figure 3. Utilization of haploid cell lines.

an intact plant could be formed from a single mutated polyploid or heteroploid cell in culture. Moreover, the techniques of mutation induction could be applied in tissue culture to establish selection systems for the identification of resistance and tolerance to disease, cold, and drought at the cellular level. Development of disease resistant breeding lines could be based on selection in the test tube, thereby enhancing the efficiency of selection.

PROPAGATION OF ELITE TREES. The culture of shoot apices can speed up propagation of elite trees, providing a large number of excellent plantings with fully developed root systems. This method has already been applied to other crops (de Fossard, 1978a,b).

PRODUCTION OF DISTANT HYBRIDS. By using the techniques of embryo culture, it is possible to recover the embryos and subsequent plants derived from distant hybridization. Recovery of distant hybrids could permit incorporation of useful characters into cultivated rubber.

LITERATURE REVIEW

As early as 1966, Chua reported the effects of osmotic concentration, carbohydrates, and pH values of culture media on the induction of callus growth in plumule tissues from *Hevea* seedlings (Chua, 1966).

Research on tissue culture of *Hevea* has been done on a large scale by the Rubber Research Institute of Malaysia. As Paranjothy and Ghandimathi (1975) have reported, the callus cultures from various explants were established and some of these were successfully maintained by repeated subculture. Embryo formation was observed in callus cultures derived from the somatic tissues of anther, and methods for culture of shoot apices and seed embryos in liquid media with the subsequent establishment of rooted plants in soil were successfully achieved.

Satchuthananthavale (1973; Satchuthananthavale and Irugulbandara, 1972) reported that callus formation from anther tissues and pollen grains was observed after 4-5 weeks in culture, and callus continued to maintain good growth on the same medium through 6 subcultures over a period of 6 months.

Wilson and colleagues (Wilson and Street, 1975; Wilson et al., 1976) observed that callus cultures initiated from stem explants of young plants of *H. brasiliensis* could be maintained over long periods by serial subculture. When the serially propagated suspensions were maintained for several months without subculture the larger cell aggregates gave rise to embryo-like structures. Attempts to promote further development of these embryo-like structures into plantlets were unsuccessful.

Toruan and Suryatmana (1977) obtained tissue from stem segments and anthers, as well as seedlings obtained from embryos without cotyledons in MS medium and on soil.

Intact plantlets have been obtained in culture from seedling stem segments with buds, as well as from embryos without cotyledons (Institute of Genetics, Academia Sinica, in cooperation with The Baoting Institute of Tropical Crops and The Hainan Institute of Rubber Tree; Chen, unpublished). Some of these plantlets were transplanted successfuly into the field. The first five pollen plantlets of *H. brasiliensis* were obtained by anther culture in 1977 (Chen et al., 1977). Subsequently,the change of ploidy of cultures and development of the microspores into embryos in the anther culture process were studied systematically (Cen et al., 1981; Chen et al.,1981a). Hundreds of pollen plantlets have already been obtained, and some of them have been cloned. With established anther culture techniques, usually 100 embryos are formed per 100 inoculated anthers, with a maximum of 149 embryos. The frequency of viable pollen plantlets has been enhanced to 3% (number of pollen plantlets per 100 anthers inoculated) (Chen et al., 1981b). Moreover, Guo et al. (1982) reported recently that two plantlets were recovered from unpollinated ovules of *H. brasiliensis*, but their ploidy has not been examined yet.

PROTOCOLS FOR ANTHER CULTURE

General Procedures

INDUCTION OF POLLEN PLANTS. The induction of pollen plants is carried out in three steps. First is the inoculation of anthers on primary medium to induce callus formation, in which the microspores grow and develop into multicellular masses, haploid embryos, or pollen calli. This step is very important because all pollen embryos and calli are formed at this step. It is noteworthy that under favorable conditions there are two different processes that exist simultaneously--the formation of somatic calli and the initiation of microspore development into small embryos or pollen calli. Upon examination of somatic calli visible to the naked eye it is still impossible to distinguish the development of microspores. Therefore, it should be emphasized that during the culture period, sufficient attention must be given to the development of microspores, as well as to the influence of various factors (including the callusing of somatic tissues) on developing microspores.

The second step of anther culture is initiated by transferring anther-derived callus to differentiation medium at about 50 days in culture. The embryos grow and differentiate into embryos that are visible to the naked eye, or the pollen calli differentiate into small embryos that subsequently develop into the embryos visible to the naked eye.

The time of transfer of callus to differentiation medium is critical. After 20-25 days in culture, the anthers become filled with meristematic parenchymatous cells derived from the somatic tissues of anthers. Cytological investigations have revealed that most of the mitotic metaphases of these callus cells are diploid (36 chromosomes). However, after 50 days of culture the somatic calli are senescing while the embryos and calli that originated from microspores begin to grow vigorously and divide rapidly. At that point, about 70% of the cells have the haploid chromosome number (18 chromosomes). This time appears to be the ideal time to transfer the callusing anthers to differentiation medium in order to promote the further development and differentiation of haploid embryos (Chen et al., 1979; Cen et al., 1981).

At the third step the well-developed microspore embryos are transferred to plantlet-forming medium on which they continue development into intact plantlets (Fig. 4).

During the process of anther culture the formation of embryos is generally asynchronous, and as a result of different conditions experienced by each embryo—e.g., hormone level, mineral and organic nutrients, or the position of the embryos--morphological diversity has been observed in their development. Round-type, rod-type, trumpet-type, and cotyledon-type have all been observed (Fig. 5). Among these types the trumpet- and cotyledon-type are more normal. Usually the plantlets are derived from these two types, while a few plants are recovered from rod-type embryos.

The embryos that are visible to the naked eye emerge about one month after callusing anthers are transferred onto the differentiation medium. At this time differentiation of the embryos has not been completed. The embryos form a terminal bud only after 2-3 months in culture; therefore they should not be transferred into plantlet-forming medium too early. Otherwise, the embryos form roots but not shoots.

COMPOSITION OF MEDIA. Dedifferentiation medium, differentiation medium, and plantlet-forming medium are all required for induction of

Figure 4. A pollen plantlet.

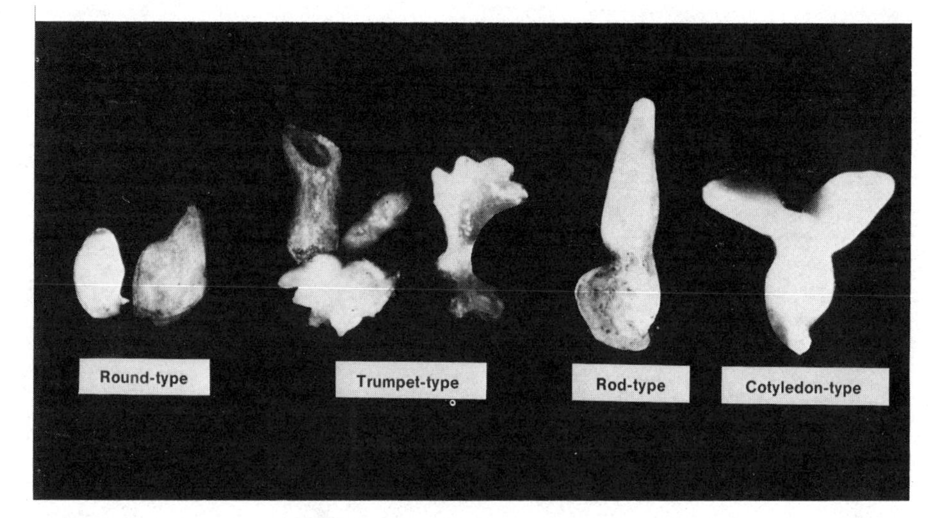

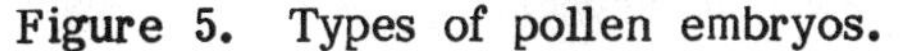

Figure 5. Types of pollen embryos.

pollen plants of *Hevea*. Two basic media are used for dedifferentiation. One is MB medium, which contains macronutrients and iron salt of MS medium, and micronutrients, vitamins, and organic supplements of the medium designed by Bourgin and Nitsch (1967) for tobacco. The other medium is a modified version of MB. Both media can induce callus development from cultured anthers and development of microspores into embryos. In general, the induction frequency of embryos in the modified MB medium is higher than that in the MB medium. The type and concentration of hormones in both media are identical. In the differentiation and plantlet-forming media, the concentration of macronutrients is reduced, but the concentration of micronutrients is higher than in MB medium (Tables 1 and 2).

EFFECTS OF SOME FACTORS ON THE DEVELOPMENT OF POLLEN EMBRYOS AND SUBSEQUENT PLANTLET FORMATION. **Mineral Salts in Dedifferentiation Medium.** Among the mineral salts in the medium, the effect of the chemicals containing nitrogen on development of microspores appears to be most significant. It has been found that there is a notable difference in the requirement for nitrogen (both chemical composition and concentration) for the dedifferentiation of microspores versus the callusing of somatic tissues of anthers. The concentration of NH_4^+ ion in MS medium appears to be adequate for the development of embryos. Moreover, the decrease of NO_3^- ion concentration in MS medium is favorable for the development of pollen embryos. On the contrary, the differentiation of somatic tissues of anthers requires a lower level of NH_4^+ ions; however, this concentration of NH_4^+ is unfavorable for development of pollen embryos, because the induction of pollen embryos requires a higher concentration of total nitrogen (50.58-60.05 mM appears to be adequate). When the total

Table 1. Composition of Media for Anther Culture of *Hevea*

COMPONENT	DIFFERENTIATION MEDIA		DIFFERENTIATION AND PLANTLET-FORMING MEDIA
	MB	MB-modified	
Macronutrients (mM)			
KNO_3	18.8	9.4	15.0
NH_4NO_3	20.6	20.6	16.5
$MgSO_4 \cdot 7H_2O$	1.5	1.5	1.5
$CaCl_2 \cdot 2H_2O$	2.3	2.3	2.3
KH_2PO_4	1.2	3.7	1.0
Iron Salts (mM)			
Na_2EDTA	0.11	0.11	0.11
$FeSO_4 \cdot 7H_2O$	0.11	0.11	0.11
Micronutrients (μM)			
$MnSO_4 \cdot 4H_2O$	11.0	4.5	11.0
$ZnSO_4 \cdot 7H_2O$	3.5	29.0	70.0
H_3BO_3	160.0	190.0	320.0
$CuSO_4 \cdot 5H_2O$	0.1	0.1	0.2
$Na_2MoO_4 \cdot 2H_2O$	1.0	1.0	2.0
KI	0	0	4.8
$CoCl_2 \cdot 2H_2O$	0	0	0.3
Vitamins and Organic Supplements (μM)			
Thiamine	1.5	1.5	1.5
Pyridoxine	2.4	2.4	2.4
Nicotinic Acid	40.0	4.0	4.0
Glycine	30.0	30.0	30.0
Folic Acid	1.1	1.1	0
Biotin	0.2	0.2	0

concentration of nitrogen decreases to 30.03 mM, the percentage of callusing anthers increases to 80%, while that of embryos decreases to 0.8% (see Fig. 6).

Our experiments also showed that both callus and embryo formation required high levels of KH_2PO_4. The greatest frequency of callus and embryo induction occurred with the highest concentration of KH_2PO_4 tested (3.75 mM). This concentration of KH_2PO_4 is superior when compared to 2.5 mM and the concentration in MS medium (1.25 mM). Moreover, cultures grown with 3.75 mM KH_2PO_4 had much greater vitality than those in media with lower concentrations of KH_2PO_4.

Hormones. The addition of 9.3 μM KIN and 9.0 μM 2,4-D to the dedifferentiation medium is sufficient to induce a considerable frequency

Table 2. Growth Regulators and Other Ingredients in Culture Medium Used for Cell Culture of *Hevea*

COMPONENT	DEDIFFERENTIATION MEDIUM (MB and MB-modified)	DIFFERENTIATION MEDIUM	PLANTLET-FORMING MEDIUM
Growth regulators (μM)			
KIN	4.6	2.3-4.6	0
2,4-D	4.5	0	0
NAA	5.7	1.1-1.6	0
GA	0	1.4	2.9-11.5
IAA	0	0	2.9-5.7
5-Bromouracil	0	0	2.6-5.2
Other ingredients[a]			
Myo-inositol (mM)	0.56	0.56	0.56
Coconut water (v/v)	5%	0	0
Sucrose (mM)	204	204-233	117-175

Agar, 7 g/l
Final pH adjusted to 5.8 with 1 M KOH or 0.2 N HCl.

[a]Added directly to media during preparation.

of callusing anthers. However, for dedifferentiation of microspores and subsequent development into embryos it is necessary to add three hormones (4.6 μM KIN, 4.5 μM 2,4-D, and 5.7 μM NAA). Experiments showed that the addition of 5.7 μM NAA in combination with 4.5 μM 2,4-D and 4.6 μM KIN had a significant positive effect on the formation of multicellular masses, doubling the induction frequency when compared with medium without NAA (Chen et al., 1981a; Table 3). Addition of 5.7 μM NAA was also superior to 2.7 μM NAA.

Out of a total of 1166 anthers inoculated on medium with 2.7 μM NAA, only 160 embryos were obtained (14.2%). At the same time, of 1073 anthers inoculated on medium with 5.7 μM NAA, 245 embryos were obtained (22.8%). In addition, we observed that most cells of the embryos induced on the medium with 5.7 μM NAA were haploid, with a chromosome number 2n = 18, while 85% of the cells of embryos induced on medium without NAA had 28-36 chromosomes (Chen et al., 1981a).

Three hormones (KIN, NAA, and GA) are required in the differentiation medium. The concentration of KIN is especially critical. In the differentiation medium without KIN the induction frequency of embryos decreased and the embryos did not develop shoots. The simultaneous addition of 1.1 μM NAA and 4.6 μM KIN had a significant positive effect on the development of small embryos into macroscopic embryos. Optimal concentrations of GA promoted further growth of the embryo and subsequent cotyledon formation. The embryos grown on differentia-

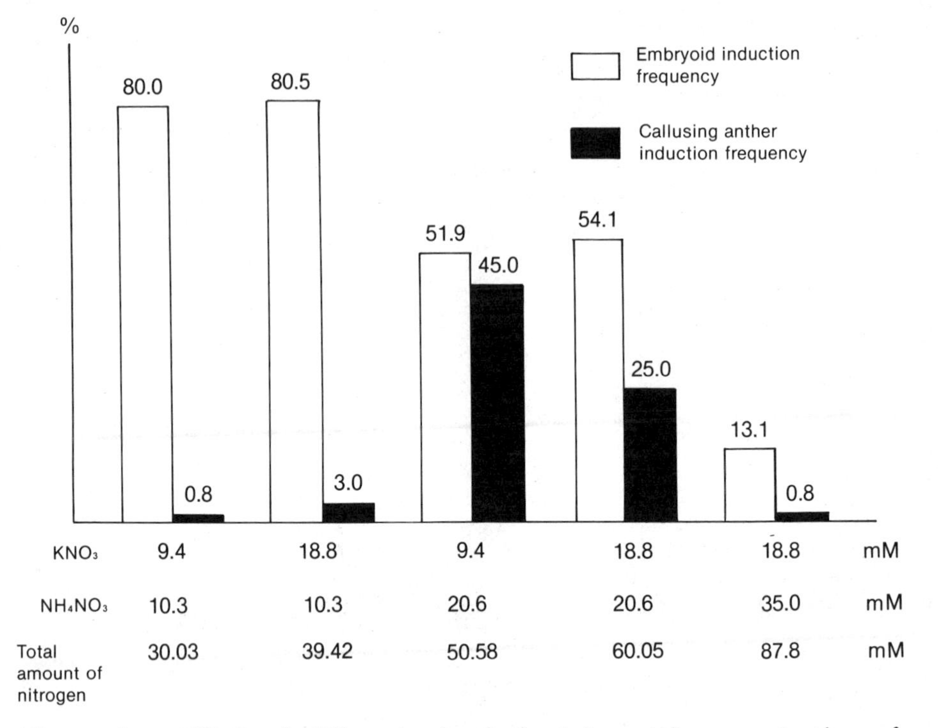

Figure 6. Effect of different chemical status and concentration of nitrogen on ions on development of embryos.

tion medium without GA were much smaller in size--many of them abnormal--and the cotyledons developed anomalously.

Sucrose. The concentration of sucrose in the dedifferentiation medium is important in the survival frequency of pollen grains and subsequent formation of pollen embryos. When the level of sucrose is increased to 10%, the somatic cells do not proliferate rapidly and the frequency of anthers forming callus is very low, though the survival frequency of pollen grains is increased. In culture media with 3% sucrose the somatic callus of the anther grows very vigorously, which inhibits formation of multicellular masses of pollen, so that the embryo induction frequency is also low. Hence, there are many undeveloped pollen grains in the anthers.

The concentration of sucrose that can induce at least 50% of the anthers to callus is 7-8%. This level of sucrose can also depress excessive callus growth and thus promote the development of pollen grains, resulting in the highest rate of embryo formation (Chen et al., 1979).

For the differentiation medium the most favorable concentration of sucrose is also 7-8%. Reducing the concentration of sucrose results in the restoration of vigorous growth of somatic cells, which inhibit the development of embryos and differentiation of terminal buds.

Table 3. Effects of NAA on Formation of Multicellular Masses from Cultured Anthers of *Hevea*

TREATMENT	NO. POLLEN OBSERVED	POLLEN MULTI-CELLULAR MASSES	
		Number	Percent
With NAA	730	149	20.4
Without NAA	795	84	10.6

The optimal concentration of sucrose in the plantlet-forming medium for germination of embryos and subsequent root and shoot growth is 4-6%.

Coconut Water. The addition of CW to the medium supplemented with KIN and 2,4-D inhibits the proliferation of somatic cells of cultured anthers. The induction frequency of callus formation from anthers when 10% CW is added is significantly lower than without CW. As a general rule, the higher the concentration of CW in the medium, the lower the induction frequency of callus formation. However, 5-10% CW added simultaneously with 4.6 μM KIN, 4.5 μM 2,4-D and 5.7 μM NAA to the medium can markedly increase the induction frequency of pollen embryos (Chen et al., 1979).

Selection of Material for Inoculation

Growth of rubber in tropical and subtropical zones is characterized by a great density of bacteria and eumycetes. Therefore, care must be taken to prevent contamination at every step of anther culture.

1. Anthers used for inoculation must have well-developed microspores. The anthers from male-sterile varieties are inappropriate for inoculation. In early spring and hot summer, some varieties often have a lot of inviable microspores in their anthers as a result of the influence of unfavorable environmental conditions. As a rule, pollen embryos would not be obtained from such anthers since one would only recover somatic callus. Hence, it is necessary to choose those varieties of which the microspores are normal and have the potential to develop. There are many floral branches on the inflorescences. Every floral branch has both pistillate and staminate flowers. The pistillate flowers are located at the top of the central axis and main lateral branches. However, there are staminate flowers on every little lateral floral branch, and their total number in an inflorescence is much greater than that of pistillate flowers.

2. There are 10 anthers in a male flower bud, with two rows of anthers found around the staminal column (5 anthers in each row) in a peeled flower bud. The uninucleate stage is optimal for the inoculation of *Hevea* anthers. When the corollas become yellow, the microspores in the anthers are at the binucleate stage. Anthers at this stage should not be used for inoculation, as the somatic tissues of these anthers do not callus and would therefore repress androgenesis.
3. Do not pick flower buds for inoculation just after rain, because anthers from such flower buds are usually hopelessly contaminated following inoculation.
4. The refrigeration of inflorescences before inoculation has a negative effect on the microspore tissue but not on somatic tissue. For example, we inoculated anthers from inflorescences stored at 3-11 C for 20-24 hr, and found that many microspores lost the tension of their exine (having a triangular shape), and the nucleus in such microspores stained indistinctly. Experiments showed that these microspores could not develop into embryos. However, the somatic cells of these anthers proliferated normally and as a result, most of the embryos that differentiated in callusing anthers were diploid (Chen et al., 1981).
5. It is imperative to examine anthers under the microscope to determine their developmental stage. Flower buds (3-3.5 mm), in which the majority of anthers are at the uninucleate stage, are excised and wrapped in a piece of gauze (3 x 3 cm).

Preparation of Medium

1. Mineral salts and organic compounds used should be of the highest grade available.
2. Water should be double-distilled. The last distillation should be performed in glass.
3. Dissolve 10 mg 2,4-D in 2 ml ethanol, heat slightly and gradually dilute to 100 ml with water. Store in refrigerator. Dissolve NAA in the same way as 2,4-D.
4. Dissolve 10 mg of KIN in a small volume of 0.5 N HCl by heating slightly and gradually diluting to 100 ml with distilled water. Store in refrigerator. Dissolve the benzylaminopurine in the same way as KIN.
5. Dissolve 10 mg of 5-Bromo-uracil in 100 ml of double-distilled water by heating slightly.
6. Dissolve folic acid in a small volume of ammonia solution and gradually dilute with double-distilled water.
7. All of the above stock solutions should be refrigerated at 4 C and used in a short time.
8. Dispense 12 ml dedifferentiation medium into 20 x 200 mm test tubes. Dispense 25-30 ml differentiation medium or plantlet-forming medium into 30 x 200 mm test tubes. Plug up the tubes with cotton and autoclave at 120 C for 20 min. Take the test tubes out of the autoclave for cooling as soon as possible.

Sterilization of Material

1. Dip tissue into 70% ethanol in a beaker for 1 min.
2. Discard the alcohol and add 0.1-0.2% $HgCl_2$ solution. Leave in the beaker for 10-12 min, or add 10% commercial bleach and leave tissue for 30 min.
3. Rinse the tissue in sterile distilled water 4-5 times.

Inoculation

1. Autoclave the various implements (containers, etc.) and dry them in the oven.
2. Heat slides before inoculation. Place an aseptic slide in a petri dish. Anthers are excised and placed onto slide. About 25-30 anthers can be inoculated in one tube.
3. The inoculated anthers should be well-distributed on the medium.
4. Incubate the material at 26 C in culture room. (NOTE: Do not injure the anthers. The operation should be quick. Each slide is used only once.)

Subculture of Callusing Anthers

1. Estimate the induction frequency of responding anthers in order to decide the amount of medium to prepare.
2. When anthers begin to form callus, transfer them to 3-4 test tubes with about 4 anthers to each tube.
3. Don't turn the anther upside down during transfer.
4. Incubate the anthers in a culture room at 26 C.

Transfer of Embryos.

1. Put an aseptic slide in a petri dish.
2. Put the callus with embryos on the slide, and softly remove well-developed embryos from the callus. Remove the callus without doing any injury to the embryos, then transfer the embryos to the plantlet-forming medium. Transfer the remaining callused anthers with small embryos to newly prepared differentiation medium.
3. Incubate the embryos under 12 hr/day illumination (more than 2400 lux) at 27-28 C.

Histological Investigation: Acetocarmine Smears

1. Add 1 g carmine into 100 ml 45% acetic acid solution, boil it for 3-4 hr in a flask mounted with a reflux condenser, then cool and filter. Store the staining solution in a refrigerator at 4 C.
2. Dip excised flower buds into vials containing fixative.

3. Excise anthers from the buds on the slide, mordant them in 4% iron alum which is blotted up after about 1 min.
4. Tear the anthers to pieces and cover them with a drop of aceto-carmine, then cover with a coverslip for microscopic examination.

Histological Investigation: Staining with PICCH

1. Add 0.5 g carmine into 100 ml propionic acid, boil it for 3-4 hr in a flask mounted with a reflux condenser, then cool and filter.
2. Add 2 g chloral hydrate to 5 ml propionic carmine solution, stirring until fully dissolved.
3. Add several drops of propionic acid solution saturated with $Fe(OH)_3$.
4. Transfer the material to vials containing fixative of 95% alcohol: chloroform:propenoic acid (6:3:1) and leave them for 12-23 hr.
5. Rinse with 95% alcohol several times.
6. Transfer the material to a slide and add a few drops of PICCH (propionic acid-iron-carmine-chloral hydrate), then cover with a coverslip for examination (Hepei Agricultural University, 1977).

Cytological Investigation

1. Take the fresh callus, embryos, or root tips of plantlets between 11:50 and 12:00 a.m. and pretreat them with 0.003 M 8-hydroxy-quinoline at room temperature for 4 hr.
2. Transfer the material into alcohol:glacial acetic acid (3:1) and fix for 24 hr.
3. Hydrolyze tissue in the hydrochloric acid:alcohol (1:1) for 40 min.
4. Transfer tissue to iron alum (4%) for half an hour.
5. Stain in hematoxylin for 2 hr.
6. Smear in 40% acetic acid for observation (Chen et al., 1979).

Transplanting Pollen Plantlets

In contrast to seedlings, the pollen plantlets of the rubber tree lack strong roots and endosperm. The first successful transplantation of rubber tree pollen plantlets in the field was accomplished by Baoting Institute of Tropical Crops in 1979 (Fig. 7). The survival frequency of transplanted plants has since reached more than 50%.

In order to have transplanted plantlets survive, the following points need attention:

1. The main roots and verticillate lateral roots must be well-developed. The survival frequency of plantlets with senescent roots or insufficient roots was very low. The addition of 1.1 μM NAA to the plantlet-forming medium helped the induction of a well-developed root system on the plantlet in the test-tube.

Figure 7. A transplanted pollen tree.

2. Before transplanting, the plantlets should be hardened. When the first and second whorls become mature, take off the plugs and expose the plantlets to the sun for 2 days. This treatment can increase the survival frequency of plantlets significantly.
3. Choose the optimum time for transplantation. The tranplantation should not be too early, but rather after the first whorl of the test-plantlet has grown up and turned into mature leaves.
4. Choose the appropriate soil for transplantation. The soil should be structurally good. Experiments have indicated that all plantlets transplanted in sterilized soil do not survive, for sterile soil lacks indispensible microorganisms. The soil should not be too sticky. It is recommended that plantlets are transplanted into a bamboo basket filled with a mixture of fertile soil and sand (4:1).
5. Soak the roots in hormones before transplantation. Rinse the roots removed from the medium and immerse them into a mixed solution of 0.188 mM NAA and 0.2 mM IAA, or 0.114 mM IAA and 0.098 mM IBA for 15-30 min. This treatment has a positive effect on the development of roots after transplantation.
6. The operation should be done quickly. Transplant the plantlets (after soaking the roots) into soil within 1-2 min.
7. After transplantation raise the temperature of the culture room to 25-30 C and maintain high humidity. Illuminate with mercury vapor lamps (500 W) for 15 hr every day.
8. When the plantlets have survived and grown new leaves, harden the plantlets outside the culture room by exposing them to the sun from 7:00 to 10:00 a.m. and from 4:00 to 6:00 p.m. for 2 weeks. After hardening, transplant the plantlets in the field.

Results of Cytological Investigations on Cultures and Transplanted Pollen Plants

EMBRYOS. Cytological investigations have confirmed that the embryos obtained by anther culture using our methods originated from pollen grains. About 80% of mitotic metaphases observed had 18 chromosomes, i.e., the chromosome number of the pollen grains. Some of them (about 16%) had 27 chromosomes, while a few of them (about 3%) had 9 chromosomes (Fig. 8). Also, about 1% of metaphases had a chromosome number of 32, 36, or 45. Most of the chromosome numbers (9, 18, 27, 36, 45) are multiples of 9.

ROOT-TIP CELLS. The cytological observations on the root-tip cells showed that 32% of metaphases had 18 chromosomes (Fig. 9), 59% had 27 chromosomes (Fig. 10) and 4% had 9 chromosomes. In addition, a few aneuploid metaphases were also observed. Thus, we observed that there was a tendency to gradually increase chromosome number during the culture of embryos into plantlets (Table 4; Cen et al., 1981).

YOUNG LEAVES. The microscopic observations carried out on the young leaves of transplanted pollen plants that were under 50 cm in height revealed that most of the metaphases still had a chromosome

Figure 8. A cell of an embryo with 9 chromosomes.

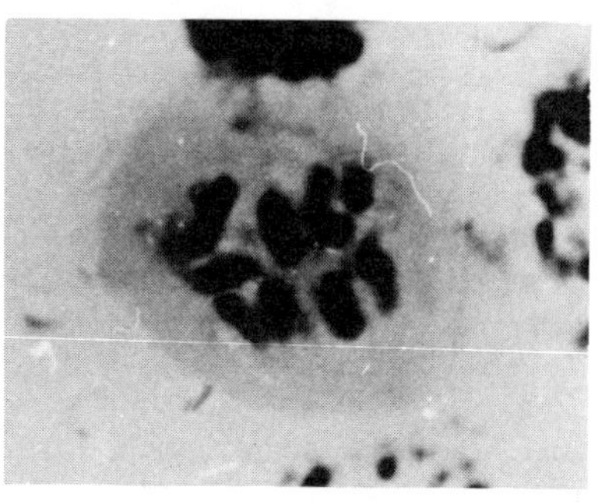

Figure 9. A root cell of a pollen plantlet with 18 chromosomes.

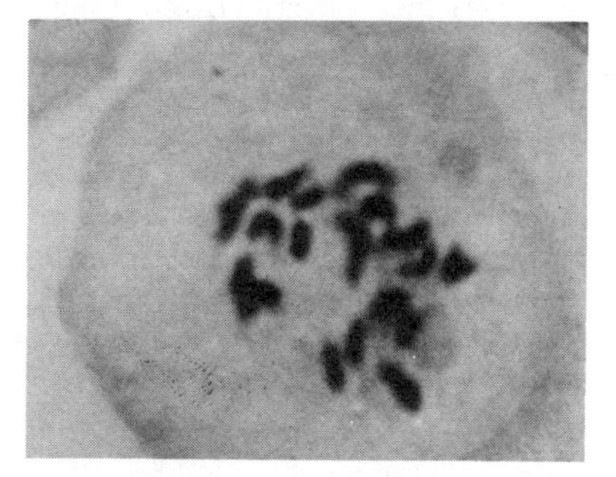

Figure 10. A root cell of a pollen plantlet with 27 chromosomes.

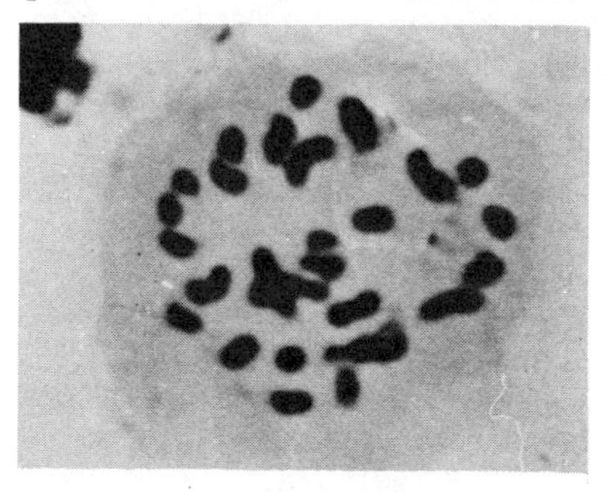

number of 18-27, while metaphases with 36 chromosomes were not seen. But in the leaves of plants more than 160 cm in height, the number of cells with more than 36 chromosomes increased. These trees had 51% leaf cells with 28-36 chromosomes and 44% cells with 18-27 chromosomes (Table 5). So this may indicate a successive increase in chromosome numbers in vivo. The frequencies of cells with different chromosome numbers appeared to continue to change in the transplanted trees, although aneuploid chromosome numbers, i.e., multiples of 9, seem to be favored. When buds from such trees can be propagated, it

Table 4. Chromosome Numbers Observed in Mitotic Metaphases of Embryos and Root Tips of Plantlets

TISSUE	NO. SPECI-MENS	NO. META-PHASES	CHROMOSOME COUNT							
			9	18	20	24	27	32	36	45
Embryo	46	238	9	190	0	0	36	1	1	1
Root tips	18	145	6	47	1	1	85	1	3	1

Table 5. Chromosome Counts in Metaphases of Young Leaf Cell from Transplanted Pollen Plants

NO. PLANTS	PLANT HEIGHT (cm)	NO. LEAVES	NO. META-PHASES	CHROMOSOME COUNT			
				9-17	18-27	28-36	36
7	50	7	245	38	196	6	5
4	100	13	10	1	44	51	4

will be possible to analyze the chromosomal variation in shoots originating from buds with various euploid and aneuploid constitutions. This chromosomal variation will be very useful in breeding work with the rubber tree.

PROTOCOLS FOR CULTURE OF OTHER EXPLANTS

Shoot Apex

As Paranjothy and Ghandimathi (1975a,b) reported, the shoot apex taken from 2-4-week-old seedlings and young clonal plants can be sterilized with commercial bleach (5%) for 5 min, followed by rinsing several times with sterilized distilled water. The scale leaves and small leaves are removed and the surface of the apex sterilized once more. The apex is grown on medium containing 0.087 M sucrose for 4-7 days to ensure sterilization, then the upper part of shoot apex (2-3 mm long) is cut and inoculated onto MS medium (0.087 M sucrose) supplemented with casein hydrolysate. Plantlets with roots form in 4 weeks, but the shoot apex from clonal young plants does not develop into plantlets by culture. There is no report yet on the successful culture of plantlets from the shoot apex of adult plants, however, contamination after inoculation makes large-scale experiments impossible.

Stem Segment

Stem segments from seedlings were regenerated into intact plantlets by us in 1975 using the following methods. Stem segments are cut into 4-cm lengths (with buds). The large leaves are removed and immersed in 10% commercial bleach for 20 min. Rinse with sterilized distilled water 4-5 times. Cut off the 0.5-1-cm portion of the stem that was damaged by the bleach, and inoculate immediately with the basal cut end in contact with the medium. If liquid medium is used, a platform made with desalted filter paper may be placed in the culture just under the level of liquid medium in the test tube (Fig. 11).

The modified Nitsch medium is optimal for stem segment culture of *Hevea* (Table 6). The addition of 3.0 mM KNO_3 to the original Nitsch medium shows a positive effect on the development of buds into plantlets. The vitamins are also essential for regeneration.

Unpollinated Ovule

The techniques used by Guo et al. (1982) for the induction of plantlets from unpollinated ovules are as follows:

Pistillate flowers with uninucleate microspores are removed from inflorescences. Flowers are sterilized with 0.1% $HgCl_2$ solution, rinsed in sterile distilled water 4-5 times, and the ovule to be used for inoculation is identified. MS or MB medium is used as basal medium with

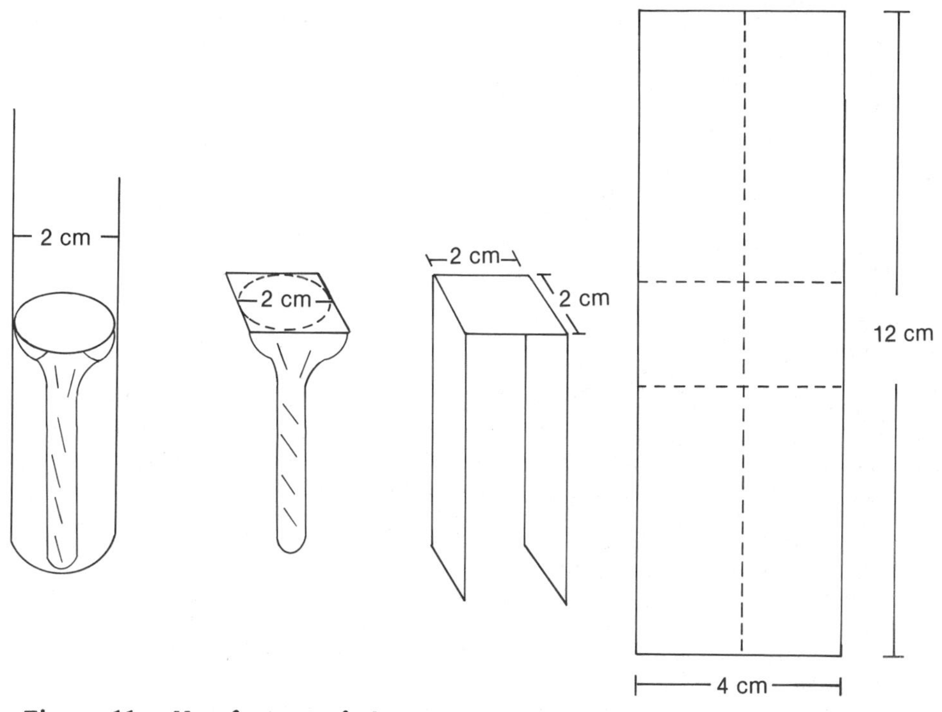

Figure 11. Manufacture of the paper tablet placed just under the level of liquid medium in test tube.

Table 6. Composition of Modified Nitsch Medium for Stem Segment Culture of *Hevea*

Component	Amount	Component	Amount
Macronutrients (mM)		Vitamins and organic supplements (μM)	
KNO_3	4.2	Glycine	100
$Ca(NO_3)_2$	3.0	Thiamine	2.96
KH_2PO_4	0.92	Pyridoxine	4.86
$MgSO_4 \cdot 7H_2O$	0.50	Nicotinic acid	20
Iron salts (mM)			
Na_2EDTA	0.11	Growth regulators (μM)	
$FeSO_4 \cdot 7H_2O$	0.10	KIN	0.9–2.3
		IAA	2.9–5.7
Micronutrients (μM)		GA	1.4–5.8
$MnSO_4 \cdot 4H_2O$	100		
$ZnSO_4 \cdot 7H_2O$	30	Coconut water	0–10%
$CuSO_4 \cdot 5H_2O$	100	Sucrose	87.7 mM
H_3BO_3	100	pH	5.8
$NaMoO_4 \cdot 2H_2O$	1		
$CoCl_2 \cdot 6H_2O$	0.1		

146–292 mM sucrose at pH 5.8. The dedifferentiation medium is supplemented with KIN (2.3–5.5 μM), 2,4-D (2.3–5.4 μM), and 5–10% CW. KIN (2.3–4.6 μM) and NAA (0–11.0 μM) are added to the differentiation medium. The medium for plantlet-formation is supplemented with 1.4–7.2 μM GA.

The ovule starts to proliferate on the sixth day after inoculation. Some ovules turn brown and die, but others break through the ovary wall and develop into callus. After 50–60 days of culture on the first medium the calli are transferred to differentiation medium. Embryos visible to the naked eye emerge after about 15 days. The embryos continue to grow on differentiation medium for 60 days and then are transferred to plantlet-forming medium. One intact plantlet has been obtained.

Embryo

Embryos from mature or immature seeds are excised and sterilized. In laminar flow cabinets or in a sterile room, excise the endosperm and part of both cotyledons, then inoculate the embryo onto medium.

The medium should contain the usual concentration of MS macronutrients, with a double dose of micronutrients and supplementary KIN (0.5–0.9 μM), NAA or IBA (1.1 μM and 2.5 μM, respectively), and GA (1.4 μM). On this medium most embryos can sprout. It is much more difficult for the embryos without cotyledons to sprout than those with a small part of cotyledon. Toruan (1977) reported that seedlings were obtained from embryos without cotyledons in MS medium and then grown on soil.

Cell Suspension

Paranjothy and Ghandimathi (1975a,b) established cell suspension cultures of somatic callus of anthers grown in 100 ml T-type tubes or nipple flasks on a disk that rotates at 2 rpm. MS medium was used, supplemented with 4.4 μM BA and 4.5 μM 2,4-D. The medium diluted twofold could promote growth of the cells, but embryo-like structures have not been obtained.

Epicotyl, Hypocotyl and Pieces of Cotyledon

Segments of epicotyl or hypocotyl and pieces of cotyledon are easy to root in culture, even with low levels of hormones in the MS medium, or in the absence of hormones. Triiodo-benzoic acid (TIBA; 0.2 μM) depresses root formation and 2,4-D promotes callus formation, while KIN in combination with 2,4-D causes vigorous growth of callus (Paranjothy and Ghandimathi, 1975a,b).

FUTURE PROSPECTS

In order to use anther culture techniques for crop improvement it is necessary to increase the number of strains from which pollen plants can be obtained and to enhance the induction frequency of pollen plantlets from different strains. Although the ploidy and origin of plantlets obtained by ovule culture still needs examination, if we could induce haploid plants from microspores in unpollinated ovules, then we could get haploid or homozygous diploid lines from male-sterile strains. The method of obtaining a large number of plantlets from the terminal apex and cambium of mature trees also would be of great use in planting and breeding. In addition, it is necessary to develop and apply the techniques of plant regeneration from cell suspensions and protoplasts, and to apply cell genetic techniques for rubber tree improvement.

KEY REFERENCES

Cen, M., Chen, Z., Qian, C., Wang, C., He, Y., and Xiao, Y. 1981. Investigation of ploidy in the process of anther culture of *Hevea brasiliensis* Muell.-Arg. Acta Genet. Sin. 8:169-174 (in Chinese, English summary).

Chen, Z., Chen, F., Chien, C., Wang, C., Chang, S., Hsu, H., Ou, H., Ho, Y., and Lu, T. 1979. A process of obtaining pollen plants of *Hevea brasiliensis* Muell.-Arg. Sci. Sin. 22:81-90.

Chen, Z., Qian, C., Qin, M., Wang, C., Suo, C. Xiao, Y., and Xu, X. 1981a. Relationship between somatic cells and microspores in the process of anther culture of *Hevea brasiliensis* Muell.-Arg. Acta Bot. Bor-Occ. Sin. 1:31-37 (in Chinese, English summary).

Chen, Z., Qian., C., Qin, M., Xu, X., Xiao, Y., Lin, M.D.Z., Wu, S., and Huan, N. 1981b. Recent advances in anther culture of rubber tree and sugarcane. In: Annual Report of the Institute of Genetics Academia Sinica, Beijing, China, p. 103.

Paranjothy, K. and Ghandimathi, H. 1975a. Tissue and organ culture of rubber tree. In: Symposium of International Rubber Conference.

REFERENCES

Bourgin, J.P. and Nitsch, J.P. 1967. Obtention de *Nicotiana* haploides as partir d'etamines cultivees in vitro. Ann. Physiol. Veg. 9:377-382.

Chen, C., Chen, F., Chien, C., Wang, C., Chang, S., Hsu, H., Ou, H., Ho, Y., and Lu, T. 1977. Induction of pollen plants of *Hevea brasiliensis* Muell.-Arg. In: Proceedings of Symposium on Anther Culture, pp. 3-8. Science Press, Beijing (in Chinese, English summary).

Chua, E.S. 1966. Studies on tissue culture of *Hevea brasiliensis*. J. Rubber Res. Inst. Malays. 19:292.

College of Tropical Crops in South China, 1980. Principles of Crop Breeding and Breeding of Rubber Tree. Forestry Press, Beijing (in Chinese).

de Fossard, R.A. 1978a. Tissue culture propagation of *Eucalyptus focifolia* F. Muell. In: Proceedings of Symposium on Plant Tissue Culture, pp. 425-438. Science Press, Beijing.

______ 1978b. Nuclear stocks multiplication rates and economic consideration of tissue culture propagation of horticultural species. In: Proceedings of Symposium on Plant Tissue Culture, pp. 439-447. Science Press, Beijing.

Edgar, A.T. 1958. Manual of Rubber Planting (Malaya). The Incorporated Society of Planters, Kuala Lumpur, Federation of Malaya.

Guo, G., Jia, X., and Chen, L. 1982. Induction of plantlet from ovule in vitro of *Hevea brasiliensis*. Hereditas 4(1):27-28.

Hepei Agricultural University, Division of Plant Breeding, Department of Horticulture. 1977. A new technique of using propionic-iron-carmine-chloral hydrate (PICCH) for chromosome staining in vegetable plants. Acta Genet. Sin. 4:82-84 (in Chinese, English summary).

Hu, H., Hsi, T., and Chia, S. 1978. Chromosome variation of somatic cells of pollen calli and plants in wheat (*Triticum aestivum* L.). Acta Genet. Sin. 5:23-30 (in Chinese, English summary).

Imle, E.P. 1978. *Hevea* Rubber: Past and Future. Econ. Bot. 32:264-277.

Ong, S.H. 1975. Chromosome morphology at the pachytene stage in *Hevea brasiliensis*: A preliminary report. In: International Rubber Conference, pp. 1-19.

Paranjothy, K. and Ghandimathi, H. 1975b. Morphogenesis in callus cultures of *Hevea brasiliensis*. Proc. Natl. Plant Tissue Cult. Symp. pp. 19-25.

Perry, B.A. 1943. Chromosome number and phylogenetic relationships in the Euphorbiaceae. Am. J. Bot. 30:527

Satchuthananthavale, R. 1973. *Hevea* tissue culture. Q. J. Rubber Res. Inst. Ceylon 50:91-97.

______ and Irugulbandara, Z.E. 1972. Propagation of callus from *Hevea* anthers. Q. J. Rubber Res. Inst. Ceylon 49:65–68.

Toraun, N.L. and Saryatmana, N. 1977. Tissue culture of *Hevea brasiliensis* Muell.-Arg. Menara Perkebunan 45:17-21.

Wang, T. and Chang, C. 1978. Triploid citrus plantlets from endosperm culture. In: Proceedings of Symposium on Plant Tissue Culture, pp. 463–467. Science Press, Beijing.

Wastie, R.L. 1975. Diseases of rubber and their control. Pest Articles and News Summaries 21:268–288.

Wilson, H.M. and Street, H.E. 1975. The growth, anatomy and suspension cultures of *Hevea brasiliensis*. Ann. Bot. 39:617–682.

Wilson, H.M., Eisa, M., and Irwin, S.W.B. 1976. The effects of agitated liquid medium on in vitro cultures of *Hevea brasiliensis*. Physiol. Plant. 36:399–402.

CHAPTER 20
Sugarcane

M.-C. Liu

HISTORY

Sugarcane is a member of the genus *Saccharum*, tribe Andropogoneae, family Gramineae. *Saccharum* species are usually highly polyploid, with no known diploids (2n = 2x = 20). Interspecific variability of chromosome number and maintenance of aneuploids are characteristic of sugarcane (Price, 1963, 1965). There are five species within the genus *Saccharum*.

First, *S. officinarum* L., the famous "noble" cane, has a chromosome number of 2n = 80. It originated in New Guinea and is now distributed throughout the tropics. There are six natural commercial varieties within this species: Batjan, Black Cheribon, Lahaina, Loether, Preanger, and Badila.

Second, *S. sinense* Roxb. and Jeswiet has a chromosome number of 2n = 106-120 and was called "Chinese cane" because it was collected by Roxburgh in 1796 in Canton, China. It was subsequently distributed and became established in India under the name Chinea, and was considered one of the best commercial cane varieties of the 1850s in that area. There are two natural varieties within this species, Uba and Tekcha.

Third, *S. barberi* Jeswiet originated in northern India. Barber (1918) recognized five major groups in this species: Sunnabile (2n = 82-116), Mungo (2n = 82), Nargori (2n = 107-124), Saretha (2n = 90-92), and Pansahi. Pansahi, not exclusive to India, was found also in Indochina, southern China, and Taiwan. Therefore, Jeswiet (1925) included it in the species *S. sinense*. There are two natural varieties in this species, Chunnee and Saretha.

Fourth, *S. spontaneum* L., with a chromosome number range of 2n = 40-128, is widely distributed in tropical southern Asia, Africa, and Taiwan. Its center of diversity is India. The famous natural variety in this species is Glagah. The varieties commonly grown in Taiwan have a chromosome number of 2n = 112.

Fifth, *S. robustum* Jeswiet, was collected in 1928 by Jeswiet and Brandes in New Guinea. It was distributed from Borneo to the New Hebrides with much variability in New Guinea. It has a chromosome number of 2n = 60-80. This species along with two others, *S. officinarum* and *S. spontaneum*, constitute the parental lines used to make modern commercial sugarcane varieties.

Each of the *Saccharum* species has its own origin, and hence the central origin of sugarcane is controversial. Brandes' proposition (Brandes, 1956) seems most reasonable. He proposed that Melanesia (New Guinea area) was the primary center of diversity of the 80-chromosome *S. officinarum*. From here, the *S. officinarum* canes migrated northwest to continental Asia where they were hybridized with local *S. spontaneum* to produce hybrid canes, resulting in the species *S. sinense*. These, the "thin" canes, survived seasonal monsoon climates of the northeastern Indian, southern Chinese area and became the local cultivated canes. The noble canes also migrated eastward across the Pacific to Polynesia and the adjacent islands, where somatic mutation has contributed to the diversity of the cultivated canes. For a more detailed account of the origin, evolution, and breeding history of sugarcane, see Stevenson (1965) and Daniels et al. (1974, 1975).

Most comercial sugarcane varieties now in use are descendants of interspecific hybrids within the genus *Saccharum* (Arcenaux, 1967). Many intergeneric hybrids have also been recorded between *Saccharum* and the genera *Erianthus*, *Miscanthus*, *Narenga*, and *Zea*. However, these hybrids have not had significant agricultural use.

Table 1 is a list of the major commercial sugarcane varieties (indicated by commonly used call letters) and the breeding stations responsible for their development.

ECONOMIC IMPORTANCE

Sugar is an important agricultural commodity for both export and domestic consumption in many countries of the world. World sugar output in 1980-81 was 86.97 million MT. Of this total, 61% was produced from sugarcane (*Saccharum* spp. hybrid) (Lichts, 1981). Brazil, Cuba, and India are the major producers. Table 2 is a list of the cane-sugar producing countries in the world based on relative importance in 1980-81.

KEY BREEDING AND IMPROVEMENT PROBLEMS

With the increase in the human population, sugar consumption is also surging. The breeding of high-sucrose cane varieties is certainly a major priority for sugarcane improvement. Because sugarcane is grown in tropical regions, it is vulnerable to attack by several major diseases

Table 1. Major Commercial Sugarcane Varieties

CANE VARIETY[a]	BREEDING STATION	COUNTRY
A	Antigua	West Indies
B	Central Breeding Station	Barbados
BO	Bihar, Orissa	India
CB	Campos	Brazil
CL	Clewiston (US Sugar Corp.), Florida	USA
Co	Coimbatore Breeding Station	India
CoL	Coimbatore, crossing; Lyallpnr, selection	India
CoS	Coimbatore, crossing; Shahja- hanpur, selection	India
CP	Canal Point, Florida	USA
D	Demerara	English Guyana
EK	(Bred by Mr. E. Karrhaus)	Java
F	Formosa (Taiwan Sugar Research Institute)	Taiwan
H	Hawaiian Sugar Planters' Associ- ation, Hawaii	USA
HQ	Hambled, Queensland	Australia
IAC	Instituto Agronomico, Campinas	Brazil
L	USDA Agriculture Experiment Station, Louisiana	USA
M	Mauritius Agric Experiment Station	Mauritius
Maur.	Mauritius Agric Experiment Station	Mauritius
MPR	Mayaguez	Puerto Rico
ML	Media Luna	Cuba
N:Co	Coimbatore, crossing; Natal, selection	South Africa
Phil	Philippine Sugar Institute	Philippines
POJ	Proefstation Oost Java	Java
PR	Puerto Rico Agriculture Experi- ment Station	Puerto Rico
R	Reunion	Reunion
Q	Queensland	Australia
ROC	Taiwan Sugar Research Institute	Taiwan
Tuc	Tucuman Agriculture Experiment Station	Argentina

[a]Designated by call number.

Table 2. World Cane-Sugar Production 1980-1981[a]

COUNTRY	PRODUCTION[b]
Brazil	8,350
Cuba	6,000
India	5,600
Australia	3,430
Mexico	2,790
United States	2,653
Philippines	2,500
China	2,250
South Africa	1,800
Argentina	1,700
Thailand	1,550
Indonesia	1,440
Colombia	1,265
Dominican Republic	1,050
Hawaii	950
Taiwan	757
Egypt	675
Pakistan	570
Mauritius	550
Peru	520
Guatemala	480
Fiji	425
Other Countries	5,938
Total	53,243

[a]Lichts, 1981.

[b]Raw value 1000 MT.

and insects. In the history of sugarcane cultivation, some cases of disease calamities, e.g., sereh and downy mildew, have threatened the existence of the industry (Stevenson, 1965). Therefore, breeding sugarcane for disease and insect resistance is a must. In reviewing the past 80 years of cane breeding, a cane-variety survey (Arcenaux, 1967) shows that the modern commercial canes were derived mostly from the germplasm of 20 nobles and fewer than 10 *S. spontaneum* derivatives of that species. Hence sugarcane breeders feel that the genetic base of commercial canes is too narrow, and urgently seek to increase genetic variability.

By taking advantage of the capacity of callus cells that undergo genetic changes in culture, many agriculturally useful callus-derived plants have been obtained (Larkin and Scowcroft, 1981b). In sugarcane, for instance, plants resistant to Fiji virus (Krishnamrthi, 1974) and smut fungus (*Ustilago scitaminea*) (Liu, 1981), and high-yielding, high-sucrose callus derivatives (Liu and Chen, 1978a,b) have been obtained. These results provide an argument for the use of tissue and cell culture techniques as tools for sugarcane improvement.

The purposes of this discussion are to illustrate the basic techniques of sugarcane tissue and cell cultures, the nature of shoot and root regeneration in vitro, the various strategies for use of in vitro culture methods for the modification of sugarcane genetic constitution, and the processes for obtaining agriculturally useful variants.

LITERATURE REVIEW

Research on sugarcane tissue and cell culture was started in Hawaii in 1961 by Nickell (1964), and its history has been reviewed by Heinz et al. (1977) and Liu (1981). The Taiwan Sugar Research Institute (TSRI) has been carrying out a program for the application of tissue culture techniques in sugarcane improvement since 1970 (Liu, 1971). Considerable work, especially on selecting variants resistant to Fiji virus and eyespot fungus, has been undertaken in Fiji (Krishnamurthi, 1974) and Australia (Larkin and Scowcroft, 1981a). Other research institutes in Florida (Lyrene, 1976; Vasil et al., 1979), the Philippines (Lat and Lantin, 1976), Brazil (Evans et al., 1980), and France (Sauvaire and Galzy, 1980) have also investigated sugarcane tissue and cell culture.

Callus induction and subsequent shoot differentiation were first reported by Heinz and Mee (1968). Suspension culture was initiated by Nickell and Maretzki (1969), and several nutrient and biochemical studies using suspension cultures have also been completed by the Hawaiian group (Maretzki et al., 1974). The initial site of callus cells in young leaf and subapical meristem explants was first identified by Liu et al. (1982). The configuration of an apical meristem surrounded by leaf primordia as a shoot differentiation pattern in callus of perhaps all graminaceous crops was first discovered by Liu and Chen (1974). Although Nadar and Heinz (1977) have proposed that the presence of 26.9 μM NAA in MS medium would stimulate the formation of roots in regenerated shoots, their finding has not been confirmed by other sugarcane tissue culture laboratories. Alternately, we have used SH medium devoid of auxin to induce root formation from differentiated shoots (Liu, 1981). Nevertheless, we (Liu and Chen, 1974; Nadar and Heinz, 1977) do agree that the shoot and root are developed independently and have different requirements in nutrient medium. Maretzki and Nickell (1973) were the first to isolate sugarcane protoplasts and to induce the formation of cell clusters. Later, Chen and Shih (1983) were able to obtain callus from protoplasts derived from suspension cells. For physiological stress studies, Liu and Yeh (1982) were the first workers to select a 0.26 M NaCl-tolerant cell line by using a plating technique. Vasil et al. (1979) was the first group to study nitrogen fixation in test tubes by establishing an association between sugarcane callus and the nitrogen-fixing bacterium *Azospirillum*. Chinese scientists (Z.H. Chen et al., 1979) have reported the regeneration of green plantlets from sugarcane anther culture. It should be mentioned that Krishnamurthi (1974) was the first to obtain a disease-resistant plant for crop improvement using cell culture methods. More importantly, Liu and Chen (1978a,b) were the first among sugarcane cell culturists to attain several high-yielding and high-sucrose callus-derived clones. Sugarcane cell culturists have contributed significantly to the basic and applied research in this area. Table 3 summarizes the most important results during the past two decades.

Broadening Genetic Variability By Using Cell and Callus Cultures

INDUCTION OF GENETIC VARIABILITY. Callus culture is mutagenic in the broadest sense, as callus cells maintained in vitro over a long period are usually cytologically unstable and give rise, after regeneration, to plantlets that are often characterized by genetic variability. Investigation into genetic variability occurring in cultured cells of differentiated plants has been concentrated on the analysis of chromosome number, morphological characters, and isoenzyme patterns. The results that have been published are summarized below.

Suspension Culture. Heinz et al. (1969) have reported that sugarcane suspension cells were variable in chromosome number such that most cells were aneuploid. Liu et al. (1977) found that chromosome numbers varied from 2n = 92 to 191 with a mean of 111 in the suspension-cultured cells of cv. F164 when compared with the donor plant chromosome number of 108. Other cultivars such as F156, F166, and F167, had a similar change of chromosomal variation (Liu and Shih, 1982).

Regenerated Plants. By studying morphological characters of eight populations of somaclones (a general term for separate plants regenerated from tissue culture), Liu and Chen (1976a) found that the most conspicuous change occurred in auricle length, which accounted for 8.6% of the change when regenerated plants were compared with donor clones. Dewlap shape ranked next (6.5%), followed by hair group (6.2%) and top leaf carriage (1.9%). The frequencies of morphological changes were dependent upon genotypes. Similar observations were also made by Heinz and Mee (1971).

Chromosome numbers varied from 2n = 86 to 126 in the F156 somaclones, representing greater variability than the donor chromosome number of 114. Somaclones of F164 varied from 2n = 88 to 108 as compared with their donor number of 108. F146-derived plants had chromosome numbers ranging from 2n = 104 to 118 contrary to the donor numer of 110 (Liu and Chen, 1976a). Heinz and Mee (1971) also found a chromosomal range between 2n = 94 and 120 among the 37 somaclones derived from H50-7209.

The isoenzyme patterns observed following polyacrylamide electrophoresis have been used to study the genetic constitution of regenerated plants (Thom and Maretzki, 1970). The somaclonal line 70-6132,

Table 3. Current Successes in Sugarcane Cell Culture Research

SUCCESS	EXPLANT	REFERENCE
Callus induction and shoot differentiation	Young internode, subapical meristem, young leaf, unemerged inflorescence	Heinz & Mee, 1968, 1969; Barba & Nickell, 1969; Liu, 1971; Lat and Lantin, 1976; Sauvaire & Galzy, 1980
Root regeneration	Neoformed shoots	Liu et al., 1972; Nadar & Heinz, 1977; Liu, 1981
Suspension culture	Young internode, subapical meristem, young leaf	Nickell & Maretzki, 1969; Heinz & Mee, 1970; Chen, 1978; Chen & Liu, 1982; Maretzki et al., 1974
Histological examination of callus origin	Subapical meristem, young leaf	Liu et al., 1982
Histological studies on organogenesis	Unemerged inflorescence, young leaf	Liu & Chen, 1974; Nadar et al., 1978
Genetic variability		
Suspension cells	Subapical meristem, young leaf	Heinz et al., 1969; Heinz & Mee, 1970; Liu et al., 1977
Regenerated plants		
Morphological characters	Subapical meristem, young leaf	Heinz & Mee, 1971; Liu & Chen, 1976a; Lat & Lantin, 1976
Chromosome number	Subapical meristem, young leaf	Heinz & Mee, 1971; Liu & Chen, 1976a
Isoenzyme patterns	Subapical meristem, young leaf	Heinz & Mee, 1971; Liu et al., 1977; Liu & Chen, 1978a
Nitrogen fixation in callus		Vasil et al., 1979
Induced mutations		
Colchicine	Subapical meristem, young leaf	Heinz & Mee, 1970; Liu et al., 1977
Gamma-ray	Subapical meristem, young leaf	Heinz, 1973; Liu & Chen, 1981
Protoplast		
Isolation	Young leaf	Chen & Liu, 1974; Krishnamurthi, 1976; Evans et al., 1980
	Suspension cells	Maretzki & Nickell, 1973; Chen & Shih, 1983

Table 3. Cont.

SUCCESS	EXPLANT	REFERENCE
Agriculturally useful somaclone		
Disease resistance		
Fiji disease	Subapical meristem, young leaf	Krishnamurthi, 1974; Krishnamurthi & Tlaskal, 1974
Downy mildew	Subapical meristem, young leaf	Krishnamurthi, 1974; Chen et al., 1979b
Eyespot disease	Subapical meristem, young leaf	Heinz et al., 1977; Larkin & Scowcroft, 1981a
Smut disease	Subapical meristem, young leaf	Liu, 1981
High yielding	Subapical meristem, young leaf	Liu & Chen, 1978a,b; 1979; Liu, 1981
High sucrose	Subapical meristem, young leaf	Liu & Chen, 1978b, 1980
Salt-tolerant cell line	Subapical meristem, young leaf	Liu & Yeh, 1982; Fitch & Morre, 1981a
Anther culture	Anther	Moore & Maretzki, 1974; Chen et al., 1979; Liu et al., 1980; Fitch & Moore, 1981b

derived from cv. F164, was devoid of two esterase bands when compared with F164, whereas other somaclones had the same bands as the donor tissue (Liu and Chen, 1978a). Three variant plants, C_1, C_2, and C_3, derived from 61-1248 (a breeding line), had two bands, three bands, and one band less, respectively, for esterase when compared with the donor tissue that had eleven bands (Liu et al., 1977).

OBTAINING AGRICULTURALLY USEFUL VARIANTS: High-Yielding Variants. To screen high-yielding lines, replicated field tests for selected somaclones were completed each year since 1974.

The results obtained in 1974-1975 indicated that 70-6132 had a higher cane yield (32%), sugar yield (34%), and stalk number (6%) than the donor, F164. Differences between 70-6132 and the donor for cane and sugar yield were significant at the 5% probability level, as shown in Table 4 (Liu and Chen, 1978a).

Among the F160 derivatives, 74-3216 performed well and has been released in regional field trials. Yield data collected from ten locations for new planting canes are presented in Table 4. Derivative 74-

Table 4. Comparisons Between Callus-Derived Plants and Their Donor Plants

	CANE YIELD			SUGAR YIELD		STALK NUMBER	
DERIVED PLANT AND VARIETY	Mean (t/ha)	CD/D[b] (%)	SUCROSE (%)	Mean (t/ha)	CD/D (%)	Mean (1000/ha)	CD/D (%)
F164[c]-70[d]-6132[e]	130.8[a]	132	13.53	17.7	133[a]	62.7	106
F160[c]-74[d]-3216[f]	147.1	106	12.81	18.8	110	63.3	106
F160[c]-75[d]-3070[g]	167.0	108	12.65[a]	21.1	115	58.0	97
F164 (donor)[e]	99.1		13.42	13.3		59.2	
F160 (donor)[f]	139.2		12.28	17.0		59.6	
F160 (donor)[g]	154.0		11.88	18.3		60.0	

[a]The difference between callus-derived plant and its donor is significant at the 5% level.

[b]CD/D = callus derived plant per donor.

[c]The donor.

[d]Year in which the callus-derived plants were transplanted into the field from a sterilized medium.

[e]Derived plants (assigned by a serial number) and its donor tested in 1974/1975 at one location (Liu and Chen, 1978a).

[f]Derived plant and its donor tested in 1981/1982 at ten locations (Liu and Chen, 1982b).

[g]Derived plant and its donor tested in 1978/1979 at one location (Liu and Chen, 1979).

3216 was higher in cane yield (6%), sugar yield (10%), and stalk number (6%) than its donor (Liu and Chen, 1982). Another of the somaclones from F160, 75-3070, was 8% and 15% higher in cane yield and sugar yield, respectively, than its donor (Liu and Chen, 1979). Although these two somaclones, 74-3216 and 75-3070, were only 10-15% higher in sugar yield, this achievement is probably the first of its kind and suggests that agriculturally useful, callus derived plants can be selected using somaclonal variation (Larkin and Scowcroft, 1981b).

High-Sucrose Variants. The sucrose contents of selected high-sugar somaclones and their donors were compared in paired groupings at each analyzed data (Table 5). Among these lines, 71-4829 is an example whose percentage sucrose has been consistently higher than its donor, H37-1933, since it was first established. The difference is significant at the 1% probability level, suggesting that 71-4829's high-sucrose level is genetically stable. The same results were obtained when 75-3070 was compared to F160. Unfortunately, 75-3070 was sensitive to sclerotic disease and therefore was not entered into regional field tests (Liu and Chen, 1980).

Table 5. High-Sucrose Lines Selected from Callus-Derived Plants

DATE OF ANALYSIS	AVAILABLE SUCROSE (%)[a]			
	71-4829	Donor (H37-1933)	75-3070	Donor (F160)
October	11.12	9.12	9.50	8.92
November	13.11	11.18	11.96	10.44
December	13.25	10.92	13.16	11.88
January	15.33	11.99	14.16	13.89
February	14.64	12.51	14.46	14.20
Average	13.49	11.14	12.65	11.87
Mean difference	2.35[c]		0.78[b]	

[a] Average of data collected from 3-4 consecutive years.

[b,c] Significant at the 5% and 1% probability levels, respectively, according to paired t-tests.

Disease-Resistant Variants. Beginning in 1970, Krishnamurthi (1974) and Krishnamurthi and Tlaskal (1974) began screening the somaclones of a number of varieties for reaction to Fiji disease (a leafhopper-transmitted virus) and downy mildew (*Sclerospora sacchari*). Among the 300 somaclones derived from varieties susceptible to Fiji disease, four were resistant and one had intermediate resistance. Similar screening for downy mildew resistance using susceptible varieties produced some variants that were more resistant than donor clones. Some of the Pindar somaclones resistant to both Fiji and downy mildew diseases have been tested for yield in Fiji (Krishnamurthi, 1974) and tested independently in Australia (Larkin and Scowcroft, 1981b). These lines did not have reduced sucrose yields when compared to Pindar; in fact, some lines had shown a slightly increased yield, though this was not statistically significant.

Larkin and Scowcroft (1981a) tested the 260 somaclones derived from a clone susceptible to an eyespot disease (*Helminthosporium sacchari*), viz. Q101, for disease resistance. They found that 8.9% of the somaclones were highly resistant or nearly immune. These clones are otherwise indistinguishable for the donor lines in gross morphology.

Cultivar F177 was the No. 2 commercial variety during 1980-81 in Taiwan. Its major limitation is high susceptibility to smut disease. Among the nine somaclones tested, one of them, 76-5530, was found to have only a 43.8% infection rate when compared with its donor's 88.2%. This line also showed a slight increase in sucrose content (Liu, 1981; Liu and Chen, 1982b).

Salt-Tolerant Cell Lines. Suspended cells of cultivar F164 were cultured in a serial liquid MS-C medium salinized with increasing the NaCl concentration by 0.02 M during each subculture for 4-8 passages. All the cultures, no matter what NaCl concentration was initially used or how many salinized passages were used, could only tolerate 0.16 M NaCl. When the cells were inhibited from further growth they were plated onto agar-solidified MS-C media containing 0.17 and 0.26 M NaCl. Within 50 days of cultivation, more than 99.9% of plated cells were killed by the high concentrations of salt. However, a few cell aggregates survived and grew into small colonies. Colonies grown from the 0.26 M NaCl plate are referred to as the 0.26 M NaCl-selected line.

There were two notable characteristics concerning the salt-selected cell line. First, the salt-selected cell line grew faster in high salt medium. The growth of the 0.26 M NaCl-selected cell line and nonselected cell lines from the original population were tested on MS-C agar media containing a gradient of NaCl concentrations (0, 0.086 M, 0.17 M, and 0.26 M). After 40 days of culture, the growth index of these cultures was measured and is summarized in Table 6. The results suggest that the salt-selected cell line grew slower than the nonselected line with no added NaCl, but grew faster in 0.17 M and 0.26 M NaCl. This suggests that the salt-selected line has become tolerant to a high-salt environment. Second, the salt-selected line absorbed higher concentrations of Na^+ and Cl^-. The salt-selected callus absorbed less K^+ than its counterpart, however, this line contained much more Na^+ and Cl^- than the control.

From these two characteristics it is suggested that a general shift has occurred toward the halophytic mode of salt tolerance within the salt-selected cell line (Liu and Yeh, 1982).

Induction of Genetic Variability through Chemical and Physical Mutagens

COLCHICINE. By treating suspension cells of cv. F164 with 0.25 mM colchicine, the chromosome number increased to as many as 309

Table 6. Salt-selected and Nonselected Cell Lines' Growth Index and the Ratio of Callus Dry Weight to Fresh Weight After 40 Days Cultivation

	FINAL FRESH CALLUS GROWTH INDEX[a]		RATIO: FINAL DRY WT/FINAL FRESH WT	
NaCl CONCENTRATION IN MEDIUM (%)	Nonselected cell line	Salt-selected cell line	Nonselected cell line	Salt-selected cell line
0	9.6[+]a[*]	8.8 a	0.051	0.184
0.5	3.8 b	4.7 b	0.278	0.279
1.0	1.4 bc	2.9 c	0.343	0.335
1.5	0.2 c	1.0 d	0.327	0.318

[a]Growth index = [final callus fresh weight (F.W.) - initial callus F.W.]/initial callus F.W.

[+]Values are means of five replications.

[*]Means followed by one or more letters in common are not significantly different at the 5% probabilty level.

per cell (see Fig. 22), almost tripling the original donor chromosome number of 108. The same induction procedure was applied to clone 61-1248 (a breeding line). Three variant plants, viz. 61-1248-C_1, -C_2, and -C_3, have been regenerated from colchicine-treated cells. Their chromosome number increased to 2n = 156, which is aneuploid when compared with the donor chromosome number of 104. Plants having such high chromosome numbers were stunted and lacked vigor (Liu et al., 1977). Heinz and Mee (1970) also obtained, by using a similar induction procedure, 46 variant plants among which 22 were polyploids, 2 were mixoploids, and 22 were diploids.

GAMMA-RAY. The calli of cv. F177 that contained numerous green shoot primordia were irradiated by gamma-rays (Union Industrial Research Laboratories, Hsinchu, Taiwan) with various dosages. The calli irradiated with 8, 12, 14, and 20 Krad doses were transferred to MS-C medium, whereas those treated with 4, 6, 8, and 10 Krad doses were transferred to MS-D_{56} medium for further growth. All treated calli were cut into 0.5-cm^2 pieces and inoculated at a spacing of 2 x 2 cm on a thin-layer (3-4 mm) MS-C medium in a 90-cm petri dish. At this time there were approximately 10-15 green nodules (small shoot buds) on each piece of callus. After one month of cultivation, these green-buds were examined and the data is summarized in Table 7.

The shoot-buds grown on MS-C and MS-D_{56} media exhibited different growth responses. Because MS-C medium contained a higher level of auxin (13.6 μM 2,4-D), the shoot-buds grew normally even though they received as high as a 16 Krad dose. Apparently the dose at 8 Krads had a stimulating effect on the growth of the shoot-buds. While examining shoot-buds grown on MS-D_{56} medium, it was observed that growth was both poor and vulnerable to irradiation damage. The LD_{50} for shoot-buds on MS-D_{56} was 10 Krad in contrast to the little damage caused by 20 Krad on the shoot-buds grown on MS-C (Liu and Chen, 1981).

Surviving shoots on both media may be raised to maturity according to procedures described in the protocol section of this chapter. Selection against the donor's major defect—smut disease susceptibility—will be practiced in later cane generations.

So far we have not found another report that successfully used gamma-rays to induce higher mutation frequency in callus culture (cf. Heinz, 1973).

PROTOCOLS FOR CALLUS CULTURES

The basic equipment and procedures for conducting plant cell culture research have been described in Vol. 1 of this series. The following methods and procedures have been successfully applied to sugarcane.

Production of Callus-Derived Plants

SHOOT REGENERATION. Tissue Used. The following three tissues are most frequently used: (a) subapical meristematic tissue including the

Table 7. Growth Response of Irradiated F177 Shoot-buds on Different Media[a]

DOSE (Krad)	MS-C MEDIA			MS-D56 MEDIA		
	Poorly growing[b] (%)	Normally growing[c] (%)	Callus-bud growth vigor[d]	Poorly growing[b] (%)	Normally growing[c] (%)	Callus-bud growth vigor[d]
0	2	98	+	5	95	+++
4	0	0		27	73	++
6	0	0		25	75	++
8	8	92	++	30	70	+
10	0	0		43	57	+
12	10	90	+	0	0	
16	12	88	+	0	0	
20	17	83	-	0	0	

[a]Callus bud per 0.5 cm^2.

[b]Poorly growing defined as survival rate at 0-30%.

[c]Normally growing defined as survival rate at 31-100%.

[d] - indicates no growth; + to +++, weak to vigorous growth of shoot-buds.

first to eighth internodes (node designation is according to Artschwager, 1925), (b) the second innermost young leaves surrounding the shoot apex in lengths of 2.0-2.5 cm, and (c) unemerged inflorescences.

Culture Medium. The nutrient medium contains the salts according to Murashige and Skoog supplemented with various vitamins and plant growth regulators as summarized in Table 8. Callus was initiated and shoots differentiated using variations of MS medium. Roots formed from regenerated shoots using a modified SH medium (Schenk and Hildebrandt, 1972).

Excision Procedures. Because the apical and the rolled young leaf tissues are wrapped deep in leaf sheaths, they are generally pathogen-free, and thus it is not necessary to disinfect these explants. The outer leaf sheaths are stripped off approximately to the tenth node (Fig. 1). The other tender sheaths are twisted carefully until they are broken off at the layer of insertion of the leaf sheath on the nodes until only the first and second innermost leaf sheaths remain attached to the shoot apex. The apical dome (Fig. 2) and the 2-cm portion of the young leaf (Fig. 3) are separately cut off and immediately inoculated onto the MS-C medium. A brown exudate is frequently obtained from the apical dome. Control of this phenomenon has been discussed in a previous paper (Liu, 1981).

Culture Conditions. Cultures were maintained at 26-28 C under cool white fluorescent light of an intensity of about 3800 lux for a 12-hr photoperiod.

Callus Initiation. As early as 4 days after explantation, callus is initiated. The callus cells from the leaf explant originate from the primary phloem parenchyma, whereas those from the apical dome arise from the primary phloem as reported by Liu et al. (1982). After 6 weeks cultivation, a considerable mass of callus can be accumulated as shown in Fig. 2 (apical dome), Fig. 4 (young leaf), and Fig. 5 (young inflorescence). Among the three explants used, the highest frequency of both producing callus and regenerating plants is obtained using the inflorescence.

Utilization and Maintenance of Callus. After 4-5 weeks of growth, each piece of callus is subdivided into two separate culture tubes—one containing MS-C medium for callus proliferation and one with MS-D_{56} medium for plant differentiation.

Sequential Processes of Shoot Regeneration. The current concept regarding in vitro plant regeneration suggests two pathways: (a) organogenesis resulting in shoot formation, and (b) somatic embryogenesis (Kohlenbach, 1977). Histological examinations of the processes of shoot and root differentiation from sugarcane callus have been reported (Liu and Chen, 1974; Nadar et al., 1978). Based upon the distinction between organogenesis and embryogenesis, it seems inappropriate to apply the term embryogenesis to the sugarcane morphogenesis reported to date. In addition to justifying the term, this section will describe in steps the process of shoot differentiation from a sugarcane callus mass.

Table 8. Media for Callus Induction, and Shoot and Root Differentiation[a]

INGREDIENT	CALLUS FORMATION MS-C[b] (mM)	SHOOT DIFFEREN-TIATION MS-C[b] (mM)	ROOT FORMATION Modified SH[c] (mM)
NH_4NO_3	20.6	20.6	-
KNO_3	18.8	18.8	24.7
$NH_4H_2PO_4$	-	7	300
KH_2PO_4	1.25	1.25	-
$CaCl_2 \cdot 2H_2O$	2.99	2.99	1.36
$MgSO_4 \cdot 7H_2O$	1.5	1.5	1.62
H_3BO_3	0.1	0.1	0.08
$MnSO_4 \cdot 4H_2O$	0.1	0.1	-
$MnSO_4 \cdot H_2O$	-	-	0.06
$ZnSO_4 \cdot 7H_2O$	0.03	0.03	0.003
KI	0.005	0.005	0.006
$Na_2MoO_4 \cdot 2H_2O$	0.001	0.001	40.0
$CuSO_4 \cdot 5H_2O$	10.0	10.0	40.0
$CoCl_2 \cdot 6H_2O$	10.0	10.0	40.0
Na_2 DTA	0.1	0.1	0.1
$FeSO_4 \cdot 7H_2O$	0.1	0.1	0.1
Myo-inositol	0.55	0.55	5.5
Thiamine•HCl	300.0	300.0	148.0
Pyridoxine•HCl	-	-	240.0
Nicotinic acid	-	-	0.4
Casein hydrolysate	-	400.0 mg/l	-
CW (v/v)	10%	10%	10%
2,4-D	13.5 μM	-	-
α-Naphthalene-acetic acid	-	5.4 μM	-
Kinetin	-	5.4 μM	-
Sucrose	5.8	5.8	5.8
Agar	9000 mg/l	8000 mg/l	8000 mg/l
pH (adjusted with 1 N NaOH)	5.7	5.7	5.7

[a]Used by Taiwan Sugar Research Institute.

[b]Modified Murashige & Skoog (1962) Medium.

[c]Modified Schenk & Hildebrandt (1972) Medium.

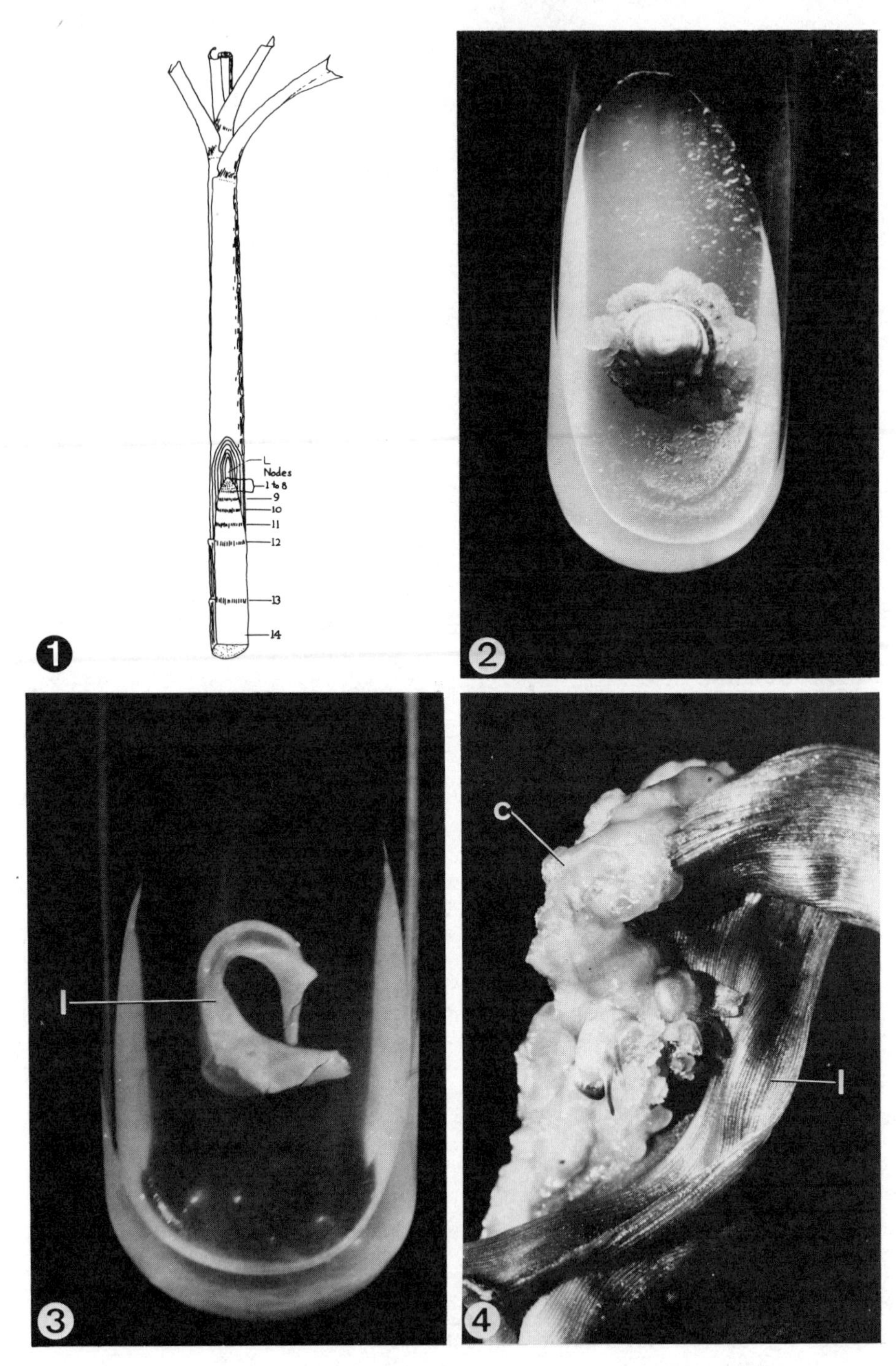
L
Nodes
1 to 8
9
10
11
12
13
14
1
2
I
3
C
I
4

Figure 1. Top portion of a sugarcane stem with leaf sheaths removed and stem tip split open to expose the two tissues to be excised for explantation: 1—second innermost young leaf (L); 2—stem-tip tissue, nodes 1-8.

Figure 2. A stem-tip tissue and the callus mass produced from it. Six weeks after explantation.

Figure 3. A piece of meristematic leaf (1) excised from just above a stem tip. One day after explantation.

Figure 4. Callus mass (c) developed from a piece of meristematic leaf (l). Six weeks after explantation.

Calli in the third or fourth passages derived from the immature inflorescence of cv. F162 were used as the study material. The method for preparation of paraffin sections is similar to that previously described (Liu and Chen, 1974).

Callus from MS-C medium (containing 13.6 μM 2,4-D) usually has relatively large parenchymous cells in the central portion, and compact, small cells, which are characterized by isodiametric shapes and densely staining nuclei, in the peripheral zone (Fig. 6). This configuration agrees well with the observations of Gautheret (1959). Occasionally, some part of the peripheral region develops into meristematic centers as shown in the upper left corner of Fig. 6. When callus grown on MS-C is transferred to MS-D_1 medium (MS-C medium minus 2,4-D plus 0.35 mM dalapon) for a further 16 days of growth, numerous nodule-like meristemoids can be formed at the peripheral region (Fig. 7). These meristemoids are presumably developed from the preexisting meristematic centers. With further growth (23 days after subculture), these meristemoids can progressively develop into shoot apices, each with an apical meristem in the center and leaf primordia surrounding it (Fig. 8). This configuration is very similar to the sugarcane stem apex (Artschwager, 1925). At this stage, numerous green dots emerge on the callus surface. With further growth of 3-5 days, when the young leaves grow larger, the shoot apex is found to be covered by a coleoptile-like structure (Fig. 9). The shoot apex develops into a shoot-bud 30-35 days after subculture (Figs. 10 and 11). The shoot-bud looks leafy and succulent. Its appearance is quite different from a shoot derived from sugarcane seed. The shoot-buds gradually grow larger and eventually cover the whole surface of the callus (Fig. 12). When the shoots grow to 1.5-2.5 cm, the cluster of shoots (Fig. 13) is transferred to a modified SH medium (Fig. 18) for root induction. Usually, there are 20-35 plantlets per culture tube (30 x 150 mm, containing 13 ml of agar medium).

Enhancement of Shoot Regeneration Frequency. In the past 10 years, we have studied methods for increasing the frequency of plant regeneration from sugarcane callus. In the early years of our studies, the differentiation medium used was MS-D_1 medium (Liu et al., 1972). This medium results in a regeneration rate of only 13.5% per passage (4

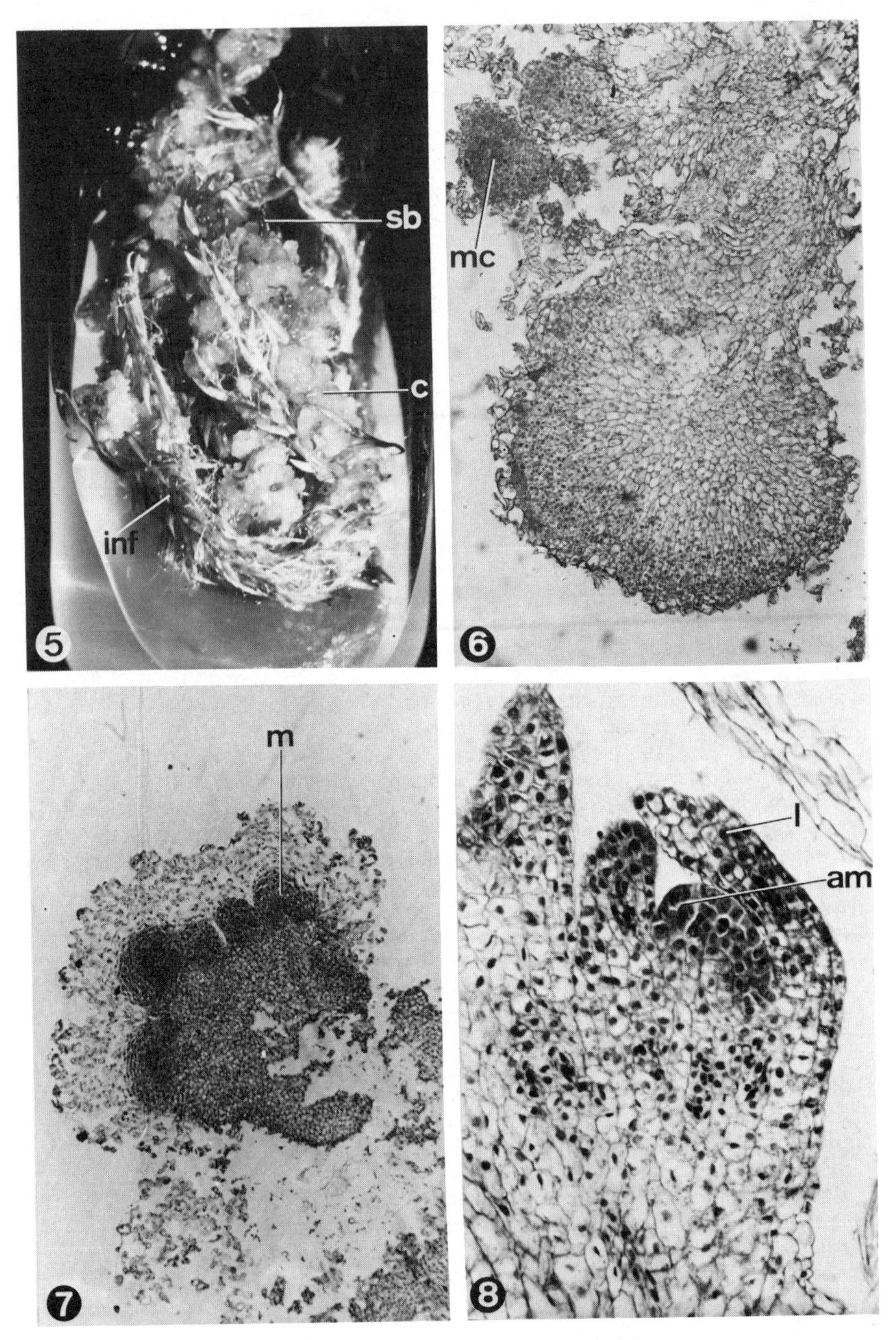
sb
mc
c
inf
5
6
m
l
am
7
8

Figure 5. Callus mass (c) proliferating vigorously from young inflorescence tissue (inf) (mostly from rachis). Note numerous shoot-buds (sb) differentiated directly from the callus. Five weeks after explantation.

Figure 6. A transverse section of a callus grown on 2,4-D-containing MS medium. Note that some meristematic centers (mc) are present in the callus.

Figure 7. A transverse section of a callus grown on 2,4-D-free MS medium. Note that numerous nodule-like meristemoids have been developed at the peripheral region. Sixteen days after subculture.

Figure 8. Longitudinal section of a neoformed shoot apex from callus, twenty-three days after subculture; am—apical meristem; l—leaf primordia.

weeks per passage) (Fig. 14). Figure 14 represents the average of 3500 units counted, representing 50 cane cultivars. The highest regeneration frequency is obtained at the second passage when the initial explant culture is considered the first passage. The shoot differentiation frequency decreases stepwise and usually terminates at the eighth passage. In order to increase the regeneration frequency, the MS medium must be supplemented with various plant growth regulators. Among them, MS-D_{56} raises the regeneration frequency from 13.5% to 24.2% (Fig. 14). The composition of MS-D_{56} is identical to that of MS-D_1 but minus dalapon, and with the addition of 5.3 μM NAA, 4.65 μM KIN and 400 mg/l CH. The regeneration capacity has also been extended to the eleventh passage; however, after that only rarely are shoots regenerated.

ROOT REGENERATION. During sugarcane regeneration the shoots and roots initiate independently. Twenty-seven days after transferring to MS-D_1 medium, some callus regenerates (Fig. 15, top). When this callus is examined microscopically, it is found that root primordia form as early as 14 days after subculture (Fig. 16). However, callus with root primordia rarely differentiate into shoots. Therefore, no structures such as embryo-radicle axes can be found in a regenerating callus. On the other hand, regenerated shoots of most cane cultivars have difficulty forming roots when grown on MS-D_1 medium. Thus, a special effort has to be made to stimulate root initiation.

When the shoot-regeneration-promoting capacity of a modified SH medium (Table 8) was tested, it was incidentally found that numerous roots appear on the surface of the callus (Fig. 15, bottom). A microscopic section of this callus shows that as early as 13 days after subculture, numerous tender roots have been developed from the callus (Fig. 17). This suggests that auxin-free SH medium might be effective for root induction. Thus, the neoformed shoots, when grown to the size shown in Fig. 13, are cultured on modified SH medium. Fifty days later, profuse roots are induced from the shoot-crowns (Fig. 18). Since then, the modified SH medium has been routinely used for rooting the shoots.

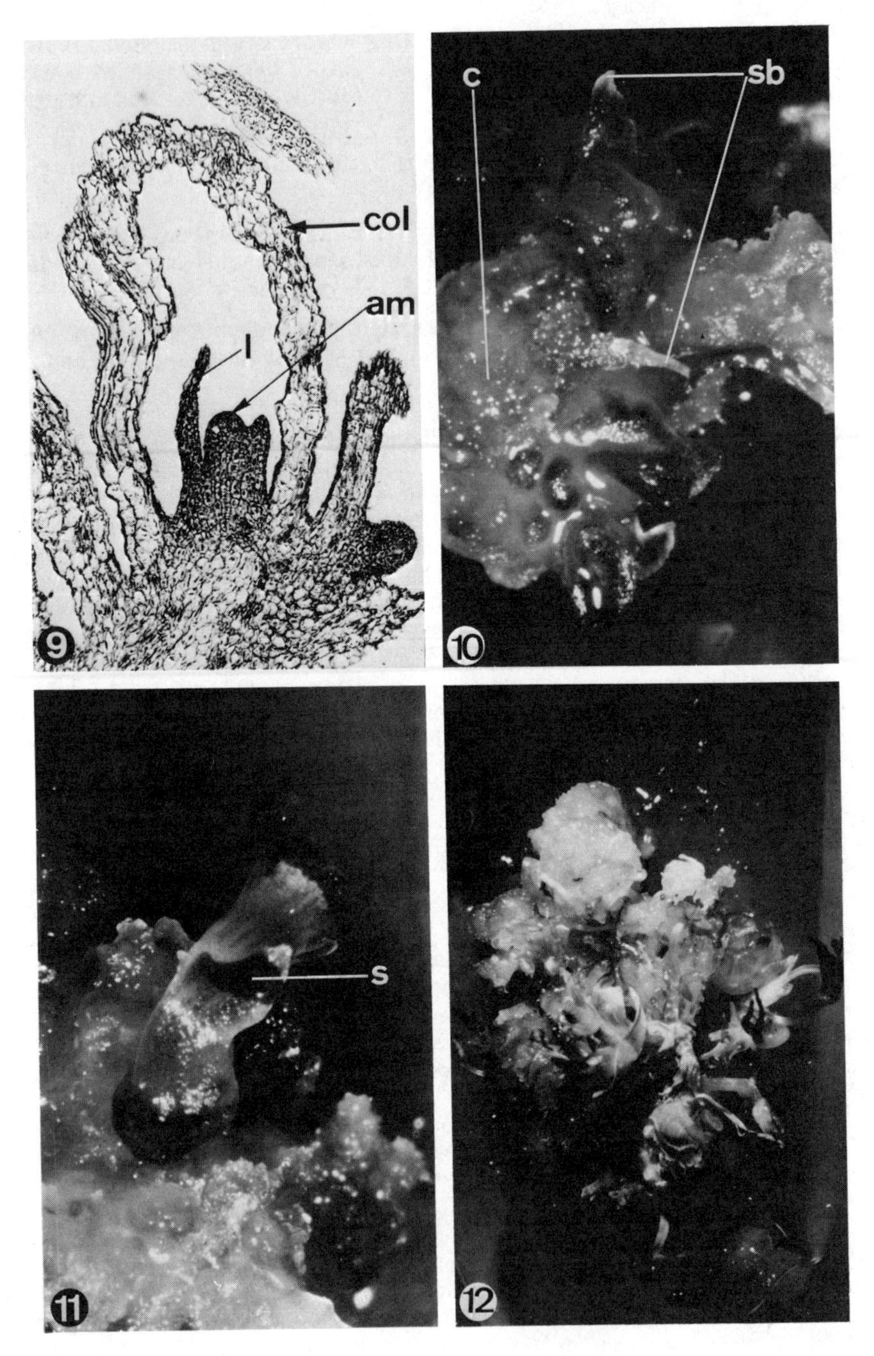
col
am
l
9
c
sb
10
s
11
12

Figure 9. Longitudinal section of a neoformed shoot apex from callus twenty-eight days after subculture; am—apical meristem; l—leaf primordia, col—coleoptile-like structure.

Figure 10. Numerous shoot-buds (sb) differentiated from the surface of callus, thirty-one days after subculture.

Figure 11. The shoot-bud gradually develops into a leafy and succulent shoot (s).

Figure 12. Larger and vigorously growing shoots derived from a callus mass, forty days after subculture.

SUCCESSFUL REARING OF SEEDLINGS. When the plantlets grow to a height of 6-9 cm in flasks (Fig. 18), they are transferred to vermiculite in pots (Fig. 19) and watered with Hoagland solution every other day. When reaching 15-18 cm, they are transferred to an intermediate bed filled with soil. With further growth for 4-6 weeks, they are ready to be transplanted into the field under regular sugarcane spacing: 35 cm between plants in rows with 1.37 m between each row. Donor plants should be planted in adjacent rows (Fig. 20).

Selection of Callus-Derived Plants

CHROMOSOME NUMBER DETERMINATION. To count the chromosomes in pollen mother cells, young spikelets from callus-derived plants are fixed in Farmer's fixative (95% ethanol:acetic acid = 3:1; Evans and Reed, 1981), and squashed in 4% iron aceto-carmine. Somatic cells are examined by the leaf-squash method of Price (1962). Five to ten cells per plant should be examined.

YIELD TEST. Cuttings excised from the second year, asexually propagated, callus-derived plants and their donors are planted (0.35 m apart) in 8-row plots in autumn, each row having a size of 6 x 1.37 m. The experimental array should be a randomized complete block design with 2-3 replicates. Canes are harvested about 12-16 months after planting. Data are collected at harvest and subjected to analysis of variance according to the field design.

SUCROSE ANALYSIS. Every promising callus-derived plant is first checked for field Brix by means of a hand refractometer. If Brix values are high, lines are subjected to detailed chemical analyses using procedures described in the Cane Sugar Handbook (Taiwan Sugar Corporation, 1956). Since sugar content is influenced by environmental factors such as temperature, soil moisture, age, and sampling technique, callus-derived plants and their donors are always grown side by side and their sugar data, either field Brix or percent sucrose, are collected at the same time to insure reliable comparison. To determine the

13

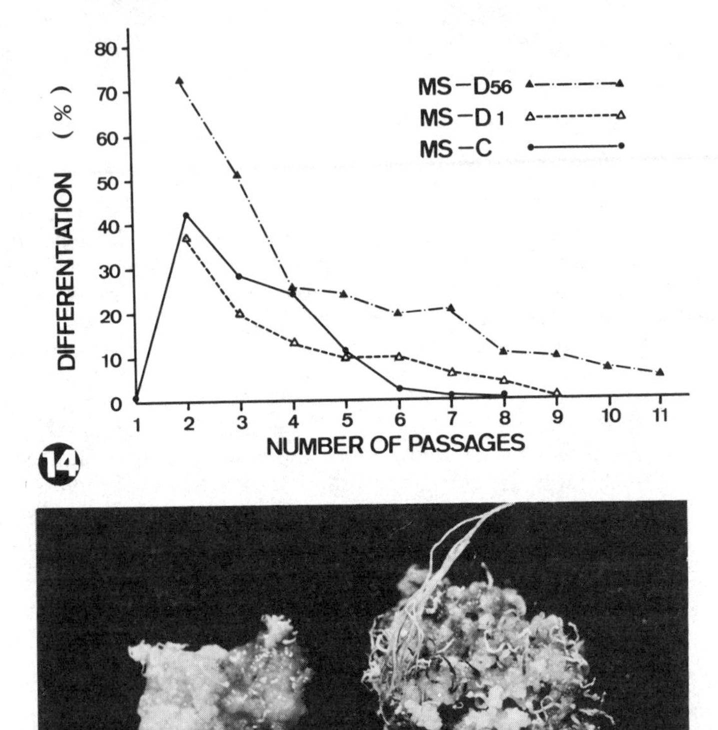
DIFFERENTIATION (%)
0
10
20
30
40
50
60
70
80
MS−D56
MS−D1
MS−C
1
2
3
4
5
6
7
8
9
10
11
NUMBER OF PASSAGES
14
15

Figure 13. Fifty-day-old plantlets which are at the correct stage to be transferred to SH medium for root initiation.

Figure 14. Differentiation frequency of callus-shoots as a function of the number of passages (explant itself as the 1st passage) grown on MS-D_{56}, MS-D_1, and MS-C media. Callus sources involve stem tip tissues and 2nd innermost leaf only. Note that the MS-C medium can give rise to 0.9% regeneration rate at the first passage.

Figure 15. Top: Some tender roots differentiated from a MS-D_1-medium-growing callus. Bottom: Numerous roots differentiated from the callus grown on a modified SH medium.

average sucrose content of callus-derived plants and donor plants, a pair of tests is run of all high-sucrose lines (Steel and Torrie, 1960).

DISEASE RESISTANCE TEST. At the present time, disease resistance tests have been concentrated on smut disease. The reaction of neoformed plants derived from a highly susceptible clone, e.g., F177, is tested by the wound-paste method described by Leu and Teng (1974). Data is analyzed using a chi-square test to determine if there is a significant difference in smut infection between callus-derived lines and respective donor clones.

PROTOCOL FOR SUSPENSION CULTURES

Cell suspension cultures have been used to study the plant cell's physiological, biochemical, and metabolic characteristics. Recently emphasis has been placed on the production of primary and secondary plant products using cell cultures. Detailed discussions of plant suspension culture techniques have been published (Street, 1973, 1977; Fowler, 1977; Kibler and Neumann, 1980). The following procedures have been applied to sugarcane suspension culture.

1. Select friable callus for inoculation, usually at the fifth to seventh passage.
2. Inoculate callus (ca. 1.0 g) into 50 ml of liquid MS-C medium (containing 13.6 µM 2,4-D and 87.6 mM sucrose) in 250 Erlenmeyer flasks.
3. The cultures are agitated on a gyrotory shaker (120 rpm) for 7 days to release single cells and aggregates.
4. The cell culture is passed through a filter (300-µm pore diameter). Residual cells on the filter are discarded and the filtrate is collected.
5. Pool the several (three) filtrates together and centrifuge at 100 x g for 5 min.
6. Collect the pellet and inoculate into freshly prepared MS-C medium contained in Nephelo flasks with side arms (Manasse, 1972).
7. The growth of the cell cultures is measured every other day by swirling the culture, tilting toward the sidearm, and reading the height of the settled cell volume (SCV) after 10-15 min.

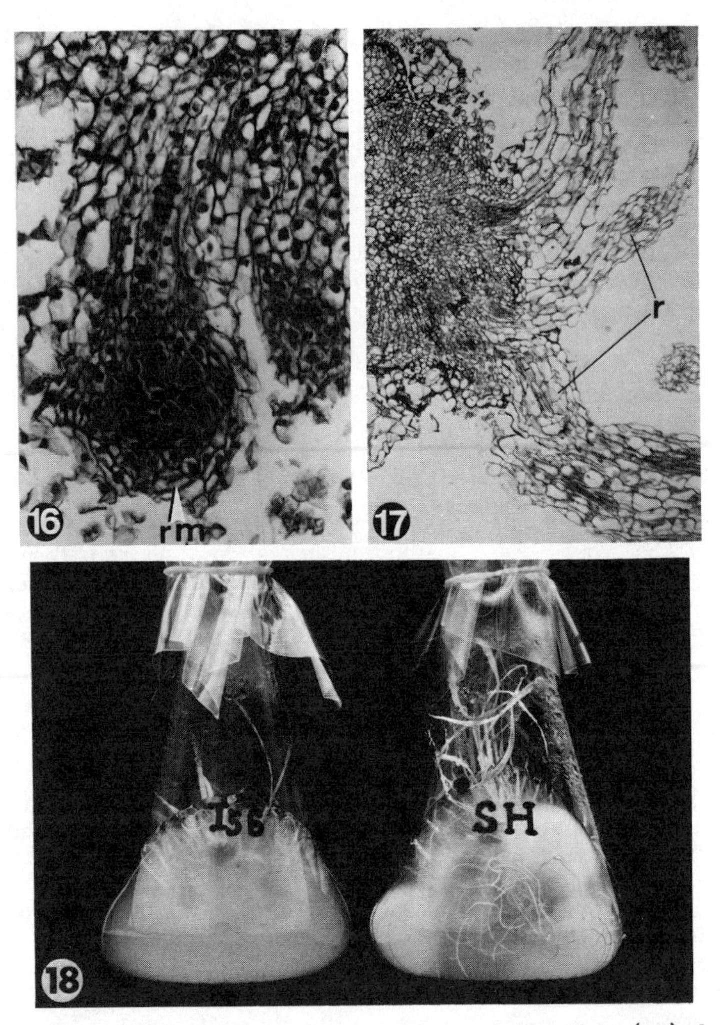

Figure 16. A longitudinal section of a root primordium (rm) developed from a MS-D_1-medium-growing callus, fourteen days after subculture.

Figure 17. A longitudinal section of root system (r) developed from a SH-medium-growing callus, thirteen days after subculture.

Figure 18. Modified SH medium is effective in stimulating rooting of neoformed shoots as compared with I_{56} which is identical to MS-D_{56} in Table 8.

After 3 months of medium-specific adaptation, the culture should be composed of smaller aggregates containing 20–50 cells that appear white in color (Fig. 21). Whether the degree of dispersion is complete or not depends on the genotype of the cane. It is often difficult to initiate a cell culture for cultivars that contain large amounts of phenolic compounds. In such cases, the composition of the medium must be

Figure 19. Fully developed plants growing in a pot of vermiculite.

Figure 20. Callus-derived plants undergoing field testing along with their donor clones.

varied. The MS-C medium with CW deleted and supplemented with 0.29-0.57 mM L-cysteine·HCl works well for the cultivars that produce phenolics. More detailed discussions on factors controlling the establishment and development of plant suspension cultures have been discussed by Dougall (1980).

PROTOCOLS FOR PROTOPLAST CULTURE

Since the development of enzymatic procedures to isolate protoplasts from plants by Cocking (1960), plant protoplasts have become important experimental tools for studying parasexual hybridization, genetic transformation, and mutant selection (Cocking, 1980; Gamborg et al., 1981). We have isolated and cultured protoplasts of sugarcane to develop a system applicable to sugarcane.

From Young Leaves of Sugarcane

The second innermost leaves just above the shoot apex are used as donor material. The leaves are cut into 1-mm square pieces and immersed in an enzyme mixture that is agitated on a reciprocal shaker (120 rpm) at 32 C. At 1-hr intervals a small sample of leaf-enzyme mixture is taken and the protoplasts are counted using a haemocytometer. The enzyme mixture of Onozuka and Macerozyme in a ratio of 4:3 in 0.7 M mannitol gives the highest yield of protoplasts (22 x 10 /g

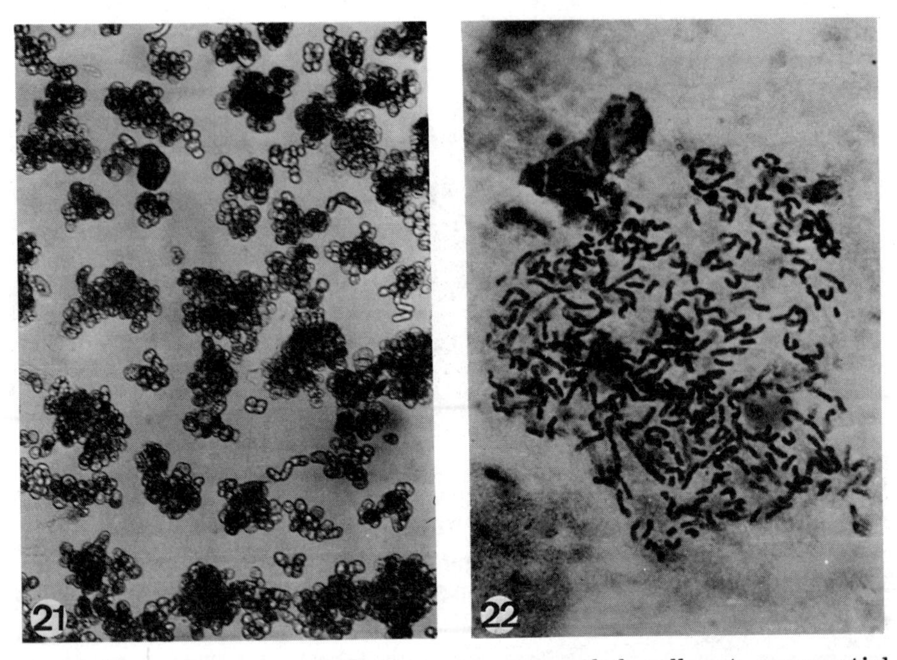

Figure 21. Appearance of sugarcane suspended cells at exponential phase of growth (Liu and Yeh, 1982).

Figure 22. A cell can have as many as 309 chromosomes as the result of treatment with 250 μM colchicine (Liu et al., 1977).

fresh weight) after 2 hr incubation (Chen and Liu, 1974). About 7% of the freshly isolated protoplasts have more than one nucleus, presumably the result of spontaneous fusion. After culturing for 24 hr, new cell wall synthesis can be observed (Chen and Liu, 1976). The protoplasts undergo first cell division within 2-5 days if cultured using a hanging drop method. Cell clusters are formed after 3 weeks cultivation (Liu and Chen, 1976b).

From Sugarcane Suspension Culture

PROTOPLAST ISOLATION. One ml of 2-5-day-old cells are collected from suspension cultures of sugarcane cv. F164 and mixed with 10 ml of a filter-sterilized enzyme solution—4.0 % cellulase Onozuka R10, 1.0% glusulase (Endo Lab. Inc., New York), 0.35 M D-mannitol, 0.35 M sorbitol, 5 mM $CaCl_2 \cdot 2H_2O$, and 1 mM $Ca(H_2PO_4)_2 \cdot H_2O$. The cell-enzyme mixture is incubated on a gyrotary shaker (100 rpm) at ca. 28 C. After 2-3 hr incubation, about 27% of the cells are converted into protoplasts. Treatment times shorter than 1 hr or longer than 4 hr decrease the quality of protoplast production. The growth stage of cell culture is critical for optimum protoplast release. Usually, cells at stationary growth phase fail to liberate protoplasts. The pH of the enzyme solution also influences the rate of protoplast release. The lower the pH, the higher the protoplast yield; pH 4.0 is optimum. After release, the protoplasts are washed three times with P_6 medium by low speed centrifugation (90 x g, 10 min). The P_6 medium is composed of MS salts and the following: 0.3 M D-mannitol, 0.3 M D-glucose, 58.4 mM sucrose, 80.3 mM $CaCl_2 \cdot 2H_2O$, 5 mM L-glutamine, 0.5 mM $Ca(H_2PO_4)_2 \cdot H_2O$, 3 μM thiamine·HCl, 0.5 mM myo-inositol, 13.6 μM 2,4-D, 5.4 μM NAA, 4.4 μM BA, and 10% (v/v) CW (Chen and Shih, 1982).

USE OF CELL-FEEDER TECHNIQUE. Fine cells of sugarcane cv. F164 are plated in a 6-cm petri dish in a thin layer, as previously described (Chen, 1978). After 7-14 days incubation at 30 C, numerous small cell colonies are formed. A piece of polyester monofilament fabric (Swiss Silk Bolting Cloth Mfg. Co. Ltd., Zurich) with mesh size of 10 μm is placed on the cell layer, followed by spreading 3 ml of agar P_1 medium onto the fabric. The P_1 medium is P_6 medium minus 73.5 mM $CaCl_2 \cdot 2H_2O$. Dishes are sealed with parafilm and incubated for 10-15 days at 26 C. A 0.2 ml fraction of the washed protoplasts (ca. 10^5 cells/ml) is layered onto the medium surface, and the whole dish is incubated at 26 C. The growth of protoplasts is routinely monitored.

Three days after incubation, it is found that most protoplasts increase in size by 2-3 times. Some of them bud or burst. Five days later, some of the surviving protoplasts regenerate cell walls and undergo first cell divison. Approximately 10% of the protoplasts divide into 4-8 cell clusters by the tenth day of incubation. The growth of cell clusters is enhanced by adding several drops of P_6 liquid medium with reduced osmoticum. Twenty-five days later, cell clusters develop into 0.1-1.0-mm colonies. The colonies are then isolated and cultured on MS-C medium (Liu, 1981) after an additional 10 days of incubation. The isolated colonies rapidly grow into callus. At the present time, no plantlet regeneration has been observed in protoplast-derived callus. More research of this sort is needed (Chen and Shih, 1983).

FUTURE PROSPECTS

Plant tissue culture technology has been available to the plant breeder for more than two decades. Too little progress has been made with some techniques for sugarcane improvement, such as fusion of isolated protoplasts to achieve somatic hybridization, genetic transformation through gene DNA exchange, and culture of anther or pollen to produce haploid plants for raising homozygous lines. Nonetheless, sugarcane tissue culturists are ahead of researchers working on other crops in the following areas: (a) investigation of the genetic variability occurring in cultured cells and the regenerated plants (Heinz and Mee, 1971; Liu and Chen, 1976a; Lat and Lantin, 1976), (b) fully exploiting the genetic variability to attain disease-resistant (Krishnamurthi, 1974; Liu, 1981), high-yielding (Liu and Chen, 1978a,b) and high-sucrose (Liu and Chen, 1980) somaclones for commercial use, (c) examining organogenesis and embryogenesis in callus masses (Liu and Chen, 1974; Nadar et al., 1978).

From a plant breeder's perspective, the production of agriculturally useful variants has already had some impact on the improvement of sugarcane. The purpose is to take advantage of the capacity of callus cells to undergo genetic change in culture particularly when only a few agronomic traits are altered. However, these genetic changes are undirected. Therefore, difficulties confronting the tissue culturist in screening a cane variety to be improved for one or two characters while still holding its high-yielding ability are the same as those faced by conventional plant breeders. To raise the frequency of success in selection, the callus-derived plant populations must be large enough and more efficient techniques must be developed to detect the changes that alter the agronomic characters.

ACKNOWLEDGMENTS

The author acknowledges with gratitude the constant encouragement and support of Mr. I-Sun Shen and Dr. Soh-Chao Shih, Advisor to TSC and Director of the Taiwan Sugar Research Institute, respectively. Special thanks are due to his colleagues, research assistants, past and present. The financial support from the National Science Council, Republic of China, is also gratefully acknowledged.

KEY REFERENCES

Heinz, D.J., Krishnamurthi, M., Nickell, L.G., and Maretzki, A. 1977. Cell, tissue and organ culture in sugarcane improvement. In: Applied and Fundamental Aspects of Plant Cell, Tissue and Organ Culture (J. Reinert and Y.P.S. Bajaj, eds.) pp. 3-17. Springer-Verlag, Berlin.

Larkin, P.J. and Scowcroft, W.R. 1981b. Somaclonal variation: A novel source of variability from cell cultures for plant improvement. Theor. Appl. Genet. 60:197-214.

Liu, M.C. 1981. In vitro methods applied to sugarcane improvement. In: Plant Tissue Culture: Methods and Applications in Agriculture (T.A. Thorpe, ed.) pp. 299-323. Academic Press, New York.

Reinert, J. and Bajaj, Y.P.S. (eds.). 1977. Applied and Fundamental Aspects of Plant Cell, Tissue and Organ Cultures. Springer-Verlag, Berlin.

Stevenson, G.C. 1965. Genetics and Breeding of Sugarcane. Longmans, Green and Co., London.

Thorpe, T.A. (ed.). 1981. Plant Tissue Culture: Methods and Applications in Agriculture. Academic Press, New York.

REFERENCES

Arcenaux, G. 1967. Cultivated sugarcanes of the world and their botanical derivation. Proc. Congr. Int. Soc. Sugar Cane Technol. (Puerto Rico) 12:844-854.

Artschwager, E. 1925. Anatomy of the vegetative organs of sugarcane. J. Agric. Res. 30:197-221.

Barba, R. and Nickell, L.G. 1969. Nutrition and organ differentiation in tissue cultures of sugarcane, a monocotyledon. Planta 89:299-302.

Barber, C.A. 1918. Studies in Indian sugarcanes, No. 3. The classification of Indian canes with special reference to the Saretha and Sunnabile groups. Memorandum Department of Agriculture, India, Botany Series 9.

Brandes, E.W. 1956. Origin, dispersal and use in breeding of the Melanesian garden sugarcane and their derivatives, *Saccharum officinarum* L. Proc. Congr. Int. Soc. Sugar Cane Technol. (India) 9:709-750.

Chen, W.H. 1978. Tissue and cell culture as a tool in sugarcane breeding. M. Phil. Thesis, Univ. of Leicester, England.

______ and Liu, M.C. 1974. Isolation of protoplasts from sugarcane young leaves. Rep. Taiwan Sugar Res. Inst. 64:1-10.

______ and Liu, M.C. 1976. Nuclear behavior and cell wall regeneration in protoplasts from sugarcane young leaves. Rep. Taiwan Sugar Res. Inst. 71:1-8.

______ and Liu, M.C. 1982. Utilization of sugar and nitrogen by sugarcane cells in suspension cultures. Taiwan Sugar 29:93-98.

______ and Shih, S.C. 1983. Isolation of protoplasts from cell culture and subsequent callus regeneration in sugarcane. Proc. Congr. Int. Soc. Sugar Cane Technol. (Cuba) 18:

______, Liu, M.C., and Chao, C.Y. 1979. The growth of sugarcane downy mildew fungus in tissue culture. Can. J. Bot. 57:528-533.

Chen, Z.H., Qian, C.F., Qin, M., Wang, C.H., Sui, C.J., Chen, F.Z., and Deng, Z.T. 1979. The induction of pollen plants of sugarcane. In: Annual Report Institute of Genetics, Academia Sinica, pp. 91-93. Beijing, China.

Cocking, E.C. 1960. A method for the isolation of plant protoplasts and vacuoles. Nature 187:927-929.

______ 1980. Protoplasts: Past and present. In: Advances in Protoplast Research (L. Ferenczy and G.L. Farkas, eds.) pp. 3-15. Pergamon Press, Oxford.

Daniels, J., Smith, P., and Paton, N.H. 1974. Prehistory and origin of sugarcane. ISSCT Sugarcane Breed. Newsl. 34:21-26.

______, and Williams, C.A. 1975. The origin of the genus *Saccharum*. ISSCT Sugarcane Breed. Newsl. 36:24-39.

Dougall, D.K. 1980. Nutrition and metabolism. In: Plant Tissue Culture as a Source of Biochemicals (E.J. Staba, ed.) pp. 21-58. CRC Press, Boca Raton, Florida.

Evans, D.A. and Reed, S.M. 1981. Cytogenetic techniques. In: Plant Tissue Culture: Methods and Applications in Agriculture (T.A. Thorpe, ed.) pp. 213-240. Academic Press, New York.

Evans, D.A., Crocomo, O.J., and de Carvalho, M.T.V. 1980. Protoplast isolation and subsequent callus regeneration in sugarcane. Z. Pflanzenphysiol. 98:355-358.

Fitch, M.M. and Moore, P.H. 1981a. The selection for salt resistance in sugarcane (*Saccharum* spp. hybrids) tissue culture. Plant Physiol. 67(No. 6, Suppl.):26 (Abstr.).

______ 1981b. Anther culture production of haploids from *Saccharum spontaneum*, a tropical grass related to sugarcane. Plant Physiol. 67 (No. 6, Suppl.):26 (Abstr.).

Fowler, M.W. 1977. Growth of cell cultures under chemostat conditions. In: Plant Tissue Culture and Its Biotechnological Application (W. Barz, E. Reinhard, and M.H. Zenk, eds.) pp. 253-265. Springer-Verlag, Berlin.

Gamborg, O.L., Shyluk, J.P., and Shahin, E.A. 1981. Isolation, fusion and culture of plant protoplasts. In: Plant Tissue Culture: Methods and Applications in Agriculture (T.A. Thorpe, ed.) pp. 115-153. Academic Press, New York.

Gautheret, R.J. 1959. La culture des tissu végétaux, techniques et realisations. Masson, Paris.

Glenn, E.P. 1981. Growth effects and metabolism of arginine in stationary and rapidly dividing stage sugarcane cell suspensions. Physiol. Plant. 52:59-64.

Heinz, D.J. 1973. Sugarcane improvement through induced mutations using vegetative propagules and cell culture techniques. In: Induced Mutations in Vegetatively Propagated Plants, pp. 53-59. International Atomic Energy Agency, Vienna.

______ and Mee, G.W.P. 1968. Tissue callus differentiation and regeneration of plants in *Saccharum* spp. Agron. Abstr., p. 10.

______ and Mee, G.W.P. 1969. Plant differentiation from callus tissue of *Saccharum* species. Crop Sci. 9:346-348.

______ and Mee, G.W.P. 1970. Colchicine-induced polyploids from cell suspension cultures of sugarcane. Crop Sci. 10:696-699.

______ and Mee, G.W.P. 1971. Morphologic, cytogenetic and enzymatic variation in *Saccharum* species hybrid clones derived from callus tissue. Am. J. Bot. 58:257-262.

______, Mee, G.W.P., and Nickell, L.G. 1969. Chromosome numbers of some *Saccharum* species hybrids and their cell suspension cultures. Am. J. Bot. 56:450-454.

Jeswiet, J. 1925. Beschrijving der Soorten van het Suikerriet, llde Bijdrage. Bijdrage tot de Systematiek van het Geslacht *Saccharum*. Arch. Suikerind. Ned. Indie, Meded. Proefsta. Java Suiker Ind.

Kibler, R. and Neumann, K.H. 1980. On cytogenetic stability of cultured tissue and cell suspensions of haploid and diploid origin. In: Plant Cell Cultures: Results and Perspectives (F. Sala, B. Parisi, R. Cella, and O. Ciferri, eds.) pp. 59-65. Elsevier/North-Holland Biomedical Press, Amsterdam.

Kohlenbach, H.W. 1977. Basic aspects of differentiation and plant regeneration from cell and tissue culture. In: Plant Tissue Culture and Its Biotechnological Application (W. Barz, E. Reinhard, and M.H. Zenk, eds.) pp. 355-366. Springer-Verlag, Berlin.

Krishnamurthi, M. 1974. Notes on disease resistance of tissue culture subclones and fusion of sugarcane protoplasts. ISSCT Sugarcane Breed. Newsl. 35:24-26.

_____ 1976. Isolation, fusion and multiplication of sugarcane protoplast and comparison of sexual and parasexual hybridization. Euphytica 25:145-150.

_____ and Tlaskal, J. 1974. Fiji disease resistant *Saccharum officinarum* var. Pindar subclones from tissue cultures. Proc. Congr. Int. Soc. Sugar Cane Technol. (South Africa) 15:130-137.

Larkin, P.J. and Scowcroft, W.R. 1981a. Eyespot disease of sugarcane: Host-specific toxin induction and its interaction with leaf cells. Plant Physiol. 67:408-414.

Lat, J.B. and Lantin, M.M. 1976. Agronomic performance of sugarcane clones derived from callus tissue. Philipp. J. Crop Sci. 1:117-123.

Leu, L.S. and Teng, W.S. 1974. Culmicolous smut of sugarcane in Taiwan. V. Two pathogenic strains of *Ustilago scitaminea* Sydow. Proc. Congr. Int. Soc. Sugar Cane Technol. (South Africa) 15:275-279.

Lichts, F.O. 1981. International Sugar Report, Feb. 2, 1981.

Liu, M.C. 1971. A new method for sugarcane breeding: Tissue culture technique. Taiwan Sugar 18(1):8-10.

_____ and Chen, W.H. 1973. Effect of organic nitrogen on growth, and patterns of growth and division in suspension cultures of sugarcane. J. Agric. Assoc. China 81:20-28.

_____ and Chen, W.H. 1974. Histological studies on the origin and process of plantlet differentiation in sugarcane callus mass. Proc. Congr. Int. Soc. Sugar Cane Technol. (South Africa) 15:118-128.

_____ and Chen, W.H. 1976a. Tissue and cell culture as aids to sugarcane breeding. I. Creation of genetic variability through callus culture. Euphytica 25:393-403.

_____ and Chen, W.H. 1976b. Fusion and cell clump formation in sugarcane protoplasts. ISSCT Sugarcane Breed. Newsl. 37:39-46.

_____ and Chen, W.H. 1978a. Tissue and cell culture as aids to sugarcane breeding. II. Performance and yield potential of sugarcane callus derived lines. Euphytica 27:273-282.

_____ and Chen, W.H. 1978b. Significant improvement in sugarcane by using tissue culture methods. In: Fourth International Congress Plant Tissue and Cell Culture (Abstr.) p. 163. Univ. Calgary, Alberta, Canada.

_____ and Chen, W.H. 1979. Application of tissue and cell culture technique for sugarcane improvement. Annual Report Research Development Council, pp. 34-37. Taiwan Sugar Corp., Taiwan (in Chinese).

_____ and Chen, W.H. 1980. Application of tissue and cell culture technique for sugarcane improvement. Annual Report Research Development Council, pp. 26-29. Taiwan Sugar Corp., Taiwan (in Chinese).

_____ and Chen, W.H. 1981. Application of tissue and cell culture technique for sugarcane improvement. Annual Report Research Development Council, pp. 18-21. Taiwan Sugar Corp., Taiwan (in Chinese).

_____ and Chen, W.H. 1982a. The structure and protoplast isolation of soybean root nodule. In: Monograph No. 4: Plant Tissue and Cell Culture (W.C. Chang, ed.) pp. 132-140. Institute of Botany, Academia Sinica, Taipei, Taiwan.

______ and Chen, W.H. 1982b. Application of tissue and cell culture technique for sugarcane improvement. Annual Report Research Development Council, pp. 14-15. Taiwan Sugar Corp., Taiwan (in Chinese).

______ and Shih, S.C. 1982. Chromosomal variations in suspension cells of sugarcane. Proc. Congr. Int. Soc. Sugar Cane Technol. (Cuba) 18:

______ and Yeh, H.S. 1982. Selection of NaCl tolerant line through stepwise salinized sugarcane cell cultures. In: Proceedings Fifth International Congres Plant Tissue and Cell Culture 1982 (A. Fujiwara, ed.) pp. 477-478. Japanese Association Plant Tissue Culture, Tokyo.

______, Huang, Y.J., and Shih, S.L. 1972. The in vitro production of plants from several tissues of *Saccharum* species. J. Agric. Assoc. China 77:52-58.

______, Shang, K.C., Chen, W.H., and Shih, S.C. 1977. Tissue and cell culture as aids to sugarcane breeding. III. Aneuploid cells and plants induced by treatment of cell suspension culture with colchicine. Proc. Congr. Int. Soc. Sugar Cane Technol. (Brazil) 16:29-42.

______, Chen, W.H., and Yang, L.S. 1980. Anther culture in sugarcane. I. Structure of anther and its pollen grain developmental stages. Taiwan Sugar 27:86-91.

______, Chen, W.H., and Shih, S.C. 1982. Histogenesis of sugarcane callus originating from young leaf and stem tip explants. Can. J. Bot.

Lyrene, P.M. 1976. Tissue culture and mutations in sugarcane. ISSCT Sugarcane Breed. Newsl. 38:61-62.

Manasse, R. 1972. A new culture flask for plant suspension cultures. Experientia 28:723-724.

Maretzki, A. and Nickell, L.G. 1973. Formation of protoplasts from sugarcane cell suspensions and the regeneration of cell cultures from protoplasts. In: Protoplastes et Fusion de Cellules Somatiques Vegetales. Colloq. Int. C. N. R. S. 212:51-63.

Maretzki, A., Thom, M., and Nickell, L.G. 1974. Utilization and metabolism of carbohydrates in cell and callus cultures. In: Tissue Culture and Plant Science 1974 (H.E. Street, ed.) pp. 329-361. Academic Press, London.

Moore, P. and Maretzki, A. 1974. Anther cultures. Annual Report Experiment Station, Hawaiian Sugar Planters' Association (1974) pp. 31-32.

Nadar, H.M. and Heinz, D.J. 1977. Root and shoot development from sugarcane callus tissue. Crop Sci. 17:814-816.

Nadar, H.M., Soepraptopo, S., Heinz, D.J., and Ladd, S.L. 1978. Fine structure of sugarcane (*Saccharum* spp.) callus and the role of auxin in embryogenesis. Crop Sci. 18:210-216.

Nickell, L.G. 1964. Tissue and cell cultures of sugarcane: Another research tool. Hawaii. Plant. Rec. 57:223-229.

______ and Maretzki, A. 1969. Growth of suspension cultures of sugarcane cells in chemically defined media. Physiol. Plant. 22:117-125.

Price, S. 1962. A modified leaf-squash technique for counting chromosomes in somatic cells of *Saccharum* and related grasses. Proc. Congr. Int. Soc. Sugar Cane Technol. (Puerto Rico) 12:583-585.

______ 1963. Cytogenetics of modern sugarcane. Econ. Bot. 17:97-106.

______ 1965. Cytology of *Saccharum robustum* and related sympatric species and natural hybrids. USDA/ARS Technical Bulletin 1337:47.

Sauvaire, D. and Galzy, R. 1980. Une methode de planification experimentale appliquee aux cultures de tissus vegetaux. Exemple de la canne a sucre (*Saccharum* spp.). Can. J. Bot. 58:264-269.

Schenk, R.R. and Hildebrandt, A.C. 1972. Medium and techniques for induction and growth of monocotyledonous and dicotyledonous plant cell cultures. Can. J. Bot. 50:199-204.

Steel, R.G.D. and Torrie, J.H. 1960. Principles and Procedures of Statisitics, pp. 78-79, 132-156, 370-371. McGraw-Hill, New York.

Street, H.E. 1973. Cell (suspension) cultures: Techniques. In: Plant Tissue and Cell Culture (H.E. Street, ed.) pp. 59-99. Univ. California Press, Berkeley.

______ 1977. Applications of cell suspension cultures. In: Applied and Fundamental Aspects of Plant Cell, Tissue and Organ Culture (J. Reinert and Y.P.S. Bajaj, eds.) pp. 649-667. Springer-Verlag, Berlin.

Taiwan Sugar Corporation. 1956. Cane Sugar Handbook. Taiwan Sugar Corporation, Taipei, Taiwan (in Chinese).

Thom, M. and Maretzki, A. 1970. Peroxidase and esterase isozymes in Hawaiian sugarcane. Hawaii. Plant. Rec. 58:81-94.

Vasil, V., Vasil, I.K., Zuberer, D.A., and Hubbell, D.A. 1979. The biology of *Azospirillum*-sugarcane association. I. Establishment of the association. Z. Pflanzenphysiol. 95:141-147.

CHAPTER 21
Tobacco

C.E. Flick and *D.A. Evans*

The genus *Nicotiana* consists of 66 species found predominantly throughout South America and Australia. *N. tabacum*, commercial tobacco, is an amphidiploid species, not found in the wild state, that presumably arose as a natural hybrid between *N. sylvestris* and *N. tomentosiformis*. A great range of types are represented in this single species. Other *Nicotiana* species are used for purposes similar to *N. tabacum*, but to a limited extent. *N. rustica* plays a minor role in world tobacco production, being primarily of local importance in Russia and India. Prior to the introduction of *N. tabacum*, Australian aborigines chewed the leaves of *N. suaveolens*.

HISTORY AND ECONOMIC IMPORTANCE

Although commercial tobacco, *N. tabacum*, originated in the New World, wild species from Australia as well as America represent a source for important agronomic attributes; e.g., *N. debneyi*, an Australian species, is resistant to many of the diseases of commercial tobacco. Wild tobacco species have been used to incorporate desirable traits into tobacco through conventional breeding, e.g., tobacco mosaic virus (TMV) resistance from *N. glutinosa* has been incorporated into some varieties of *N. tabacum*. *N. longiflora* has been used as a source of wildfire disease resistance in *N. tabacum*. The initial cross *N. tabacum* x *N. longiflora* produced the breeding line TL 106. Resistance was stabilized by backcrossing to *N. tabacum* White Burley to produce the Burley 21 variety. Burley 21 has been subsequently used to transfer wildfire resistance into other tobacco varieties. Burley 49 carries disease resistance factors derived from three wild species: black root rot from *N. debneyi*, wildfire disease from *N. longiflora*, and TMV from *N. glutinosa*.

Successful cultivation of tobacco poses several problems. Seedlings of tobacco species are very small. Direct seeding in the field is usually unsuccessful. This necessitates the use of seedbeds to raise the seedlings to transplant size. Tobacco needs a longer growing season than is available in most temperate climates. Indeed, for cultivation in most northern climates, e.g., Canada, seedlings must be raised in a greenhouse. For the 3 months from seed sowing to field transplantation, the seedling must be closely supervised in the seedbed. This early supervision considerably reduces the labor needed to produce a crop. Even in some tropical climates, a seedbed is used. For example, in Africa, because the length of the wet season does not span an entire tobacco growing season, seedbeds are used. Associated with the use of seedbed and field for cultivation of tobacco are a distinct set of diseases that attack tobacco at each growth stage. Despite these problems cultivation of tobacco has spread worldwide, probably because tobacco can be grown in poor soil that is not suited for the cultivation of other crops and the profits of tobacco cultivation are large.

Tobacco is grown worldwide. North America leads the world in production, followed by China, India and Pakistan combined, and Brazil. In the United States alone the value of the 1980 tobacco crop was about $2.5 billion. Due to price supports for tobacco established by the U.S. government, about 40% of that figure or $1 billion was profit. Whereas the value of the 1979 tomato harvest, another solanaceous crop, was about $4800/hectare and the value of the 1979 potato harvest, also a solanaceous crop, was about $2000/hectare, the value of the tobacco harvest was estimated at $7500/hectare. Despite immense profits in tobacco, yields of tobacco worldwide have not improved significantly over the past 10 years. In most agricultural areas tobacco production has remained constant while production of food crops has increased dramatically.

GENETICS AND CROP IMPROVEMENT

Much early genetic and cell culture research was conducted with tobacco. Noteworthy has been the pioneering work on interspecific hybrids by Kostoff (1943) and Goodspeed (1954). Murashige and Skoog (1962) developed a culture medium for routine propagation of tobacco. Haploid plants were produced through anther culture of both *N. tabacum* and its progenitor species *N. sylvestris* (Nitsch and Nitsch, 1969). Plant regeneration from protoplasts was first demonstrated in *N. tabacum* (Nagata and Takebe, 1971). The first somatic hybrid in higher plants was produced by protoplast fusion of *N. glauca* and *N. langsdorffii* (Carlson et al., 1972). Carlson isolated the first auxotrophs from mutagenized cell cultures of haploid *N. tabacum*. Genetic analysis of mutants isolated in vitro has been most extensively accomplished in *N. tabacum* (e.g., Chaleff and Parsons, 1978a). Most of the techniques developed in plant tissue culture have arisen in *N. tabacum* or one of the wild species of tobacco.

Although a foundation of basic research in tissue culture has been developed in *N. tabacum*, this knowledge has not been applied to crop improvement to a large extent. Several breeding problems could be addressed through cellular genetics and tissue culture. Incorporation of disease resistance into *N. tabacum* from a wild species has been achieved via tissue culture. Although *N. suaveolens* x *N. tabacum* hybrids usually die as seedlings, Lloyd (1975) regenerated plants from callus initiated from cotyledons of this hybrid and thus was able to transfer brown spot resistance from *N. suaveolens* to *N. tabacum*. Nakamura et al. (1974) developed disease resistant breeding lines of the flue-cured tobacco variety MC 1610 by anther culture of F_1 hybrids. Resistance to black shank was demonstrated in *N. nesophila* x *N. tabacum* sexual hybrids obtained by ovule culture (Reed and Collins, 1980). Resistance to TMV was observed in *N. nesophila* + *N. tabacum* somatic hybrids (Evans et al., 1981). Chaleff and Parsons (1978a) directly selected mutants in cell cultures resistant to the herbicide picloram. In addition in vitro mutant isolation was used to directly obtain *N. tabacum* lines resistant to wildfire disease (Carlson, 1973). In this instance selection in vitro for resistance to methionine sulfoximine resulted in resistant plants that overproduced methionine, an analog of the wildfire toxin.

Alkaloid concentration in *N. tabacum* may also be manipulated by cell culture. Because of somaclonal variation, it may be possible to identify clonal lines with altered alkaloid composition following routine regeneration of plants from tissue culture. In addition, both embryo rescue and protoplast fusion may produce hybrids which incorporate desirable alkaloid composition from a wild species.

The value of the wild *Nicotiana* species cannot be overemphasized with respect to breeding for both resistance to disease organisms and alkaloid concentration. The 65 wild species of tobacco represent a major genetic resource for improvement of the crop. Many of these wild species are sexually incompatible with *N. tabacum* in a unilateral or bilateral way. However, through tissue culture techniques, e.g., embryo culture and protoplast fusion, advantage can be taken of the wild species.

LITERATURE REVIEW

Plant Propagation

Tobacco has been used as a model system for in vitro studies on regeneration, since the classical studies of Skoog and Miller (1957). Species of tobacco which have been regenerated in vitro are listed in Table 1. Varieties of cultivated tobacco, *N. tabacum*, represent the easiest material to manipulate in vitro. Protocols for plant regeneration from leaf explants are given in the protocol section. The nutrient solution most often used for in vitro cultivation of plant species, MS culture medium, was formulated from results of growth experiments with *N. tabacum* (Murashige and Skoog, 1962). The effects of growth-regu-

Table 1. *Nicotiana* Species Propagated In Vitro

SPECIES	REFERENCE
N. accuminata	Helgeson, 1979
N. africana	Evans, unpublished
N. alata	Tran Thanh Van & Trinh, 1978
N. glauca	Evans, unpublished
N. glutinosa	Evans, unpublished
N. goodspeedii	Helgeson, 1979
N. longiflora	Ahuja & Hagen, 1966
N. megalosiphon	Helgeson, 1979
N. nesophila	Evans, unpublished
N. otophora	Tran Thanh Van & Trinh, 1978
N. plumbaginifolia	Tran Thanh Van & Trinh, 1978
N. repanda	Evans, unpublished
N. stocktonii	Evans, unpublished
N. suaveolens	Helgeson, 1979
N. sylvestris	Flick, unpublished
N. tabacum	Gamborg et al., 1979
N. tomentosiformis	Tran Thanh Van & Trinh, 1978

lating auxins and cytokinins are quite specific and reproducible. Callus has been initiated from leaf or stem explants of *N. tabacum*, and all *Nicotiana* species were examined using MS medium with the addition of 4.5 µM 2,4-D and 2 g/l CH (Evans, unpublished). Callus can be initiated on MS medium with other hormone concentrations, e.g., 11.4 µM IAA and 2.3 µM KIN (Murashige and Skoog, 1962), but in all cases an auxin is necessary, and in the case of 2,4-D, may be sufficient for the induction of callus. Once formed, callus can be maintained on MS medium with 4.5 µM 2,4-D, while the CH becomes unnecessary, or the callus may be shifted to an alternate medium, e.g., B5 medium with 4.5 µM 2,4-D (Gamborg et al., 1979).

In addition to direct plant regeneration, it is possible to obtain either callus-mediated plant regeneration or plant regeneration from cell suspension cultures in tobacco. If callus growth is initiated and maintained in medium with 2,4-D, the 2,4-D must be replaced with a cytokinin to achieve shoot formation. In this manner plant regeneration has been obtained from long term suspension cultures of tobacco (Gamborg et al., 1979). Numerous combinations of growth regulating auxins and cytokinins have been varied by researchers attempting to achieve plant regeneration in tobacco.

Haploids

Anther culture has been reviewed by Bajaj (1983). Species of tobacco from which haploid plants have been recovered following anther culture are listed in Table 2. Mature haploid plants have been recov-

Table 2. Anther Culture of *Nicotiana* Species

SPECIES	REFERENCE
Nicotiana alata	Nitsch, 1971
N. attenuata	Collins & Sunderland, 1974
N. clevelandii	Vyskot & Novak, 1973
N. glutinosa	Tomes & Collins, 1976
N. knightiana	Tomes & Collins, 1976
N. otophora	Collins et al., 1972
N. paniculata	Tomes & Collins, 1976
N. raimondii	Collins & Sutherland, 1974
N. rustica	Tomes & Collins, 1976
N. sanderae	Vyskot & Novak, 1976
N. sylvestris	Tomes & Collins, 1976
N. tabacum	Nitsch, 1974

ered in large quantities from cultured *Nicotiana* spp. anthers (Bourgin and Nitsch, 1967). The haploid condition was confirmed by chromosome counts and sterility of haploid flowers. Formation of plantlets occurred one month following culture of anthers containing uninucleate microspores. The haploid embryos followed the normal embryogenic process with globular, heart, and torpedo stages observed. Haploid tobacco plants are smaller than comparable diploid plants, and an abundant number of flowers are frequently produced. Atypical morphological features are usually observed in vitro, but new growth following transfer to pots becomes phenotypically normal (Sunderland, 1970).

Sunderland and Wicks (1969) determined the best developmental stage for anther culture of *N. tabacum* to obtain plantlet formation. After testing buds at different developmental stages, they concluded that the most rapid response was achieved using (a) petals of 8-16 mm with uninucleated pollen grains or microspores, then (b) petals of 17-20 mm with microspores undergoing the first pollen grain mitosis. They concluded that the first pollen grain mitosis was the critical transitional point for plantlet formation in *N. tabacum*. Unfortunately, there is no single optimum microsporogenic stage for cultured anthers, even for closely related species.

When microspores are set in culture, three routes of development have been described by Sunderland (1973, 1974). In route A, as occurs in *N. tabacum*, embryos are formed from the vegetative cell while the generative cell degenerates soon after or just following one or two mitotic divisions. Route B, as for *D. innoxia*, results in the absence of an assymetrical mitosis and leads to the formation of two equal, diffuse, vegetative-type nuclei. Route B can occur in anthers cultured at any stage prior to pollen mitosis. The frequency of routes A and B vary considerably between species and even from one anther to another within the same plant. There are some suggestions that the generative cell can follow an organogenesis pattern of development as reported by Sharp et al. (1971) and Raghavan (1976). Haploid plants are not recovered in route C in which the generative and vegetative nuclei fuse prior to nuclear division.

In *Datura* and *Nicotiana* spp. only sucrose (2%) in agar was necessary for induction of embryogenesis up to the globular stage with subsequent development occurring only when inorganic salts are added to the medium. Nitsch (1971) reported the role of iron as a key element in the complete development of embryos. Iron deficiency in higher plants has been associated with an abnormally high level of arginine as demonstrated by suppression of embryo development when arginine was added to the culture medium of tobacco anthers.

Induction of the anther is dependent on the plant age and growth environment (Engvild et al., 1972; Dunwell and Sunderland, 1973; Rashid and Street, 1973). Anther response is greater at the beginning of the flowering period than at the end (Sunderland and Wicks, 1971), and there are differences between seasonal plantings of greenhouse plants. It has been demonstrated that short days are more proper for tobacco anther development than long days (Dunwell and Perry, 1973).

Pretreatment of anthers before placing in culture has been shown to enhance the frequency of anther induction. A cold shock treatment (3-5 C for 48-72 hr) resulted in an increased number of embryos per anther in *Nicotiana* and *Datura* (Nitsch and Norrell, 1972; Nitsch, 1974). A procedure for anther culture of tobacco is summarized in the protocol section.

In several instances haploids have been used in tobacco breeding. Doubled haploids appear to be reduced in agronomic characteristics including yield (Brown and Wernsman, 1982), but the reason for this is still unclear. Although Burk and Matzinger (1976) found doubled haploids reduced for several agronomic characteristics, two traits, alkaloids and reducing sugars, were both increased. The reduction of vigor in doubled haploids is not confined to *N. tabacum* as it has also been observed in *N. sylvestris* (De Paepe et al., 1977).

Doubled haploids of F_1 hybrids have been used to develop disease resistant breeding lines in *N. tabacum.* Nakamura et al. (1974) obtained three doubled haploid breeding lines with increased resistance to bacterial wilt, black shank, and black root rot from the Japanese flue-cured variety MC 1610. Agronomic and chemical traits of the lines are similar to the parent line. Chaplin (1978) evaluated doubled haploids produced from a VY32 (potato virus Y resistant line from Rhodesia) x Coker 86 (TMV, root-rot nematode, black shank, bacterial wilts, and fusarium wilt resistant). A line has been developed that is resistant to potato virus Y and black shank and that has good agronomic characteristics. Wark (1977) transferred resistance to two strains of blue mold from *N. velutina* or *N. excelsior* and *N. goodspeedii* to flue-cured tobacco through the use of doubled haploids. In addition, *N. tabacum* lines with good agronomic characteristics and resistance to TMV and blue mold were obtained with doubled haploids.

Ovule Culture

A large number of sexual crosses between species in the genus *Nicotiana* are either unilaterally or bilaterally incompatible. In some cases this is due to mechanical causes, e.g., inability of the pollen tube to reach the ovule. However, in many instances the embryo fails to develop despite successful fertilization. Failure of the embryo to develop may be reversed by transferring the fertilized ovule to in vitro culture, i.e., optimal growth conditions. Indeed failure of an embryo to develop in some cases may be attributed to the inadequate development of the endosperm rather than the embryo itself leading to starvation of the embryo (North, 1976).

Despite the potential benefits of embryo or ovule culture for recovery of novel sexual hybrids, this technique has not been exploited for hybridization of tobacco species. Dulieu (1976) determined the optimal conditions for development and germination of *Nicotiana* embryos in vitro. Pollen can be collected and stored in a dessicator at 5 C prior to in vivo fertilization. Usually ovules are transferred to in vitro culture 5-10 days following fertilization. Hormone-free MS medium is the basis of ovule culture medium. Many organic compounds commonly added to plant tissue culture media are either toxic (e.g., CH, KIN, NAA, glycine, and asparagine) or ineffective (e.g., glutamine) for ovule culture. B vitamins are necessary for embryo development, whereas 20% CW in the culture medium enhances embryo survival. Sucrose concentration was critical for ovule culture. Although most culture media contain 2-3% sucrose, embryo development is best in 4-8% sucrose. Concentrations of sucrose greater than 8% inhibit embryo development. Dulieu (1976) proposed an optimal system for culture of *Nicotiana* ovules in which fertilized ovules are cultured on 4-8% sucrose until embryos reach a 400-500 micron size. Ovules are then transferred to a lower osmotic pressure, e.g., 0.8% agar, 2% sucrose. Under these conditions 50% germination can be achieved. Despite the detailed analysis of the conditions necessary for in vitro development of *Nicotiana* embryos, embryo and/or ovule culture have only been attempted to a very limited extent in tobacco. A protocol for ovule culture is summarized in the protocol section.

N. nesophila x *N. tabacum* and *N. stocktonii* x *N. tabacum* hybrids were produced by Reed and Collins (1978) using fertilized ovule culture. Successful rescue of fertilized ovules was dependent on the length of time following fertilization before in vitro culture. Contrary to Dulieu's work, Reed and Collins (1978) cultured ovules on culture medium containing low sucrose concentrations. Hybrid plants were obtained only when the wild *Nicotiana* species was used as the female parent. The potential usefulness of ovule culture of tobacco species was obvious as these ovule rescue hybrids were as resistant to black shank disease as the wild species parent (Reed and Collins, 1980).

Protoplast Isolation and Culture

Methods for protoplast isolation and culture including *Nicotiana* species have been recently reviewed (Evans and Bravo, 1983). Fusion of protoplasts has been discussed by Evans (1983) and Gleba and Evans (1983). As more research on protoplast isolation, culture, and fusion has been undertaken in *Nicotiana* than any other genus these reviews are extensive. Here we will only discuss techniques involved in isolation and culture of protoplasts.

Protoplasts have been isolated from at least 17 tobacco species (see Table 3). A protocol for protoplast isolation is given in Evans and Bravo (Vol. 1, Chapter 4). Conditions for protoplast isolation have varied enormously among the species, both between species and within species (Evans and Bravo, Vol. 1, Chapter 4). At least in the case of *N. tabacum*, protoplast isolation has been reported from four different cell sources, i.e., mesophyll, epidermis, cell suspension culture, and stem callus. The large amount of variability in successful techniques for preparation of tobacco protoplasts may reflect the ease with which the system can be manipulated.

The choice of osmoticum for isolation and culture of *Nicotiana* protoplasts has been quite variable. In most instances the nonmetabolized sugar D-mannitol has been used. However, concentrations of mannitol have varied from 0.23 M as reported for ornamental *Nicotianas* (Passiatore and Sink, 1981) to as high as 0.7 M for *N. tabacum* (Nagata and Takebe, 1971). *Nicotiana* protoplasts may not be highly sensitive to osmotic potential as *N. tabacum* mesophyll protoplasts have been isolated in 0.4 M mannitol (Vasil and Vasil, 1974) up to 0.7 M mannitol (Nagata and Takebe, 1971). Most commonly mannitol has been used at 0.4–0.5 M. Glucose has also been used successfully at concentrations of 0.35–0.40 M (e.g., Evans, 1979).

As with osmoticum a variety of enzymes has been used for protoplast isolation. In most instances desalting of enzymes is not reported, although the importance of this step in protoplast growth and subsequent plant regeneration has been recently described (Patnaik et al., 1981). Onozuka R10 cellulase has been universally used for isolation of *Nicotiana* protoplasts. Usually this cellulase is used in conjunction with macerozyme (macerase). In some species (e.g., *N. sylvestris*; Nagy and Maliga, 1976), this enzyme combination is sufficient. In many instances cellulase and macerase are supplemented with driselase (e.g., Bourgin et al., 1979). Cellulase is usually used at concentrations between 1.0 and 2.0%. A low concentration of macerase is required, i.e., usually about 0.5%. Driselase is used at low concentrations, e.g., 0.05% by Bourgin et al. (1979), but has been used at concentrations as high as 1% (Passiatore and Sink, 1981).

Many different media have been used to culture *Nicotiana* protoplasts, as well as a variety of plant growth regulators (see Evans and Bravo, Vol. 1, Chapter 4). In all cases protoplasts are initially cultured in a medium with a high osmotic strength. Whereas tobacco callus can be cultured with an auxin as the only growth regulator, protoplasts are generally cultured in the presence of both an auxin and a cytokinin. In addition, the culture medium is often enriched with vitamins, sugars, amino acids, CW, or other organic nutrients not usually found in callus culture medium (e.g., medium 8p; Kao and Michayluk, 1975). Three to six weeks after protoplast isolation callus growth is sufficient for transfer of cells to shoot regeneration medium. Shoots are formed on medium with high cytokinin concentration, e.g., 5.0 µM 6BA (Evans, 1979), or a high cytokinin concentration in conjunction with a weak auxin, e.g., 22.8 µM IAA and 11.9 µM KIN

Table 3. *Nicotiana* Species Capable of Plant Regeneration from Protoplasts

SPECIES	PROTOPLAST SOURCE	REFERENCE
N. acuminata	Mesophyll	Bourgin et al., 1979
N. alata	Mesophyll	Bourgin et al., 1979; Passiatore & Sink, 1981
N. alata (haploid)	Mesophyll	Bourgin & Missonier, 1978
N. alata x *N. sanderae*	Flower petals	Flick & Evans, 1983
N. debneyi	Mesophyll	Piven, 1981
N. debneyi	Mesophyll, suspension	Scowcroft & Larkin, 1980
N. forgettiana	Mesophyll	Passiatore & Sink, 1981
N. glauca	Mesophyll	Bourgin et al., 1979
N. langsdorffii	Mesophyll	Bourgin et al., 1979; Evans, 1979
N. longiflora	Mesophyll	Bourgin et al., 1979
N. nesophila	Mesophyll	Evans, 1979
N. otophora	Mesophyll	Bourgin et al., 1979; Banks & Evans, 1976
N. paniculata	Mesophyll	Bourgin et al., 1979
N. plumbaginifolia	Mesophyll	Gill et al., 1979
N. repanda	Mesophyll	Evans, 1979
N. rustica	Mesophyll	Gill et al., 1979
N. sanderae	Mesophyll	Passiatore & Sink, 1981
N. stocktonii	Mesophyll	Evans, 1979
N. sylvestris	Mesophyll	Banks & Evans, 1976; Nagy & Maliga, 1976
N. sylvestris (haploid)	Mesophyll	Facciotti & Pilet, 1979
N. sylvestris x *N. otophora*	Mesophyll	Banks & Evans, 1976
N. tabacum	Mesophyll	Banks & Evans, 1976; Nagata & Takebe, 1971
	Epidermis	Davey et al., 1974
	Suspension	Gamborg et al., 1979
	Stem callus	Vasil & Vasil, 1974
N. tabacum (haploid)	Mesophyll	Nitsch & Ohyama, 1971
N. tabacum x *N. otophora*	Mesophyll	Banks & Evans, 1976

(Nagata and Takebe, 1971). Shoots are usually rooted on hormone-free medium.

Isolated protoplasts can be induced to fuse giving rise to hybrid cells. Table 4 lists somatic hybrids between *Nicotiana* species. The

Table 4. *Nicotiana* Somatic Hybrids

HYBRID	REFERENCE
Intraspecific	
N. debneyi + *N. debneyi*	Scowcroft & Larkin, 1981
N. sylvestris + *N. sylvestris*	White & Vasil, 1979
N. tabacum + *N. tabacum*	Melchers & Labib, 1974
Interspecific	
N. glauca + *N. langsdorffii*	Carlson et al., 1972
N. sylvestris + *N. knightiana*	Maliga et al., 1977
N. tabacum + *N. glauca*	Evans et al., 1980
N. tabacum + *N. glutinosa*	Uchimiya, 1982
N. tabacum + *N. knightiana*	Maliga et al., 1981
N. tabacum + *N. nesophila*	Evans et al., 1981
N. tabacum + *N. otophora*	Evans et al., 1983
N. tabacum + *N. plumbaginifolia*	Sidorov et al., 1981
N. tabacum + *N. repanda*	Nagao, 1982
N. tabacum + *N. rustica*	Nagao, 1978
N. tabacum + *N. stocktonii*	Evans et al., 1981
N. tabacum + *N. sylvestris*	Zelcer et al., 1978; Medgyesy et al., 1980; Evans et al., 1983

method for PEG-induced fusion of tobacco protoplasts is outlined by Evans (1983). Although protoplast fusion is technically simple, identification of hybrid cell lines or plants is difficult. Selection for hybrids can be based on one or more criteria. The first somatic hybrids were selected based on different nutritional requirements of parental and hybrid cells. Certain sexual hybrids of tobacco species, e.g., *N. glauca* x *N. langsdorffii*, produce genetic tumors. Growth of cell cultures of these tumors is auxin autotrophic, whereas growth in cell cultures of either parent is auxin dependent. Carlson et al. (1972) selected for auxin autotrophy in cell cultures following fusion of mesophyll protoplasts of *N. glauca* and *N. langsdorffii*. Regenerated plants were shown to be somatic hybrids. This selection scheme is limited in that only a few interspecies hybrids in *Nicotiana* produce tumors.

The use of albino cell lines has been widely applied in selection of tobacco somatic hybrids. The semidominant sulfur mutant of *N. tabacum* (Su) has proved useful in selection of somatic hybrids between *N. tabacum* and other species of tobacco. As Su is semidominant, Su/Su is albino, Su/su is light green, and su/su is the wild-type dark green. Hybrid plants have been regenerated following fusion of protoplasts isolated from a cell suspension culture of *N. tabacum* Su/Su and mesophyll protoplasts isolated from *N. tabacum* (Evans, unpublished), *N. sylvestris* (Evans et al., 1983), *N. otophora* (Evans et al., 1983), *N. glauca* (Evans et al., 1980), *N. nesophila* (Evans et al., 1981), and *N. stocktonii* (Evans et al., 1981). All hybrids are light green and hybridity has been verified by several criteria including chromosome number, plant morphology and isoenzyme analysis. In addition complementation of two albinos to produce a green hybrid plant has been used for identification of hybrids (Melchers and Labib, 1974). In this instance both parents are limited to the albino lines, whereas with the use of a semidominant marker such as Su any species can be used as the second parent.

The use of biochemical markers, such as mutants isolated in vitro, in protoplast fusion has not been fully exploited in tobacco. Biochemical markers can be useful because they may be encoded in either the nuclear or cytoplasmic genomes. Hence either nuclear hybrids, i.e., somatic hybrids or cytoplasmic hybrids, or cybrids, can be selected. A hybrid between two cell lines that both have dominant nuclear mutations, e.g., resistance to an antimetabolite, will be resistant to both selective compounds rather than one or the other as the parental cell lines. For example, White and Vasil (1979) fused protoplasts isolated from cell lines of *N. sylvestris* resistant to the amino acid analogs 5-methyltryptophan and S-aminoethylcysteine and recovered hybrid cell lines with both resistances. Alternatively, complementation of two auxotrophic mutants can be used to isolate hybrid cells. Glimelius et al. (1978) fused protoplasts isolated from nitrate-reductase-deficient cnx cell lines with nitrate-reductase-deficient nia cell lines. Although neither parental cell line can utilize nitrate or regenerate shoots, hybrid cell lines utilize nitrate and regenerate shoots. Nitrate reductase mutants are discussed more thoroughly in Flick (1983).

Through the use of cytoplasmic selective markers, e.g., antibiotic resistance, cybrid cell lines can be selected. Medgyesy et al. (1980) selected for transfer of *N. tabacum* chloroplasts into *N. sylvestris*. The nucleus of a streptomycin resistant cell line of *N. tabacum* (SR 1) was inactivated with iodoacetate. Treated *N. tabacum* protoplasts were fused to untreated *N. sylvestris* protoplasts using PEG. Streptomycin resistant calli were selected. As *N. tabacum* protoplasts did not proliferate only cybrids were streptomycin resistant. Thus through the use of mutants isolated in vitro somatic hybrids and cybrids can be easily selected following protoplast fusion.

The protoplast isolation, culture, and fusion techniques available with tobacco species have not been exploited for crop improvement. Indeed in most instances agronomic traits have not been discussed in connection with protoplast derived plants. The potential agricultural usefulness of a somatic hybrid has been examined by Evans et al. (1981). In this instance *N. nesophila* + *N. tabacum* somatic hybrids have resistance to tobacco mosaic virus as does the wild species parent. In addition comparison of this somatic hybrid to the comparable sexual hybrid obtained through ovule rescue techniques indicates that clones of the somatic hybrids express a much greater degree of variability than the sexual hybrids (Evans et al., 1982). It is apparent from these observations that protoplast techniques may be useful for crop improvement in tobacco.

Mutant Isolation

The current literature and techniques for in vitro isolation of mutants in tobacco as well as other species are reviewed elsewhere (Flick, 1983). Table 5 lists mutants isolated in vitro from *Nicotiana*

Table 5. Mutants of *Nicotiana* Isolated In Vitro

SPECIES	PHENOTYPE	REFERENCE
N. plumbaginifolia	Isoleucine auxotroph	Sidorov et al., 1981a
	Uracil auxotroph	Sidorov et al., 1981a
N. sylvestris	Abscisic acid resistant	Wong & Sussex, 1980
	S-Aminoethylcysteine resistant	White & Vasil, 1979
	Chloramphenicol resistant	Dix, 1981
	Cold treatment	Dix & Street, 1976; Dix, 1977
	Kanamycin resistant	Dix et al., 1977
	5-Methyltryptophan resistant	White & Vasil, 1979
	Salt tolerant	Dix & Pearce, 1981
	Streptomycin resistant	Maliga, 1981
	Thienylalanine resistant	Vunsh et al., 1980
N. tabacum	Abscisic acid resistant	Wong & Sussex, 1980
	p-Aminobenzoic acid auxotroph	Carlson, 1970
	S-Aminoethylcysteine resistant	Widholm, 1976, 1977
	Amphotericin-B resistant	Chiu et al., 1980
	Arginine auxotroph	Carlson, 1970
	8-Azaguanine resistant	Lescure, 1973
	Bentazone resistant	Radin & Carlson, 1978
	Biotin auxotroph	Carlson, 1970
	Bromodeoxyuridine resistant	Maisuryan et al., 1981; Marton & Maliga, 1975
	Carboxin resistant	Polacco & Polacco, 1977
	Cycloheximide resistant	Maliga et al., 1976
	p-Fluorophenylalanine resistant	Widholm, 1977; Berlin & Vollmer, 1979; Berlin & Widholm, 1977, 1978; Flick et al., 1981
	Glycerol utilization	Chaleff & Parsons, 1978b
	Glycine hydroxamate resistant	Lawyer et al., 1980
	α-Hydroxylysine resistant	Widholm, 1976, 1977
	Hydroxyurea resistant	Chaleff & Kiel, 1981
	Hypoxanthine auxotroph	Carlson, 1970
	Isopropyl-N-phenylcarbamate resistant	Aviv & Galun, 1977

Table 5. Cont.

SPECIES	PHENOTYPE	REFERENCE
N. tabacum	Kanamycin resistant	Owens, 1981
	Lysine auxotroph	Carlson, 1970
	Methionine sulfoximine resistant	Carlson, 1973
	5-Methyltryptophan resistant	Widholm, 1972, 1977
	Paraquat resistant	Miller & Hughes, 1980
	Phenmedifarm resistant	Radin & Carlson, 1978
	Picloram resistant	Chaleff & Parsons, 1978a
	Proline auxotroph	Carlson, 1970
	Salt tolerant	Nabors et al., 1980
	Selenoamino acid resistant	Flashman & Filner, 1978
	Streptomycin resistant	Maliga et al., 1973, 1975; Yurina et al., 1978; Umiel & Goldner, 1976
	Temperature sensitive	Malmberg, 1980
	Threonine resistant	Heimer & Filner, 1970
	Valine resistant	Bourgin, 1978; Caboche & Muller, 1980

species. Techniques for isolation of mutants are discussed in detail by Flick (1983). A total of 43 mutants representing 36 different phenotypes have been isolated from three tobacco species, *N. plumbaginifolia*, *N. sylvestris*, and *N. tabacum*. However, most of these mutants (32/43) have been isolated in *N. tabacum*. The majority of mutants isolated have been dominant or semidominant. Carlson (1970) isolated recessive auxotrophic mutants from haploid *N. tabacum*, but these mutants were all leaky due to the allotetraploidy of *N. tabacum*. Both *N. plumbaginifolia* and *N. sylvestris* are diploid species. Whereas all of those mutants isolated from *N. sylvestris* are diploid, the auxotrophs of *N. plumbaginifolia* were isolated from mesophyll protoplasts of haploid plants.

It is quite apparent from Table 5 that positive selection has been used for isolation of most tobacco mutants, i.e., selection for resistance to an antimetabolite. A range of toxic compounds have been used including amino acid analogs (e.g., Flick et al., 1981), nucleic acid base analogs (e.g., Lescure, 1973), a fungicide (e.g., Polacco and Polacco,

1977), antibiotics (e.g., Maliga et al., 1973, 1975), and herbicides (e.g., Chaleff and Parsons, 1978a). Carlson (1970) used bromodeoxyuridine to enrich for auxotrophs in *N. tabacum*. Other auxotrophs, i.e., those in *N. plumbaginifolia*, were isolated by large scale screening of nutritional requirements of individual cell colonies.

Plants have been regenerated from many tobacco mutants isolated in vitro (see Flick, 1983), but genetic analysis of mutants has been rare. Chaleff and Parsons (1978a) showed that picloram resistance in *N. tabacum* was due to a dominant nuclear allele. On the other hand, the auxotrophs of *N. tabacum* isolated by Carlson (1970) were recessive. Maliga (1981) isolated a recessive nuclear streptomycin mutant. Most streptomycin mutants are cytoplasmically inherited (Maliga et al., 1975; Umiel, 1979). At least one of these streptomycin resistant mutants (SR 1) was shown to be due to an altered chloroplast ribosomal protein (Yurina et al., 1978). Thus dominant, recessive, and cytoplasmic mutations are all possible when mutants are isolated in vitro.

Few of the 43 tobacco mutants isolated in vitro have direct agricultural application. Salt-tolerant cell lines of both *N. sylvestris* (Dix and Pearce, 1981) and *N. tabacum* (Nabors et al., 1980) have been isolated. Salt tolerance in *N. tabacum* is transmitted sexually although at low frequency (Nabors et al., 1980). No plants were regenerated from salt-tolerant *N. sylvestris* (Dix and Pearce, 1981). Although cold-tolerant lines of *N. sylvestris* have been described (Dix and Street, 1976; Dix, 1977), this trait is not transmitted sexually. The toxicity of some herbicides has been used to select mutants in vitro. Plants resistant to Bentazone (Radin and Carlson, 1978), paraquat (Miller and Hughes, 1980), Phenmedifarm (Radin and Carlson, 1978), and picloram (Chaleff and Parsons, 1978a) have been regenerated from cell cultures of *N. tabacum*. Bentazone and Phenmedifarm resistance are due to recessive nuclear mutations (Radin and Carlson, 1978), whereas picloram resistance is a dominant nuclear mutation (Chaleff and Parsons, 1978a). Some of these tobacco mutants may be useful as selective markers in protoplast fusion experiments. As discussed above mutants isolated in vitro have been used to select hybrid cell lines.

PROTOCOLS

In Vitro Propagation of *Nicotiana* Species from Leaf Explant

1. Select leaf from young, preflowering plant.
2. Surface sterilize leaf. For greenhouse-grown plants 6 min in 8% commercial bleach (0.42% sodium hypochlorite) is sufficient. Use of ethanol for sterilization should be avoided as some *Nicotiana* species are ethanol-sensitive. The bleach is rinsed off in three changes of sterile distilled water. All sterilization procedures may be carried out in a 100-mm petri dish.
3. Cut leaf into 1-cm^2 sections avoiding large veins. Leaf explants are transferred aseptically to solid regeneration medium. Most tobacco species will regenerate shoots from leaf explants on MS medium with 5 μM BA.

4. Shoot proliferation should be extensive in 3-5 weeks. Shoots may be excised and rooted on MS medium with 110 μM 3-aminopyridine and without growth regulators.
5. Rooted shoots may be transferred to soil.

Anther Culture of *Nicotiana* Species

1. Use only the first flowers on a plant. These buds will have synchronized microspores and will produce better results.
2. Buds are surface sterilized in 8% commercial bleach for 6 min as described for leaf explants.
3. Following rinsing with sterile distilled water anthers are sterilely excised from the bud and transferred to culture medium. The filament should be removed and care must be taken to not damage the anther wall as it may proliferate if damaged. *Nicotiana* anthers are usually cultured on Nitsch's H medium (i.e., half concentration MS medium, but with normal levels of Fe•EDTA). No growth regulators are included in the medium.
4. In 3-4 weeks the anther dehisces and plantlets emerge from it. These plantlets can be transferred to growth regulator free MS medium for further growth. Shoot cultures can be initiated on MS medium with 1-5 μM BA, but care should be exercised as BA causes chromosome doubling in many species, e.g., *N. sylvestris* (Flick, unpublished).
5. Plants can be transferred to soil for further greenhouse growth. Haploidy should be verified by chromosome counts in cells from root tips.
6. Optional procedures to increase the frequency of androgenesis are (a) cold pretreatment of excised buds for 48 hr at 4 C, and (b) inclusion of activated charcoal in culture medium.

Ovule Rescue of *Nicotiana* Interspecific Hybrids

1. Hand-pollinate flowers by emasculating unopened flowers and pollinating with donor pollen collected on the previous day. Cover the stigma and style with a paper straw to insure no cross fertilization.
2. Remove capsules at various intervals after fertilization for in vitro culture. The optimum interval should be determined for each species combination.
3. Surface sterilize capsules as follows: 2.5 min in 70% ethanol, 5 min in 20% commercial bleach (1% sodium hypochlorite), 2 rinses in sterile water.
4. Remove ovary wall sterilely with scalpel and forceps. Transfer fertilized ovules to Nitsch's medium N.
5. Transfer germinating seedlings to half-strength Nitsch's medium.
6. If plants are chlorotic or otherwise unhealthy section into small pieces and induce callus on LS medium with 9.3 μM KIN and 5.7 μM IAA.

7. Shoots regenerated from callus are rooted on half-strength Nitsch's medium N. After rooting, plants are transferred to soil.

FUTURE PROSPECTS

The genus *Nicotiana* offers more prospects for crop improvement via in vitro techniques than any other genus of plants. Whereas in most crop species only a limited number of in vitro techniques can be utilized, tobacco is unique in that it has developed as a model system for tissue culture. A great deal of basic research has been accomplished in *Nicotiana*, but the techniques developed have not been applied to the improvement of commercial tobacco.

Routine propagation of tobacco through in vitro culture is possible. Depending on the desired results, plants can be clonally propagated, or in vitro methods can be used to increase genetic variability in a variety. Clonal propagation may be advantageous for maintenance of male steriles, other varieties with low seed production, or hybrid varieties. In this manner many plants can be routinely produced starting from a limited number of explants. Clonal propagation is usually best achieved through shoot apex culture. Plant regeneration from non-meristematic tissue may result in increased genetic variability in a given variety. Increased variability as compared to a seed population was observed in populations of *N. alata* Nicki hybrids regenerated from leaf explants by Bravo and Evans (unpublished). Some of this variability is in horticultural traits which could result in variety improvement.

Since tobacco is a sexually propagated species, a new variety or breeding line is not useful until it breeds true, i.e., is homozygous. Conventionally homozygous lines are obtained by repeated self-fertilization. This process can take years before an acceptable homozygous line is available. A large number of sexual generations can be circumvented using anther culture. Haploids obtained through anther culture may have more than one potential use. First, by doubling the chromosome number of haploids a homozygous diploid is obtained. Hence years of sexual propagation are bypassed. Second, the fusion of protoplasts isolated from haploid plants or cell cultures will produce hybrids with the same chromosome number as sexual hybrids. This may be important if protoplast fusion is being used to transfer a cytoplasmic trait, e.g., male sterility between varieties, and if increased chromosome number is not desirable. Third, recessive mutants can be isolated only from haploid cultures. Any of these applications of haploids could be utilized immediately as the basic techniques are available in tobacco.

The wild species of tobacco represent a large reservoir of genetic information that could be potentially incorporated into *N. tabacum*. However, many of the wild species are sexually incompatible with *N. tabacum*. If this incompatibility is due to a post-fertilization event, then in some cases ovule culture can be useful in obtaining novel sexual hybrids. This was demonstrated by Reed and Collins (1978) as they obtained *N. nesophila* x *N. tabacum* and *N. stocktonii* x *N. tabacum* sexual hybrids using fertilized ovule culture. These hybrids may be of agricultural interest as they incorporate resistance to black shank disease from the wild species. However, the application of ovule culture is limited by several factors. First, fertilization is required. If sexual incompatibility is based on inability to obtain fertilization, then ovule culture is not useful. Second, despite the use of in vitro culture of fertilized ovules, two species still may be only unilaterally sexually compatible. Indeed this was the case with *N. nesophila* and *N. stocktonii* x *N. tabacum* hybrids obtained through ovule culture. Successful embryo development was only possible when the wild species of tobacco was the female parent (Reed and Collins, 1978).

Protoplasts have many potential applications in crop improvement. Those problems outlined above for obtaining hybrids through fertilized ovule culture are not applicable to protoplast fusion, because somatic cells are being fused. In addition all possible nuclear-cytoplasmic combinations are possible following protoplast fusion. Through the use of appropriate selective markers cytoplasmic transfer without nuclear hybridization can be achieved via protoplast fusion. Thus a large array of novel hybrids are possible. Whereas sexual hybrids between two species are uniform, the comparable somatic hybrids demonstrate a great deal of genetic variability. This phenomenon was documented when *N. nesophila* + *N. tabacum* somatic hybrids were compared to the comparable sexual hybrids (Evans et al., 1982). Hence in addition to the availability of novel hybrids produced through protoplast fusion, more variability may be available in somatic hybrids than conventional sexual hybrids. Protoplast fusion introduces a large reservoir of genetic information for potential crop improvement.

The potential value of protoplasts for crop improvement in tobacco has not been examined. Genetic variability which is evident when plants are regenerated from nonmeristematic tissue should also be present when plants are regenerated from protoplasts. Protoplast-derived plants, in contrast to regenerates from leaf explants, are derived from single cells and thus each plant should be genetically uniform. Shepard and Totten (1977) have demonstrated in potato that there is extensive variability in plants regenerated from protoplasts.

Direct selection of mutants from cell culture offers an opportunity to select specific desirable traits, e.g., herbicide resistance, cold tolerance, or salt tolerance. Herbicide resistant mutants would permit the use of normally lethal concentrations of herbicides on crop species. Resistance to pathotoxins has been demonstrated in potato (Behnke, 1979) and maize (Gengenbach and Green, 1975), but not in tobacco. Disease resistance may indeed be one of the most prospective applications of in vitro mutant isolation. Carlson showed that plants regenerated from methionine sulfoximine resistant mutants isolated in vitro were resistant to wildfire disease (Carlson, 1973). In contrast to sexual or somatic hybrids where complete genomes are combined, mutant isolation enables one to isolate a line with only a single desired genetic modification. Hence extensive breeding to eliminate unnecessary alien germplasm is not necessary.

It is quite apparent that in vitro techniques can be of immediate use in addressing breeding problems of tobacco. It is possible to incorporate new genetic information into tobacco using each of the techniques described above. Although disease resistance has been most often addressed in tobacco these methods could equally well be applied to modification of the alkaloid content of tobacco or other agronomic characteristics of breeding interest. Tissue culture techniques will only be successful for tobacco improvement when integrated with conventional breeding programs.

KEY REFERENCES

Bajaj, Y.P.S. 1983. Haploid plant production. In: Handbook of Plant Cell Culture, Vol. 1 (D.A. Evans, P.V. Ammirato, W.R. Sharp, and Y. Yamada, eds.) pp. 228-290. Macmillan, New York.

Collins, G.B. and Legg, P.D. 1979. Use of tissue and cell culture methods in tobacco improvement. In: Plant Cell and Tissue Culture: Principles and Applications (W.R. Sharp, P.O. Larsen, E.F. Paddock, and V. Raghavan, eds.) pp. 585-614. Ohio State Univ. Press, Columbus.

Evans, D.A. 1983. Protoplast fusion. In: Handbook of Plant Cell Culture, Vol. 1 (D.A. Evans, P.V. Ammirato, W.R. Sharp, and Y. Yamada, eds.) pp. 291-321. Macmillan, New York.

______ and Bravo, J.E. 1983. Protoplast isolation and culture. In: Handbook of Plant Cell Culture, Vol. 1 (D.A. Evans, P.V. Ammirato, W.R. Sharp, and Y. Yamada, eds.) pp. 124-176. Macmillan, New York.

Flick, C.E. 1983. Isolation of mutants from cell cultures. In: Handbook of Plant Cell Culture, Vol. 1 (D.A. Evans, P.V. Ammirato, W.R. Sharp, and Y. Yamada, eds.) pp. 393-441. Macmillan, New York.

______, Evans, D.A., and Sharp, W.R. 1983. Organogenesis. In: Handbook of Plant Cell Culture, Vol. 1 (D.A. Evans, P.V. Ammirato, W.R. Sharp, and Y. Yamada, eds.) pp. 13-81. Macmillan, New York.

Gleba, Yu.Yu. and Evans, D.A. 1983. Transmission genetics of somatic hybrids. In: Handbook of Plant Cell Culture, Vol. 1 (D.A. Evans, P.V. Ammirato, W.R. Sharp, and Y. Yamada, eds.) pp. 322-357. Macmillan, New York.

REFERENCES

Ahuja, M.R. and Hagen, G.L. 1966. Morphogenesis in *Nicotiana debneyi, N. tabacum, N. longiflora* and their tumor-forming hybrid derivatives in vitro. Dev. Biol. 13:408-423.

Aviv, D. and Galun, E. 1977. Isolation of tobacco protoplasts in the presence of isopropyl N-phenylcarbamate and their culture and regeneration into plants. Z. Pflanzenphysiol. 83:267-273.

Banks, M.S. and Evans, P.K. 1976. A comparison of the isolation and culture of mesophyll protoplast from several *Nicotiana* species and their hybrids. Plant Sci. Lett. 7:409-416.

Behnke, M. 1979. Selection of potato callus for resistance to culture filtrates of *Phytophthora infestans* and regeneration of resistant plants. Theor. Appl. Genet. 55:69–71.

Berlin, J. and Vollmer, B. 1979. Effect of α-aminooxy-β-phenylpropionic acid on phenylalanine metabolism in p-fluorophenylalanine sensitive and resistant tobacco cells. Z. Naturforsch. 34c:770–775.

Berlin, J. and Widholm, J.M. 1977. Correlation between phenylalanine ammonia lyase activity and phenolic biosynthesis in p-fluorophenylalanine-sensitive and -resistant tobacco and carrot tissue cultures. Plant Physiol. 59:550–553.

______ 1978. Metabolism of phenylalanine and tyrosine in tobacco cell lines resistant and sensitive to p-fluorophenylalanine. Phytochemistry 17:65–68.

Bourgin, J.P. 1978. Valine-resistant plants from in vitro selected tobacco cells. Mol. Gen. Genet. 161:225–230.

______ and Missonier, C. 1978. Culture of haploid mesophyll protoplasts from *Nicotiana alata*. Z. Pflanzenphysiol. 87:55–64.

______ and Nitsch, J.P. 1967. Obtention de *Nicotiana* haploides au partir d'etamines cultivees in vitro. Ann. Physiol. Veg. 9:377-382.

______, Chupeau, Y., and Missonier, C. 1979. Plant regeneration from mesophyll protoplasts of several *Nicotiana* species. Physiol. Plant. 45:288–292.

Brown, J.S. and Wernsman, E.A. 1982. Nature of reduced productivity of anther-derived dihaploid lines of flue-cured tobacco. Crop Sci. 22:1-5.

Burk, L.G. and Matzinger, D.F. 1976. Variation among anther derived doubled haploids from an in-bred line of tobacco. J. Hered. 67:382-384.

Caboche, M. and Muller, J.F. 1980. Use of a medium allowing low cell density growth for in vitro selection experiments: Isolation of valine-resistant clones from nitrosoguanidine-mutagenized cells and gamma-irradiated tobacco plants. In: Plant Cell Cultures: Results and Perspectives (F. Sala, B. Parisi, R. Cella, and O. Ciferri, eds.). Elsevier/North-Holland Biomedical Press, Amsterdam.

Carlson, P.S. 1970. Induction and isolation of auxotrophic mutants in somatic cell cultures of *Nicotiana tabacum*. Science 168:487-489.

______ 1973. Methionine sulfoximine-resistant mutants of tobacco. Science 180:1366-1368.

______, Smith, H.H., and Dearing, R.D. 1972. Parasexual interspecific plant hybridization. Proc. Natl. Acad. Sci. USA 69:2292–2294.

Chaleff, R.S. and Keil, R.L. 1981. Genetic and physiological variability among cultured cells and regenerated plants of *Nicotiana tabacum*. Molec. Gen. Genet. 181:254–258.

Chaleff, R.S. and Parsons, M.F. 1978a. Direct selection in vitro for herbicide-resistant mutants of *Nicotiana tabacum*. Proc. Natl. Acad. Sci. USA 75:5104-5107.

______ 1978b. Isolation of a glycerol-utilizing mutant of *Nicotiana tabacum*. Genetics 89:723-728.

Chaplin, J.F. 1978. Genetic manipulation for tailoring the tobacco plant to meet the requirements of the grower, manufacturer and consumer. CORESTA Information Bulletin. International Tobacco Scientific Symposium, 1978, pp. 17-32. Sofia, Bulgaria.

Chiu, P.L., Bottino, P.J., and Patterson, G.W. 1980. Sterol composition of nystatin and amphotericin B resistant tobacco calluses. Lipids 15:50-54.

Collins, G.B. and Sunderland, N. 1974. Pollen derived haploids of *Nicotiana knightiana, N. raimondii,* and *N. attenuata.* J. Exp. Bot. 25: 1030-1039.

Collins, G.B., Legg, P.D., and Kasperbauer, M.J. 1972. Chromosome numbers in anther derived haploids of two *Nicotiana* species. J. Hered. 63:113-118.

Davey, M.R., Frearson, E.M., Withers, L.A., and Power, J.B. 1974. Observations on the morphology, ultrastructure and regeneration of tobacco leaf epidermal protoplasts. Plant Sci. Lett. 2:23-27.

DePaepe, C., Nitsch, C., Godard, M., and Pernes, J. 1977. Potential for haploids and possible uses in agriculture. In: Plant Tissue Culture and Its Biotechnological Application (N. Barz, E. Reinhard, and M.H. Zenk, eds.) pp. 341-352. Springer-Verlag, Berlin.

Dix, P.J. 1977. Chilling resistance is not transmitted sexually in plants regenerated from *Nicotiana sylvestris* cell lines. Z. Pflanzenphysiol. 84:223-226.

______ 1981. Inheritance of chloramphenicol resistance, a trait selected in cell cultures of *Nicotiana sylvestris* Speg. and Comes. Ann. Bot. 48:321-325.

______ and Pearce, R.S. 1981. Proline accumulation in NaCl-resistant and -sensitive cell lines of *Nicotiana sylvestris.* Z. Pflanzenphysiol. 102:243-248.

______ and Street, H.E. 1976. Selection of plant cell lines with enhanced chilling resistance. Ann. Bot. 40:903-910.

______, Joo, F., and Maliga, P. 1977. A cell line of *Nicotiana sylvestris* with resistance to kanamycin and streptomycin. Molec. Gen. Genet. 157:285-290.

Dulieu, H.L. 1976. Pollination of excised ovaries and culture of ovules of *Nicotiana tabacum* L. Phytomorphology 26:69-75.

Dunwell, J.M. and Perry, M.E. 1973. The influence of in vivo growth conditions for *N. tabacum* plants on the in vitro embryogenic potential of their anthers. In: John Innes Annual Report No. 64. John Innes Institute. Norwich, England.

Dunwell, J.M. and Sunderland, N. 1973. Anther culture of *Solanum tuberosum* L. Euphytica 22:317-323.

Engvild, K.C., Linde-Laursen, I.B., and Lundgvist, A. 1972. Anther cultures of *Datura innoxia*: Flower bud stage and embryoid level of ploidy. Hereditas 72:331-332.

Evans, D.A. 1979. Chromosome stability of plants regenerated from mesophyll protoplasts of *Nicotiana* species. Z. Pflanzenphysiol. 95:459-463.

______, Wetter, L.R., and Gamborg, O.L. 1980. Somatic hybrid plants of *Nicotiana glauca* and *Nicotiana tabacum* obtained by protoplast fusion. Physiol. Plant. 48:225-230.

______, Flick, C.E., and Jensen, R.A. 1981. Disease resistance: Incorporation into sexually incompatible somatic hybrids of the genus *Nicotiana.* Science 213:907-909.

______, Flick, C.E., Kut, S.A., and Reed, S.M. 1982. Comparison of *Nicotiana tabacum* and *Nicotiana nesophila* hybrids produced by ovule culture and protoplast fusion. Theor. Appl. Genet. 62:193-198.

______, Bravo, J.E., Kut, S.A., and Flick, C.E. 1983. Genetic behavior of somatic hybrids in the genus *Nicotiana: N. otophora + N. tabacum* and *N. sylvestris + N. tabacum.* Theor. Appl. Genet. 65:93-101.

Faccioti, D. and Pilet, P.E. 1979. Plants and embryoids from haploid *Nicotiana sylvestris* protoplasts. Plant Sci. Lett. 15:1-6.

Flashman, S.M. and Filner, P. 1978. Selection of tobacco cell lines resistant to selenoamino acids. Plant Sci. Lett. 13:219-229.

Flick, C.E. and Evans, D.A. 1983. Isolation, culture and plant regeneration from protoplasts isolated from flower petals of ornamental *Nicotiana* species. Z. Pflanzenphysiol. 109:379-384.

Flick, C.E., Jensen, R.A., and Evans, D.A. 1981. Isolation, protoplast culture, and plant regeneration of PFP-resistant variants of *Nicotiana tabacum* Su/Su. Z. Pflanzenphysiol. 103:239-245.

Gamborg, O.L., Shyluk, J.P., Fowke, L.C., Wetter, L.R., and Evans, D.A. 1979. Plant regeneration from protoplasts and cell cultures of *N. tabacum* sulfur mutant (Su/Su). Z. Pflanzenphysiol. 95:255-264.

Gengenbach, B.G. and Green, C.E. 1975. Selection of T-cytoplasm maize callus cultures resistant to *Helminthosporium maydis* race T pathotoxin. Crop Sci. 15:645-649.

Gill, R., Rashid, A., and Maheshwari, S.C. 1979. Isolation of mesophyll protoplasts of *Nicotiana rustica* and their regeneration into plants flowering in vitro. Physiol. Plant. 47:7-10.

Glimelius, K., Eriksson, T., Grafe, R., and Muller, A.J. 1978. Somatic hybridization of nitrate-deficient mutants of *Nicotiana tabacum* by protoplast fusion. Physiol. Plant. 44:273-277.

Goodspeed, T.H. 1954. The Genus *Nicotiana*: Chronica Botanica. Waltham, Massachusetts.

Heimer, Y.M. and Filner, P. 1970. Regulation of the nitrate assimilation pathway of cultured tobacco cells. II. Properties of a variant cell line. Biochim. Biophys. Acta 215:152-165.

Helgeson, J.P. 1979. Tissue and cell suspension culture. In: *Nicotiana*: Procedures for Experimental Use (R.D. Durbin, ed.) pp. 52-59. USDA, Washington, D.C.

Kao, K.N. and Michayluk, M.R. 1975. Nutritional requirements for growth of *Vicia hajastana* cells and protoplasts at a very low population density in liquid media. Planta 126:105-110.

Kostoff, D. 1943. Cytogenetics of the Genus *Nicotiana*. State Printing House, Sofia.

Lawyer, A.L., Berlyn, M.B., and Zelitch, I. 1980. Isolation and characterization of glycine hydroxamate-resistant cell lines of *Nicotiana tabacum*. Plant Physiol. 66:334-341.

Lescure, A.M. 1973. Selection of markers of resistance to base-analogues in somatic cell cultures of *Nicotiana tabacum*. Plant Sci. Lett. 1:375-383.

Lloyd, R. 1975. Tissue culture as a means of circumventing lethality in an interspecific *Nicotiana* hybrid. Tob. Sci. 19:4-6.

Maisuryan, A.N., Khadeeva, N.V., and Pogosov, V.Z. 1981. Isolation of 5-bromodeoxyuridine resistant cell lines of haploid tobacco. Sov. Plant Physiol. 28:395-400.

Maliga, P. 1981. Streptomycin resistance is inherited as a recessive Mendelian trait in a *Nicotiana sylvestris* line. Theor. Appl. Genet. 60:1-3.

______, Breznovits, A.S., and Marton, L. 1973. Streptomycin-resistant plants from callus culture of haploid tobacco. Nature 244:29-30.

______, Breznovits, A.S., Marton, L., and Joo, F. 1975. Non-mendelian streptomycin-resistant tobacco mutant with altered chloroplasts and mitochondria. Nature 255:401-402.

______, Lazar, G., Svab, Z., and Nagy, F. 1976. Transient cycloheximide resistance in a tobacco cell line. Molec. Gen. Genet. 149:267-271.

______, Lazar, G., Joo, F., Nagy, A.H., and Menczel, L. 1977. Restoration of morphogenetic potential of *Nicotiana* by somatic hybridization. Molec. Gen. Genet. 157:291-296.

______, Kiss, Z.R., Nagy, A.H., and Lazar, G. 1978. Genetic instability in somatic hybrids of *Nicotiana tabacum* and *Nicotiana knightiana*. Molec. Gen. Genet. 163:145-151.

Malmberg, R.L. 1980. Biochemical, cellular and developmental characterization of a temperature-sensitive mutant of *Nicotiana tabacum* and its second site revertant. Cell 22:603-609.

Marton, L. and Maliga, P. 1975. Control of resistance in tobacco cells to 5-bromodeoxyuridine by a simple Mendelian factor. Plant Sci. Lett. 5:77-81.

Medgyesy, P., Menczel, L., and Maliga, P. 1980. The use of cytoplasmic streptomycin resistance: Chloroplast transfer from *Nicotiana tabacum* into *Nicotiana sylvestris*, and isolation of their somatic hybrids. Molec. Gen. Genet. 179:693-698.

Melchers, G. and Labib, G. 1974. Somatic hybridization of plants by fusion of protoplasts. I. Selection of light resistant hybrids of "haploid" light sensitive varieties of tobacco. Molec. Gen. Genet. 135: 277-294.

Miller, O.K. and Hughes, K.W. 1980. Selection of paraquat-resistant variants of tobacco from cell cultures. In Vitro 16:1085-1091.

Murashige, T. and Skoog, F. 1962. A revised medium for rapid growth and bioassays with tobacco tissue cultures. Physiol. Plant. 15:473-497.

Nabors, M.W., Gibbs, S.E., Bernstein, C.S., and Meis, M.E. 1980. NaCl-tolerant tobacco plants from cultured cells. Z. Pflanzenphysiol. 97: 13-17.

Nagao, T. 1978. Somatic hybridization by fusion of protoplasts. I. The combination of *Nicotiana tabacum* and *Nicotiana rustica*. Jpn. J. Crop Sci. 47:491-498.

______ 1979. Somatic hybridization by fusion of protoplasts. II. The combinations of *Nicotiana tabacum* and *N. glutinosa* and of *N. tabacum* and *N. alata*. Jpn. J. Crop Sci. 48:385-392.

______ 1982. Somatic hybridization by fusion of protoplasts. III. Somatic hybrids of sexually incompatible combinations *Nicotiana tabacum* + *Nicotiana repanda* and *Nicotiana tabacum* + *Salpiglossis sinuata*. Jpn. J. Crop Sci. 51:35-42.

Nagata, T. and Takebe, I. 1971. Plating of isolated tobacco mesophyll protoplasts on agar medium. Planta 99:12-20.

Nagy, J.I. and Maliga, P. 1976. Callus induction and plant regeneration from mesophyll protoplasts of *Nicotiana sylvestris*. Z. Pflanzenphysiol. 78:453-455.

Nakamura, A., Yamada, T., Kadotani, N., and Itagaki, R. 1974. Improvement of flue-cured tobacco variety MC 1610 by means of haploid breeding methods and some problems of this method. In: Haploids in Higher Plants: Advances and Potential (K.J. Kasha, ed.) pp. 277-278. Univ. Guelph Press, Guelph, Ontario.

Nitsch, C. 1974. Pollen culture: A new technique for mass production of haploid and homozygous plants. In: Haploids in Higher Plants: Advances and Potential (K.J. Kasha, ed.) pp. 123-135. Univ. Guelph Press, Guelph, Ontario.

______ and Norrell, B. 1972. Factors favoring the formation of androgenetic embryos in anther cultures. In: Genes, Enzymes, and Populations, Vol. 2 (A. Srb, ed.) pp. 128-144. Plenum Press, New York.

Nitsch, J.P. 1971. La production in vitro d'embryons haploides: Resultats et perspectives. In: Colloque internationaux C.N.R.S. 193, Les Cultures de Tissus de Plantes, pp. 282-294. C.N.R.S. Paris.

______ and Nitsch, C. 1969. Haploid plants from pollen grains. Science 163:85-87.

______ and Ohyama, K. 1971. Obtention de plantes a partir de protoplastes haploides cultivees in vitro. C.R. Acad. Sci. Paris 273:801-804.

North, C. 1976. In vitro culture of plant material as an aid to hybridization. Acta Hortic. 63:67-75.

Owens, L.D. 1981. Characterization of kanamycin-resistant cell lines of *Nicotiana tabacum*. Plant Physiol. 67:1166-1168.

Passiatore, J.E. and Sink, K.C. 1981. Plant regeneration from leaf mesophyll protoplasts of selected ornamental *Nicotiana* species. J. Am. Soc. Hortic. Sci. 106:799-803.

Patnaik, G., Wilson, D., and Cocking, E.C. 1981. Importance of enzyme purification for increased plating efficiency and plant regeneration from single protoplasts of *Petunia parodii*. Z. Pflanzenphysiol. 102: 199-202.

Piven, N.M. 1981. Regeneration of whole plants from isolated leaf mesophyll protoplasts of *Nicotiana debneyi* Domin. Soviet Genet. 15: 35-39.

Polacco, J.C. and Polacco, M.L. 1977. Inducing and selecting valuable mutations in plant cell culture: A tobacco mutant resistant to carboxin. Ann. N.Y. Acad. Sci. 287:385-400.

Radin, D.N. and Carlson, P.S. 1978. Herbicide-tolerant tobacco mutants selected in situ and recovered via regeneration from cell culture. Genet. Res. 32:85-89.

Raghavan, V. 1976. Role of the generative cell in androgenesis of henbane. Science 191:388-389.

Rashid, A. and Street, H.E. 1973. The development of haploid embryoids from anther cultures of *Atropa belladona* L. Planta 113:263-270.

Reed, S.M. and Collins, G.B. 1978. Interspecific hybrids in *Nicotiana* through in vitro culture of fertilized ovules. J. Hered. 69:311-315.

______ 1980. Chromosome pairing relationships and black shank resistance in three *Nicotiana* interspecific hybrids. J. Hered. 71:423-426.

Scowcroft, W.R. and Larkin, P.J. 1980. Isolation, culture and plant regeneration from protoplasts of *Nicotiana debneyi*. Aust. J. Plant Physiol. 7:635-644.

______ 1981. Chloroplast DNA assorts randomly in intraspecific somatic hybrids of *Nicotiana debneyi*. Theor. Appl. Genet. 60:179-184.

Sharp, W.R., Raskin, R.S., and Sommer, H.E. 1971. Haploids in *Lilium*. Phytomorphology 21:334-337.

Shepard, J.F. and Totten, R.E. 1977. Mesophyll cell protoplasts of potato: Isolation, proliferation, and plant regeneration. Plant Physiol. 60:313-316.

Sidorov, V.A., Menczel, L., and Maliga, P. 1981a. Isoleucine-requiring *Nicotiana* plant deficient in threonine deaminase. Nature 294:87-88.

Sidorov, V.A., Menczel, L., Nagy, F., and Maliga, P. 1981b. Chloroplast transfer in *Nicotiana* based on metabolic complementation between irradiated iodoacetate treated protoplasts. Planta 152:341-345.

Skoog, F. and Miller, C.O. 1957. Chemical regulation of growth and organ formation in plant tissues cultivated in vitro. In: Biological Action of Growth Substances. Symp. Soc. Exp. Biol. 11:118-131.

Sunderland, N. 1970. Pollen plants and their significance. New Sci. 16:142-143.

______ 1973. Pollen and anther culture. In: Plant Tissue and Cell Culture (H.E. Street, ed.) pp. 205-238. Univ. California Press, Berkeley.

______ 1974. Anther culture as a means of haploid production. In: Haploids in Higher Plants: Advances and Potential (K.J. Kasha, ed.) pp. 91-122. Univ. Guelph Press, Guelph, Ontario.

______ and Wicks, F.M. 1969. Cultivation of haploid plants from tobacco pollen. Nature 224:1227-1229.

______ and Wicks, F.M. 1971. Embryoid formation in pollen grains of *Nicotiana tabacum*. J. Exp. Bot. 22:213-226.

Tomes, D.T. and Collins, G.B. 1976. Factors affecting haploid plant production from in vitro anther cultures of *Nicotiana* species. Crop Sci. 16:837-840.

Tran Thanh Van, K. and Trinh, H. 1978. Morphogenesis in thin cell layers: Concept, methodology and results. In: Frontiers of Plant Tissue Culture (T.A. Thorpe, ed.) pp. 37-48. Univ. Calgary Press, Calgary.

Uchimiya, H. 1982. Somatic hybridization between male sterile *Nicotiana tabacum* and *N. glutinosa* through protoplast fusion. Theor. Appl. Genet. 61:69-72.

Umiel, N. 1979. Streptomycin resistance in tobacco. III. A test on germinating seedlings indicates cytoplasmic inheritance in the St-R701 mutant. Z. Pflanzenphysiol. 92:295-301.

______ and Goldner, R. 1976. Effects of streptomycin on diploid tobacco callus cultures and the isolation of resistant mutants. Protoplasma 89:1031-1037.

Vasil, I.K. and Hildebrandt, A.C. 1965. Differentiation of tobacco plants from single, isolated cells in micro cultures. Science 150:889-892.

Vasil, V. and Vasil, I.K. 1974. Regeneration of tobacco and petunia plants from protoplasts and culture of corn protoplasts. In Vitro 10: 83-96.

Vuush, R., Aviv, D., and Galun, E. 1980. An analogue-resistant cell line derived from haploid *Nicotiana sylvestris*. In: Plant Cell Cultures: Results and Perspectives (F. Sala, B. Parisi, R. Cella, and O. Ciferri, eds.) pp. 145-150. Elsevier/North-Holland Biomedical Press, Amsterdam.

Vyskot, B. and Novak, F.J. 1973. Experimental androgenesis in vitro in *Nicotiana clevelandii* Gray and *N. sanderae* Hort. Theor. Appl. Genet. 44:138-140.

Wark, D.C. 1977. Doubled haploids in Australian tobacco breeding. Proceedings of the 3rd International Congress of SABRAO 2:19-23.

White, D.W.R. and Vasil, I.K. 1979. Use of amino acid analogue-resistant cell lines for selection of *Nicotiana sylvestris* somatic cell hybrids. Theor. Appl. Genet. 55:107-112.

Widholm, J.M. 1972. Cultured *Nicotiana tabacum* cells with an altered anthranilate synthetase which is less sensitive to feedback inhibition. Biochim. Biophys. Acta 261:52-58.

______ 1976. Selection and characterization of cultured carrot and tobacco cells resistant to lysine, methionine, and proline analogues. Can. J. Bot. 54:1523-1529.

______ 1977. Selection and characterization of amino acid analog resistant cell cultures. Crop Sci. 17:597-600.

Wong, J.R. and Sussex, I.M. 1980. Isolation of abscisic acid-resistant variants from tobacco cell cultures. II. Selection and characterization of variants. Planta 148:103-107.

Yurina, N.P., Odintsova, M.S., and Maliga, P. 1978. An altered chloroplast ribosomal protein in a streptomycin resistant tobacco mutant. Theor. Appl. Genet. 52:125-128.

Zelcer, A., Aviv, D., and Galun, E. 1978. Interspecific transfer of cytoplasmic male sterility by fusion between protoplasts of normal *Nicotiana sylvestris* and X-ray irradiated protoplasts of male-sterile *N. tabacum*. Z. Pflanzenphysiol. 90:397-407.

Species Index

Subject Index

ABBREVIATIONS

Growth Regulators

IAA	Indole-3-acetic acid
IBA	Indole-3-butyric acid
NAA	1-Naphthaleneacetic acid
2,4-D	(2,4-Dichlorophenoxy)acetic acid
2,4,5-T	(2,4,5-Trichlorophenoxy)acetic acid
CPA	(4-Chlorophenoxy)acetic acid
PIC	Picloram (4-amino-3,5,6-trichloropicolinic acid)
NOA	2-Naphthoxyacetic acid
BTOA	2-Benzothiazoleacetic acid
BA or 6BA	6-Benzylaminopurine
ZEA	Zeatin
KIN	Kinetin
2iP	(2-Isopentenyl)adenine

Additives

CH	Casein hydrolysate
CW	Coconut water
EDTA	(Ethylenedinitrilo)tetraacetic acid
GA	Gibberellic acid (Gibberellin A_3)
ABA	Abscisic acid

Macro- and Micronutrient Formulations

ADE	Adenine
MS	Murashige and Skoog (1962)
B5	Gamborg et al. (1968)
ER	Erikkson (1965)
WH	White (1963)
SH	Schenk and Hildebrandt (1972)